S0-BIX-340

Fatigue and Tribological Properties

Of Plastics And Elastomers

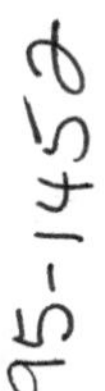

Plastics Design Library

ISBN 1-884207-15-4
Library of Congress Card Number 94-066603

Published in the United States of America, Norwich, NY by Plastics Design Library a division of William Andrew Inc.

Comments, criticisms and suggestions are invited, and should be forwarded to Plastics Design Library.

Manufactured in the United States of America.

Plastics Design Library, P.O. Box 443, Morris, NY 13808 Tel: 607/263-2318 Fax: 607/263-2446

Table of Contents

Fluoroplastic

Polyamide

Polycarbonate

Polyester

Polyurethane

Styrenic Resin

Vinyl Resin

Thermoplastic Alloys

Plastic Alloy

Thermosets

Thermoplastic Elastomers

Section Two - Tribological Properties

Thermoplastics

Polyester

Polyimide

Polyketone

Polyolefin

Polyphenylene Sulfide

Polysulfone

Styrenic Resin

Thermoplastic Alloys

Thermosets

Thermoplastic Elastomers

Rubbers

Appendices

Indices

Introduction

Plastics Design Library is pleased to introduce ***Fatigue and Tribological Properties Of Plastics And Elastomers***, a unique reference and data bank on the fatigue behavior and friction and wear characteristics of polymeric materials. The basic physical characteristics of polymers are generally well defined by manufacturers. However, data on the more capricious phenomenological issues, such as fatigue, are difficult to find, especially in a comprehensive compilation and in a manner which shows behavior under different test variables. This volume serves to turn disparate information from wide ranging sources (i.e. conference proceedings, test laboratories, materials suppliers, monographs, trade and technical journals) into useful engineering knowledge.

The information provided ranges from a general overview of the fatigue and tribological properties of plastics and elastomers to detailed discussions and test results. For users to whom the study and use of this data are relatively new, the primer provided in the introduction and detailed glossary of terms, including descriptions of test methods, will prove useful. For those who wish to delve beyond the data presented, source documentation is presented in detail.

Data presented in these pages detail differences in behavior between generic families of plastic and rubber materials. Also covered are differences within the same generic family due to factors such as temperature, test frequency and mating surface details (in the case of tribological properties) or material characteristics such as sample preparation and material composition. This data serves as an indication of how one material is likely to behave relative to another material or relative to the same material tested under different conditions.

In compiling data, the philosophy of Plastics Design Library is to provide as much information as is available. This means that complete information corresponding to each test result (often referred to as metadata) is provided. At the same time, an effort is made to provide information for as many tests, conditions, and materials combinations as possible. Therefore, even if detailed test metadata are not available, information is still provided. The belief is that some limited information serves as a reference point and is better than no information. In all cases, we undertake to provide information in as complete and detailed a form as it was presented in the source document. Flexibility and ease of use are also carefully considered in designing the layout of the book.

How a material performs in its end use environment is a critical consideration and the information here gives useful guidelines. However, this or any other information resource should not serve as a substitute for actual testing in determining the applicability of a particular part or material in a given end use environment.

We trust you will greet this reference publication with the same enthusiasm as other Plastics Design Library titles and that it will be a useful tool in your work. As always, your feedback on improving this volume or others in the PDL Handbook series is appreciated and encouraged.

Plastics Design Library
13 Eaton Avenue
Norwich, NY 13815
Tel: 607-337-5080 Fax: 607-337-5090

Some Notes About The Information In This Book

This publication contains data and information from many disparate sources. In order to make the product most useful to users, Plastics Design Library arranges information to be easily accessible in consistent formats. Although substantial effort is exerted throughout the editorial process to maintain accuracy and consistency in unit conversion and presentation of information, possibility for error exists. Often these errors occur due to insufficient or inaccurate information in the source document.

As with all PDL products, complete information (as it was presented in the source document) is provided. This includes details of test methods, test conditions, mating surface, sample size, material composition and other factors which may affect the resulting value. Therefore, the user has all available information on which to make a judgment or comparison.

How The Book Is Organized

This publication is divided into ***two major sections***. Section I, which covers fatigue of plastics and elastomers, includes Chapters One (1) through sixty-eight (68). Within each chapter in Section I, information is presented as combinations of text and graphs. Textual data are presented as conscise discussions of topics relating to the fatigue behavior of the material of interest in the chapter. Graphs include S-N curves (Woehler Diagrams), fatigue crack propagation curves and other curves which provide insight into the fatigue behavior of a material. Figures referenced in the legends of the graphs can be found in Appenix V - Figures. This reference is designed to give more information on the data contained in the graphs.

Section II, which covers the tribological (friction and wear) properties of plastics and elastomers, includes Chapters sixty-nine (69) through ninety-nine (99) as well as Appendix I through IV. Within the chapters in Section II, information is presented as combinations of text and graphs. Textual data are presented as conscise discussions of topics relating to the friction and wear behavior of the material of interest in the chapter. Graphs show how variables affect the friction and wear properties. Appendix I covers coefficient of friction data, Appendix II wear factor (K) and wear rate, Appendix III covers PV Limits and Appendix IV details abrasion resistance data.

In addition, a glossary of terms, including test methods used in the publication is included. A trade name index and complete source documentation are also included for reference.

FATIGUE PROPERTIES

The fracture or decrease in load bearing capability of structural components under cyclic or intermittent application of stress or strain is known as dynamic fatigue. The term static fatigue or creep rupture applies to the effect of continuously applied loads. The subject of this book and the use of the term fatigue throughout apply to dynamic fatigue. Static fatigue is covered in the Plastics Design Library publication ***The Effect of Creep and Other Time Related Factors on Plastics.***

Examples of materials subject to fatigue loading include a snap action plastic latch which is constantly opened and closed, a reciprocating mechanical part on a machine, a gear tooth, a bearing, any structural component subjected to vibration or any part which will be subjected to repeated impacts. Less obvious examples of components subject to fatigue loading include an air line filter bowl subjected to on off pressurizations of less than 10 times per day or a hot water kettle which is subjected to the stresses of heating and cooling at the rate of perhaps two times per day.[414]

Estimates in the literature indicate that between 20% and 80% of plastic part service failures can be attributed to fatigue. Despite the importance of dynamic fatigue failure in plastics, very little has been done to standardize test methods, thereby leading to many different approaches. The problem is that there are many variables which affect test results and to cover them all in a comprehensive test program for even one material is an impossible task.[461]

In fatigue testing, a specimen of the material being tested is subjected to repeated cycles of short term stress or deformation. Eventually, micro-cracks or defects form in the specimen's structure, causing decreased toughness, impact strength, and tensile elongation - and the likelihood of failure at stress levels considerably lower than the material's original ultimate tensile strength. The number of cycles to failure at any given stress level depends on the inherent strength of the resin, the size and number of defects induced at that stress level, the environment of the test specimen and the conditions under which the material was tested.[78]

Typical fatigue tests are conducted on a machine which subjects a cantilever beam to reverse flexural loading cycles at different maximum stress levels. Fatigue tests are also made in tension, compression, alternating tension and compression, cycling around zero stress, cycling superimposed on a static preload and at constant deformation or constant load. The number of cycles to failure is recorded for each stress level. The data are generally presented in a plot of log stress versus log cycles called an S-N curve. Significant differences in the S-N curve can be produced by testing at different frequencies, different mean stresses, different waveforms, and different test methods, i.e., tension rather than bending. Therefore, appropriate test conditions should be reported with the curve.

Although fatigue test data give some indication of the relative ability of plastic materials to survive failure, the designer must be aware of the above variables. The tests are run on specially prepared samples in a test environment which never resembles the actual loading and environment of the actual parts. Therefore, it is essential that tests be run on actual injection molded parts under end use operating conditions to determine the true fatigue endurance of any part subjected to cyclic loading.

Nevertheless fatigue test data resembling end use conditions are helpful in understanding plastics fatigue performance, ranking materials and qualitatively guiding design. Fatigue data are also useful in measuring the effects of the many variables that affect fatigue performance of plastics such as frequency, temperature and loading conditions.

Fatigue Crack Initiation and Fatigue Crack Growth

Heterogenities inherent in the microstructure of most materials result in a random field of defects whose geometry, size, and orientation are also random. Such a random field of defects, influenced by the imposed stress, gives rise to a complex process of growth and interaction of defects, which ultimately leads to the initiation of macroscopic cracks. A crack propagates first in a stable manner to a stage at which it undergoes a transition to unstable (uncontrolled) propagation. Although failure as a result of externally applied forces is essentially a continuous process from beginning to actual, phenomenologically it can be split into two main phases, crack initiation and crack propagation.. Depending on the severity of defects, crack initiation may comprise 20 to 80% of the total lifetime. Therefore, the fundamental importance of crack initiation cannot be overemphasized and the exclusive examination of crack propagation, and a comparison of material fatigue resistance simply on this count, can be misleading. The reason the crack propagation is often of particular interest is because it responds to mathematical treatment and analysis.[355, 462]

When compared on a total time under load basis, materials under intermittently or cyclically applied loads will fail sooner than materials under continuously applied loads. An explanation for this is that under cyclic or intermittent loading, the plastic deformation formed at the tip of the crack relaxes during unload periods. By this mechanism, the growth of plastic deformation is controlled and the onset of premature failure is accelerated. Even at such a low frequency as that of diurnal application of the load, reductions occur in load bearing lifetime.[355]

Ductile to Brittle Transition

Under continuously applied loads, failure is controlled by two main mechanisms, yielding and cracking. If the yield processes are dominant, then ductile failure occurs. At some stage the influence of cracking begins to take over and brittle fracture intervenes.

A dramatic reduction in fatigue resistance coincides with this transition from ductile failure to brittle fracture. The position of this transition will be accelerated to lower critical cycles by: lower frequencies, higher temperatures, stress concentrations, chemical attack, environmental stress cracking, thermal degradation, thicker products (suppression of yielding at crack tips), larger products (high levels of strain energy for crack propagation) and complex stress fields.[414] Although the position on the time scale and the severity of the transition depends on these factors, it is important to note that it is materials based.[355]

It should be noted that much fatigue data is shown with a minimum of 100,000 cycles. Since the ductile to brittle transition usually occurs at less than 100,000 cycles, representation of this dramatic transition is not reflected in much of the most easily obtained data. Furthermore, much fatigue data is generated using high test frequencies (i.e. 30 Hz) and high stress amplitudes. This causes the material to fail through thermal degradation. Therefore, often failures at less than 100,000 cycles can be legitimately eliminated as not being true fatigue failures.[414]

One explanation for testing at higher frequencies is that it is cheaper. Another explanation may be that ASTM D671 - the test method commonly employed for fatigue testing of plastics is inconvenient. The apparatus used in this test is designed to operate at 30 Hz and requires major modifications and efforts for operation at other frequencies.

Factors Affecting Fatigue Performance

Depending on the stress amplitude and the frequency of load application, fatigue failure of polymers can occur by two means, thermal fatigue failure and mechanical fatigue failure. Thermal fatigue failure involves thermal softening (or yielding), which precedes crack propagation leading to ultimate failure. This mechanism dominates in certain materials at large stress amplitudes within a particular range of frequency of load applications. At a lower stress amplitude, on the other hand, a conventional form of fatigue crack propagation (FCP) mechanism is generally observed. Low frequency is also found to cause fatigue fracture by conventional crack propagation at high stress amplitude.[462]

Polymer fatigue behavior is generally sensitive to temperature, frequency, and environment, as well as molecular weight, molecular weight density and aging. S-N curves that do not account for these effects should be used with caution.[463] Fatigue resistance will also be affected by processing history since ease of crack initiation depends upon whether flaws or imperfections are present in the material. Good particulate dispersion or avoidance of residual stresses/ strains by correct processing effectively improves the resistance to fatigue failure of the resulting components.[355]

The high damping and low thermal conductivity of polymers cause a strong dependency of temperature rise on the rate of load application (frequency) and on the deformation level (stress or strain amplitude). From a thermodynamic point of view, part of the mechanical work done during cyclic loading is spent on irreversible molecular processes, leading to microscopic deformations such as crazes, shear bands, voids, and microcracks. The other part of the mechanical work evolves as heat. Both processes are obviously interdependent. [462]

The effect of moisture on fatigue of plastics has not been widely studied. In the few studies reported, the effect of absorbed moisture is not uniformly bad, but in fact sometimes acts to improve performance. This was true in the case of polycarbonate and polyester materials. However, the effects of moisture must be taken into consideration with hygroscopic polymers such as nylon. Absorbed moisture plasticizes the nylon and may weaken the fiber-resin bond resulting in lowered fatigue endurance limits.

Sinusoidal waveforms are commonly used in plastics fatigue testing in emulation of the practices for metals testing. However, other waveforms are being increasingly used ; the most useful are square waves and repeated gate functions, the response to which can readily be linked to the creep rupture data. (Dynamic fatigue data using sinusoidal waveforms cannot be directly compared because if the excitation is sinusoidal, neither the stress amplitude nor the time under load can be converted quantitatively.)[464]

In the case where the mean stress $\neq 0$, the presence of such a stress can well dominate the strain history resulting in accumulated increasing creep strain as the number of cycles increases. As a result, the nature of failure is also affected.

Important Points about Fatigue in Plastics [355, 414, 462, 464]

- An intermittent stress or a periodic stress usually constitutes a more severe condition than a continuous stress of the same magnitude.

- Sharp notches or any equivalent feature should never be stressed intermittently at levels greater than the fatigue limit.

- Low frequencies and, particularly, long rest periods often induce brittle failure.

- High stress amplitudes and high frequencies can cause thermal failure.

- In general, crystalline polymers are less prone to fatigue than amorphous polymers.

- Polymers which are resistant to failure under impact loads, are not necessarily also resistant to fatigue. The mechanisms involved in crack initiation are not the same for very short term and very long-term failure.

- Fatigue crack initiation is dominated by time under load, while fatigue crack propagation is controlled by the number of cycles.

- Failures most often occur at very low frequencies and after less than 50,000 cycles.

- It is common for fatigue lifetime data from well-controlled samples to spread over a few orders of magnitude. This, in fact, reflects the complex nature of the fracture processes involved.

Fatigue Crack Propagation

Fatigue Crack Propagation (FCP), based on the principles of fracture mechanics, is an alternative approach to fatigue testing. The analysis assumes that all real structures contain one or more sharp cracks, either as a result of fabrication defects or material flaws. Thus the problem becomes one of defining exactly the level of stress that may be applied to this flaw before it becomes a moving brittle crack.

The FCP test begins with a notched, or precracked, sample. The sample is subjected to repeated loading and the macroscopic crack propagation is monitored. Typical data obtained from an FCP test show that crack length, *a*, advances with the number of cycles, *N*. The rate the crack grows per cycle, *da/DN*, is determined from this raw data. The rate of growth becomes increasingly greater as the crack grows longer. Also, for a given number of cycles, a higher level of stress produces a higher rate of growth. The crack growth rate is therefore a function of the crack length and the applied stress (or stress amplitude). Several relationships were postulated to describe the crack propagation in metals, applying fracture mechanics principles. The model proposed by Paris has gained the widest acceptance.

Paris found that the elastic stress field in the vicinity of a crack tip can be described by a single term parameter designated as the stress intensity factor, *K*. It is a function of the flaw geometry and the nominal stress acting in the region in which the flaw resides. Therefore, if the relationship between the stress intensity factor, *K*, and the pertinent external variables (applied stress and flaw size) are known for a given structural geometry containing a particular defect, the stress intensity in the region of the crack tip can be established from knowledge of the applied stress and flaw size alone. Since fatigue is cyclic loading, there is a maximum stress and a minimum stress applied to the sample. Likewise, there is a K_{max} and a K_{min}, and the difference, ΔK, is the stress intensity factor range. The Paris model describes the relationship between crack growth rate and the stress intensity factor range as:

$$\frac{\mathbf{da}}{\mathbf{DN}} = A(\Delta K)^m$$

where *da/dN* is the cyclic crack growth rate, ΔK is the stress intensity factor range, and *A* and *m* are material and loading dependent constants.

As shown in the diagram, typical FCP behavior falls into three distinct regions - slow crack growth, stable crack growth and rapid, unstable crack growth. It is the stable crack growth or linear portion of the FCP curve which is used to interpret the fatigue resistance. A decrease in the slope of the linear region indicates improved fatigue resistance (decreased crack growth rate at the same stress intensity level). Similarly, a shift in the curve to the right indicates improved fatigue resistance (higher stress intensity to promote the same rate of crack growth).

A critical value of the stress intensity factor, conventionally designated K_c, can be used to define the critical crack tip stress condition for failure. The critical stress intensity factor for fracture instability is designated as K_{Ic}..[465]

Commonly used geometries include single edge notched (SEN) and compact tension (CT) specimens. Although a variety of loading cycles may be applied, it is common to study FCP under tension loading programs of different waveforms, such as sinusoidal, triangular, or rectangular. The majority of FCP experiments, however, are conducted under tensile sinusoidal loads. The frequency of load applications, the load amplitude, and the stress level determined by its maximum or mean values represent the basic loading variables.[462]

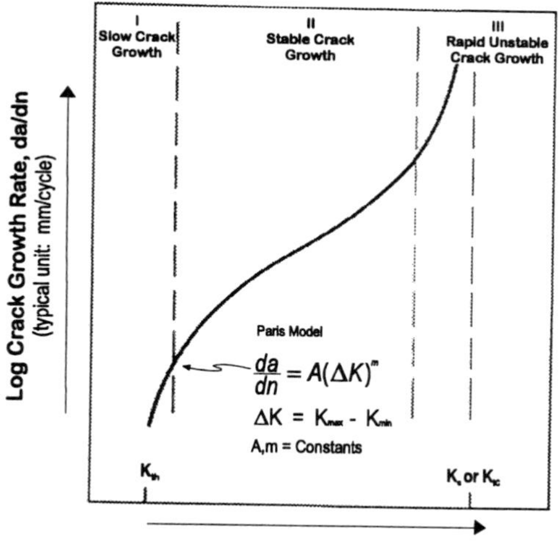

This diagram shows the three distinct regions of typical fatigue crack propagation (FCP) behavior. ***Region I***: Region I starts with a threshold value of the stress intensity factor range (ΔK_{th}), below which propagation of the crack is not observed. The value of ΔK_{th} has been attributed to the attainment of a sufficient level of activity in the notch tip region to cause its propagation. The initial slope of region I is usually very steep. ***Region II***: As the crack becomes longer, that is, as ΔK becomes larger, reduced crack acceleration occurs, leading to region II. The FCP curve is effectively linear in region II in the majority of cases. This commonly observed linearity of the FCP rate within region II promoted the general acceptance of the Paris model to describe the phenomenon. A lack of linearity in some polymers is immediately obvious when the test is conducted over a wide range of ΔK. Nevertheless, the Paris model can still be used to evaluate the relative resistance of materials to FCP. this is achieved by examining the rate of FCP at a particular value of ΔK. The higher the *da/dN*, the lower the FCP resistance. Alternatively, the higher the ΔK for a particular *da/dN*, the more resistant the material is supposed to be. ***Region III***: The rate of FCP approaches its asymptotic value at $K = K_c$, where a transition from a stable condition to rapid unstable crack growth occurs.

TRIBOLOGICAL PROPERTIES

Tribology is the science that deals with the design, friction, wear, and lubrication of interacting surfaces in relative motion. All materials used in bearing applications must have a suitable combination of mechanical and tribological properties under the conditions experienced in use. The three main performance areas that need to be examined in the study of tribology are wear, friction and limiting PV. An ideal bearing material will have a very low coefficient of friction and wear factor (or rate) and a very high limiting PV over a wide range of operating conditions.[392]

The phenomenon of wear occurs when two solid surfaces in contact are sliding relative to each other. It is sometimes common to call erosion of the surface of interest "wear" and erosion of the counter surface "abrasion" or "galling." Wear processes have been classified into four principle categories: abrasion, adhesion, surface fatigue and tribochemical wear. Adhesive wear occurs when opposing/ mating surfaces slide against each other, and fragments of one surface pull off and adhere to the other. In a lubricated material, these fragments form a very fine powder on the mating surface, indicating proper wear. Abrasive wear occurs when harder surfaces scrape or abrade away the mating part. This type of wear is characterized by grooves or gouges cut into the part surface. Fatigue and pressure wear occur when the pressure and/or velocity exceed the capabilities of the mating surfaces, resulting in a melted, deformed, embrittled or fractured part. Tribochemical wear deals with wear in which chemical reaction with the environment is significant. Combined, these effects lead to rapid part failure.[457]

The primary wear mechanism for thermoplastics is adhesive wear. Adhesive wear is characterized by the formation of a coherent transferred layer on the mating counterpart. The adhesive forces between the polymer and the counterpart are sufficient to inhibit sliding at the original interface. The adhesive junctions which form at the real points of contact rupture within the polymer itself and a layer of polymer is deposited on the counterpart. The counterpart topography can become effectively smoother leading to a reduction in the rate of wear. PTFE has been shown to be very effective at forming such a transfer film. The tribological properties of most neat polymers (without additives) are relatively poor since adhesion of most polymeric transfer layers to metal counterparts is very weak.[382]

In addition to wear, friction generates heat because work is done at the contact surface. If the thermoplastic is sensitive to temperature change (typified by low heat deflection temperature or low melting point), this frictional heating may cause the polymer to soften or even melt. This means the wear mode (thus the wear rate) and part shape are changed rapidly and the wear part can no longer function adequately due to large dimensional deformation. High friction coefficient is usually indicative of such a softening of the thermoplastic part.

If the wear particles are hard and abrasive (i.e. glass fiber), they will roll between the surfaces and scratch the counter surface, causing severe abrasion. This roller-bearing type action also creates a cooling effect because there is now a small air gap between the surfaces. Whether cooling helps lower the wear rate depends largely on which factors will dominate in this complicated happening.[454]

Different wear mechanisms can occur at the same time. This makes interpretation of laboratory wear data and the prediction of actual wear difficult. An understanding of wear dynamics as well as the development of materials that offer improved physical properties, tailored to the application, are necessary to combat wear. Formulated materials such as thermosets, thermoplastics, and ceramics are increasingly being used in cost effective industrial applications where wear resistance is a major concern.[387] By compounding wear additives with virtually any thermoplastic, it is possible to improve abrasion resistance and coefficients of friction, even in materials that already have good wear properties. Additives like polytetrafluoroethylene (PTFE), molybdenum disulfide, graphite powder, silicone fluid, glass fiber, carbon fiber and aramid fiber serve to make plastics self lubricating and reduce stress on mating parts.

Besides improved wear properties, cost effectiveness is another primary advantage in selecting wear resistant plastics over metals. The manufacturer can save money and energy by using an injection molded thermoplastic impeller rather than a machined metal impeller. Start-up torque is significantly lower compared to its heavier metal counterpart.[457]

Wear Factor

The wear resistance of a bearing operating at a temperature below its limit can be predicted from an experimentally determined wear factor, K. The wear factor is derived from an equation relating the volume of material removed by wear in a given time per unit of load and surface velocity.

$$\mathbf{W = KFVT}$$

where: W = wear volume, cm^3 (in^3); K = wear factor, cm^3-min/m-kg-h (in^3-min/ft-lb-hr); F = supported load, kg (lb); T = time, h; V = velocity, m/sec (ft/min or fpm). For flat surfaces the equation is modified so that:

$$\mathbf{X = KPVT \quad or \quad K = X/PVT}$$

where: X = wear depth, cm (in); P = pressure, MPa (psi).

The wear factor K, although shown here as a proportional constant, in fact does vary with the PV multiplier. But at a given PV condition, a lower wear rate factor also indicates lower wear rate. The PV multiplier also represents the work done per unit area per unit time at the contact surface. A part of this work is then transformed into heat as follows:

$$\textbf{Heat generation} = \mathbf{(PV) \times f}$$

where f is the coefficient of friction.

For a given wear system where heat transfer rate is defined, heat generation can be measured through temperature rise as follows:

$$\mathbf{\Delta T \sim (PV) \times f}$$

This equation indicates that in selecting a wear material, one should consider the temperature capability of the polymer.[454]

Once a K factor is established, it can be used to calculate wear rates of such components as bearings and gears. However, the engineer must bear in mind that the wear rate of the plastic is affected by test PV, plastic material finish, part geometry, ambient temperature, mating surface finish, mating surface hardness and mating surface thermal conductivity and a prototype should therefore be constructed and tested under actual end use conditions. Nonetheless, as a relative measure of one material versus another under the same operating conditions, K factors have proved to be highly reliable.[453]

Typically materials are ranked based on their equilibrium wear rates. These numbers are very important in selecting materials for specific tribological applications. However, the behavior of a material prior to the establishment of equilibrium or steady state can also reveal valuable mechanistic information about wearing systems. The time necessary to achieve equilibrium wear, and whether the wear rate increases or decreases prior to equilibrium are important.[382]

Coefficient of Friction

Friction is a force of resistance to the relative motion of two contacting surfaces, so that a low coefficient of friction is required for a bearing material. There are two coefficients of friction. In a stationary specimen, the static coefficient of friction applies: The static coefficient of friction is the force required to create motion divided by the force pressing mating surfaces together. With a specimen in motion, the dynamic coefficient of friction applies: The dynamic coefficient of friction is the force required to sustain motion at a specified surface velocity divided by the force pressing mating surfaces together. The lower the coefficient of friction, the easier the two surfaces slide over each other. The coefficient of friction varies with applied load, velocity and temperature.

Frictional properties of plastics differ markedly from those of metals. The rigidity of even the highly reinforced resins is low compared to that of metals; therefore, plastics do not behave according to the classic laws of friction. Metal to plastic friction is characterized by adhesion and deformation of the plastic, resulting in frictional forces that are proportional to velocity rather than load. In thermoplastics, friction actually decreases as load increases.

A unique characteristic of most thermoplastics is that the static coefficient of friction is typically less than the dynamic coefficient of friction. This accounts for the slip/ stick sliding motion associated with many plastics on metal and with plastics on plastics.[457]

PV Limit

In addition to the wear factor and coefficient of friction, another key parameter that is often used to select a material for parts requiring excellent resistance to the effects of wear is the PV limit. By definition, the PV limit is simply a PV multiplier above which the material can no longer function as a wear part due to softening, melting, and deformation. But in reality, the PV limit remains more a concept than a clear cut number that one can determine experimentally.[454]

The PV limit for a material is the product of limiting bearing pressure MPa (psi) and peripheral velocity m/min (fpm), or bearing pressure and limiting velocity, in a given dynamic system. It describes a critical, easily recognizable change in the bearing performance of the material in the given system. When the PV limit is exceeded, one of the following manifestations may occur: (1) melting, (2) cold flow or creep, (3) unstable friction, (4) transition from mild to severe wear. PV limit is generally related to rubbing surface temperature limit. As such, PV limit decreases with increasing ambient temperature. The PV limits determined on any given tester geometry and ambient temperature can rank materials, but translation of test PV limits to other geometries is difficult.

For a given bearing application, the product of pressure and velocity (PV) is independent of the bearing material. Wear is dependent on PV for any material.

The use of experimentally determined PV limits in specific applications should be considered approximate, since all pertinent factors are not easily defined. This means that a generous safety factor is an important consideration in bearing design. As long as the mechanical strength of the bearing material is not exceeded, the temperature of the bearing surface is generally the most important factor in determining PV limit. Factors known to affect PV limits are: (1) absolute pressure, (2) velocity, (3) lubrication, (4) ambient temperature, (5) clearance, (6) type of mating materials and (7) surface roughness.[347]

A high value of limiting PV (LPV) demonstrates an ability to operate under high applied loads and surface velocities. In general, when the operating conditions exceed approximately one-half of the LPV, wear accelerates rapidly. Thus, the design PV for a composite may be determined by dividing the LPV by two.[453]

Thrust Washer Test

The most common means of evaluating the tribological characteristics of plastics is the thrust washer testing machine or wear tester. Normally it is operated according to ASTM test D3702. The wear tester rotates the plastic test specimen against a thrust washer. In these tests, the stationary specimen (wear ring) is mounted in an antifriction bearing equipped with a torque arm. The moving specimen (thrust washer), which is mounted in the upper sample holder, bears against the stationary specimen. The tests, to be considered valid, are run until an equilibrium condition is reached.[453] The thrust washer test generates wear information for a material based on area contact, not line or point contact as needed in some bearing applications.

The thrust washer test is capable of generating wear data for plastic against metal, plastic against plastic, or against virtually any mating surface. The equipment generates data from which wear factors, coefficients of friction and PV ratings can be calculated. Testing can be done for a wide temperature range and/ or submerged in various fluids.[457]

Wear Resistant Additives

Even among plastic materials with excellent natural lubricity, wear characteristics between two thermoplastics differ greatly. When an application calls for plastic on plastic, dissimilar polymers should be used and incorporated with one or more wear resistant additives. Reinforcements such as glass, carbon, and aramid fibers enhance wear resistance by increasing the thermal conductivity and creep resistance, thus improving the LPV and working PV of the part.[457]

Polytetrafluoroethylene (PTFE) has the lowest coefficient of friction of any internal lubricant. Its particles shear during operation to form a lubricous film on the part surface. Often referred to as the best lubricant for metal mating surfaces, PTFE modifies the mating surface after an initial break-in period. PTFE goes an extra step in lessening wear and fatigue failure by actually cushioning shock. What is most important about PTFE is its distribution throughout the thermoplastic compound. PTFE has a typical optimum loading of 15% in amorphous thermoplastic resins and 20% in crystalline resins. However, there is a price performance limit at which PTFE can actually begin to demonstrate diminishing returns.

Molybdenum disulfide (MoS_2), otherwise known as "moly," is a solid lubricant usually used in nylon composites to reduce wear rates and increase PV limits. Acting as a nucleating agent, MoS_2 creates a better wearing surface by changing the structure of nylons to become more crystalline, creating a harder and more wear resistant surface.. MoS_2 will not lower the coefficients of friction like other modifiers, and its use is therefore confined to nylons where it has this crystallizing effect on the nylon molecular structure.

MoS_2 also has a high affinity for metal. Once attracted to the metal, it fills the metal's microscopic pores, making the metal surface slippery. This makes MoS_2 the ideal lubricant for applications in which nylon wears against metal, such as industrial bushings, cam components and ball joints. Two added benefits occur during molding: fast injection molding times which lower per part costs; and lower, uniform shrinkage.

Graphite's unique chemical lattice structure allows its molecules to slide easily over one another with little friction. This is especially true in an aqueous environment and makes graphite powder an ideal lubricant for many underwater applications such as water meter housings, impellers and valve seals.

Silicone or Polysiloxane fluid is a migratory lubricant. A particular silicone fluid is chosen that is compatible enough with the base resin to allow compounding, yet incompatible enough to migrate to the surface of the compound to continuously regenerate the wear surface.

Silicone offers engineers several unique advantages based on its ability to be both a boundary lubricant and an alloying partner with the base resin. Silicone acts as a boundary lubricant because silicone moves or migrates to the surface of a part over time, by both diffusion as a result of random molecular movement, and by its exclusion from the resin matrix which is a result of migration. As a partial alloying material with the base resin, silicone remains in the component over its service lifetime, but because silicone is incompatible enough, the silicone is constantly moving from the matrix to the surface. This continuous secretion eases friction and wear at start-up and when high speed lubricity is necessary. Silicone is excellent for start-up, high speed, and low pressure wear applications such as keyboard keycap receptacles and high speed printer components.

Silicone fluid is available in a wide range of viscosities. The lower the viscosity, the more fluid the additive is, and the quicker it will migrate to the surface and provide lubrication. This is particularly important in wear applications that require numerous start and stop actions. However, if the additive's viscosity is too low, the silicone can vaporize during processing, or migrate too quickly from the molded part.

Silicone and PTFE will work together to create a high temperature grease which will create better wearing characteristics and lower friction, particularly at high speeds and during start-ups. When used together, PTFE acts as a thickening agent as well as an extreme pressure additive to make the grease at the surface. Because the silicone is constantly moving to the surface, this provides the added lubricity necessary during start-ups and at high speeds. Since failure at high speeds is more dependent on wear than failure at low speeds, and because the benefits of the silicone/PTFE synergy are most evident at these higher speeds, this combination should not be considered for low-speed components. In these cases, usually a PTFE-only compound is needed.

Glass fibers are mainly added to resins to improve both short term mechanical and thermal performance properties, particularly strength, creep resistance, hardness, and heat distortion. Wear resistance can also be improved with the addition of glass fibers, but the improvement is directly correlated to the efficiency of the glass sizing system which bonds the resins and fibers together. Glass reinforcement results in a marked improvement of the resins limiting PV by enhancing creep resistance, thermal conductivity, and heat distortion.

Glass fiber reinforcement often leads to increased coefficient of friction and mating surface wear. This can be ameliorated with the addition of an internal lubricant.

Carbon fibers are added to engineering resins to produce high strength, heat distortion temperatures, and modulus as well as creep and fatigue resistance. Often referred to as the perfect additive for wear and friction resins, carbon fibers also greatly increase thermal conductivity and lower coefficients of friction and wear rates. In fact, the strengthened compound may have lower friction coefficients than the base resin.

Carbon fibers should be considered as replacements or alternatives for glass fiber when wear and friction are not sufficiently addressed in glass fiber reinforced components. Unlike glass, carbon is a softer and less abrasive fiber. It will not score the surface of iron or steel. Most resins which are reinforced with 10% or more of carbon fibers will dissipate static electricity and overcome problems with static buildup on moving parts. This can be extremely important for business machine, textile equipment and other electronic components.

Aromatic Polyamide Fiber, commonly known as aramid fiber or Kevlar, is one of the latest wear resistant additives to be used in thermoplastic composites. Unlike the traditional fiber reinforcements of glass and carbon, aramid is the softest and least abrasive fiber. This is a major advantage in wear applications, particularly if the mating surface is sensitive to abrasion.

Chapter 1

Acetal Resin

Fatigue Properties

DuPont: Delrin 100 (features: surface lubricity, low flow; melt flow index: 1.0 g/10 min.); **Delrin 100ST** (features: high impact, low flow, surface lubricity; melt flow index: 1.0 g/10 min.); **Delrin 500** (features: general purpose grade, surface lubricity; melt flow index: 6.0 g/10 min.); **Delrin 500T** (features: impact modified, surface lubricity; melt flow index: 6.0 g/10 min.)

Delrin acetal resins have extremely high resistance to fatigue failure from -40°C to 82°C (-40°F to 180°F). Furthermore, their resistance to fatigue is little affected by water, solvents, neutral oils, and greases. For highest fatigue endurance select Delrin 100, 100ST and 500T. For example, in gear tests, Delrin 100 exhibits approximately 40% higher fatigue endurance than Delrin 500.

Reference: *Delrin Design Handbook For Du Pont Engineering Plastics,* supplier design guide (E-62619) - Du Pont Company, 1987.

GRAPH 01: Fatigue Cycles to Failure vs. Stress in Flexure for 30% Glass Fiber Reinforced LNP Thermocomp KF1006 Acetal Resin.

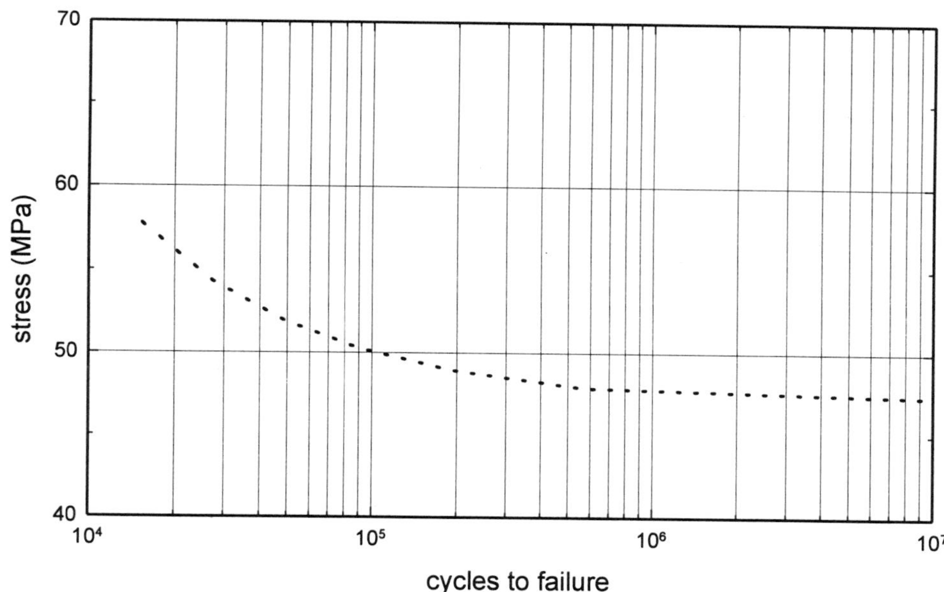

....	LNP Thermocomp KF1006 Acetal (30% glass fiber); flexure; 30 Hz; 23°C
Reference No.	394

GRAPH 02: Fatigue Cycles to Failure vs. Stress in Tension at Different Temperatures for DuPont Delrin 500 Acetal Resin.

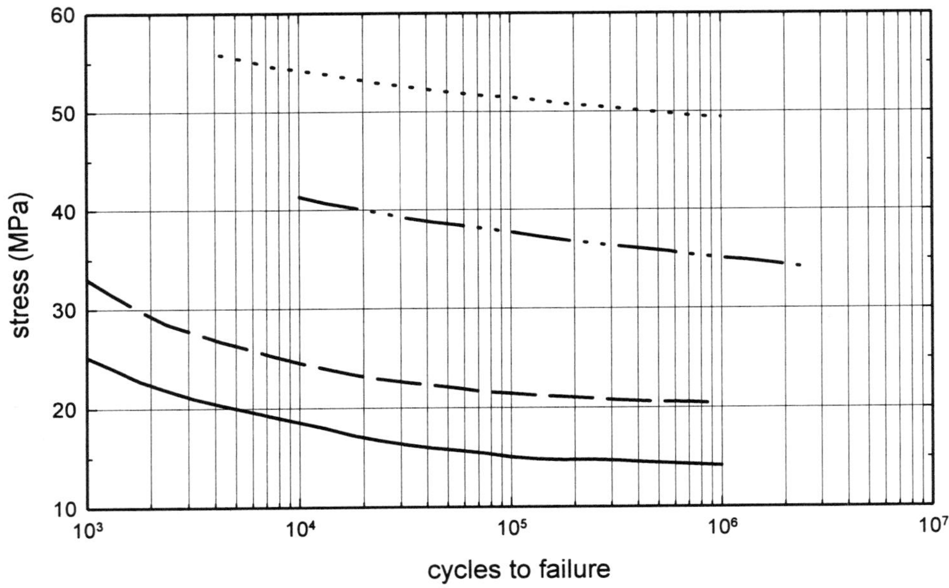

···············	DuPont Delrin 500 Acetal (gen. purp. grade, 94 Rockwell M, surface lubricity, 120 Rockwell R; 6.0 g/10min. MFI); tension-zero; 30 Hz; 23°C [fig f]
—··—··—··	DuPont Delrin 500 Acetal; tension- compression; 30 Hz; 23°C [fig d]
— — — —	DuPont Delrin 500 Acetal tension- compression; 30 Hz; 66°C [fig d]
———	DuPont Delrin 500 Acetal tension- compression; 30 Hz; 100°C [fig d]
Reference No.	201

GRAPH 03: Fatigue Cycles to Failure vs. Stress at Different Frequencies for Acetal Resin.

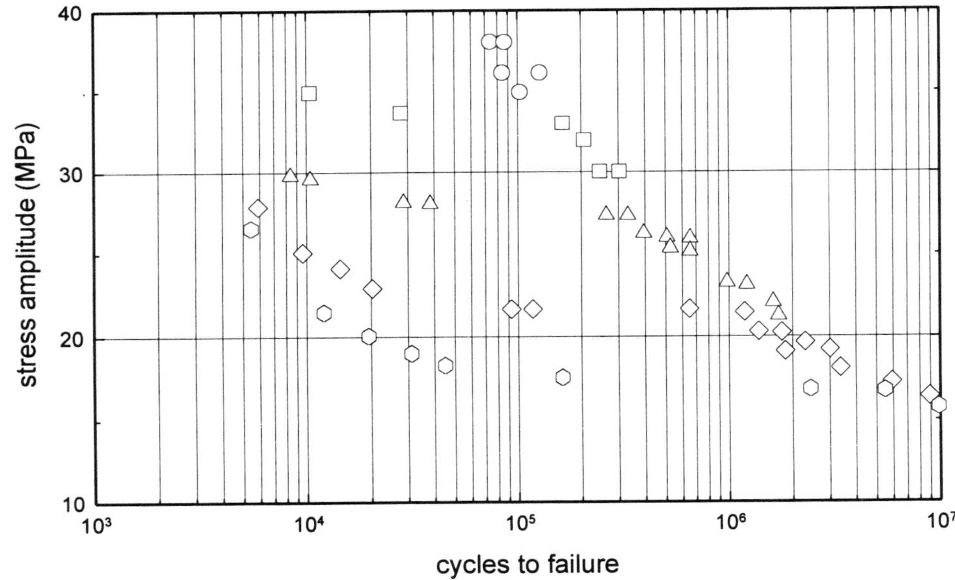

○	Acetal; tension-compression; 0.167 Hz; 23°C
□	Acetal; tension-compression; 0.5 Hz; 23°C
△	Acetal; tension-compression; 1.67 Hz; 23°C
◇	Acetal; tension-compression; 5.0 Hz; 23°C
⬡	Acetal; tension-compression; 10 Hz; 23°C
Reference No.	369

Chapter 2

Acetal Copolymer

Fatigue Properties

Hoechst Celanese: Celcon M90

Results show that polyacetal will either stress soften, or harden, with accompanying changes in the dynamic viscoelastic behavior and energy densities, depending on whether it is fatigued in the thermally or mechanically dominated regimes, respectively. The data suggest that the transition knee occurs at approximately 50 MPa above which the fatigue life is dominated by thermal effects and below which the life is dominated by other mechanisms. It should be mentioned that the location of this knee as well as the S-N data in the thermally dominated region are strongly affected by the frequency of the applied load and the size of the specimen tested. In contrast, the data in the mechanically dominated region are relatively insensitive to loading frequency and specimen size and represent the material's intrinsic response to fatigue loading. One mechanism presumed dominant in this regime is that which relates to fatigue crack growth in this material.

Reference: Lesser, Alan J., *High-Cycle Fatigue Behavior Of Engineering Thermoplastics,* ANTEC 1995, conference proceedings - Society of Plastics Engineers, 1995.

Hoechst Celanese: Celcon

Because of its highly crystalline polymer structure, Celcon acetal copolymer has exceptional resistance to tensile and flexural fatigue stress. However, fatigue from repeated loading can eventually cause failure of a part in service. It is important in designing gears, hinges, and other reciprocating mechanical parts to design for sufficient strength to prevent fatigue failure.

Under metal fatigue endurance limits, which are defined as stress at which specimens fail at 10^7 cycles, Celcon acetal copolymer can be said to have a flexural fatigue endurance limit of 22.7 MPa (3300 psi) at 23°C (73°F) and 50% RH with a cantilevered specimen. With center loaded beams, flexural fatigue is 29.6 MPa (4300 psi). Flexural fatigue data were obtained on a Budd fatigue tester. The tensile fatigue endurance limit of Celcon acetal copolymer at similar conditions is 28.9 MPa (4200 psi). Tensile fatigue life was determined on a Sonntag Universal Fatigue Tester.

Reference: *Celcon Acetal Copolymer,* supplier design guide (90-350 7.5M/490) - Hoechst Celanese Corporation, 1990.

GRAPH 04: **Fatigue Cycles to Failure vs. Stress in Flexure for Hoechst Hostaform Acetal Copolymer.**

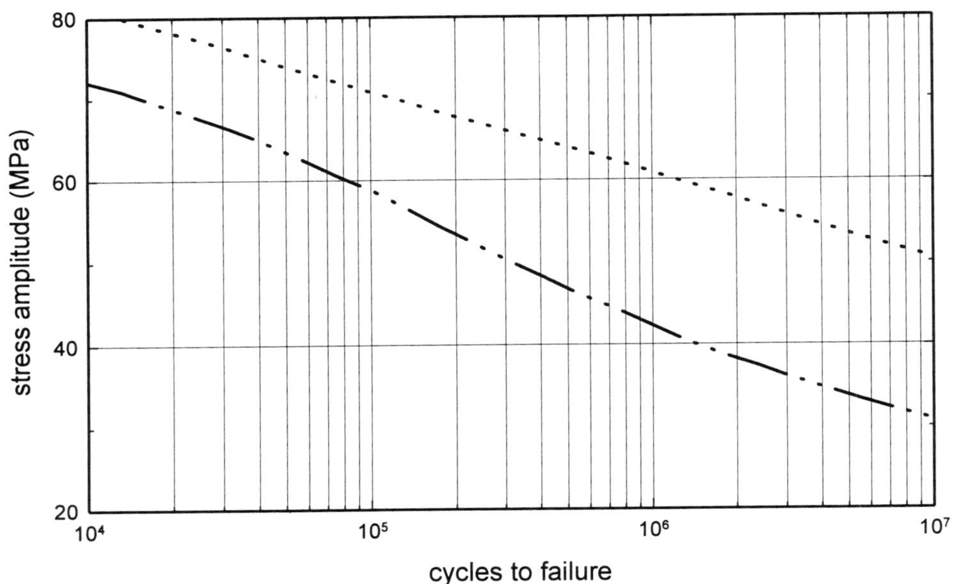

....	Hoechst AG Hostaform C9021GV1/30 Acetal Copol.; (glass reinforced; 5.0 g/10 min. MFI); flexure (alternating); 10 Hz; 23°C [fig d]
—·—·—	Hoechst AG Hostaform C9021 Acetal Copol. (gen. purp. grade; 9.0 g/10 min. MFI); flexure (alternating); 10 Hz; 23°C [fig d]
Reference No.	323

GRAPH 05: **Fatigue Cycles to Failure vs. Stress in Flexure for BASF Ultraform N2320 Acetal Copolymer.**

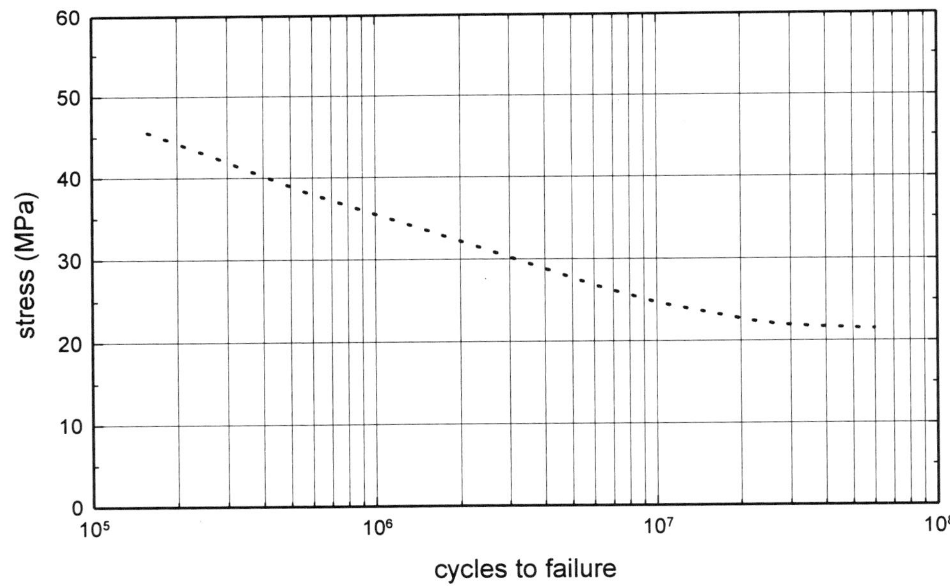

....	BASF AG Ultraform N2320 Acetal Copol. (moderate flow, gen. purp. grade); DIN 53442; 10 Hz; 23°C; 50% RH; laboratory atmosphere
Reference No.	181

GRAPH 06: **Fatigue Cycles to Failure vs. Stress in Flexure for Mitsubishi Gas Chemical Iupital Acetal Copolymer.**

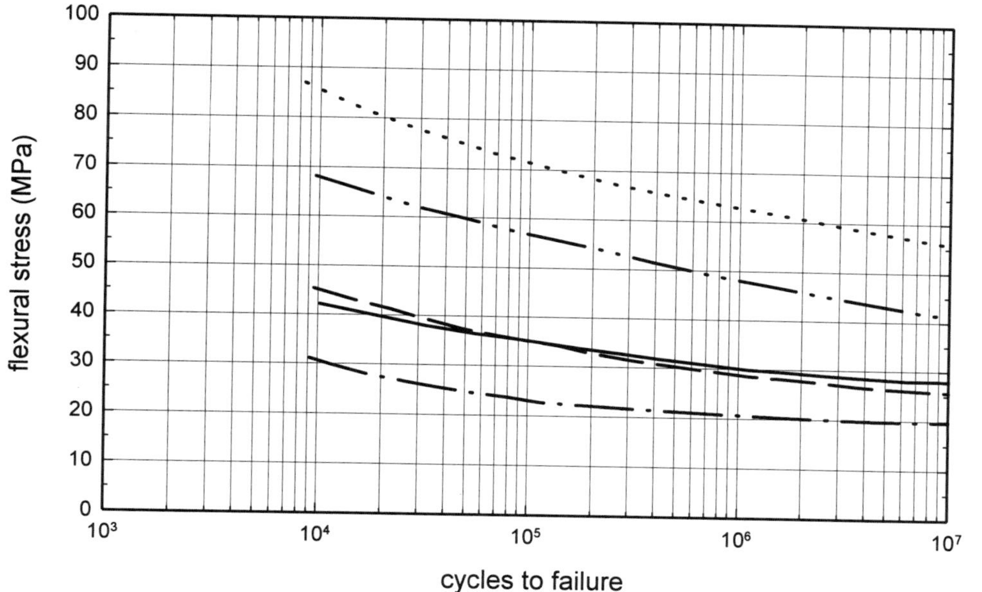

···············	Mitsubishi Iupital FC2020H Acetal Copol. (high modulus; 20% carbon fiber); flexure; ASTM D671-71B; 23°C
—··—··	Mitsubishi Iupital FG2025 Acetal Copol. (25% glass fiber); flexure; ASTM D671-71B; 23°C
— — —	Mitsubishi Iupital F30 Acetal Copol. (unfilled, mod.-high flow); flexure; ASTM D671-71B; 23°C
———	Mitsubishi Iupital F10 Acetal Copol. (low flow, unfilled); flexure; ASTM D671-71B; 23°C
—·—·—	Mitsubishi Iupital FL2020 Acetal Copol. (low COF; 20% fluorocarbon); flexure; ASTM D671-71B; 23°C
Reference No.	364

GRAPH 07: **Fatigue Cycles to Failure vs. Stress in Flexure for Hoechst Hostaform Acetal Copolymer.**

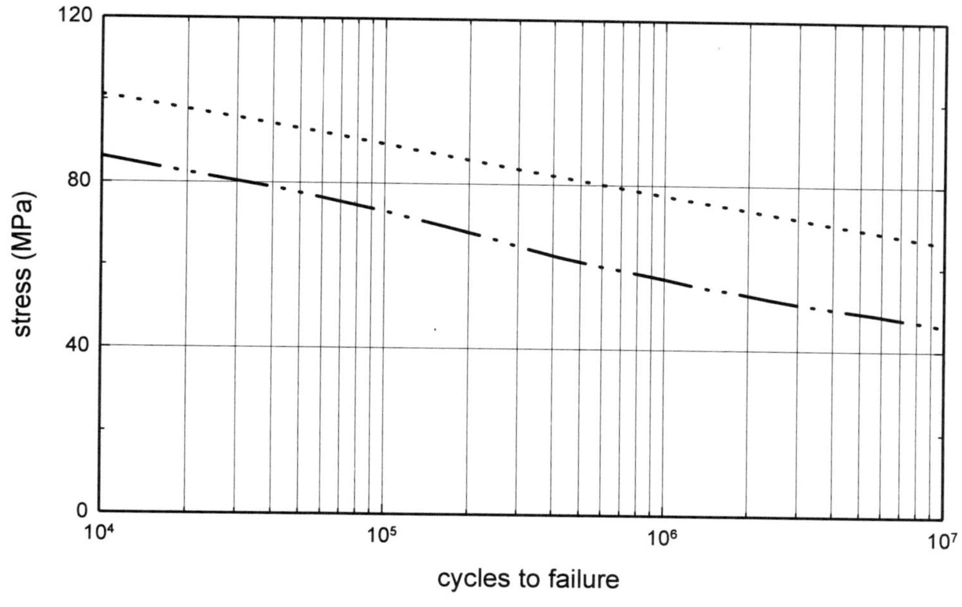

···············	Hoechst AG Hostaform C9021GV1/30 Acetal Copol. (glass reinforced; 5.0 g/10 min. MFI); flexure (pulsating); 10 Hz; 23°C [fig f]
—··—··	Hoechst AG Hostaform C9021 Acetal Copol. (gen. purp. grade; 9.0 g/10 min. MFI); flexure (pulsating); 10 Hz; 23°C [fig f]
Reference No.	323

GRAPH 08: **Fatigue Cycles to Failure vs. Stress in Flexure for General Purpose Hoechst Celanese Celcon M90 Acetal Copolymer.**

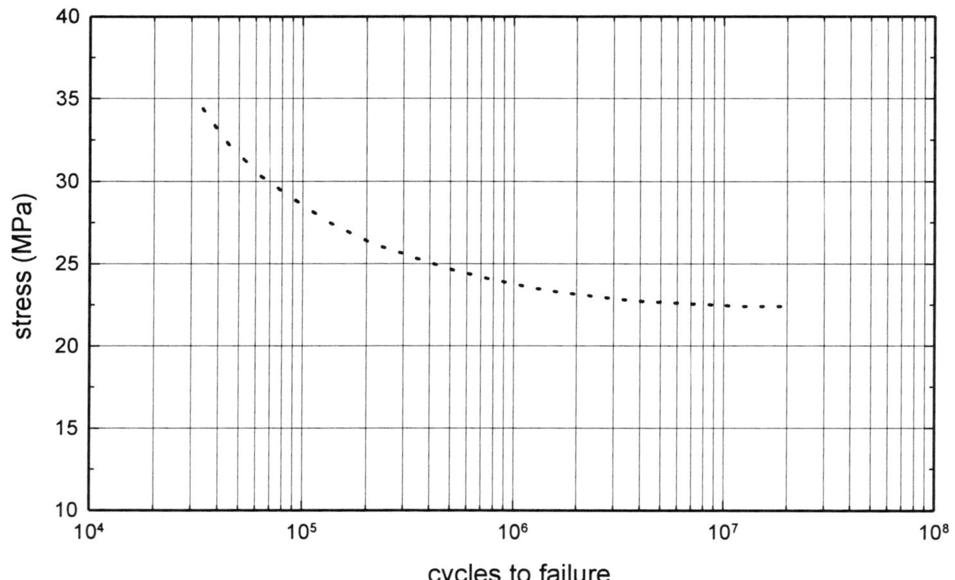

....	Hoechst Cel. Celcon M90 Acetal Copol. (gen. purp. grade, 80 Rockwell M; 9.0 g/10 min. MFI); flexure (alternating, cantilevered); ASTM D671-61; 30 Hz; 23°C [fig j]
Reference No.	210

GRAPH 09: **Fatigue Cycles to Failure vs. Stress in Tension for Hoechst Celanese Celcon M90 Acetal Copolymer**

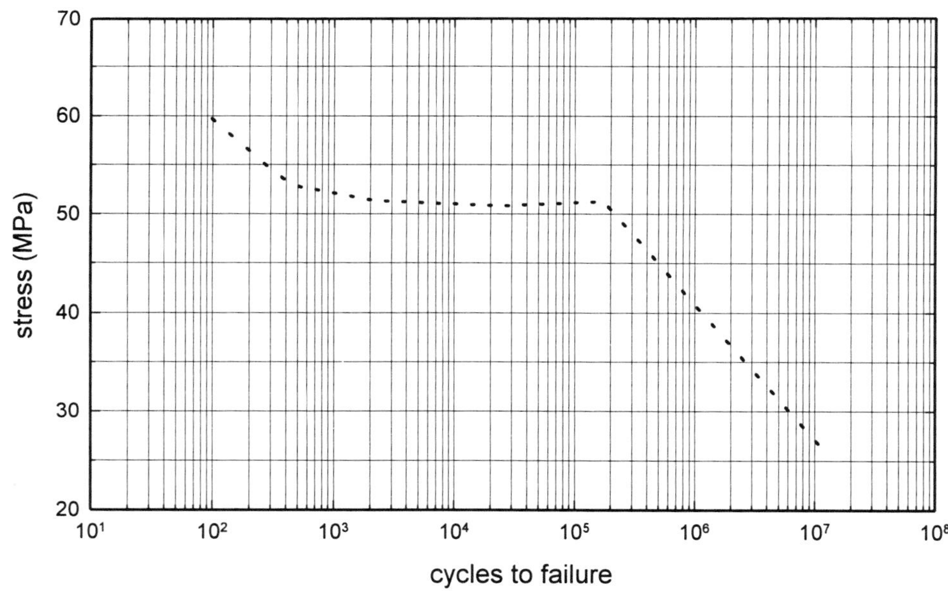

....	Hoechst Celanese Celcon M90; tension-tension; 2 Hz; 23°C; ASTM D638 Type I tensile bar
Reference No.	370

GRAPH 10: **Fatigue Cycles to Failure vs. Stress in Flexure at Low Test Frequency for Acetal Copolymer.**

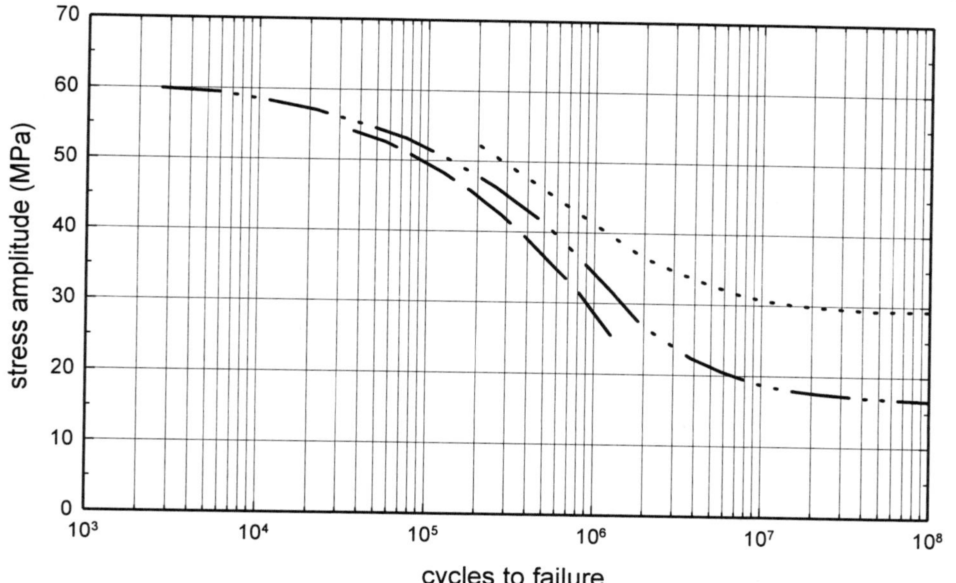

....	Acetal Copol. (high molecular weight, extrusion grade); flexure (alternating); 0.5 Hz; 20°C; mean stress = 0
—··—··	Acetal Copol. (inj. mold. grade, moderate mol. wgt.); flexure (alternating); 0.5 Hz; 20°C; mean stress = 0
— — — —	Acetal Copol. (inj. mold. grade, low molecular weight, high flow); flexure (alternating); 0.5 Hz; 20°C; mean stress = 0
Reference No.	355

GRAPH 11: **Fatigue Cycles to Failure vs. Stress in Flexure at Two Temperatures for General Purpose Hoechst Celanese Celcon M90 Acetal Copolymer.**

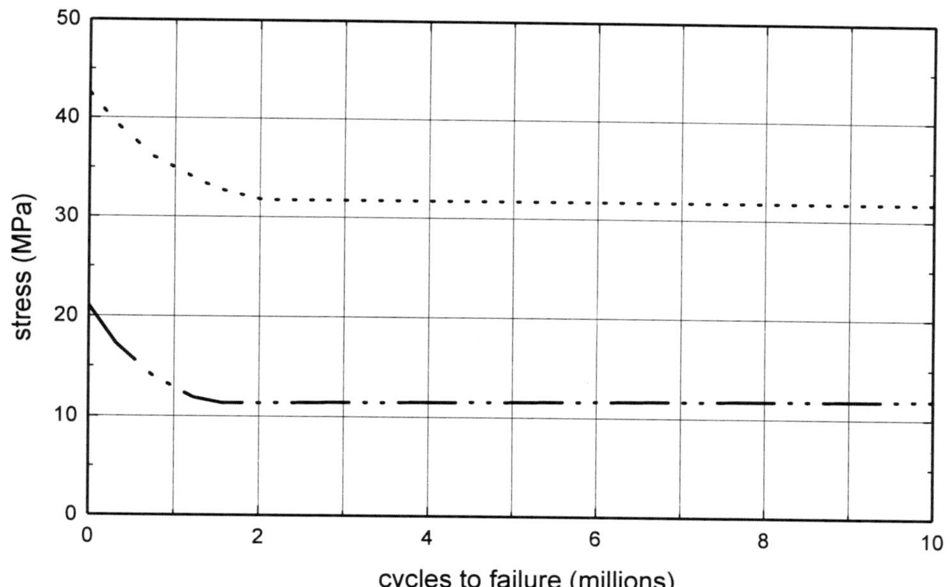

....	Hoechst Cel. Celcon M90 Acetal Copol. (gen. purp. grade, 80 Rockwell M; 9.0 g/10 min. MFI); flexure (center loaded); 23°C [fig k, m]
—··—··	Hoechst Cel. Celcon M90 Acetal Copol. (gen. purp. grade, 80 Rockwell M; 9.0 g/10 min. MFI); flexure (center loaded); 104°C [fig k, m]
Reference No.	210

GRAPH 12: **Fatigue Cycles to Failure vs. Stress in Flexure at Various Temperatures for 20% Carbon Fiber Filled Mitsubishi Gas Chemical Iupital Acetal Copolymer.**

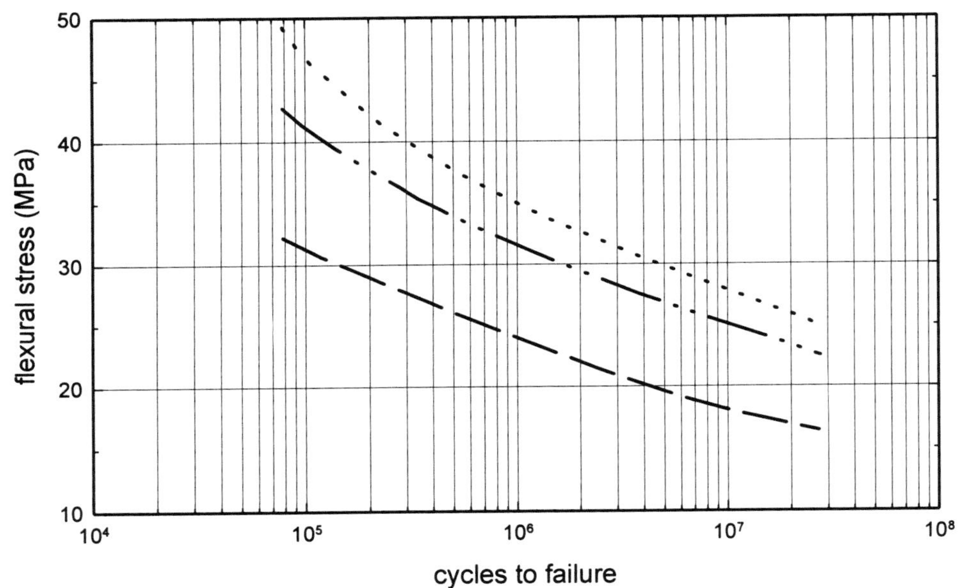

....	Mitsubishi Iupital FC2020H Acetal Copol. (high modulus; 20% carbon fiber); flexure; ASTM D671-71B; 23°C
—··—··	Mitsubishi Iupital FC2020H Acetal Copol. (high modulus; 20% carbon fiber); flexure; ASTM D671-71B; 50°C
– – – –	Mitsubishi Iupital FC2020H Acetal Copol. (high modulus; 20% carbon fiber); flexure; ASTM D671-71B; 70°C
Reference No.	364

GRAPH 13: **Fatigue Cycles to Failure vs. Stress in Tension for Hoechst Hostaform C9021 Acetal Copolymer.**

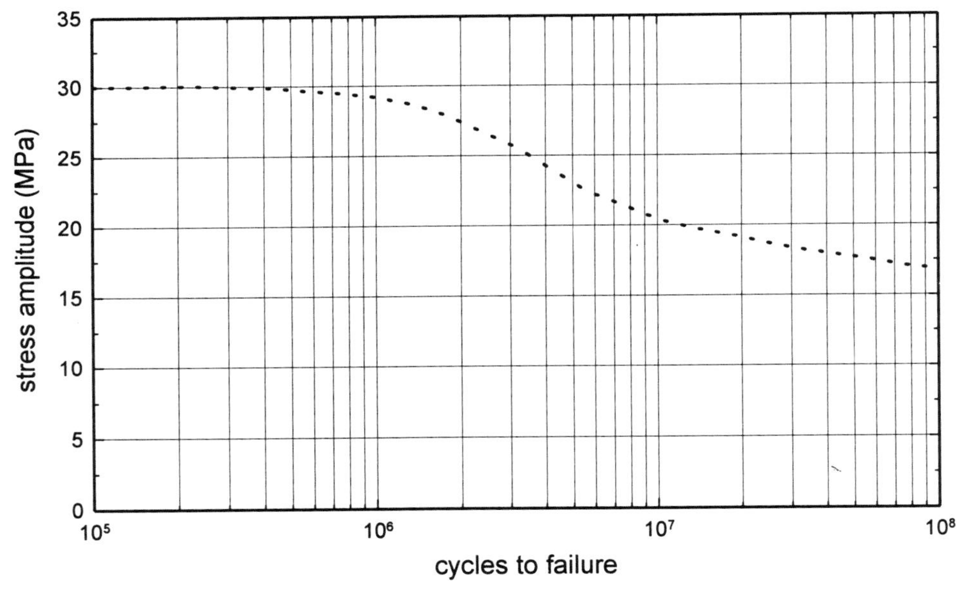

....	Hoechst AG Hostaform C9021 Acetal Copol. (gen. purp. grade; 9.0 g/10 min. MFI); tension-compression; DIN 53455; 10 Hz; 23°C; DIN 53455 specimen [fig d]
Reference No.	323

GRAPH 14: **Fatigue Cycles to Failure vs. Stress in Tension for Hoechst Hostaform C9021 Acetal Copolymer.**

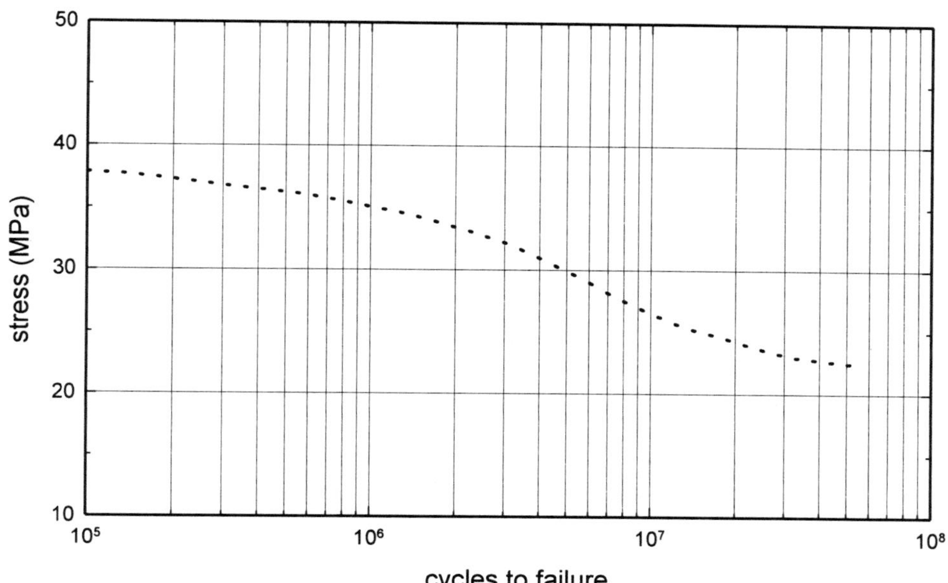

....	Hoechst AG Hostaform C9021 Acetal Copol. (gen. purp. grade; 9.0 g/10 min. MFI); tension- zero; DIN 53455; 10 Hz; 23°C; DIN 53455 specimen [fig f]
Reference No.	323

GRAPH 15: **Fatigue Cycles to Failure vs. Stress in Tension at Two Temperatures for General Purpose Hoechst Celanese Celcon M90 Acetal Copolymer.**

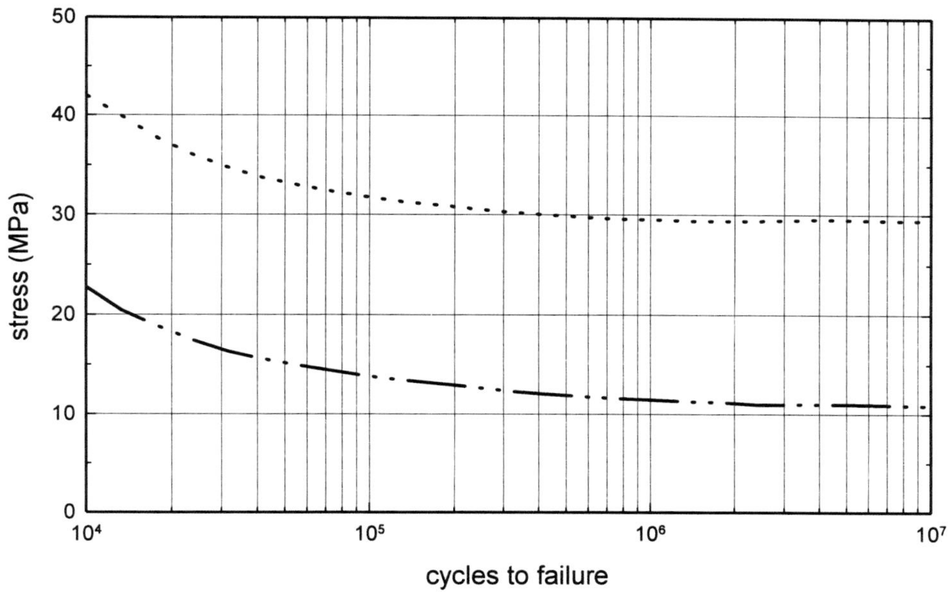

....	Hoechst Cel. Celcon M90 Acetal Copol. (gen. purp. grade, 80 Rockwell M; 9.0 g/10 min. MFI); tension; 50% RH; 30 Hz; 23°C; [fig o]
—··— ··	Hoechst Cel. Celcon M90 Acetal Copol. (gen. purp. grade, 80 Rockwell M; 9.0 g/10 min. MFI); tension; 30 Hz; 104°C; [fig o]
Reference No.	210

GRAPH 16: **Fatigue Cycles to Failure vs. Stress in Torsion for Hoechst Hostaform C9021 Acetal Copolymer.**

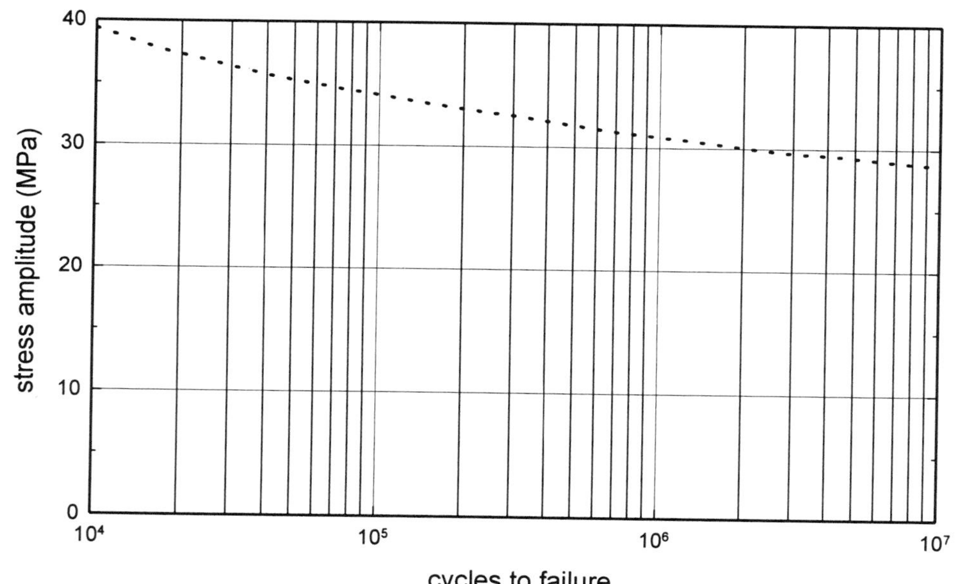

....	Hoechst AG Hostaform C9021 Acetal Copol. (gen. purp. grade; 9.0 g/10 min. MFI); torsion (alternating); 10 Hz; 23°C [fig d]
Reference No.	323

GRAPH 17: **Fatigue Cycles to Failure vs. Stress in Torsion for Hoechst Hostaform C9021 Acetal Copolymer.**

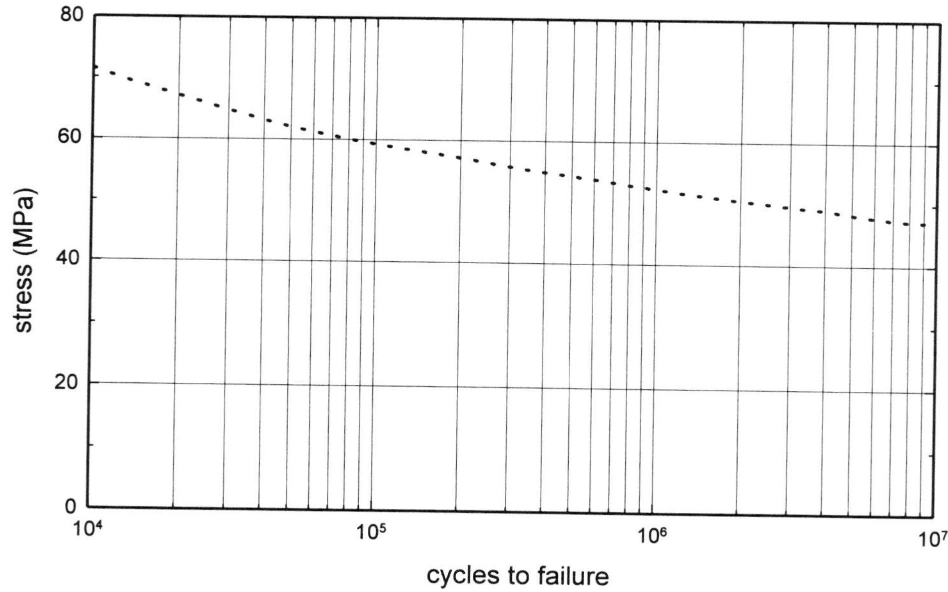

....	Hoechst AG Hostaform C9021 Acetal Copol. (gen. purp. grade; 144 (358/30) Ball ind. hardness; 9.0 g/10 min. MFI); torsion (pulsating); 10 Hz; 23°C [fig f]
Reference No.	323

Chapter 3

Polymethyl Methacrylate

GRAPH 18: **Fatigue Cycles to Failure vs. Stress in Flexure for Polymethyl Methacrylate.**

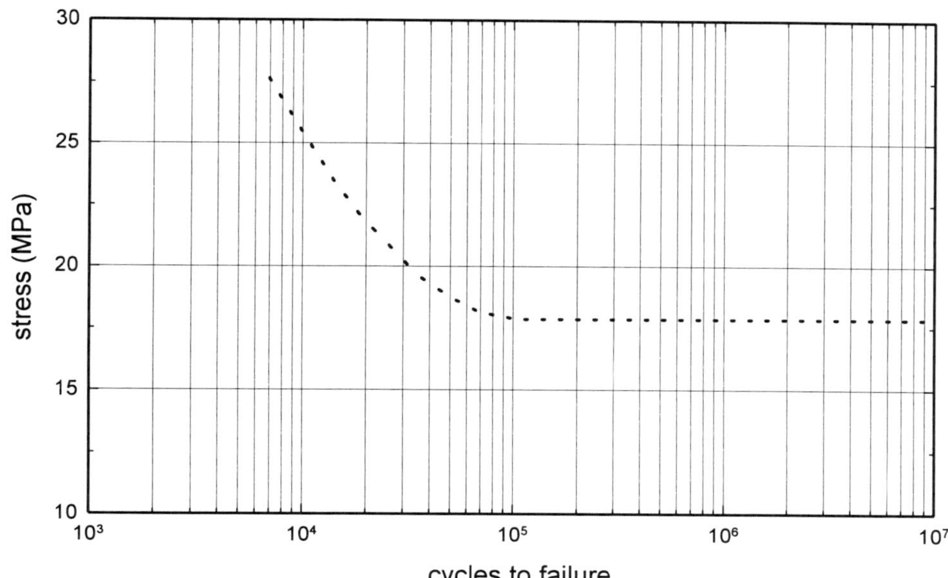

···············	PMMA (6.4 mm thick); flexure (alternating); 30 Hz; 23°C; mean stress= 0
Reference No.	365

GRAPH 19: **Fatigue Cycles to Failure vs. Stress in Tension for Bulk Cast Polymethyl Methacrylate Showing the Effect of Notch Tip Radius.**

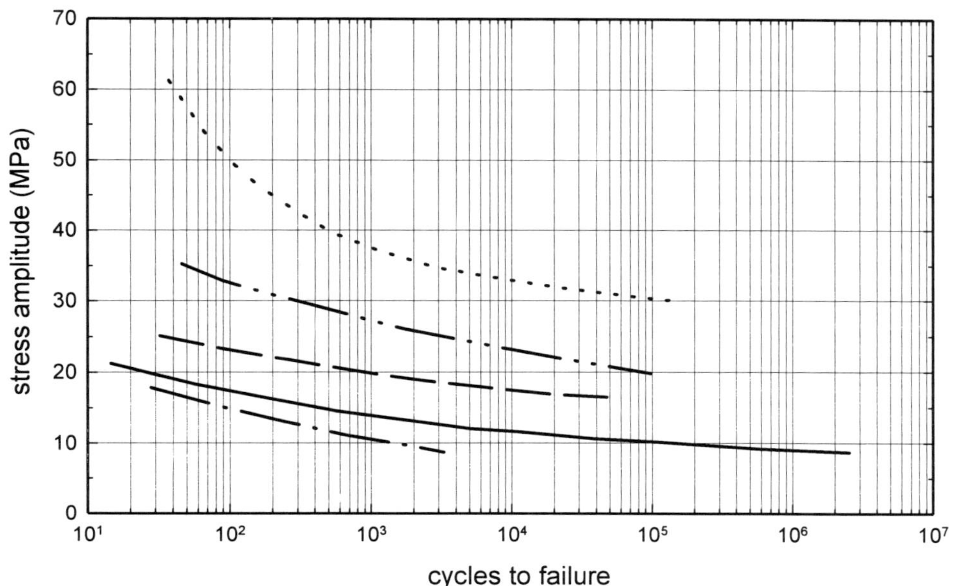

···············	PMMA (transparent; bulk cast); tension-compression; 0.5 Hz; 20°C; means stress = 0; plain specimen [fig d]
—··—··	PMMA (transparent; bulk cast); means stress = 0; tip radius = 1 mm
– – – –	PMMA (transparent; bulk cast); means stress = 0; tip radius = 0.5 mm
———	PMMA (transparent; bulk cast); means stress = 0; tip radius = 0.25 mm
—·—·—	PMMA (transparent; bulk cast); means stress = 0; tip radius = 10 μm
Reference No.	355

GRAPH 20: **Increase in Surface Temperature vs. Stress in Flexural Fatigue With and Without Cooling for Polymethyl Methacrylate.**

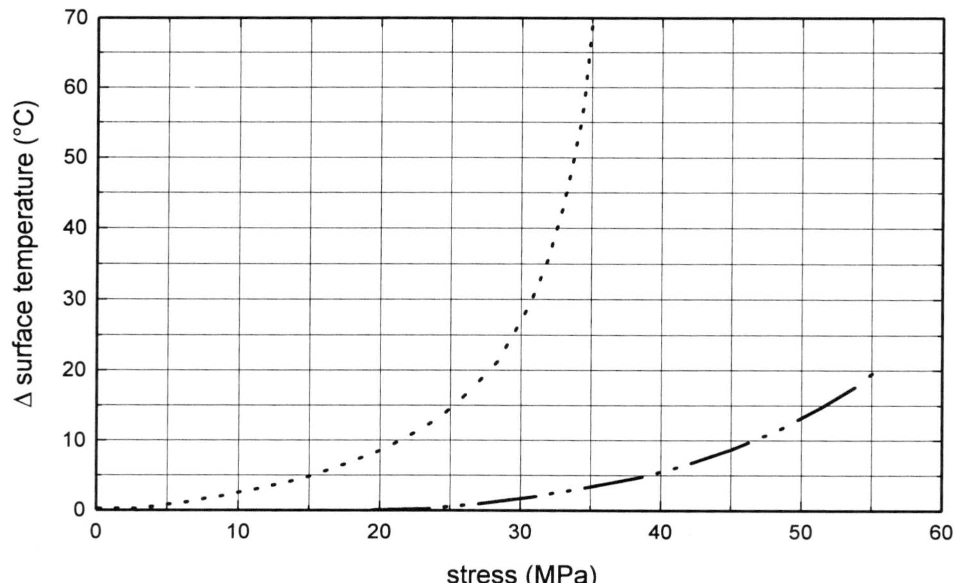

....	PMMA (transparent, gen. purp. grade); flexure; 2.5 Hz; 20°C; 65% RH; square wave; no cooling
—··—··	PMMA (transparent, gen. purp. grade); flexure; 2.5 Hz; 20°C; 65% RH; square wave; forced air cooling
Reference No.	355

GRAPH 21: **Fatigue Crack Propagation (FCP) Rate vs. Stress Intensity Range for Polymethyl Methacrylate.**

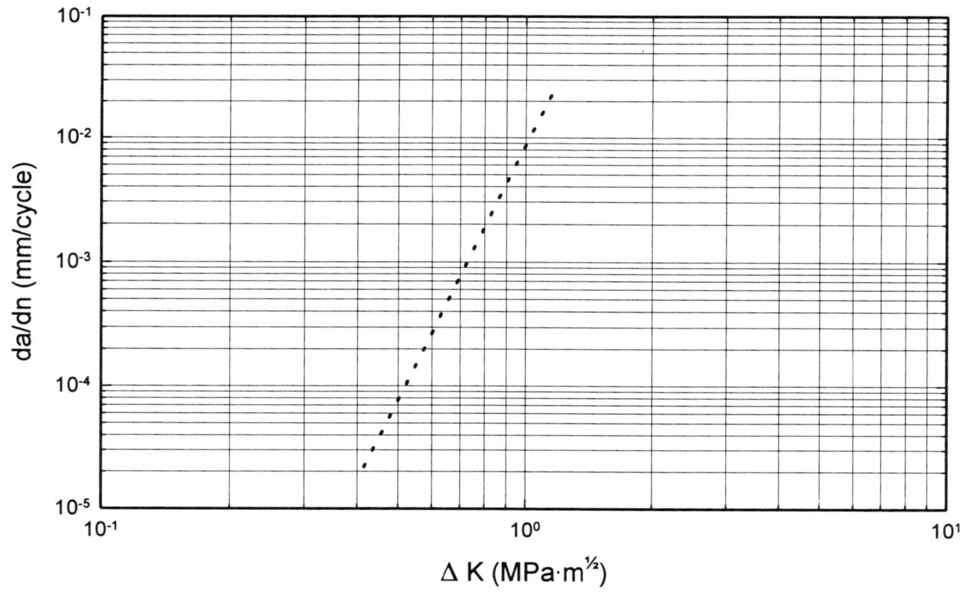

....	PMMA (transparent, gen. purp. grade)
Reference No.	355

Chapter 4

Ethylene-Tetrafluoroethylene Copolymer

Fatigue Properties

DuPont: Tefzel 200 (features: general purpose grade, 50 Rockwell R hardness, 75 Shore D hardness); **Tefzel HT-2004** (features: 74 Rockwell R hardness; material compostion: 25% glass fiber reinforcement)

Tefzel HT-2004 has a greater fatigue limit than Tefzel 200, but both are quite sensitive to stress levels. For best fatigue performance, stress levels below 20.7 MPa (3,000 psi) and 12.1 MPa (1,750 psi) are suggested for Tefzel HT-2004 and Tefzel 200 respectively.

Reference: *Tefzel Fluoropolymer Design Handbook,* supplier design guide (E-31301-1) - Du Pont Company, 1973.

GRAPH 22: Fatigue Cycles to Failure vs. Stress in Flexure for Carbon and Glass Reinforced Ethylene Tetrafluoroethylene Copolymer.

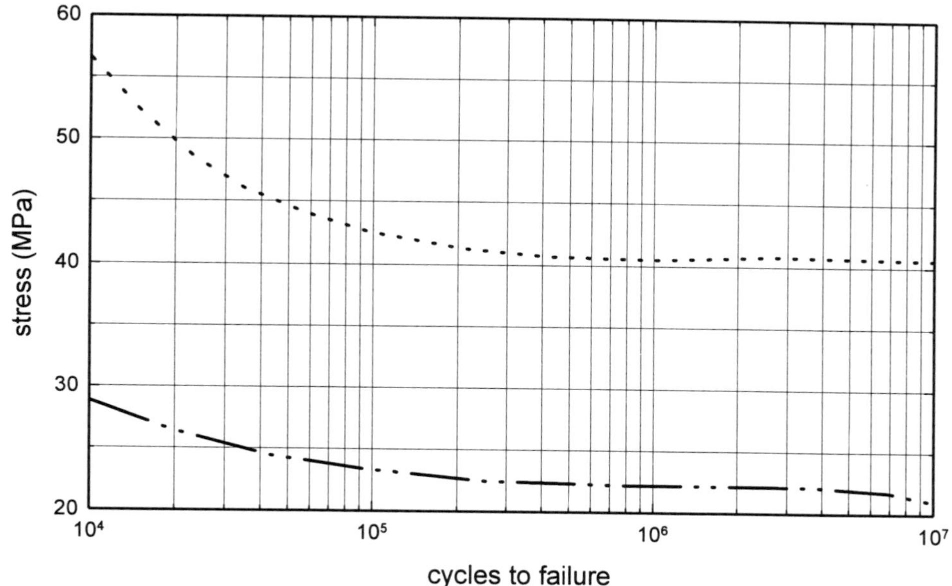

....	LNP ETFE (30% carbon fiber); flexure; ASTM D671; 30 Hz; 23°C [fig d]
—··—··	LNP ETFE; (30% glass fiber); flexure; ASTM D671; 30 Hz; 23°C [fig d]
Reference No.	395

GRAPH 23: **Fatigue Cycles to Failure vs. Stress in Flexure for DuPont Tefzel Ethylene Tetrafluoroethylene Copolymer.**

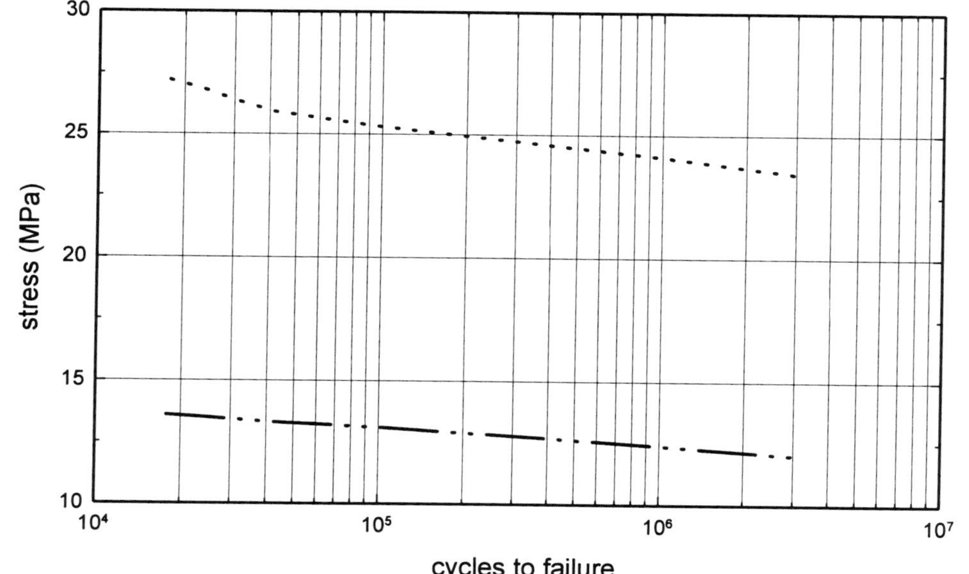

....	DuPont Tefzel HT-2004 ETFE (74 Rockwell R; 25% glass fiber); flexure (alternating); ASTM D671; 30 Hz; 23°C; 50% RH; sample type I, small [fig d]
—··—··	DuPont Tefzel 200 ETFE (gen. purp. grade, 50 Rockwell R, 75 Shore D); flexure (alternating); ASTM D671; 30 Hz; 23°C; 50% RH; sample type I, small [fig d]
Reference No.	205

Chapter 5

Polytetrafluoroethylene

GRAPH 24: **Fatigue Cycles to Failure vs. Stress in Flexure at Various Thicknesses for Unfilled Polytetrafluoroethylene.**

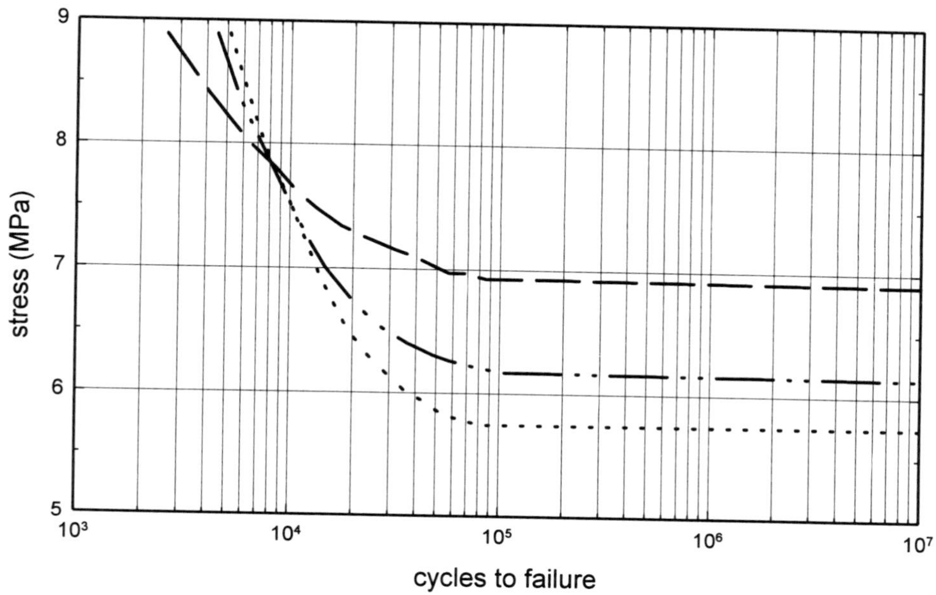

....	TFE (10.7 mm thick; unfilled); flexure (alternating); 30 Hz; 23°C; mean stress= 0
—··—··	TFE (6.6 mm thick; unfilled); flexure (alternating); 30 Hz; 23°C; mean stress= 0
– – – –	TFE (3.6 mm thick; unfilled); flexure (alternating); 30 Hz; 23°C; mean stress= 0
Reference No.	365

GRAPH 25: **Fatigue Cycles to Failure vs. Stress At Different Test Frequencies for Polytetrafluoroethylene.**

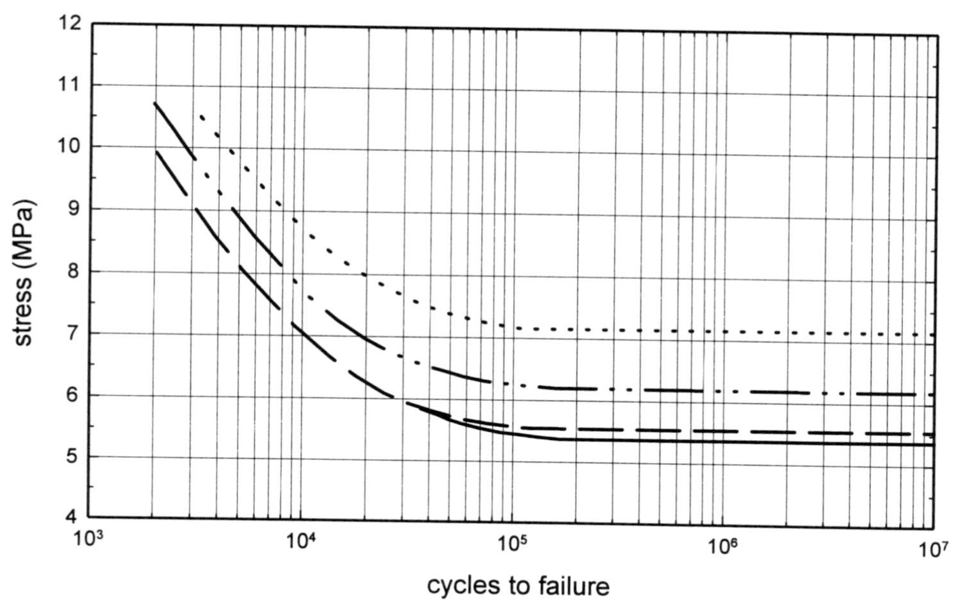

....	TFE; 20 Hz; 23°C
—··—··	TFE; 30 Hz; 23°C
– – – –	TFE; 40 Hz; 23°C
———	TFE; 60 Hz; 23°C
Reference No.	368

Chapter 6

Polyvinyl Fluoride

GRAPH 26: Fatigue Crack Propagation (FCP) Rate vs. Stress Intensity Range for Polyvinyl Fluoride.

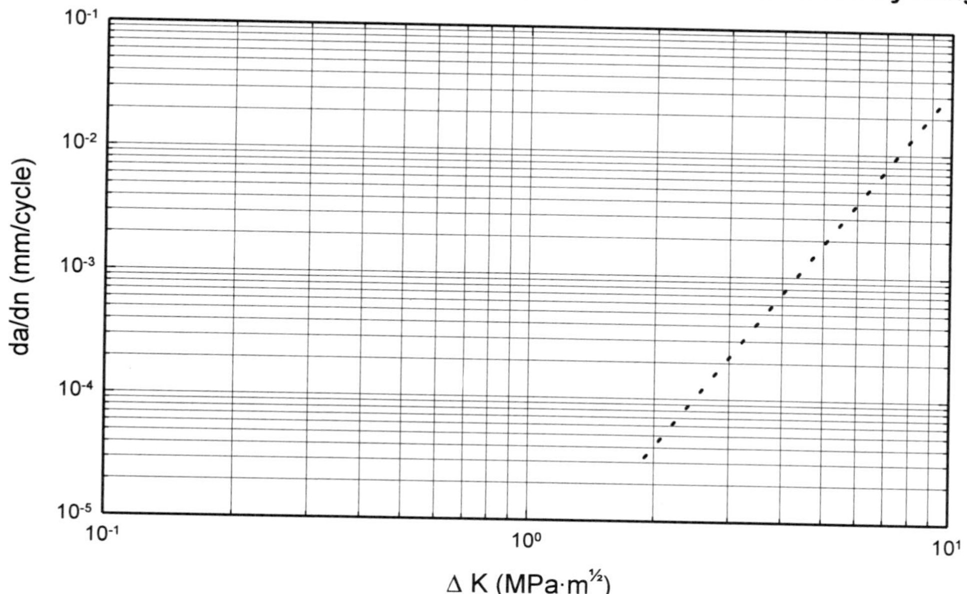

....	PVF
Reference No.	355

Chapter 7

Amorphous Nylon

Fatigue Crack Propagation

DuPont: Bexloy APC-803 (note: rubber toughened); **Huls America: Trogamid T**

The data for unmodified amorphous nylon clearly indicate that this material is inferior to nylon 66 in resistance to fatigue crack propagation. The results for toughened amorphous nylon (Bexloy AP C 803) show that it too exhibits poorer fatigue resistance than that of the pure nylon 66 despite its far superior impact strength. This is a clear indication that fatigue and impact fracture mechanisms are distinct and that toughening and alloying systems may not necessarily be optimized for both failure modes. Different results for the two amorphous nylons were suspected to be due to the presence of the rubber-toughening agent in the Bexloy material rather than to differences in chemical structure between the Bexloy and the Trogamid T.

Reference: Wyzgoski, M.G. (General Motors), Novak, G.E. (General Motors), *Fatigue-Resistant Nylon Alloys,* Journal of Applied Polymer Science (1994), technical journal (Vol. 51; CCC 0021-8995/94/050873-13) - John Wiley & Sons, Inc., 1994.

GRAPH 27: Fatigue Cycles to Failure vs. Stress in Tension for General Purpose Huls Trogamid T5000 Amorphous Nylon.

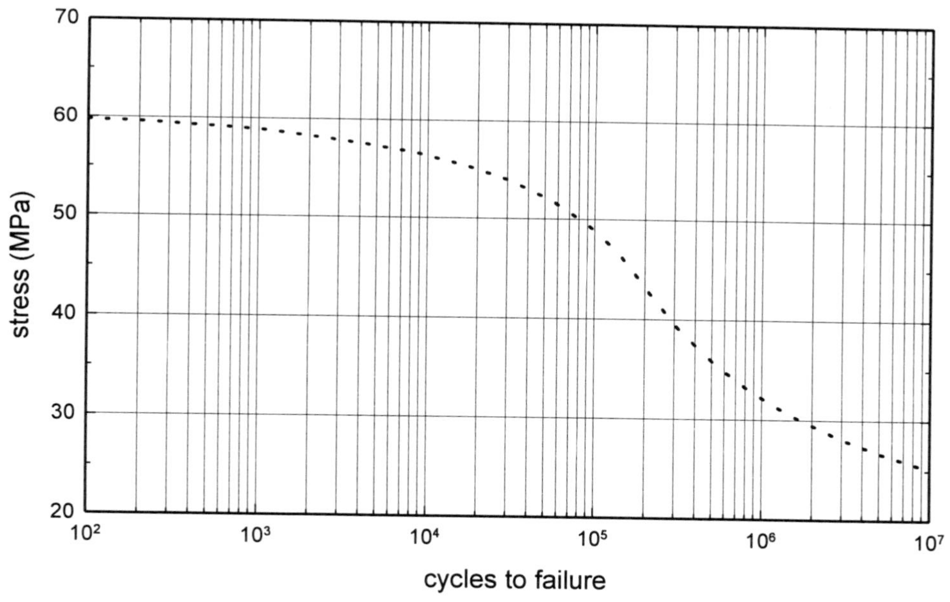

....	Huls Trogamid T5000 Amorphous Nylon (transparent, gen. purp. grade); tension-compression; 5 Hz; 23°C
Reference No.	415

GRAPH 28: **Fatigue Crack Propagation (FCP) Rate vs. Stress Intensity Range for Amorphous Nylon.**

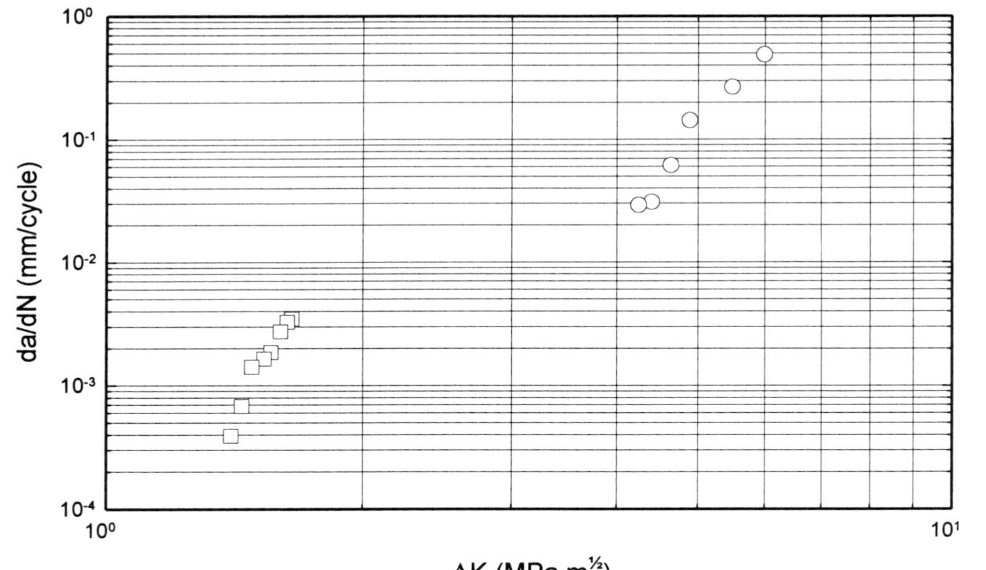

○	DuPont Bexloy APC-803 Amorphous Nylon (rubber toughened); tension; 0.5 Hz; 23°C; compact tension specimen, precracked, parallel to flow
□	Huls Trogamid T Amorphous Nylon; tension; 0.5 Hz; 23°C; compact tension specimen, precracked, parallel to flow
Reference No.	401

Chapter 8

Nylon 46

Fatigue Properties

DSM: Stanyl

The high level of crystallinity and small spherulite structure of Stanyl nylon 46 lead to a fatigue resistance which is superior to ordinary nylons. Compared to nylon 66, Stanyl nylon 46 can be subjected either to a higher stress level with the same lifetime, or to many more cycles at a comparable stress level. At high temperatures, Stanyl nylon 46 retains its excellent fatigue behavior when compared to nylon 6 and nylon 66.

Reference: *Stanyl 46 Nylon General Information,* supplier design guide (MBC-PP-492-5M) - DSM, 1992.

GRAPH 29: **Fatigue Cycles to Failure vs. Stress in Flexure for 30% Glass Fiber Reinforced DSM Stanyl TW200F6 Nylon 46.**

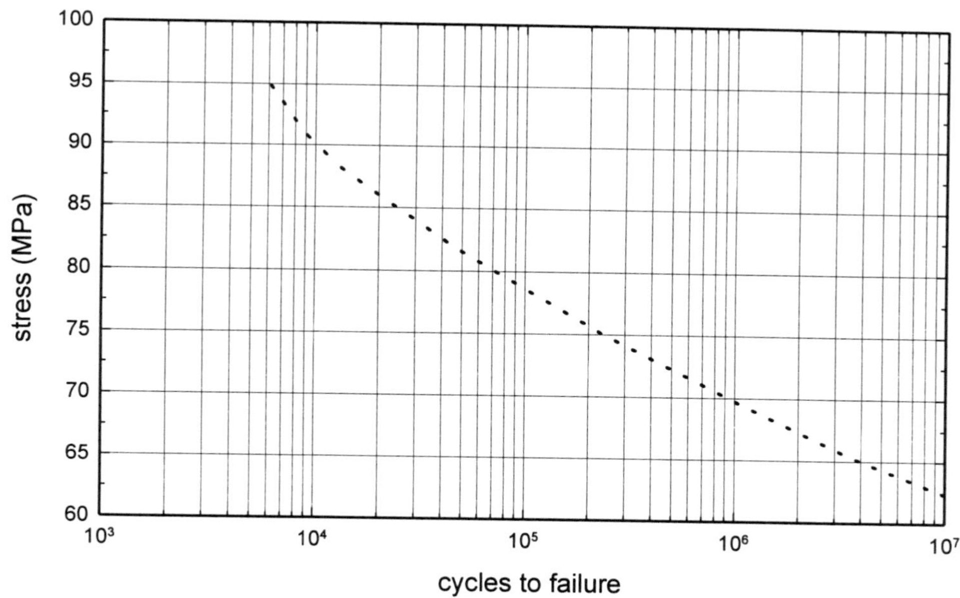

....	DSM Stanyl TW200F6 Nylon 46 (30% glass fiber); flexure; ASTM D671; 20 Hz; 23°C
Reference No.	419

GRAPH 30: **Fatigue Cycles to Failure vs. Stress in Flexure at 160°C for Glass Fiber Reinforced DSM Stanyl Nylon 46.**

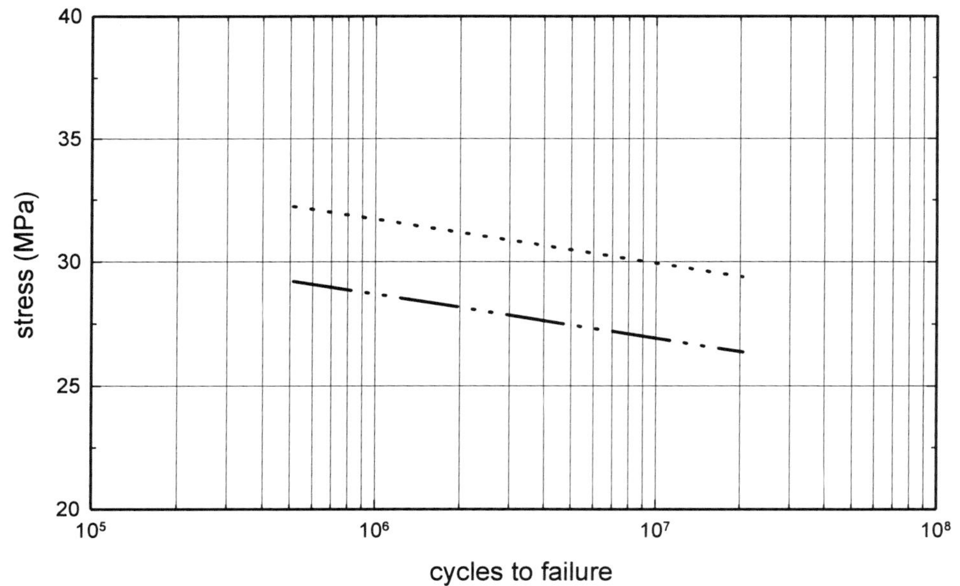

···············	DSM Stanyl TW200F9 Nylon 46 (45% glass fiber); flexure; ASTM D671; 20 Hz; 160°C
—··—··	DSM Stanyl TW200F6 Nylon 46 (30% glass fiber); flexure; ASTM D671; 20 Hz; 160°C
Reference No.	419

GRAPH 31: **Fatigue Cycles to Failure vs. Stress in Tension for DSM Stanyl TW300 Nylon 46.**

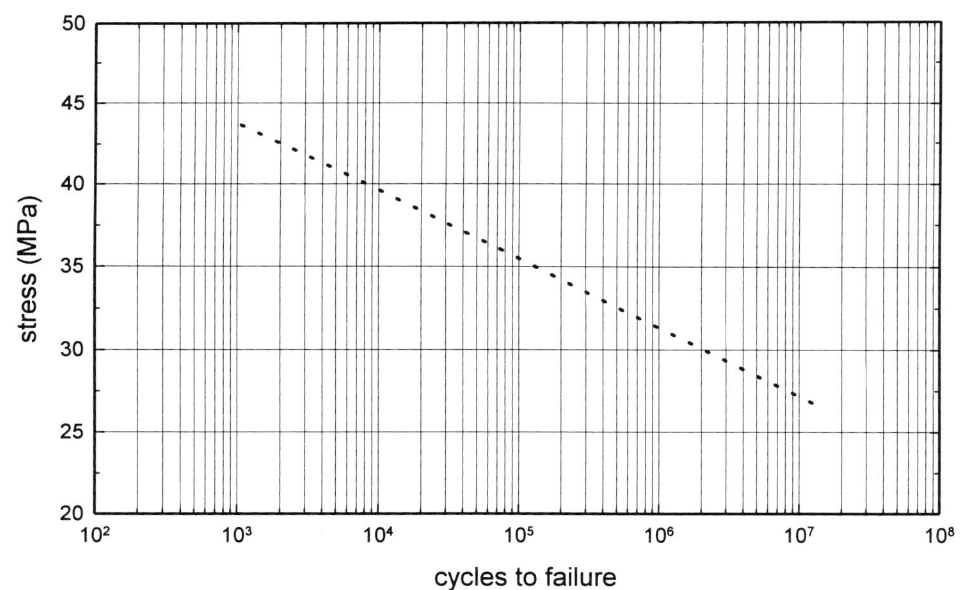

···············	DSM Stanyl TW300 Nylon 46 (gen. purp. grade, unfilled); tension; 20 Hz; 23°C; DAM; prestressed @ 20.7 MPa [fig g]
Reference No.	419

GRAPH 32: **Fatigue Cycles to Failure vs. Stress in Flexure for Nylon 6.**

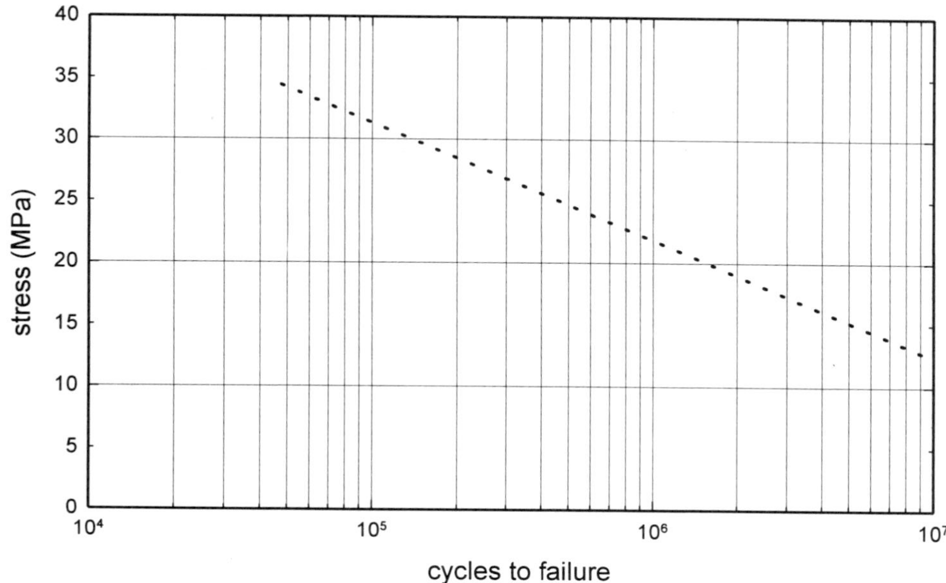

....	Nylon 6 (4.6 mm thick); flexure (alternating); 30 Hz; 23°C; DAM; mean stress= 0
Reference No.	365

GRAPH 33: **Fatigue Cycles to Failure vs. Stress in Flexure for 30% Glass Fiber Reinforced LNP Thermocomp PF1006 Nylon 6.**

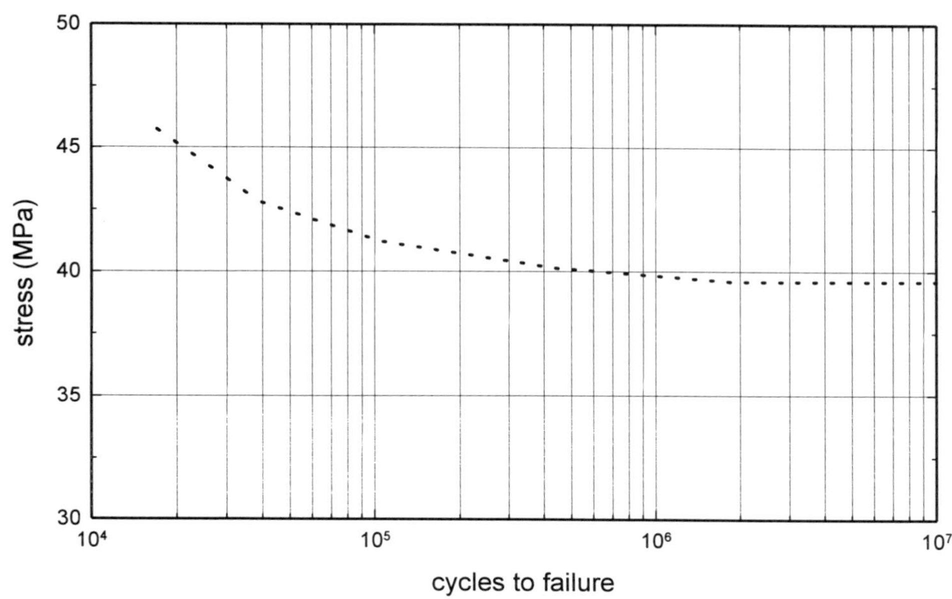

....	LNP Thermocomp PF1006 Nylon 6 (30% glass fiber); flexure; 30 Hz; 23°C
Reference No.	394

GRAPH 34: Fatigue Cycles to Failure vs. Stress in Flexure at 160°C for 45% Glass Fiber Reinforced Nylon 6.

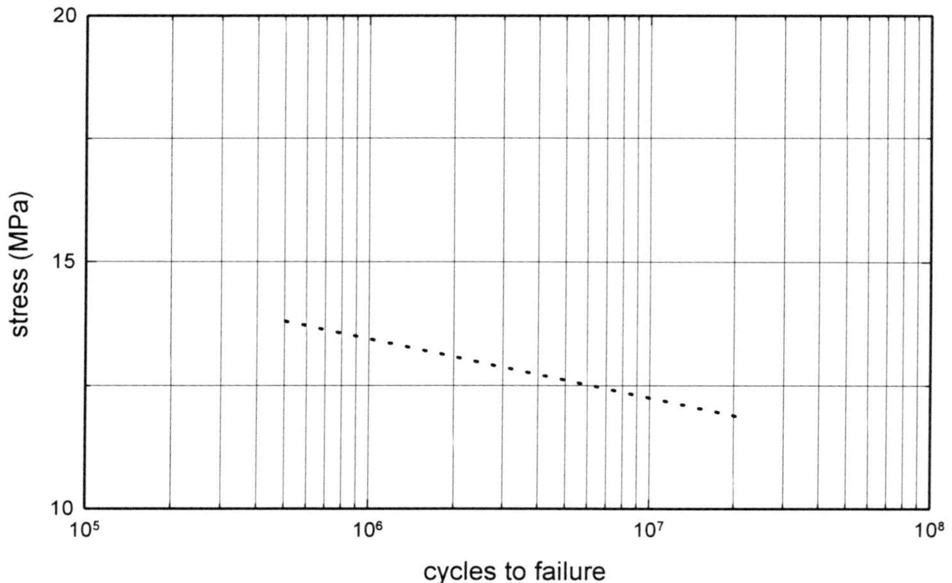

....	Nylon 6 (45% glass fiber); flexure; ASTM D671; 20 Hz; 160°C
Reference No.	419

GRAPH 35: Fatigue Cycles to Failure vs. Stress in Tension-Compression for Ube Nylon 6.

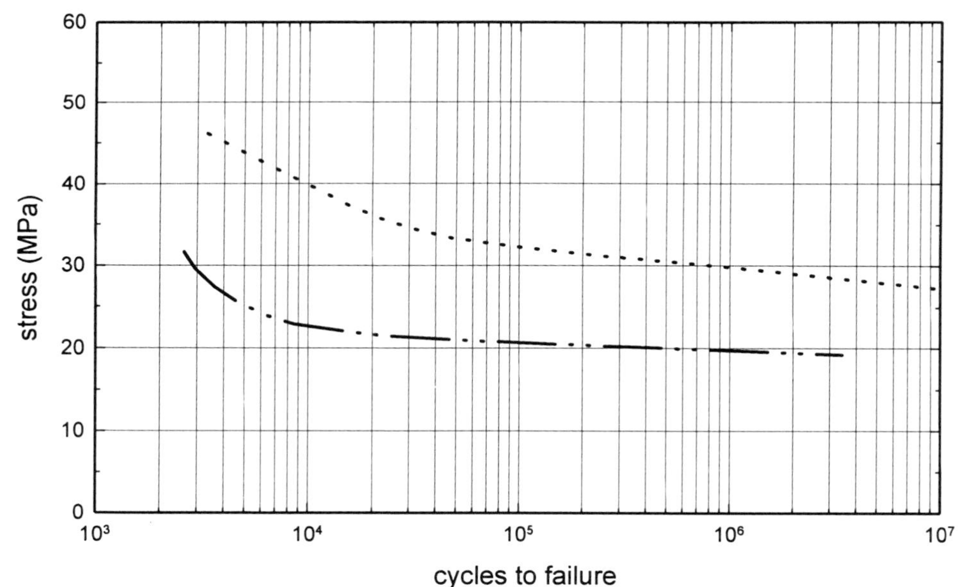

....	Ube Ube 1011GC4 Nylon 6 (20% glass fiber); tension-compression; 16.7 Hz; DAM; 4 mm thick
—··—··	Ube Ube 1013B Nylon 6 (moderate flow, gen. purp. grade); tension-compression; 16.7 Hz; DAM; 4 mm thick
Reference No.	238

GRAPH 36: **Fatigue Cycles to Failure vs. Stress for BASF Ultramid B Nylon 6.**

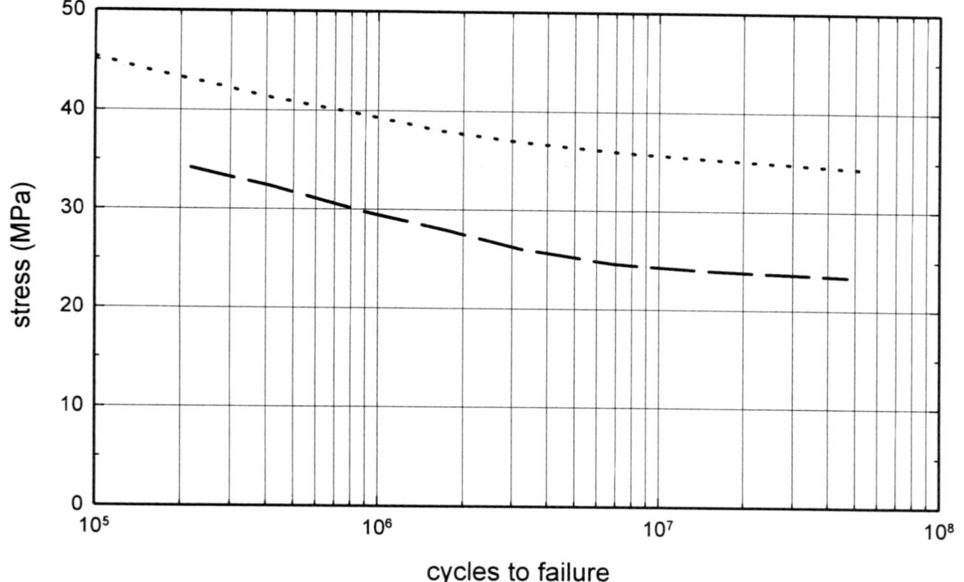

....	BASF Ultramid BG5 Nylon 6 (25% glass fiber); rotating beam; DIN 53442; 7.5 Hz; 23°C; 50% RH; laboratory atmosphere [fig. u]
....	BASF Ultramid BG7 Nylon 6 (35% glass fiber); rotating beam; DIN 53442; 7.5 Hz; 23°C; 50% RH; laboratory atmosphere [fig. u]
– – – –	BASF Ultramid BG5 Nylon 6 (25% glass fiber); tension- compression; DIN 53442; 7.5 Hz; 23°C; 50% RH; laboratory atmosphere [fig. v]
– – – –	BASF Ultramid BG7 Nylon 6 (35% glass fiber); tension- compression; DIN 53442; 7.5 Hz; 23°C; 50% RH; laboratory atmosphere [fig. v]
Reference No.	93

Chapter 10

Nylon 610

GRAPH 37: Fatigue Cycles to Failure vs. Stress in Flexure for Glass Fiber Reinforced LNP Nylon 610.

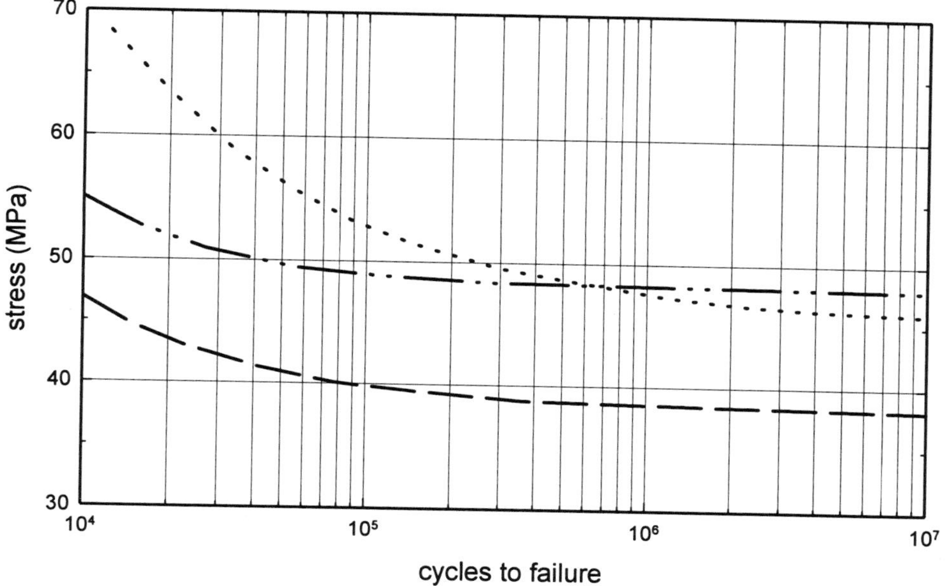

···············	LNP Lubricomp QFL4017ERHS Nylon 610 (lubricated, heat stabilized, mold release; 35% glass fiber, 5% PTFE modified); flexure; 30 Hz; 23°C
—··—··	LNP Thermocomp QF1008 Nylon 610 (40% glass fiber); flexure; 30 Hz; 23°C
– – – –	LNP Thermocomp QF1006 Nylon 610 (30% glass fiber); flexure; 30 Hz; 23°C
Reference No.	394

Chapter 11

Nylon 612

GRAPH 38: Fatigue Cycles to Failure vs. Stress in Tension for DuPont Zytel 158L Nylon 612.

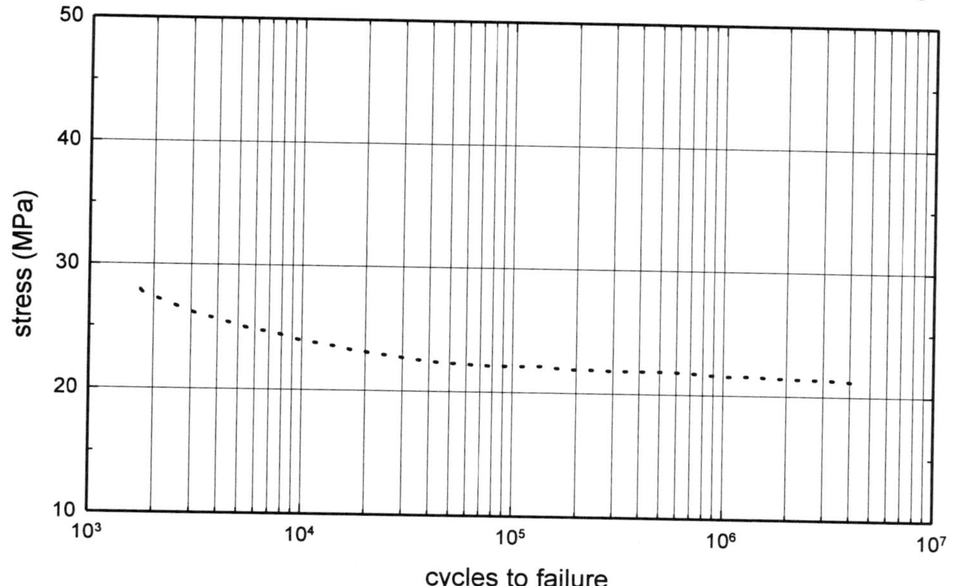

....	DuPont Zytel 158L Nylon 612 (lubricated, gen. purp. grade); tension-compression; 30 Hz; 23°C; 50%RH [fig d]
Reference No.	68

Chapter 12

Nylon 66

Fatigue Properties

DuPont: Zytel (reinforcement: glass fiber)

Elevated temperatures and the presence of oils, greases, gasolines and detergents can affect the fatigue resistance of some plastics. Zytel nylon resins, however, exhibit good fatigue and vibration resistance under these conditions, by showing only moderate effects from elevated temperatures, and virtually no effect from prolonged exposure to gasoline vapors.

Glass reinforced GRZ resins provide excellent fatigue resistance at high stress levels and in situations where repeated load variations are encountered. In applications such as gears where rubbing occurs, unreinforced resins give much longer wear. GRZ gears have been used in high stress, limited duty applications.

To rate the family of nylon resins in terms of fatigue endurance at one million cycles can be misleading. The GRZ resins show higher values with test specimens stressed in the direction of fiber orientation. In an actual part with more random fiber distribution, the fatigue endurance could be substantially lower. Moreover, in a part subject to vibration, the high flexural modulus of the GRZ resins would result in a much higher induced stress than would be experienced by the same part made of an unreinforced nylon. In other words, fatigue life with constant strain will be much higher with the unreinforced nylons. The use of Zytel tubing instead of metal tubing for hydraulic lines to vibrating machinery is one example.

In general, it can be said that the 66 nylons, reinforced and unreinforced, will exhibit better fatigue endurance than the 6 or 612 nylons. The Minlon resins exhibit somewhat lower fatigue endurance than the unreinforced 66 nylons.

Reference: *Design Handbook For Du Pont Engineering Plastics - Module II,* supplier design guide (E-42267) - Du Pont Engineering Polymers.

Fatigue Crack Propagation

DuPont: Zytel ST 801 (features: impact modified; note: Zytel ST801 contains 20% EPDM rubber (grafted))

Pure Zytel 801 supertough nylon 66 is highly resistant to fatigue crack initiation, and stable fatigue crack propagation could only be produced at very high ΔK levels. In fact, at the ΔK levels employed for the supertough nylon, most nylons would have fractured immediately since the maximum stress intensity level in the fatigue experiment would have exceeded the fracture toughness of the materials.

Reference: Wyzgoski, M.G. (General Motors), Novak, G.E. (General Motors), *Fatigue-Resistant Nylon Alloys,* Journal of Applied Polymer Science (1994), technical journal (Vol. 51; CCC 0021-8995/94/050873-13) - John Wiley & Sons, Inc., 1994.

Nylon 66 (note: 90% Zytel 122L/ 10% Zytel ST 801); **Nylon 66** (note: 80% Zytel 122L/ 20% Zytel ST 801)

Measurable improvements in fatigue crack propagation are noted for blends containing 10 or 20% of toughened nylon 66 (Zytel ST 801). Interestingly, these blends are not as efficient as equivalent blends in which the same amount of rubber is grafted onto the amorphous nylon. Available data suggest that the amorphous nylon is the preferred polymer for the preblend.

Reference: Wyzgoski, M.G. (General Motors), Novak, G.E. (General Motors), *Fatigue-Resistant Nylon Alloys,* Journal of Applied Polymer Science (1994), technical journal (Vol. 51; CCC 0021-8995/94/050873-13) - John Wiley & Sons, Inc., 1994.

Nylon 66 (note: 96% Zytel 122L/ 4% Uniroyal X465 (EPDM rubber)); **Nylon 66** (note: 98% Zytel 122L/ 2% Uniroyal X465 (EPDM rubber))

Rubber alone is effective in reducing fatigue crack growth rates; however, it is not as effective as the alloys also containing amorphous nylon. These results suggest that the miscible amorphous nylon does play a key role in imparting improved fatigue resistance to the crystalline nylon 66.

Reference: Wyzgoski, M.G. (General Motors) Vak, G.E. (General Motors), *Fatigue-Resistant Nylon Alloys,* Journal of Applied Polymer Science (1994), technical journal (Vol. 51; CCC 0021-8995/94/050873-13) - John Wiley & Sons, Inc., 1994.

Effect of Filler Content on Fatigue Behavior

LNP Engineering Plastics: Thermocomp RF-100-10 HS (material compostion: 50% glass fiber reinforcement; note: short glass fiber (average length 160-190 microns)); **Verton RF-700-10 HS** (material compostion: 50% glass fiber reinforcement; note: long glass fibers (average fiber length of 1180-1290 microns))

In the case of nylon composites, better fatigue performance was observed at higher glass contents since the crack initiation mode did not involve the interface, and since the cracks would travel more slowly through the higher glass content composites.

Reference: Grove, D.A. (LNP)' Kim, H.C. (LNP), *Effect of Constituents on the Fatigue Behavior of Long Fiber Reinforced Thermoplastics,* ANTEC 1995, conference proceedings - Society of Plastics Engineers, 1995.

Nylon 66

The S-N curves for a series of glass reinforced, filled and neat resin nylon 66 indicate the positive effect of increasing glass fiber content on fatigue endurance limit within a specific resin family. A moisture conditioned (50% RH), 40% glass fiber reinforced nylon 66 has a 17% higher cyclic failure stress at 10^{6} cycles than a 30% glass fiber reinforced nylon 66 under the same conditions. The addition of 30% glass beads to nylon 66 provides a lower endurance limit than its glass fiber analog; however, it exceeds the fatigue resistance of the unfilled resin.

The addition of carbon fibers to thermoplastics improves fatigue endurance limits more than does glass fiber reinforcement. For example, a comparison of 40% carbon fiber and a 40% glass fiber reinforced nylon 66 shows an improvement of 21% (57 MPa vs. 48 MPa) for the carbon reinforced compound. Other resin systems are also improved with carbon fibers, but the amount of improvement varies with different base resins.

Reference: Newby, G.B. (LNP) Theberge, J.E. (LNP, *Long-Term Behavior of Reinforced Thermoplastics,* Machine Design, trade journal - Penton Publishing, 1984.

Effect of Glass Reinforcement on Fatigue Behavior

LNP Engineering Plastics: Thermocomp RF-100-10 HS (material compostion: 50% glass fiber reinforcement; note: short glass fiber (average length 160-190 microns)); **Verton RF-700-10 HS** (material compostion: 50% glass fiber reinforcement; note: long glass fibers (average fiber length of 1180-1290 microns))

Unlike tensile fatigue, where a uniform stress is applied across the sample, in flexural fatigue the stress varies from a large tensile stress on one side of the specimen to a large compressive stress on the opposite side. The cyclic tensile and compression loads in the outer regions lead to various levels of fiber-resin debonding coupled with some fiber breakage. In the case of the various glass reinforced nylon materials, the initial cracks emanated from fiber surface flaws. The initial fiber cracks were probably formed during the compression part of the stress cycle since most fibers have the tendency to buckle and break under compression loading conditions.

Reference: Grove, D.A. (LNP), Kim, H.C. (LNP), *Effect of Constituents on the Fatigue Behavior of Long Fiber Reinforced Thermoplastics,* ANTEC 1995, conference proceedings - Society of Plastics Engineers, 1995.

Effect of Moisture Content on Fatigue Behavior

Nylon 66

Moisture can have a significant effect on hygroscopic resins, specifically nylons. The absorption of moisture can cause plasticization and swelling of the resin matrix, weakening the fiber/ matrix interface, which reduces both static loading capacity and fatigue performance.

Reference: Newby, G.B. (LNP), Theberge, J.E. (LNP), *Long-Term Behavior of Reinforced Thermoplastics,* Machine Design, trade journal - Penton Publishing, 1984.

GRAPH 39: **Fatigue Cycles to Failure vs. Stress in Flexure for Hoechst Celanese Nylon 1500 Nylon 66 at different moisture contents.**

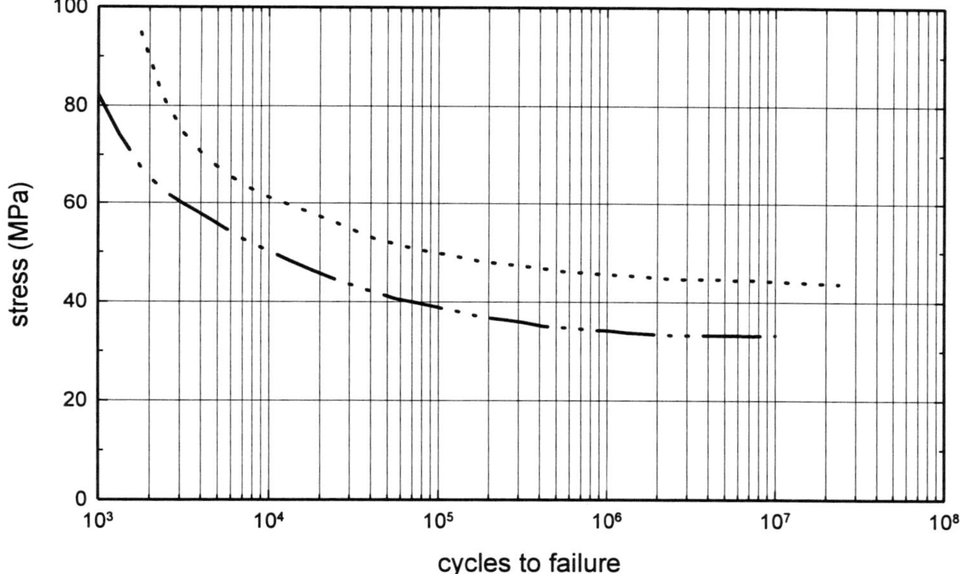

....	Hoechst Cel. Nylon 1500 Nylon 66 (33% glass fiber); flexure (alternating); 30 Hz; 23°C; DAM [fig h, i]
—··—··	Hoechst Cel. Nylon 1500 Nylon 66 (33% glass fiber); flexure (alternating); 30 Hz; 23°C; 50% RH; 1.7% moisture content [fig h, i]
Reference No.	317

<u>GRAPH 40:</u> Fatigue Cycles to Failure vs. Stress in Flexure for Carbon Fiber Reinforced LNP Stat-Kon Nylon 66.

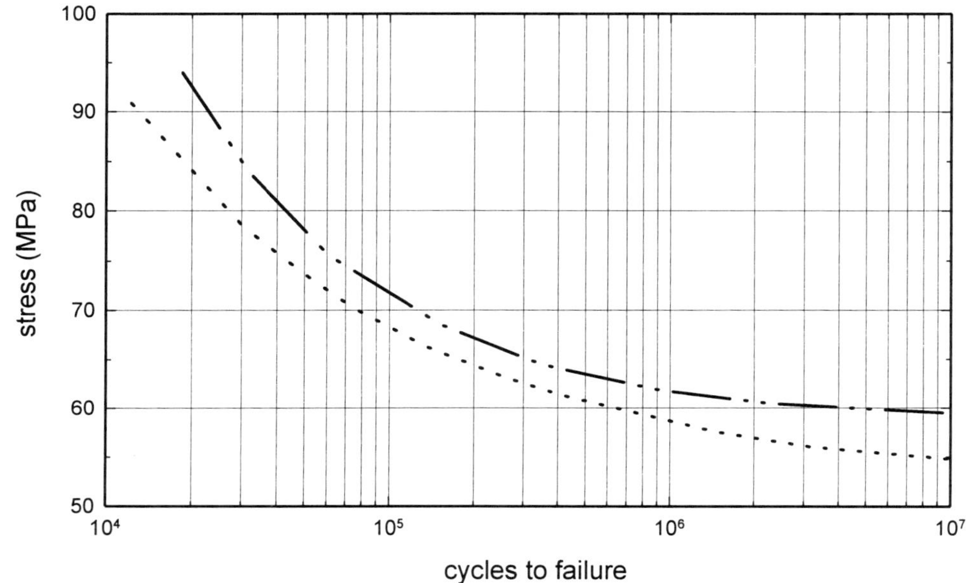

....	LNP Stat-Kon RC1006 Nylon 66 (static dissipative; 30% carbon fiber); flexure; 30 Hz; 23°C
—··—··	LNP Stat-Kon RC1008 Nylon 66 (static dissipative; 40% carbon fiber); flexure; 30 Hz; 23°C
Reference No.	394

<u>GRAPH 41:</u> Fatigue Cycles to Failure vs. Stress in Flexure for Long Glass Fiber Reinforced LNP Verton Nylon 66.

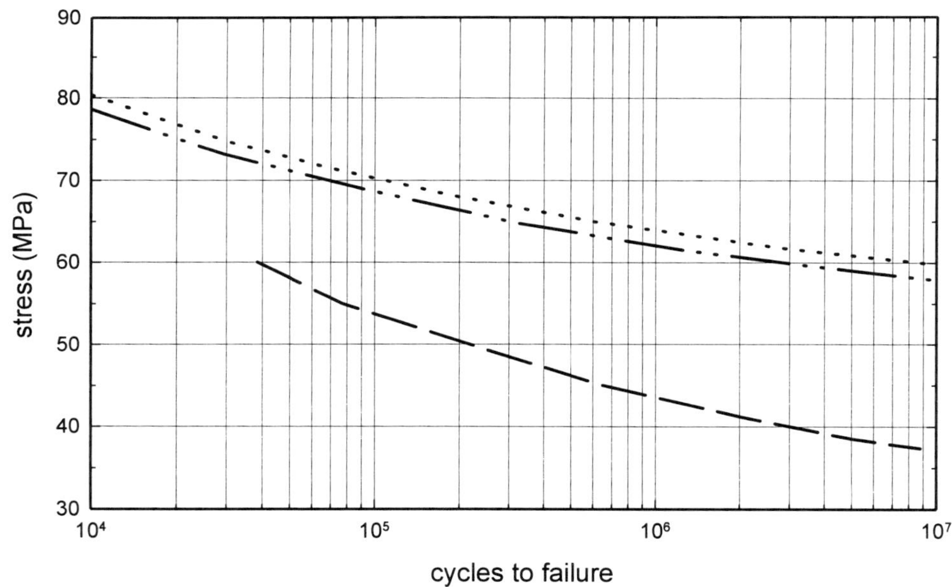

....	LNP Verton RF70012 Nylon 66 (60% glass fiber; long glass fibers); flexure; 30 Hz; 23°C
—··—··	LNP Verton RF70010 Nylon 66 (50% glass fiber; long glass fibers); flexure; 30 Hz; 23°C
— — — —	LNP Verton RF7007 Nylon 66 (35% glass fiber; long glass fibers); flexure; 30 Hz; 23°C
Reference No.	394

GRAPH 42: Fatigue Cycles to Failure vs. Stress in Flexure for DuPont Zytel 101 Nylon 66.

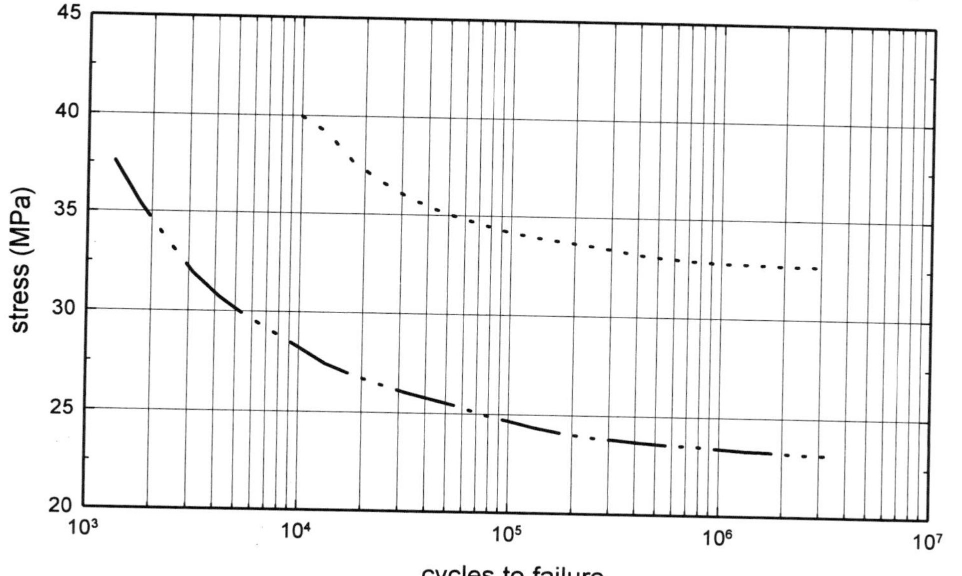

....	DuPont Zytel 101 Nylon 66 (gen. purp. grade, unlubricated); flexure; 23°C; DAM
—·· —··	DuPont Zytel 101 Nylon 66 (gen. purp. grade, unlubricated); flexure; 23°C;
Reference No.	68

GRAPH 43: Fatigue Cycles to Failure vs. Stress in Flexure for DuPont Minlon Nylon 66.

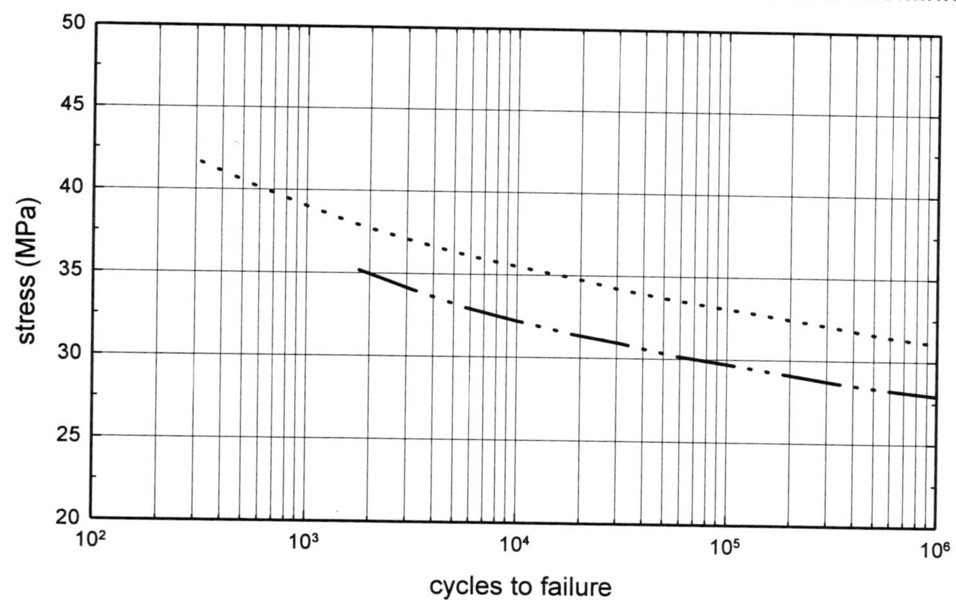

....	DuPont Minlon 20B Nylon 66 (glass/ mineral reinforced); flexure; 23°C; DAM
—·· —··	DuPont Minlon 10B Nylon 66 (mineral filled); flexure; 23°C; DAM
Reference No.	68

GRAPH 44: Fatigue Cycles to Failure vs. Stress in Flexure for Glass Reinforced LNP Nylon 66.

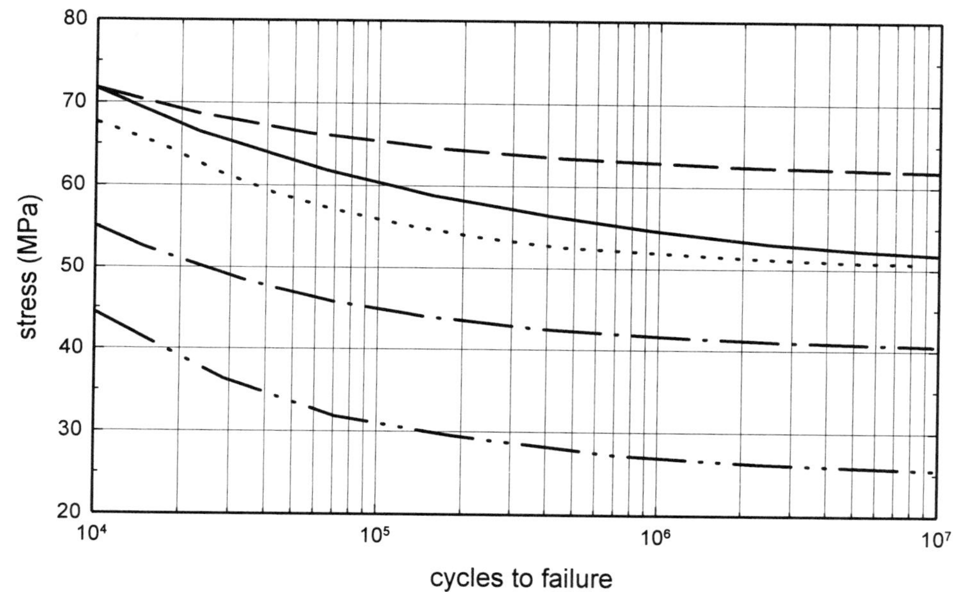

....	LNP Lubricomp RF40 Nylon 66 (lubricated; 40% glass fiber); flexure; 30 Hz; 23°C
—··—··	LNP Thermocomp RB1006 Nylon 66 (30% glass bead); flexure; 30 Hz; 23°C
— — —	LNP Thermocomp RF1008 Nylon 66 (40% glass fiber); flexure; 30 Hz; 23°C
———	LNP Thermocomp RF10010 Nylon 66 (50% glass fiber); flexure; 30 Hz; 23°C
—·—·—	LNP Thermocomp RF1006 Nylon 66 (30% glass fiber); flexure; 30 Hz; 23°C
Reference No.	394

GRAPH 45: Fatigue Cycles to Failure vs. Stress in Flexure for DuPont Minlon Nylon 66.

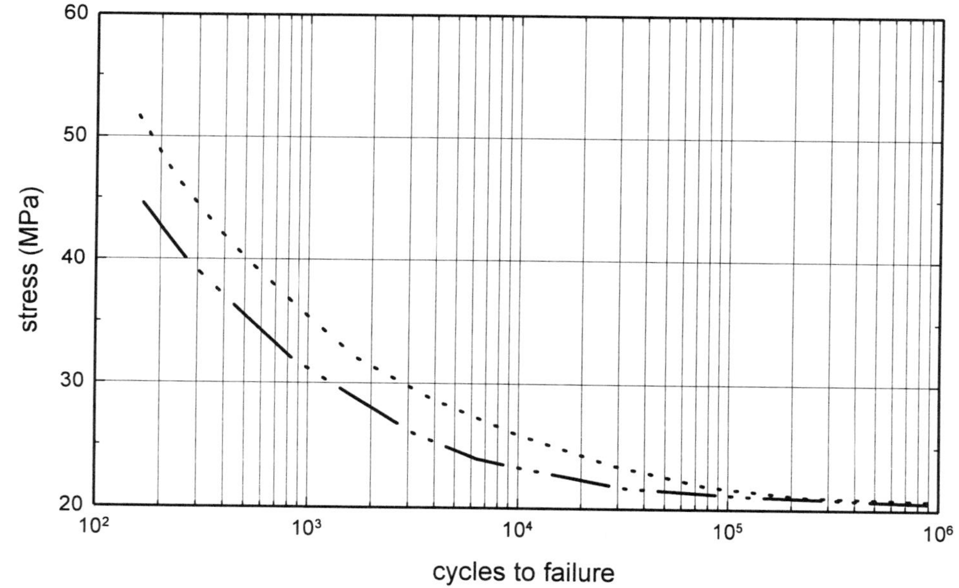

....	DuPont Minlon 11C Nylon 66 (low warp grade, high impact, chrome platable; , mineral filled); flexure; 23°C; DAM
—··—··	DuPont Minlon 12T Nylon 66 (low warp grade, high impact; , mineral filled); flexure; 23°C; DAM
Reference No.	68

GRAPH 46: Fatigue Cycles to Failure vs. Stress in Flexure for Carbon Fiber Reinforced LNP Thermocomp Nylon 66.

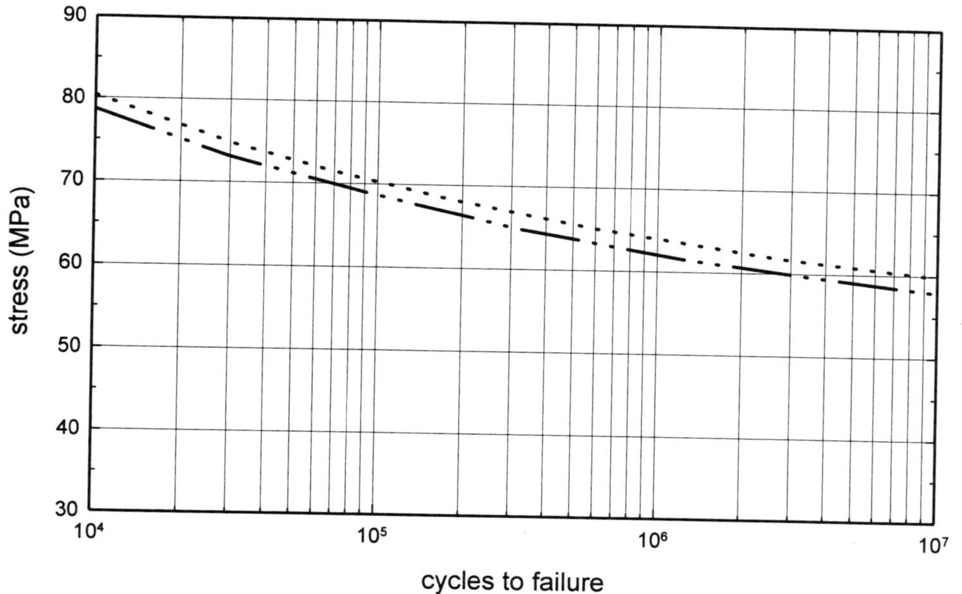

....	LNP Thermocomp RC1008 Nylon 66 (40% carbon fiber); flexure; 30 Hz; 23°C
—·-—··	LNP Thermocomp RC1006 Nylon 66 (30% carbon fiber); flexure; 30 Hz; 23°C
Reference No.	394

GRAPH 47: Fatigue Cycles to Failure vs. Stress in Flexure for Long and Short Glass Fiber Reinforced Nylon 66.

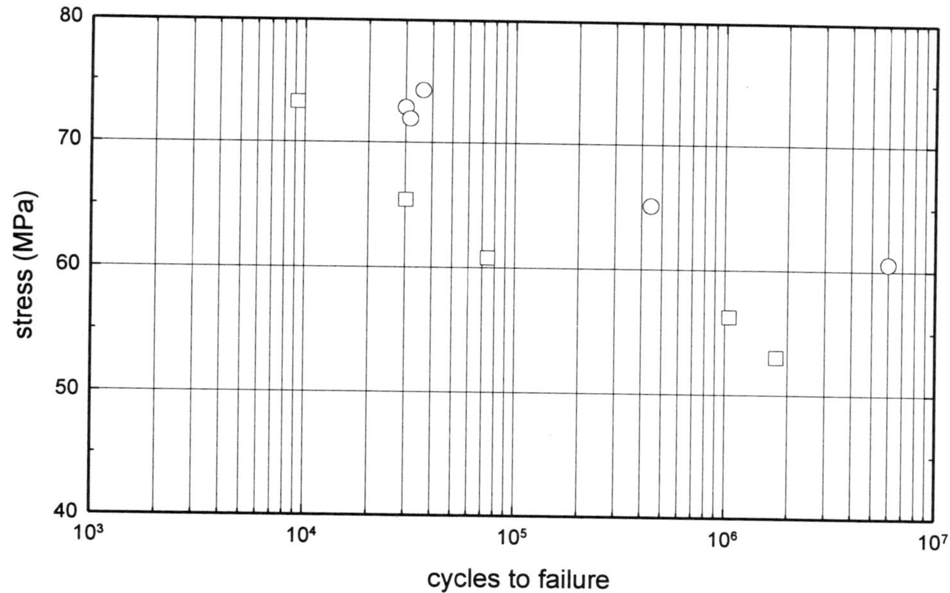

○	LNP Verton RF-700-10 HS Nylon 66 (50% glass fiber; long glass fibers (1180-1290 microns)); flexure; 30 Hz; 23°C; ASTM D671 Type B [fig d]
□	LNP Thermocomp RF-100-10 HS Nylon 66 (50% glass fiber; short glass fiber (160-190 microns)); flexure; 30 Hz; 23°C; ASTM D671 Type B [fig d]
Reference No.	391

GRAPH 48: Fatigue Cycles to Failure vs. Stress in Flexure for Filled and Unfilled BASF Ultramid A Nylon 66.

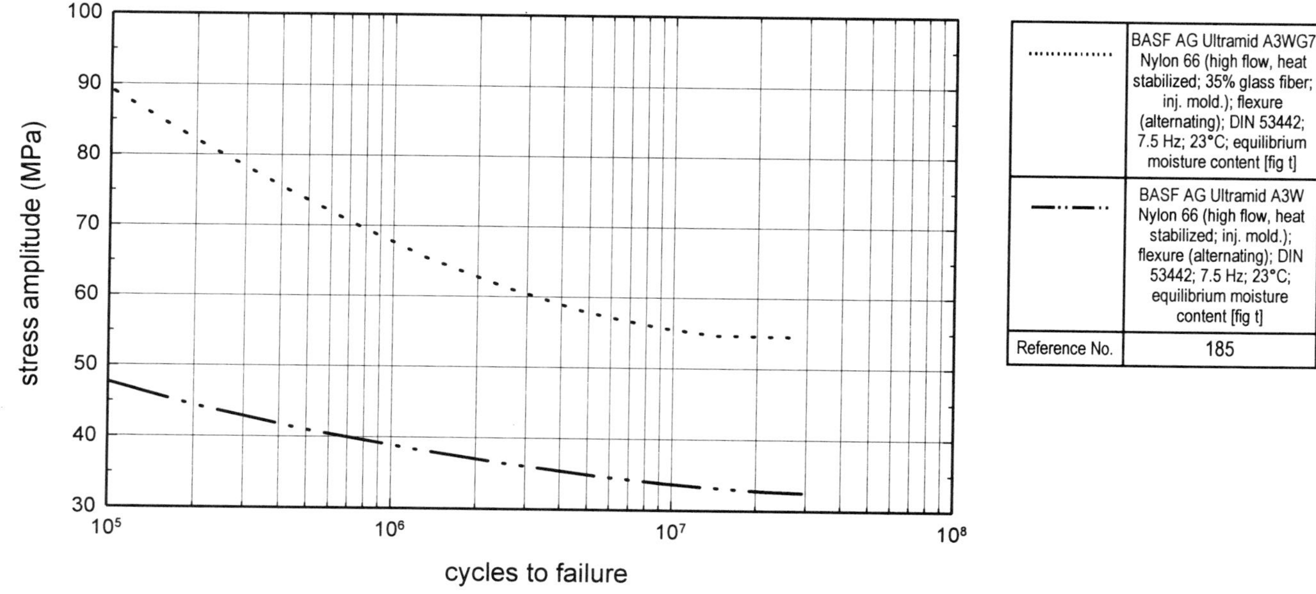

GRAPH 49: Fatigue Cycles to Failure vs. Stress in Flexure at 160°C for 45% Glass Fiber Reinforced Nylon 66.

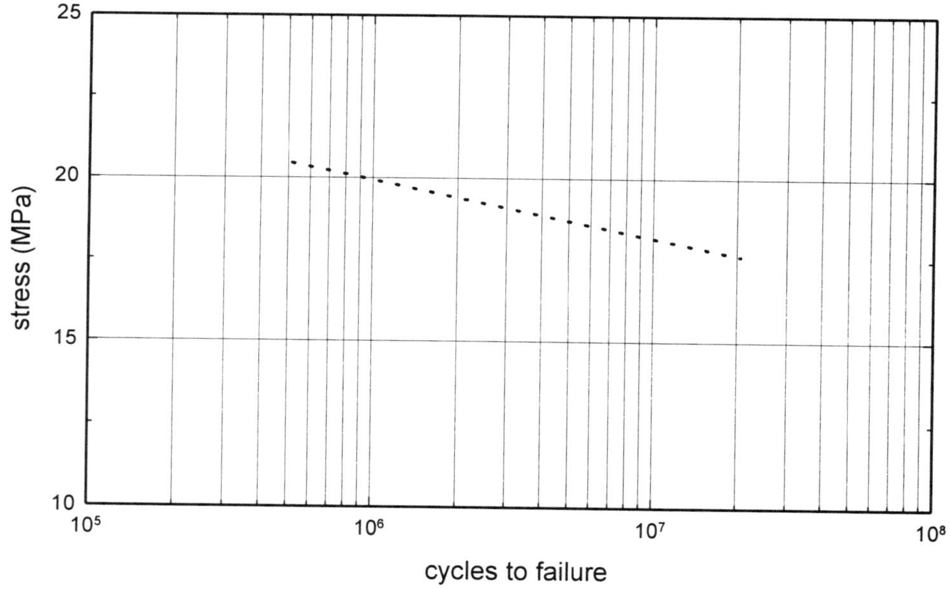

....	Nylon 66; flexure; ASTM D671; 20 Hz; 160°C
Reference No.	419

GRAPH 50: **Fatigue Cycles to Failure vs. Stress in Tension for Long and Short Glass Reinforced Nylon 66.**

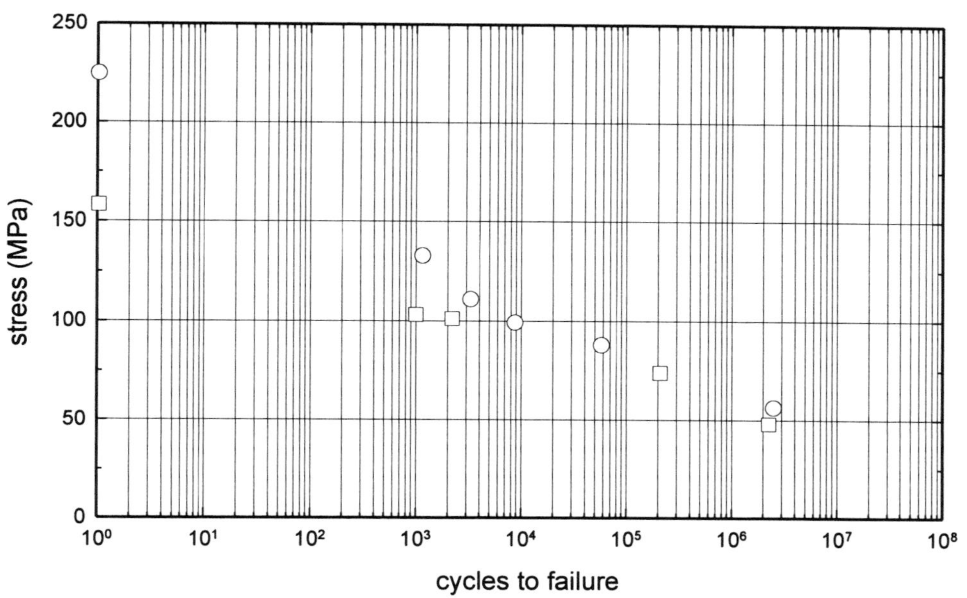

○	LNP Verton RF-700-10 HS Nylon 66 (50% glass fiber; long glass fibers (1180-1290 microns)); tension-tension; 10 Hz; 23°C; ASTM D638 Type 1 [fig g]
□	LNP Thermocomp RF-100-10 HS Nylon 66 (50% glass fiber; short glass fiber (160-190 microns)); tension- tension; 10 Hz; 23°C; ASTM D638 Type 1 [fig g]
Reference No.	391

GRAPH 51: **Fatigue Cycles to Failure vs. Stress in Tension for DuPont Zytel Nylon 66.**

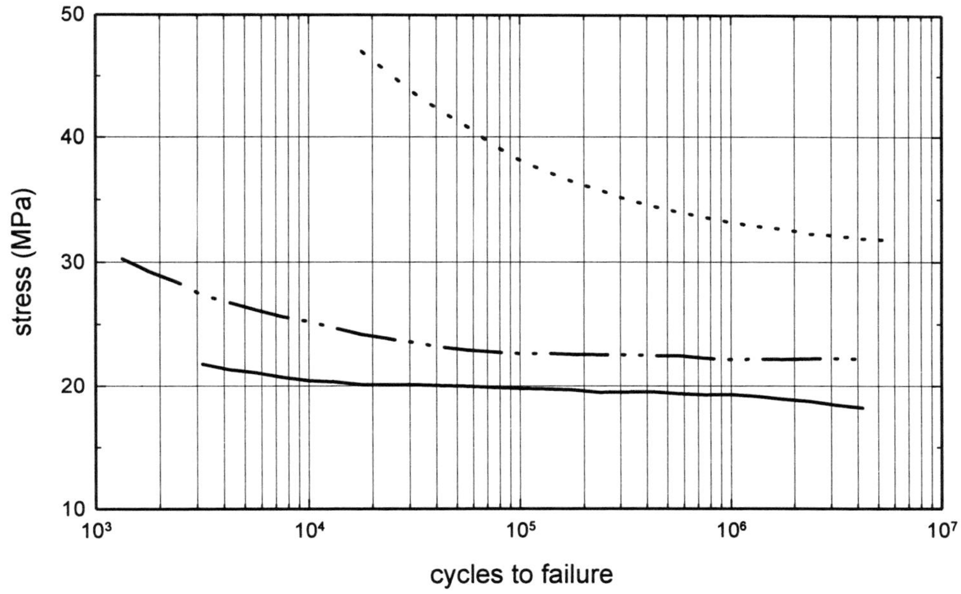

....	DuPont Zytel 101 Nylon 66 (gen. purp. grade, unlubricated); tension-compression; 30 Hz; 23°C; DAM [fig d]
—··—··	DuPont Zytel 101 Nylon 66 (gen. purp. grade, unlubricated); tension-compression; 30 Hz; 23°C; 50% RH [fig d]
———	DuPont Zytel 408 Nylon 66 (impact modified); tension-compression; 30 Hz; 23°C; 50% RH [fig d]
Reference No.	68

GRAPH 52: **Fatigue Cycles to Failure vs. Stress in Tension for Glass Fiber Reinforced DuPont Zytel Nylon 66.**

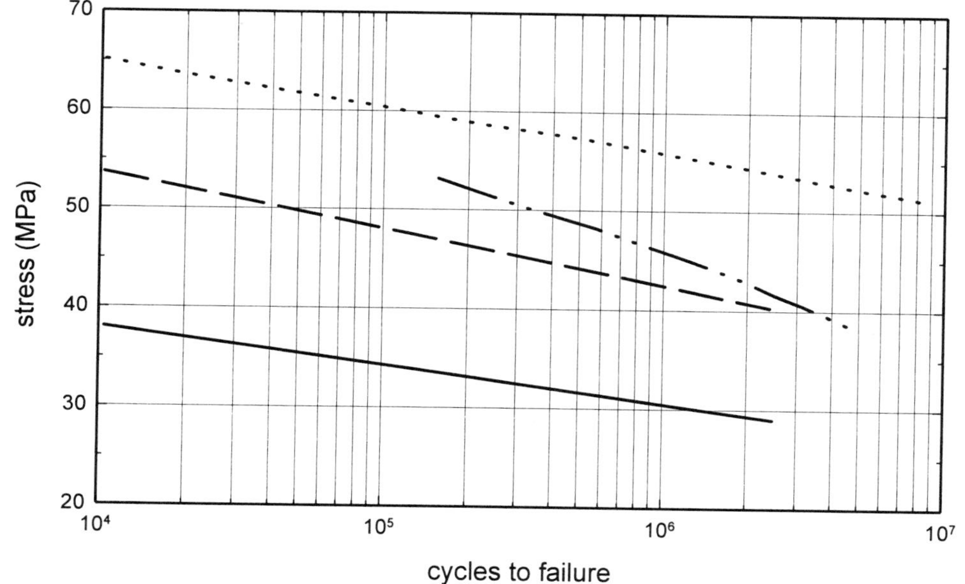

··············	DuPont Zytel 70G 33L Nylon 66 (lubricated, gen. purp. grade; 33% glass fiber); tension-compression; 30 Hz; 23°C; DAM [fig d]
—·—··	DuPont Zytel 70G 33L Nylon 66 (lubricated, gen. purp. grade; 33% glass fiber); tension-compression; 30 Hz; 23°C 50% RH; [fig d]
– – – –	DuPont Zytel 71G 33L Nylon 66 (lubricated, impact modified; 33% glass fiber); tension-compression; 30 Hz; 23°C; DAM [fig d]
———	DuPont Zytel 71G 33L Nylon 66 (lubricated, impact modified; 33% glass fiber); tension-compression; 30 Hz; 23°C; 50% RH [fig d]
Reference No.	68

GRAPH 53: **Fatigue Cycles to Failure vs. Stress in Tension for DuPont Zytel 101 Nylon 66.**

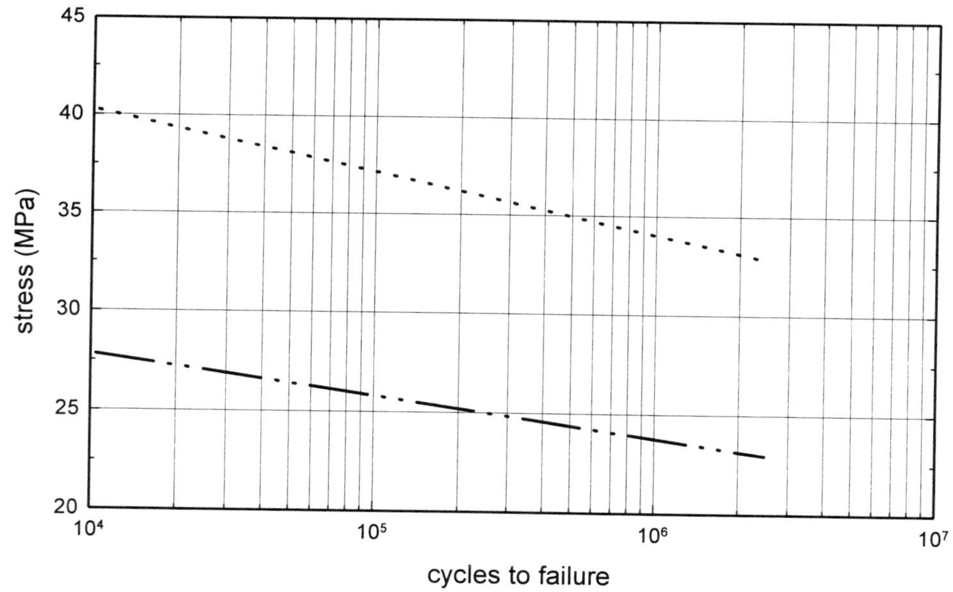

··············	DuPont Zytel 101 Nylon 66 (gen. purp. grade, unlubricated); tension-compression; 30 Hz; 23°C; DAM [fig d]
—·—··	DuPont Zytel 101 Nylon 66 (gen. purp. grade, unlubricated); tension-compression; 30 Hz; 23°C 50% RH; [fig d]
Reference No.	68

GRAPH 54: **Fatigue Cycles to Failure vs. Stress in Tension-Compression for Ube Nylon 66.**

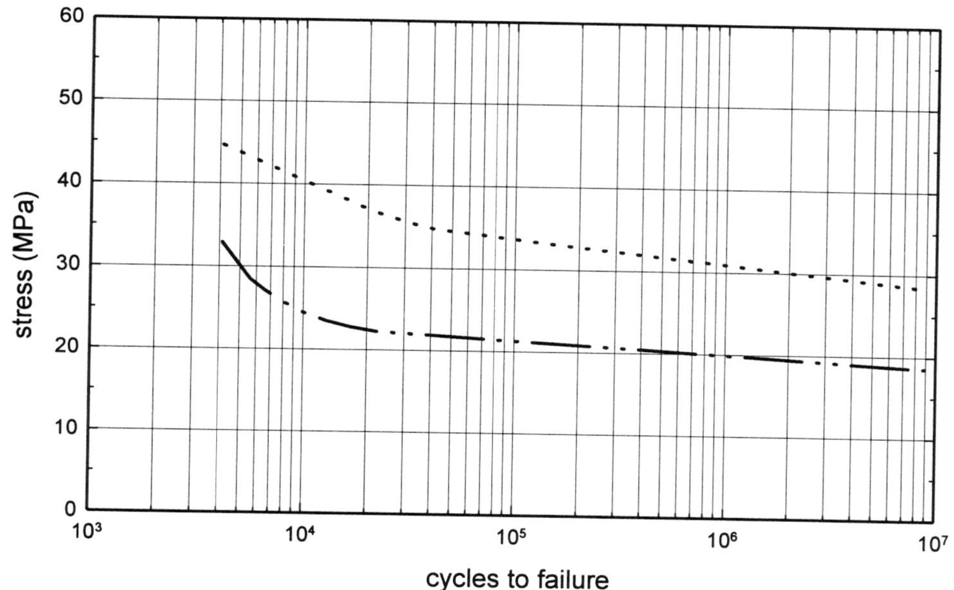

....	Ube 2020GC4 Nylon 66 (20% glass fiber); tension-compression; 16.7 Hz; DAM; 4 mm thick
—··—··	Ube 2020B Nylon 66 (moderate flow, gen. purp. grade); tension-compression; 16.7 Hz; DAM; 4 mm thick
Reference No.	238

GRAPH 55: **Fatigue Cycles to Failure vs. Stress in Tension at Different Temperatures for DuPont Zytel 101 Nylon 66.**

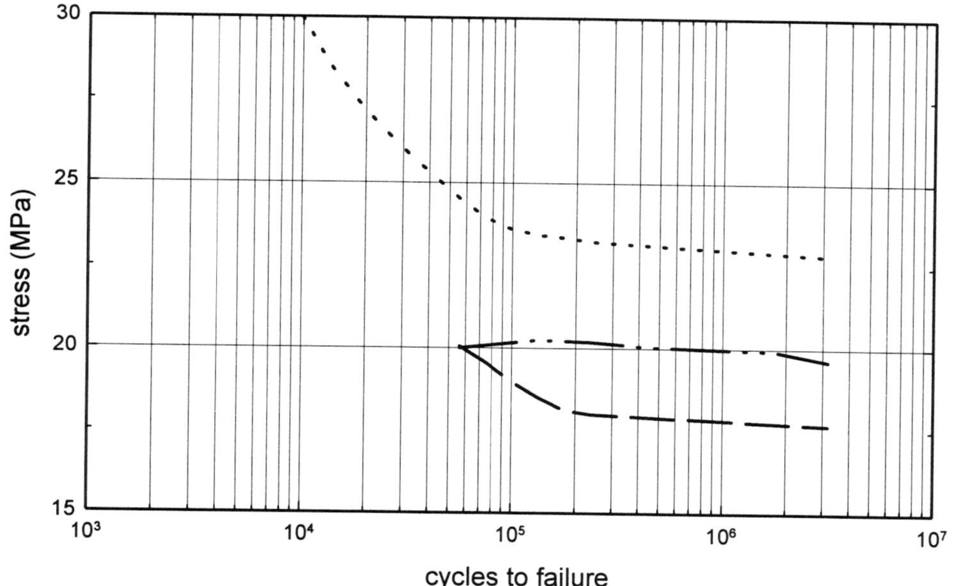

....	DuPont Zytel 101 Nylon 66 (gen. purp. grade, unlubricated); tension-compression; 30 Hz; 23°C; 50% RH [fig d]
—··—··	DuPont Zytel 101 Nylon 66 (gen. purp. grade, unlubricated); tension-compression; 30 Hz; 66°C [fig d]
– – – –	DuPont Zytel 101 Nylon 66 (gen. purp. grade, unlubricated); tension-compression; 30 Hz; 100°C [fig d]
Reference No.	68

GRAPH 56: Fatigue Cycles to Failure vs. Stress for BASF Ultramid A Nylon 66.

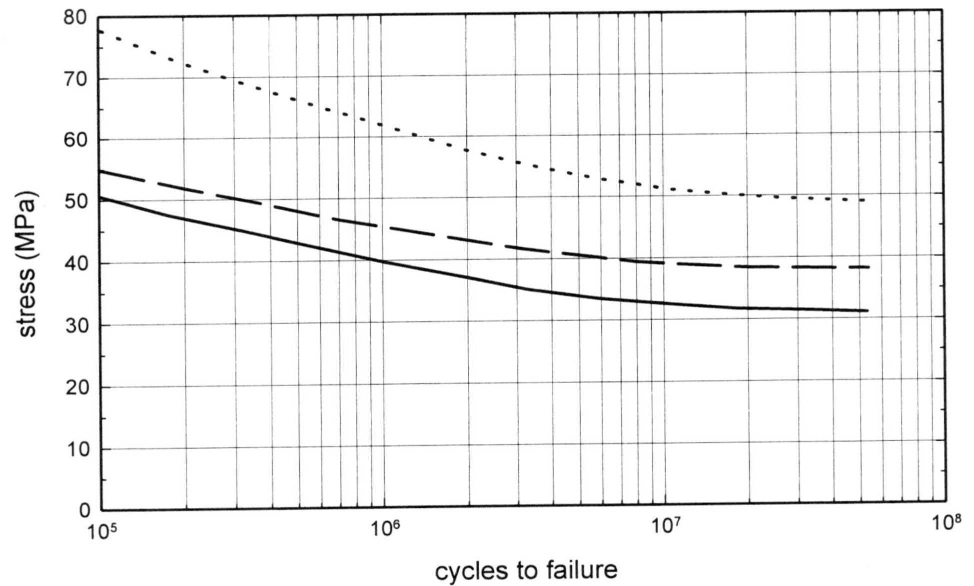

dotted	BASF Ultramid AG5 Nylon 66 (25% glass fiber); tension- compression; DIN 53442; 7.5 Hz; 23°C; 50% RH; laboratory atmosphere [fig. v]
fine dotted	BASF Ultramid AG7 Nylon 66 (35% glass fiber); tension- compression; DIN 53442; 7.5 Hz; 23°C; 50% RH; laboratory atmosphere [fig. v]
dashed	BASF Ultramid AG5 Nylon 66 (25% glass fiber); rotating beam; DIN 53442; 7.5 Hz; 23°C; 50% RH; laboratory atmosphere [fig. u]
solid	BASF Ultramid AG7 Nylon 66 (35% glass fiber); rotating beam; DIN 53442; 7.5 Hz; 23°C; 50% RH; laboratory atmosphere [fig. u]
Reference No.	93

GRAPH 57: Fatigue Crack Propagation (FCP) Rate vs. Stress Intensity Range at Different Test Frequencies for DuPont Zytel 122L Nylon 66.

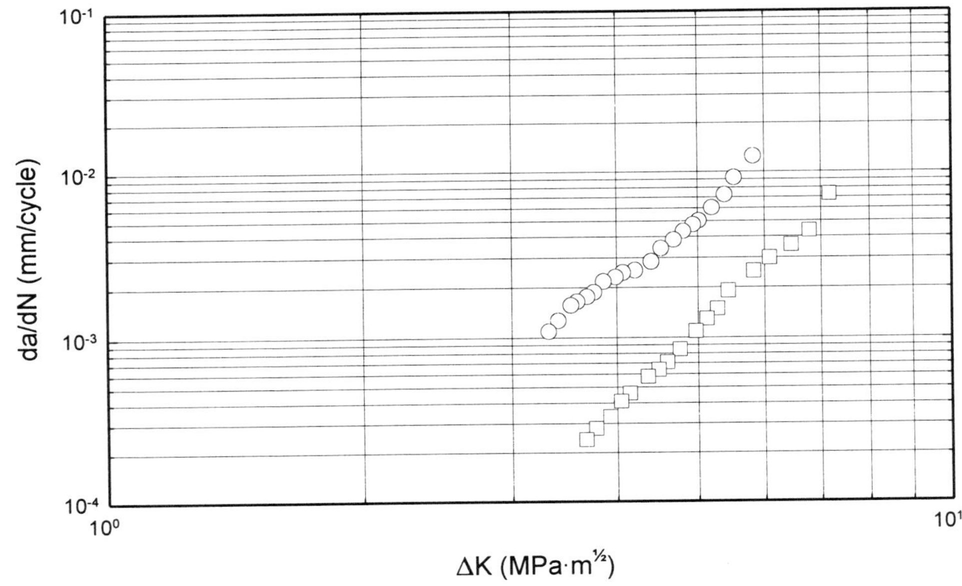

○	DuPont Zytel 122L Nylon 66; tension; 0.5 Hz; 23°C; compact tension specimen, precracked, parallel to flow
□	DuPont Zytel 122L Nylon 66; tension; 5 Hz; 23°C; compact tension specimen, precracked, parallel to flow
Reference No.	401

GRAPH 58: **Fatigue Crack Propagation (FCP) Rate vs. Stress Intensity Range for Supertough Nylon 66.**

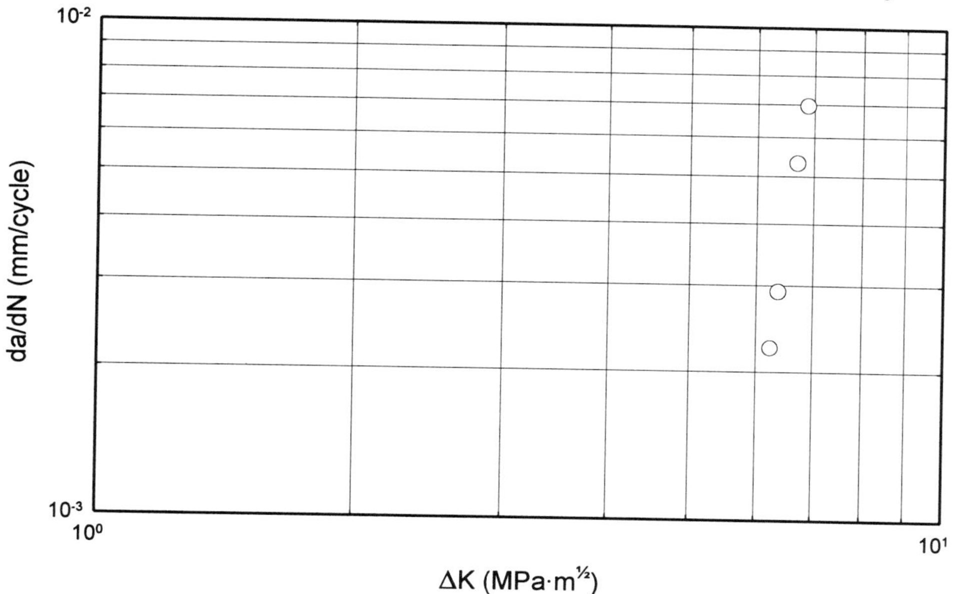

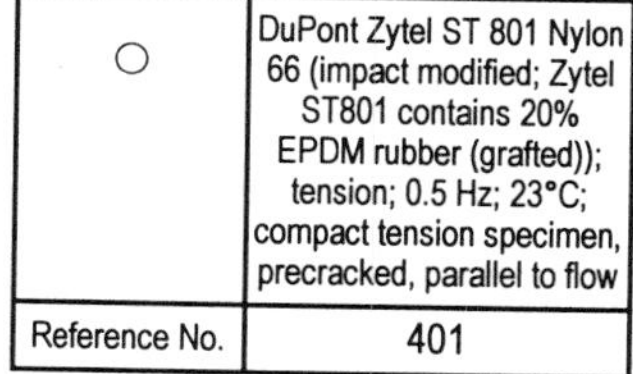

○	DuPont Zytel ST 801 Nylon 66 (impact modified; Zytel ST801 contains 20% EPDM rubber (grafted)); tension; 0.5 Hz; 23°C; compact tension specimen, precracked, parallel to flow
Reference No.	401

GRAPH 59: **Fatigue Crack Propagation (FCP) Rate vs. Stress Intensity Range for Reactive Rubber Modified Nylon 66.**

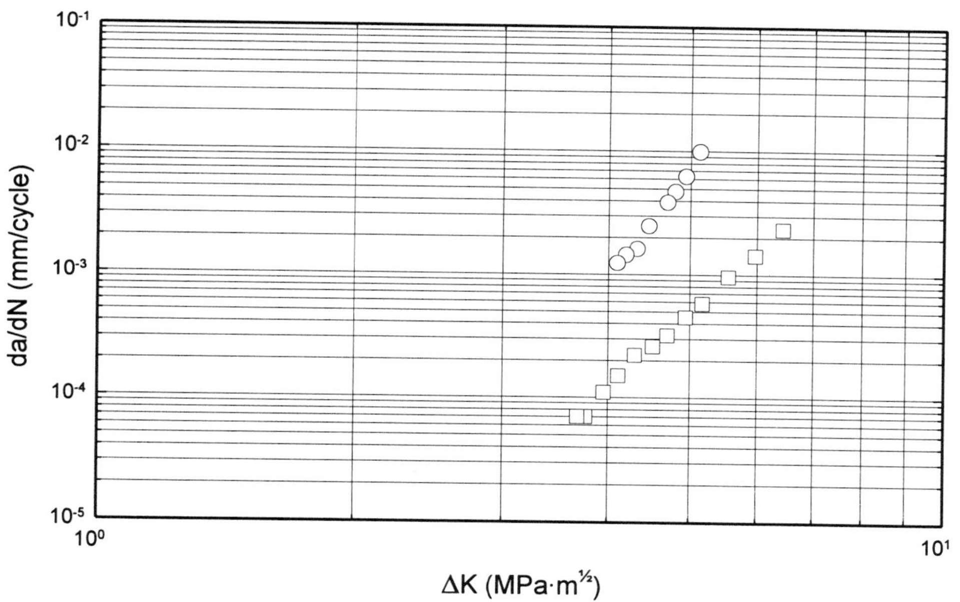

○	Nylon 66 (98% Zytel 122L/ 2% EPDM rubber; rubber toughened); tension; 0.5 Hz; 23°C; compact tension specimen, precracked, parallel to flow
□	Nylon 66 (96% Zytel 122L/ 4% EPDM rubber; rubber toughened); tension; 0.5 Hz; 23°C; compact tension specimen, precracked, parallel to flow
Reference No.	401

GRAPH 60: Fatigue Crack Propagation (FCP) Rate vs. Stress Intensity Range for Nylon 66.

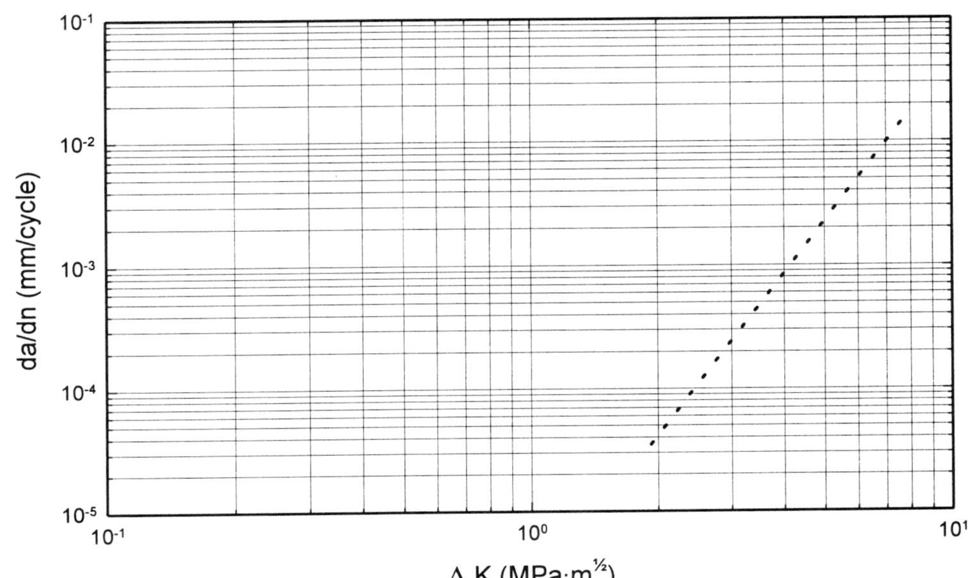

GRAPH 61: Fatigue Crack Propagation (FCP) Rate vs. Stress Intensity Range for Rubber Toughened Nylon 66 Blended with Supertough Nylon.

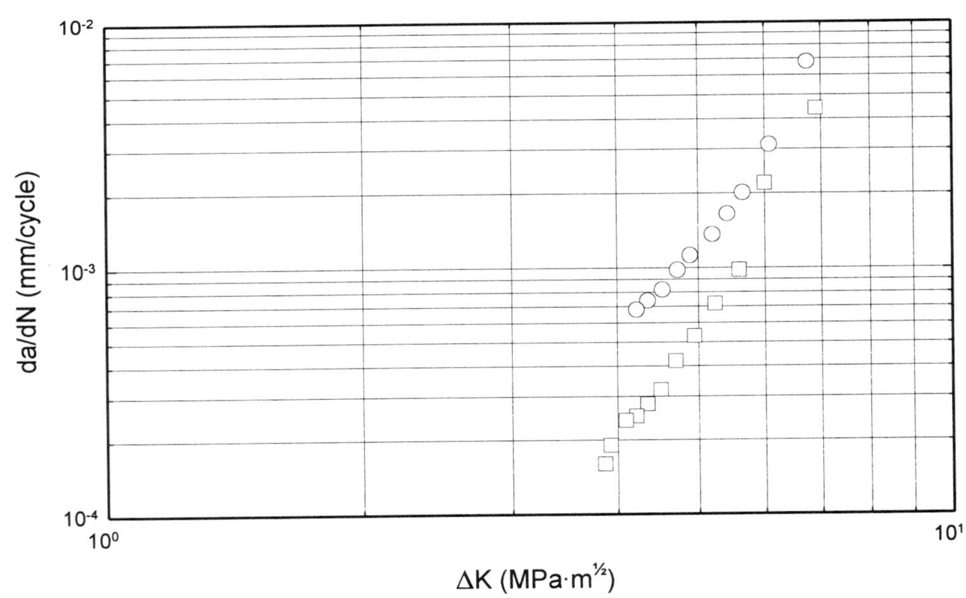

Chapter 13

Nylon MXD6

GRAPH 62: Fatigue Cycles to Failure vs. Stress in Flexure for Glass Fiber Reinforced Mitsubishi Gas Chemical Reny Nylon MXD6.

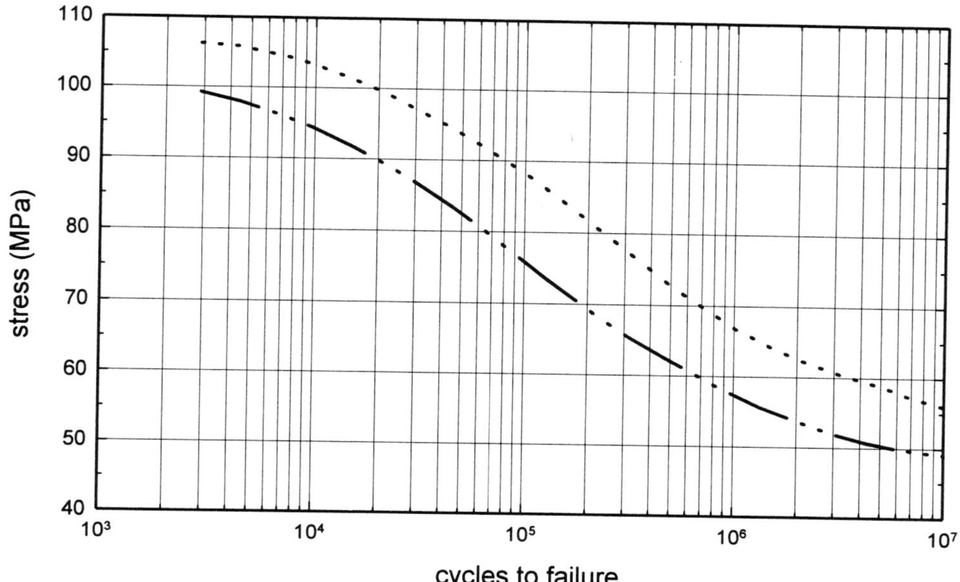

....	Mitsubishi Reny 1022 Nylon MXD6 (50% glass fiber); flexure; ASTM D671-71B; 23°C
—··—··	Mitsubishi Reny 1002 Nylon MXD6 (30% glass fiber); flexure; ASTM D671-71B; 23°C
Reference No.	236

Chapter 14

Nylon 6/6T

GRAPH 63: Fatigue Cycles to Failure vs. Stress in Flexure for Filled and Unfilled BASF Ultramid T Nylon 6/6T.

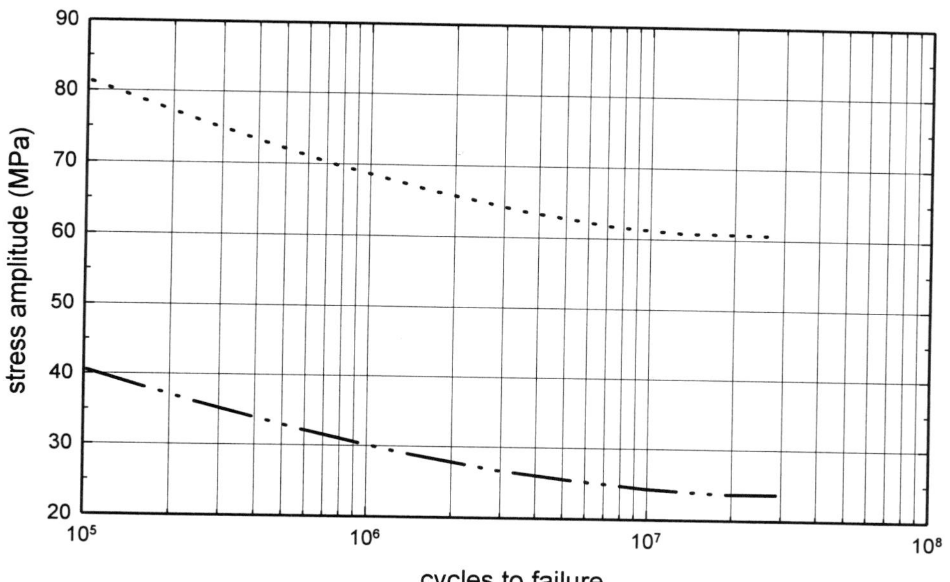

....	BASF AG Ultramid T KR4355G7 Nylon 6/6T (35% glass fiber); flexure (alternating); DIN 53442; 7.5 Hz; 23°C; equilibrium moisture content [fig t]
—··—··	BASF AG Ultramid T KR4350 Nylon 6/6T (gen. purp. grade, unfilled); flexure (alternating); DIN 53442; 7.5 Hz; 23°C; equilibrium moisture content [fig t]
Reference No.	185

Chapter 15

Polyarylamide

GRAPH 64: Fatigue Cycles to Failure vs. Stress in Flexure for 50% Glass Fiber Reinforced Solvay Ixef 1022 Polyarylamide.

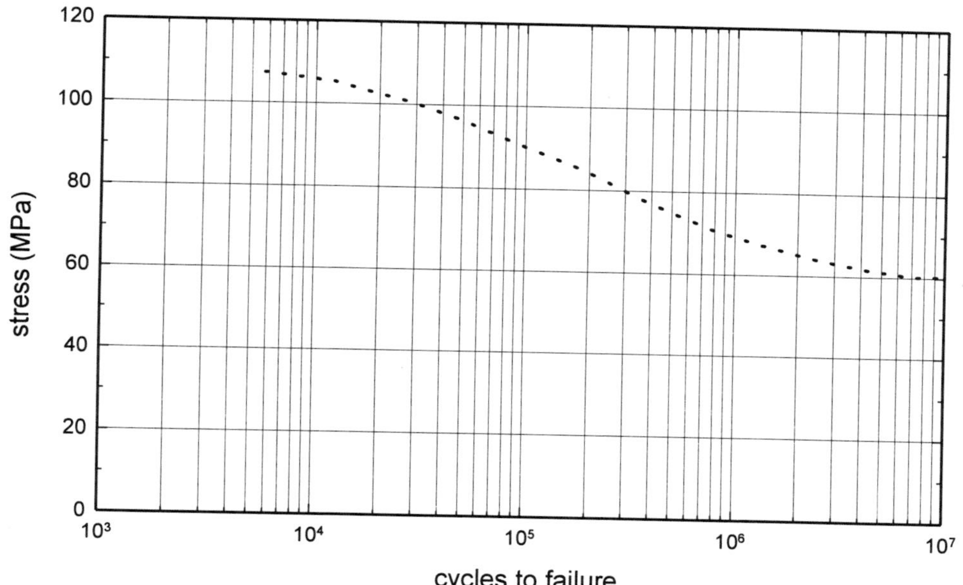

....	Solvay Ixef 1022 Polyarylamide (50% glass fiber); flexure; ASTM D671 B; 23°C
Reference No.	135

Chapter 16

Polyphthalamide

Fatigue Properties

LNP Engineering Plastics: Verton UF-700-10 (material compostion: 50% glass fiber reinforcement; note: long glass fibers)

Polyphthalamide (50% long glass fiber reinforced) composite exhibited the best fatigue performance of materials tested in both tensile and flexural fatigue tests, followed by the nylon 66 and polypropylene composites (both 50% long glass fiber reinforced). Final failure of all of the materials was characterized by a large degree of fiber pull-outs, some bare fibers, and broken filaments which presumably occur when there was insufficient resin and interface left to support the remaining fibers.

Reference: Grove, D.A. (LNP), Kim, H.C. (LNP), *Effect of Constituents on the Fatigue Behavior of Long Fiber Reinforced Thermoplastics,* ANTEC 1995, conference proceedings - Society of Plastics Engineers, 1995.

GRAPH 65: Fatigue Cycles to Failure vs. Stress in Flexure for 33% Glass Fiber Reinforced Amoco Amodel A-1133HS Polyphthalamide.

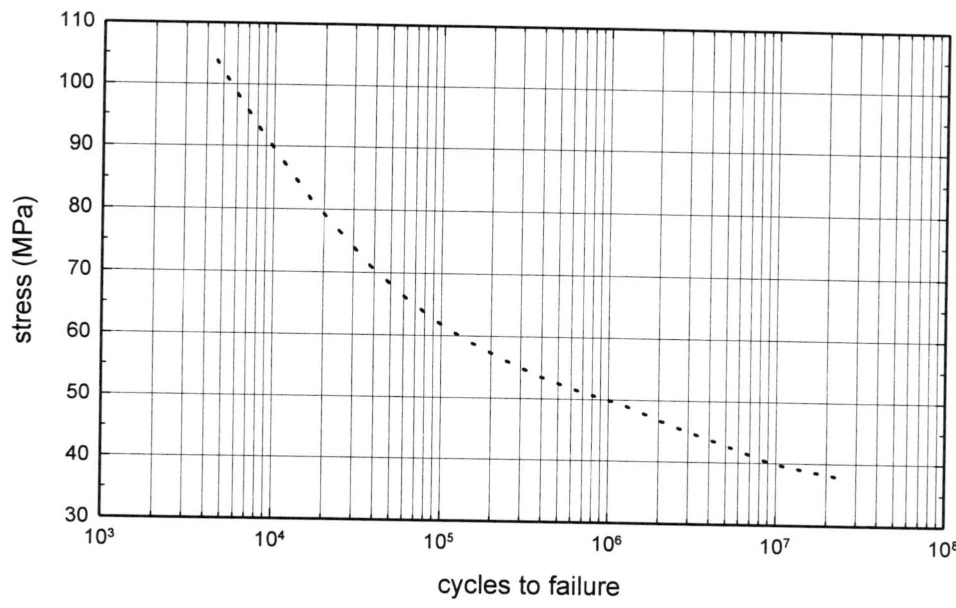

....	Amoco Amodel A1133HS PPA (33% glass fiber); flexure; 23°C
Reference No.	359

GRAPH 66: **Fatigue Cycles to Failure vs. Stress in Flexure for 50% Long Glass Fiber Reinforced Polyphthalamide.**

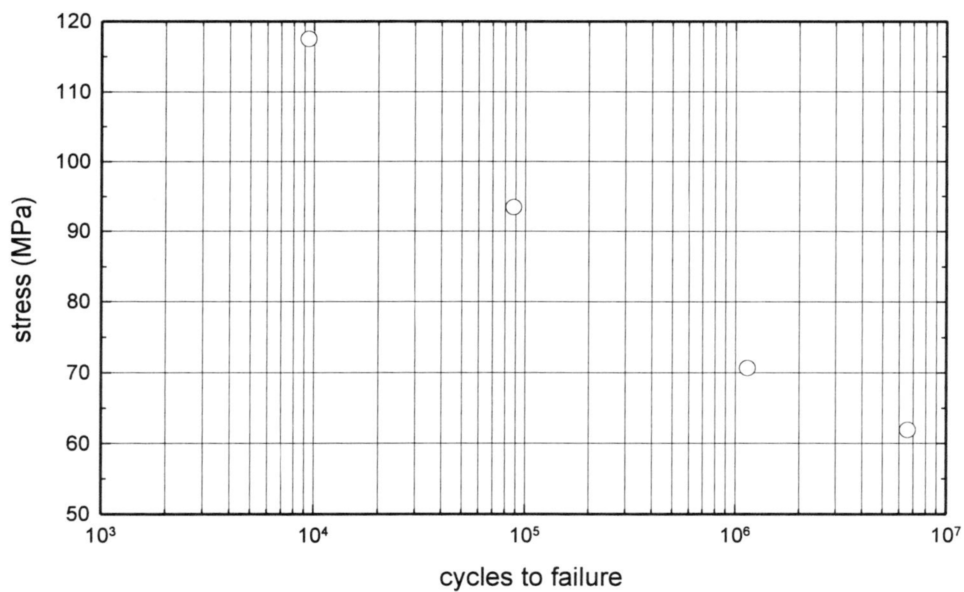

○	LNP Verton UF-700-10 PPA (50% glass fiber; long glass fibers); flexure; 30 Hz; 23°C; ASTM D671 Type B [fig d]
Reference No.	391

GRAPH 67: **Fatigue Cycles to Failure vs. Stress in Flexure for Reinforced LNP Polyphthalamide.**

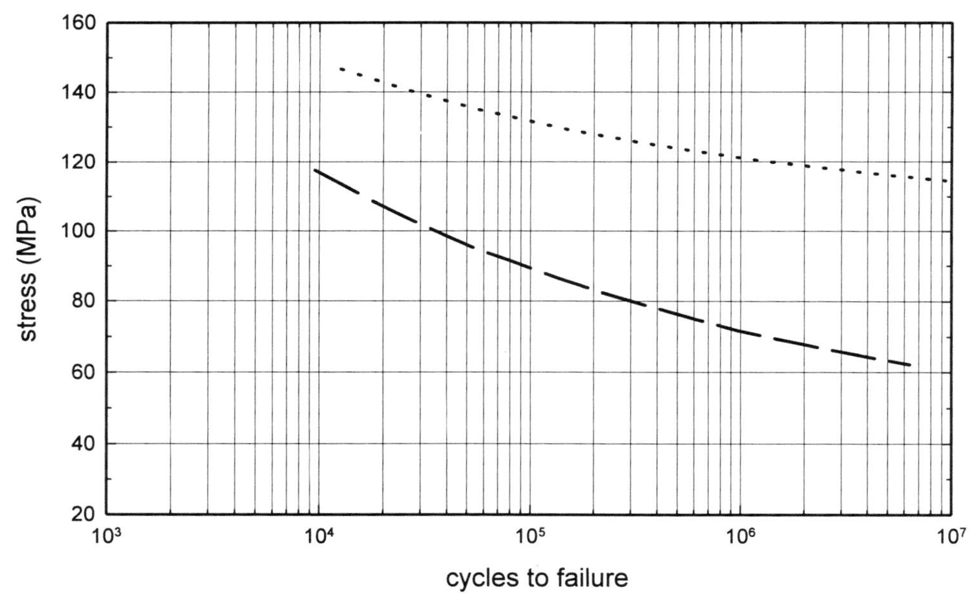

···············	LNP Stat-Kon UC1008 PPA (static dissipative; 40% carbon fiber); flexure; 30 Hz; 23°C
···············	LNP Thermocomp UC1008 PPA (40% carbon fiber); flexure; 30 Hz; 23°C
— — — —	LNP Verton UF70010 PPA (50% glass fiber; long glass fibers); flexure; 30 Hz; 23°C
Reference No.	394

GRAPH 68: Fatigue Cycles to Failure vs. Stress in Tension for 50% Long Glass Fiber Reinforced Polyphthalamide.

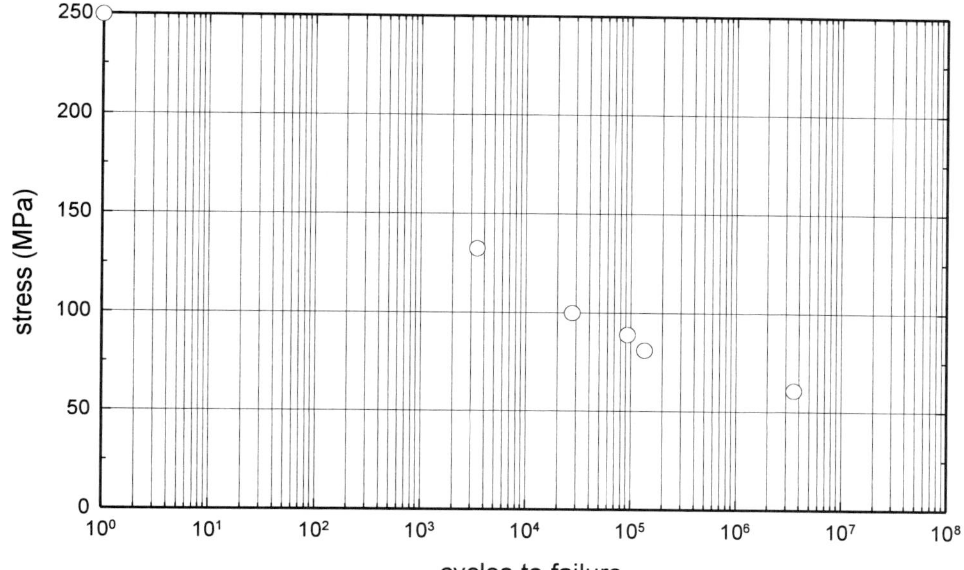

○	LNP Verton UF-700-10 PPA (50% glass fiber; long glass fibers); tension-tension; 10 Hz; 23°C; ASTM D638 Type 1 [fig g]
Reference No.	391

Chapter 17

Polycarbonate

Fatigue Properties

Bayer: Makrolon T-7855 (additives: mold release; features: moderate flow, impact modified, opaque, unfilled, easy mold release); **Makrolon**

The stress cycle diagram of Makrolon polycarbonate shows that the material only permits a moderate dynamic load at room temperature. It becomes more interesting at 100°C (212°F) if compared with other thermoplastics. At the high temperature level, however, a certain embrittlement takes place, finally reducing the ultimate tensile elongation to 40% after 10^7 cycles.

In spite of its outstanding impact resistance, the fatigue endurance of Makrolon polycarbonate grade T-7855 is only half that of normal grades and, therefore, of little interest in this respect.

Reference: *Makrolon Polycarbonate Design Guide,* supplier design guide (55-A840(10)L) - Bayer, 1988.

Dow Chemical: Calibre (features: transparent)

Because Calibre resins are viscoelastic materials, and are thus sensitive to temperature, changing any variable that affects heat transfer can cause a shift in the S-N curve. Such variables include the frequency of loading cycles, specimen thickness, and ambient temperature. Hysteretic heating (caused by constant loading-unloading) can be an important factor in fatigue failure and the designer, to minimize heating and its effects, should:

1) Reduce stress by ribbing, or by redistributing the load
2) Avoid stress concentrators such as sharp corners or abrupt changes in cross-sectional geometry.
3) Reduce the frequency of loading cycles, if possible.
4) Avoid unnecessarily thick walls which reduce heat transfer and thermal dissipation.

Reference: *Calibre Engineering Thermoplastics Basic Design Manual,* supplier design guide (301-1040-1288) - Dow Chemical Company, 1988.

Polycarbonate (features: general purpose grade)

A standard grade of Polycarbonate was tested in RAPRA Technologies laboratories under tensile/zero loading at a frequency of 0.5 Hz. The dramatic reduction in fatigue resistance after between 10,000 and 20,000 cycles coincides with a transition from ductile yielding to brittle fracture. The position of this transition is quite sensitive to those many factors that promote stress embrittlement. In particular it will be accelerated to lower critical cycles by:

> Lower frequencies; higher temperatures; stress concentrations (including notches, voids, and knit lines); chemical attack and environmental stress cracking; thermal degradation; UV degradation; larger products (high levels of strain energy for crack propagation); thicker products (suppression of yielding at crack tips); biaxial tensile stress fields

To quantify the influence of all of these factors for a given design and application would be an impossible task. However on the evidence available it would be unsafe to assume that a real product will sustain more than 5000 cycles at stress amplitudes of greater than 12 MPa.

It is not untypical for polycarbonate fatigue data to be presented at a minimum of 100,000 cycles and test frequency of 30 Hz. By doing so, the dramatic and very informative ductile to brittle transition has been avoided. This nominally very tough material will generally fail by brittle fracture under fatigue conditions, and will therefore be very sensitive to many factors.

Reference: Wright, D. (RAPRA), *Knowledge Based System for Plastics - SPE ANTEC '95,* presentation - RAPRA Technology Ltd., 1995.

Bisphenol A PC (features: 1.5 mm thick; note: specimens milled from Lexan 9030 sheet, saturation cyled in CO_2 and desorbed between 140 and 260 hours); **Bisphenol A PC** (features: 1.5 mm thick; note: specimens milled from Lexan 9030 sheet, saturation cyled in CO_2 and desorbed >1000 hours)

Saturation-cycling polycarbonate with CO_2 greatly increases the fatigue life of notched polycarbonate. The fatigue life of notched saturation-cycled specimens is approximately 70 times that of notched unprocessed specimens in tests performed with maximum cyclic stresses of approximately 20 MPa. The effect of saturation-cycling is a bulk effect, rather than a surface effect. Increased free volume is likely the mechanism. The increased fatigue life may result from dominance of the shear fatigue mechanism at lower maximum cyclic stresses than in the unprocessed polymer.

Reference: Seeler, K.A. (Lafayette College), Kumar, V. (University of Washington), *Effect of Carbon Dioxide Saturation and Desorption on the Notch Sensitivity of Polycarbonate Fatigue,* ANTEC 1994, conference proceedings - Society of Plastics Engineers, 1994.

Microcellular Polycarbonate

Microcellular polycarbonate with a relative density of 0.97 has a fatigue life four times greater than that of solid polycarbonate when tested under comparable conditions. The fatigue life of microcellular polycarbonate with relative density of 0.83 was longer than the solid polycarbonate. Microcellular polycarbonate with relative densities less than 0.83 had shorter fatigue lives than solid polycarbonate even though the average matrix stress was approximately equal to the stress in the solid specimen. Very low relative density microcellular polycarbonate foam has very short fatigue life.

It seems likely that the differences between the fatigue behavior of microcellular and solid polycarbonate result directly from the geometry of the foam blunting either the leading craze crack growth or the shear plane growth, although other factors such as non-uniform stress and strain, may also contribute.

Reference: Seeler, K.A. (Lafayette College), Kumar, V. (University of Washington, *Tension-Tension Fatigue of Microcellular Polycarbonate: Inital Results,* ANTEC 1992, conference proceedings - Society of Plastics Engineers, 1992.

Effect of Temperature on Fatigue Behavior

Bisphenol A PC (note: specimen contains a weld line)

Takemori (M.T. Takemori (1988), "Polymer Fatigue Fracture Diagrams: BPA Polycarbonate", Poly.Eng.Sci.V.28, N.10, p.641) performed sinusoidal tension-tension testing of injection molded polycarbonate specimens with a weld line in the middle of the gage length at 10 Hz and with a ratio of the minimum to maximum load of 0.1. He observed fatigue lifetime inversion in tests between -25°C (-13°F) and 50°C (122°C). Fatigue lifetime inversion did not occur in tests at 75°C (167°F) and above. Takemori attributed the fatigue lifetime inversion to competition between two fatigue fracture modes: crack growth through leading crazes and shear fracture at 45 degrees from the load direction. The shear planes develop at higher stresses, but they propagate slowly. Crack growth through leading crazes is faster, but it is blunted by shear planes. Takemori presented a fatigue fracture diagram for polycarbonate as a function of temperature and normalized stress which divides the fatigue behavior of specimens with weld lines into three regions; craze failures, shear failures and necking failures.

Reference: Seeler, K.A. (Lafayette College), Kumar, V. (University of Washington, *Tension-Tension Fatigue of Microcellular Polycarbonate: Inital Results,* ANTEC 1992, conference proceedings - Society of Plastics Engineers, 1992.

Effect of Weld Line on Fatigue Behavior

GE Plastics: Lexan 141 (features: 3.2 mm thick; note: specimen does not contain a weld line); **Lexan 141** (features: 3.2 mm thick; note: specimen contains a weld line)

When operating in the stress range (50 to 34 MPa), double gated specimens do not fail at the weld-line. The recorded lifetime was independent to whether the specimen failed at the weld-line or not. Consequently, the corresponding S-N points recorded for high stress amplitudes were used as representative of specimens containing welds as well as representative of specimens without welds. The fatigue behavior can be summarized in the following three points:

a) when the applied stress is decreased from 50 MPa to 34 MPa, no effect of weld-line seems to be noticeable;

b) when the applied stress decreases from 34 to 30 MPa the effect of the weld-line increases dramatically, resulting in an inversion in the S-N curve of the welded specimens;

c) for stresses ranging between 30 MPa and the endurance limit, the effect of the weld-line became stationary, i.e. specimens with and without weld-lines follow the same general pattern behavior.

The above presentation of the results displays the effect of the weld-line on the fatigue lifetime of polycarbonate step-by-step. In considering the whole S-N curve corresponding to each situation (with weld-lines and without weld-lines), it is interesting to note that the S-N curve for specimens with weld-lines presents a transition. The following hypotheses can be suggested for discussion.

(1) For polycarbonate, the weld-line effect acts principally via the V-notch effect. The V-notch involved is small and can be considered as a surface notch.

(2) Polycarbonate is a notch sensitive material: a notch effect will exist under some conditions and will be absent under others. These conditions include mainly: stress state, temperature, thickness and strain rate.

(3) When there is a notch effect the fatigue test performed becomes similar to a fatigue crack propagation test. When the notch effect is absent the test performed is the conventional fatigue test. Specific considerations should be undertaken for each situation.

When these hypotheses are appropriately linked to the obtained results described previously; the following qualitative explanation is proposed:

In high stress range (50 - 34 MPa) the notchy-tip is blunted within the first few cycles, consequently the notch effect is suppressed and specimens with and without weld-lines undergo the same failure mechanism. The failure is of ductile type involving extensive shear yielding. In low stress range (30 MPa - Endurance Limit), there is no notch blunting. Specimens containing weld-lines will fail earlier due to notch effect which reduces the crack initiation stage. It is worth noting that the presence of the V-notch not only shortens the crack initiation stage but also the crack propagation stage.

Reference: Boukhili, R. (Ecole Polytechnique de Montreal), Gauvin, R. (Ecole Polytechnique de Montreal), Gossel, *Fatigue Behavior of Injection Moded Polycarbonate and Polystyrene Containing Weld Lines,* ANTEC 1989, conference proceedings - Society of Plastics Engineers, 1989.

Bisphenol A PC

Matsumoto and Gifford (Matsumoto, D.S. and Gifford, S.K., 1985, "Competing Fatigue Mechanisms in BPA-Polycarbonate," Journal of Materials Science, Vol. 20, pp. 873-880) found that tests on specimens with weld-lines and specimens that were razor notched were substantially similar. They performed tests at 1 Hz and 10 Hz to demonstrate that the fatigue life inversion was not due to hysteretic heating. They further demonstrated that the increased life of notched specimens fatigued at high cyclic stress was due to strain and not heat by creep loading notched specimens at 52 MPa before cycling them at 28 MPa. Fatigue life increased by up to approximately two orders of magnitude as a function of the duration of the creep loading. Finally, they found that polished smooth bar specimens, washed to remove craze producing oils accumulated during handling, do not have fatigue life inversion. They proposed the lack of inversion was due to the time it took to initiate the fatigue crack on the smooth surface and because the specimens exhibited both discontinuous crack growth and shear band growth.

Reference: Seeler, K.A. (Lafayette College), Kumar, V. (University of Washington), *Effect of Carbon Dioxide Saturation and Desorption on the Notch Sensitivity of Polycarbonate Fatigue,* ANTEC 1994, conference proceedings - Society of Plastics Engineers, 1994.

GRAPH 69: **Fatigue Cycles to Failure vs. Stress in Flexure for Bayer Makrolon Polycarbonate.**

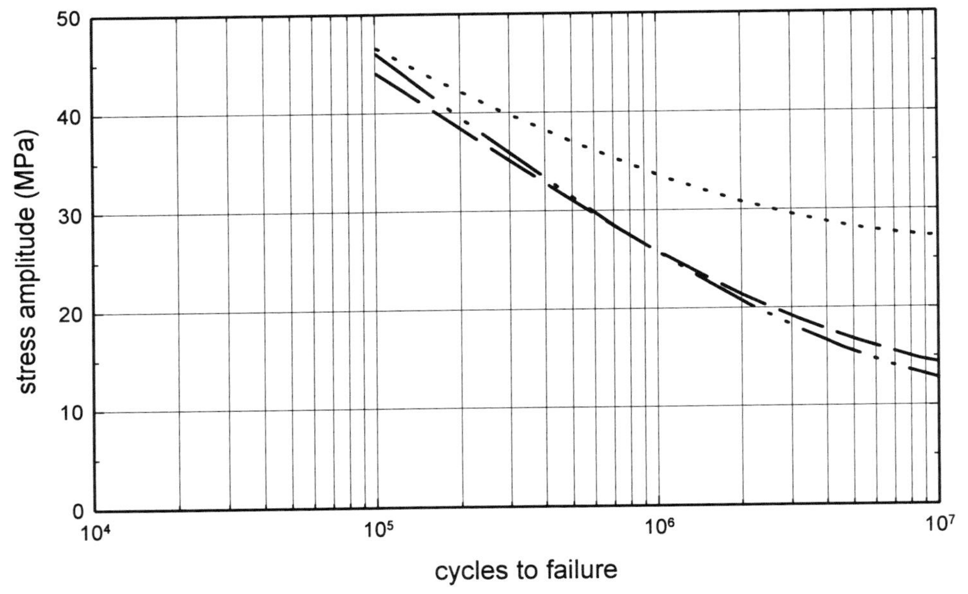

....	Bayer Makrolon 9415 Polycarbonate (moderate flow, opaque, flame retardant; 10% glass fiber; bromine free FR; w/ mold release); flexure; 7 Hz; 23°C; 2x stress amplitude = fatigue limit [fig d]
—··—··	Bayer Makrolon 3200 Polycarbonate (transparent, gen. purp. grade); flexure; 7 Hz; 23°C; 2x stress amplitude = fatigue limit [fig d]
— — — —	Bayer Makrolon 6455 Polycarbonate (moderate flow, flame retardant, unfilled; bromine free FR); flexure; 7 Hz; 23°C; 2x stress amplitude = fatigue limit [fig d]
Reference No.	416

GRAPH 70: **Fatigue Cycles to Failure vs. Stress in Flexure for Reinforced LNP Stat-Kon Polycarbonate.**

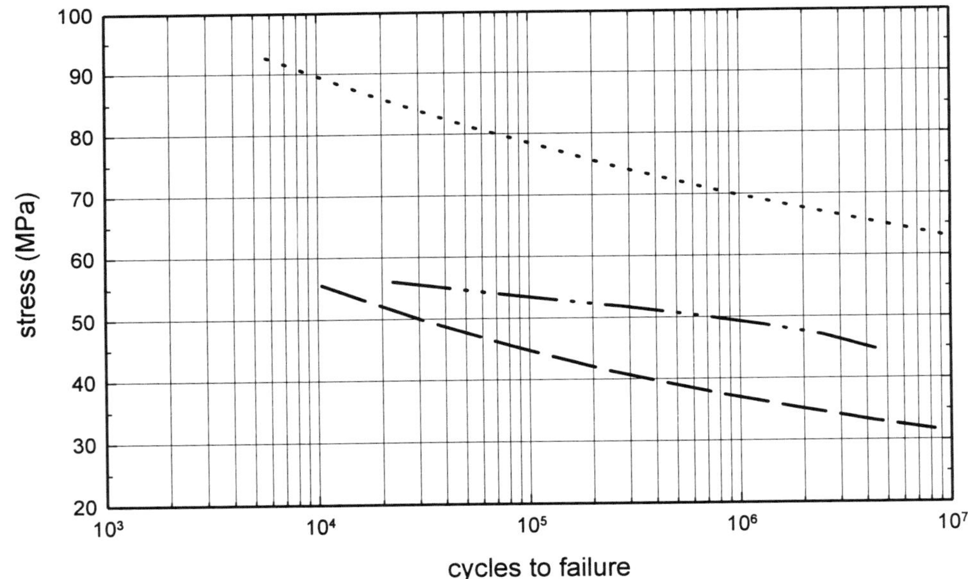

....	LNP Stat-Kon DC1006 Polycarbonate (static dissipative; 30% carbon fiber); flexure; 30 Hz; 23°C
—··—··	LNP Stat-Kon DC1003 Polycarbonate (static dissipative; 15% carbon fiber); flexure; 30 Hz; 23°C
— — — —	LNP Stat-Kon DCL4032FR Polycarbonate (flame retardant, static dissipative; 15% PTFE modified, 10% carbon fiber); flexure; 30 Hz; 23°C
Reference No.	394

GRAPH 71: Fatigue Cycles to Failure vs. Stress in Flexure for Reinforced LNP Thermocomp Polycarbonate.

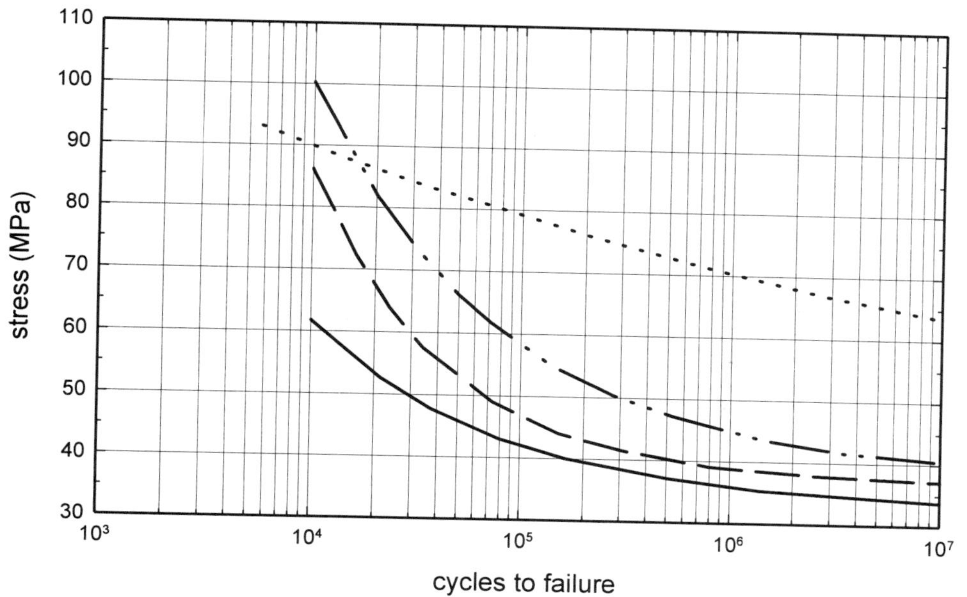

....	LNP Thermocomp DC1006 Polycarbonate (30% carbon fiber); flexure; 30 Hz; 23°C
—··—··	LNP Thermocomp DF1008 Polycarbonate (40% glass fiber); flexure; 30 Hz; 23°C
— — — —	LNP Thermocomp DF1006 Polycarbonate (30% glass fiber); flexure; 30 Hz; 23°C
————	LNP Thermocomp DF1004 Polycarbonate (20% glass fiber); flexure; 30 Hz; 23°C
Reference No.	394

GRAPH 72: Fatigue Cycles to Failure vs. Stress in Flexure for Reinforced LNP Lubricomp Polycarbonate.

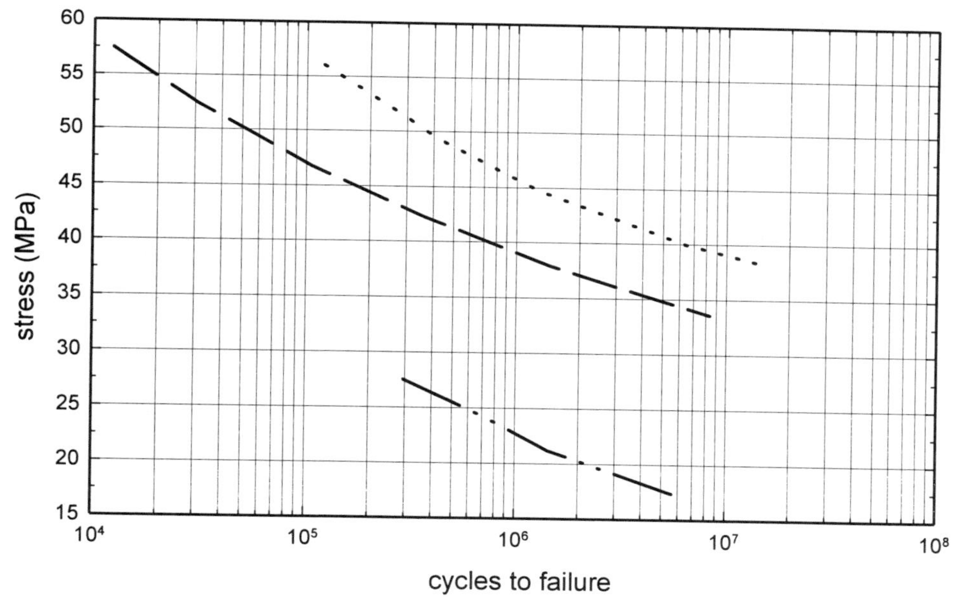

....	LNP Lubricomp DCL4023 Polycarbonate (lubricated; 15% carbon fiber, 10% PTFE modified); flexure; 30 Hz; 23°C
—··—··	LNP Lubricomp DAL4022 Polycarbonate (lubricated; 10% PTFE modified, 10% aramid fiber); flexure; 30 Hz; 23°C
— — — —	LNP Lubricomp DCL4032FR Polycarbonate (lubricated; 10% carbon fiber, 15% PTFE modified); flexure; 30 Hz; 23°C
Reference No.	394

GRAPH 73: **Fatigue Cycles to Failure vs. Stress in Flexure for Dow Chemical Calibre Polycarbonate.**

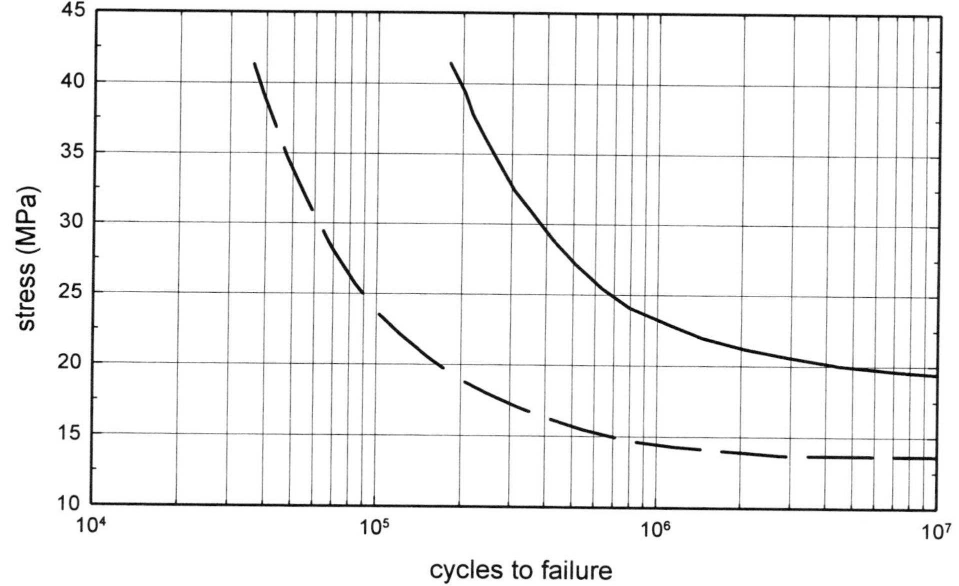

———	Dow Calibre 300-15 Polycarbonate (transparent, gen. purp. grade, 72 Rockwell M; 15 g/10 min. MFI); flexure; ASTM D671; 23°C
– – – –	Dow Calibre 300-10 Polycarbonate (transparent, gen. purp. grade, 73 Rockwell M; 10 g/10 min. MFI); flexure; ASTM D671; 23°C
Reference No.	78

GRAPH 74: **Fatigue Cycles to Failure vs. Stress in Tension for Flame Retardant Grades of General Electric Lexan Polycarbonate.**

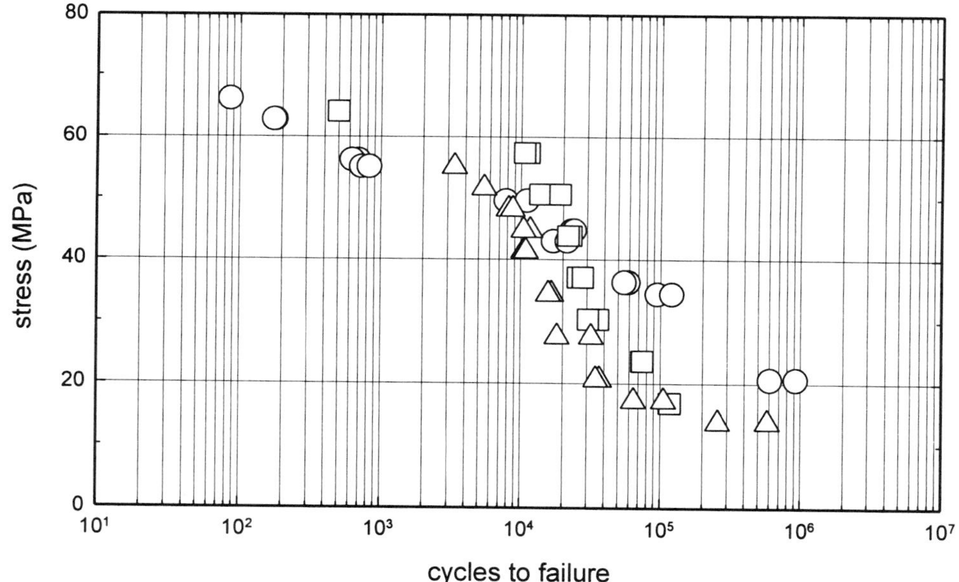

○	GE Lexan 500 Polycarbonate (flame retardant; 10% glass fiber; inj. mold.); tension-tension; GE method; 5 Hz; stress ratio: 0.1; 23°C; flow direction [fig g]
□	GE Lexan 920A Polycarbonate (transparent, appearance grade, flame retardant, unfilled; inj. mold.); tension-tension; GE method; 5 Hz; stress ratio: 0.1; 23°C; flow direction [fig g]
△	GE Lexan 940 Polycarbonate (opaque, flame retardant, unfilled; inj. mold.); tension-tension; GE method; 5 Hz; stress ratio: 0.1; 23°C; flow direction [fig g]
Reference No.	352

GRAPH 75: **Fatigue Cycles to Failure vs. Stress in Tension at Different Temperatures for Glass Fiber Reinforced General Electric Lexan Polycarbonate.**

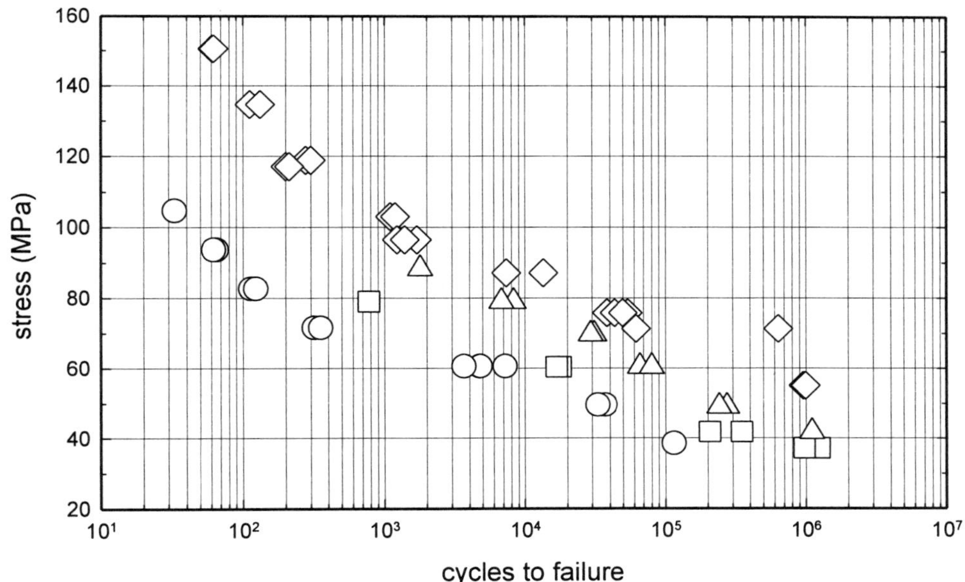

○	GE Lexan 3412 Polycarbonate (20% glass fiber); tension- tension; GE method; 5 Hz; stress ratio: 0.1; 23°C; flow direction [fig g]
□	GE Lexan 3413 Polycarbonate (30% glass fiber); tension- tension; GE method; 5 Hz; stress ratio: 0.1; 23°C; flow direction [fig g]
△	GE Lexan 3413 Polycarbonate (30% glass fiber); tension- tension; GE method; 5 Hz; stress ratio: 0.1; 66°C; flow direction [fig g]
◇	GE Lexan 3414 Polycarbonate (40% glass fiber); tension- tension; GE method; 5 Hz; stress ratio: 0.1; 23°C; flow direction [fig g]
Reference No.	352

GRAPH 76: **Fatigue Cycles to Failure vs. Stress in Tension for General Electric Lexan 141 Polycarbonate with and without Weld Line.**

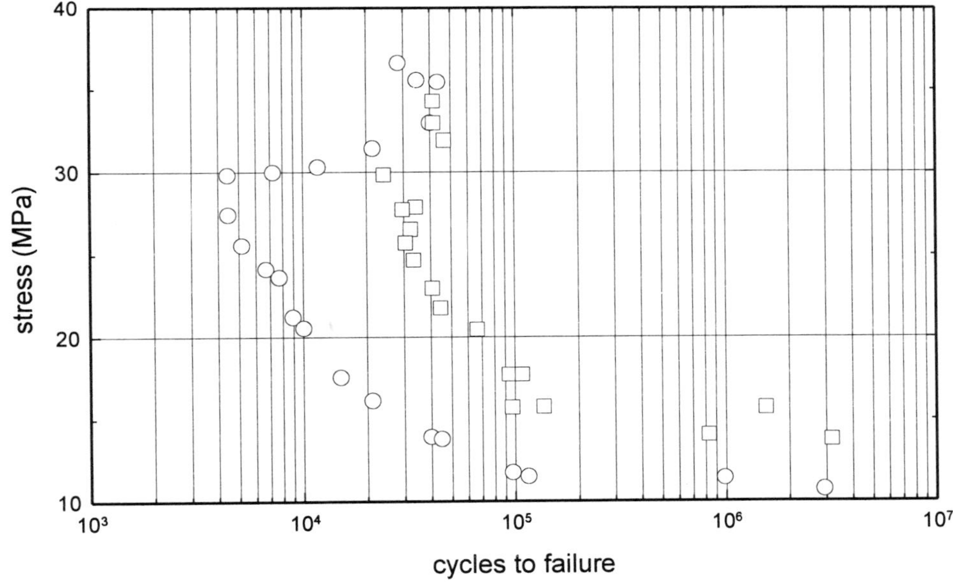

□	GE Lexan 141 Polycarbonate (3.2 mm thick; without weld line); tension- tension; 10 Hz; stress ratio: 0.1; 23°C [fig g, aa]
○	GE Lexan 141 Polycarbonate (3.2 mm thick; with weld line); tension- tension; 10 Hz; stress ratio: 0.1; 23°C [fig g, aa]
Reference No.	372

GRAPH 77: **Fatigue Cycles to Failure vs. Stress in Tension for Polycarbonate Under Uniaxial and Biaxial Stress Fields.**

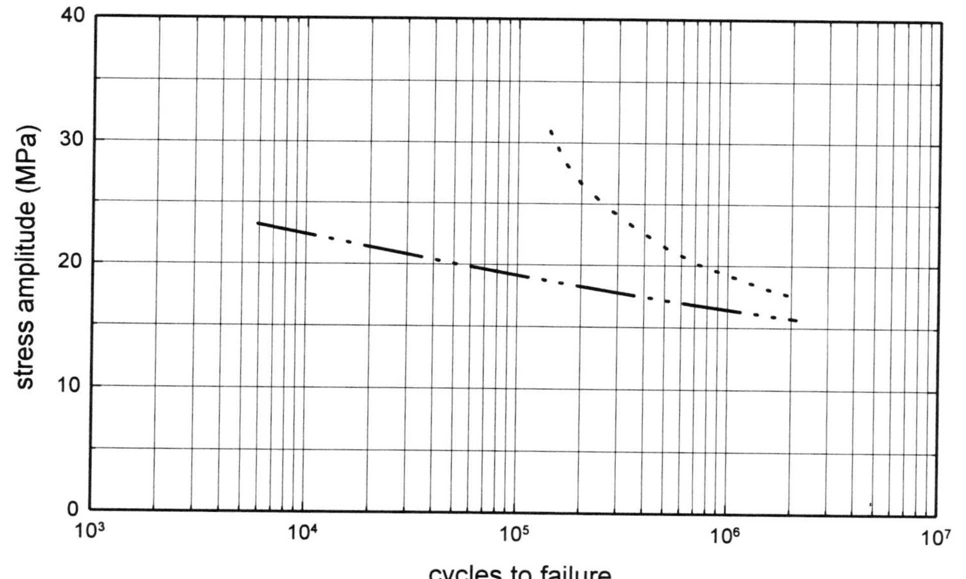

....	Polycarbonate; tension-zero; 0.5 Hz; 20°C; uniaxial loading
—··—··	Polycarbonate; tension-zero; 0.5 Hz; 20°C; biaxial loading
Reference No.	355

GRAPH 78: **Fatigue Cycles to Failure vs. Stress in Tension for General Electric Lexan 141 Polycarbonate with and without Weld Line.**

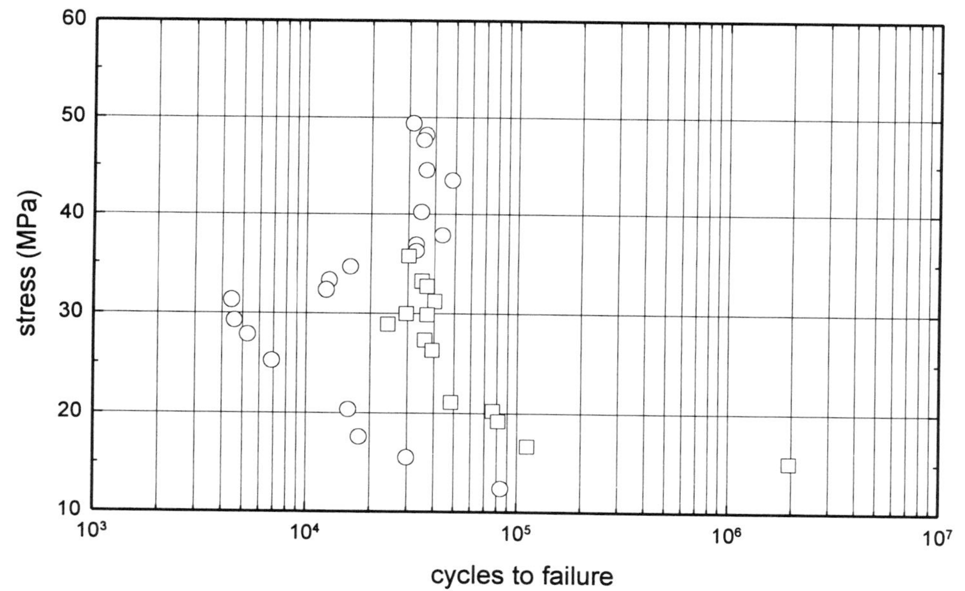

○	GE Lexan 141 Polycarbonate (3.2 mm thick; with weld line); tension- tension; 10 Hz; stress ratio: 0.1; 23°C [fig g, ab]
□	GE Lexan 141 Polycarbonate (3.2 mm thick; without weld line); tension- tension; 10 Hz; stress ratio: 0.1; 23°C [fig g, ab]
Reference No.	372

GRAPH 79: **Fatigue Cycles to Failure vs. Stress in Tension for General Purpose Grades of General Electric Lexan Polycarbonate.**

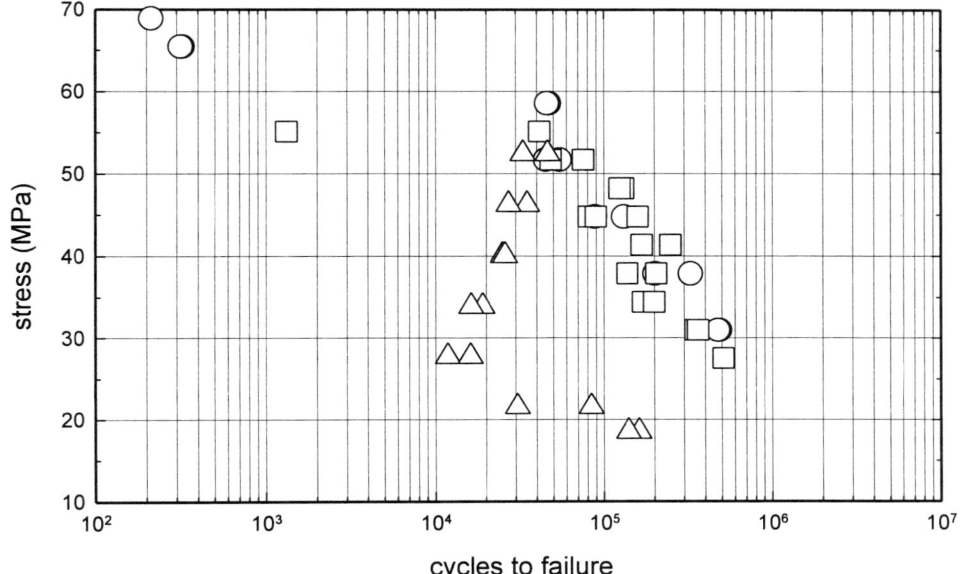

○	GE Lexan 101 Polycarbonate (transparent, flame retardant, unfilled; 7.0 g/10 min. MFI; nonhalogenated; inj. mold.); tension-tension; GE method; 5 Hz; stress ratio: 0.1; 23°C; flow direction [fig g]
□	GE Lexan 101; tension-tension; GE method; 10 Hz; stress ratio: 0.1; 23°C; flow direction [fig g]
△	GE Lexan 121 Polycarbonate (transparent, flame retardant, unfilled; 17.5 g/10 min. MFI; nonhalogenated; inj. mold.); tension- tension; GE method; 5 Hz; stress ratio: 0.1; 23°C; flow direction [fig g]
Reference No.	352

GRAPH 80: **Fatigue Cycles to Failure vs. Stress in Tension at Different Temperatures for Bayer Makrolon Polycarbonate.**

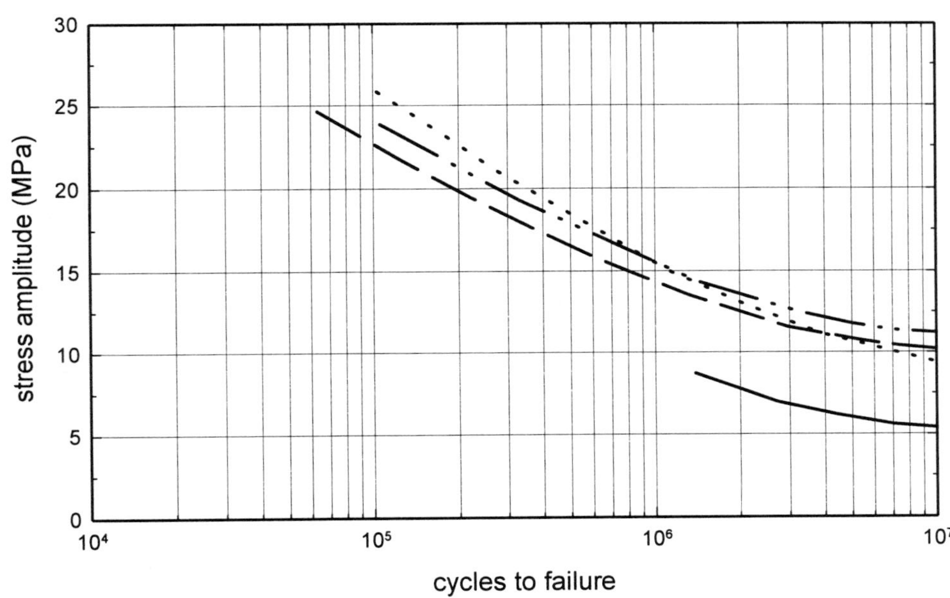

···············	Bayer Makrolon 9415 Polycarbonate (moderate flow, opaque, flame retardant; 10% glass fiber; bromine free FR; w/ mold release); tension; 7 Hz; 23°C; 2x stress amplitude = fatigue limit [fig d]
—··—··	Bayer Makrolon 6455 Polycarbonate (moderate flow, flame retardant, unfilled; bromine free FR); tension; 7 Hz; 23°C; 2x stress amplitude = fatigue limit [fig d]
– – – –	Bayer Makrolon 3200 Polycarbonate (transparent, gen. purp. grade); tension; 7 Hz; 23°C; 2x stress amplitude = fatigue limit [fig d]
———	Bayer Makrolon 3200 Polycarbonate (transparent, gen. purp. grade); tension; 7 Hz; 100°C; 2x stress amplitude = fatigue limit [fig d]
Reference No.	416

GRAPH 81: **Fatigue Cycles to Failure vs. Stress in Tension for TFE/ Glass Filled and Wear Resistant Grades of General Electric Lexan Polycarbonate.**

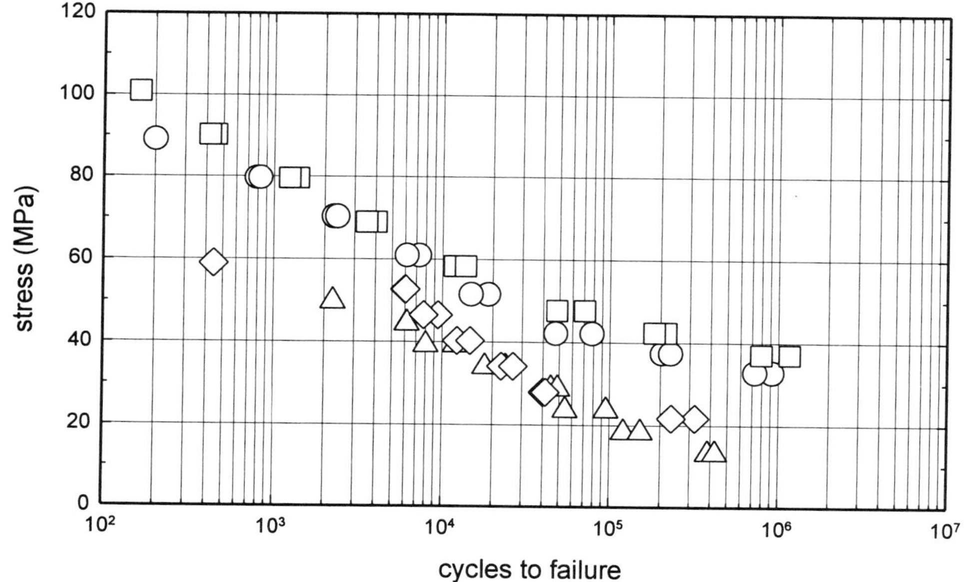

○	GE Lexan LF1520 PC (lubricated, wear resistant, flame retardant; 20% glass fiber, 15% PTFE modified; inj. mold.); tension-tension; GE method; 5 Hz; stress ratio: 0.1; 23°C; flow direction [fig g]
□	GE Lexan LF1530 PC (lubricated, wear resistant, flame retardant; 30% glass fiber, 15% PTFE modified; inj. mold.); tension-tension; GE method; 5 Hz; stress ratio: 0.1; 23°C; flow direction [fig g]
△	GE Lexan WR1210 PC (wear resistant; PTFE modified; inj. mold.); tension- tension; GE method; 5 Hz; stress ratio: 0.1; 23°C; flow direction [fig g]
◇	GE Lexan WR2310 PC (low wear; proprietary modifier); tension- tension; GE method; 5 Hz; stress ratio: 0.1; 23°C; flow direction [fig g]
Reference No.	352

GRAPH 82: **Fatigue Cycles to Failure vs. Stress in Tension for Healthcare and Houseware Grades of General Electric Lexan Polycarbonate.**

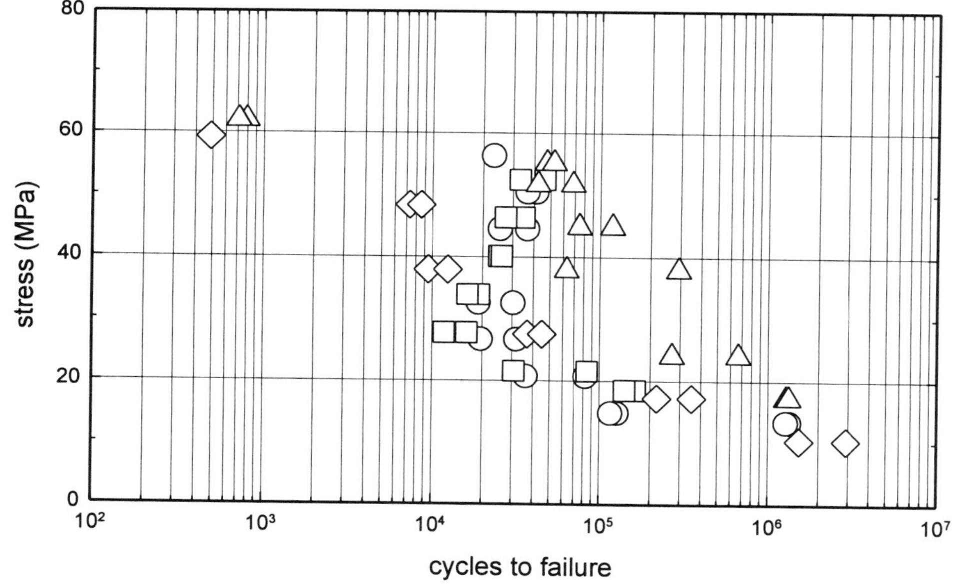

○	GE Lexan HP1 PC (transparent, FDA grade, flame retardant, unfilled; inj. mold.); tension-tension; GE method; 5 Hz; stress ratio: 0.1; 23°C; flow direction [fig g]
□	GE Lexan HP2 PC (transparent, FDA grade, flame retardant, unfilled; inj. mold.); tension-tension; GE method; 5 Hz; stress ratio: 0.1; 23°C; flow direction [fig g]
△	GE Lexan HP4 PC (transparent, FDA grade, flame retardant, unfilled; inj. mold.); tension-tension; GE method; 5 Hz; stress ratio: 0.1; 23°C; flow direction [fig g]
◇	GE Lexan HW1210 PC (housewares; high flow, flame retardant, unfilled, thin wall applic.; inj. mold.); tension- tension; GE method; 5 Hz; stress ratio: 0.1; 23°C; flow direction [fig g]
Reference No.	352

GRAPH 83: **Fatigue Cycles to Failure vs. Stress in Tension at Low Test Frequency for General Purpose Polycarbonate.**

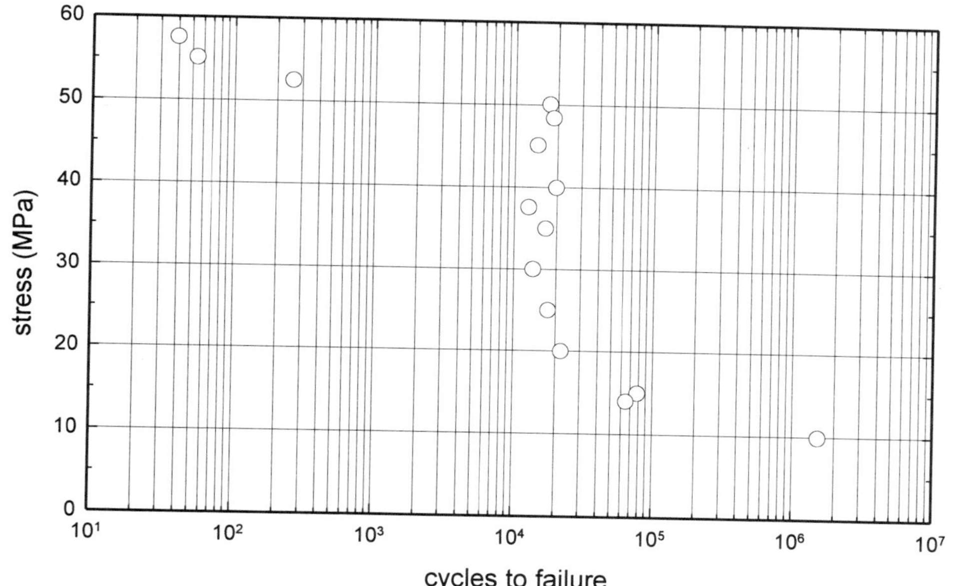

○	Polycarbonate (gen. purp. grade); tension- zero; 0.5 Hz; 23°C [fig f]
Reference No.	414

GRAPH 84: **Fatigue Cycles to Failure vs. Stress in Tension and Immersed in Water for Bayer Makrolon Polycarbonate.**

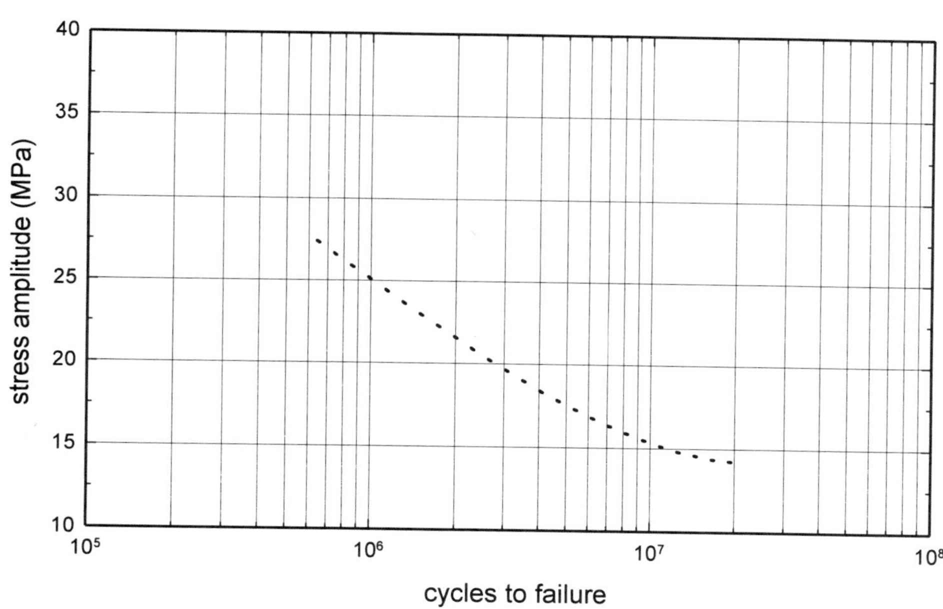

....	Bayer Makrolon Polycarbonate; tension; 7 Hz; 23°C; immersed in water; 2x stress amplitude = fatigue limit [fig d]
Reference No.	416

GRAPH 85: **Fatigue Cycles to Failure vs. Stress in Tension for Saturation Cycled and Unprocessed Bisphenol A Polycarbonate.**

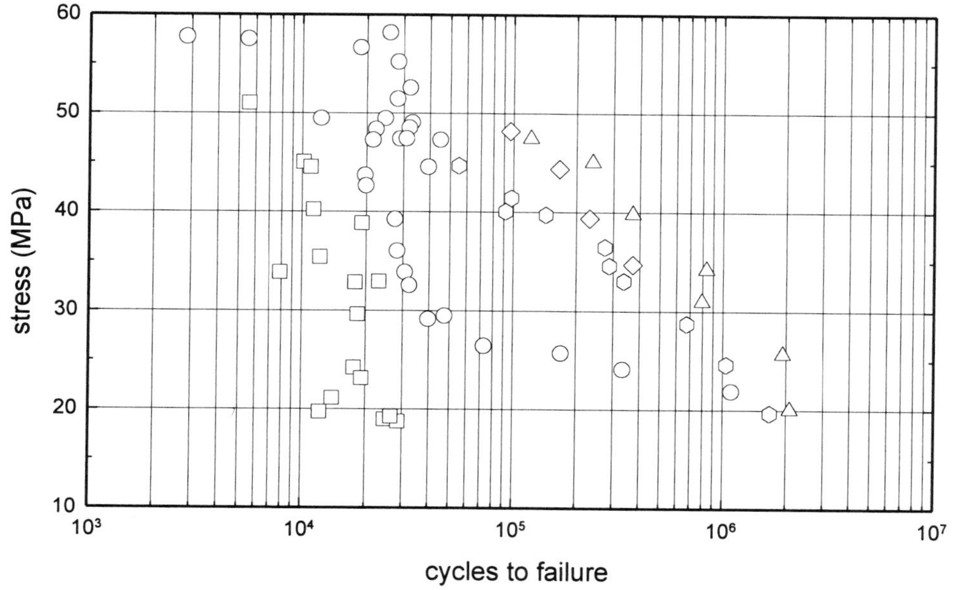

○	GE Lexan 9030 Bisphenol A PC (1.5 mm thick); tension- tension; 10 Hz; unnotched specimen; stress ratio: 0.1; 23°C
□	GE Lexan 9030 Bisphenol A PC (1.5 mm thick); tension- tension; 10 Hz; notched specimen; stress ratio: 0.1; 23°C
△	Bisphenol A PC (1.5 mm thick; milled from sheet, CO_2 saturation cyled & desorbed >1000 hours); tension- tension; 10 Hz; unnotched specimen; stress ratio: 0.1; 23°C
◇	Bisphenol A PC (1.5 mm thick; milled from sheet, CO_2 saturation cyled & desorbed 140-260 hours); tension- tension; 10 Hz; unnotched specimen; stress ratio: 0.1; 23°C
⬡	Bisphenol A PC (1.5 mm thick; milled from sheet, CO_2 saturation cyled & desorbed >1000 hours); tension- tension; 10 Hz; notched specimen; stress ratio: 0.1; 23°C
Reference No.	386

GRAPH 86: **Fatigue Cycles to Failure vs. Stress in Tension for Bisphenol A Polycarbonate with Weld Line.**

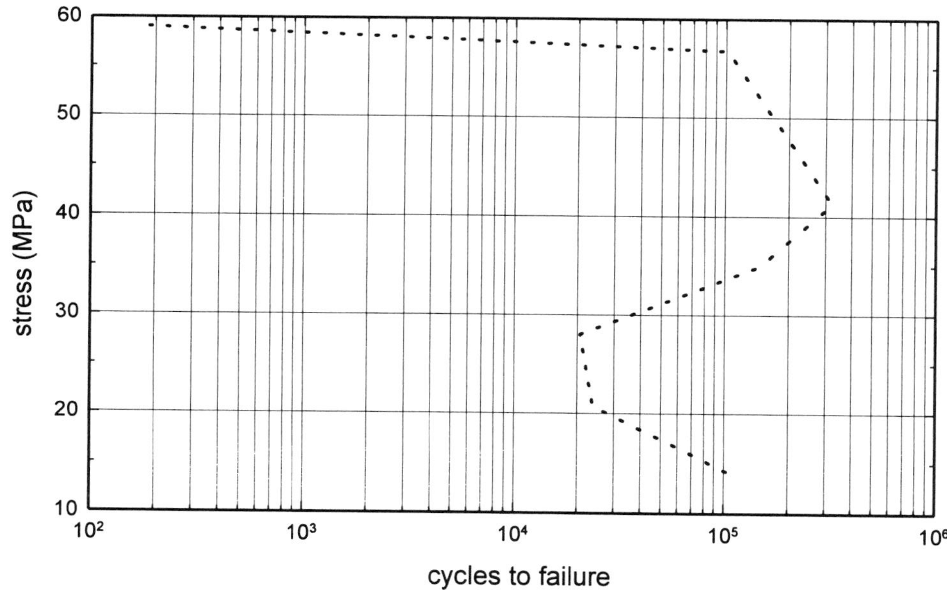

....	Bisphenol A PC (with weld line); tension- tension; 10 Hz; 25°C
Reference No.	381

GRAPH 87: Fatigue Cycles to Failure vs. Stress in Tension for Cellular Bisphenol A Polycarbonate.

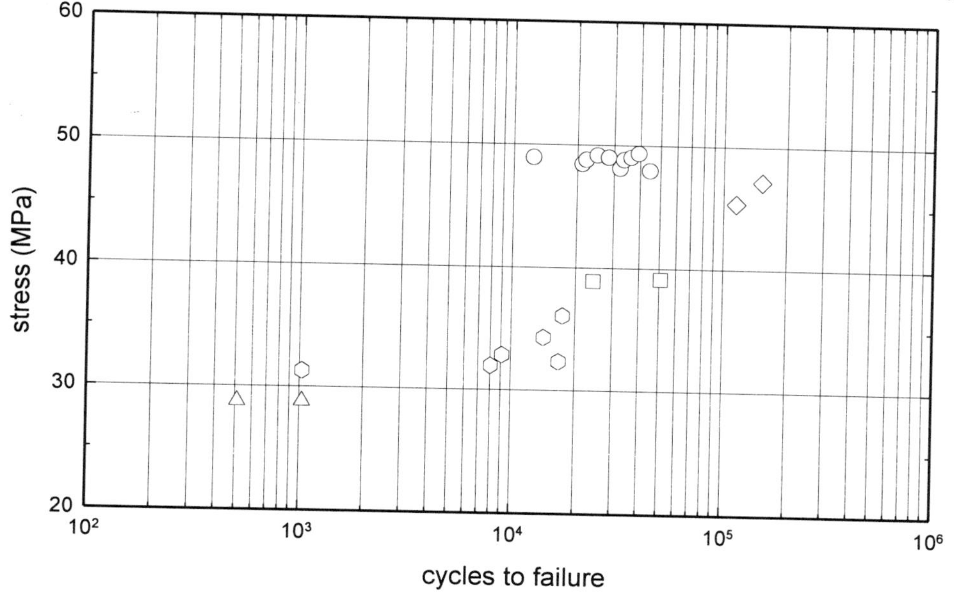

◇	Polycarbonate, Cellular (97% of solid PC density; 1.5 mm thick; milled from sheet and foamed)
⬡	Polycarbonate, Cellular (68% of solid PC density; 1.5 mm thick; milled from sheet and foamed)
○	GE Lexan 9030 Bisphenol A PC (1.5 mm thick); tension- tension; 10 Hz; 23°C; ASTM D638 type IV [fig g, ac]
□	Polycarbonate, Cellular (83% of solid PC density; 1.5 mm thick; milled from sheet and foamed); tension- tension; 10 Hz; 23°C; ASTM D638 type IV [fig g, ac]
△	Polycarbonate, Cellular (52% of solid PC density; 1.5 mm thick; milled from sheet and foamed); tension- tension; 10 Hz; 23°C; ASTM D638 type IV [fig g, ac]
Reference No.	381

GRAPH 88: Fatigue Crack Propagation (FCP) Rate vs. Stress Intensity Range for Polycarbonate.

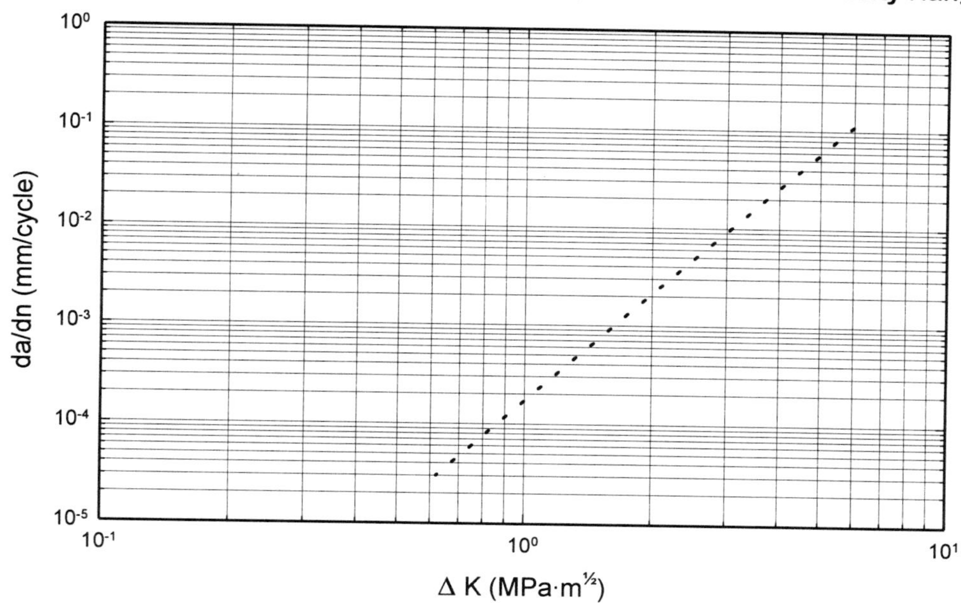

....	Polycarbonate
Reference No.	355

Chapter 18

Polyphthalate Carbonate Copolymer

GRAPH 89: Fatigue Cycles to Failure vs. Stress in Tension for General Electric Lexan PPC Polyphthalate Carbonate Copolymer.

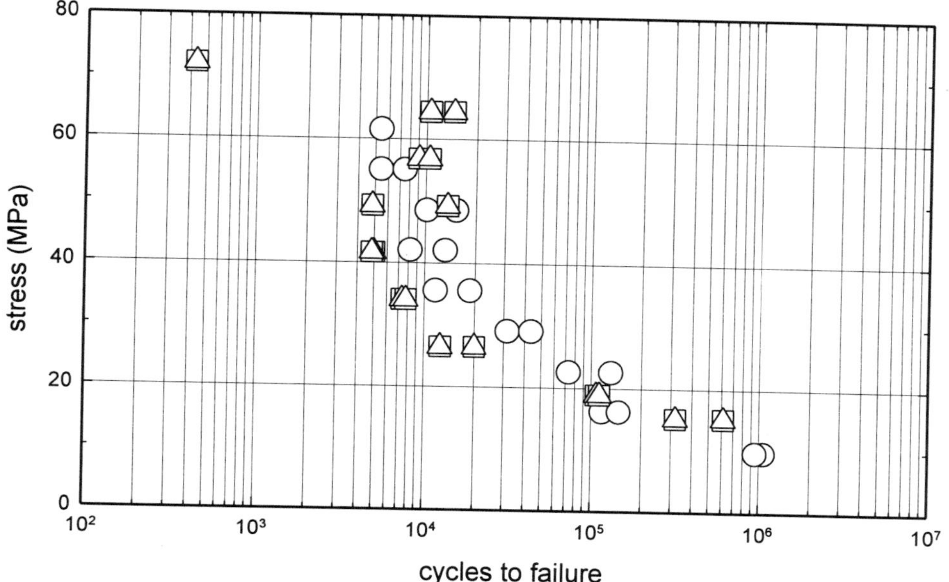

○	GE Lexan PPC4501 PPC (high heat grade, unfilled; inj. mold.); tension-tension; GE method; 5 Hz; stress ratio: 0.1; 23°C; flow direction [fig g]
□	GE Lexan PPC4701 PPC (high heat grade, unfilled; inj. mold.); tension-tension; GE method; 5 Hz; stress ratio: 0.1; 23°C; flow direction [fig g]
△	GE Lexan PPC4704 PPC (high heat grade, FDA grade, unfilled; inj. mold.); tension- tension; GE method; 5 Hz; stress ratio: 0.1; 23°C; flow direction [fig g]
Reference No.	352

Chapter 19

Polybutylene Terephthalate

Fatigue Properties

Bayer: Pocan B1505 (features: general purpose grade, unfilled); **Pocan B3235** (features: natural resin; material compostion: 30% glass fiber reinforcement); **Pocan B4235** (features: flame retardant; material compostion: 30% glass fiber reinforcement)

Pocan B 1505 fails in the dynamic fatigue test at high stress amplitudes and correspondingly low load cycles, as a result of the short-term high-temperature strength being exceeded. So the shape of the Woehler curve is to a very great extent dependent on the test conditions, e.g. load cycle frequency, shape of the stress curve, the ratio of the surface area to the cross-section of the test specimen, the heat transfer coefficients, and the ambient temperature. If an increase in temperature during the test is avoided, a stress amplitude of ± 13.8 MPa (2000 psi) at 10^7 cycles can be endured without rupture; this corresponds to a fatigue strength under pulsating stresses of 27.6 MPa (4000 psi).

With glass-fiber-reinforced Pocan B 3235 and B 4235 the relationships are similar. The temperature increases more slowly, however, and the high surface temperatures occurring with unreinforced material are not reached. With high loads and few stress cycles, fractures are produced here, too, but again the shape of the curves is to a large extent dependent on the test conditions.

Reference: *Pocan Thermoplastic PBT Polyester - A General Reference Manual,* supplier design guide (55-B635(7.5)J) - Mobay Corporation, 1985.

Effect of Glass Reinforcement on Fatigue Behavior

PBT (material compostion: 45% glass fiber reinforcement; note: 0.171 mm fiber length); **PBT** (material compostion: 30% glass fiber reinforcement; note: 0.165 mm fiber length); **PBT** (material compostion: 15% glass fiber reinforcement; note: 0.171 mm fiber length)

High stress/low cycle fatigue strength is increased with higher glass fiber contents of 15 wt%, 30 wt% and 45 wt%. Low stress/high cycle fatigue resistence is improved by increasing the fiber content from 15 wt% to 30 wt%. The highest fiber content of 45 wt% does not improve low stress/high cycle fatigue behavior. The fatigue limits for 10^6 cycles to fracture of GF-PBT modifications are approximately 40 MPa for material with 15 wt%, 45 MPa with 30 wt% and 46 MPa with 45 wt% glass fiber fraction.

Reference: Boukhili, R. (Ecole Polytechnique de Montreal), Gauvin, R. (Ecole Polytechnique de Montreal), Gossel, *Fatigue Behavior of Injection Moded Polycarbonate and Polystyrene Containing Weld Lines,* ANTEC 1989, conference proceedings - Society of Plastics Engineers, 1989.

GRAPH 90: Fatigue Cycles to Failure vs. Stress in Flexure for Hoechst Celanese Celanex Polybutylene Terephthalate Polyester.

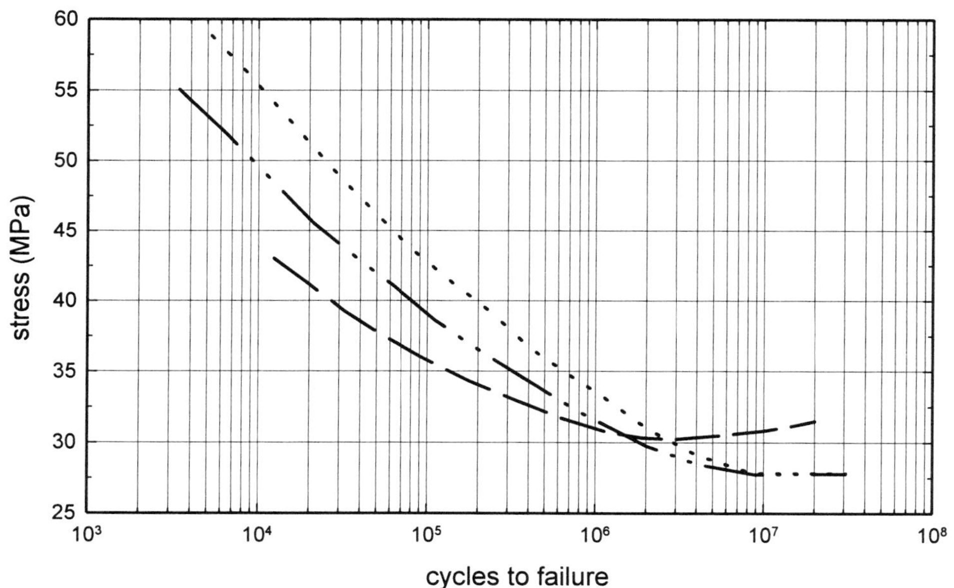

··············	Hoechst Cel. Celanex 3310 PBT (flame retardant, 90 Rockwell M; 30% glass fiber); flexure
—··—··	Hoechst Cel. Celanex 3210 PBT (flame retardant, 90 Rockwell M; 18% glass fiber); flexure
— — — —	Hoechst Cel. Celanex 3300 PBT (gen. purp. grade, 90 Rockwell M; 30% glass fiber); flexure
Reference No.	122

GRAPH 91: Fatigue Cycles to Failure vs. Stress in Flexure for Hoechst Celanex 2300 GV 1/30 Polybutylene Terephthalate Polyester.

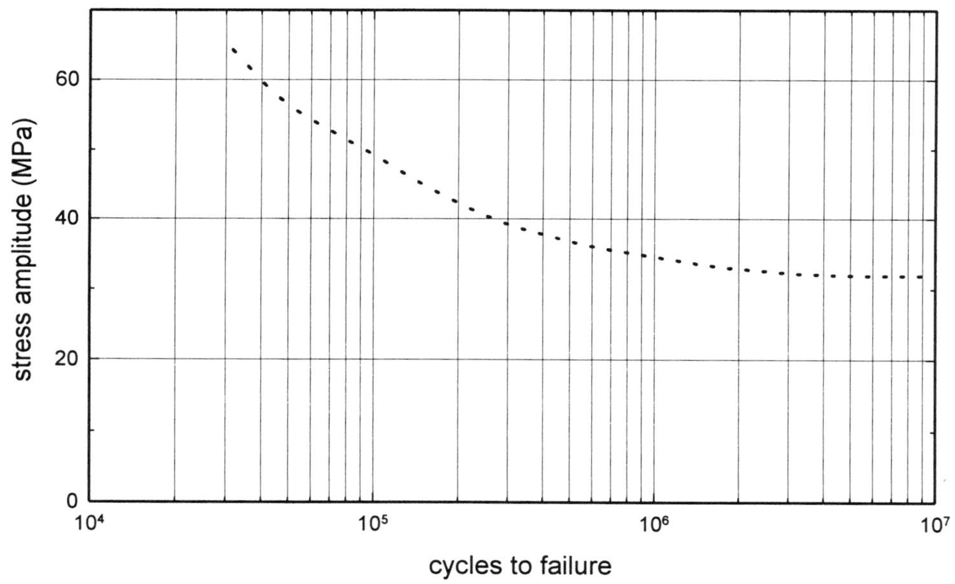

··············	Hoechst AG Celanex 2300 GV 1/30 PBT (gen. purp. grade; 30% glass fiber); flexure (alternating); 10 Hz; 23°C [fig d]
Reference No.	316

GRAPH 92: Fatigue Cycles to Failure vs. Stress in Flexure for Reinforced LNP Polybutylene Terephthalate Polyester.

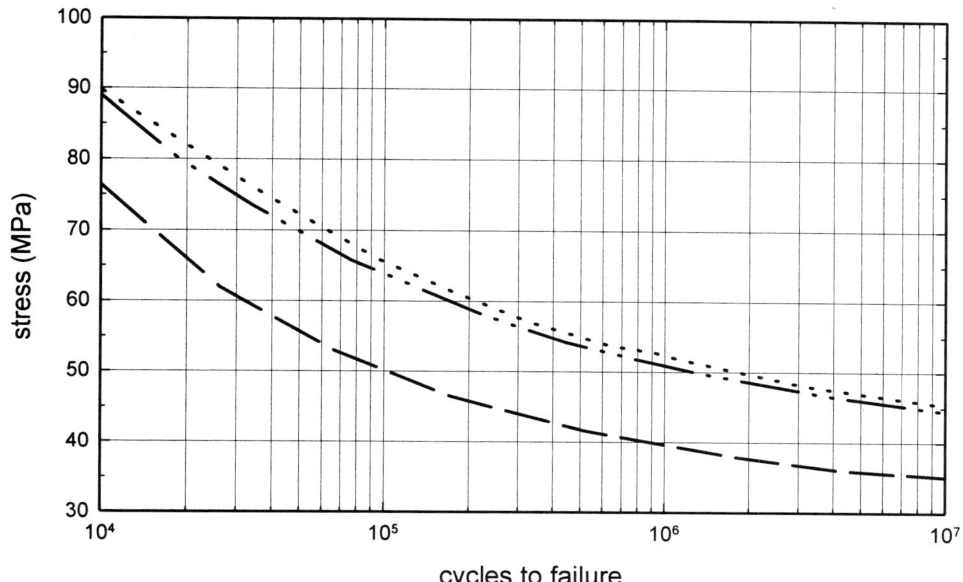

....	LNP Stat-Kon WC1006 PBT (static dissipative; 30% carbon fiber); flexure; 30 Hz; 23°C
—··—··	LNP Thermocomp WC1006 PBT (30% carbon fiber); flexure; 30 Hz; 23°C
— — — —	LNP Thermocomp WF1006 PBT (30% glass fiber); flexure; 30 Hz; 23°C
Reference No.	394

GRAPH 93: Fatigue Cycles to Failure vs. Stress in Tension for Glass Fiber Reinforced General Electric Valox Polybutylene Terephthalate Polyester.

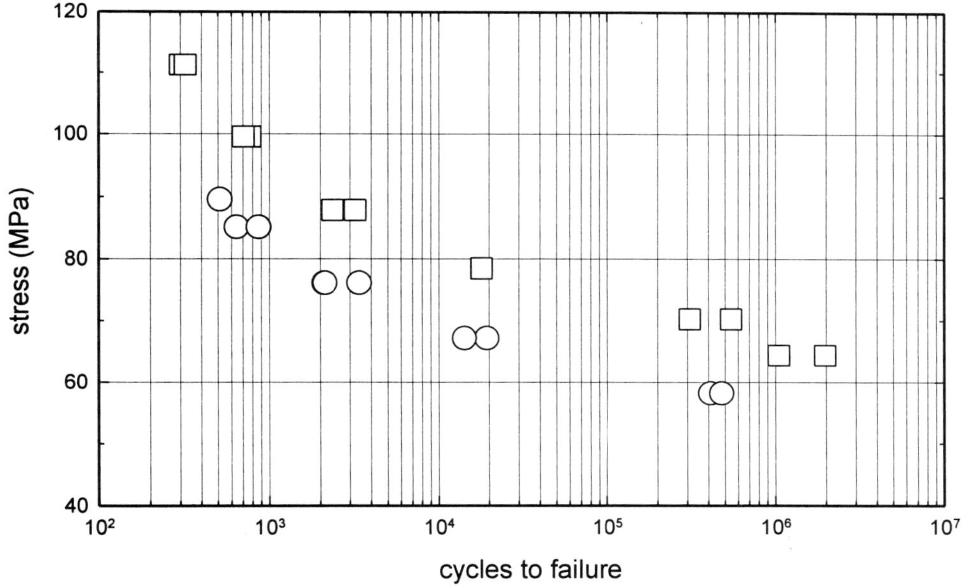

○	GE Valox 850 PBT (15% glass fiber; inj. mold.); tension- tension; GE method; 5 Hz; stress ratio: 0.1; 23°C; flow direction [fig g]
□	GE Valox 865 PBT (flame retardant; 30% glass fiber; inj. mold.); tension-tension; GE method; 5 Hz; stress ratio: 0.1; 23°C; flow direction [fig g]
Reference No.	352

GRAPH 94: **Fatigue Cycles to Failure vs. Stress in Tension for Various Grades of General Electric Valox Polybutylene Terephthalate Polyester.**

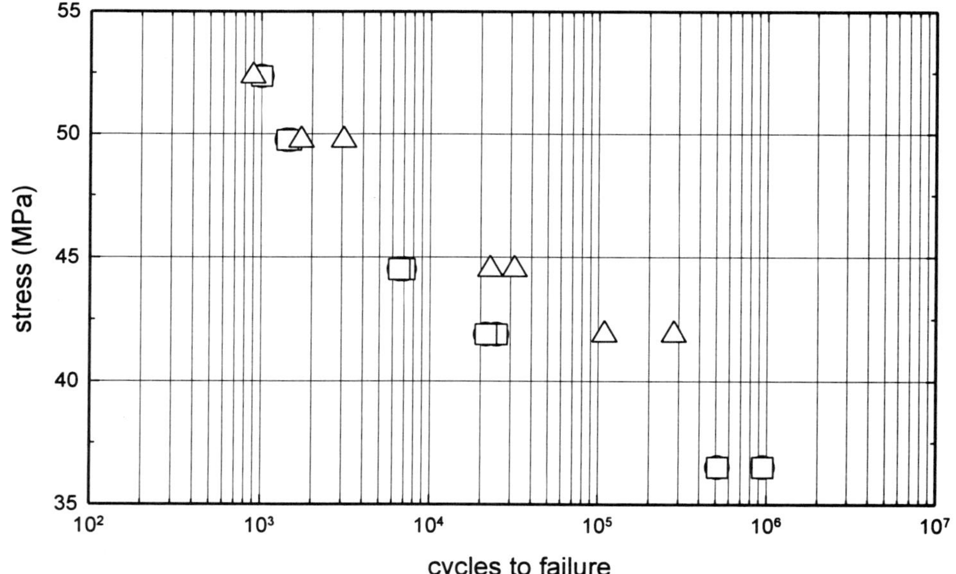

○	GE Valox 210HP PBT (medical grade, high flow, gamma rad. stable; inj. mold.); tension- tension; GE method; 5 Hz; stress ratio: 0.1; 23°C; flow direction [fig g]
□	GE Valox 295 PBT; tension- tension; GE method; 5 Hz; stress ratio: 0.1; 23°C; flow direction [fig g]
△	GE Valox 300 PBT (unfilled; inj. mold.); tension- tension; GE method; 5 Hz; stress ratio: 0.1; 23°C; flow direction [fig g]
Reference No.	352

GRAPH 95: **Fatigue Cycles to Failure vs. Stress in Tension for 300 Series Grades of General Electric Valox Polybutylene Terephthalate Polyester.**

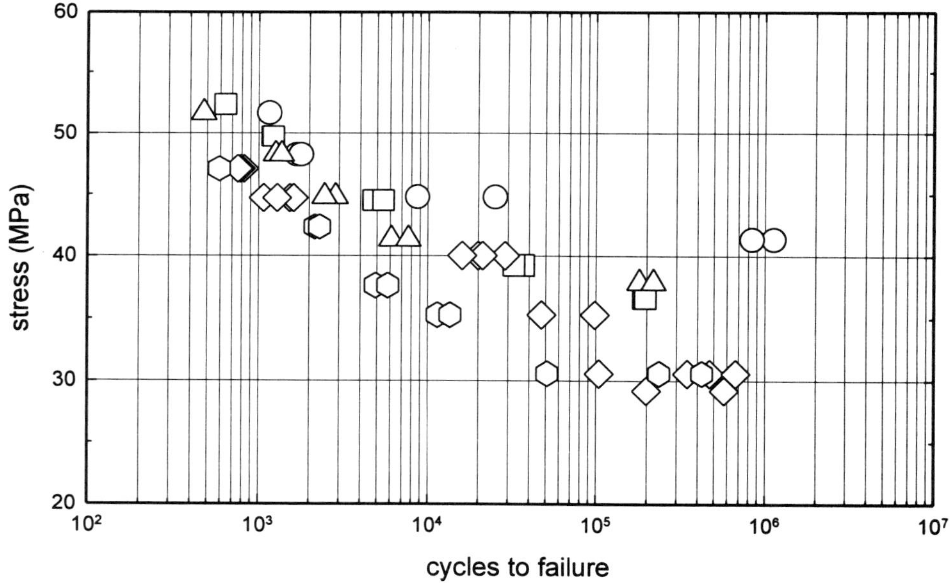

○	GE Valox 310SE0 PBT (flame retardant, unfilled; inj. mold.); tension- tension; GE method; 5 Hz; stress ratio: 0.1; 23°C; flow direction [fig g]
□	GE Valox 317 PBT (unfilled; inj. mold.); tension- tension; GE method; 5 Hz; stress ratio: 0.1; 23°C; flow direction [fig g]
△	GE Valox 325 PBT (gen. purp. grade, unfilled; inj. mold.); tension- tension; GE method; 5 Hz; stress ratio: 0.1; 23°C; flow direction [fig g]
◇	GE Valox 357 PBT (flame retardant, unfilled; inj. mold.); tension- tension; GE method; 5 Hz; stress ratio: 0.1; 23°C; flow direction [fig g]
⬡	GE Valox 357 PBT (flame retardant, unfilled; inj. mold.); tension- tension; GE method; 20 Hz; stress ratio: 0.1; 23°C; flow direction [fig g]
Reference No.	352

GRAPH 96: **Fatigue Cycles to Failure vs. Stress in Tension for Glass Fiber Reinforced Polybutylene Terephthalate Polyester.**

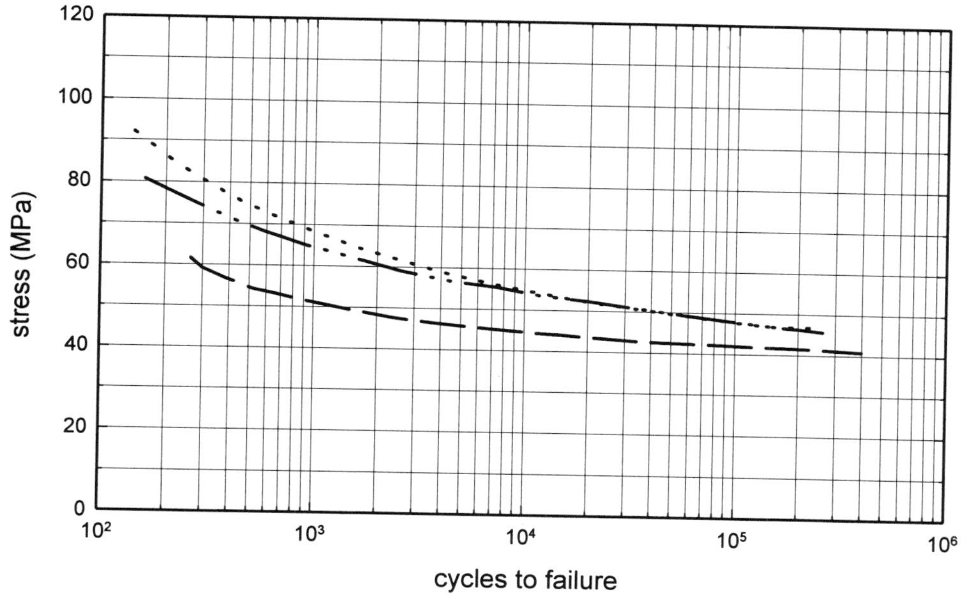

···············	PBT (45% glass fiber, 0.171 mm fiber length); tension- tension; 1 Hz; stress ratio: 0.01; 23°C;
—··—··	PBT (30% glass fiber, 0.165 mm fiber length); tension- tension; 1 Hz; stress ratio: 0.01; 23°C;
— — — —	PBT (15% glass fiber, 0.171 mm fiber length); tension- tension; 1 Hz; stress ratio: 0.01; 23°C;
Reference No.	372

GRAPH 97: **Fatigue Cycles to Failure vs. Stress in Tension for 700 Series Glass/ Mineral Reinforced General Electric Valox Polybutylene Terephthalate Polyester.**

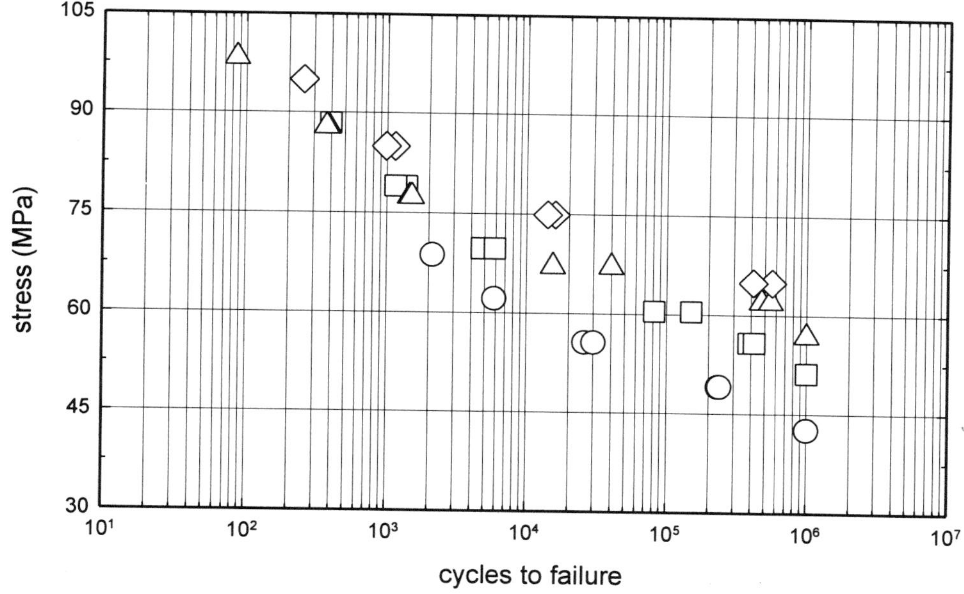

○	GE Valox 732E PBT (low warp grade, high flow; 30% mineral/ glass; inj. mold.); tension- tension; GE method; 5 Hz; stress ratio: 0.1; 23°C; flow direction [fig g]
□	GE Valox 735 PBT (40% mineral/ glass; inj. mold.); tension- tension; GE method; 5 Hz; stress ratio: 0.1; 23°C; flow direction [fig g]
△	GE Valox 736 PBT (45% mineral/ glass; inj. mold.); tension- tension; GE method; 5 Hz; stress ratio: 0.1; 23°C; flow direction [fig g]
◇	GE Valox 780 PBT (flame retardant; 40% mineral/ glass; non-blooming; inj. mold.); tension- tension; GE method; 5 Hz; stress ratio: 0.1; 23°C; flow direction [fig g]
Reference No.	352

GRAPH 98: **Fatigue Cycles to Failure vs. Stress in Tension for Bayer Pocan Polybutylene Terephthalate Polyester.**

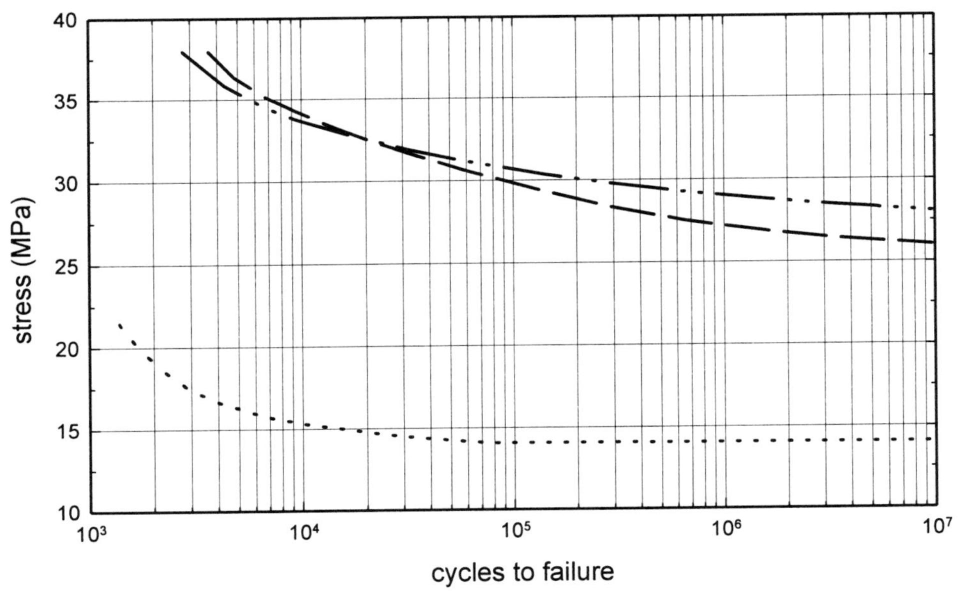

....	Bayer Pocan B1505 PBT (gen. purp. grade, unfilled); tension; 23°C; 2x stress amplitude = fatigue limit
—··—	Bayer Pocan B3235 PBT (natural resin; 30% glass fiber); tension; 23°C; 2x stress amplitude = fatigue limit
– – – –	Bayer Pocan B4235 PBT (flame retardant; 30% glass fiber); tension; 23°C; 2x stress amplitude = fatigue limit
Reference No.	227

GRAPH 99: **Fatigue Cycles to Failure vs. Stress in Tension at Various Test Conditions for Glass Fiber Reinforced General Electric Valox Polybutylene Terephthalate Polyester.**

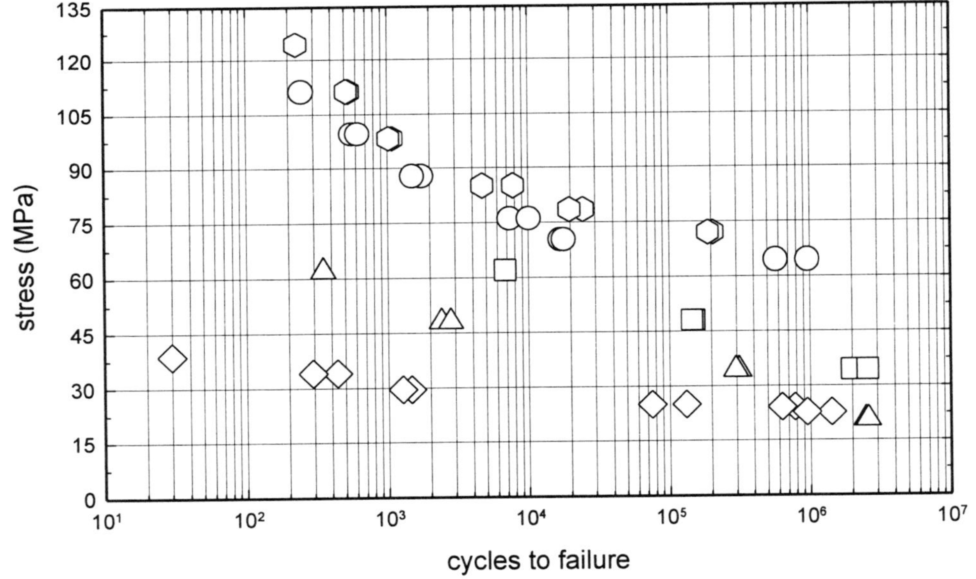

○	GE Valox 420 PBT (30% glass fiber; inj. mold.); tension- tension; GE method; 5 Hz; stress ratio: 0.1; 23°C; flow direction [fig g]
□	GE Valox 420 PBT (30% glass fiber; inj. mold.); tension- tension; GE method; 30 Hz; stress ratio: 0.1; 23°C; flow direction [fig g]
△	GE Valox 420 PBT (30% glass fiber; inj. mold.); tension- tension; GE method; 30 Hz; stress ratio: 0.1; 66°C; flow direction [fig g]
◇	GE Valox 420 PBT (30% glass fiber; inj. mold.); tension- tension; GE method; 5 Hz; stress ratio: 0.1; 149°C; flow direction [fig g]
⬡	GE Valox 430 PBT (impact modified; 33% glass fiber, inj. mold.); tension- tension; GE method; 5 Hz; stress ratio: 0.1; 23°C; flow direction [fig g]
Reference No.	352

GRAPH 100: **Fatigue Cycles to Failure vs. Stress Under Different Loading Conditions for BASF Ultradur B 4300G6 Polybutylene Terephthalate Polyester.**

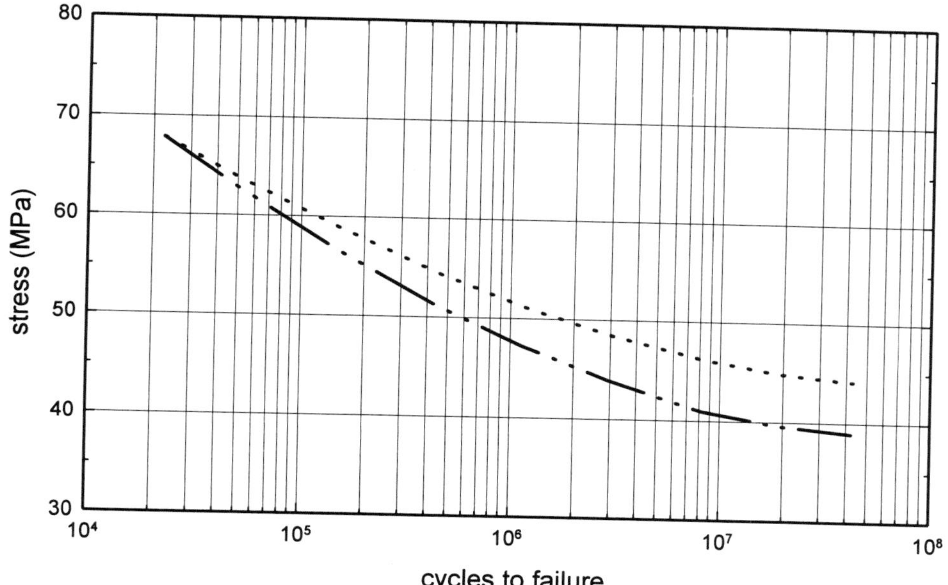

....	BASF AG Ultradur B4300G6 PBT (30% glass fiber); rotating beam; DIN 53442; 25 Hz; 23°C; [fig u]
—··—··	BASF AG Ultradur B4300G6 PBT (30% glass fiber); flexure (alternating); DIN 53442; 15 Hz; 23°C; [fig v]
Reference No.	180

Chapter 20

Polyethylene Terephthalate

GRAPH 101: **Fatigue Cycles to Failure vs. Stress in Flexure for Toughned DuPont Rynite 430 Polyethylene Terephthalate Polyester.**

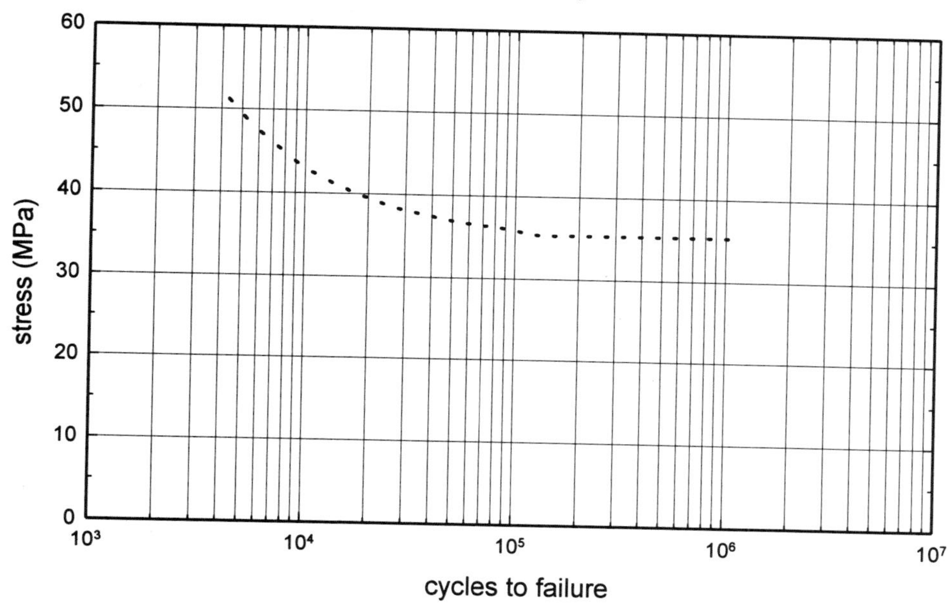

....	DuPont Rynite 430 PET (impact modified, 114 Rockwell R, 82 Rockwell M; 30% glass fiber); flexure; ASTM D671; 60 Hz; 23°C
Reference No.	200

GRAPH 102: **Fatigue Cycles to Failure vs. Stress in Flexure for General Purpose DuPont Rynite Polyethylene Terephthalate Polyester.**

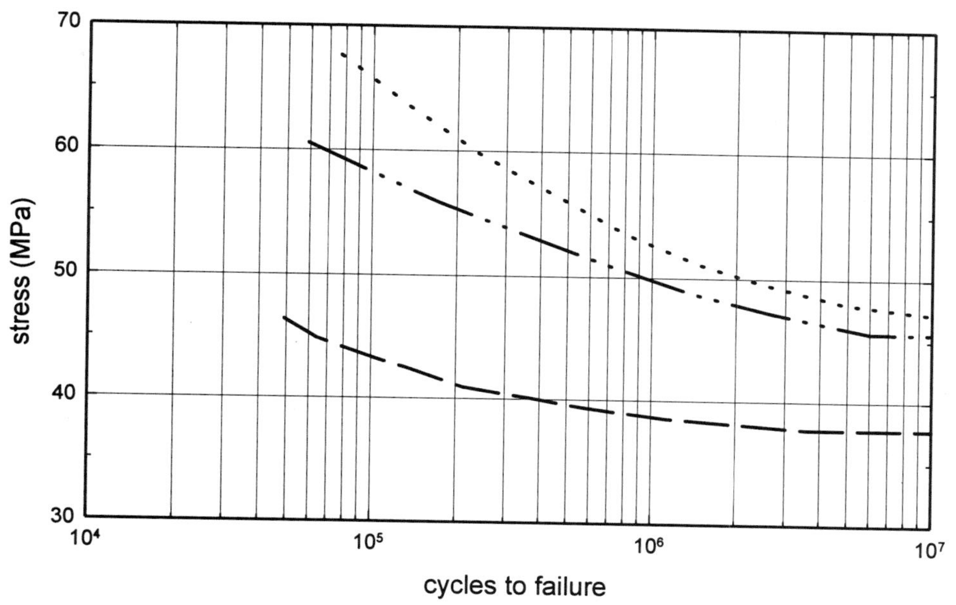

....	DuPont Rynite 555 PET (121 Rockwell R, 105 Rockwell M; 55% glass fiber); flexure; ASTM D671; 60 Hz; 23°C
—··—··	DuPont Rynite 545 PET (100 Rockwell M, 120 Rockwell R; 45% glass fiber); flexure; ASTM D671; 60 Hz; 23°C
— — — —	DuPont Rynite 530 PET (120 Rockwell R, 100 Rockwell M; 30% glass fiber); flexure; ASTM D671; 60 Hz; 23°C
Reference No.	200

GRAPH 103: **Fatigue Cycles to Failure vs. Stress in Flexure for Flame Retardant DuPont Rynite Polyethylene Terephthalate Polyester.**

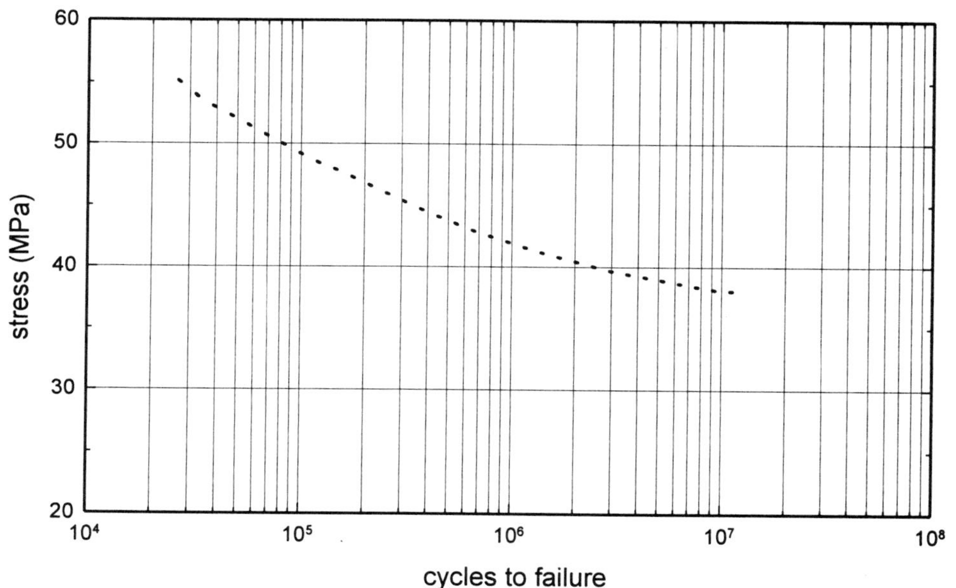

....	DuPont Rynite FR530 PET (flame retardant, 100 Rockwell M, 120 Rockwell R; 30% glass fiber); flexure; ASTM D671; 60 Hz; 23°C
Reference No.	200

GRAPH 104: **Fatigue Cycles to Failure vs. Stress in Flexure for Low Warp DuPont Rynite Polyethylene Terephthalate Polyester.**

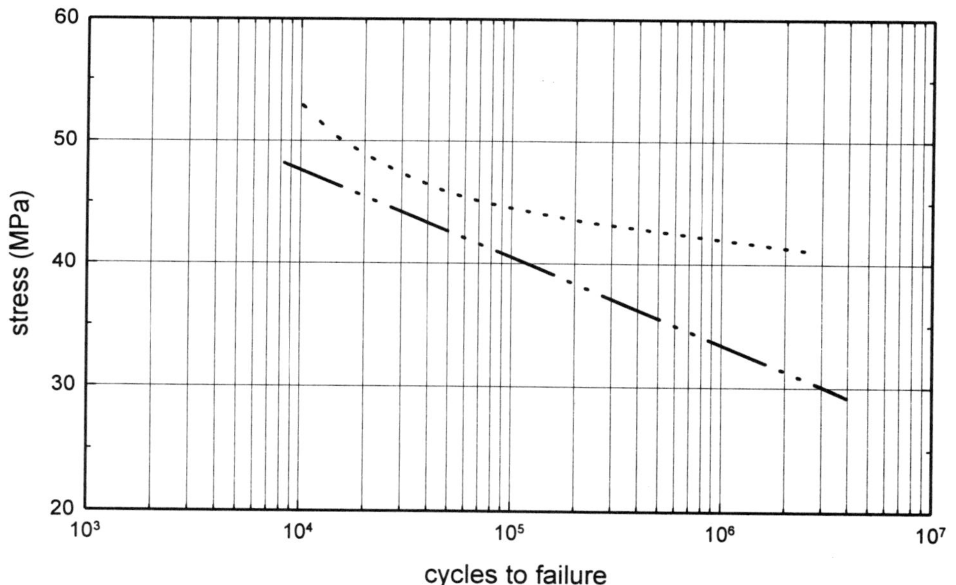

....	DuPont Rynite 940 PET (low warp grade, 118 Rockwell R; 40% mica/ glass); flexure; ASTM D671; 60 Hz; 23°C
—··—··	DuPont Rynite 935 PET (low warp grade, 119 Rockwell R, 85 Rockwell M; 35% mica/ glass); flexure; ASTM D671; 60 Hz; 23°C
Reference No.	200

GRAPH 105: Fatigue Cycles to Failure vs. Stress in Tension for General Electric Valox Polyethylene Terephthalate Polyester.

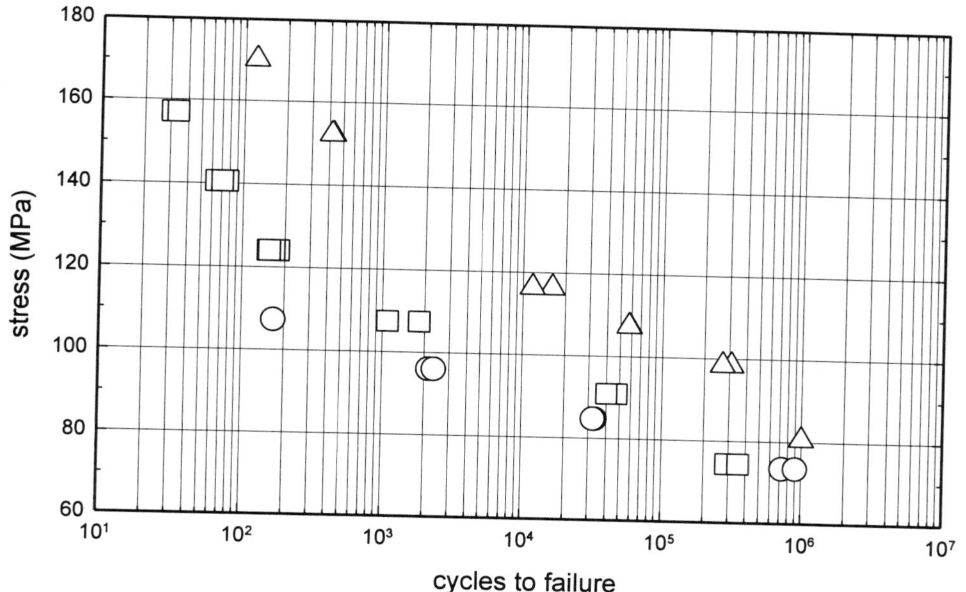

○	GE Valox 9215 PET (15% glass fiber; inj. mold.); tension- tension; GE method; 5 Hz; stress ratio: 0.1; 23°C; flow direction [fig g]
□	GE Valox 9230 PET (30% glass fiber; inj. mold.); tension- tension; GE method; 5 Hz; stress ratio: 0.1; 23°C; flow direction [fig g]
△	GE Valox 9245 PET (45% glass fiber; inj. mold.); tension- tension; GE method; 5 Hz; stress ratio: 0.1; 23°C; flow direction [fig g]
Reference No.	352

Chapter 21

Polycyclohexylenedimethylene Terephthalate

GRAPH 106: **Fatigue Cycles to Failure vs. Stress in Tension for PDR Series General Electric Valox Polycyclohexylenedimethylene Terephthalate Polyester.**

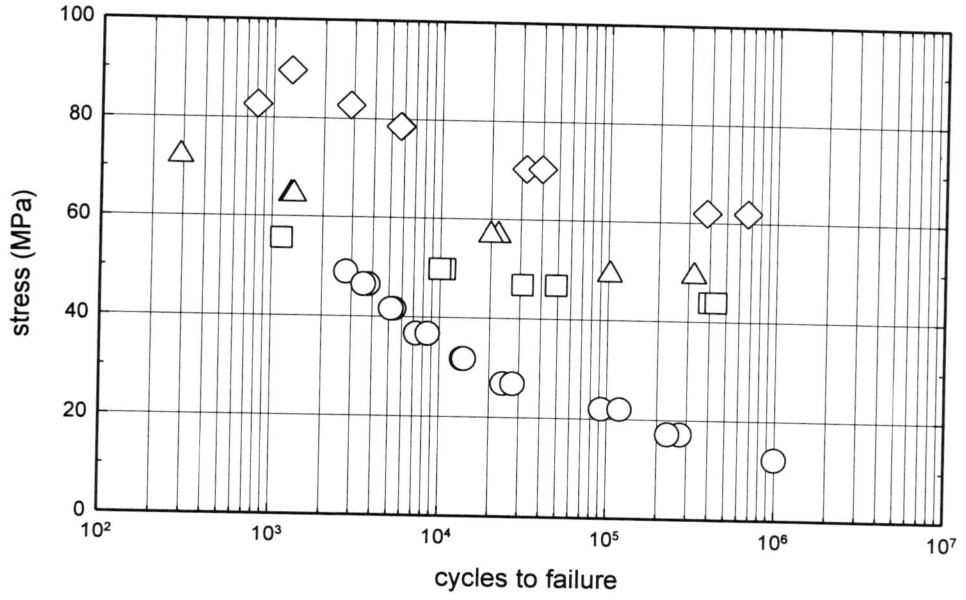

○	GE Valox PDR365 PCT; tension- tension; GE method; 5 Hz; stress ratio: 0.1; 23°C; flow direction [fig g]
□	GE Valox PDR726R PCT; tension- tension; GE method; 5 Hz; stress ratio: 0.1; 23°C; flow direction [fig g]
△	GE Valox PDR751 PCT; tension- tension; GE method; 5 Hz; stress ratio: 0.1; 23°C; flow direction [fig g]
◇	GE Valox PDR830U PCT; tension- tension; GE method; 5 Hz; stress ratio: 0.1; 23°C; flow direction [fig g]
Reference No.	352

GRAPH 107: **Fatigue Cycles to Failure vs. Stress in Tension for DR and EF Series General Electric Valox Polycyclohexylenedimethylene Terephthalate Polyester.**

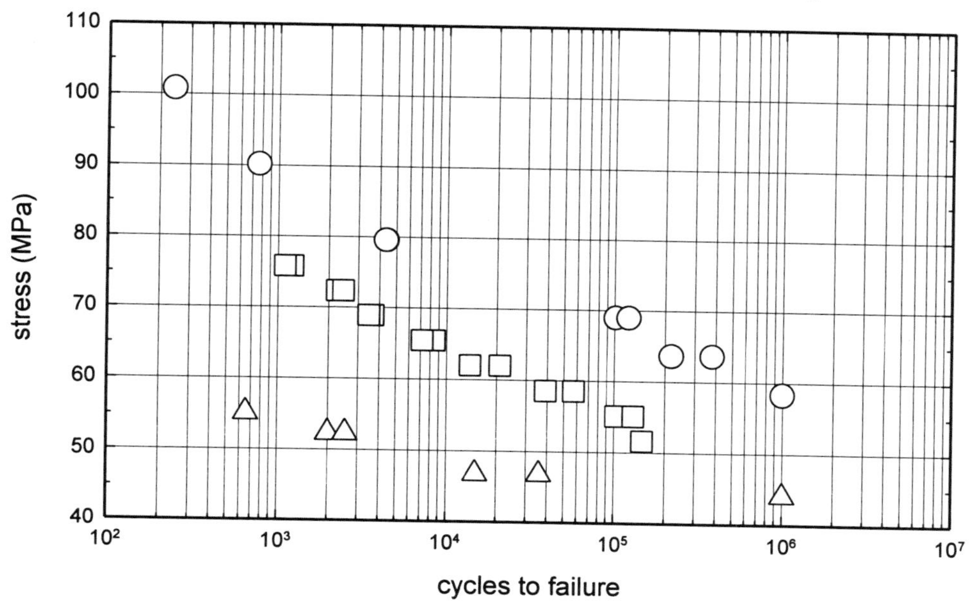

○	GE Valox DR48 PCT (flame retardant; 17% glass fiber; inj. mold.); tension- tension; GE method; 5 Hz; stress ratio: 0.1; 23°C; flow direction [fig g]
□	GE Valox DR51 PCT (15% glass fiber; inj. mold.); tension- tension; GE method; 5 Hz; stress ratio: 0.1; 23°C; flow direction [fig g]
△	GE Valox EF3500 PCT (high flow, flame retardant, unfilled; inj. mold.); tension- tension; GE method; 5 Hz; stress ratio: 0.1; 23°C; flow direction [fig g]
Reference No.	352

GRAPH 108: **Fatigue Cycles to Failure vs. Stress in Tension for General Electric Valox Polycyclohexylenedimethylene Terephthalate Polyester.**

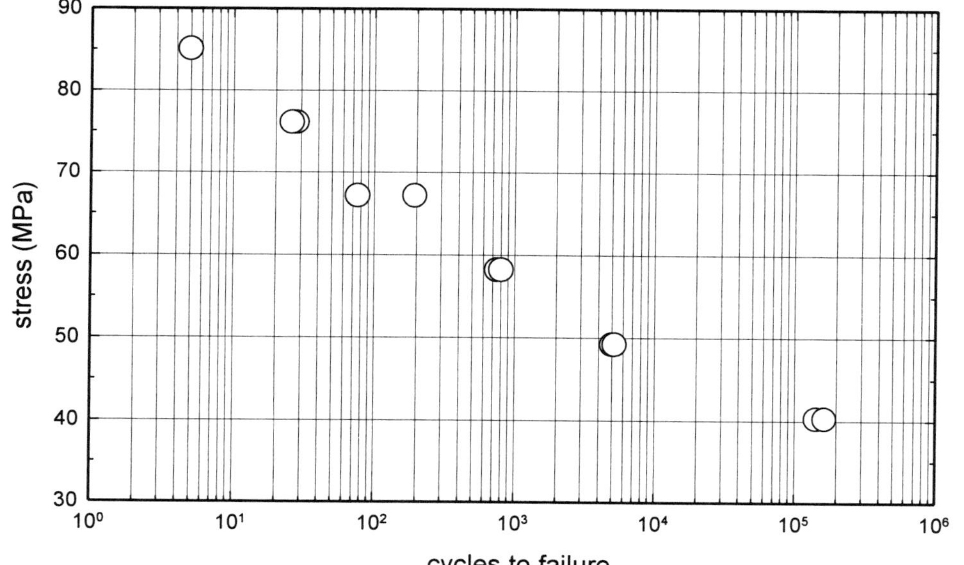

○	GE Valox 9715 PCT (15% glass fiber; inj. mold.); tension- tension; GE method; 5 Hz; stress ratio: 0.1; 23°C; flow direction [fig g]
Reference No.	352

Chapter 22

Liquid Crystal Polymer

GRAPH 109: Fatigue Cycles to Failure vs. Stress in Flexure for Hoechst Vectra Liquid Crystal Polyester.

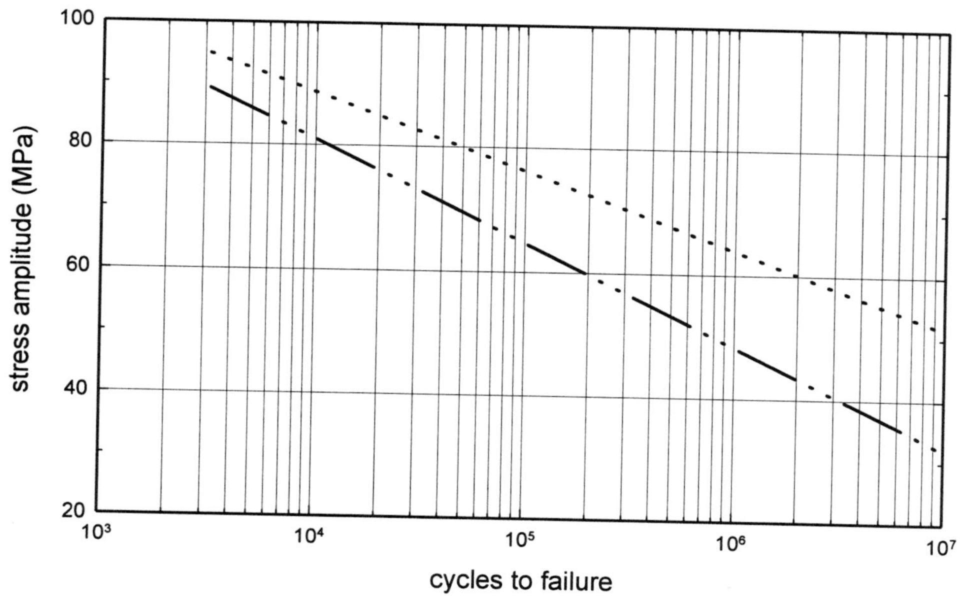

···············	Hoechst AG Vectra A130 LCP (80 Rockwell M; 30% glass fiber, Vectra A950 matrix); flexure (alternating); 10 Hz; 23°C; mean stress = 0
—··—··	Hoechst AG Vectra A530 LCP (65 Rockwell M; Vectra A950 matrix, 30% mineral filler); flexure (alternating); 10 Hz; 23°C; mean stress = 0
Reference No.	70

GRAPH 110: Fatigue Cycles to Failure vs. Stress in Flexure for Liquid Crystal Polyester.

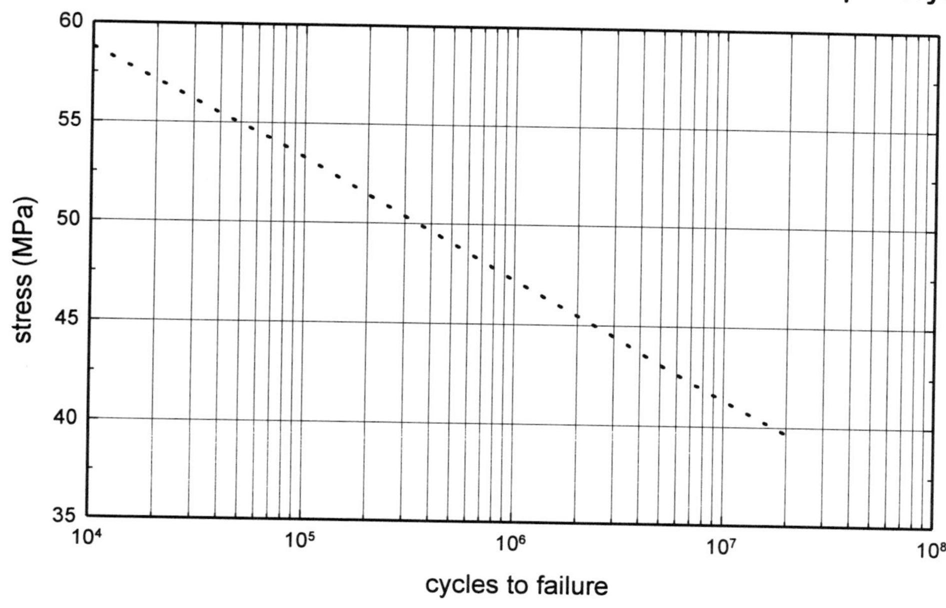

···············	DuPont Zenite 6130 LCP (3.2 mm thick, 61 Rockwell M, 108 Rockwell R; 30% glass fiber); flexure
Reference No.	354

Chapter 23

Polyamideimide

Fatigue Properties

Amoco Performance Products: Torlon 7130 (features: fatigue resistance; material composition: 1% fluorocarbon, 30% graphite); **Torlon**

Torlon parts resist cyclic stress. Torlon 7130, a graphite fiber reinforced grade, has exceptional fatigue strength, and is superior to competitive engineering resins. The S-N curves for selected Torlon grades show that even after 10,000,000 cycles, Torlon poly(amide-imide) has excellent resistance to cyclical stress in the flexural mode, and demonstrate the integrity of Torlon 7130 under tension/tension cyclical stress. At lower frequencies, the fatigue strength of Torlon 7130 is even higher.

Even at high temperature, Torlon polymers maintain strength under cyclical stress. Flexural fatigue tests were run at 177°C (350°F) on specimens pre-conditioned at that temperature. The results suggest Torlon polymers are suitable for applications requiring fatigue resistance at high temperature.

Reference: *Torlon Engineering Polymers / Design Manual,* supplier design guide (F-49893) - Amoco Performance Products.

Fracture Toughness

Amoco Performance Products: Torlon

Fracture toughness can be assessed by measuring the fracture energy (G_I) of a polymer. The Naval Research Laboratory (NRL) uses a compact tension specimen to determine G_I, a measure of a polymer's ability to absorb and dissipate impact energy without fracturing - larger values correspond to higher fracture toughness.

	Fracture Energy		T_g	
	kJ/m²	ft.lb/in²	°C	°F
Thermosets				
Polyimide-1	0.20	0.095	350	662
Polyimide-2	0.12	0.057	360	680
Tetrafunctional epoxy	0.076	0.036	260	500
Thermoplastics				
Poly(amide-imide)	3.4	1.6	275	527
Polysulfone	3.1	1.5	174	345
Polyethersulfone	2.6	1.2	230	446
Polyimide-4	2.1	1.0	365	689
Polyimide-3	0.81	0.38	326	619
Polyphenylene sulfide	0.21	0.10	-	-

As expected, thermosetting polymers cannot absorb and dissipate impact energy as well as thermoplastics and consequently have lower fracture energies. Torlon poly(amide-imide) exhibits outstanding fracture toughness, with a G_I of 3.4 kJ/m^2 (1.6 ft.lb/in^2). Glass transition temperatures are included to indicate the trade-off between fracture toughness and useful temperature range. Poly(amide-imide) is characterized by a balance of toughness and high T_g.

Reference: *Torlon Engineering Polymers / Design Manual,* supplier design guide (F-49893) - Amoco Performance Products.

GRAPH 111: **Fatigue Cycles to Failure vs. Stress in Flexure for Amoco Performance Products Torlon Polyamideimide.**

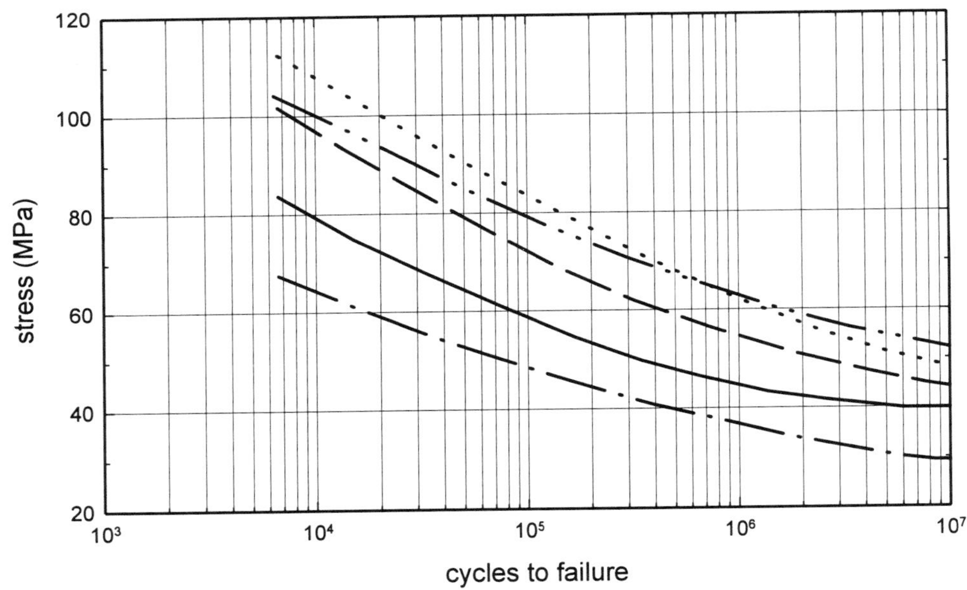

···············	Amoco Torlon 9040 PAI (40% glass fiber, 1% fluorocarbon); flexure; 30 Hz; 23°C
—··—··	Amoco Torlon 7130 PAI (fatigue resist.; 1% fluorocarbon, 30% graphite); flexure; 30 Hz; 23°C
— — — —	Amoco Torlon 5030 PAI (30% glass fiber, 1% fluorocarbon); flexure; 30 Hz; 23°C
———	Amoco Torlon 4203L PAI (3% TiO_2, 0.5% fluorocarbon); flexure; 30 Hz; 23°C
—·—·—	Amoco Torlon 4275 PAI (wear resistant; 3% fluorocarbon, 30% graphite powd.); flexure; 30 Hz; 23°C
Reference No.	20

GRAPH 112: **Fatigue Cycles to Failure vs. Stress in Flexure at High Temperature for Amoco Performance Products Torlon Polyamideimide.**

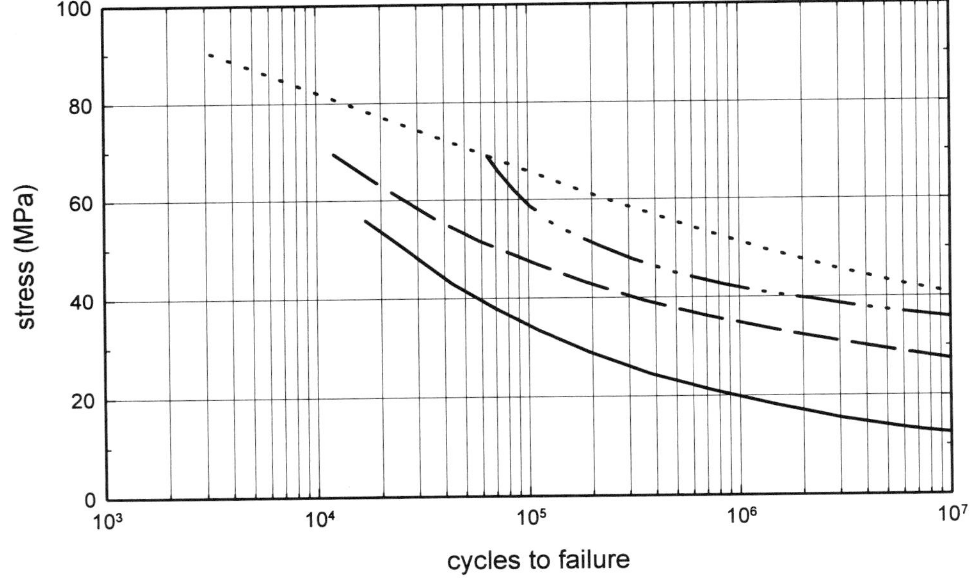

···············	Amoco Torlon 7130 PAI (fatigue resist.; 1% fluorocarbon, 30% graphite); flexure; 30 Hz; 177°C
—··—··	Amoco Torlon 5030 PAI (30% glass fiber, 1% fluorocarbon); flexure; 30 Hz; 177°C
— — — —	Amoco Torlon 9040 PAI (40% glass fiber, 1% fluorocarbon); flexure; 30 Hz; 177°C
———	Amoco Torlon 4203L PAI (3% TiO_2, 0.5% fluorocarbon); flexure; 30 Hz; 177°C
Reference No.	20

GRAPH 113: Fatigue Cycles to Failure vs. Stress in Tension for Amoco Performance Products Torlon Polyamideimide.

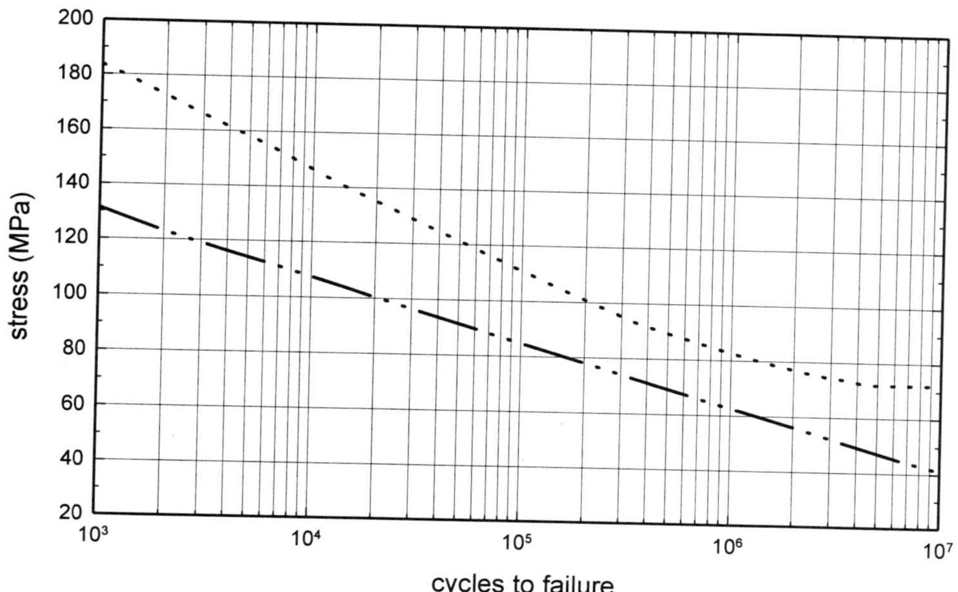

....	Amoco Torlon 7130 PAI (fatigue resist.; 1% fluorocarbon, 30% graphite); tension- tension; 30 Hz; stress ratio: 0.9; 23°C [fig g]
—··—··	Amoco Torlon 4203L PAI (3% TiO_2, 0.5% fluorocarbon); tension-tension; 30 Hz; stress ratio: 0.9; 23°C [fig g]
Reference No.	20

GRAPH 114: Fatigue Cycles to Failure vs. Stress in Tension at Low Test Frequency for Amoco Performance Products Torlon 7130 Polyamideimide.

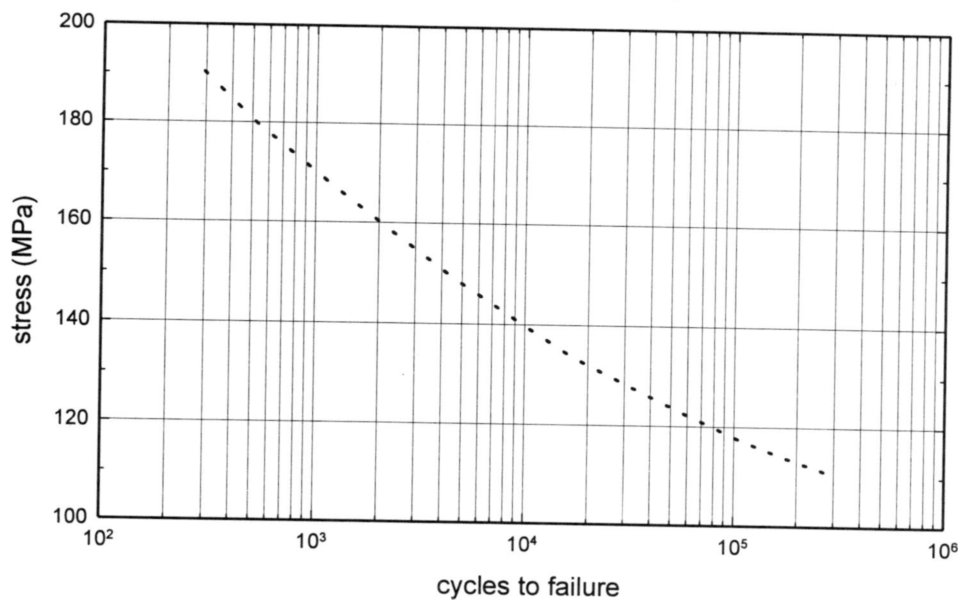

....	Amoco Torlon 7130 PAI (fatigue resist.; 1% fluorocarbon, 30% graphite); tension- tension; 2 Hz; stress ratio: 0.9; 23°C [fig g]
Reference No.	20

Chapter 24

Polyetherimide

GRAPH 115: **Fatigue Cycles to Failure vs. Stress in Flexure for Reinforced LNP Polyetherimide.**

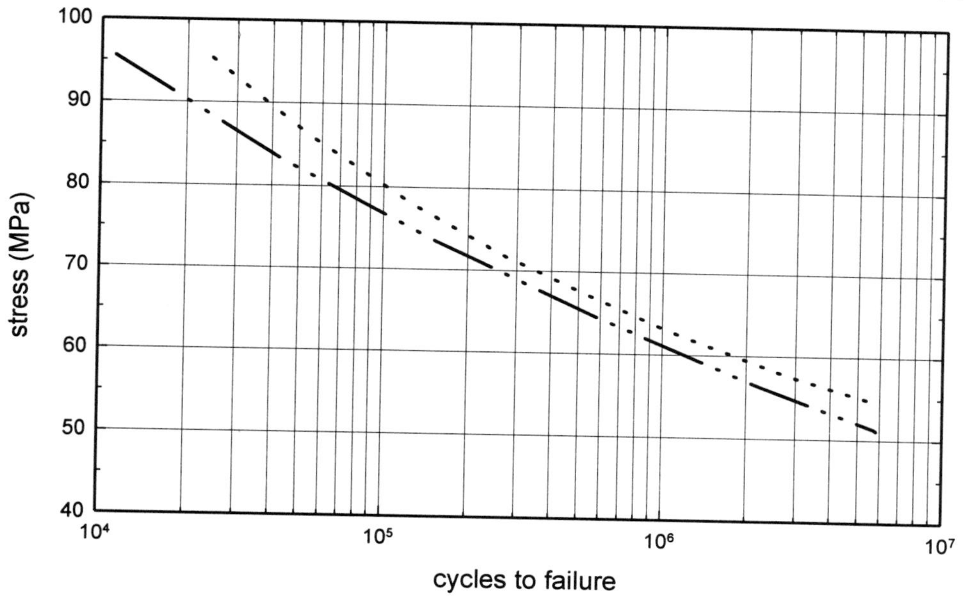

....	LNP Lubricomp ECL4036 PEI (lubricated; 15% PTFE modified, 30% carbon fiber); flexure; 30 Hz; 23°C
—··—··	LNP Thermocomp EF1006 PEI (30% glass fiber); flexure; 30 Hz; 23°C
Reference No.	394

GRAPH 116: **Fatigue Cycles to Failure vs. Stress in Tension for Glass Fiber Reinforced General Electric Ultem Polyetherimide.**

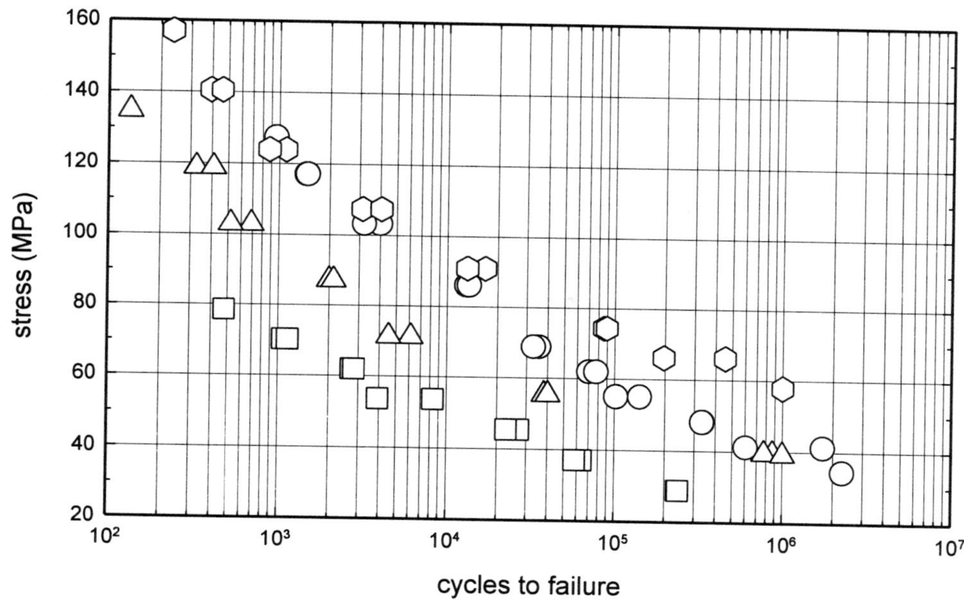

○	GE Ultem 2200 PEI (flame retardant; 20% glass fiber; inj. mold.); tension-tension; GE method; 5 Hz; stress ratio: 0.1; 23°C; flow direction [fig g]
□	GE Ultem 2212 PEI (high flow, flame retardant; 20% glass fiber; milled glass; inj. mold.); tension-tension; GE method; 5 Hz; stress ratio: 0.1; 23°C; flow direction [fig g]
△	GE Ultem 2240 PEI (20% glass fiber; inj. mold.); tension- tension; GE method; 5 Hz; stress ratio: 0.1; 23°C; flow direction [fig g]
⬡	GE Ultem 2440 PEI (40% glass fiber; inj. mold.); tension- tension; GE method; 5 Hz; stress ratio: 0.1; 23°C; flow direction [fig g]
Reference No.	352

GRAPH 117: Fatigue Cycles to Failure vs. Stress in Tension for High Temperature and FDA Compliant General Electric Ultem Polyetherimide.

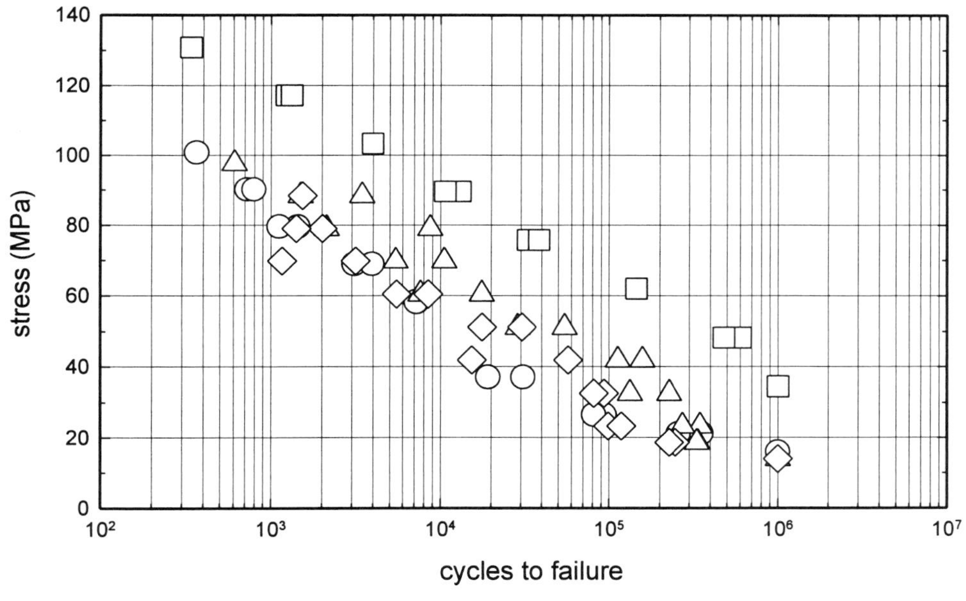

○	GE Noryl 6000 PEI (flame retardant, unfilled; inj. mold.); tension- tension; GE method; 5 Hz; stress ratio: 0.1; 23°C; flow direction [fig g]
□	GE Noryl 6200 PEI (flame retardant; 20% glass fiber; inj. mold.); tension-tension; GE method; 5 Hz; stress ratio: 0.1; 23°C; flow direction [fig g]
△	GE Ultem CRS5001 PEI (transparent, FDA grade, flame retardant, unfilled; extrus., inj. mold.); tension-tension; GE method; 5 Hz; stress ratio: 0.1; 23°C; flow direction [fig g]
◇	GE Ultem CR55011 PEI (transparent, high flow, FDA grade, flame retardant, unfilled; extrus., inj. mold.); tension-tension; GE method; 5 Hz; stress ratio: 0.1; 23°C; flow direction [fig g]
Reference No.	352

GRAPH 118: Fatigue Cycles to Failure vs. Stress in Tension for Glass Fiber/ Mica Reinforced and Wear Resistant General Electric Ultem Polyetherimide.

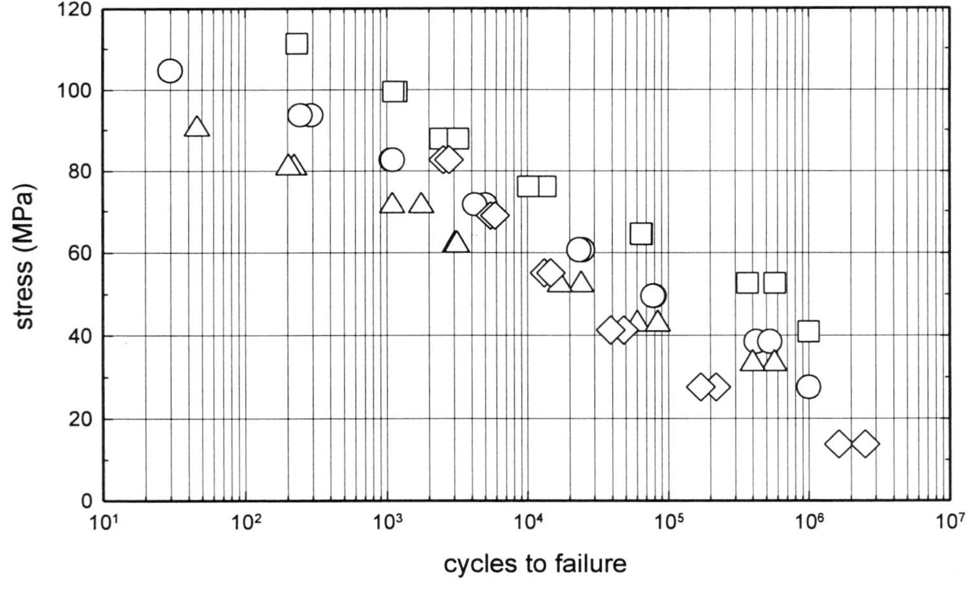

○	GE Ultem 3453 PEI (flame retardant; 20% glass fiber, 25% mineral filler; inj. mold.); tension- tension; GE method; 5 Hz; stress ratio: 0.1; 23°C; flow direction [fig g]
□	GE Ultem 3455 PEI (high flow, flame retardant; 20% glass fiber, 25% mineral filler; inj. mold.); tension-tension; GE method; 5 Hz; stress ratio: 0.1; 23°C; flow direction [fig g]
△	GE Ultem 4000 PEI (wear resistant, flame retardant, unfilled; inj. mold.); tension-tension; GE method; 5 Hz; stress ratio: 0.1; 23°C; flow direction [fig g]
◇	GE Ultem 4001 PEI (lubricated, wear resistant, flame retardant, unfilled; inj. mold.); tension-tension; GE method; 5 Hz; stress ratio: 0.1; 23°C; flow direction [fig g]
Reference No.	352

GRAPH 119: **Fatigue Cycles to Failure vs. Stress in Tension for Glass Fiber Reinforced General Electric Ultem Polyetherimide.**

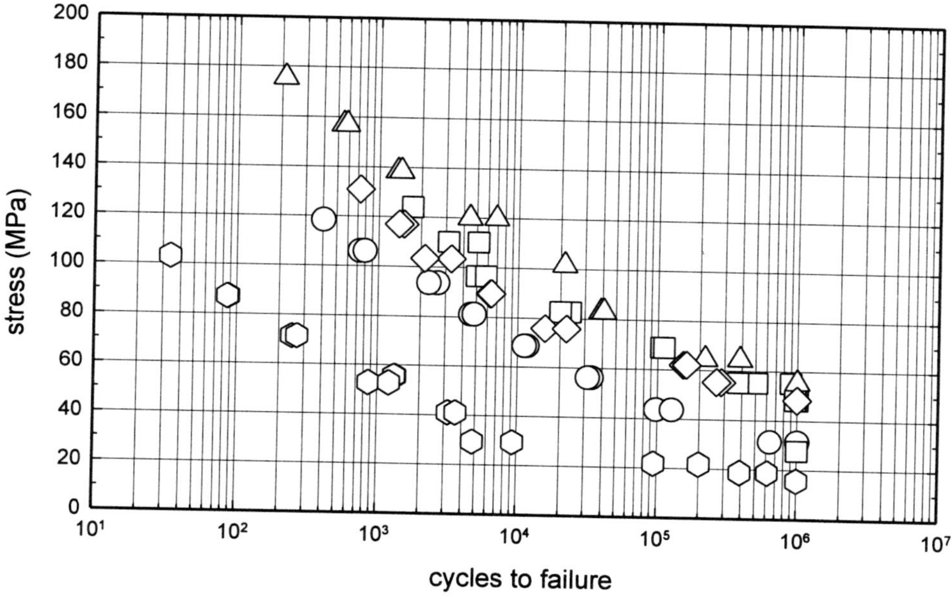

○	GE Ultem 2100 PEI (flame retardant; 10% glass fiber; inj. mold.); tension-tension; GE method; 5 Hz; stress ratio: 0.1; 23°C; flow direction [fig g]
□	GE Ultem 2300 PEI (flame retardant; 30% glass fiber; inj. mold.); tension-tension; GE method; 10 Hz; stress ratio: 0.1; 23°C; flow direction [fig g]
△	GE Ultem 2310 PEI (high flow, flame retardant; 30% glass fiber; inj. mold.); tension- tension; GE method; 5 Hz; stress ratio: 0.1; 23°C; flow direction [fig g]
◇	GE Ultem 2340 PEI (30% glass fiber; inj. mold.); tension- tension; GE method; 5 Hz; stress ratio: 0.1; 23°C; flow direction [fig g]
⬡	GE Ultem 2342 PEI (30% glass fiber; inj. mold.); tension- tension; GE method; 5 Hz; stress ratio: 0.1; 23°C; flow direction [fig g]
Reference No.	352

GRAPH 120: **Fatigue Cycles to Failure vs. Stress in Tension at Various Test Conditions for 1000 Series General Electric Ultem Polyetherimide.**

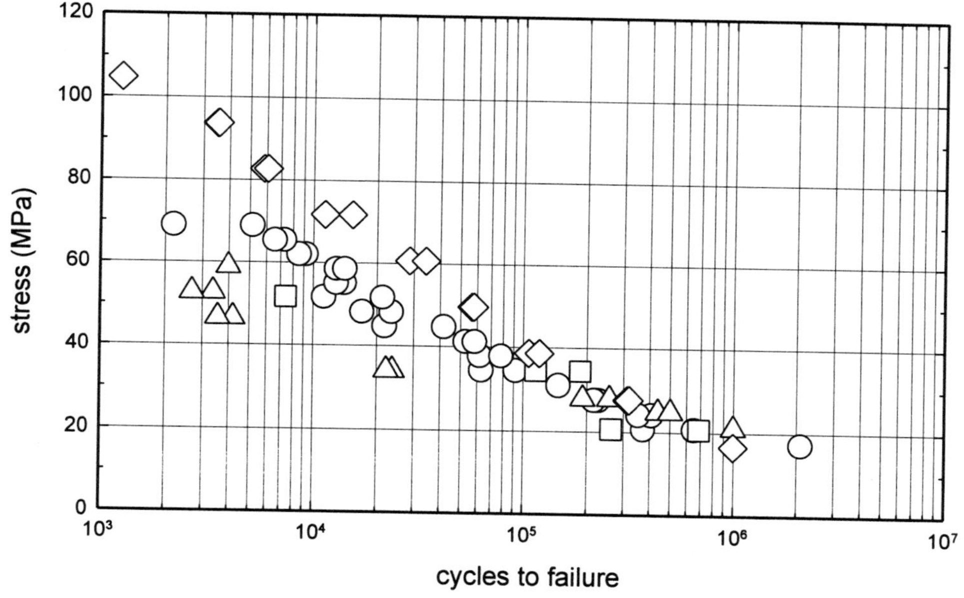

○	GE Ultem 1000 PEI (transparent, flame retardant, unfilled; extrus., inj. mold.); tension-tension; GE method; 5 Hz; stress ratio: 0.1; 23°C; flow direction [fig g]
□	GE Ultem 1000 PEI (transparent, flame retardant, unfilled; extrus., inj. mold.); tension-tension; GE method; 30 Hz; stress ratio: 0.1; 23°C; flow direction [fig g]
△	GE Ultem 1000 PEI (transparent, flame retardant, unfilled; extrus., inj. mold.); tension-tension; GE method; 5 Hz; stress ratio: 0.1; 93°C; flow direction [fig g]
◇	GE Ultem 1020 PEI (transparent, high flow, unfilled; inj. mold.); tension-tension; GE method; 5 Hz; stress ratio: 0.1; 23°C; flow direction [fig g]
Reference No.	352

Chapter 25

Polyaryletherketone

GRAPH 121: Fatigue Cycles to Failure vs. Stress in Flexure for BASF Ultrapek A Polyaryletherketone.

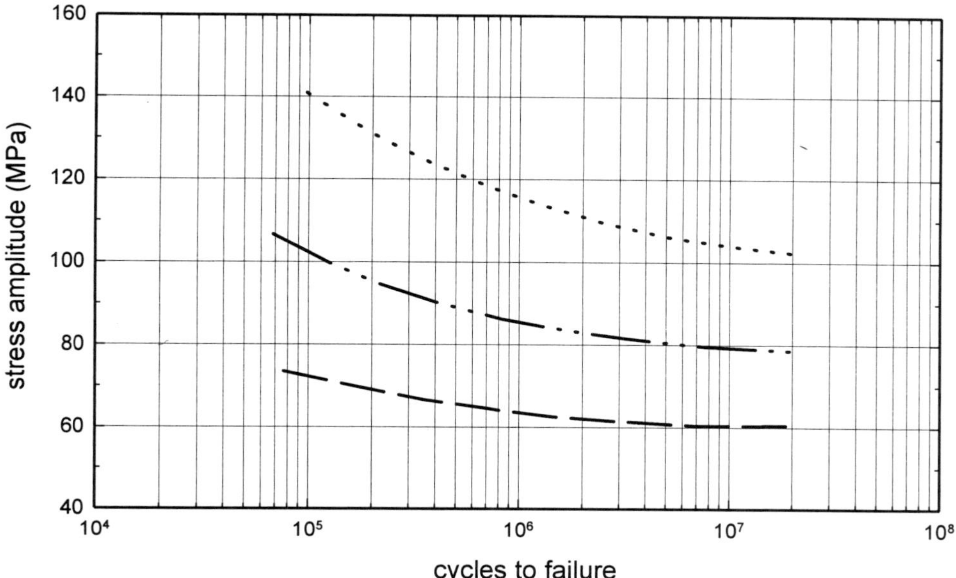

··············	BASF AG Ultrapek A2000C6 PAEK (moderate flow; 30% carbon fiber); DIN 53442; 15 Hz; 23°C; 50% RH; laboratory atmosphere
—··—··	BASF AG Ultrapek A2000G6 PAEK (moderate flow; 30% glass fiber); DIN 53442; 15 Hz; 23°C; 50% RH; laboratory atmosphere
— — — —	BASF AG Ultrapek A3000 PAEK (low flow); DIN 53442; 15 Hz; 23°C; 50% RH; laboratory atmosphere
Reference No.	27

Chapter 26

Polyetheretherketone

Fatigue Properties

Victrex USA: Victrex PEEK

'Victrex' PEEK shows good fatigue performance. As with constant load data, operation above the line should be avoided. In common with all plastics 'Victrex' PEEK is sensitive to the presence of notches. In general, the smaller the notch radius or the sharper the notch the greater will be the reduction in fatigue performance.

Reference: *Victrex PEEK The High Temperature Engineering Thermoplastic - Data For Design,* supplier design guide (VK4/2/1290) - ICI Advanced Materials, 1990.

GRAPH 122: Fatigue Cycles to Failure vs. Stress in Flexure for Reinforced LNP Polyetheretherketone.

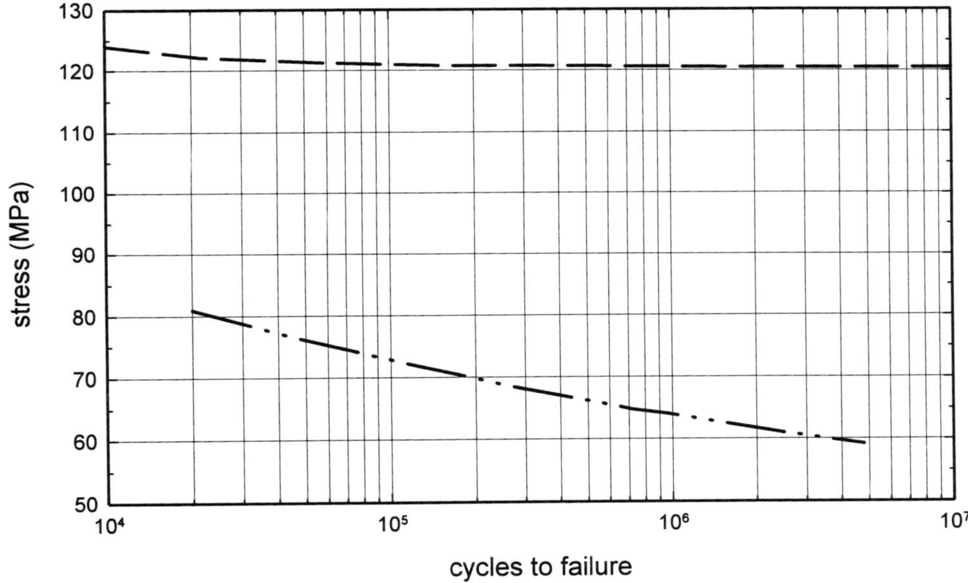

– – – –	LNP Stat-Kon LC1006 PEEK (static dissipative; 30% carbon fiber); flexure; 30 Hz; 23°C
—··—··	LNP Lubricomp LCL4033EM PEEK (lubricated; 15% carbon fiber, 15% PTFE modified); flexure; 30 Hz; 23°C
– – – –	LNP Thermocomp LC1006 PEEK (30% carbon fiber); flexure; 30 Hz; 23°C
Reference No.	394

GRAPH 123: **Fatigue Cycles to Failure vs. Stress in Tension at Low Test Frequency for Victrex PEEK Polyetheretherketone.**

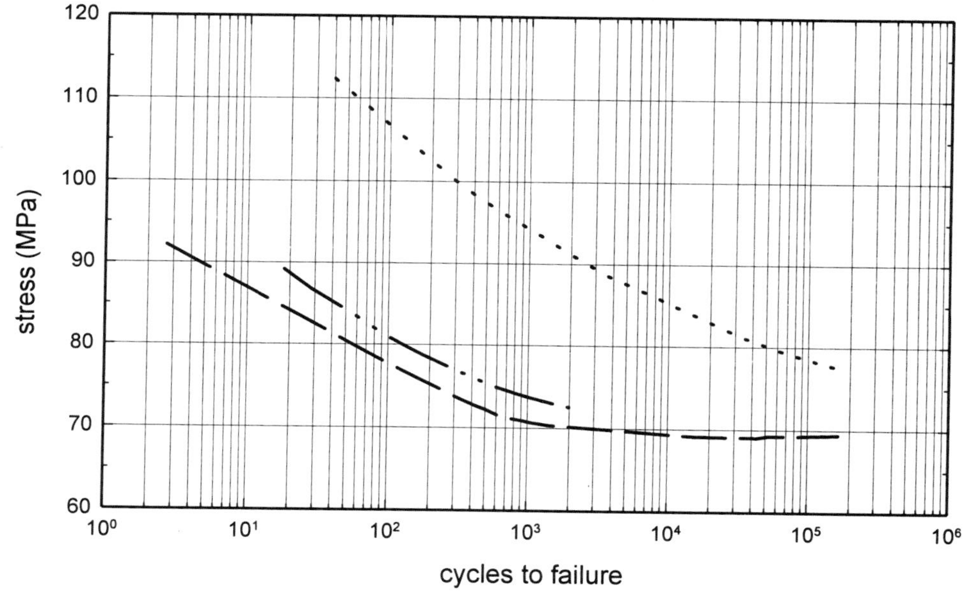

....	Victrex PEEK 450CA30 PEEK (30% carbon fiber); tension- zero; 0.5 Hz; 23°C; perpendicular to flow; square wave [fig f, y]
—··—··	Victrex PEEK 450GL30 PEEK (30% glass fiber); tension- zero; 0.5 Hz; 23°C; perpendicular to flow; square wave [fig f, y]
— — — —	Victrex PEEK 450G (gen. purp. grade); tension- zero; 0.5 Hz; 23°C; perpendicular to flow; square wave [fig f, y]
Reference No.	337

GRAPH 124: **Fatigue Cycles to Failure vs. Stress in Tension at Low Test Frequency for Victrex PEEK 450G Polyetheretherketone.**

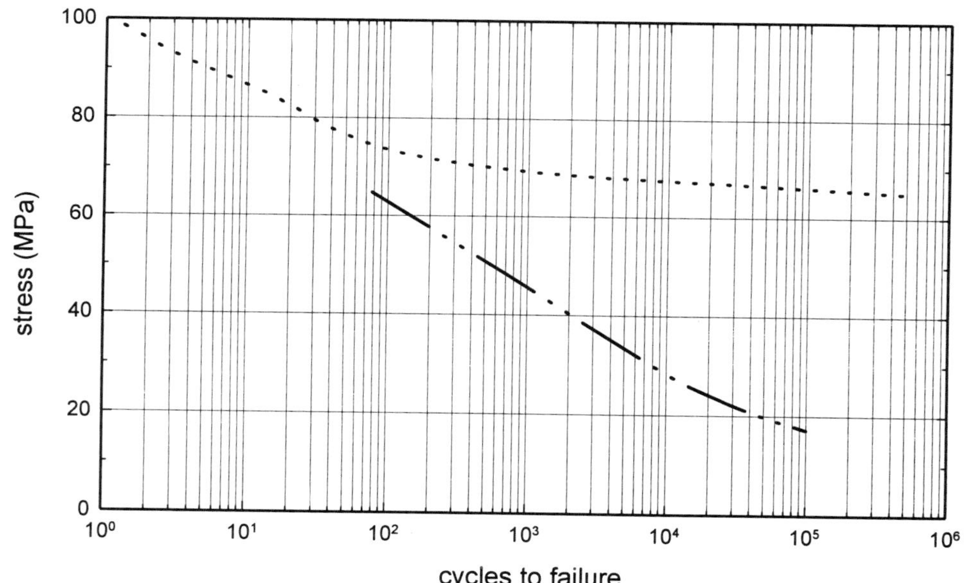

....	Victrex PEEK 450G (gen. purp. grade); tension- zero; 0.5 Hz; 23°C; square wave; unnotched [fig f, w]
—··—··	Victrex PEEK 450G (gen. purp. grade); tension- zero; 0.5 Hz; 23°C; square wave; notched; 0.25 mm radius [fig f, x]
Reference No.	337

GRAPH 125: **Fatigue Cycles to Failure vs. Initial Strain in Tension at Two Temperatures for Polyetheretherketone.**

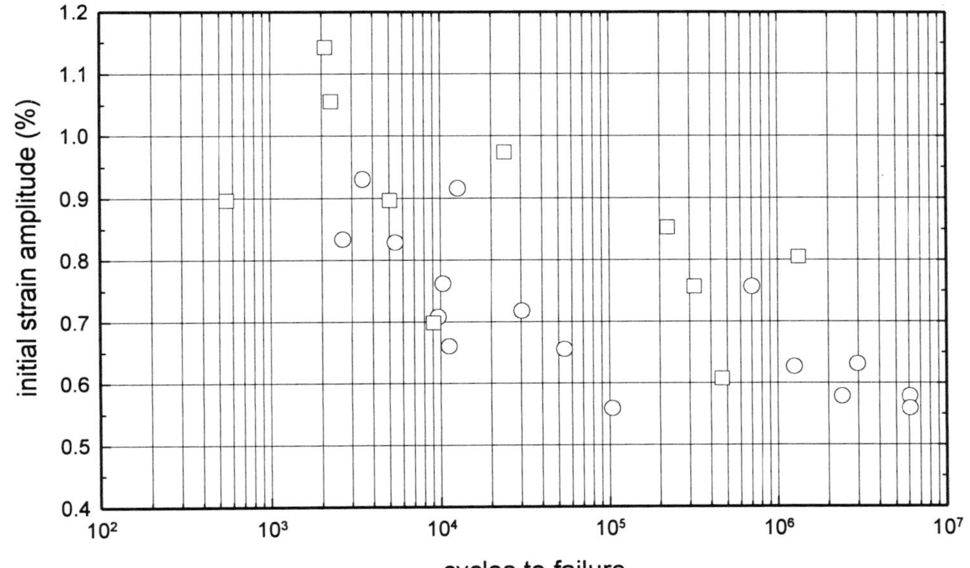

○	PEEK; tension- zero; 0.5 Hz; 20°C [fig d]
□	PEEK; tension- zero; 0.5 Hz; 120°C [fig d]
Reference No.	355

Chapter 27

Polyetherketone

Fatigue Properties

Victrex USA: Victrex PEK

Victrex PEK is particularly useful where vibrational or cyclic loading is likely to occur, and also where components are required to withstand high mechanical loads for long periods, even at very high temperatures. Victrex PEK offers excellent fatigue properties coupled with good creep resistance - a combination rarely found in crystalline polymers.

Reference: *Victrex PEK Properties And Processing,* supplier design guide (VP2/ October 1987) - ICI Advanced Materials, 1987.

Chapter 28

High Density Polyethylene

Fatigue Properties

Paxon HDPE: Virgin HDPE (features: virgin resin); **Waste Alternatives Co.: Recycled HDPE** (features: recycled resin)

At flexing frequencies approximately equal to and less than 330 cpm both recycled and virgin HDPE exhibit brittle failure after high cycle counts. For this frequency and below, the flexural fatigue failure is caused by crack propagation. A macro crack is initiated at the edge of the sample then propagates transversely across the specimen at three different rates during the lifetime of the specimen. Recycled HDPE has a shorter fatigue lifetime than virgin HDPE due to its lower resistance to crack initiation and higher crack propagation rate. The nature of flexural fatigue behavior of HDPE materials at frequencies higher than 330 cpm and at elevated temperatures requires further investigation.

Reference: Yang, S.G. (University of Florida), Bennett, D. (University of Florida), Beatty, C. (University of F, *Flexural Fatigue Resistance of Recycled HDPE,* ANTEC 1994, conference proceedings - Society of Plastics Engineers, 1994.

GRAPH 126: Fatigue Cycles to Failure vs. Stress in Flexure for Hoechst Hostalen GM 5010T2 High Density Polyethylene.

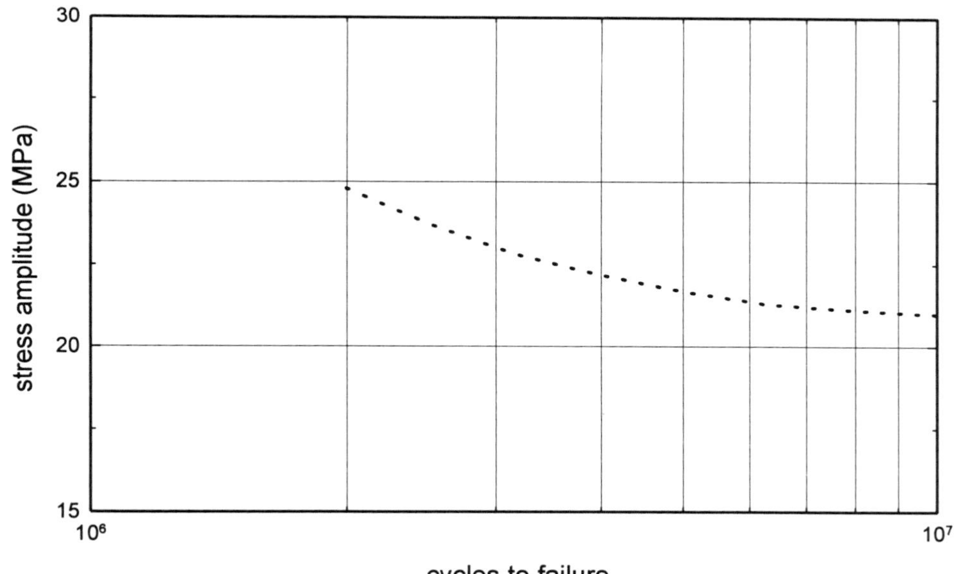

....	Hoechst AG Hostalen GM5010T2 HDPE (pressure pipe; black, 60 Shore D; 11-17 g/ 10 min. MFI); flexure (alternating); DIN 53442; 10 Hz; 23°C; mean stress = 0
Reference No.	94

GRAPH 127: Fatigue Cycles to Failure vs. Stress in Flexure at Different Test Frequencies for Recycled High Density Polyethylene.

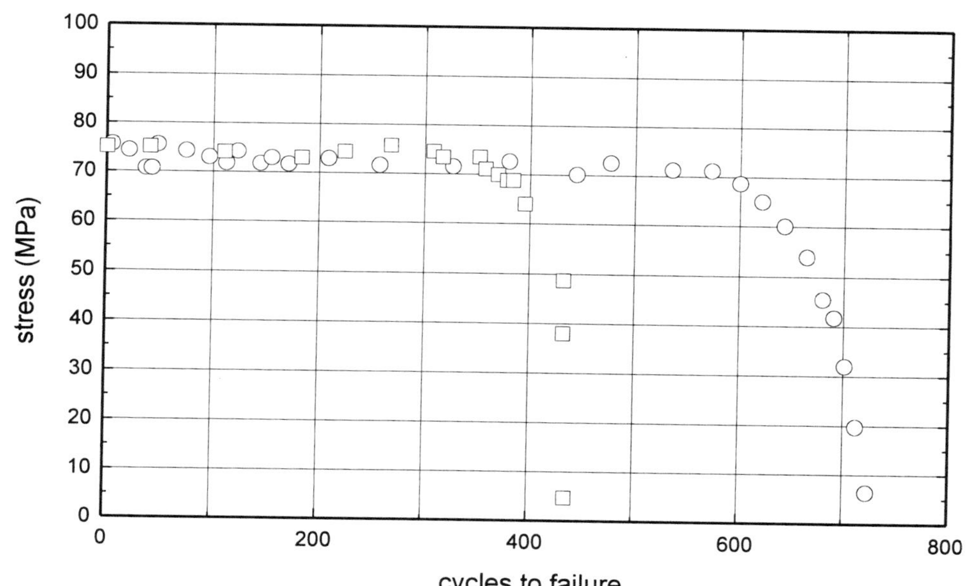

GRAPH 128: Cycles to Failure vs. Deflection in Flexure for Recycled and Virgin High Density Polyethylene.

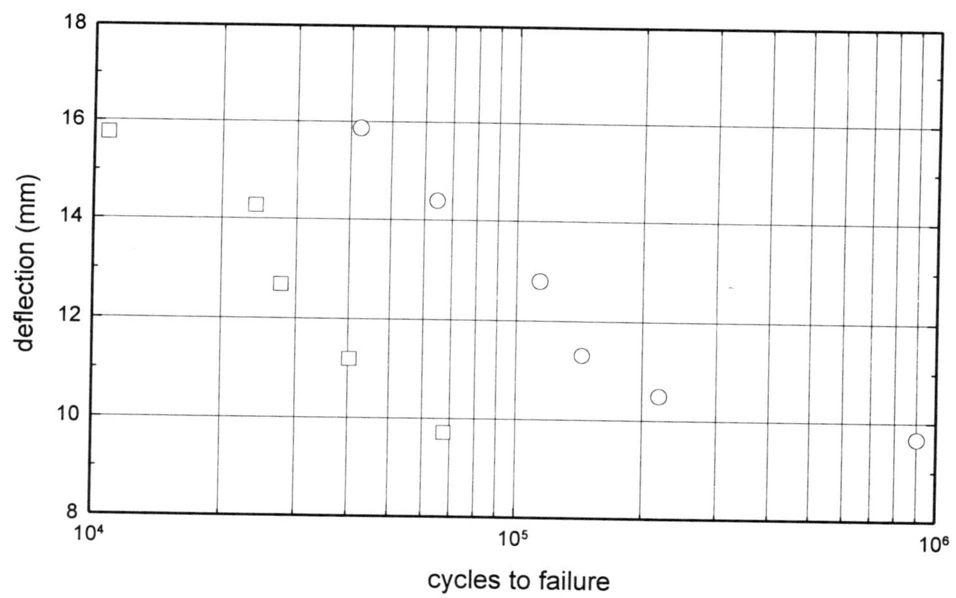

GRAPH 129: **Fatigue Crack Propagation (FCP) Rate vs. Stress Intensity Range for 0.59 Melt Index High Density Polyethylene.**

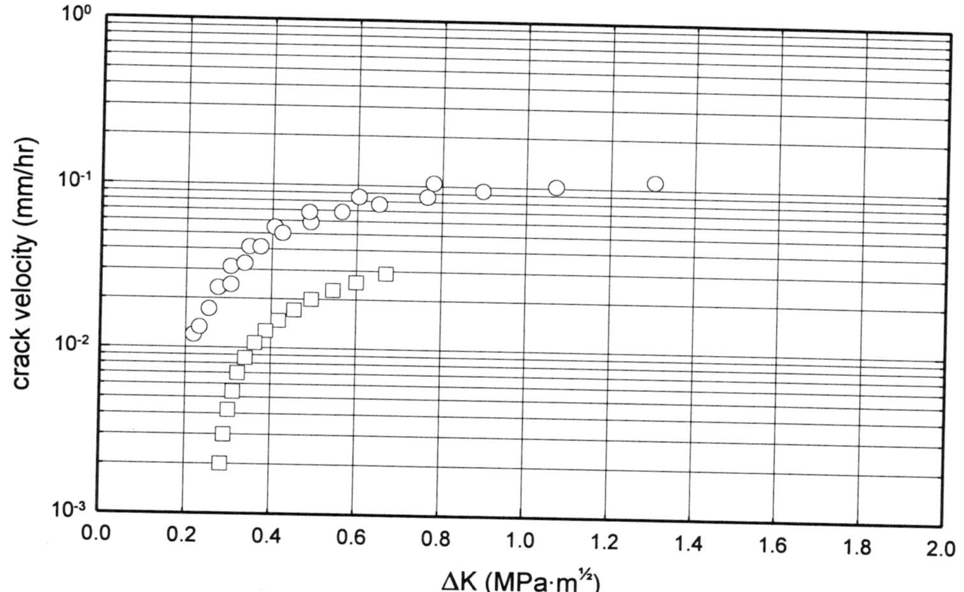

○	HDPE (pipe grade; 0.59 g/10 min. MFI; prepared w/ pipes previously utilized in natural gas system); tension; 50°C; single edge notched specimen
□	HDPE (pipe grade; 0.59 g/10 min. MFI; prepared w/ pipes previously utilized in natural gas system); tension; 40°C; single edge notched specimen
Reference No.	376

Chapter 29

Polypropylene

Fatigue Properties

Eastman: Tenite

Polypropylene possesses a unique characteristic - the ability to be molded or formed into a tough and highly fatigue resistant one piece hinge. A properly formed integral hinge in polypropylene is very strong, even though the hinge area is usually the thinnest part of the entire article. The superior strength results from the molecular orientation induced in the hinge by flexing. Flexing the hinge causes the molecules to align themselves perpendicular to the hinge, concentrating the molecular chain strength in the direction of alignment. Unoriented molecular chains have a random configuration which distributes the chain strength in all directions. If practical, the hinge should be flexed as soon after forming as possible since the molecules can be oriented much easier, and the possibility of breakage is reduced, if the hinge is flexed while it retains heat imparted to it during the forming operation.

Reference: *Utilizing Inherent Hinge Properties of Tenite Polypropylene and Tenite Polyallomer,* supplier technical report (TR-14E) - Eastman Chemical, 1987.

Effect of Glass Reinforcement on Fatigue Behavior

LNP Engineering Plastics: Verton MFX-700-10 HS (material compostion: 50% glass fiber reinforcement; note: long glass fibers (average fiber length of 1330-1370 microns), chemically coupled glass fibers); **Verton MFX-7006 HS** (material compostion: 30% glass fiber reinforcement; note: long glass fibers, chemically coupled glass fibers); **Verton MFX-7008 HS** (material compostion: 40% glass fiber reinforcement; note: chemically coupled glass fibers, long glass fibers)

The flexural fatigue behavior of 30%, 40%, and 50% long glass-reinforced polypropylene composites was tested. Under static testing conditions there is a pronounced difference in tensile properties with numbers ranging from 107 MPA to 128 MPA, but as the fatigue cycles increase to 1000-10,000 cycles, all of the data converges to similar values. As the cycles increase into the million cycle region, slight differences are observed which favor the higher glass content material.

Interfacial crack initiation points are more abundant in the higher glass content material than the lower glass content material due to the greater interface area in the 50% glass-reinforced composite. Once the cracks are initiated, the crack propagation rate will be lower under higher fiber density conditions. Thus, the fatigue behavior of a 30% glass-reinforced material is characterized by fewer initiation points coupled with a faster rate of crack propagation, whereas a 50% glass reinforced composite has more initiation points along with a slower crack propagation rate. As the cycles increase further, the crack propagation rate begins to have a more significant influence over the interfacial effect, and then a slightly better flexural fatigue resistance is observed in the higher glass content materials.

Reference: Grove, D.A. (LNP) m, H.C. (LNP), *Effect of Constituents on the Fatigue Behavior of Long Fiber Reinforced Thermoplastics,* ANTEC 1995, conference proceedings - Society of Plastics Engineers, 1995.

LNP Engineering Plastics: Thermocomp MFX-100-10 HS (note: chemically coupled glass fibers, short glass fiber (average length 230-280 microns)); **Verton MFX-700-10 HS** (material compostion: 50% glass fiber reinforcement; note: long glass fibers (average fiber length of 1330-1370 microns), chemically coupled glass fibers)

Unlike tensile fatigue, where a uniform stress is applied across the sample, in flexural fatigue the stress varies from a large tensile stress on one side of the specimen to a large compressive stress on the opposite side. The cyclic tensile and compression loads in the outer regions lead to various levels of fiber-resin debonding coupled with some fiber breakage.

Crack initiation in glass-reinforced polypropylene composites occurred mostly due to resin-fiber debonding coupled with some fiber breakage. After the initial cracks were created, the cracks progressed towards the center of the part. The cracks would preferentially move around or pull-out the long fibers; whereas the cracks were more likely to cause pull-outs in the short fiber material. Evidence from previous papers clearly showed that fiber pull-out, fiber breakage, etc. are part of the dominant flexural fatigue crack propagation behavior. Since longer perpendicular fibers are known to slow the progress of cracks through some pull-out, fiber breakage, and crack opening displacement criteria, the longer fiber material is expected to exhibit superior longitudinal fatigue resistance. This behavior was indeed observed in the long glass- reinforced composites.

In the case of polypropylene, the significant resin-glass debonding behavior was the main cause for the poorer fatigue behavior of this material versus nylon and polyphthalamide. The much greater number of initial cracks formed from debonding fibers was the main reason for the poorer performance of the polypropylene material versus other composite systems.

Reference: Grove, D.A. (LNP) m, H.C. (LNP), *Effect of Constituents on the Fatigue Behavior of Long Fiber Reinforced Thermoplastics,* ANTEC 1995, conference proceedings - Society of Plastics Engineers, 1995.

Effect of Molecular Weight on Fatigue Behavior

LNP Engineering Plastics: PP (material compostion: 50% glass fiber reinforcement; molecular weight: 150,000; note: long glass fibers); **PP** (material compostion: 50% glass fiber reinforcement; molecular weight: 210,000; note: long glass fibers)

The molecular weight influence on flexural fatigue behavior was investigated by comparing the fatigue behavior of two 50% glass reinforced polypropylene composites formed from 210,000 and 150,000 weight average molecular weight polypropylene resins. Despite the employment of a lower molecular weight material, which may have adversely affected the crack propagation rate through the more brittle matrix, similar flexural fatigue behavior was observed. Perhaps, lower molecular weight polymer chains participate better in interfacial bonding to offset the increase in the expected crack propagation rate.

Reference: Grove, D.A. (LNP), Kim, H.C. (LNP), *Effect of Constituents on the Fatigue Behavior of Long Fiber Reinforced Thermoplastics,* ANTEC 1995, conference proceedings - Society of Plastics Engineers, 1995.

GRAPH 130: Fatigue Cycles to Failure vs. Stress in Flexure for 50% Glass Fiber Reinforced Polypropylene with Different Molecular Weights.

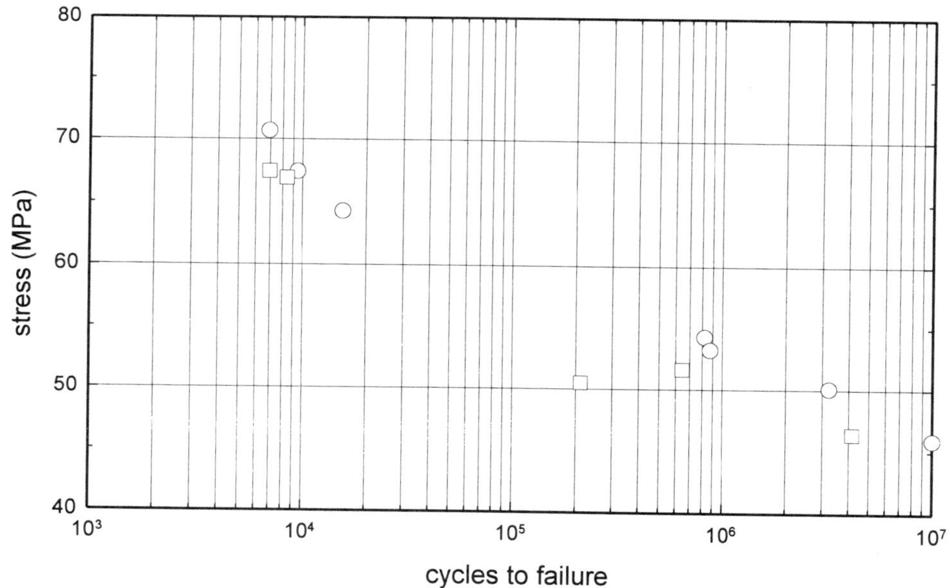

○	LNP PP (50% glass fiber; 210,000 mol. wgt.; long glass fibers); flexure; 30 Hz; 23°C; ASTM D671 Type B [fig d]
□	LNP PP (50% glass fiber; 150,000 mol. wgt.; long glass fibers); flexure; 30 Hz; 23°C; ASTM D671 Type B [fig d]
Reference No.	391

GRAPH 131: **Fatigue Cycles to Failure vs. Stress in Flexure for Hoechst Hostalen PPN 7180TV20 Polypropylene.**

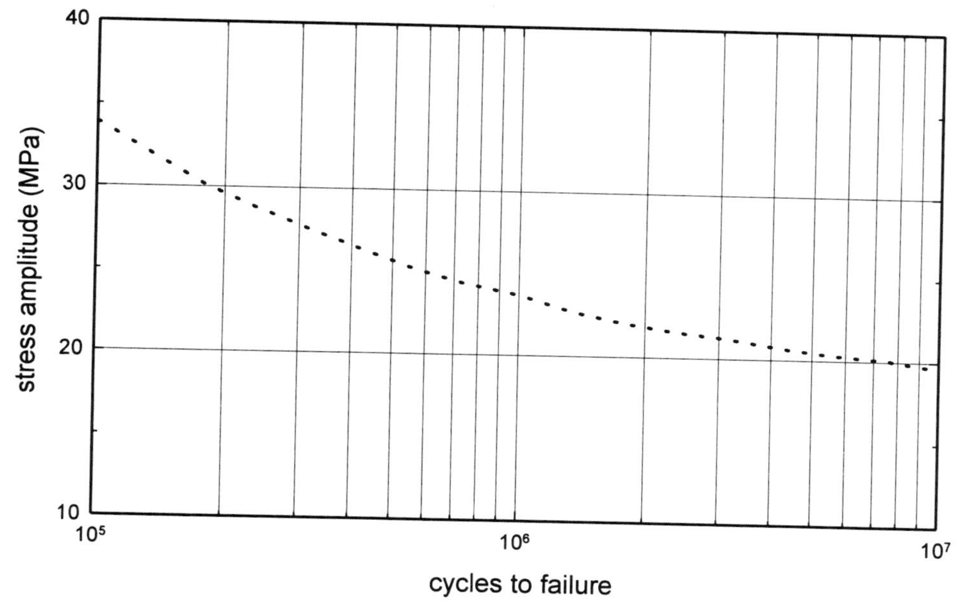

....	Hoechst AG Hostalen PPN 7180TV20 PP; flexure (alternating); 10 Hz; 23°C; mean stress = 0 [fig d]
Reference No.	325

GRAPH 132: **Fatigue Cycles to Failure vs. Stress in Flexure for Hoechst Hostalen PPN 7790 GV2/30 Polypropylene.**

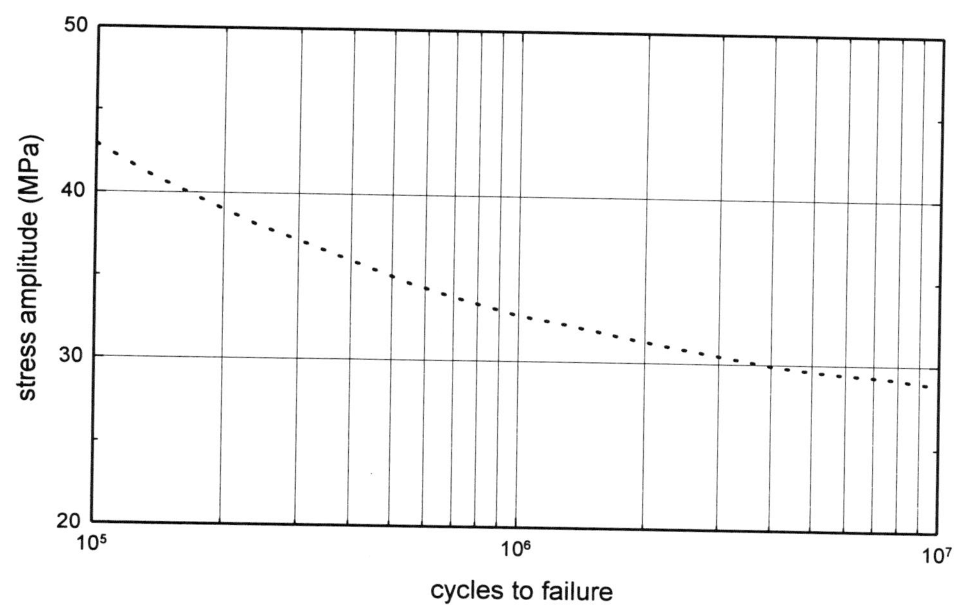

....	Hoechst AG Hostalen PPN 7790GV2/30 PP; flexure (alternating); 10 Hz; 23°C; mean stress = 0 [fig d]
Reference No.	325

GRAPH 133: Fatigue Cycles to Failure vs. Stress in Flexure for Hoechst Hostalen PPN 7790 GV2/30 Polypropylene.

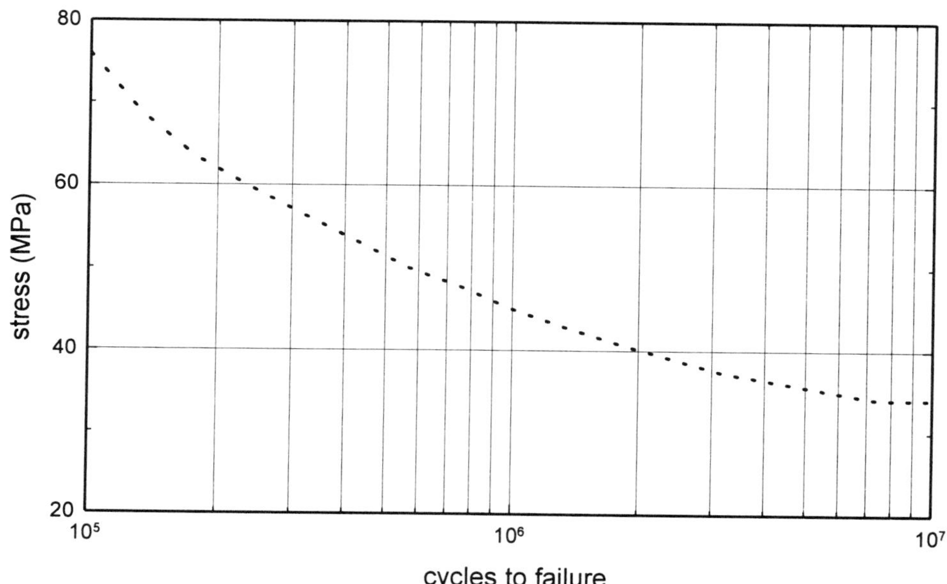

....	Hoechst AG Hostalen PPN 7790GV2/30 PP; flexure; 10 Hz; 23°C [fig f]
Reference No.	325

GRAPH 134: Fatigue Cycles to Failure vs. Stress in Flexure for Hoechst Hostacom G3N01 Polypropylene.

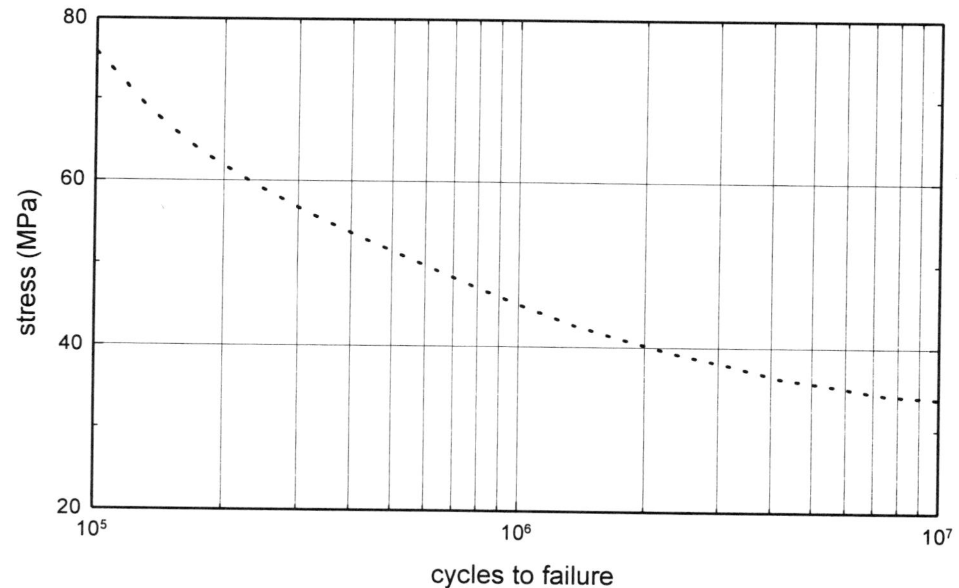

....	Hoechst AG Hostacom G3N01 PP (30% glass fiber, chemically coupled; 1 g/10 min. MFI); flexure (pulsating); 10 Hz; 23°C [fig f]
Reference No.	326

GRAPH 135: Fatigue Cycles to Failure vs. Stress in Flexure for Long Glass Fiber Reinforced Polypropylene.

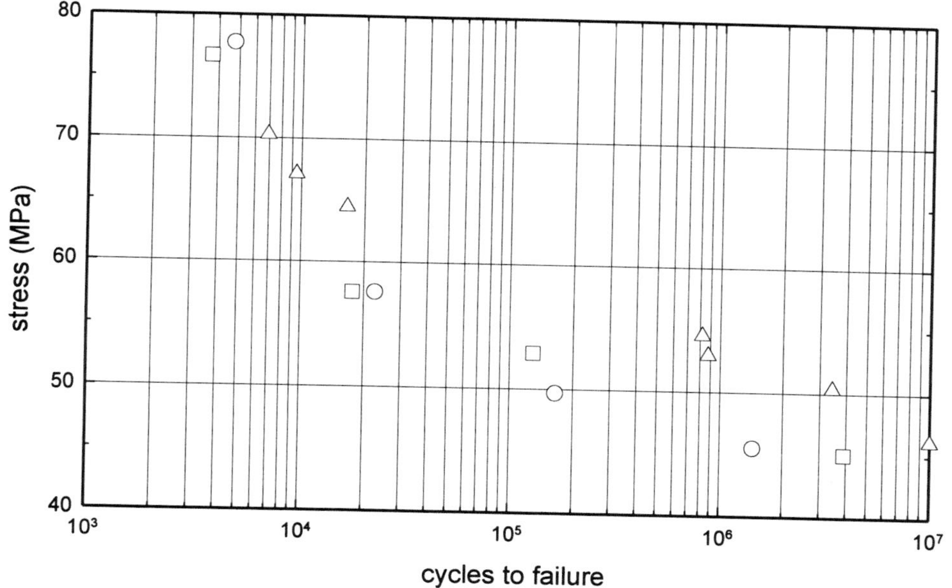

○	LNP Verton MFX-7006 HS PP (30% glass fiber, chemically coupled; long glass fibers); flexure; 30 Hz; 23°C; ASTM D671 Type B [fig d]
□	LNP Verton MFX-7008 HS PP (40% glass fiber, chemically coupled; long glass fibers); flexure; 30 Hz; 23°C; ASTM D671 Type B [fig d]
△	LNP Verton MFX-700-10 HS PP (50% glass fiber, chemically coupled; long glass fibers (1330-1370 microns)); flexure; 30 Hz; 23°C; ASTM D671 Type B [fig d]
Reference No.	391

GRAPH 136: Fatigue Cycles to Failure vs. Stress in Flexure for Long and Short Glass Fiber Reinforced Polypropylene.

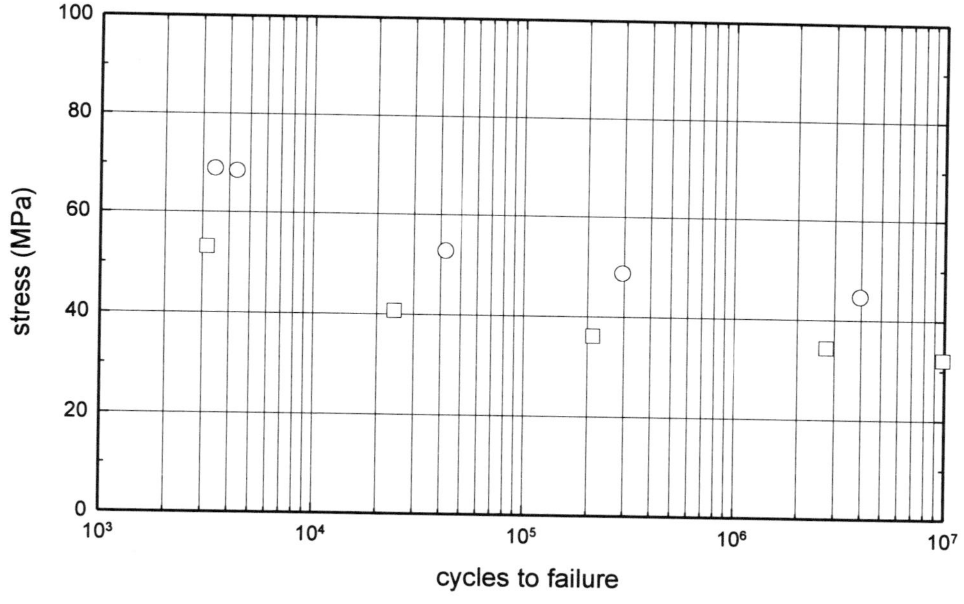

○	LNP Verton MFX-700-10 HS PP (50% glass fiber, chemically coupled; long glass fibers (1330-1370 microns)); flexure; 30 Hz; 23°C; ASTM D671 Type B [fig d]
□	LNP Thermocomp MFX-100-10 HS PP (chemically coupled; short glass fiber (230-280 microns)); flexure; 30 Hz; 23°C; ASTM D671 Type B [fig d]
Reference No.	391

GRAPH 137: Fatigue Cycles to Failure vs. Stress in Flexure for Hoechst Hostacom G3N01 Polypropylene.

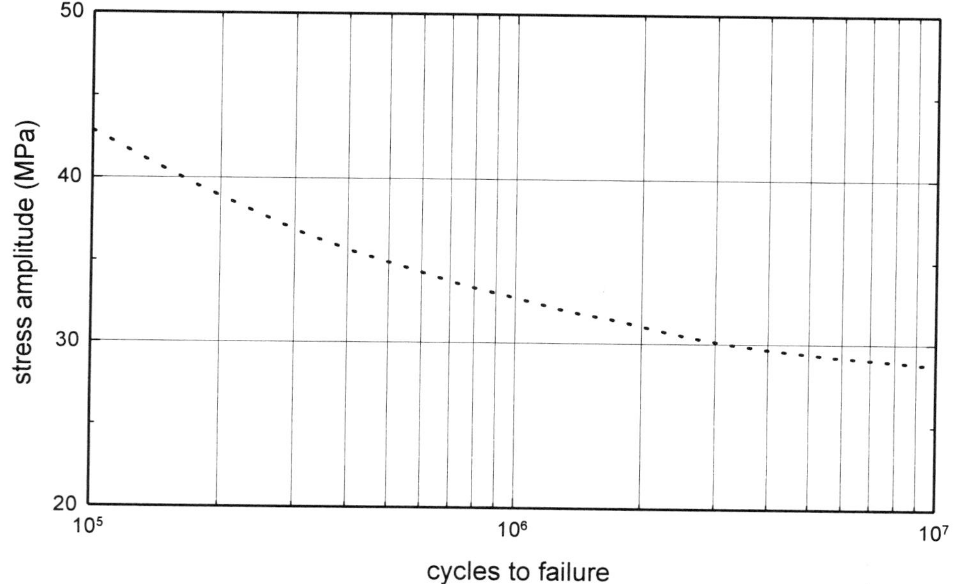

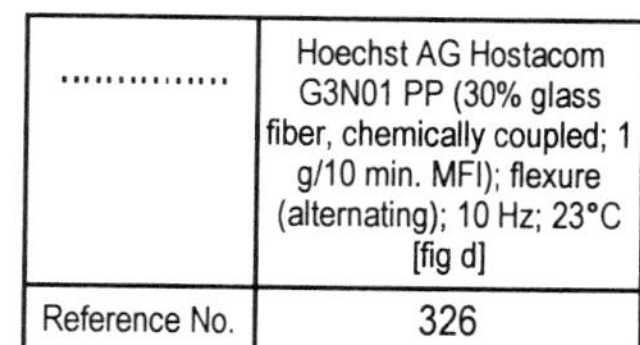

GRAPH 138: Fatigue Cycles to Failure vs. Stress in Flexure for Hoechst Hostacom M2N01 Polypropylene.

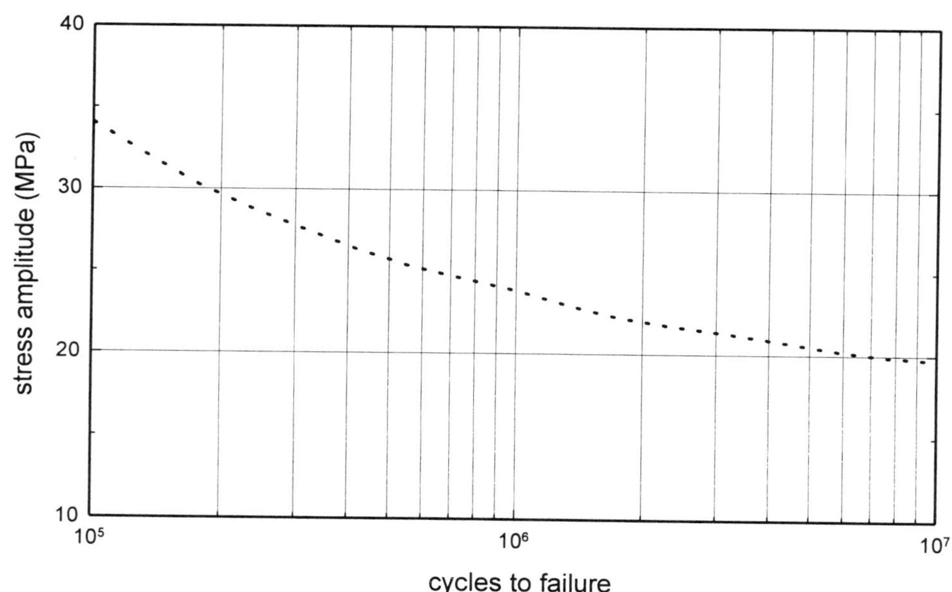

....	Hoechst AG Hostacom M2N01 PP (low warp grade, good surface; 20% talc; 2 g/10 min. MFI); flexure (alternating); 10 Hz; 23°C [fig d]
Reference No.	326

GRAPH 139: **Fatigue Cycles to Failure vs. Stress in Flexure for Glass Fiber Reinforced LNP Polypropylene.**

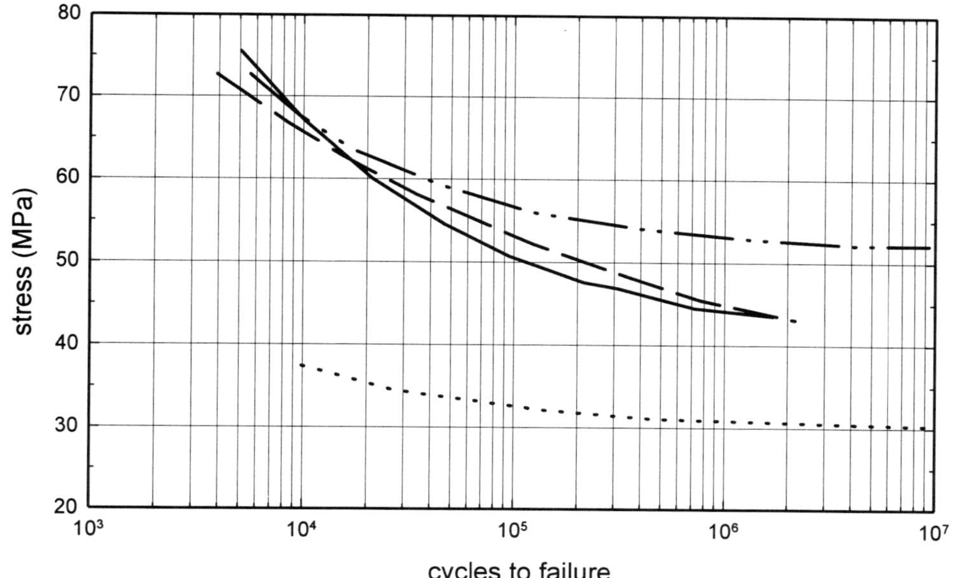

....	LNP Thermocomp MF1006 PP (30% glass fiber); flexure; 30 Hz; 23°C
—··—··	LNP Verton MFX70010 PP (50% glass fiber, chemically coupled; long glass fibers); flexure; 30 Hz; 23°C
— — —	LNP Verton MFX7008 PP (40% glass fiber, chemically coupled; long glass fibers); flexure; 30 Hz; 23°C
———	LNP Verton MFX7006 PP (30% glass fiber, chemically coupled; long glass fibers); flexure; 30 Hz; 23°C
Reference No.	394

GRAPH 140: **Fatigue Cycles to Failure vs. Stress in Tension for 25% Glass Fiber Reinforced Polypropylene.**

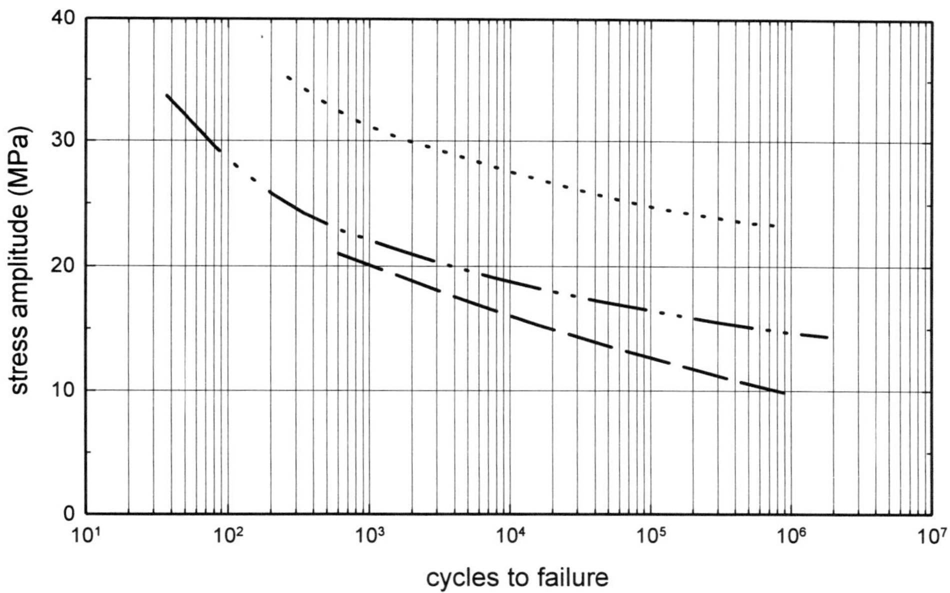

....	PP (25% glass fiber); tension- compression; 0.5 Hz; 20°C; 6 mm thick; perpendicular to flow; mean stress = 0 [fig d]
—··—··	PP (25% glass fiber); tension- compression; 0.5 Hz; 20°C; 6 mm thick; parallel to flow; mean stress = 0 [fig d]
— — —	PP; tension- compression; 0.5 Hz; 20°C; mean stress = 0 [fig d]
Reference No.	355

GRAPH 141: Fatigue Cycles to Failure vs. Stress in Tension for Long and Short Glass Reinforced Polypropylene.

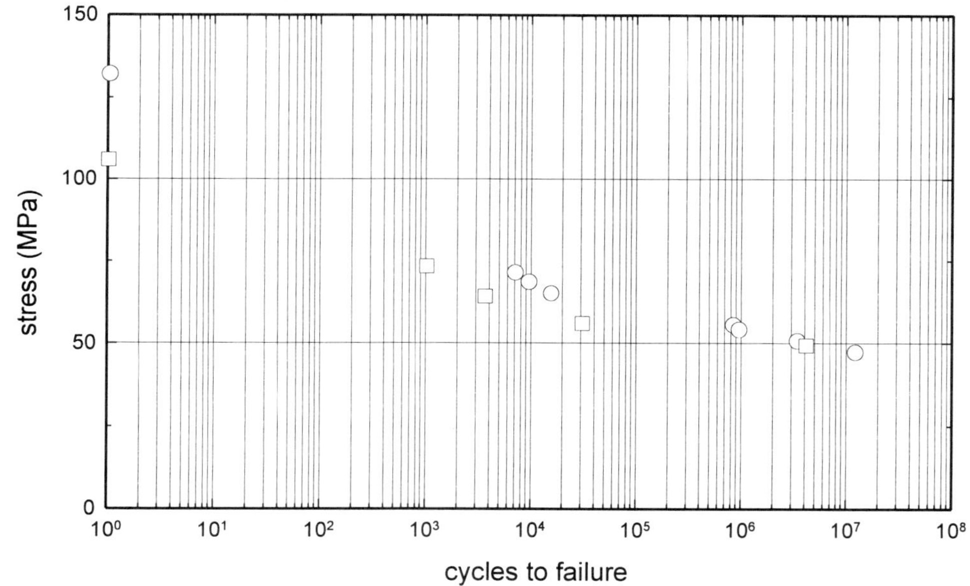

○	LNP Verton MFX-700-10 HS PP (50% glass fiber, chemically coupled; long glass fibers (1330-1370 microns)); tension- tension; 10 Hz; 23°C; ASTM D638 Type 1 [fig g]
□	LNP Thermocomp MFX-100-10 HS PP (50% glass fiber; chemically coupled; short glass fiber (230-280 microns)); tension- tension; 10 Hz; 23°C; ASTM D638 Type 1 [fig g]
Reference No.	391

GRAPH 142: Fatigue Cycles to Failure vs. Stress in Tension at Low Test Frequency for Polypropylene.

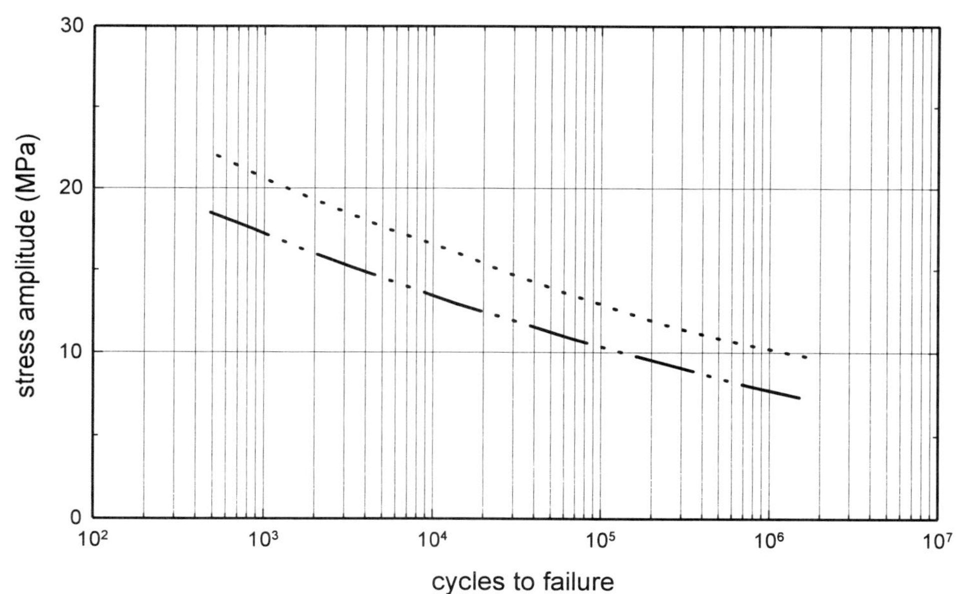

....	PP; tension- compression; 0.5 Hz; 20°C; mean stress = 0 [fig d]
—··—··	PP Copolymer; tension- compression; 0.5 Hz; 20°C; mean stress = 0 [fig d]
Reference No.	355

GRAPH 143: **Fatigue Cycles to Failure vs. Initial Strain in Tension at Different Test Frequencies for Unreinforced and 25% Glass Fiber Reinforced Polypropylene.**

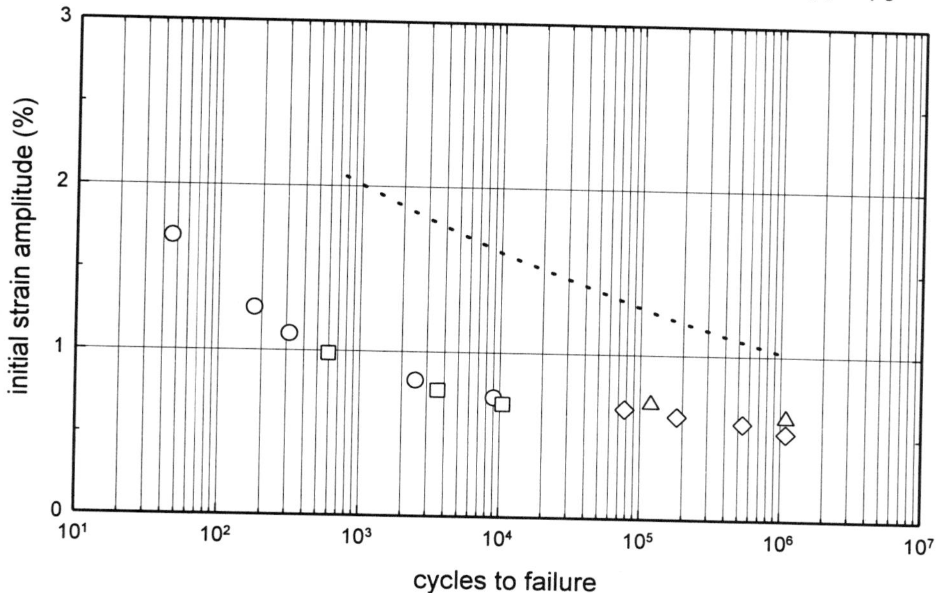

···············	PP; tension- compression; 0.5 Hz; 20°C; mean stress = 0 [fig d]
○	PP (25% glass fiber); tension- compression; 0.5 Hz; 20°C; 6 mm thick; parallel to flow; mean stress = 0 [fig d]
□	PP (25% glass fiber); tension- compression; 0.5 Hz; 20°C; 6 mm thick; perpendicular to flow; mean stress = 0 [fig d]
△	PP (25% glass fiber); tension- compression; 2.5 Hz; 20°C; 6 mm thick; perpendicular to flow; mean stress = 0 [fig d]
◇	PP (25% glass fiber); tension- compression; 2.5 Hz; 20°C; 6 mm thick; parallel to flow; mean stress = 0 [fig d]
Reference No.	355

GRAPH 144: **Fatigue Cycles to Failure vs. Initial Strain in Tension at Low Test Frequency for Polypropylene.**

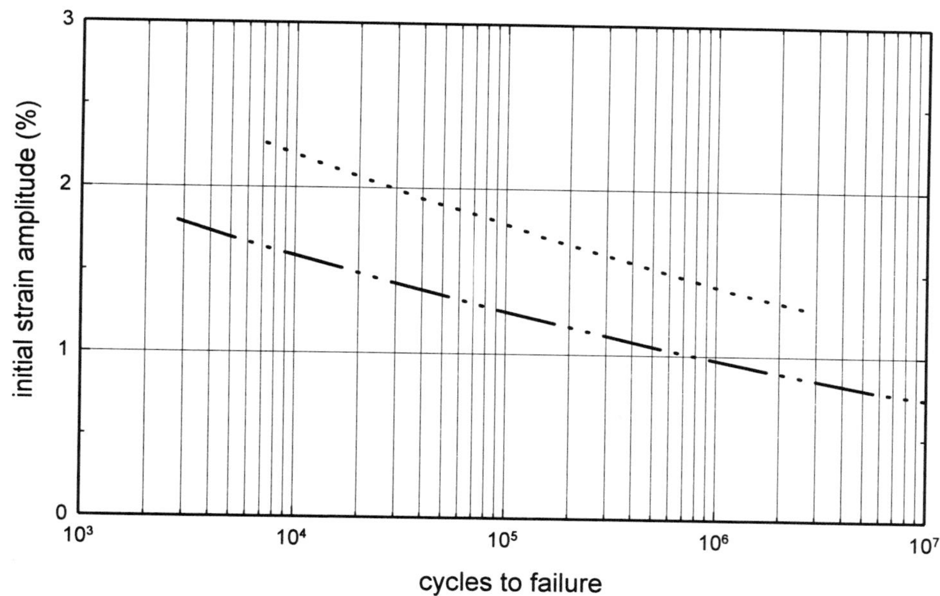

···············	PP Copolymer; tension- compression; 0.5 Hz; 20°C; mean stress = 0 [fig d]
—··—··	PP; tension- compression; 0.5 Hz; 20°C; mean stress = 0 [fig d]
Reference No.	355

GRAPH 145: Fatigue Cycles to Failure vs. Stress in Tension at Low Test Frequency for Polypropylene.

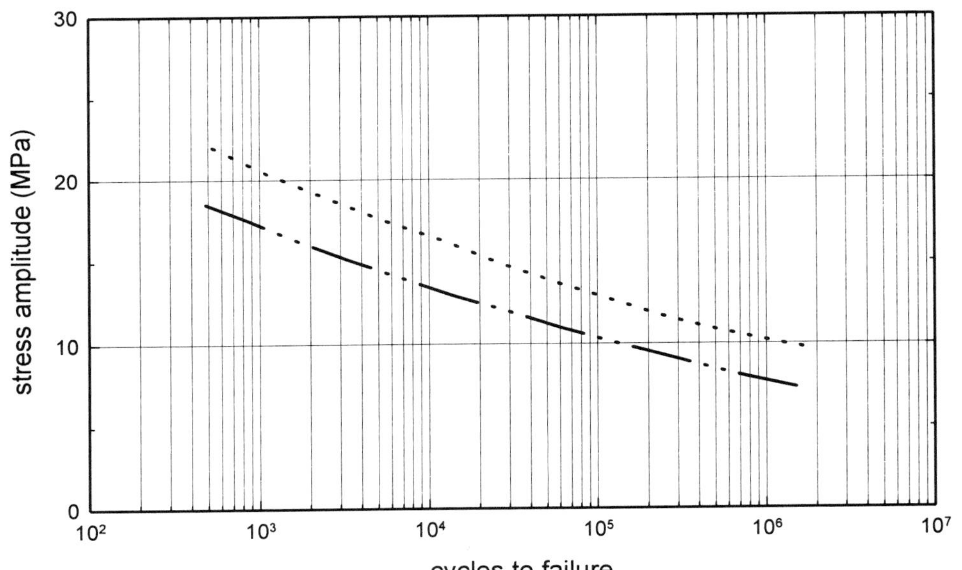

....	PP; tension- compression; 0.5 Hz; 20°C; mean stress = 0 [fig d]
—·· —··	PP Copolymer; tension-compression; 0.5 Hz; 20°C; mean stress = 0 [fig d]
Reference No.	355

Chapter 30

Polystyrene Modified Polyphenylene Ether

GRAPH 146: **Fatigue Cycles to Failure vs. Stress in Flexure for Mitsubishi Gas Chemical Iupiace Modified Polyphenylene Ether.**

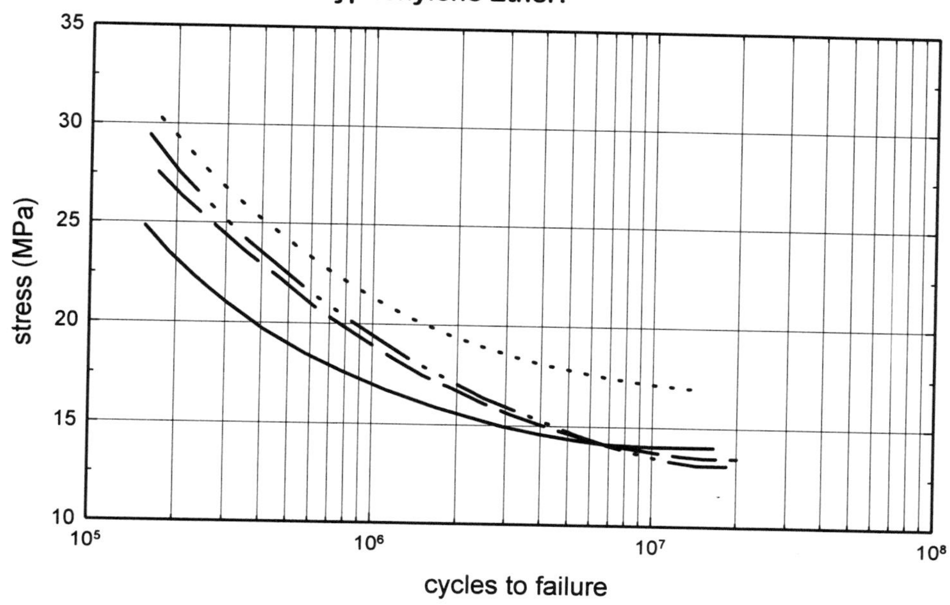

....	Mitsubishi Iupiace AV40 Modified PPE; flexure; 30 Hz; 23°C
—··—··	Mitsubishi Iupiace AV60 Modified PPE; flexure; 30 Hz; 23°C
– – – –	Mitsubishi Iupiace AH60 Modified PPE; flexure; 30 Hz; 23°C
———	Mitsubishi Iupiace AN30 Modified PPE (flame retardant); flexure; 30 Hz; 23°C
Reference No.	363

GRAPH 147: **Fatigue Cycles to Failure vs. Stress in Flexure for 30% Glass Fiber Reinforced LNP Thermocomp ZF1006 Modified Polyphenylene Ether.**

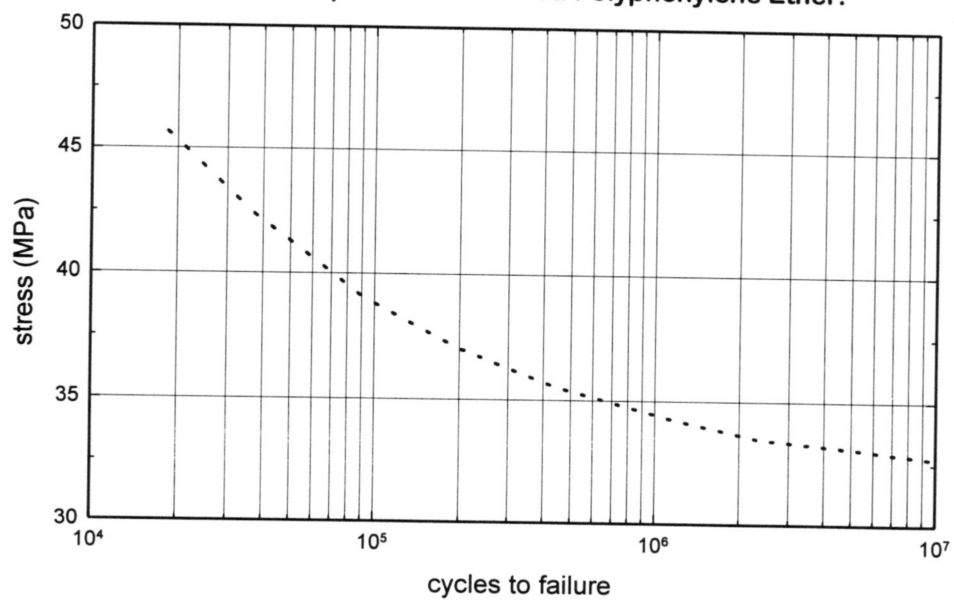

....	LNP Thermocomp ZF1006 Modified PPE (30% glass fiber); flexure; 30 Hz; 23°C
Reference No.	394

GRAPH 148: **Fatigue Cycles to Failure vs. Stress in Flexure at Two Temperatures for Mitsubishi Gas Chemical Iupiace EV80 Modified Polyphenylene Ether.**

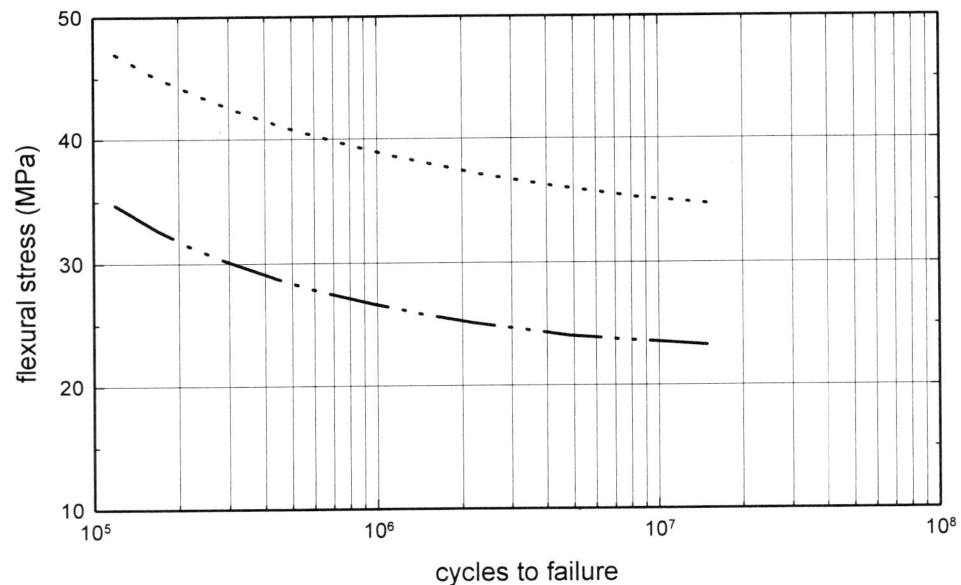

....	Mitsubishi Iupiace EV80 Modified PPE; flexure; 3 Hz; 23°C
—··—··	Mitsubishi Iupiace EV80 Modified PPE; flexure; 3 Hz; 100°C
Reference No.	362

GRAPH 149: **Fatigue Cycles to Failure vs. Stress in Tension for Flame Retardant Grades of General Electric Noryl Modified Polyphenylene Oxide.**

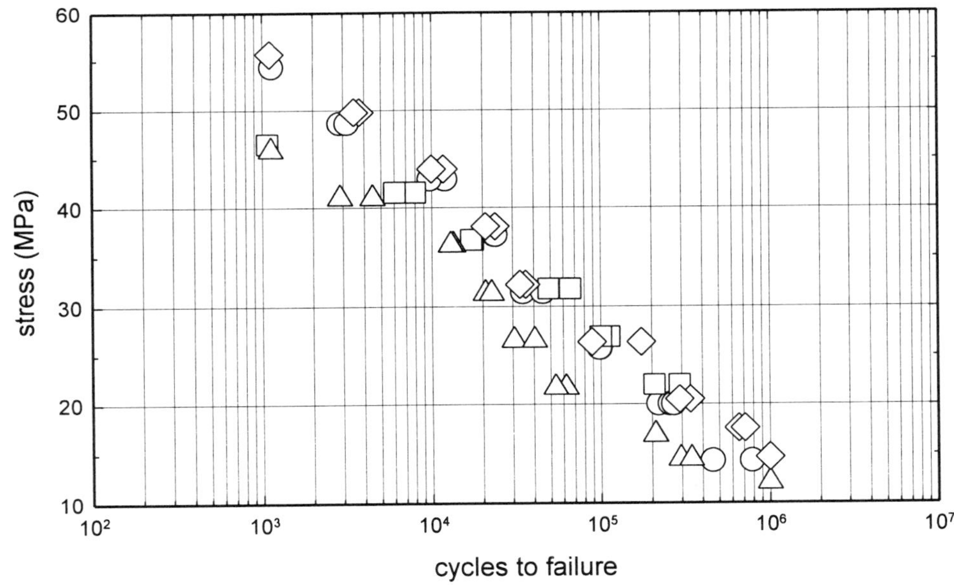

○	GE Noryl SE1 Modified PPE (flame retardant, unfilled; inj. mold.); tension-tension; GE method; 5 Hz; stress ratio: 0.1; 23°C; flow direction [fig g]
□	GE Noryl SE100 Modified PPE (flame retardant, unfilled; inj. mold.); tension-tension; GE method; 5 Hz; stress ratio: 0.1; 23°C; flow direction [fig g]
△	GE Noryl SE100X Modified PPE (flame retardant, unfilled; no brominated or chlorinated additives; inj. mold.); tension- tension; GE method; 5 Hz; stress ratio: 0.1; 23°C; flow direction [fig g]
◇	GE Noryl SE1X Modified PPE (flame retardant, unfilled; no brominated or chlorinated additives; inj. mold.); tension- tension; GE method; 5 Hz; stress ratio: 0.1; 23°C; flow direction [fig g]
Reference No.	352

GRAPH 150: Fatigue Cycles to Failure vs. Stress in Tension for High Heat Grades of General Electric Noryl Modified Polyphenylene Oxide.

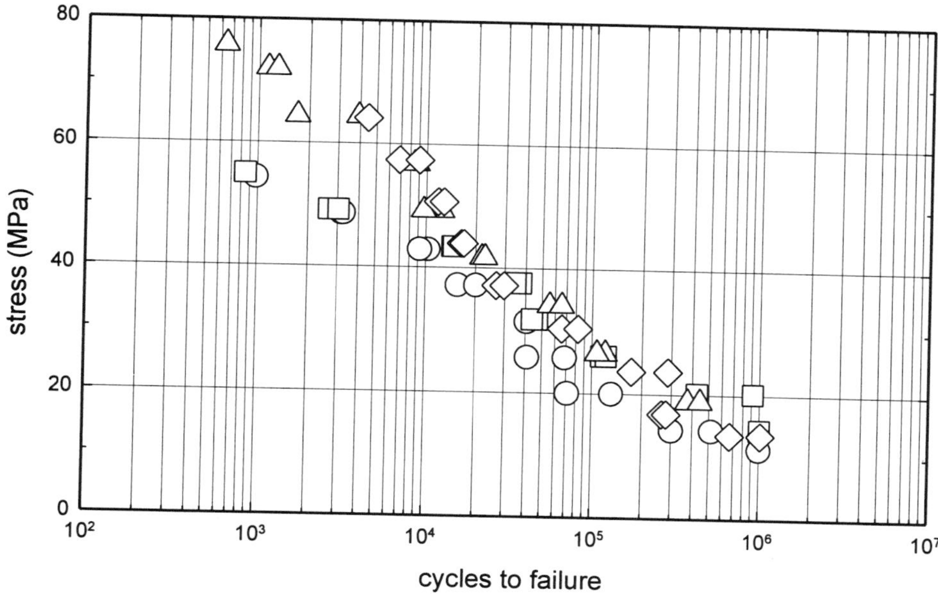

○	GE Noryl N225 Modified PPE (high heat grade, flame retardant, unfilled; inj. mold.); tension-tension; GE method; 5 Hz; stress ratio: 0.1; 23°C; flow direction [fig g]
□	GE Noryl N225X Modified PPE (high heat grade, flame retardant, flame retardant, unfilled; inj. mold., inj. mold.); tension-tension; GE method; 5 Hz; stress ratio: 0.1; 23°C; flow direction [fig g]
△	GE Noryl N300 Modified PPE (flame retardant); tension- tension; GE method; 5 Hz; stress ratio: 0.1; 23°C; flow direction [fig g]
◇	GE Noryl N300X Modified PPE (high heat grade, flame retardant, unfilled; inj. mold.); tension-tension; GE method; 5 Hz; stress ratio: 0.1; 23°C; flow direction [fig g]
Reference No.	352

GRAPH 151: Fatigue Cycles to Failure vs. Stress in Tension for Glass/ Mineral Filled Grades of General Electric Noryl Modified Polyphenylene Oxide.

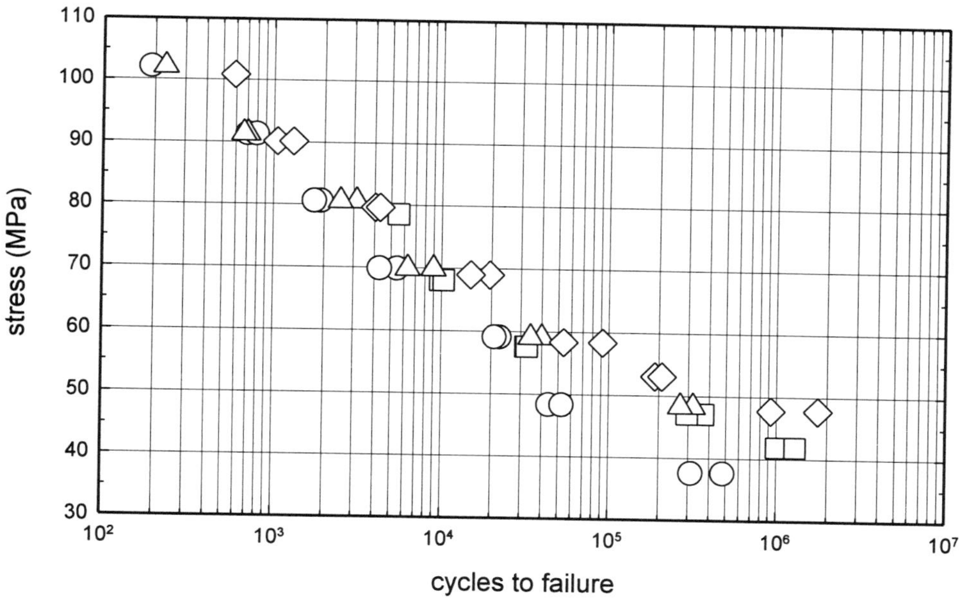

○	GE Noryl HM3020 Modified PPE (flame retardant; 30% mineral/ glass; inj. mold.); tension- tension; GE method; 5 Hz; stress ratio: 0.1; 23°C; flow direction [fig g]
□	GE Noryl HM3020X Modified PPE (30% mineral/ glass; inj. mold.); tension- tension; GE method; 5 Hz; stress ratio: 0.1; 23°C; flow direction [fig g]
△	GE Noryl HM4025 Modified PPE (flame retardant; 40% mineral/ glass; inj. mold.); tension- tension; GE method; 5 Hz; stress ratio: 0.1; 23°C; flow direction [fig g]
◇	GE Noryl HM4025X Modified PPE (40% mineral/ glass; inj. mold.); tension- tension; GE method; 5 Hz; stress ratio: 0.1; 23°C; flow direction [fig g]
Reference No.	352

GRAPH 152: **Fatigue Cycles to Failure vs. Stress in Tension for Flame Retardant and Filled Grades of General Electric Noryl Modified Polyphenylene Oxide.**

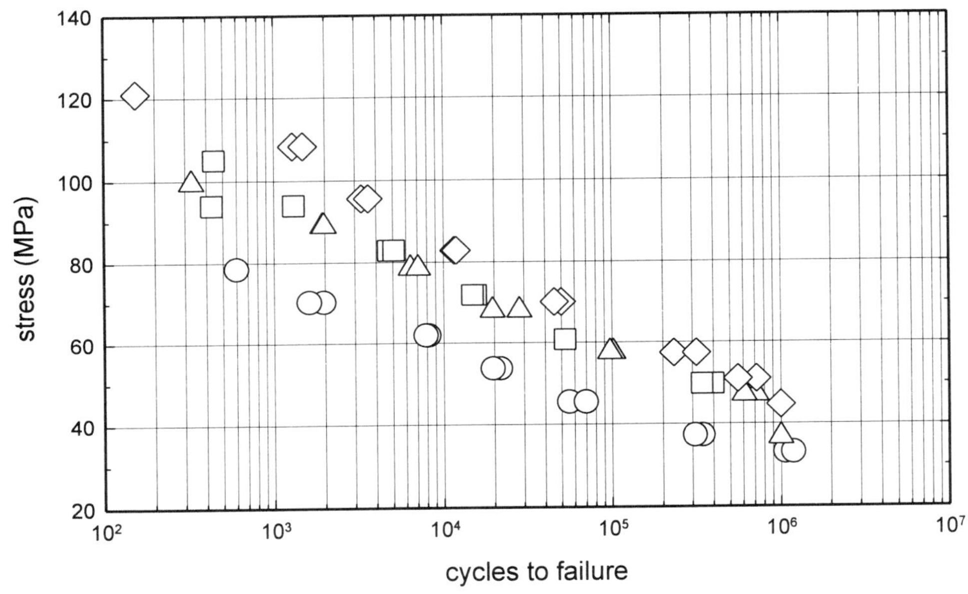

○	GE Noryl SE1GFN1X Modified PPE (flame retardant; 10% glass fiber; no brominated or chlorinated additives; inj. mold.); tension- tension; GE method; 5 Hz; stress ratio: 0.1; 23°C; flow direction [fig g]
□	GE Noryl SE1GFN2 Modified PPE (flame retardant; 20% glass fiber; nonhalogenated; inj. mold.); tension- tension; GE method; 5 Hz; stress ratio: 0.1; 23°C; flow direction [fig g]
△	GE Noryl SE1GFN2X Modified PPE (flame retardant; 20% glass fiber; no brominated or chlorinated additives; inj. mold.); tension- tension; GE method; 5 Hz; stress ratio: 0.1; 23°C; flow direction [fig g]
◇	GE Noryl SE1GFN3X Modified PPE (flame retardant; 30% glass fiber; no brominated or chlorinated additives; inj. mold.); tension- tension; GE method; 5 Hz; stress ratio: 0.1; 23°C; flow direction [fig g]
Reference No.	352

GRAPH 153: **Fatigue Cycles to Failure vs. Stress in Tension for General Electric Prevex Modified Polyphenylene Ether.**

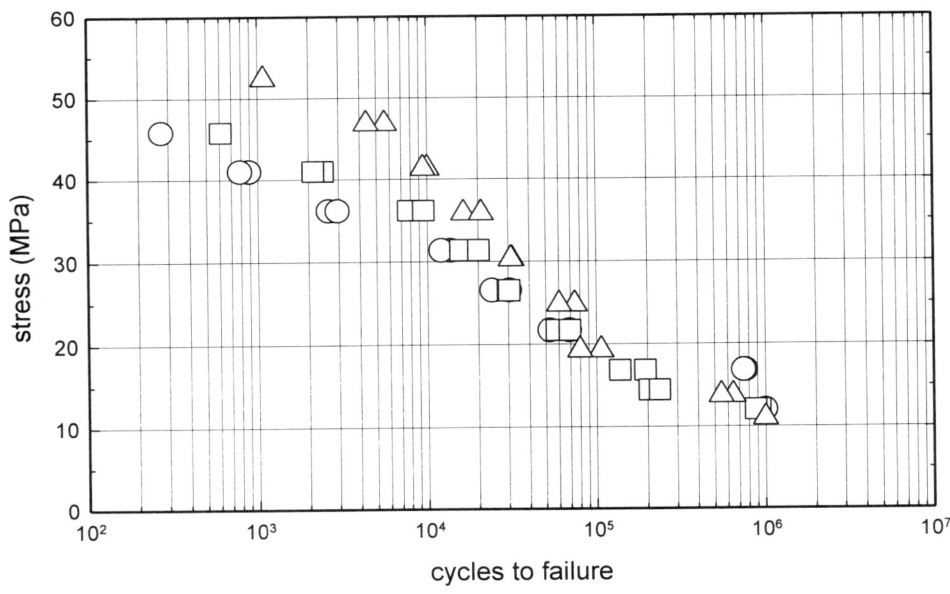

○	GE Prevex VFAX Modified PPE (high impact, unfilled; inj. mold.); tension- tension; GE method; 5 Hz; stress ratio: 0.1; 23°C; flow direction [fig g]
□	GE Prevex VGAX Modified PPE (flame retardant, unfilled; no brominated or chlorinated additives; inj. mold.); tension- tension; GE method; 5 Hz; stress ratio: 0.1; 23°C; flow direction [fig g]
△	GE Prevex VKAX Modified PPE (flame retardant, unfilled; no brominated or chlorinated additives; inj. mold.); tension- tension; GE method; 5 Hz; stress ratio: 0.1; 23°C; flow direction [fig g]
Reference No.	352

GRAPH 154: **Fatigue Cycles to Failure vs. Stress in Tension for Mineral Filled Grades of General Electric Noryl Modified Polyphenylene Oxide.**

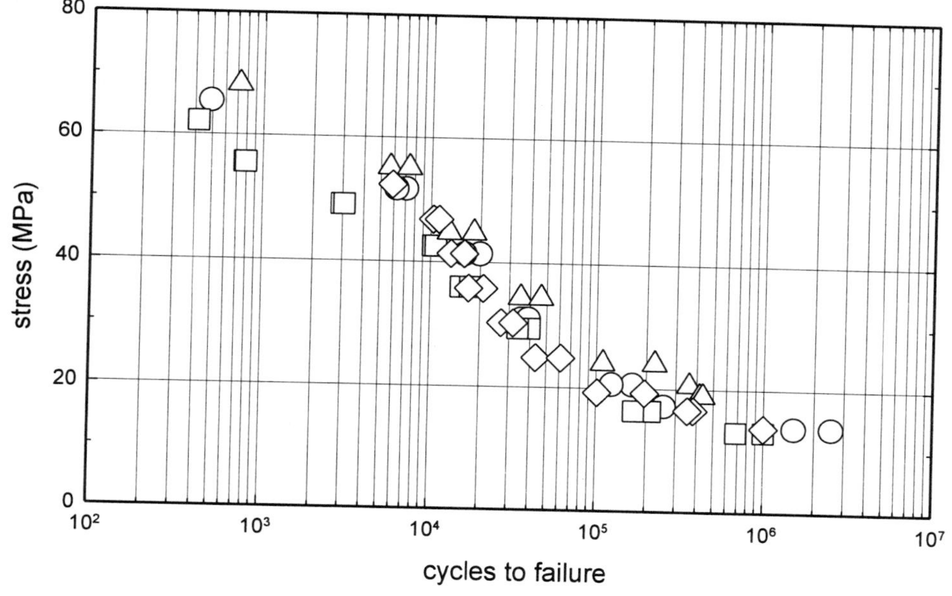

○	GE Noryl HS1000 Modified PPE (10% mineral filler; inj. mold.); tension- tension; GE method; 5 Hz; stress ratio: 0.1; 23°C; flow direction [fig g]
□	GE Noryl HS1000X Modified PPE (flame retardant; 14% mineral filler; inj. mold.); tension-tension; GE method; 5 Hz; stress ratio: 0.1; 23°C; flow direction [fig g]
△	GE Noryl HS2000 Modified PPE (10% mineral filler; inj. mold.); tension- tension; GE method; 5 Hz; stress ratio: 0.1; 23°C; flow direction [fig g]
◇	GE Noryl HS2000X Modified PPE (flame retardant; 18% mineral filler; inj. mold.); tension-tension; GE method; 5 Hz; stress ratio: 0.1; 23°C; flow direction [fig g]
Reference No.	352

GRAPH 155: **Fatigue Cycles to Failure vs. Stress in Tension for Flame Retardant Grades of General Electric Noryl Modified Polyphenylene Oxide.**

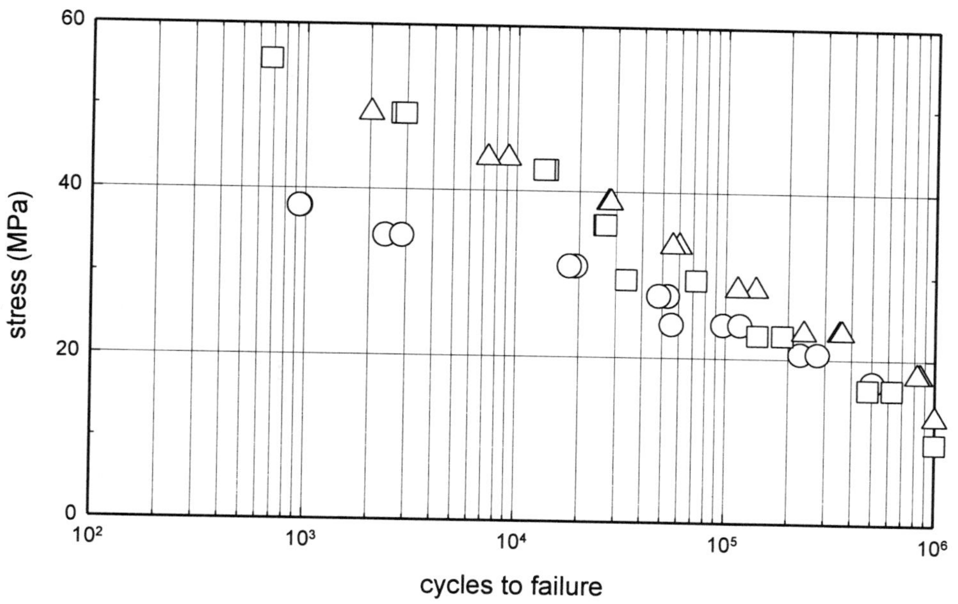

○	GE Noryl N190 Modified PPE (high impact, flame retardant; inj. mold.); tension- tension; GE method; 5 Hz; stress ratio: 0.1; 23°C; flow direction [fig g]
□	GE Noryl N190HX Modified PPE (high gloss, flame retardant, unfilled; inj. mold.); tension- tension; GE method; 5 Hz; stress ratio: 0.1; 23°C; flow direction [fig g]
△	GE Noryl N190X Modified PPE (high impact, flame retardant, unfilled; no brominated or chlorinated additives; inj. mold.); tension- tension; GE method; 5 Hz; stress ratio: 0.1; 23°C; flow direction [fig g]
Reference No.	352

GRAPH 156: **Fatigue Cycles to Failure vs. Stress in Tension at Different Temperatures for Filled Grades of General Electric Noryl Modified Polyphenylene Oxide.**

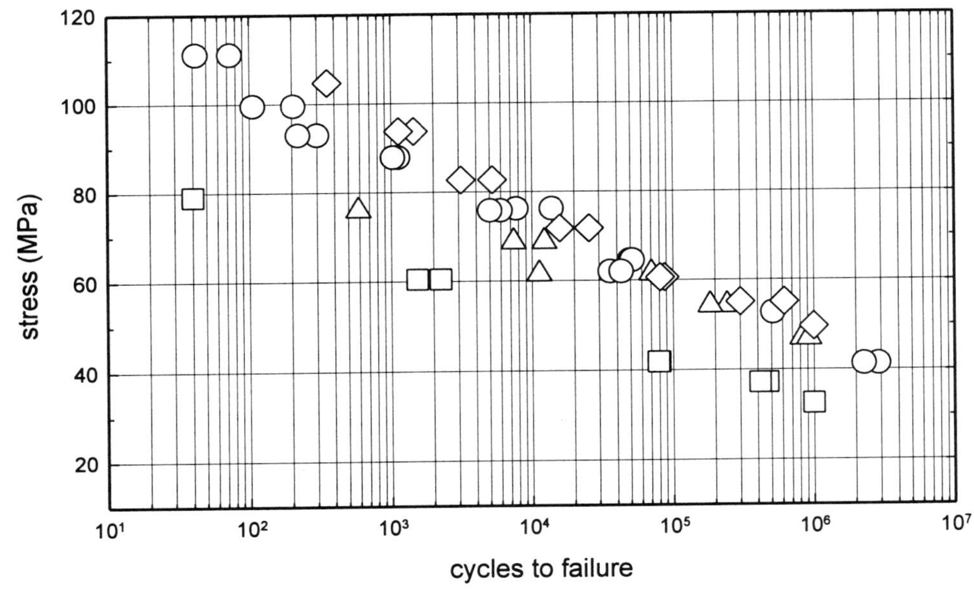

○	GE Noryl GFN3 Modified PPE (FDA grade; 30% glass fiber; inj. mold.); tension- tension; GE method; 5 Hz; stress ratio: 0.1; 23°C; flow direction [fig g]
□	GE Noryl GFN3 Modified PPE (FDA grade; 30% glass fiber; inj. mold.); tension- tension; GE method; 5 Hz; stress ratio: 0.1; 66°C; flow direction [fig g]
△	GE Noryl HMC1010 Modified PPE (flame retardant, static dissipative; 10% carbon fiber; inj. mold.); tension- tension; GE method; 5 Hz; stress ratio: 0.1; 23°C; flow direction [fig g]
◇	GE Noryl HMC3008A Modified PPE (flame retardant; 30% carbon/ glass/ mineral reinforced; inj. mold.); tension- tension; GE method; 5 Hz; stress ratio: 0.1; 23°C; flow direction [fig g]
Reference No.	352

GRAPH 157: **Fatigue Cycles to Failure vs. Stress in Tension for General Purpose and Business Equipment Grades of General Electric Noryl Modified Polyphenylene Oxide.**

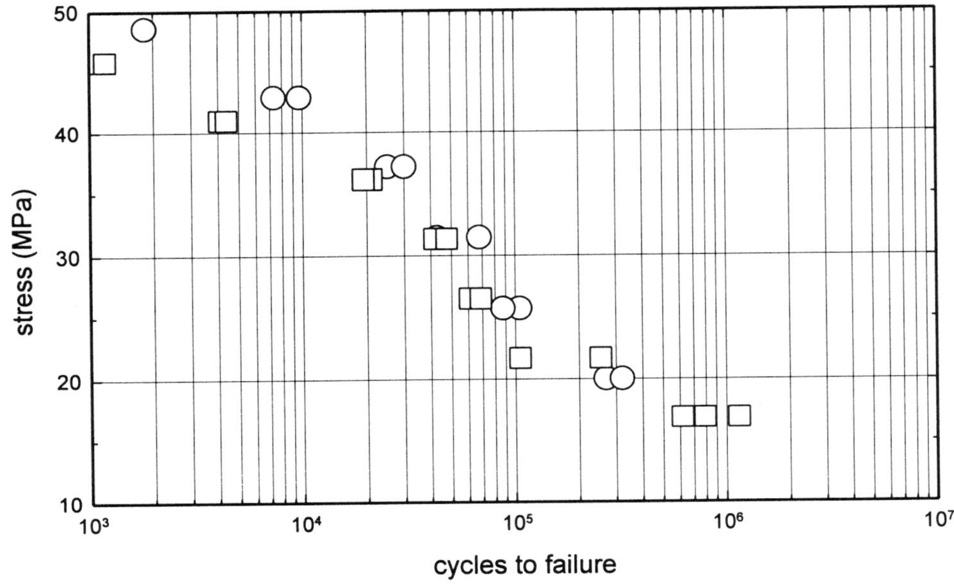

○	GE Noryl 731 Modified PPE (gen. purp. grade, unfilled; inj. mold.); tension-tension; GE method; 5 Hz; stress ratio: 0.1; 23°C; flow direction [fig g]
□	GE Noryl CRT200 Modified PPE (gen. purp. grade, flame retardant); tension-tension; GE method; 5 Hz; stress ratio: 0.1; 23°C; flow direction [fig g]
Reference No.	352

GRAPH 158: Fatigue Cycles to Failure vs. Stress in Tension for Structural Foam Grades of General Electric Noryl Modified Polyphenylene Oxide.

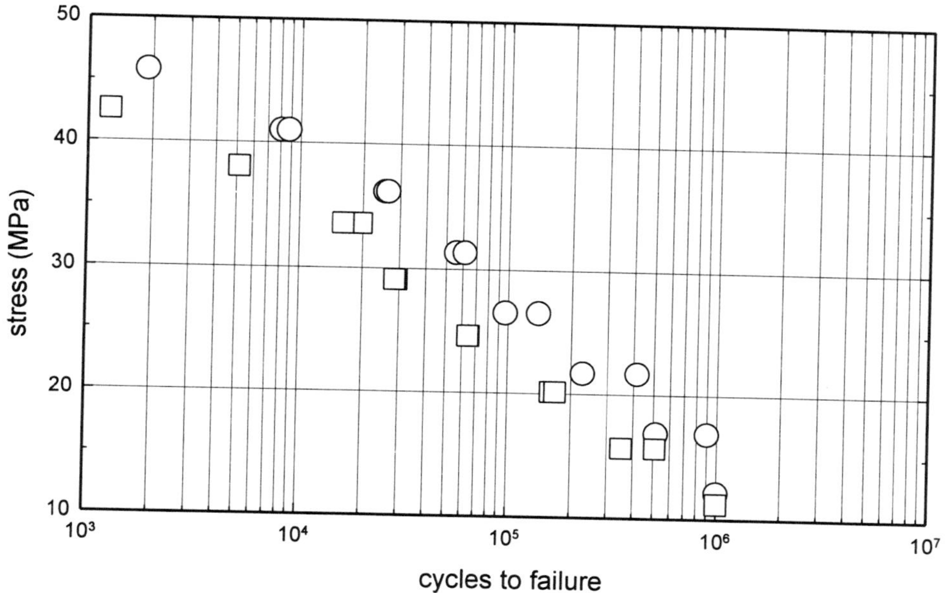

○	GE Noryl FN150X Modified PPE (flame retardant, unfilled; 20% wgt. reduction; structual foam); tension- tension; GE method; 5 Hz; stress ratio: 0.1; 23°C; flow direction [fig g]
□	GE Noryl FN215X Modified PPE (flame retardant, unfilled; 20% wgt. reduction; structual foam); tension- tension; GE method; 5 Hz; stress ratio: 0.1; 23°C; flow direction [fig g]
Reference No.	352

GRAPH 159: Fatigue Cycles to Failure vs. Stress in Tension for Blow Molding and Extrusion Grades of General Electric Noryl Modified Polyphenylene Oxide.

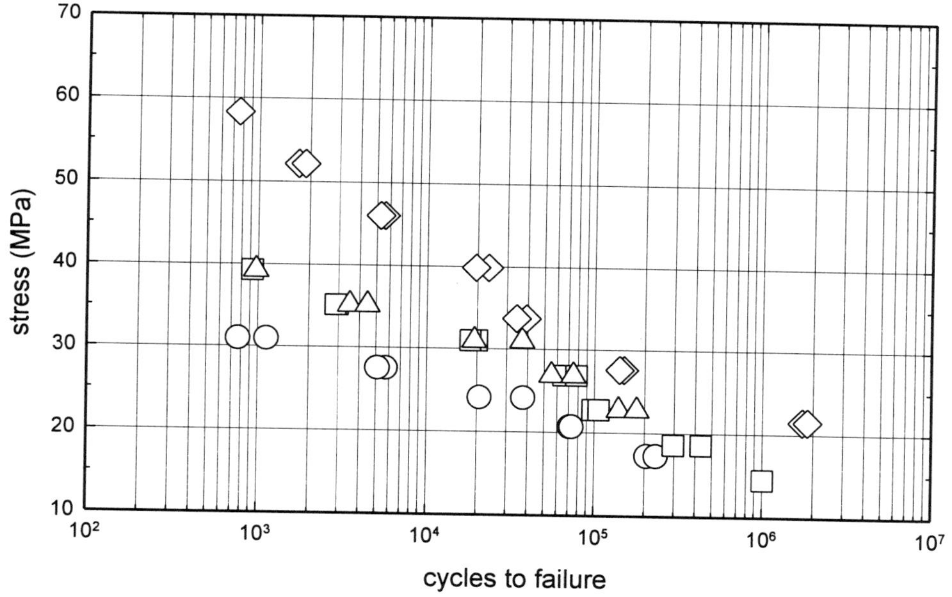

○	GE Noryl BN11 Modified PPE (unfilled; blow molding); tension- tension; GE method; 5 Hz; stress ratio: 0.1; 23°C; flow direction [fig g]
□	GE Noryl BN31 Modified PPE (flame retardant, unfilled, unfilled; blow molding, blow molding); tension- tension; GE method; 5 Hz; stress ratio: 0.1; 23°C; flow direction [fig g]
△	GE Noryl BN41 Modified PPE; tension- tension; GE method; 5 Hz; stress ratio: 0.1; 23°C; flow direction [fig g]
◇	GE Noryl EN265 Modified PPE (flame retardant, unfilled; extrus.); tension-tension; GE method; 5 Hz; stress ratio: 0.1; 23°C; flow direction [fig g]
Reference No.	352

GRAPH 160: **Fatigue Crack Propagation (FCP) Rate vs. Stress Intensity Range for Modified Polyphenylene Oxide.**

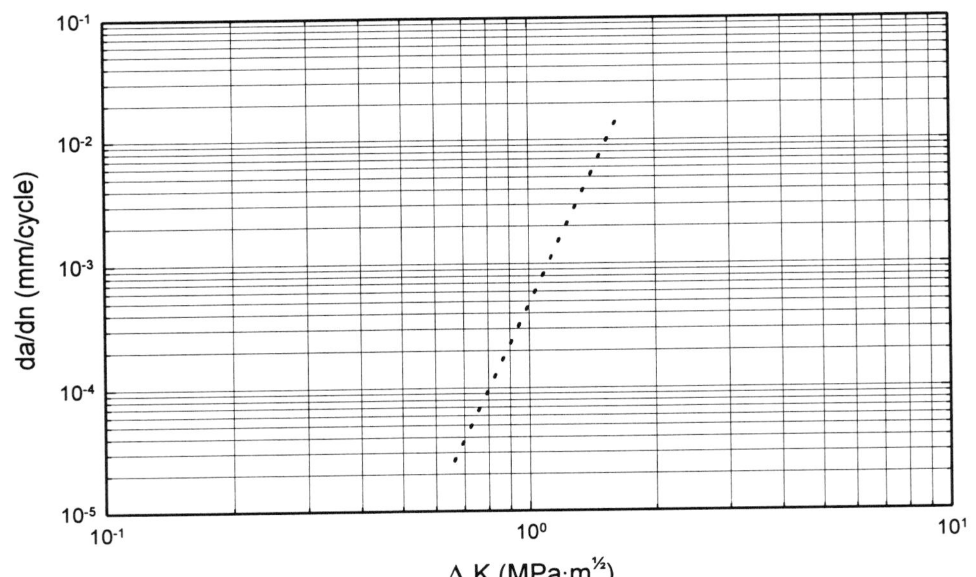

....	Modified PPE
Reference No.	355

Chapter 31

Polyphenylene Sulfide

GRAPH 161: **Fatigue Cycles to Failure vs. Stress in Flexure for Glass Fiber Reinforced Phillips Ryton R-4 Polyphenylene Sulfide.**

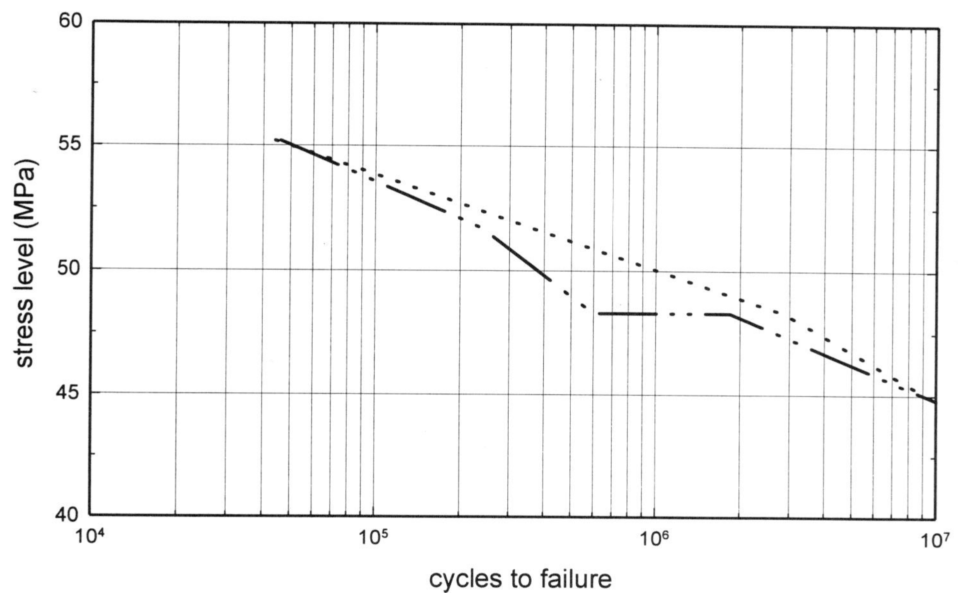

....	Phillips Ryton R4 PPS (unstabilized, low crystallinity; 40% glass fiber); flexure; 23°C; 6.3 mm thick
—··—··	Phillips Ryton R4 PPS (unstabilized, high crsytallinity; 40% glass fiber); flexure; 23°C; 6.3 mm thick
Reference No.	140

GRAPH 162: **Fatigue Cycles to Failure vs. Stress in Flexure for Glass Fiber and Mineral Reinforced Phillips Ryton R-8 Polyphenylene Sulfide.**

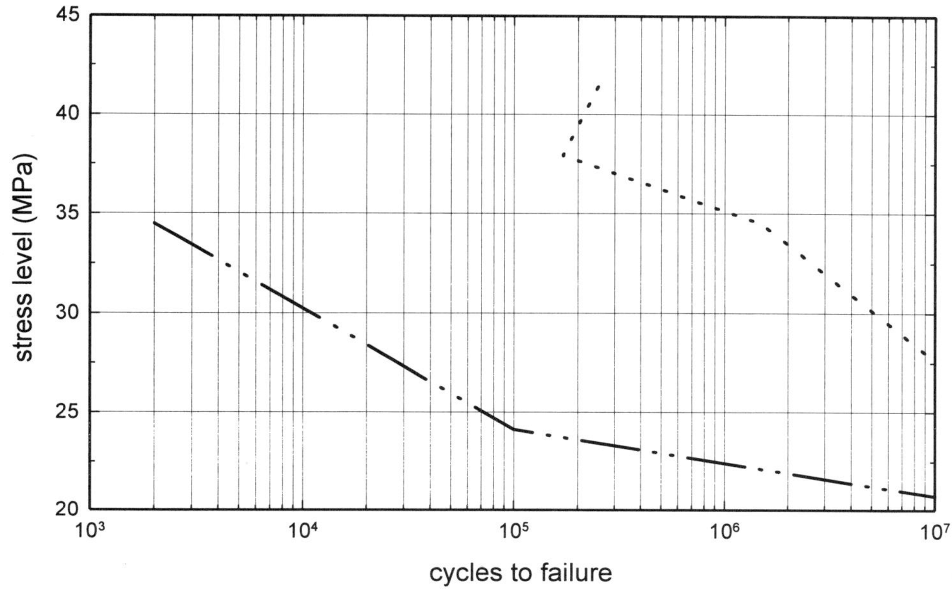

....	Phillips Ryton R8 PPS (low crystallinity; glass/ mineral reinforced); flexure; 23°C; 6.3 mm thick
—··—··	Phillips Ryton R8 PPS (high crystallinity; glass/ mineral reinforced); flexure; 23°C; 6.3 mm thick
Reference No.	140

GRAPH 163: **Fatigue Cycles to Failure vs. Stress in Flexure for Carbon Fiber Reinforced LNP Polyphenylene Sulfide.**

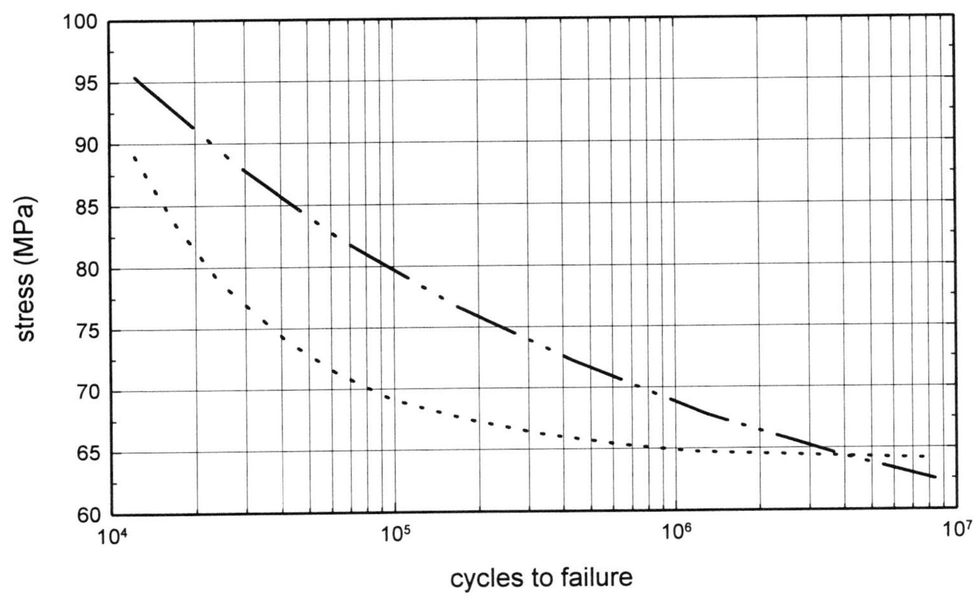

···············	LNP Stat-Kon OC1006 PPS (static dissipative; 30% carbon fiber); flexure; 30 Hz; 23°C
—··—··	LNP Lubricomp OCL4036 PPS (lubricated; 30% carbon fiber, 15% PTFE modified); flexure; 30 Hz; 23°C
···············	LNP Thermocomp OC1006 PPS (30% carbon fiber); flexure; 30 Hz; 23°C
Reference No.	394

GRAPH 164: **Fatigue Cycles to Failure vs. Stress in Tension for General Electric Supec Polyphenylene Sulfide.**

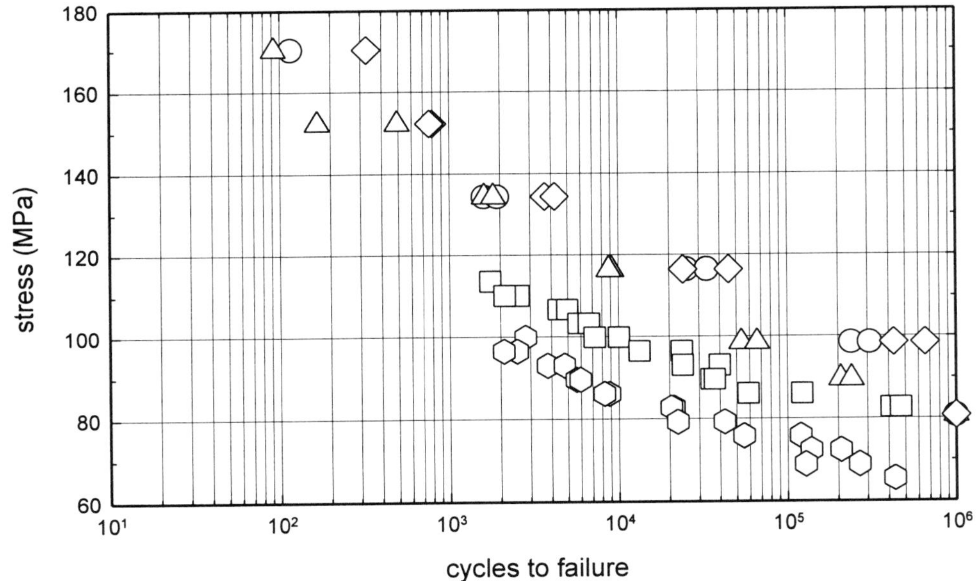

○	GE Supec G301T PPS (flame retardant; 30% glass fiber; inj. mold.); tension- tension; GE method; 5 Hz; stress ratio: 0.1; 23°C; flow direction [fig g]
□	GE Supec G401 PPS (flame retardant; 40% glass fiber; inj. mold.); tension- tension; GE method; 5 Hz; stress ratio: 0.1; 23°C; flow direction [fig g]
△	GE Supec G401M PPS (40% glass fiber; inj. mold.); tension- tension; GE method; 5 Hz; stress ratio: 0.1; 23°C; flow direction [fig g]
◇	GE Supec G401T PPS (flame retardant; 40% glass fiber; inj. mold.); tension- tension; GE method; 5 Hz; stress ratio: 0.1; 23°C; flow direction [fig g]
⬡	GE Supec G402 PPS (high flow; 40% glass fiber; inj. mold.); tension- tension; GE method; 5 Hz; stress ratio: 0.1; 23°C; flow direction [fig g]
Reference No.	352

GRAPH 165: **Fatigue Cycles to Failure vs. Initial Strain in Tension at Two Temperatures for Polyphenylene Sulfide.**

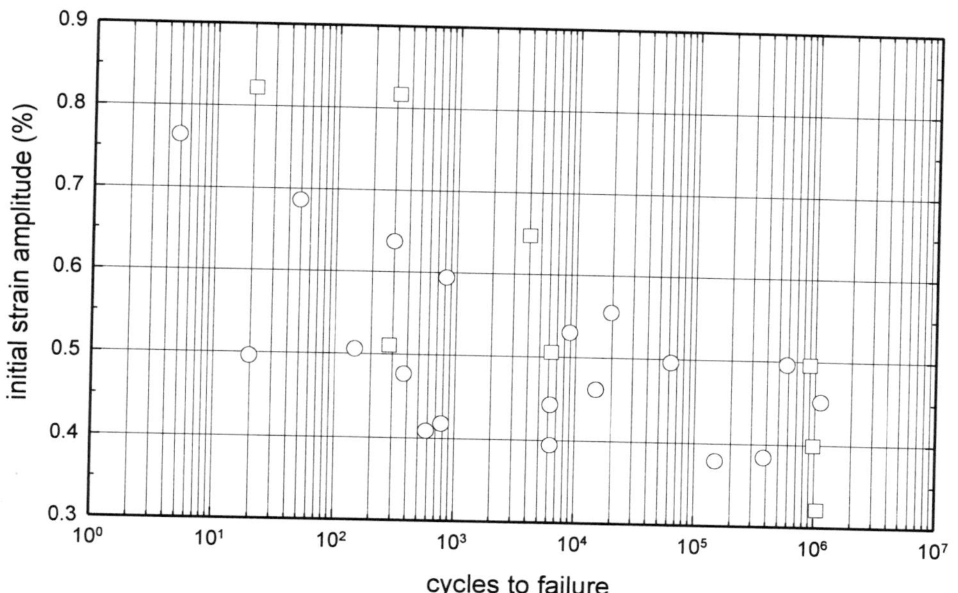

○	PPS; tension- zero; 0.5 Hz; 20°C [fig d]
□	PPS; tension- zero; 0.5 Hz; 120°C [fig d]
Reference No.	355

GRAPH 166: **Fatigue Cycles vs. Percent Tensile Strength Retained for Phillips Ryton Polyphenylene Sulfide.**

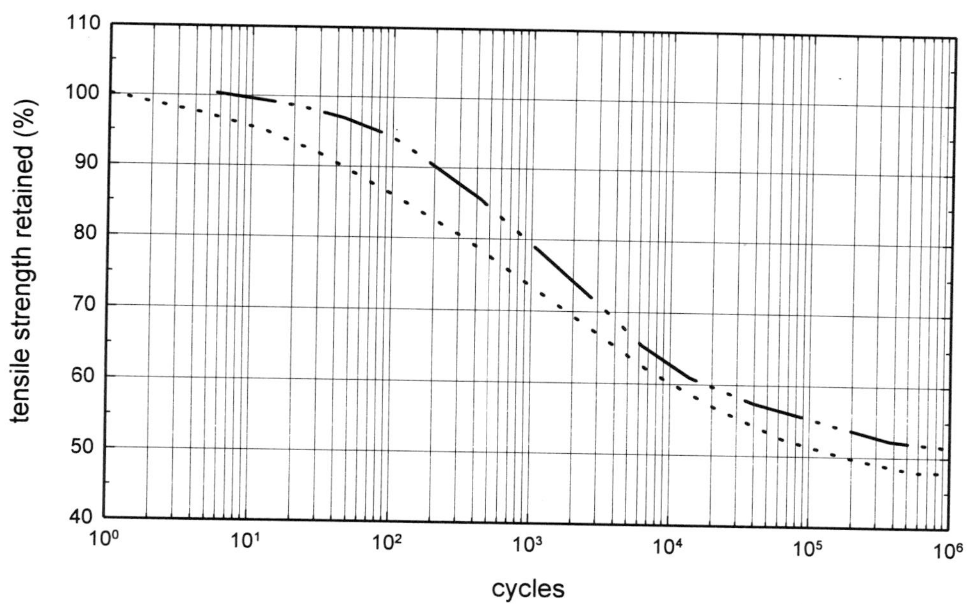

···············	Phillips Ryton R7 PPS (121 Rockwell R; 60% mineral/ glass); tension; 10 Hz; 23°C
—··—··	Phillips Ryton A200 PPS (120 Rockwell R; 40% glass fiber); tension; 10 Hz; 23°C
Reference No.	102

GRAPH 167: Fatigue Cycles vs. Percent Tensile Strength Retained for Phillips Ryton Polyphenylene Sulfide.

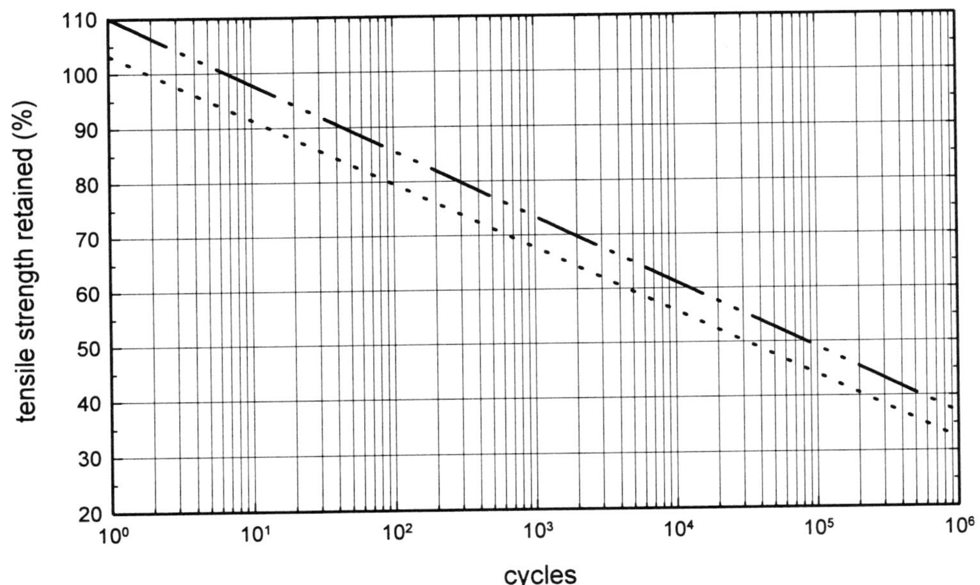

—·-—··	Phillips Ryton R4 PPS (natural resin, 123 Rockwell R; 40% glass fiber); tension; 10 Hz; 23°C
....	Phillips Ryton R4 04 PPS (natural resin, 123 Rockwell R; 40% glass fiber); tension; 10 Hz; 23°C
Reference No.	102

Chapter 32

Polysulfone

Fatigue Properties

Amoco Performance Products: Udel (features: transparent, amber tint)

The fatigue endurance limit (10^6 cycles at 1,800 cycles/min.) is approximately 1,000 psi (96.9 MPa). This is about equivalent to the performance of polycarbonate resin and lower than the fatigue endurance of acetal or nylon resins.

Reference: *Udel Polysulfone Design Engineering Handbook,* supplier design guide (F-47178) - Amoco Performance Products, Inc., 1988.

GRAPH 168: Fatigue Cycles to Failure vs. Stress in Flexure for Glass Fiber Reinforced LNP Thermocomp Polysulfone.

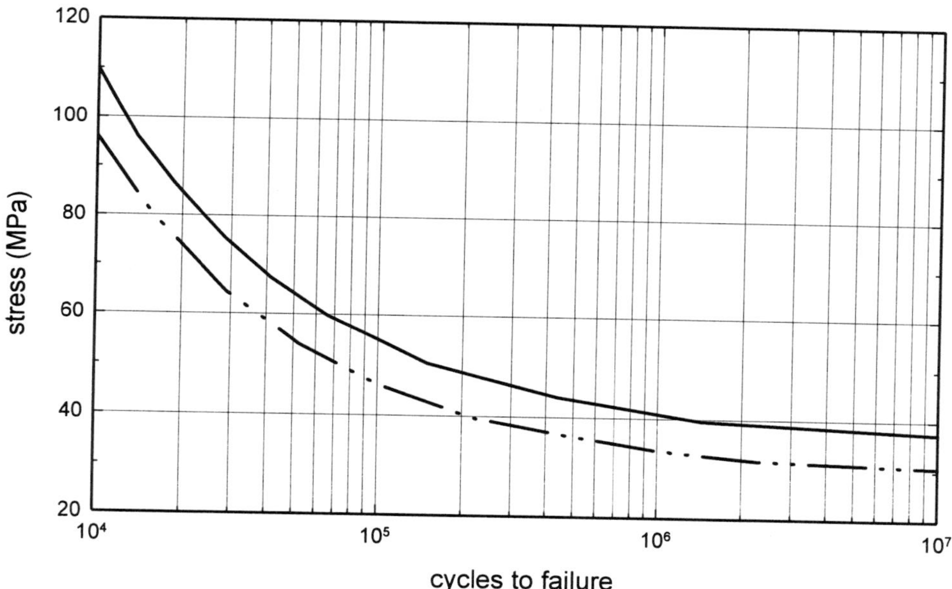

———	LNP Thermocomp GF1008 Polysulfone (40% glass fiber); flexure; 30 Hz; 23°C
—··—··	LNP Thermocomp GF1006 Polysulfone (30% glass fiber); flexure; 30 Hz; 23°C
Reference No.	394

GRAPH 169: **Fatigue Cycles to Failure vs. Stress in Flexure for Filled and Unfilled BASF Ultrason S Polysulfone.**

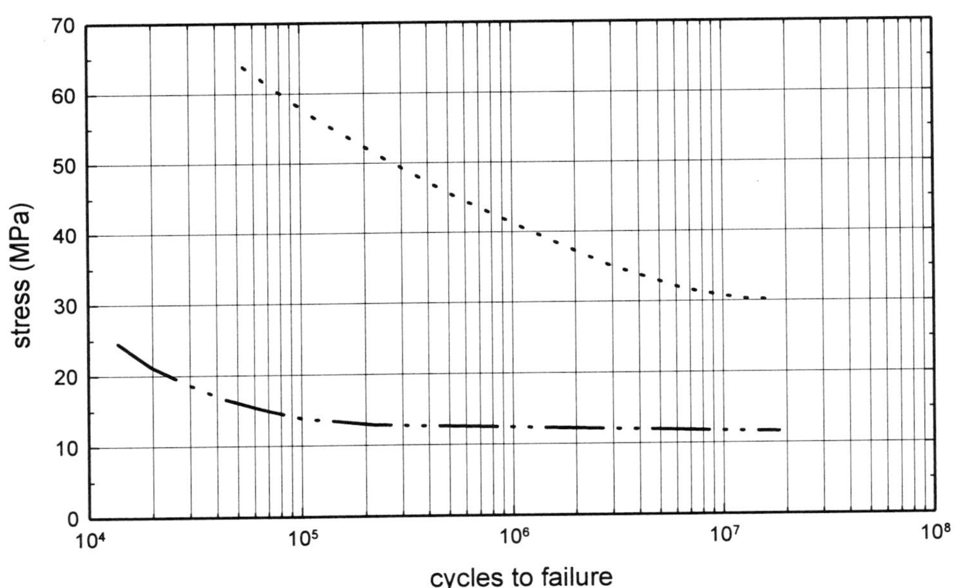

···············	BASF AG Ultrason S 2010G4 Polysulfone (moderate flow; 20% glass fiber); flexure (alternating); DIN 53442; 15 Hz; 23°C; 50% RH; laboratory atmosphere
—··—··	BASF AG Ultrason S 2010 Polysulfone (moderate flow, gen. purp. grade); flexure (alternating); DIN 53442; 15 Hz; 23°C; 50% RH; laboratory atmosphere
Reference No.	28

GRAPH 170: **Fatigue Cycles to Failure vs. Stress in Tension for Amoco Performance Products Udel Polysulfone.**

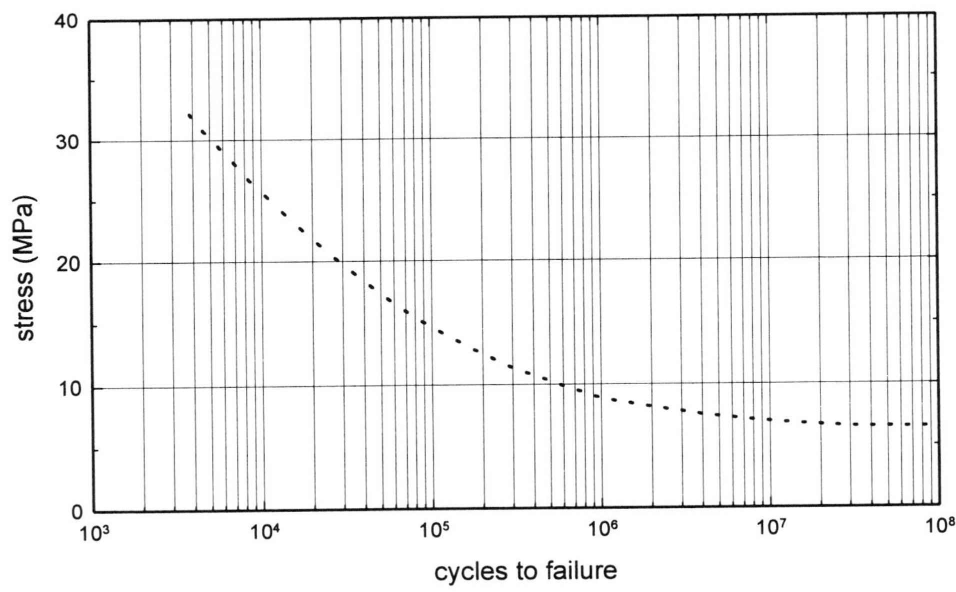

···············	Amoco Udel Polysulfone (transparent, amber tint); tension; 30 Hz; 22°C; interpolated after 10^7 cycles
Reference No.	15

Chapter 33

Polyethersulfone

GRAPH 171: Fatigue Cycles to Failure vs. Stress in Flexure for Filled and Unfilled BASF Ultrason E Polyethersulfone.

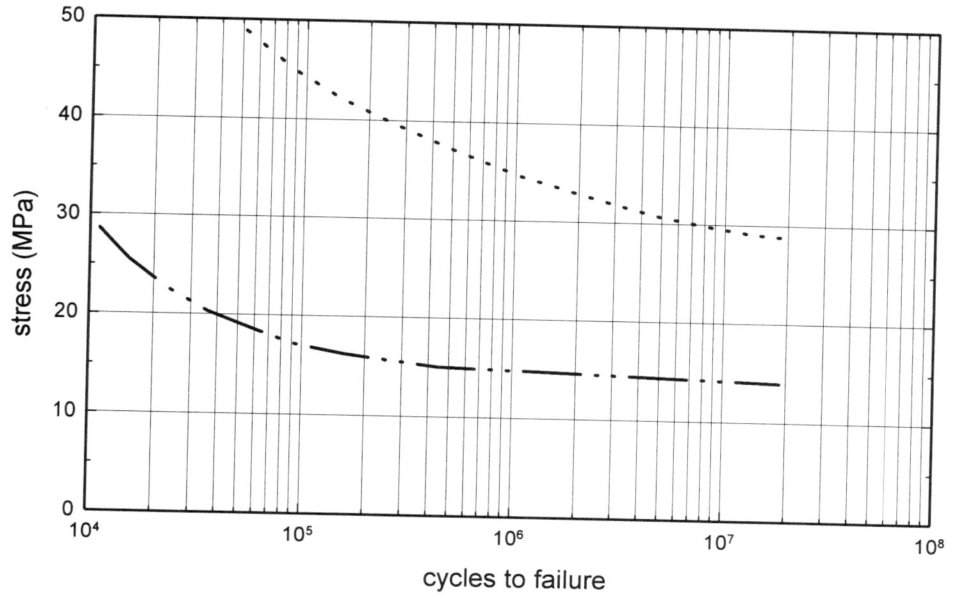

....	BASF AG Ultrason E 2010G4 PES (moderate flow; 20% glass fiber); flexure (alternating); DIN 53442; 15 Hz; 23°C; 50% RH; laboratory atmosphere
—·-—··	BASF AG Ultrason E 2010 PES (moderate flow, gen. purp. grade); flexure (alternating); DIN 53442; 15 Hz; 23°C; 50% RH; laboratory atmosphere
Reference No.	28

GRAPH 172: Fatigue Cycles to Failure vs. Stress in Flexure for Reinforced LNP Polyethersulfone.

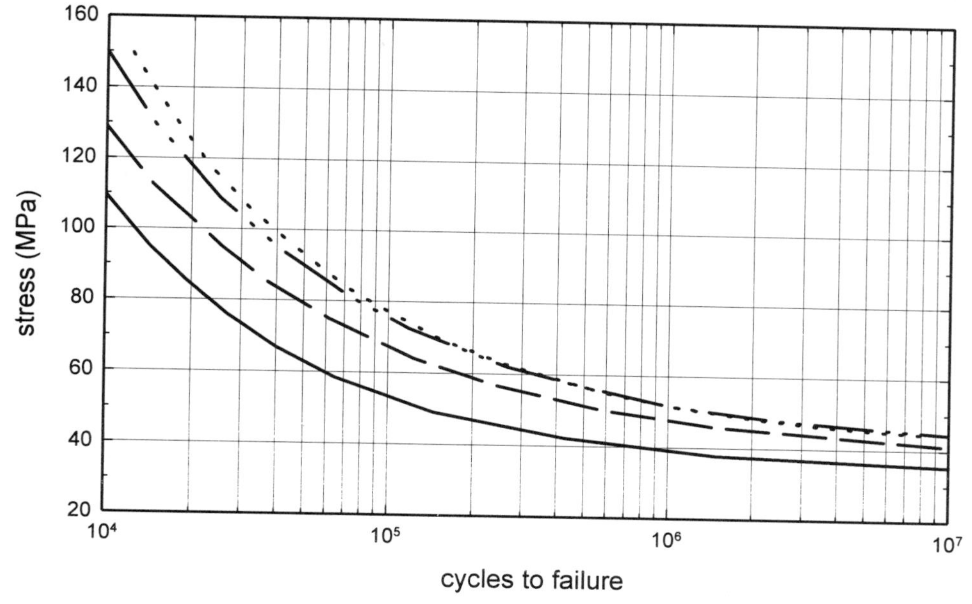

....	LNP Stat-Kon JC1006 PES (static dissipative; 30% carbon fiber); flexure; 30 Hz; 23°C
—·-—··	LNP Thermocomp JC1006 PES (30% carbon fiber); flexure; 30 Hz; 23°C
— — —	LNP Thermocomp JF1008 PES (40% glass fiber); flexure; 30 Hz; 23°C
———	LNP Thermocomp JF1006 PES (30% glass fiber); flexure; 30 Hz; 23°C
Reference No.	394

GRAPH 173: **Fatigue Cycles to Failure vs. Stress in Tension at Low Test Frequency for Filled and Unfilled Victrex PES Polyethersulfone.**

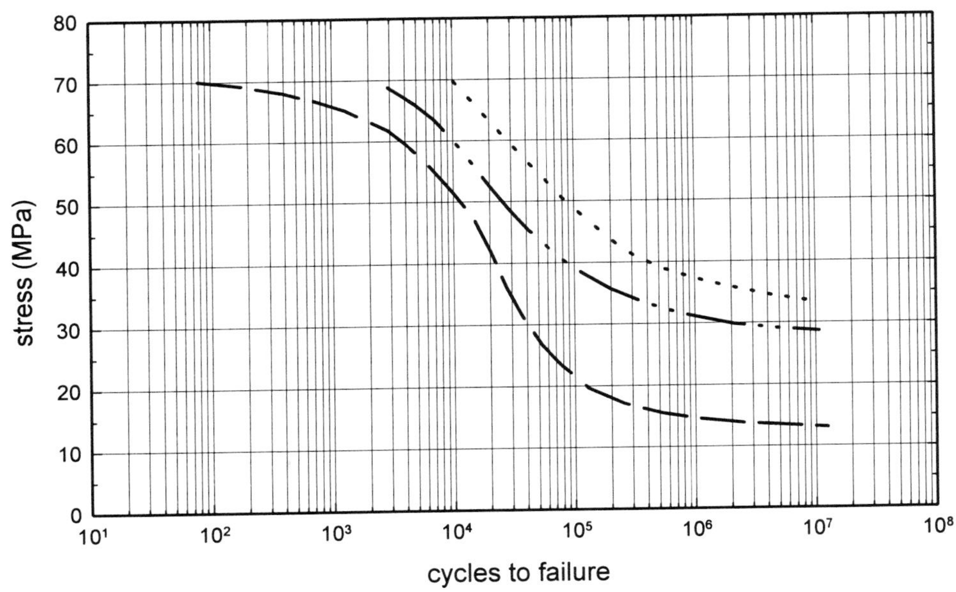

....	ICI Victrex PES 4101GL30 PES (gen. purp. grade; 30% glass fiber); tension; 0.5 Hz; 23°C; parallel to flow; square wave
—··—··	ICI Victrex PES 4101GL20 PES (gen. purp. grade; 20% glass fiber; modified grade); tension; 0.5 Hz; 23°C; parallel to flow; square wave
– – – –	ICI Victrex PES 4100G PES (gen. purp. grade); tension; 0.5 Hz; 23°C; parallel to flow; square wave
Reference No.	393

GRAPH 174: **Fatigue Cycles to Failure vs. Stress in Tension at Low Test Frequency for Victrex Polyethersulfone.**

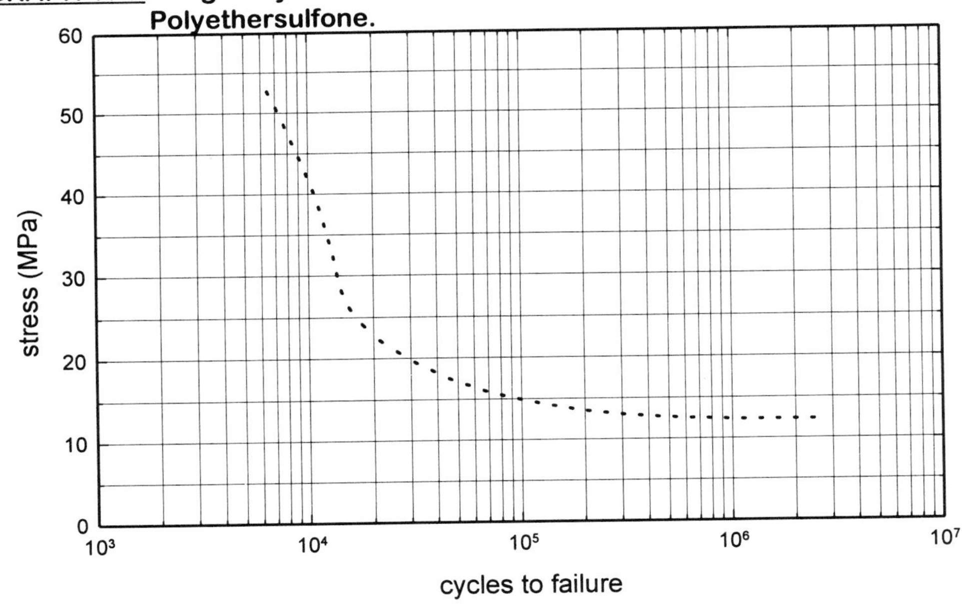

....	ICI Victrex PES 4800P PES (low flow); tension-compression; 0.5 Hz; 20°C; square wave; mean stress = 0
Reference No.	76

GRAPH 175: Fatigue Cycles to Failure vs. Stress in Tension at Two Temperatures for Glass Fiber Reinforced Polyethersulfone.

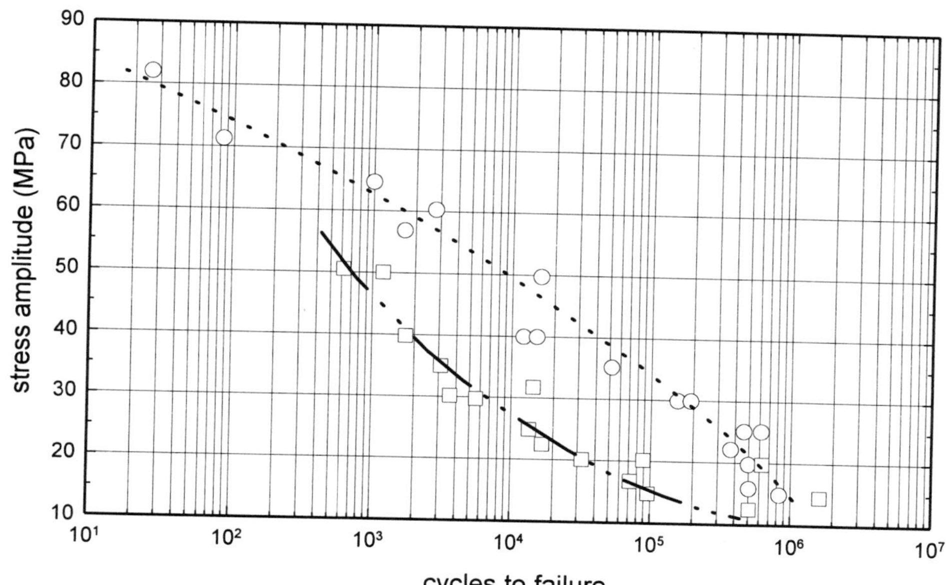

○	PES (30% glass fiber); tension- zero; 0.5 Hz; 20°C [fig d]
□	PES (30% glass fiber); tension- zero; 0.5 Hz; 120°C [fig d]
Reference No.	355

GRAPH 176: Fatigue Cycles to Failure vs. Initial Strain in Tension at Two Temperatures for Glass Fiber Reinforced Polyethersulfone.

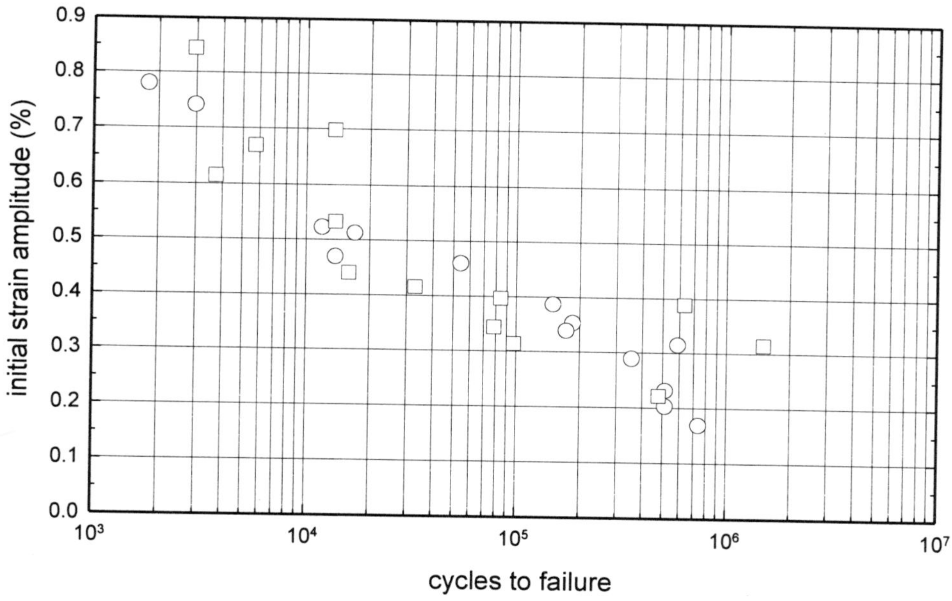

○	PES (30% glass fiber); tension- zero; 0.5 Hz; 20°C [fig d]
□	PES (30% glass fiber); tension- zero; 0.5 Hz; 120°C [fig d]
Reference No.	355

Chapter 34

Rigid Thermoplastic Urethane

Fatigue Properties

Dow Chemical: Isoplast 300 (note: w/ MDI, hexanediol based, linear RTPU); **Isoplast 301** (note: mixed RTPU, mixture of 2 diols); **Isoplast 302** (note: cyclic RTPU, 4,4' cyclohexanedimethanol based)

At 34% of yield stress hexanediol-MDI RTPU performs the longest fatigue lifetime (about 8500 cycles), whereas 4,4' cyclohexanedimethanol - MDI RTPU (Isoplast 302) shows the shortest lifetime, (about 700 cycles). Crack propogation resistance accounts for 60% of fatigue lifetime for RTPU's.

Reference: Tanrattanakul, V. (Case Western Reserve University), Moet, A. (Case Western Reserve University), Ban, *Fatigue Behavior of Rigid Thermoplastic Polyurethanes,* ANTEC 1993, conference proceedings - Communication Channels, Inc., 1993.

GRAPH 177: Fatigue Crack Propagation (FCP) Rate vs. Stress Intensity Range for Dow Chemical Isoplast Rigid Thermoplastic Urethane.

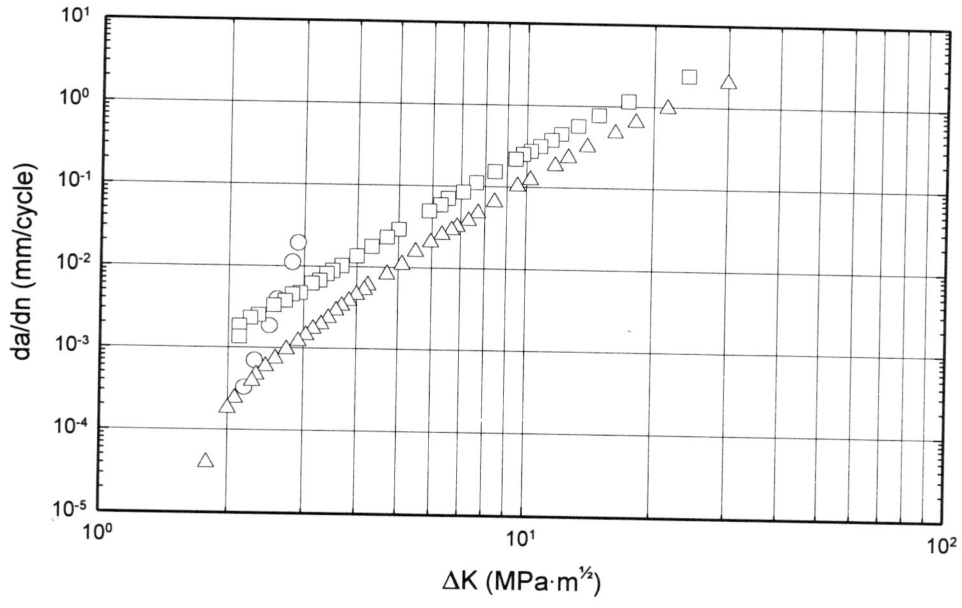

○	Dow Isoplast 302 RTPU (4,4' cyclohexane-dimethanol based, cyclic RTPU); tension- tension; 0.5 Hz; stress ratio: 0.1; 2 mm notch
□	Dow Isoplast 301 RTPU (mixture of 2 diols, mixed RTPU); tension- tension; 0.5 Hz; stress ratio: 0.1; 2 mm notch
△	Dow Isoplast 300 RTPU (linear RTPU, w/ MDI, hexanediol based); tension- tension; 0.5 Hz; stress ratio: 0.1; 2 mm notch
Reference No.	375

Chapter 35

Acrylonitrile-Butadiene-Styrene Copolymer

Fatigue Properties

ABS

The fatigue behavior of ABS is a function of the cycling rate of the test. At high rates (>1 Hz), the presence of loose chain ends in the rubber and the poor thermal conductivity of both components results in severe hysteric heating which reduces the modulus and causes failure by ductile tearing. At low cycling rates (much less than 1 Hz), fatigue is isothermal and the rubber increases the fatigue life by increasing the fracture toughness of the material through crack blunting.

Reference: Adams, M.E., Buckley, D.J., Colborn, R.E., England, W.P., Schissel, D.N., *Acrylonitrile Butadiene Styrene Polymers - Rapra Review Report 70,* review report - RAPRA Technology Ltd., 1993.

GRAPH 178: Fatigue Cycles to Failure vs. Stress in Tension for General Purpose Grades of General Electric Cycolac Acrylonitrile Butadiene Styrene.

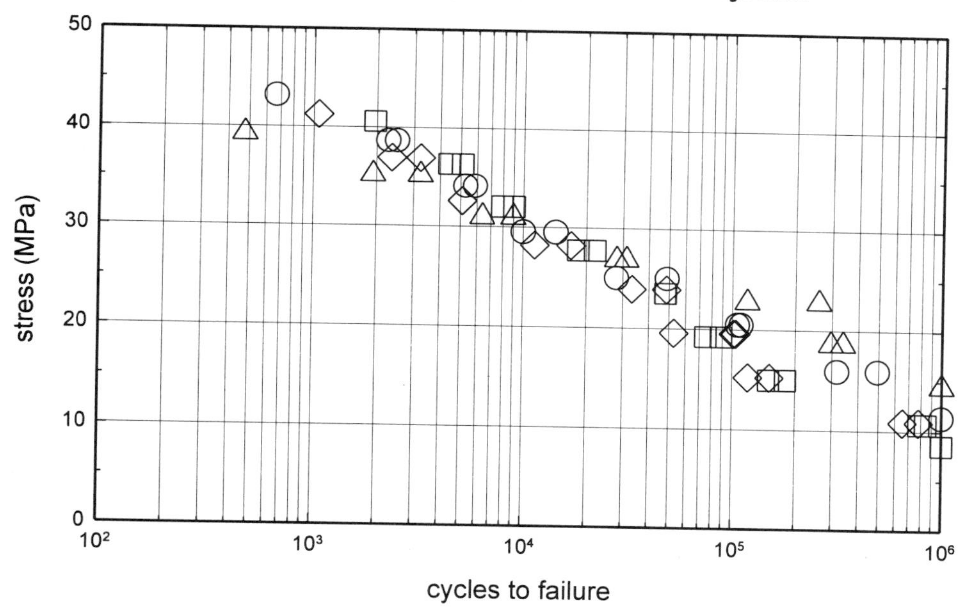

○	GE Cycolac DFA-R ABS (high gloss, high flow; 3.9 g/10 min. MFI; extrus., inj. mold.); tension- tension; GE method; 5 Hz; stress ratio: 0.1; 23°C; flow direction [fig g]
□	GE Cycolac GSM ABS (high impact, appearance grade, FDA grade, unfilled; inj. mold.); tension-tension; GE method; 5 Hz; stress ratio: 0.1; 23°C; flow direction [fig g]
△	GE Cycolac L ABS (FDA grade, unfilled; inj. mold.); tension- tension; GE method; 5 Hz; stress ratio: 0.1; 23°C; flow direction [fig g]
◇	GE Cycolac T ABS (FDA grade, unfilled; inj. mold.); tension- tension; GE method; 5 Hz; stress ratio: 0.1; 23°C; flow direction [fig g]
Reference No.	352

GRAPH 179: **Fatigue Cycles to Failure vs. Stress in Tension for Flame Retardant Grades of General Electric Cycolac Acrylonitrile Butadiene Styrene.**

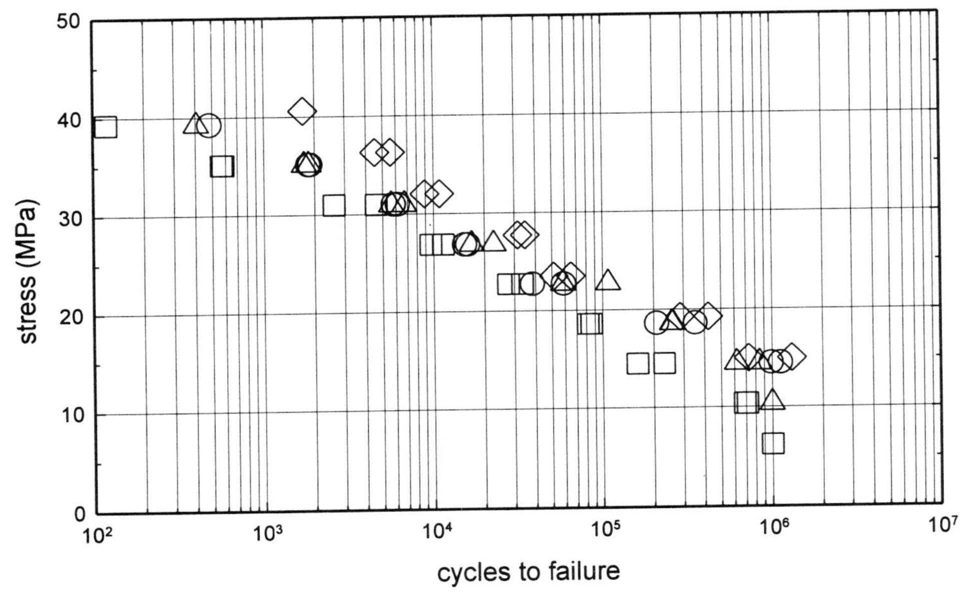

○	GE Cycolac CKM1 ABS (flame retardant; inj. mold.); tension- tension; GE method; 5 Hz; stress ratio: 0.1; 23°C; flow direction [fig g]
□	GE Cycolac CKM2 ABS (flame retardant; extrus., inj. mold.); tension- tension; GE method; 5 Hz; stress ratio: 0.1; 23°C; flow direction [fig g]
△	GE Cycolac KJB ABS (flame retardant, unfilled; inj. mold.); tension- tension; GE method; 5 Hz; stress ratio: 0.1; 23°C; flow direction [fig g]
◇	GE Cycolac KJW ABS (business equip.; flame retardant, unfilled; inj. mold.); tension- tension; GE method; 5 Hz; stress ratio: 0.1; 23°C; flow direction [fig g]
Reference No.	352

GRAPH 180: **Fatigue Cycles to Failure vs. Stress in Tension for Transparent Grades of General Electric Cycolac Acrylonitrile Butadiene Styrene.**

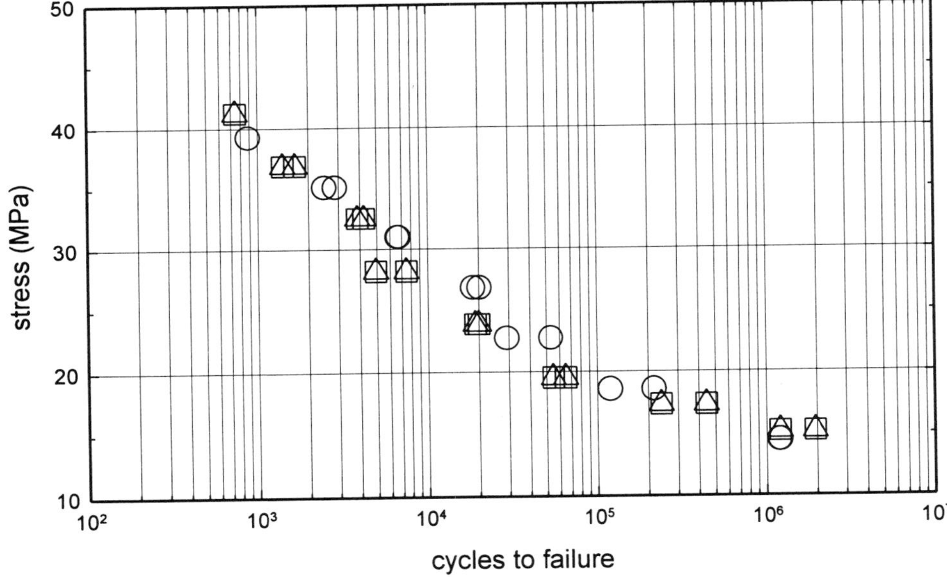

○	GE Cycolac CTB ABS (transparent; inj. mold.); tension- tension; GE method; 5 Hz; stress ratio: 0.1; 23°C; flow direction [fig g]
□	GE Cycolac GTM5300 ABS (transparent; inj. mold.); tension- tension; GE method; 5 Hz; stress ratio: 0.1; 23°C; flow direction [fig g]
△	GE Cycolac HP10 ABS (gamma rad. stable; inj. mold.); tension- tension; GE method; 5 Hz; stress ratio: 0.1; 23°C; flow direction [fig g]
Reference No.	352

GRAPH 181: Fatigue Cycles to Failure vs. Stress in Tension for Flame Retardant Grades of General Electric Cycolac Acrylonitrile Butadiene Styrene.

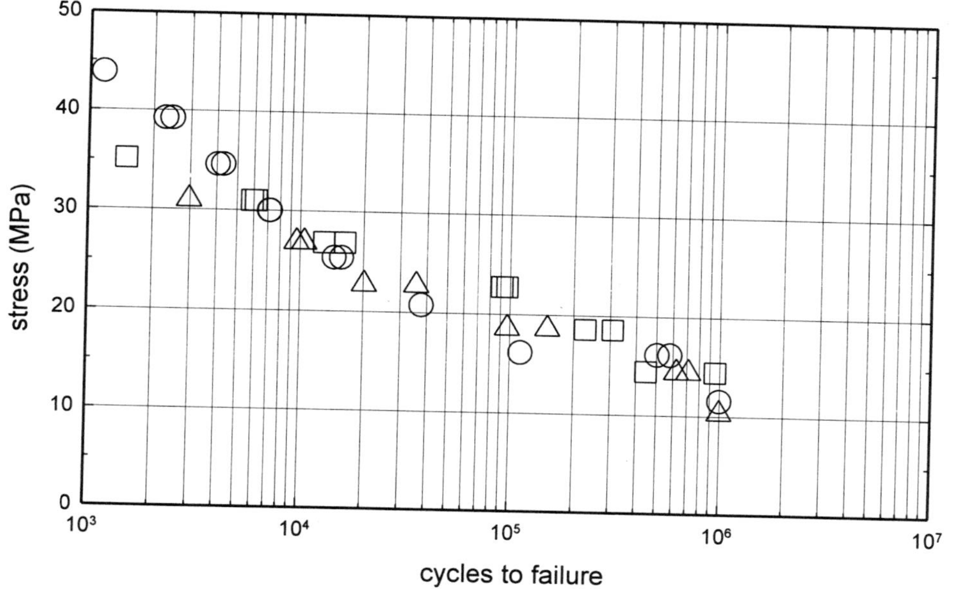

○	GE Cycolac V100 ABS (flame retardant, unfilled; non-PBBE additives; inj. mold.); tension- tension; GE method; 5 Hz; stress ratio: 0.1; 23°C; flow direction [fig g]
□	GE Cycolac V200 ABS (indoor UV stable, flame retardant; non-PBBE additives; inj. mold.); tension- tension; GE method; 5 Hz; stress ratio: 0.1; 23°C; flow direction [fig g]
△	GE Cycolac VW300 ABS (indoor UV stable, flame retardant, unfilled; inj. mold.); tension- tension; GE method; 5 Hz; stress ratio: 0.1; 23°C; flow direction [fig g]
Reference No.	352

GRAPH 182: Fatigue Cycles to Failure vs. Stress in Tension for Electroplating Grades of General Electric Cycolac Acrylonitrile Butadiene Styrene.

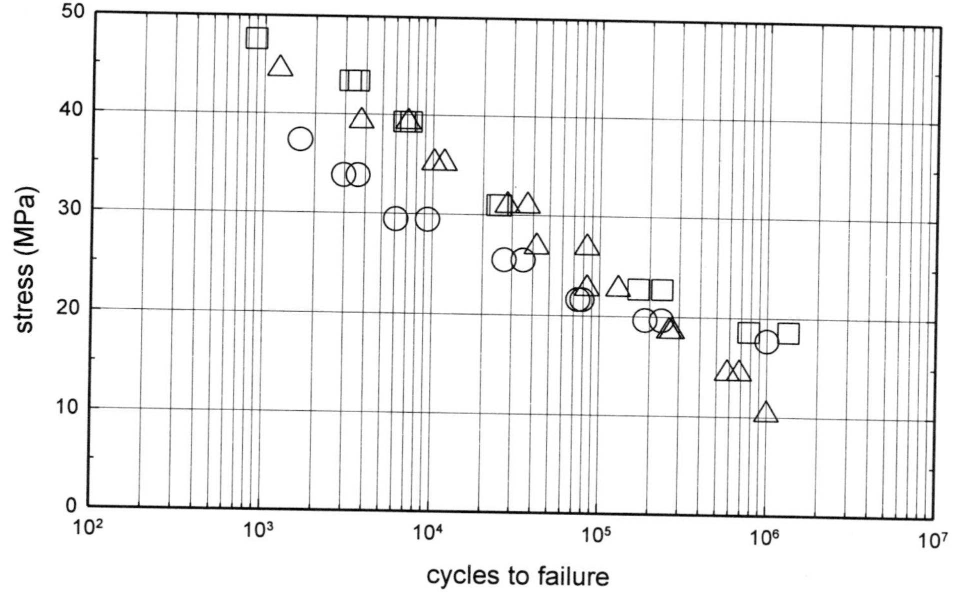

○	GE Cycolac EP ABS (plating grade; inj. mold.); tension- tension; GE method; 5 Hz; stress ratio: 0.1; 23°C; flow direction [fig g]
□	GE Cycolac EPB ABS (plating grade; 3.9 g/10 min. MFI; inj. mold.); tension- tension; GE method; 5 Hz; stress ratio: 0.1; 23°C; flow direction [fig g]
△	GE Cycolac EPBM ABS (plating grade; inj. mold.); tension- tension; GE method; 5 Hz; stress ratio: 0.1; 23°C; flow direction [fig g]
Reference No.	352

GRAPH 183: **Fatigue Cycles to Failure vs. Stress for Toray Toyolac Acrylonitrile Butadiene Styrene.**

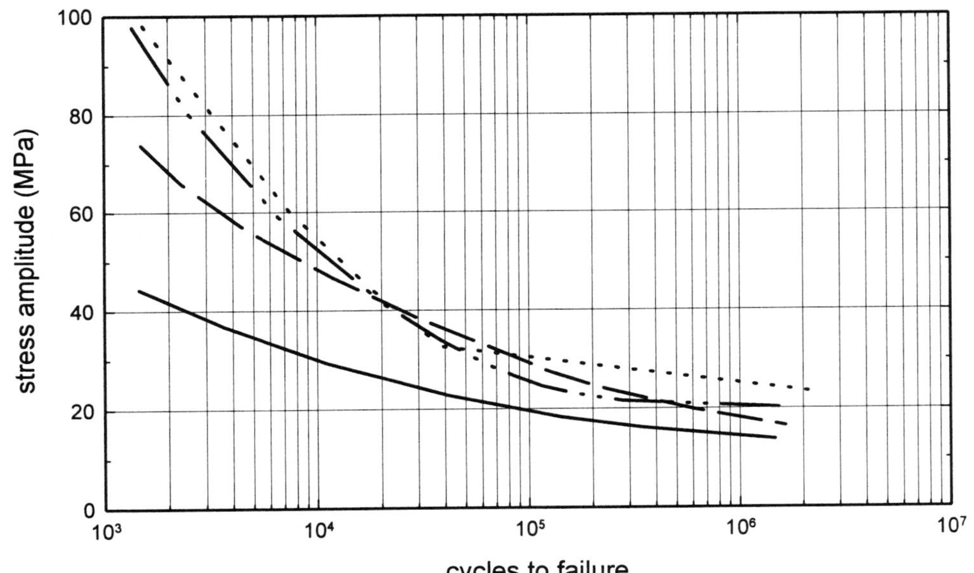

···············	Toray Toyolac 100 ABS (high impact; extrus., inj. mold.); flexure; 5 Hz; 23°C;
—··—··	Toray Toyolac 300 ABS (high impact, super tough; inj. mold.); flexure; 5 Hz; 23°C;
— — —	Toray Toyolac 500 ABS (high modulus; inj. mold.); flexure; 5 Hz; 23°C;
———	Toray Toyolac 600 ABS (high impact; extrus.); flexure; 5 Hz; 23°C;
Reference No.	407

Chapter 36

Acrylonitrile-Styrene-Acrylate Copolymer

GRAPH 184: **Fatigue Cycles to Failure vs. Stress in Flexure for BASF Luran S Acrylate Styrene Acrylonitrile Polymer.**

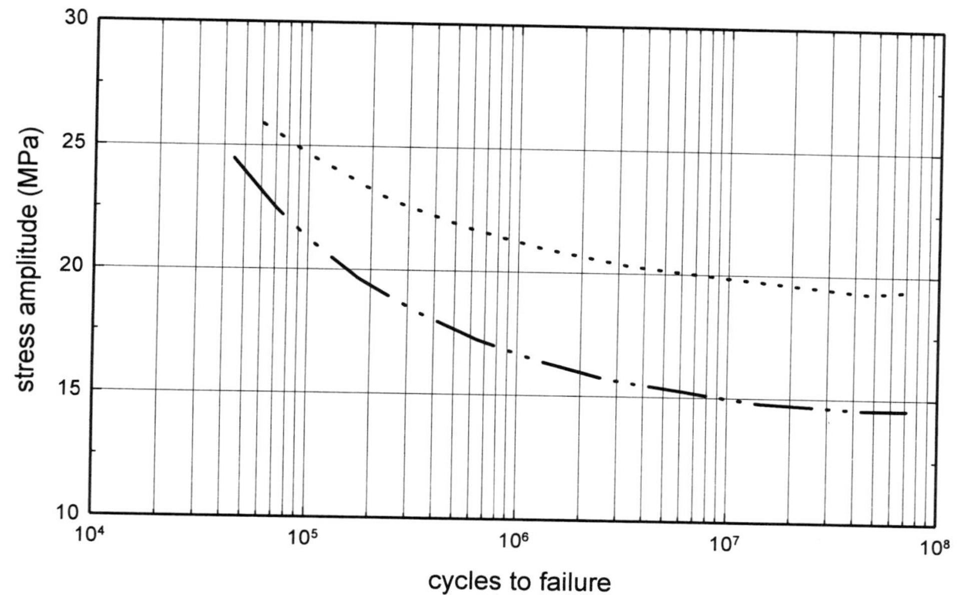

....	BASF AG Luran S 776 S ASA (moderate flow, high impact); flexure (alternating); DIN 53442; 10 Hz; 23°C
—··—··	BASF AG Luran S 757 R ASA (gen. purp. grade); flexure (alternating); DIN 53442; 10 Hz; 23°C
Reference No.	143

GRAPH 185: **Fatigue Cycles to Failure vs. Stress in Tension for General Electric Geloy GY1120 Acrylate Styrene Acrylonitrile Polymer.**

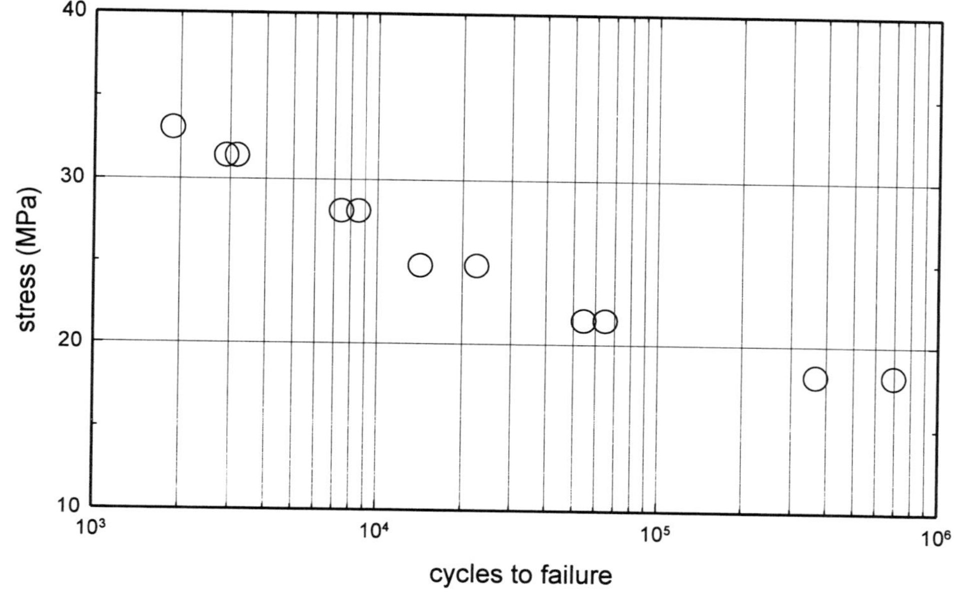

○	GE Geloy GY1120 ASA (profile extrusion; lubricated, white, UV stabilized, unfilled; extrus., blow molding); tension-tension; GE method; 5 Hz; stress ratio: 0.1; 23°C; flow direction [fig g]
Reference No.	352

Chapter 37

Polystyrene

GRAPH 186: **Fatigue Crack Propagation (FCP) Rate vs. Stress Intensity Range for Polystyrene.**

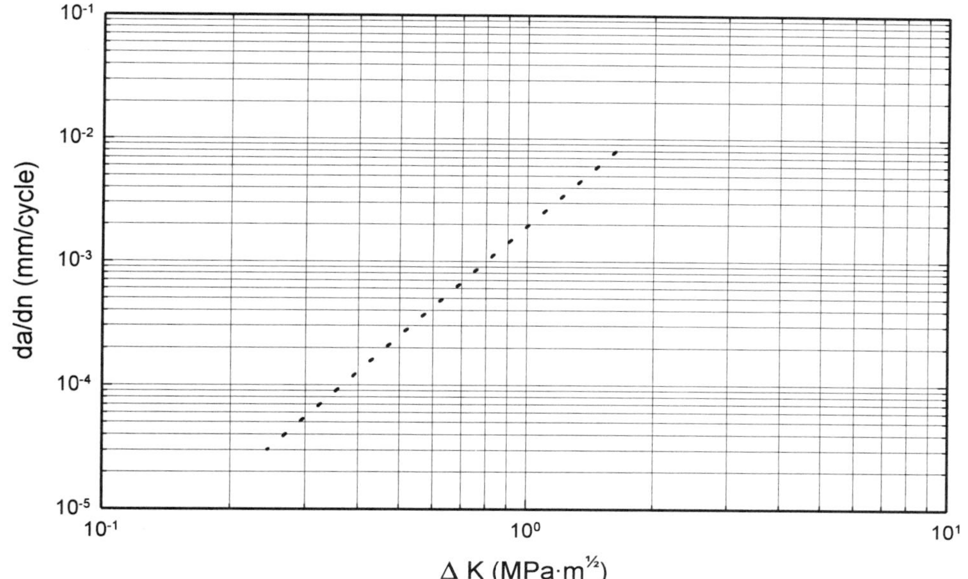

....	PS
Reference No.	355

Chapter 38

General Purpose Polystyrene

Effect of Weld Line on Fatigue Behavior

Polysar GPPS: (features: 3.2 mm thick; note: specimen does not contain a weld line); **GPPS** (note: specimen contains a weld line)

The S-N curves of welded and non-welded polystyrene have the same general pattern behavior. As expected the fatigue lifetime is decreased by the presence of the weld-line. In the case of specimens containing a weld-line the crack initiates and propagates from the V-notch. In specimens without a weld-line, the crack initiates at the corner and tends to propagate through the thickness. In both cases the specimens undergo a brittle failure at any applied stress level. Observation of the fracture surfaces shows that the fracture mechanisms involved are the same.

Reference: Boukhili, R. (Ecole Polytechnique de Montreal), Gauvin, R. (Ecole Polytechnique de Montreal), Gossel, *Fatigue Behavior of Injection Moded Polycarbonate and Polystyrene Containing Weld Lines,* ANTEC 1989, conference proceedings - Society of Plastics Engineers, 1989.

GRAPH 187: Fatigue Cycles to Failure vs. Stress in Flexure for Glass Fiber Reinforced LNP Thermocomp General Purpose Polystyrene.

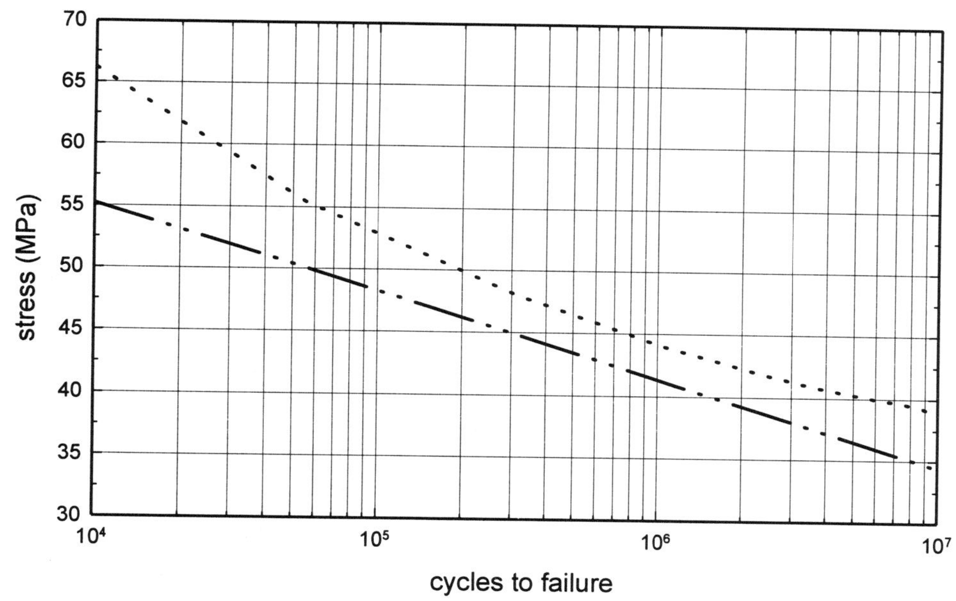

··············	LNP Thermocomp CF1008 GPPS (40% glass fiber); flexure; 30 Hz; 23°C
—··—··	LNP Thermocomp CF1006 GPPS (30% glass fiber); flexure; 30 Hz; 23°C
Reference No.	394

GRAPH 188: **Fatigue Cycles to Failure vs. Stress in Tension for Polysar General Purpose Polystyrene With and Without Weld Line.**

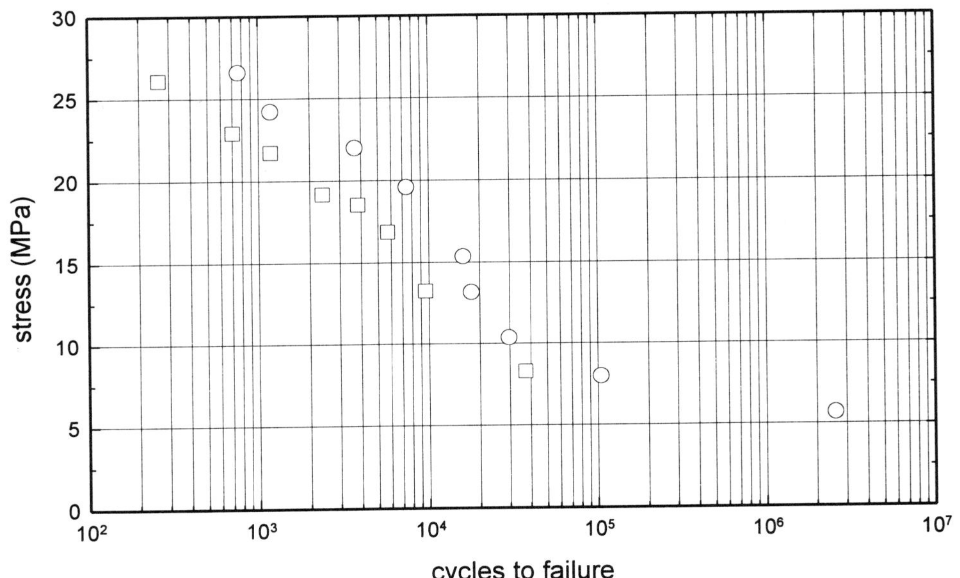

○	Polysar GPPS (with weld line); tension- tension; 10 Hz; stress ratio: 0.1; 23°C [fig g, aa]
□	Polysar GPPS (3.2 mm thick; without weld line); tension- tension; 10 Hz; stress ratio: 0.1; 23°C [fig g, aa]
Reference No.	372

GRAPH 189: **Fatigue Cycles to Failure vs. Stress for General Purpose Polystyrene.**

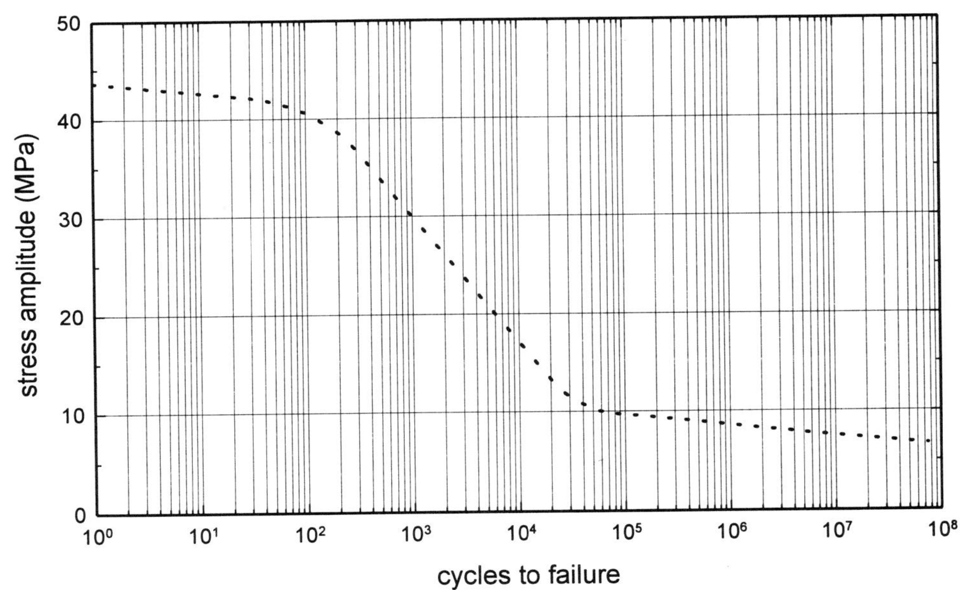

....	Dow Styron GPPS (injection molding); 23°C
Reference No.	263

Chapter 39

Impact Resistant Polystyrene

GRAPH 190: Fatigue Cycles to Failure vs. Stress in Flexure for Mobil PS 4600 Impact Polystyrene.

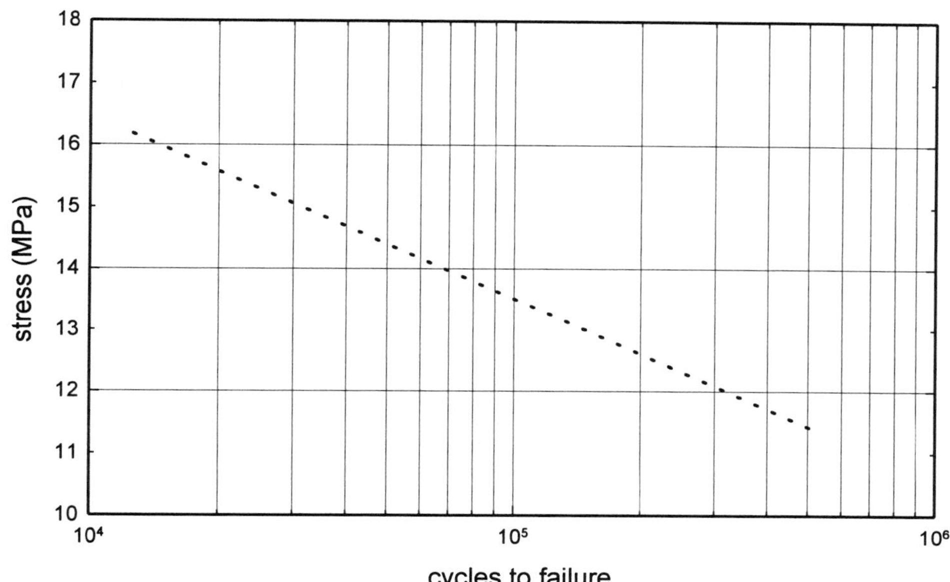

....	Mobil PS 4600 IPS; flexure; ASTM D671; 30 Hz; 23°C
Reference No.	366

GRAPH 191: Fatigue Cycles to Failure vs. Stress for Notched Dow Chemical Styron Impact Polystyrene.

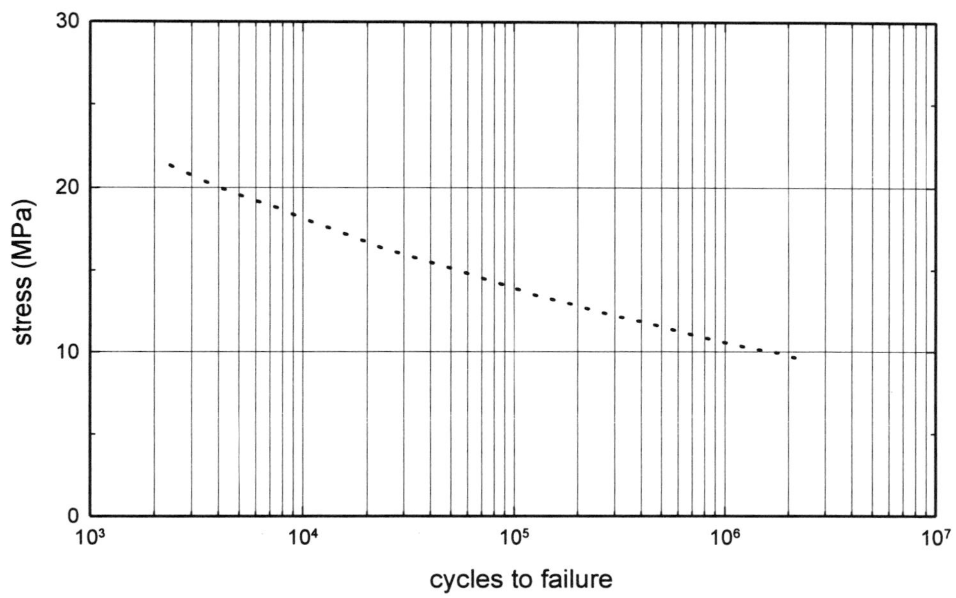

....	Dow Styron IPS (injection molding); 30 Hz; 22°C; notched; 6.3 mm radius notch
Reference No.	262

Chapter 40

Styrene-Acrylonitrile Copolymer

GRAPH 192: **Fatigue Cycles to Failure vs. Stress in Flexure for Injection and Compression Molded BASF Luran 368R Styrene Acrylonitrile Copolymer.**

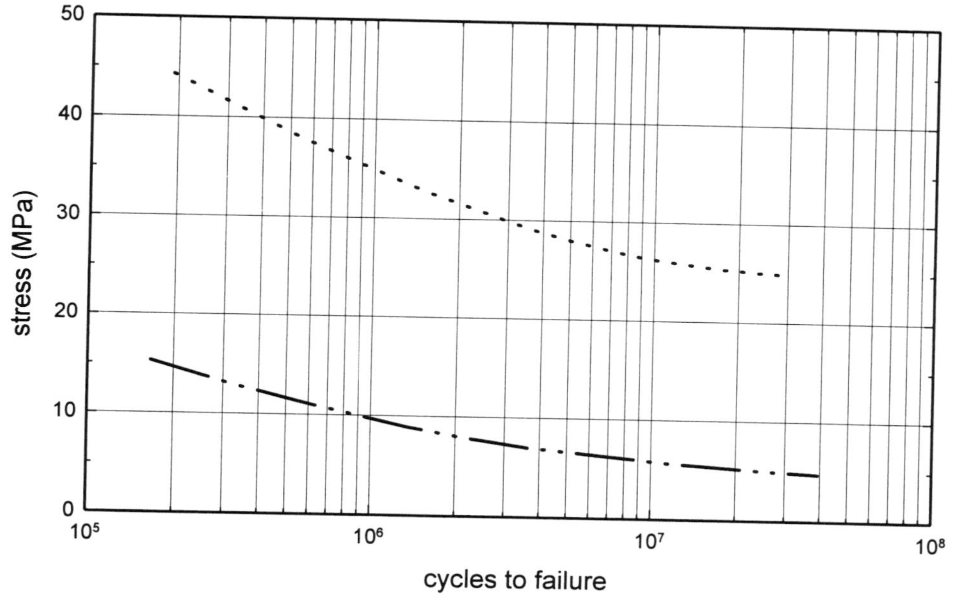

···············	BASF AG Luran 368R SAN (injection molding); flexure (alternating); DIN 53442; 23°C
—··—··	BASF AG Luran 368R SAN (compression molded, machined spec.); flexure (alternating); DIN 53442; 23°C
Reference No.	30

GRAPH 193: **Fatigue Cycles to Failure vs. Stress in Flexure for 30% Glass Fiber Reinforced LNP Thermocomp BF1006 Styrene Acrylonitrile Copolymer.**

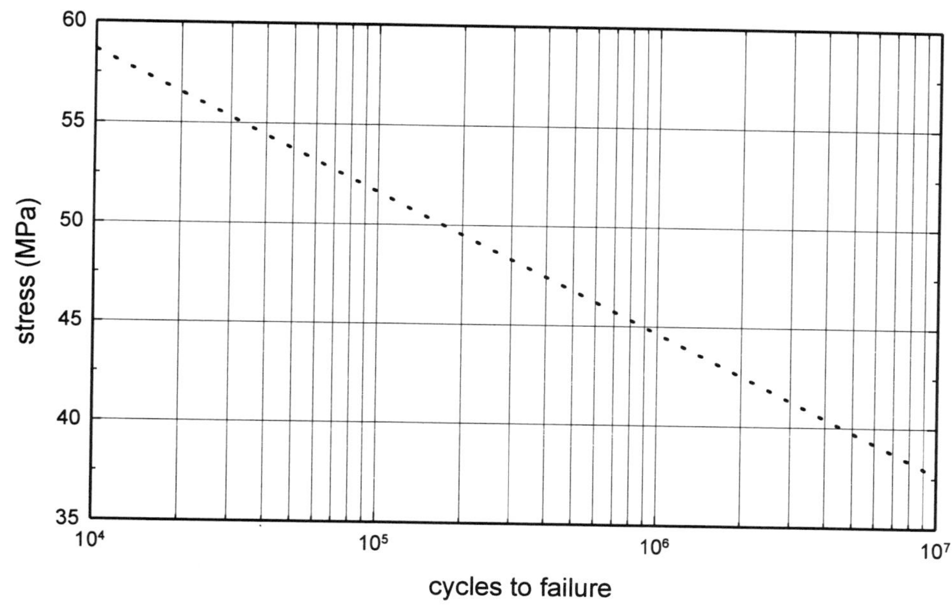

···············	LNP Thermocomp BF1006 SAN (30% glass fiber); flexure; 30 Hz; 23°C
Reference No.	394

Chapter 41

Polyvinyl Chloride

Fatigue Properties

PVC Geon Company:

Fatigue data for some model glass reinforced PVC compounds show the existence of a fatigue plateau at approximately 35% of the ultimate tensile strength of the materials below which failure did not occur. As is well known, this fatigue plateau is common for metals, but is not common for plastics, having been noted only for nylon, PET and PEEK.

Reference: *BFGoodrich Fiberloc Polymer Composites Engineering Design Data,* supplier design guide (FL-0101) - BFGoodrich Geon Vinyl Division, 1989.

GRAPH 194: **Fatigue Cycles to Failure vs. Stress in Tension for Filled and Unfilled Polyvinyl Chloride.**

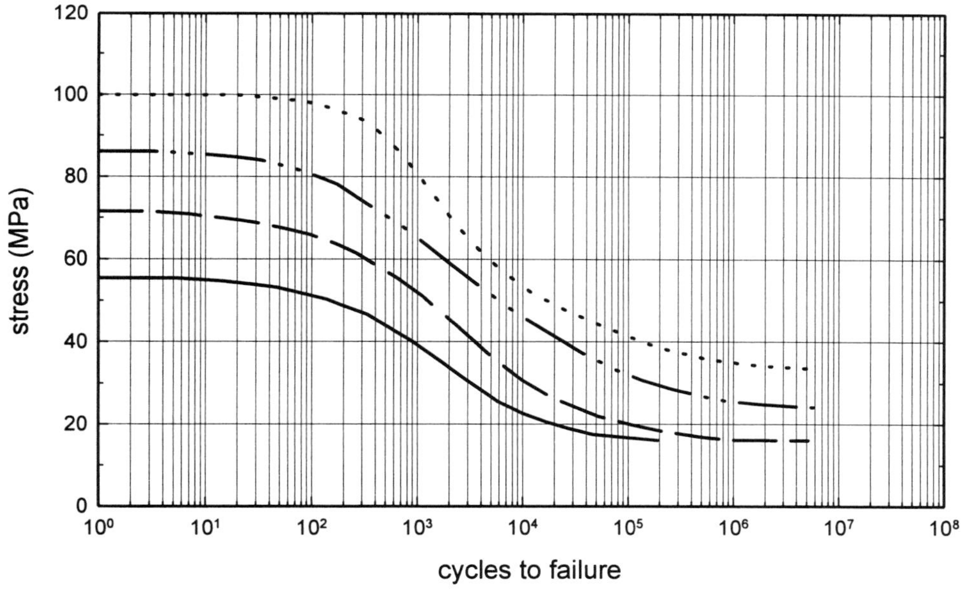

....	Geon Co. Fiberloc PVC (30% glass fiber); tension; 10 Hz; stress ratio: 0.1; 25°C [fig g]
—··—··	Geon Co. Fiberloc PVC (20% glass fiber); tension; 10 Hz; stress ratio: 0.1; 25°C [fig g]
– – – –	Geon Co. Fiberloc PVC (10% glass fiber); tension; 10 Hz; stress ratio: 0.1; 25°C [fig g]
———	Geon Co. PVC; tension; 10 Hz; stress ratio: 0.1; 25°C [fig g]
Reference No.	190

GRAPH 195: **Fatigue Cycles to Failure vs. Stress in Tension at Low Test Frequency for Pipe Grade Polyvinyl Chloride.**

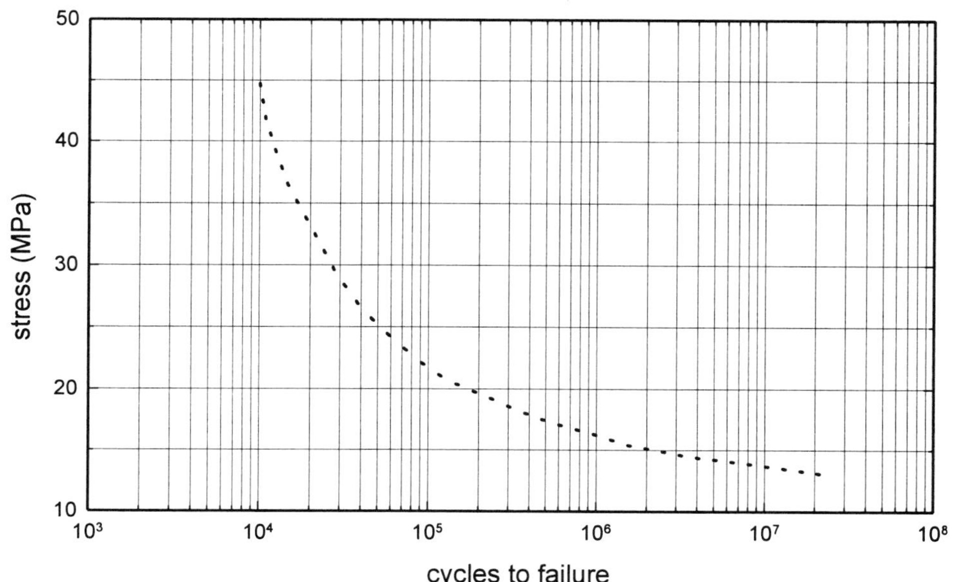

....	PVC (pipe grade; unplasticized); tension-zero; 0.5 Hz; 20°C; square wave
Reference No.	355

GRAPH 196: **Fatigue Cycles to Failure vs. Stress in Tension at Low Test Frequency for Poorly Processed and Well Processed Polyvinyl Chloride.**

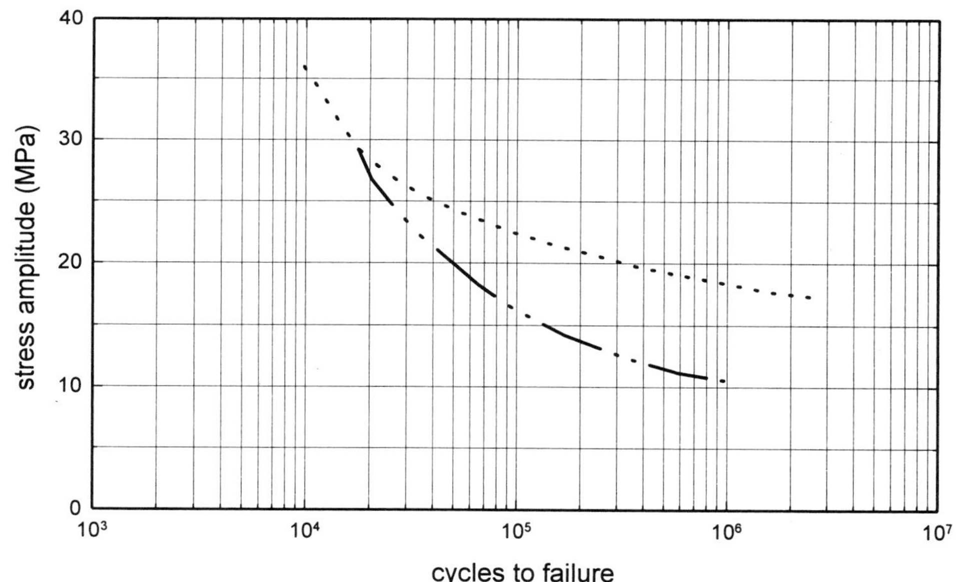

....	PVC (pipe grade; unplasticized); tension-zero; 0.5 Hz; 20°C; square wave; well processed sample [fig f]
—··—··	PVC (pipe grade; unplasticized); tension-zero; 0.5 Hz; 20°C; square wave; poorly processed sample [fig f]
Reference No.	355

GRAPH 197: Fatigue Cycles to Failure vs. Stress for Polyvinyl Chloride Showing the Influence of Waveform Shape.

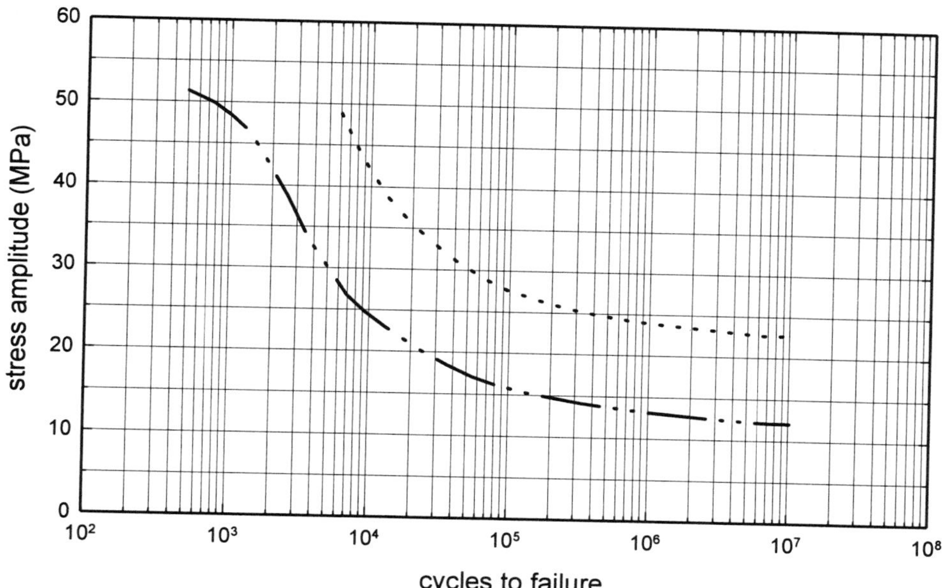

GRAPH 198: Fatigue Cycles to Failure vs. Strain in Flexure for Geon Polyvinyl Chloride.

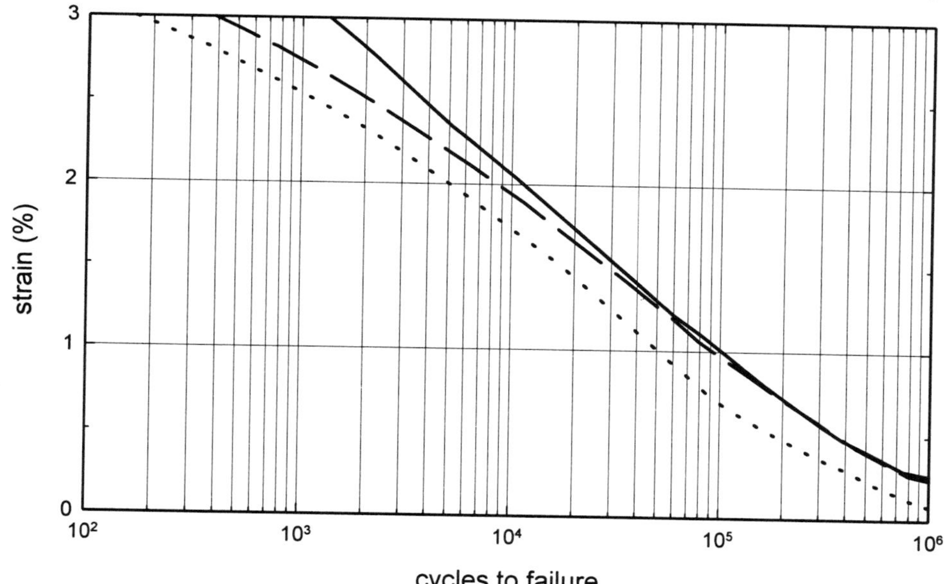

GRAPH 199: **Total Time Under Load vs. Stress in Tension with Continuous and Cyclic Loading Conditions for Polyvinyl Chloride.**

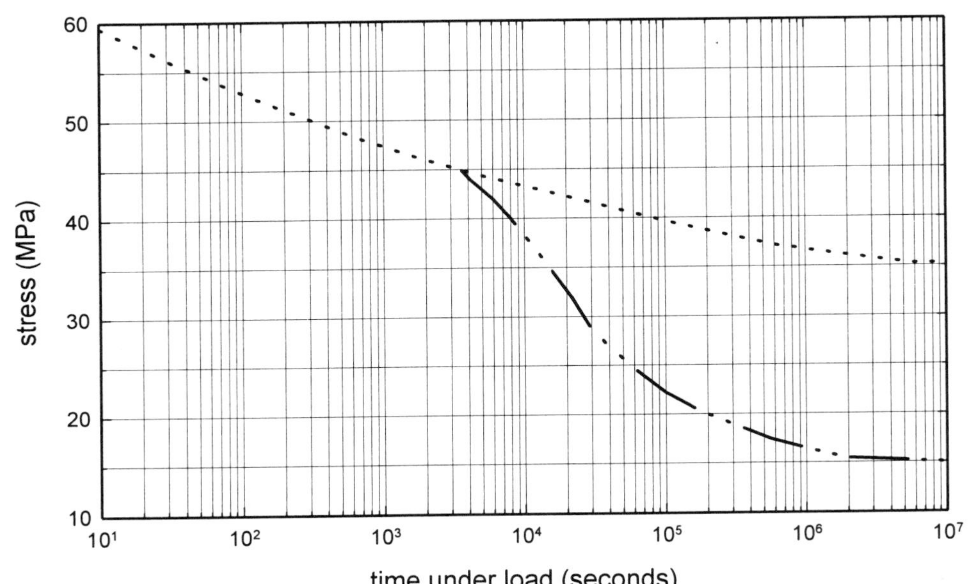

....	PVC (unplasticized); tension; 20°C; continuous load
—··—··	PVC (unplasticized); tension; 20°C; cyclic loading; max. time represents failure[1]
Reference No.	355

GRAPH 200: **Total Time Under Load vs. Stress in Tension with Continuous and Cyclic Loading Conditions for Polyvinyl Chloride, Pipe Grade.**

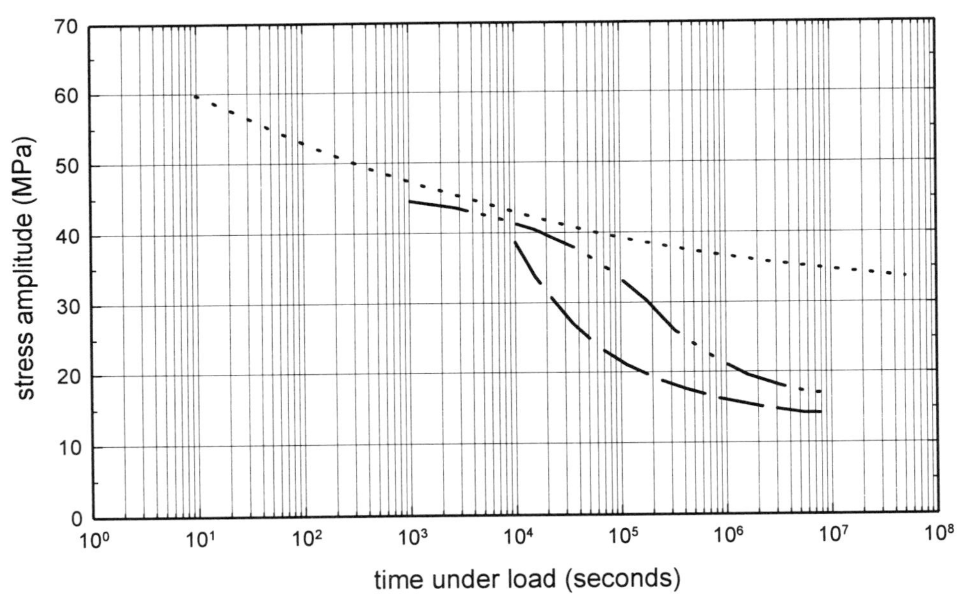

....	PVC (pipe grade; unplasticized); tension; 20°C; continuous load
—··—··	PVC (pipe grade; unplasticized); tension; 0.005 Hz; 20°C; cyclic load[1]
— — — —	PVC (pipe grade; unplasticized); tension; 0.5 Hz; 20°C; cyclic load[1]
Reference No.	355

[1] Although fatigue data are usually presented in their simplest and most easily recognized form, i.e. as conventional S-N curves, in the case of square wave loading the data transpose to real time characteristics. This is achieved by summing the time under tensile load over the N_f cycles to failure. The resulting stress (or strain) vs. time characterisitc can then be compared with the corresponding creep rupture characteristics for the material.

GRAPH 201: **Total Time Under Load vs. Strain in Tension with Continuous and Cyclic Loading Conditions for Transparent Polyvinyl Chloride.**

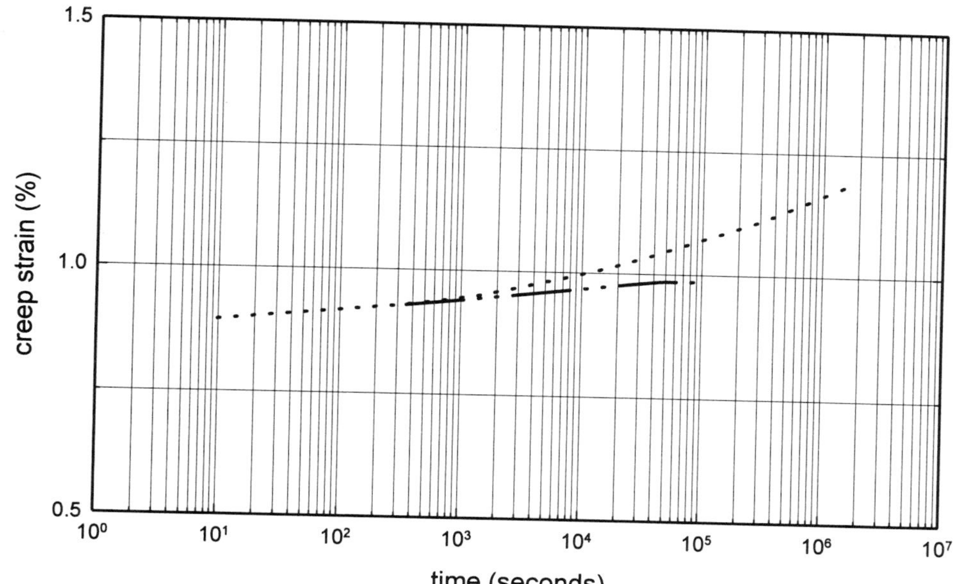

....	PVC (transparent; unplasticized); tension; 20°C; continuous load
—··—··	PVC (transparent; unplasticized); tension; 20°C; cyclic loading; max. time represents failure[2]
Reference No.	355

GRAPH 202: **Total Time Under Load to Failure vs. Frequency in Cyclic Loading Conditions for Polyvinyl Chloride.**

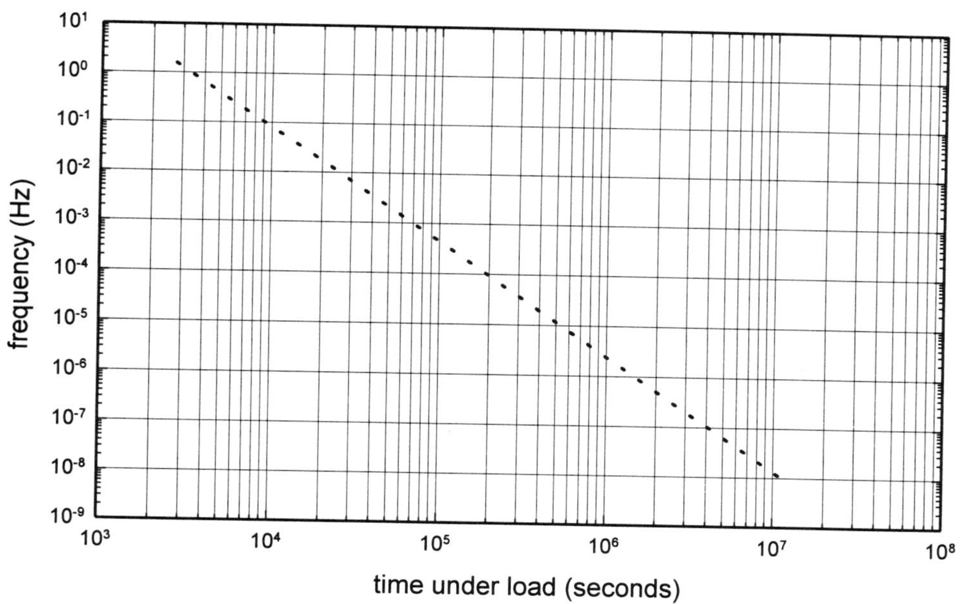

....	PVC (pipe grade; unplasticized); 20°C; stress level: 35 MPa; (The relationship between frequency and lifetimes, specified as real time and plotted on log/log axes, approximates to being linear, with time to failure increasing as frequency decreases.)
Reference No.	355

[2] Although fatigue data are usually presented in their simplest and most easily recognized form, i.e. as conventional S-N curves, in the case of square wave loading the data transpose to real time characteristics. This is achieved by summing the time under tensile load over the N_f cycles to failure. The resulting stress (or strain) vs. time characterisitc can then be compared with the corresponding creep rupture characteristics for the material.

GRAPH 203: Fatigue Crack Propagation (FCP) Rate vs. Stress Intensity Range for Polyvinyl Chloride.

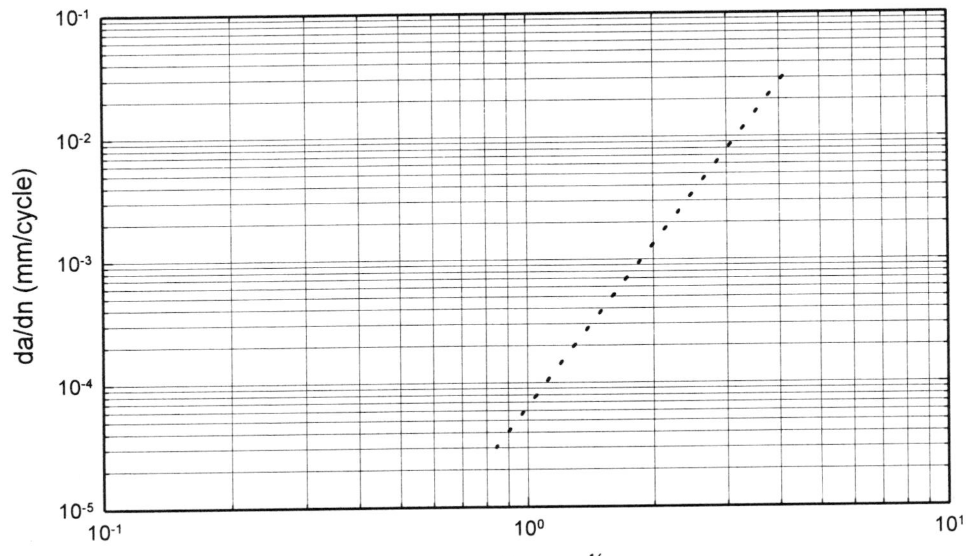

....	PVC (unplasticized)
Reference No.	355

Chapter 42

Acrylonitrile-Butadiene-Styrene Copolymer/ Polyvinyl Chloride Alloy

GRAPH 204: **Fatigue Cycles to Failure vs. Stress in Tension for General Electric Cycovin K29 ABS/ Polyvinyl Chloride Alloy.**

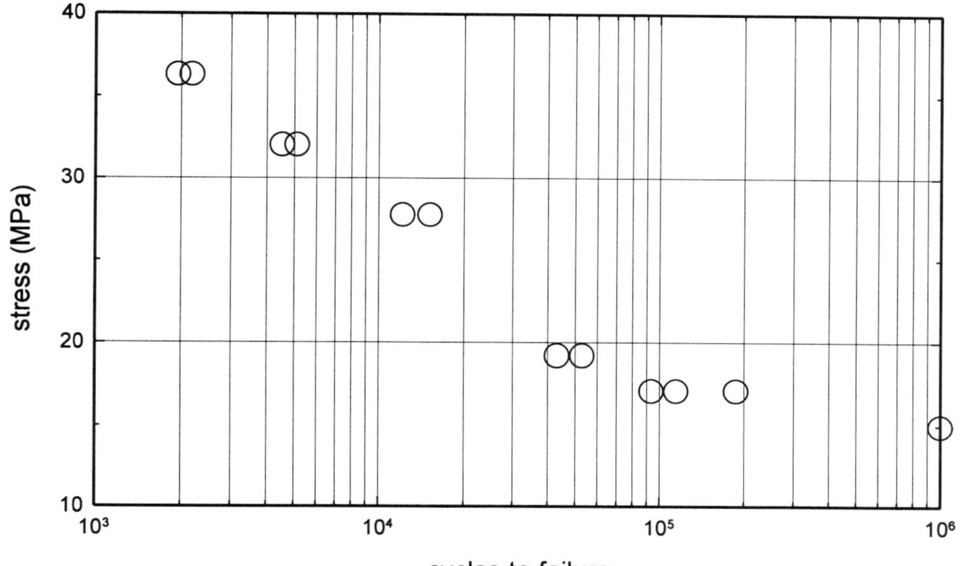

○	GE Cycovin K29 ABS/PVC; tension-tension; GE method; 5 Hz; stress ratio: 0.1; 23°C; flow direction [fig g]
Reference No.	352

Chapter 43

Acrylonitrile-Styrene-Acrylate Copolymer/ Polycarbonate Alloy

GRAPH 205: Fatigue Cycles to Failure vs. Stress in Tension for General Electric Geloy Acrylate Styrene Acrylonitrile Polycarbonate Alloy.

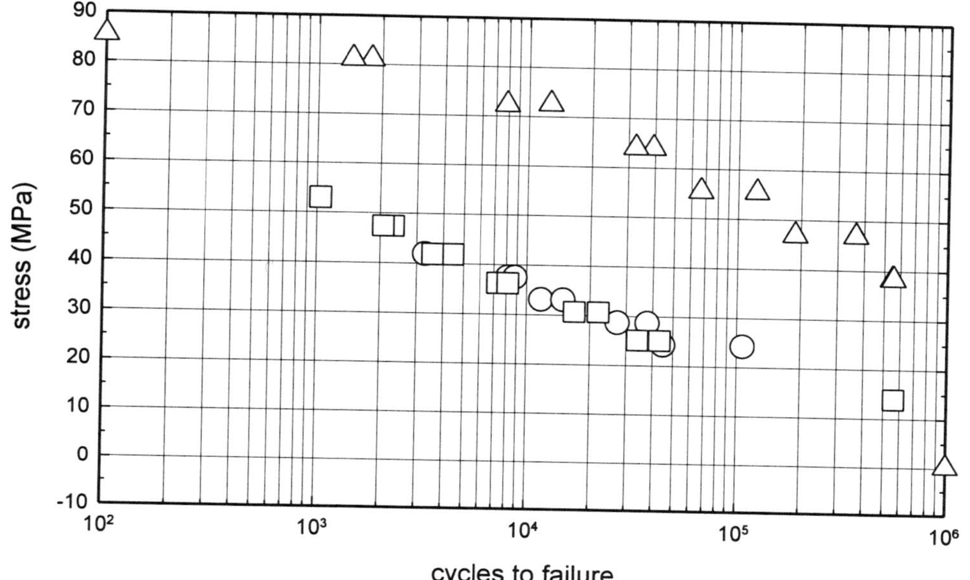

○	GE Geloy XP4001 ASA/PC (inj. mold.); tension-tension; GE method; 5 Hz; stress ratio: 0.1; 23°C; flow direction [fig g]
□	GE Geloy XP4034 ASA/PC (UV stabilized, black, limited colors, auto. grade, unfilled; inj. mold.); tension-tension; GE method; 5 Hz; stress ratio: 0.1; 23°C; flow direction [fig g]
△	GE Geloy XP4001 ASA/PC (inj. mold.); tension-tension; GE method; 5 Hz; stress ratio: 0.1; 23°C; flow direction [fig g]
Reference No.	352

Chapter 44

Acrylonitrile-Styrene-Acrylate Copolymer/ Polyvinyl Chloride Alloy

GRAPH 206: **Fatigue Cycles to Failure vs. Stress in Tension for General Electric Geloy GY1220 Acrylate Styrene Acrylonitrile Polyvinyl Chloride Alloy.**

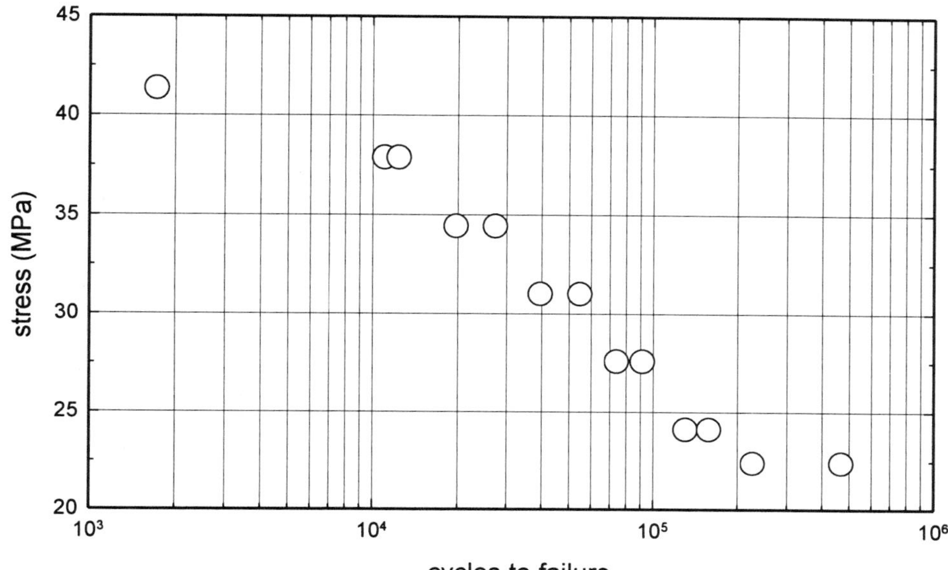

○	GE Geloy GY1220 ASA/PVC (lubricated, UV stabilized, flame retardant, unfilled, dark colors; extrus.); tension- tension; GE method; 5 Hz; stress ratio: 0.1; 23°C; flow direction [fig g]
Reference No.	352

Chapter 45

Polycarbonate/ Acrylonitrile-Butadiene-Styrene Copolymer Alloy

Fracture Toughness

Dow Chemical: PC/ABS (note: 65% polycarbonate/ 35% ABS); **PC/ABS** (material compostion: 8% talc; note: 65% polycarbonate/ 35 % ABS)

Incorporating 8% talc into a 65% PC/35% ABS blend increases the fatigue lifetime of the material. This is due to a 200% increase in the initiation time, as well as a modest increase in propagation. The increased fatigue resistance of the filled material is explained by examining the energetics of crack growth. The rate of work dissipated on crack tip plasticity is equal for both the unfilled and filled materials. However, talc increases the amount of damage formed prior to crack advance which slows the rate of crack growth.

Reference: Seibel, S.R. (Case Western), Moet, A. (Case Western), Bank, D.H. (Dow), Nichols, K. (Dow), *Effect Of Filler On the Fatigue Performance of a 65 PC/ 35 ABS Blend,* ANTEC 1993, conference proceedings - Society of Plastics Engineers, 1993.

GRAPH 207: Fatigue Cycles to Failure vs. Stress in Tension for Flame Retardant Grades of General Electric Cycoloy Polycarbonate/ ABS Alloy.

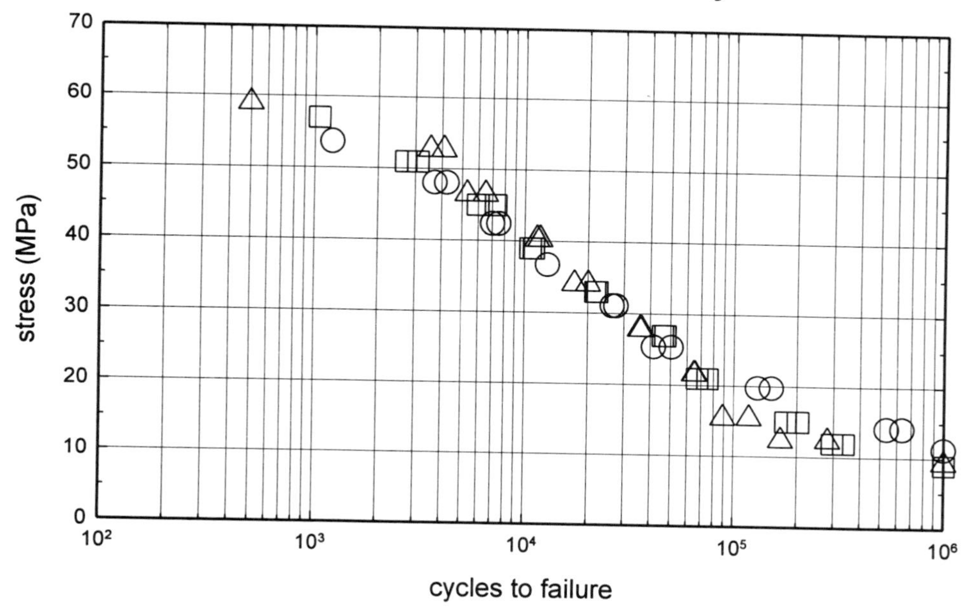

○	GE Cycoloy C2800 PC/ABS (high flow, flame retardant, unfilled; no brominated or chlorinated additives; inj. mold.); tension- tension; GE method; 5 Hz; stress ratio: 0.1; 23°C; flow direction [fig g]
□	GE Cycoloy C2950 PC/ABS (high heat grade, flame retardant, unfilled; no brominated or chlorinated additives; inj. mold.); tension- tension; GE method; 5 Hz; stress ratio: 0.1; 23°C; flow direction [fig g]
△	GE Cycoloy C2950HF PC/ABS (high heat grade, high flow, flame retardant, unfilled; no brominated or chlorinated additives; inj. mold.); tension- tension; GE method; 5 Hz; stress ratio: 0.1; 23°C; flow direction [fig g]
Reference No.	352

GRAPH 208: **Fatigue Cycles to Failure vs. Stress in Tension for Injection Molding and Blow Molding Grades of General Electric Cycoloy Polycarbonate/ ABS Alloy.**

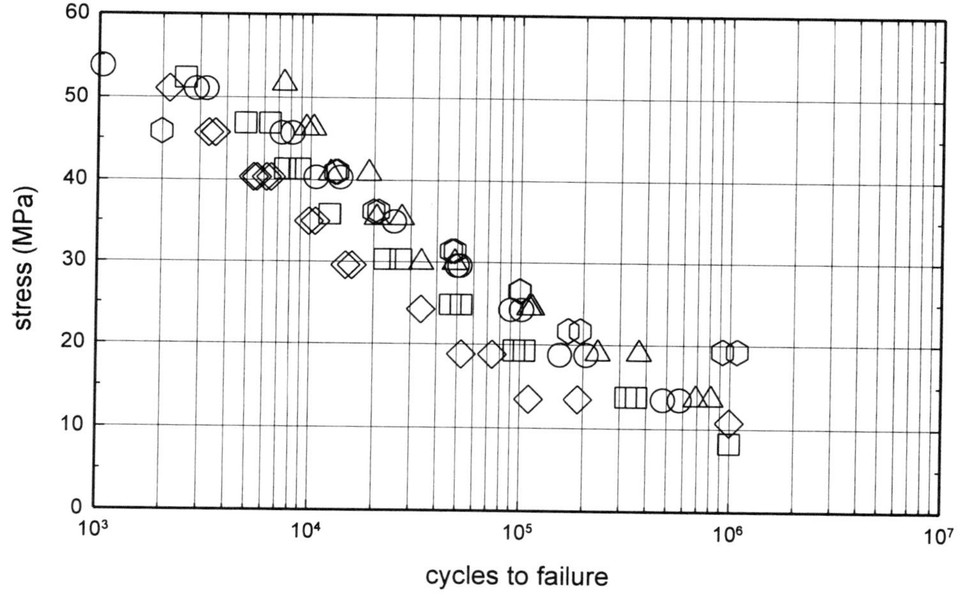

○	GE Cycoloy C1110 PC/ABS (unfilled; inj. mold.); tension- tension; GE method; 5 Hz; stress ratio: 0.1; 23°C; flow direction [fig g]
□	GE Cycoloy C1110HF PC/ABS (high flow, unfilled; inj. mold.); tension-tension; GE method; 5 Hz; stress ratio: 0.1; 23°C; flow direction [fig g]
△	GE Cycoloy C1200HF PC/ABS (high flow, unfilled; inj. mold.); tension-tension; GE method; 5 Hz; stress ratio: 0.1; 23°C; flow direction [fig g]
◇	GE Cycoloy C1950 PC/ABS (high flow, unfilled; inj. mold.); tension-tension; GE method; 5 Hz; stress ratio: 0.1; 23°C; flow direction [fig g]
⬡	GE Cycoloy MC8100 PC/ABS (unfilled; blow molding); tension- tension; GE method; 5 Hz; stress ratio: 0.1; 23°C; flow direction [fig g]
Reference No.	352

GRAPH 209: **Fatigue Crack Propagation (FCP) Rate vs. Stress Intensity Range for Mineral Filled and Unfilled Polycarbonate/ ABS Alloy.**

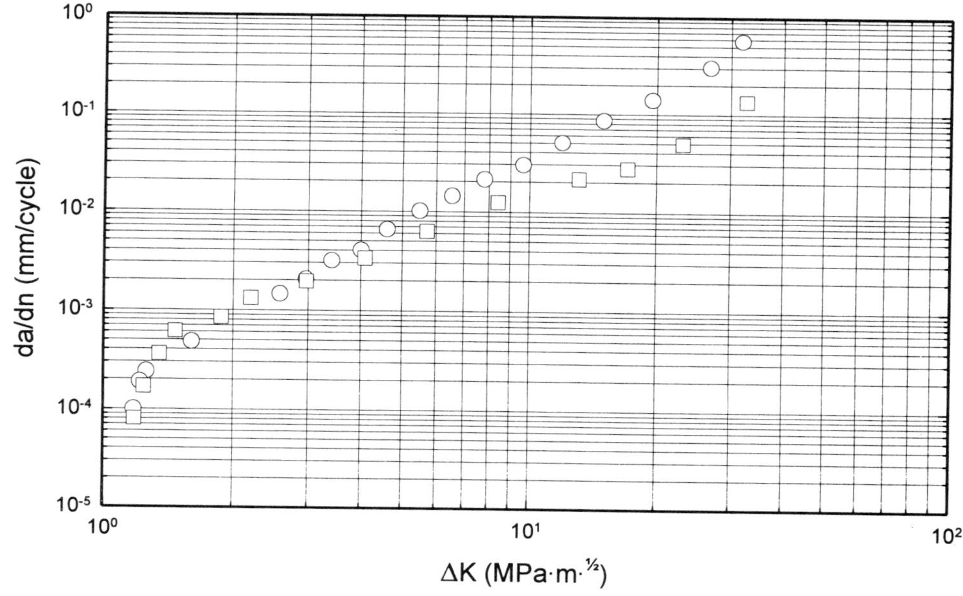

○	Dow PC/ABS (65% PC/ 35 % ABS); tension- tension; 1.0 Hz; stress ratio: 0.1; 23°C; single edge notched; parallel to flow
□	Dow PC/ABS (8% talc, 65% PC/ 35 % ABS); tension- tension; 1.0 Hz; stress ratio: 0.1; 23°C; single edge notched; parallel to flow
Reference No.	371

Chapter 46

Polycarbonate/ Polyethylene Terephthalate Alloy

Effect of Filler Content on Fatigue Behavior

PC/PET (features: impact modified); **PC/PET** (5% mineral filled; features: impact modified); **PC/PET** (9% mineral filled; features: impact modified)

Data clearly indicate that the fatigue lifetime decreases as the maximum remote stress increases. Furthermore, the fatigue resistance as measured in the perpendicular to flow direction is consistently lower in comparison to the results obtained in the parallel to flow direction. It should also be recognized that the general trend of the data indicates that the fatigue performance of the resin containing 5.5% mineral filler is consistently equal to or better than the unfilled system, while the resin containing 9.0% mineral filler has lower fatigue resistance in all cases. These observations suggest that an optimum level of filler may be attainable which allows the production of mineral filled systems with no degradation in fatigue performance.

Reference: Bank, D.H. (Dow), Seibel, S. (Case Western Reserve University), Moet, A. (Case Western Reserve Unive, *Fatigue and Fracture Toughness Evaluation of Mineral Filled Polycarbonate/ Polyethylene Terephthalate Blends,* ANTEC 1992, conference proceedings - Society of Plastics Engineers, 1992.

GRAPH 210: Fatigue Crack Propagation (FCP) Cycles vs. Crack Length Tested Parallel to Flow at 11.0 MPa Stress Amplitude for Polycarbonate/ Polyethylene Terephthalate Alloy.

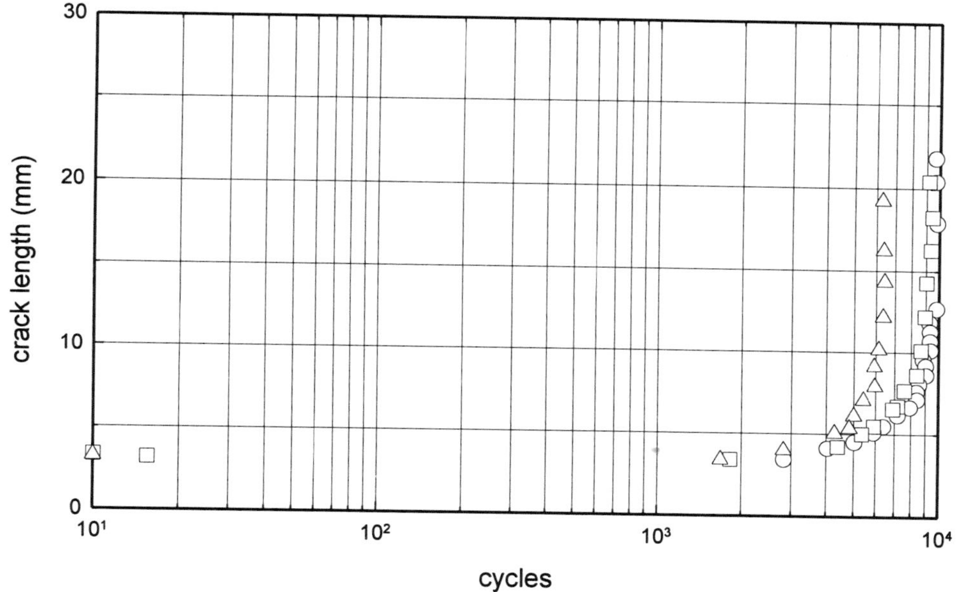

○	PC/PET (impact modified); tension- tension; 1 Hz; 5.5% mineral filled; stress ratio: 0.1; 23°C; single edge notched specimen; stress amplitude: 11.0 MPa [fig g]
□	PC/PET (impact modified); tension- tension; 1 Hz; stress ratio: 0.1; 23°C; single edge notched specimen; stress amplitude: 11.0 MPa [fig g]
△	PC/PET (impact modified; 9% mineral filled); tension-tension; 1 Hz; stress ratio: 0.1; 23°C; single edge notched specimen; stress amplitude: 11.0 MPa [fig g]
Reference No.	400

GRAPH 211: **Fatigue Crack Propagation (FCP) Cycles vs. Crack Length Tested Perpendicular to Flow at 11.0 MPa Stress Amplitude for Polycarbonate/ Polyethylene Terephthalate Alloy.**

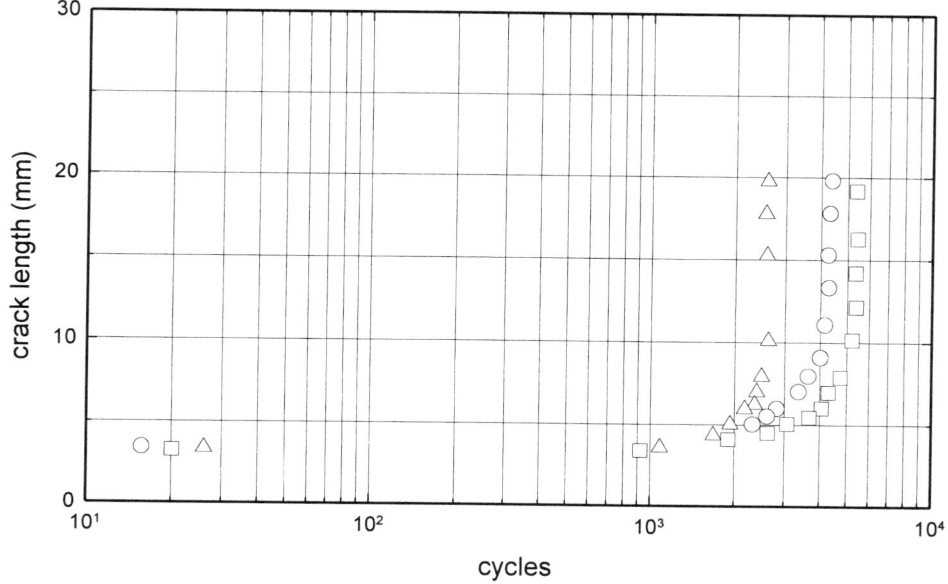

○	PC/PET (impact modified); tension- tension; 1 Hz; stress ratio: 0.1; 23°C; single edge notched specimen; stress amplitude: 11.0 MPa
□	PC/PET (impact modified, 5.5% mineral filled); tension- tension; 1 Hz; stress ratio: 0.1; 23°C; single edge notched specimen; stress amplitude: 11.0 MPa
△	PC/PET (impact modified, 9% mineral filled); tension-tension; 1 Hz; stress ratio: 0.1; 23°C; single edge notched specimen; stress amplitude: 11.0 MPa
Reference No.	400

GRAPH 212: **Fatigue Crack Propagation (FCP) Cycles vs. Crack Length Tested Perpendicular to Flow at 25.0 MPa Stress Amplitude for Polycarbonate/ Polyethylene Terephthalate Alloy.**

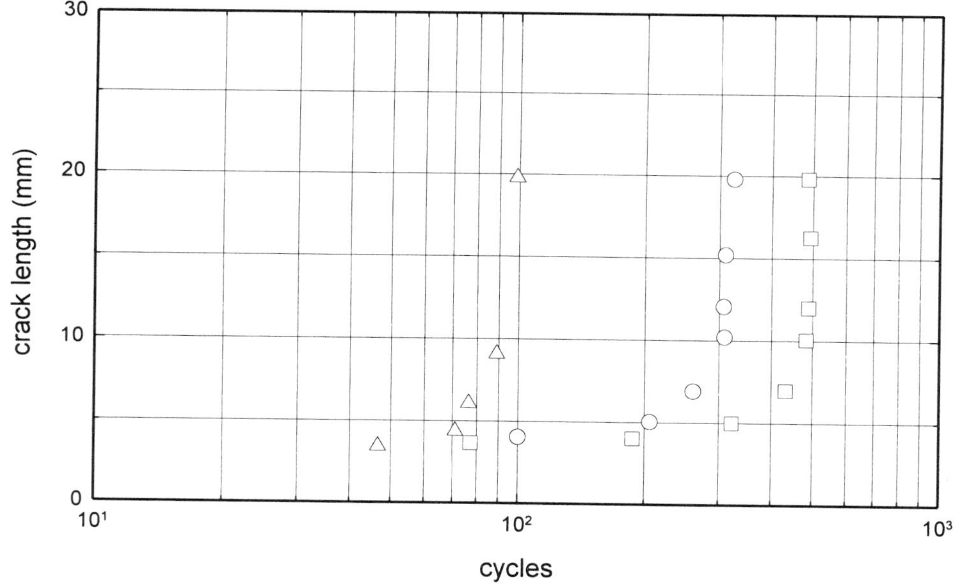

○	PC/PET (impact modified); tension- tension; 1 Hz; stress ratio: 0.1; 23°C; single edge notched specimen; stress amplitude: 25 MPa
□	PC/PET (impact modified, 5.5% mineral filled); tension- tension; 1 Hz; stress ratio: 0.1; 23°C; single edge notched specimen; stress amplitude: 25 MPa
△	PC/PET (impact modified, 9% mineral filled); tension-tension; 1 Hz; stress ratio: 0.1; 23°C; single edge notched specimen; stress amplitude: 25 MPa
Reference No.	400

GRAPH 213: **Fatigue Crack Propagation (FCP) Cycles vs. Crack Length Tested Parallel to Flow at 25 MPa Stress Amplitude for Polycarbonate/ Polyethylene Terephthalate Alloy.**

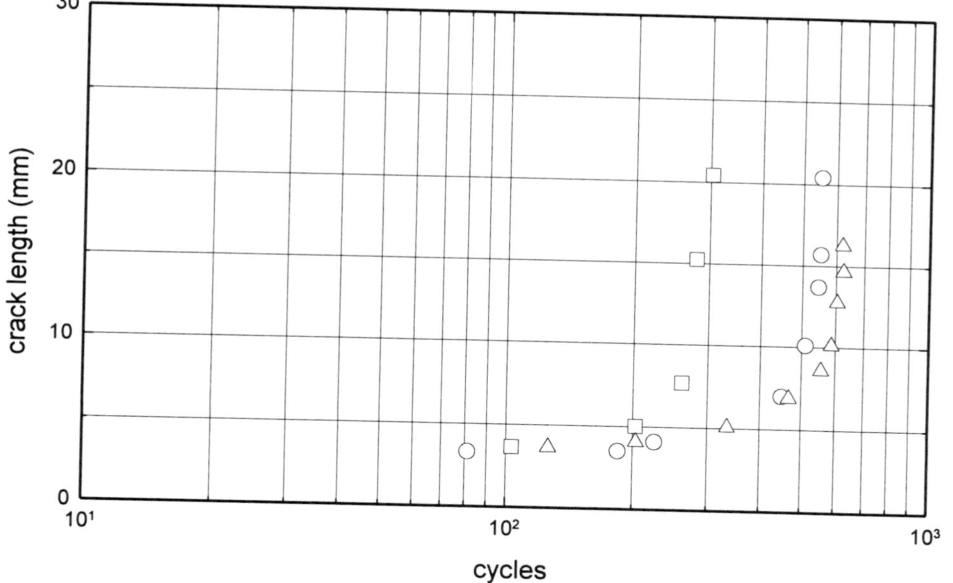

○	PC/PET (impact modified); tension- tension; 1 Hz; stress ratio: 0.1; 23°C; single edge notched specimen; stress amplitude: 25 MPa [fig g]
□	PC/PET (impact modified, 9% mineral filled); tension-tension; 1 Hz; stress ratio: 0.1; 23°C; single edge notched specimen; stress amplitude: 25 MPa
△	PC/PET (impact modified, 5.5% mineral filled); tension- tension; 1 Hz; stress ratio: 0.1; 23°C; single edge notched specimen; stress amplitude: 25 MPa
Reference No.	400

Chapter 47

Polycarbonate/ Polbutylene Terephthalate Alloy

GRAPH 214: Fatigue Cycles to Failure vs. Stress in Tension for General Electric Xenoy Polycarbonate/ Polybutylene Terephthalate Alloy.

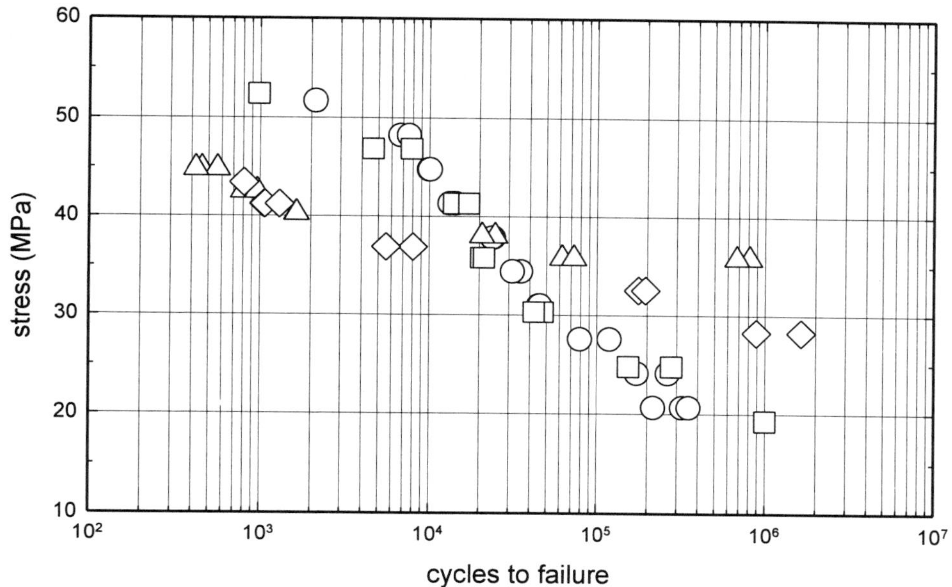

○	GE Xenoy 2230 PC/PBT (unfilled; inj. mold.); tension- tension; GE method; 5 Hz; stress ratio: 0.1; 23°C; flow direction [fig g]
□	GE Xenoy 5220 PC/PBT (unfilled; inj. mold.); tension- tension; GE method; 5 Hz; stress ratio: 0.1; 23°C; flow direction [fig g]
△	GE Xenoy 6120 PC/PBT (impact modified; inj. mold.); tension- tension; GE method; 5 Hz; stress ratio: 0.1; 23°C; flow direction [fig g]
◇	GE Xenoy 6620 PC/PBT (impact modified, unfilled; inj. mold.); tension-tension; GE method; 5 Hz; stress ratio: 0.1; 23°C; flow direction [fig g]
Reference No.	352

GRAPH 215: Fatigue Cycles to Failure vs. Stress in Tension for Glass Reinforced General Electric Xenoy Polycarbonate Polybutylene Terephthalate Alloy.

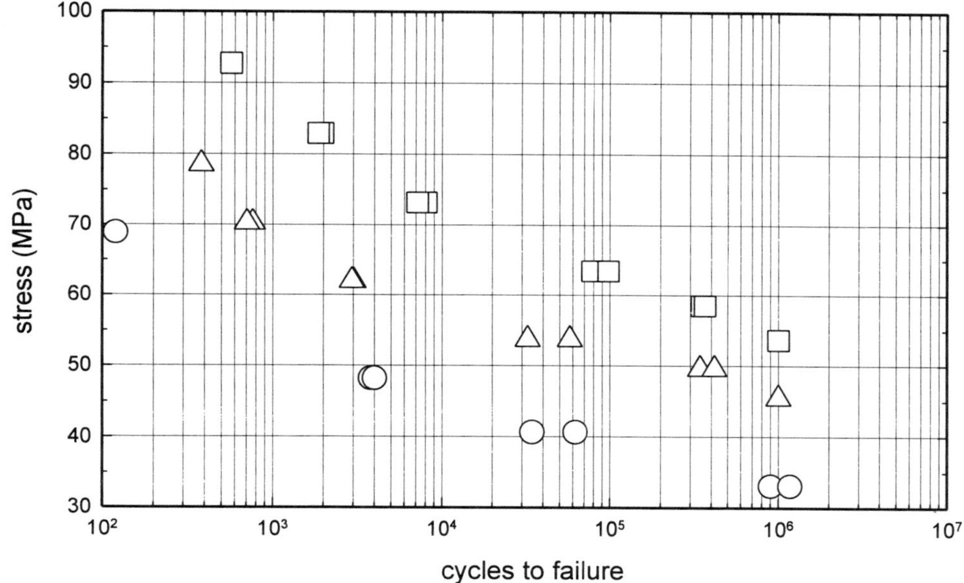

○	GE Xenoy 6240 PC/PBT (impact modified; 10% glass fiber; inj. mold.); tension- tension; GE method; 10 Hz; stress ratio: 0.1; 23°C; flow direction [fig g]
□	GE Xenoy 6370 PC/PBT (30% glass fiber; inj. mold.); tension- tension; GE method; 5 Hz; stress ratio: 0.1; 23°C; flow direction [fig g]
△	GE Xenoy 6380 PC/PBT (high gloss; 30% glass fiber; inj. mold.); tension-tension; GE method; 5 Hz; stress ratio: 0.1; 23°C; flow direction [fig g]
Reference No.	352

GRAPH 216: Fatigue Cycles to Failure vs. Stress in Tension for General Electric Xenoy Polycarbonate Polybutylene/ Terephthalate Alloy.

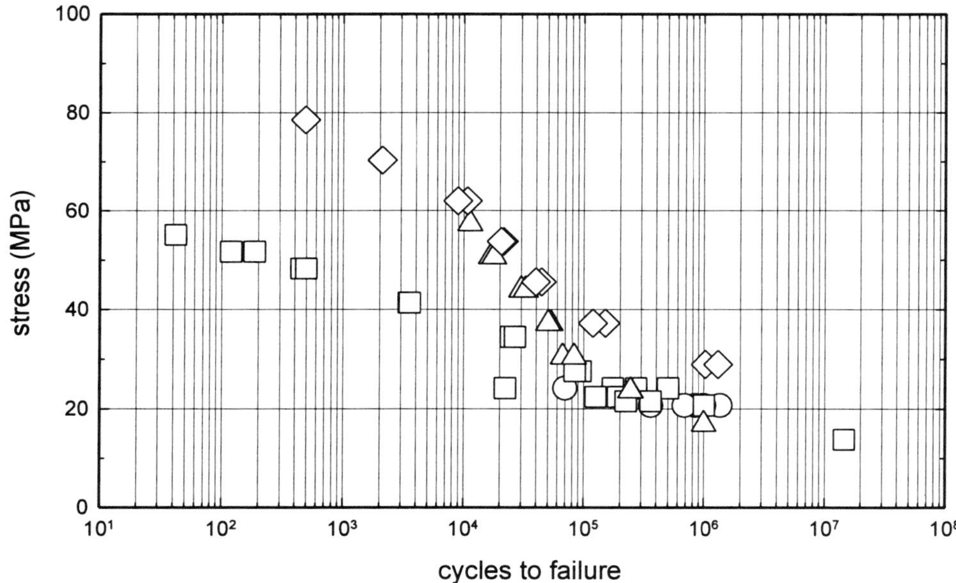

○	GE Xenoy 1102 PC/PBT (UV stabilized, unfilled; inj. mold.); tension- tension; GE method; 5 Hz; stress ratio: 0.1; 23°C; flow direction [fig g]
□	GE Xenoy 1102 PC/PBT (UV stabilized, unfilled; inj. mold.); tension- tension; GE method; 10 Hz; stress ratio: 0.1; 23°C; flow direction [fig g]
△	GE Xenoy 1731 PC/PBT (inj. mold.); tension-tension; GE method; 5 Hz; stress ratio: 0.1; 23°C; flow direction [fig g]
◇	GE Xenoy 1760 PC/PBT (unfilled; inj. mold.); tension- tension; GE method; 5 Hz; stress ratio: 0.1; 23°C; flow direction [fig g]
Reference No.	352

GRAPH 217: Fatigue Cycles to Failure vs. Stress in Tension for Glass Reinforced General Electric Valox Polycarbonate/ Polybutylene Terephthalate Alloy.

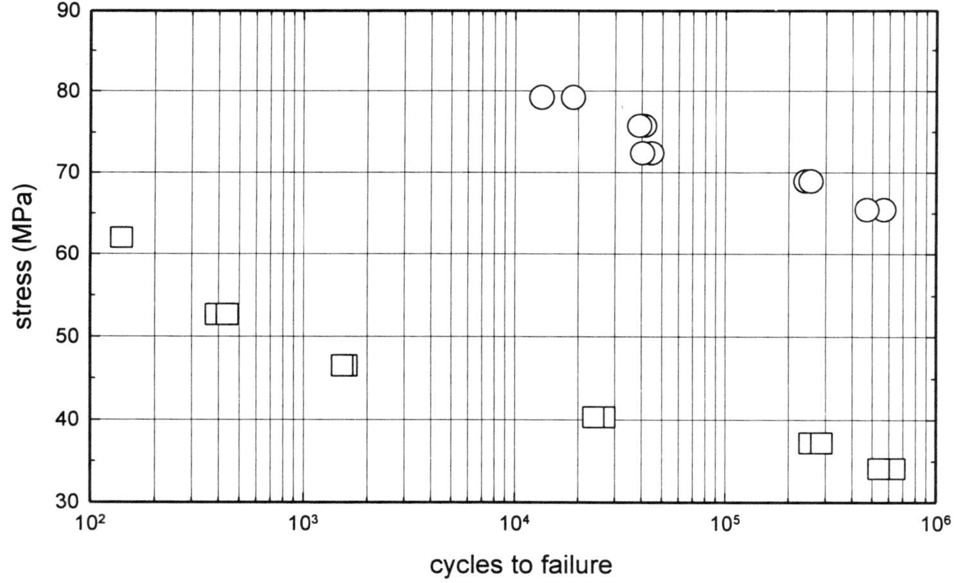

○	GE Valox 508 PC/PBT (30% glass fiber; inj. mold.); tension- tension; GE method; 5 Hz; stress ratio: 0.1; 23°C; flow direction [fig g]
□	GE Valox 508 PC/PBT (30% glass fiber; inj. mold.); tension- tension; GE method; 5 Hz; stress ratio: 0.1; 82°C; flow direction [fig g]
Reference No.	352

Chapter 48

Polyethylene Terephthalate/ Polbutylene Terephthalate Alloy

GRAPH 218: Fatigue Cycles to Failure vs. Stress in Tension for Mineral Filled General Electric "Heavy" Valox Polyethylene Terephthalate/ Polybutylene Terephthalate Alloy.

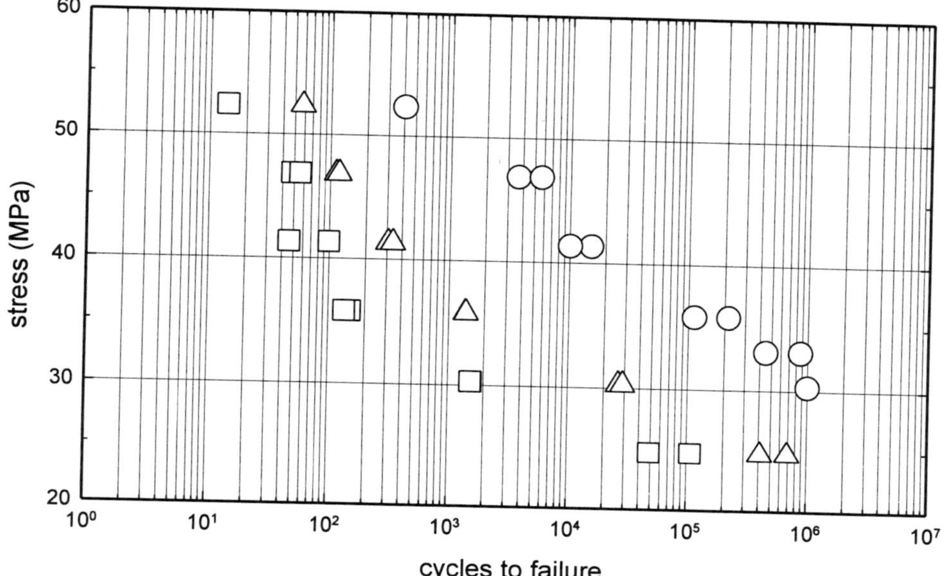

○	GE Valox HV7065 PET/PBT (ceramic, ivory replacement; high gloss; 63% mineral filler; inj. mold.); tension- tension; GE method; 5 Hz; stress ratio: 0.1; 23°C; flow direction [fig g]
□	GE Valox HV7075 PET/PBT (ceramic, ivory replacement; 68% mineral filler; inj. mold.); tension- tension; GE method; 5 Hz; stress ratio: 0.1; 23°C; flow direction [fig g]
△	GE Valox HV7085 PET/PBT (impact modified; 68% mineral filler; 3.9 g/10 min. MFI; inj. mold.); tension- tension; GE method; 5 Hz; stress ratio: 0.1; 23°C; flow direction [fig g]
Reference No.	352

Chapter 49

Polyetherimide/ Polycarbonate Alloy

GRAPH 219: **Fatigue Cycles to Failure vs. Stress in Tension for General Electric Ultem LTX100A Polyetherimide/ Polycarbonate Alloy.**

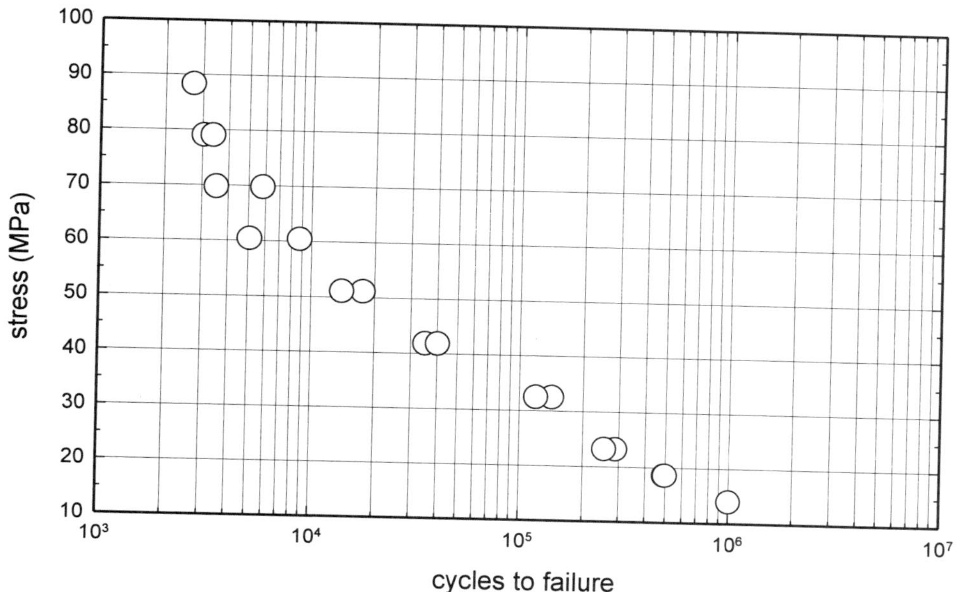

○	GE Ultem LTX100A PEI/PC (FDA grade, unfilled; inj. mold.); tension-tension; GE method; 5 Hz; stress ratio: 0.1; 23°C; flow direction [fig g]
Reference No.	352

Chapter 50

Polystyrene Modified Polyphenylene Ether/ Nylon Alloy

Fatigue Crack Propagation

GE Plastics: Noryl GTX900 (note: Nylon 66 based alloy)

It has become evident that alloys or blends that exhibit increased toughness may not necessarily show improved fatigue resistance. An example of this behavior was observed while examining the propagation of fatigue cracks in Noryl GTX, a nylon 66 based alloy. This material is a complex mixture of a rubber-toughened immiscible amorphous poly(phenylene oxide) phase that is dispersed in a crystalline nylon 66 matrix. A polymeric compatibilizer is also added to the blend. Not only is it well established that this material has improved impact strength compared to nylon 66, but the toughening mechanism has also been extensively described in the literature and used as a model system. The Noryl GTX exhibits significantly increased crack growth rates at all ΔK levels, indicating that it is inferior to the unmodified nylon (Zytel 211) in spite of its toughness under impact-loading conditions.

Reference: Wyzgoski, M.G. (General Motors), Novak, G.E. (General Motors), *Fatigue-Resistant Nylon Alloys,* Journal of Applied Polymer Science (1994), technical journal (Vol. 51; CCC 0021-8995/94/050873-13) - John Wiley & Sons, Inc., 1994.

GRAPH 220: Fatigue Cycles to Failure vs. Stress in Tension for General Electric Noryl GTX Modified Polyphenylene Ether/ Nylon Alloy.

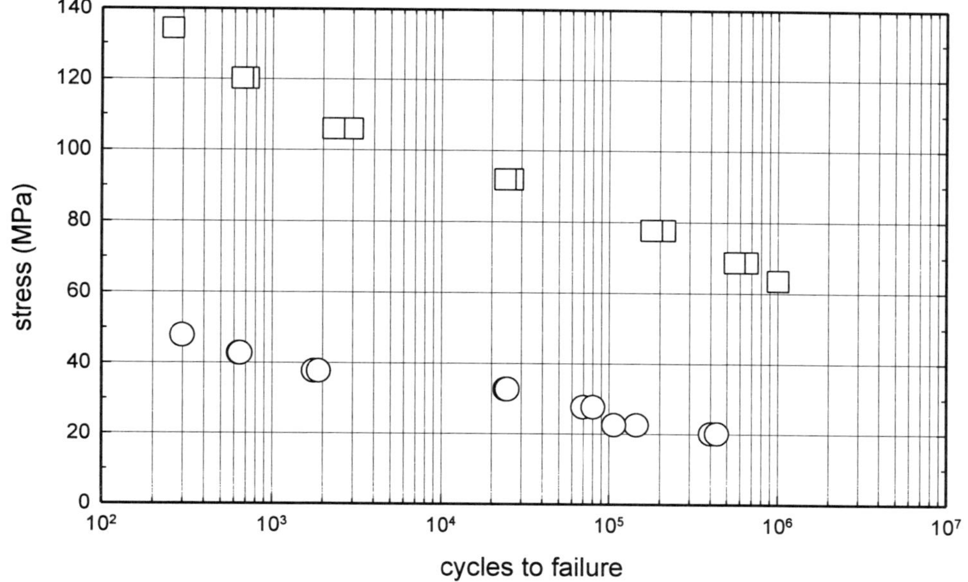

○	GE Noryl GTX625 Modified PPE/PA (extrus., blow molding); tension- tension; GE method; 5 Hz; stress ratio: 0.1; 23°C; flow direction [fig g]
□	GE Noryl GTX830 Modified PPE/PA (under the bonnet; 30% glass fiber; inj. mold.); tension- tension; GE method; 5 Hz; stress ratio: 0.1; 23°C; flow direction [fig g]
Reference No.	352

GRAPH 221: **Fatigue Cycles to Failure vs. Stress in Tension for General Electric Noryl GTX 900 Series Modified Polyphenylene Ether Nylon Alloy.**

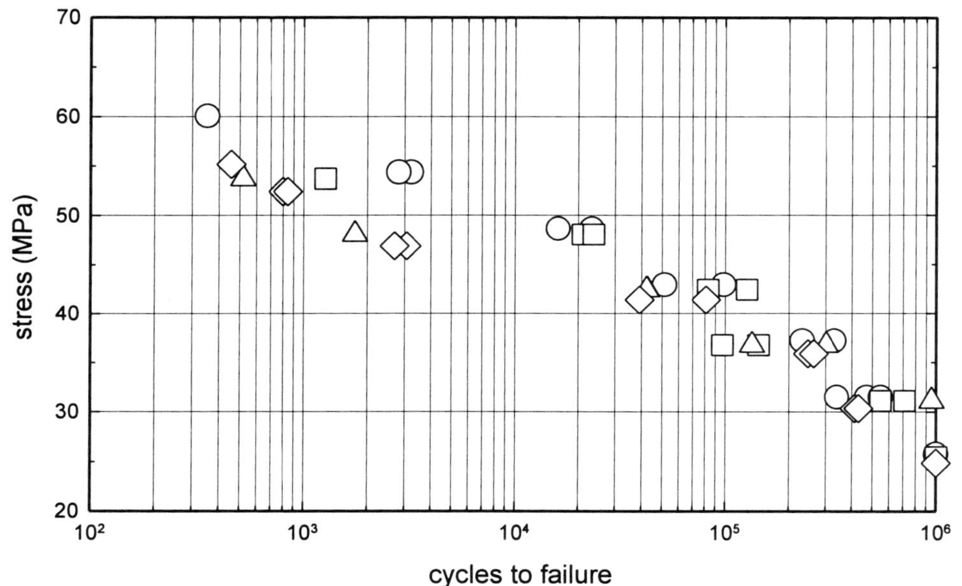

○	GE Noryl GTX902 Modified PPE/PA (unfilled, paintable; inj. mold.); tension- tension; GE method; 5 Hz; stress ratio: 0.1; 23°C; flow direction [fig g]
□	GE Noryl GTX904 Modified PPE/PA (unfilled; inj. mold.); tension- tension; GE method; 5 Hz; stress ratio: 0.1; 23°C; flow direction [fig g]
△	GE Noryl GTX909 Modified PPE/PA (unfilled; inj. mold.); tension- tension; GE method; 5 Hz; stress ratio: 0.1; 23°C; flow direction [fig g]
◇	GE Noryl GTX910 Modified PPE/PA (automotive on-line painted components; appearance grade, unfilled; inj. mold.); tension-tension; GE method; 5 Hz; stress ratio: 0.1; 23°C; flow direction [fig g]
Reference No.	352

GRAPH 222: **Fatigue Crack Propagation (FCP) Rate vs. Stress Intensity Range for General Electric Noryl GTX900 Modified Polyphenylene Ether Nylon Alloy.**

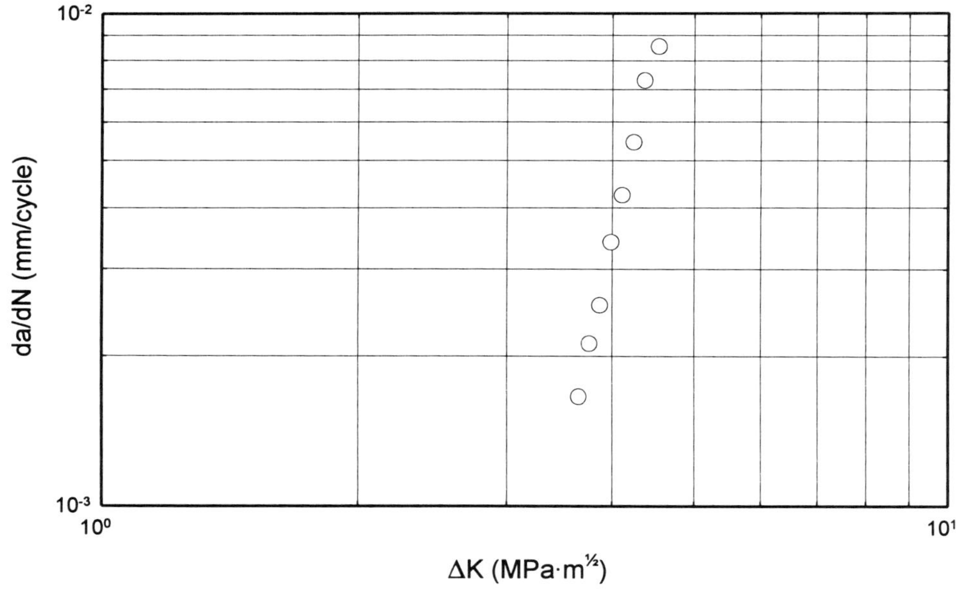

○	GE Noryl GTX900 Modified PPE/PA (Nylon 66 based); tension; 5 Hz; 23°C; compact tension specimen, precracked, parallel to flow
Reference No.	401

Chapter 51

Nylon 66/ Amorphous Nylon Alloy

Fatigue Crack Propagation

Nylon Alloy (note: 90% Zytel 122L/ 6% Trogamid T/ 4% Uniroyal X465 (EPDM rubber))

Rubber alone is effective in reducing fatigue crack growth rates; however, it is not as effective as the alloys also containing amorphous nylon. These results suggest that the miscible amorphous nylon does play a key role in imparting improved fatigue resistance to the crystalline nylon 66. Though the role of the rigid glassy amorphous nylon polymer is not clear, a possible explanation is that it acts to more uniformly disperse the rubber phase at relatively low concentrations.

Reference: Wyzgoski, M.G. (General Motors) vak, G.E. (General Motors), *Fatigue-Resistant Nylon Alloys,* Journal of Applied Polymer Science (1994), technical journal (Vol. 51; CCC 0021-8995/94/050873-13) - John Wiley & Sons, Inc., 1994.

Nylon Alloy (note: 90% Zytel 122L/ 10% Bexloy APC-803); **Nylon Alloy** (note: 90% Zytel 122L/ 10% Trogamid T)

There is a marginal improvement in fatigue crack propagation (lower crack growth rates at a given ΔK level) for the alloy containing an amorphous nylon (Trogamid T); however, the blend containing the toughened amorphous nylon (Bexloy AP C-803) shows a much more significant decrease in fatigue crack growth rates. Even the addition of only 5% of this material resulted in a significant improvement. A blend containing 15% did not exhibit a much greater effect.

Reference: Wyzgoski, M.G. (General Motors), Novak, G.E. (General Motors), *Fatigue-Resistant Nylon Alloys,* Journal of Applied Polymer Science (1994), technical journal (Vol. 51; CCC 0021-8995/94/050873-13) - John Wiley & Sons, Inc., 1994.

GRAPH 223: Fatigue Crack Propagation (FCP) Rate vs. Stress Intensity Range for Rubber Modified Nylon 66/ Amorphous Nylon Alloy.

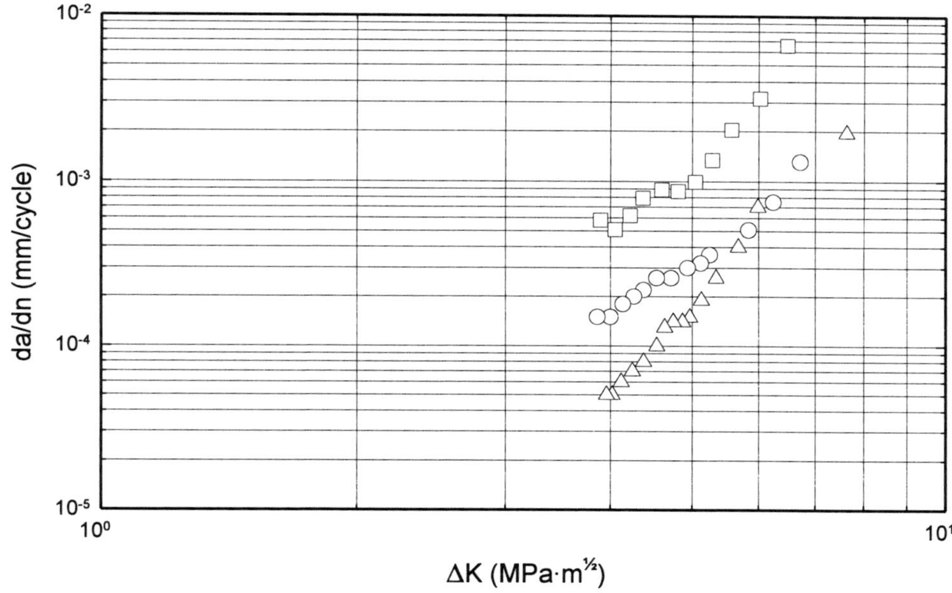

○	Nylon Alloy (90% Zytel 122L/ 8% Trogamid T/ 2% EPDM rubber); tension; 0.5 Hz; 23°C; compact tension specimen, precracked, parallel to flow
□	Nylon Alloy (90% Zytel 122L/ 6% Trogamid T/ 4% butyl acrylate rubber); tension; 0.5 Hz; 23°C; compact tension specimen, precracked, parallel to flow
△	Nylon Alloy (90% Zytel 122L/ 6% Trogamid T/ 4% EPDM); tension; 0.5 Hz; 23°C; compact tension specimen, precracked, parallel to flow
Reference No.	401

GRAPH 224: **Fatigue Crack Propagation (FCP) Rate vs. Stress Intensity Range for Nylon 66/ Amorphous Nylon Alloy.**

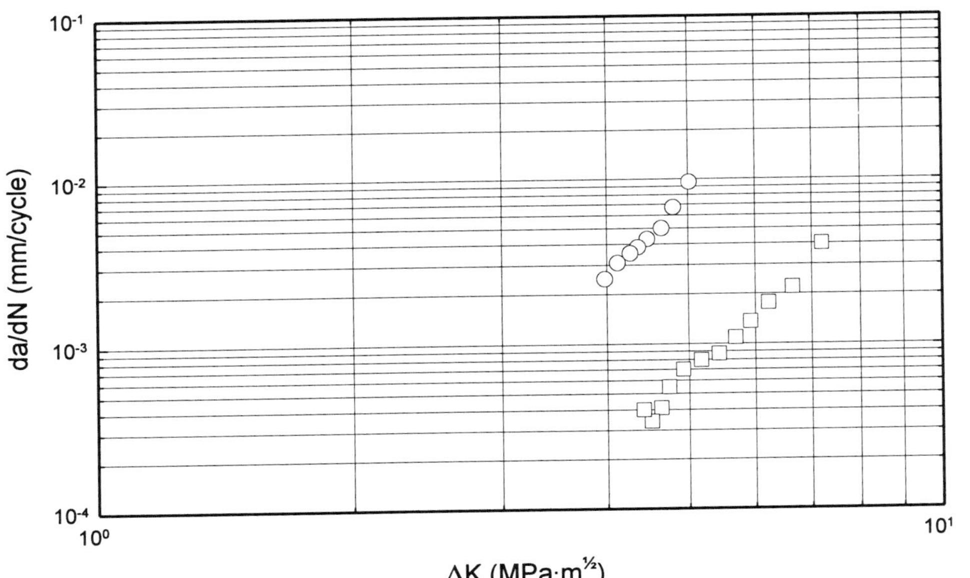

○	Nylon Alloy (90% Zytel 122L/ 10% Trogamid T); tension; 0.5 Hz; 23°C; compact tension specimen, precracked, parallel to flow
□	Nylon Alloy (90% Zytel 122L/ 10% Bexloy APC-803); tension; 0.5 Hz; 23°C; compact tension specimen, precracked, parallel to flow
Reference No.	401

Chapter 52

Polystyrene/ Polycarbonate Alloy

Fatigue Properties

PS/PC (compatibilizer: 5 wt.% PC-PS graft copolymer; material compostion: 50% glass fiber reinforcement; note: 90% PC/ 10% PS); **PS/PC** (note: 80% PC/ 20% PS); **PS/PC** (compatibilizer: 5 wt.% PC- PS graft copolymer; note: 80% PC/ 20% PS); **PS/PC** (note: 70% PC/ 30 % PS); **PS/PC** (compatibilizer: 5 wt.% PC-PS graft copolymer; note: 70% PC/ 30 % PS); **PS/PC** (note: 90% PC/ 10 % PS)

Lifetime inversion of the S-N curve was observed for unnotched pure PC specimens. By blending in 10 wt% PS, lifetime inversion disappeared. This is because the shear dominated area has disappeared. At 10 and 20 wt% PS compositions, fatigue lifetimes of systems with compatibilizer are longer than those of systems without compatibilizer, regardless of the stress amplitude. At 30 wt% PS composition and stress amplitudes higher than 35 MPa, differences between the system with compatibilizer and the system without compatibilizer disappear, but the system with compatibilizer has longer fatigue lifetimes than the systems without compatibilizer from 10 to 30 wt% PS compositions. The systems with compatibilizer have almost the same fatigue lifetimes as that of PC at 25 MPa. This is an interesting result, because a rubber toughened polymer usually exhibits shorter fatigue lifetime than a pure polymer. However, the difference between the system with compatibilizer and the system without compatibilizer at 30 wt% PS decreases. This difference might disappear all together at 40 wt% PS composition.

Reference: Mimura, K. (The University of Michigan), Hristov, H. (The University of Michigan), Yee, A.F. (The Un, *Effect of Compatibilizer on Fatigue Properties of PC/PS Blend,* ANTEC 1994, conference proceedings - Society of Plastics Engineers, 1994.

GRAPH 225: **Fatigue Cycles to Failure vs. Stress in Tension for 30% Polystyrene content Polystyrene/ Polycarbonate Alloy with and without Compatibilizer.**

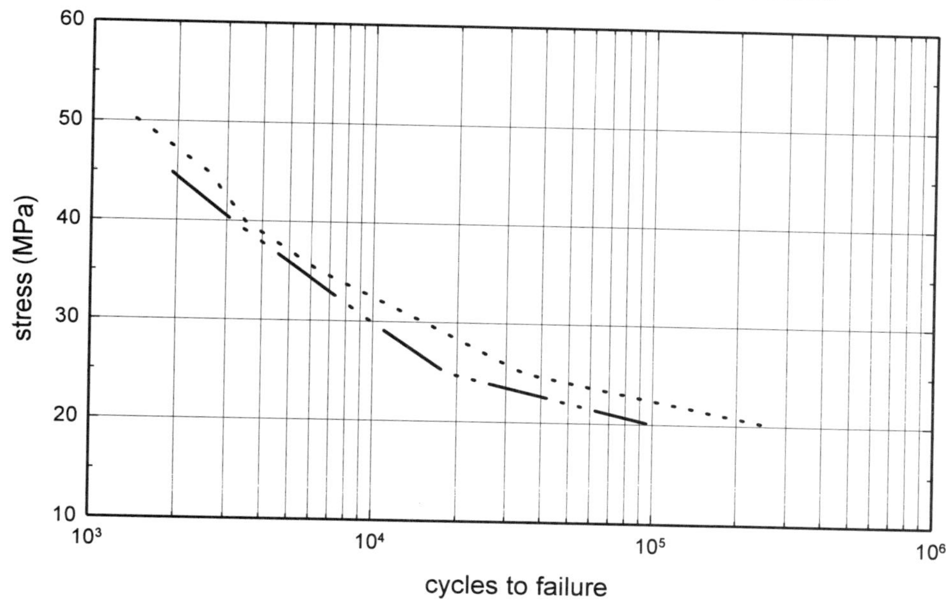

....	PS/PC (5% PC-PS graft copolymer; 70% PC/ 30 % PS); tension- tension; 5 Hz; stress ratio: 0.05; 35°C [fig g]
—··—··	PS/PC (70% PC/ 30 % PS); tension- tension; 5 Hz; stress ratio: 0.05; 35°C [fig g]
Reference No.	388

GRAPH 226: **Fatigue Cycles to Failure vs. Stress in Tension for 10% Polystyrene content Polystyrene/ Polycarbonate Alloy with and without Compatibilizer.**

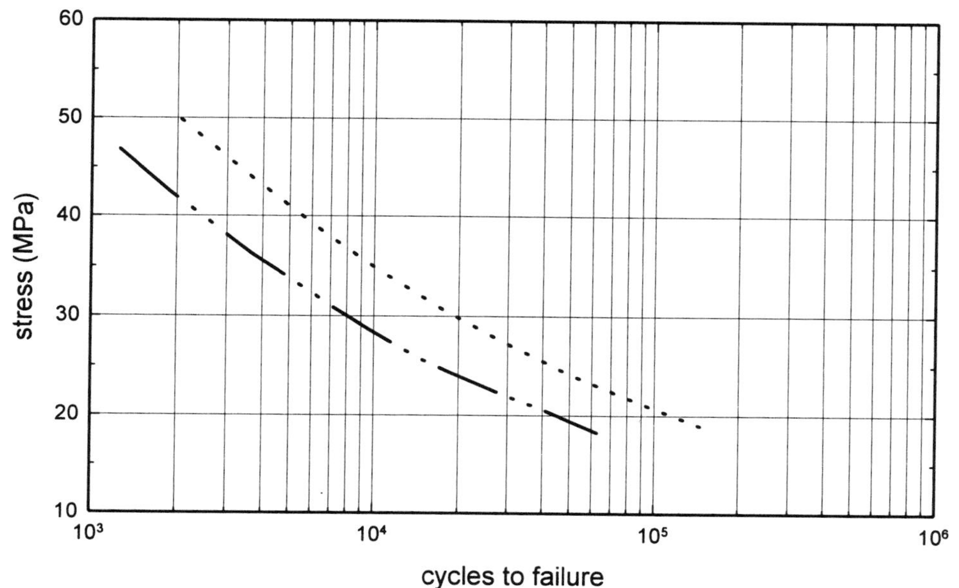

···············	PS/PC (5% PC-PS graft copolymer; 90% PC/ 10% PS); tension- tension; 5 Hz; stress ratio: 0.05; 35°C [fig g]
—··—··	PS/PC (90% PC/ 10% PS); tension- tension; 5 Hz; stress ratio: 0.05; 35°C [fig g]
Reference No.	388

GRAPH 227: **Fatigue Cycles to Failure vs. Stress in Tension for 20% Polystyrene content Polystyrene/ Polycarbonate Alloy with and without Compatibilizer.**

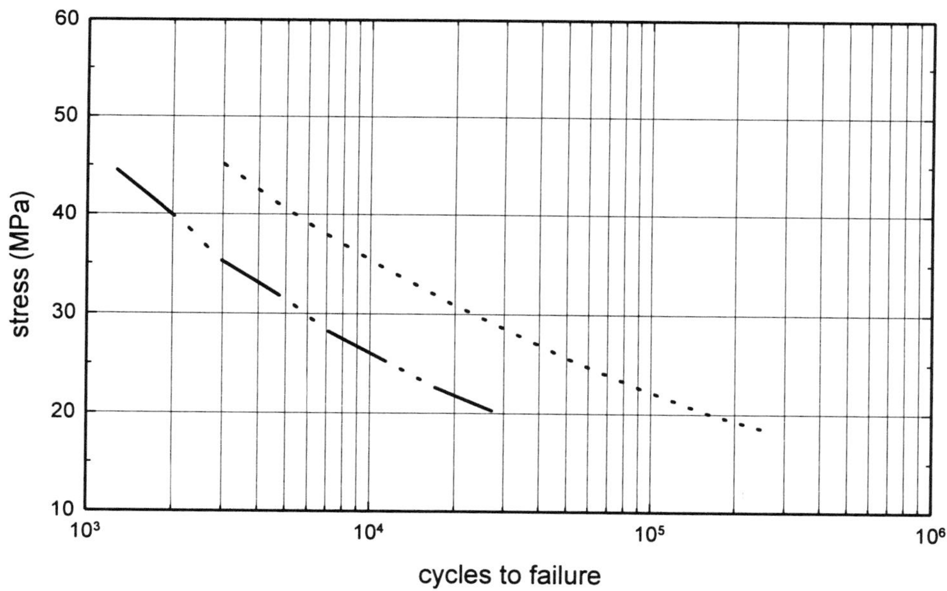

···············	PS/PC (5% PC-PS graft copolymer; 80% PC/ 20% PS); tension- tension; 5 Hz; stress ratio: 0.05; 35°C [fig g]
—··—··	PS/PC (80% PC/ 20% PS); tension- tension; 5 Hz; stress ratio: 0.05; 35°C [fig g]
Reference No.	388

Chapter 53

Epoxy Resin

Fatigue Crack Propagation

Epoxy (features: epoxy equivalent weight of 187 g/mole; modifier: 10% CTBN rubber (1.5 micron diameter); note: DGEBA resin; cure: piperidine); **Epoxy** (features: epoxy equivalent weight of 187 g/mole; modifier: 10% MBS rubber (0.2 micron diameter); note: DGEBA resin; cure: piperidine); **Epoxy** (features: epoxy equivalent weight of 187 g/mole; note: DGEBA resin; cure: piperidine)

McGarry and co-workers were among the first to show that the fracture resistance of epoxy polymers could be enhanced by incorporating rubber particles. Crack-tip plastic zone-rubber particle interactions affect the FCP behavior of rubber-modified epoxy polymers. Consequently, rubber cavitation/shear banding and plastic void growth mechanisms become active when the size of the plastic zone becomes large enough compared to the size of the rubber particles. As a result of these interactions, the use of smaller size, commercially available, 0.2 μm structured core-shell MBS rubber particles in place of 1.5 μm CTBN particles results in more than one order of magnitude improvement in FCP resistance of the rubber-modified system. In addition, results indicate that the volume fraction of the rubber particles affects the slope of the Paris regime and has no effect on the threshold behavior. On the other hand, it is the size of the rubber particle that affects the near-threshold behavior.

Reference: Azimi, H.R. (Lehigh University), Pearson, R.A. (Lehigh University), Hertzberg, R.W. (Lehigh Universi, *A Mechanistic Understanding of Fatigue Crack Propagation Behavior of Rubber-Modified Epoxy Polymers,* ANTEC 1995, conference proceedings - Society of Plastics Engineers, 1995.

GRAPH 228: Fatigue Cycles to Failure vs. Stress in Flexure for Short Glass Fiber Filled Epoxy Resin.

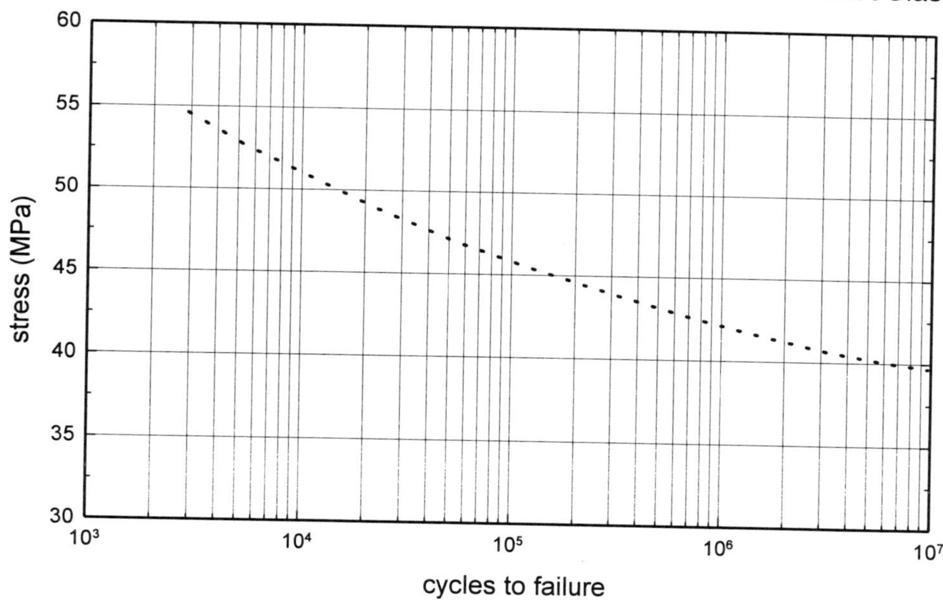

....	Epoxy (low elev. temp. outgassing; short glass reinforced); flexure; 30 Hz; 23°C
Reference No.	417

GRAPH 229: **Fatigue Cycles to Failure vs. Stress in Flexure for Epoxy Resin.**

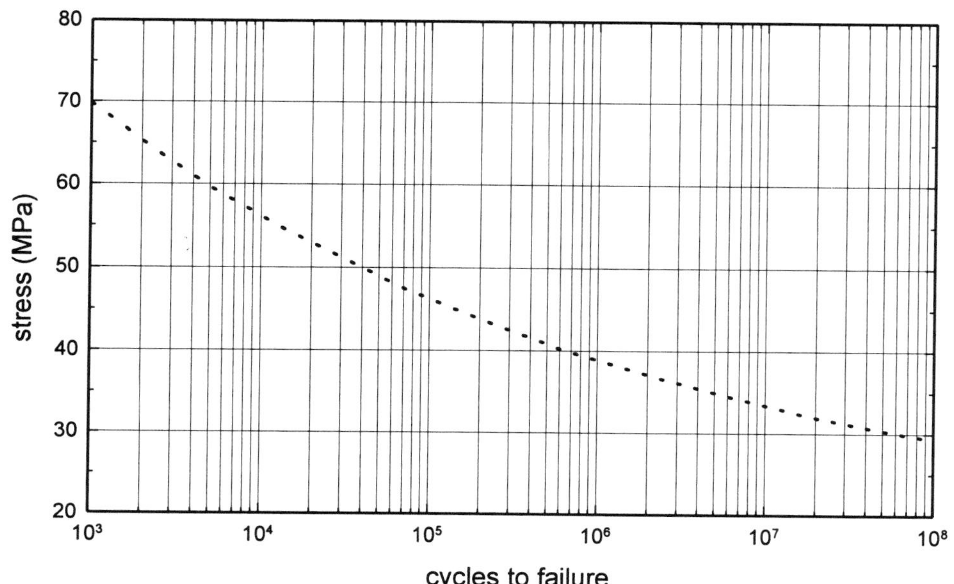

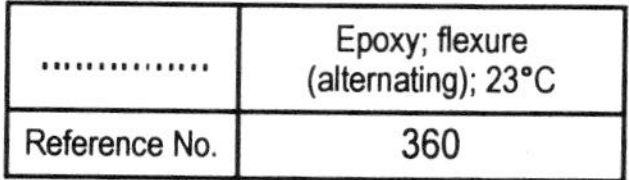

GRAPH 230: **Fatigue Crack Propagation (FCP) Rate vs. Stress Intensity Range for Unmodified and Rubber Modified Epoxy Resin.**

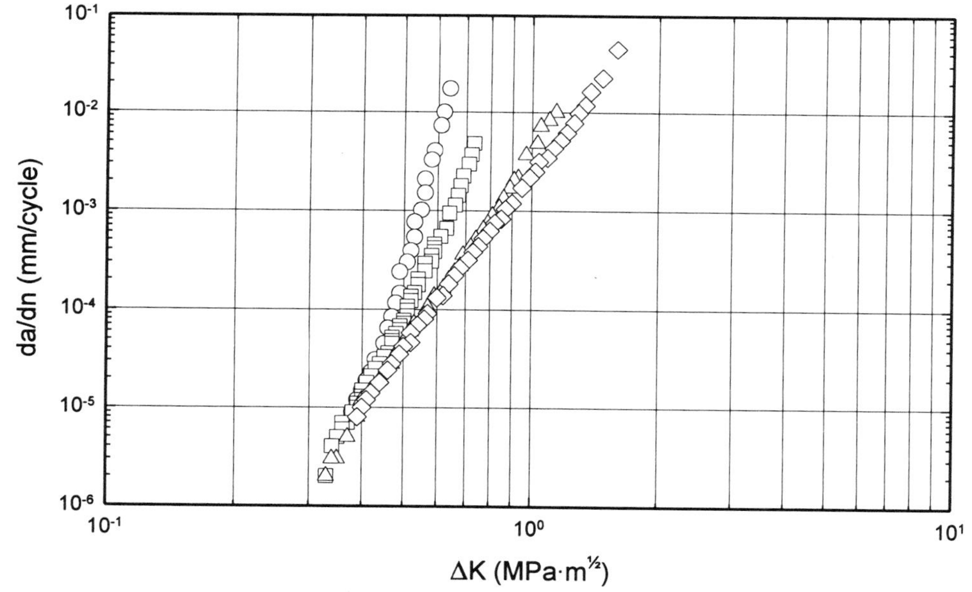

○	Epoxy (piperidine cure; epoxy equiv. wgt. - 187 g/mole; DGEBA resin)
□	Epoxy (piperidine cure; epoxy equiv. wgt. - 187 g/mole; 1% CTBN rubber modified; DGEBA resin)
△	Epoxy (piperidine cure; epoxy equiv. wgt. - 187 g/mole; 5% CTBN rubber modified; DGEBA resin)
◇	Epoxy (piperidine cure; epoxy equiv. wgt. - 187 g/mole; 10% MBS rubber modified; DGEBA resin)
Reference No.	390

GRAPH 231: **Fatigue Crack Propagation (FCP) Rate vs. Stress Intensity Range for Unmodified and Rubber Modified Epoxy Resin.**

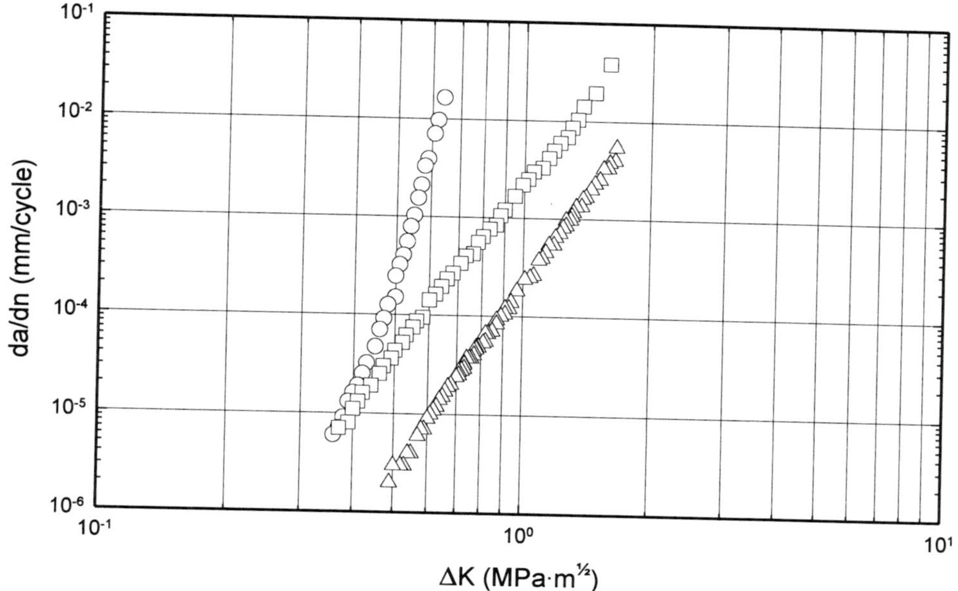

○	Epoxy (piperidine cure; epoxy equiv. wgt. - 187 g/mole; DGEBA resin)
□	Epoxy (piperidine cure; epoxy equiv. wgt. - 187 g/mole; 10% CTBN rubber modified; DGEBA resin)
△	Epoxy (piperidine cure; epoxy equiv. wgt. - 187 g/mole; 10% MBS rubber modified; DGEBA resin)
Reference No.	390

GRAPH 232: **Fatigue Crack Propagation (FCP) Rate vs. Stress Intensity Range for Unmodified and Rubber Modified Epoxy Resin.**

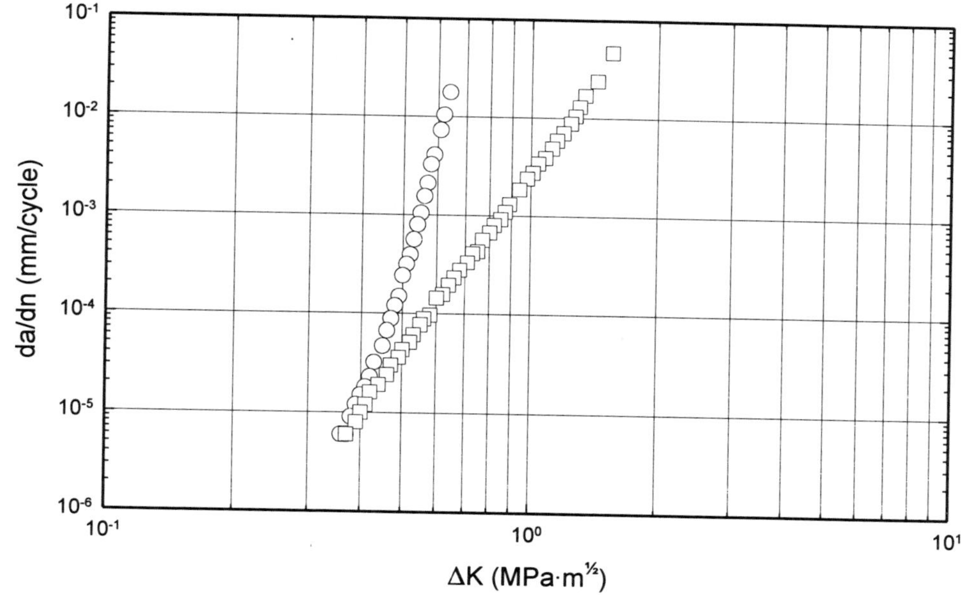

○	Epoxy (piperidine cure; epoxy equiv. wgt. - 187 g/mole; DGEBA resin)
□	Epoxy (piperidine cure; epoxy equiv. wgt. - 187 g/mole; 10% CTBN rubber modified; DGEBA resin)
Reference No.	390

Chapter 54

Phenol-Formaldehyde Copolymer

GRAPH 233: Fatigue Cycles to Failure vs. Stress in Flexure for Short Glass Fiber Filled Phenolic Resin.

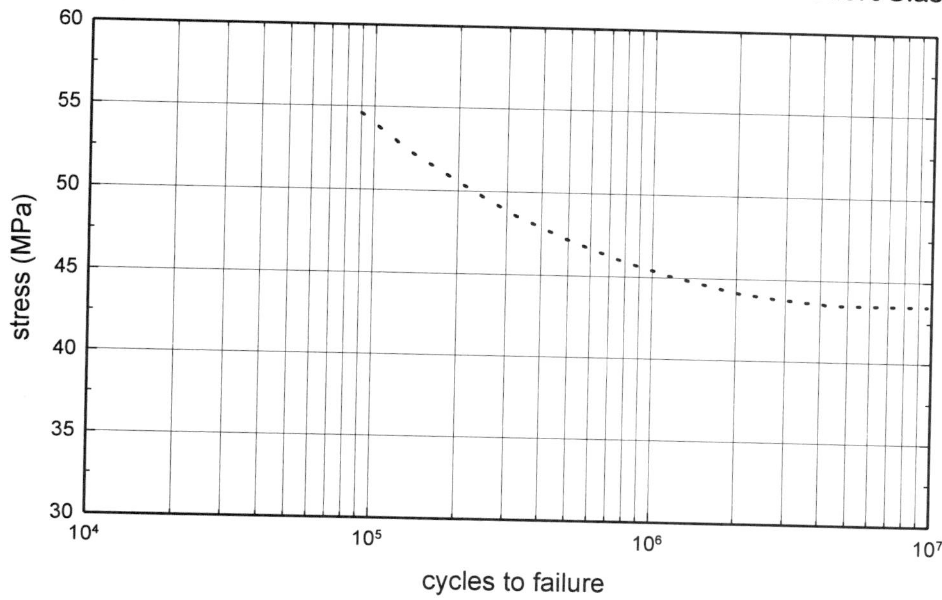

....	Phenolic (short glass reinforced); flexure; 30 Hz; 23°C
Reference No.	417

GRAPH 234: Fatigue Cycles to Failure vs. Stress at Different Temperatures for 50% Glass Fiber Reinforced Rogers RX 630 Resorcinol Modified Phenolic.

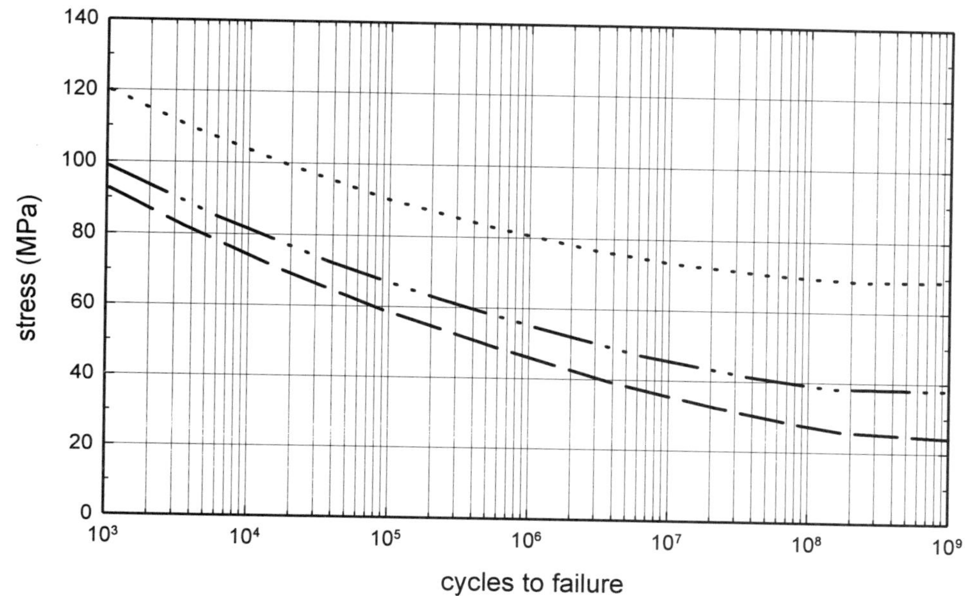

....	Rogers Rogers RX630 Phenolic (matte finish; 50% glass fiber); flexure (alternating); ASTM D671; 23°C
—··—··	Rogers Rogers RX630 Phenolic (matte finish; 50% glass fiber); flexure (alternating); ASTM D671; 100°C
– – – –	Rogers Rogers RX630 Phenolic (matte finish; 50% glass fiber); flexure (alternating); ASTM D671; 150°C
Reference No.	404

Chapter 55

Thermoset Polyester

GRAPH 235: **Fatigue Crack Propagation (FCP) Rate vs. Stress Intensity Range at 100°C for Thermoset Polyester.**

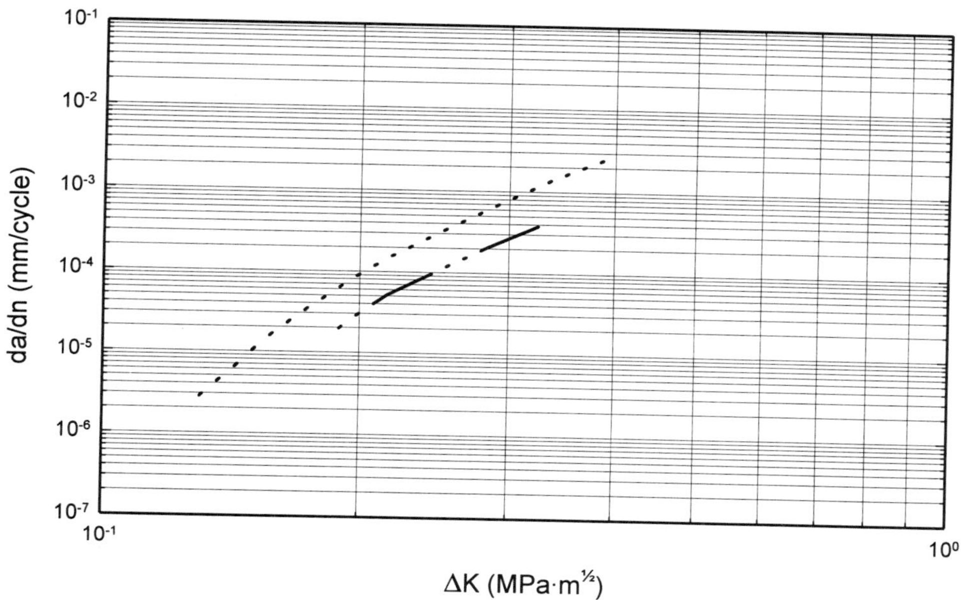

....	Reichhold Atlac 711-05A Brominated Bisphenol A Fumarate Polyester; tension- tension; 20 Hz; stress ratio: 0.6; 100°C [fig s]
—··—··	Ashland Hetron 197 Chlorendic Anhydride Polyester; tension- tension; 20 Hz; stress ratio: 0.6; 100°C [fig s]
Reference No.	397

GRAPH 236: **Fatigue Crack Propagation (FCP) Rate vs. Stress Intensity Range at Room Temperature for Thermoset Polyester.**

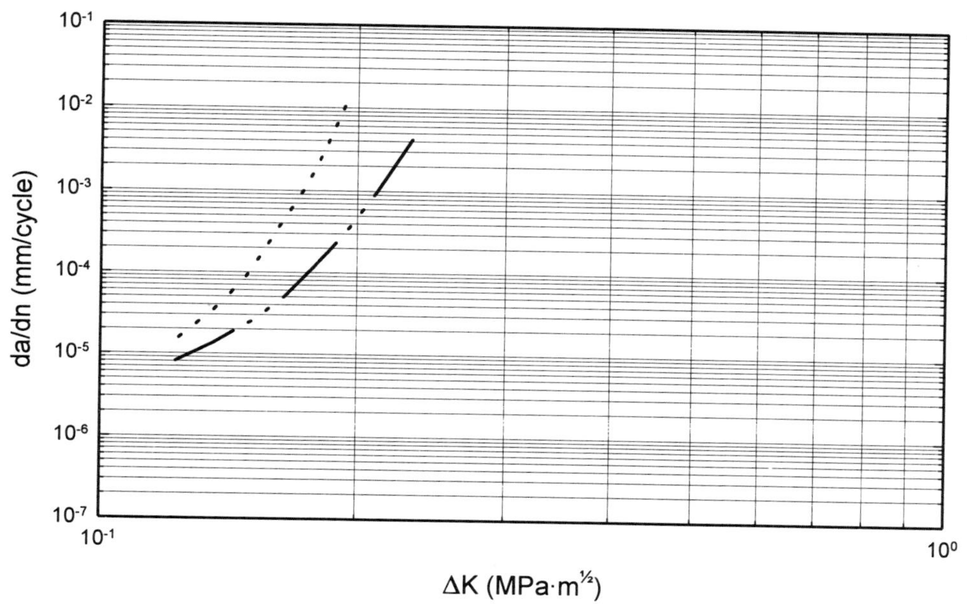

....	Reichhold Atlac 711-05A Brominated Bisphenol A Fumarate Polyester; tension- tension; 20 Hz; stress ratio: 0.6; 23°C [fig s]
—··—··	Ashland Hetron 197 Chlorendic Anhydride Polyester; tension- tension; 20 Hz; stress ratio: 0.6; 23°C [fig s]
Reference No.	397

Chapter 56

Polyimide

GRAPH 237: **Fatigue Cycles to Failure vs. Stress in Flexure for Kinel 5504 Polyimide.**

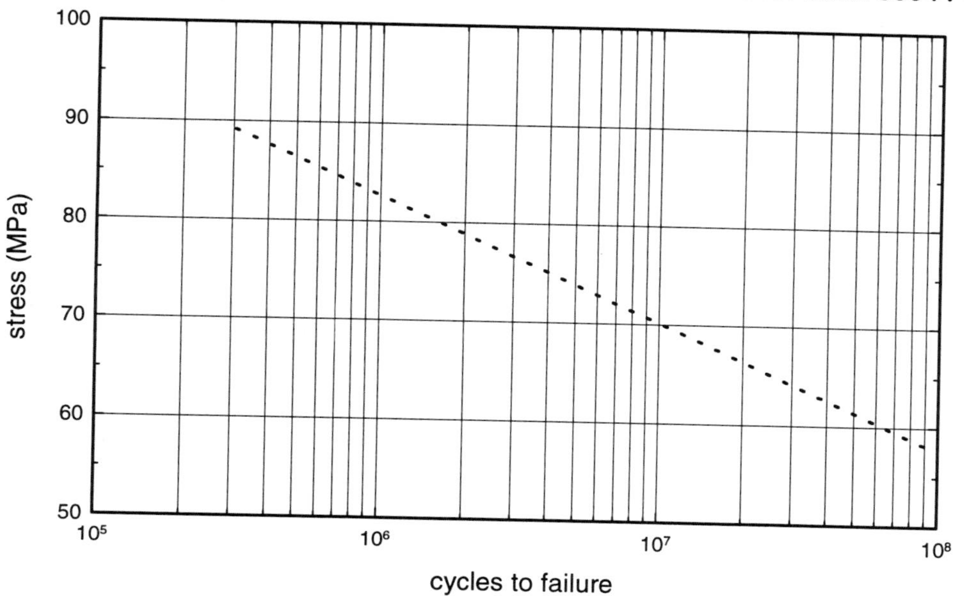

....	Rhone Pou. Kinel 5504 Polyimide (120 Rockwell M; 65% glass fiber; compression; , 6 mm long glass fibers); flexure (alternating); 23°C
Reference No.	360

GRAPH 238: **Fatigue Stress vs. Temperature at Various Cycles to Failure for Machined DuPont Vespel SP-1 Polyimide.**

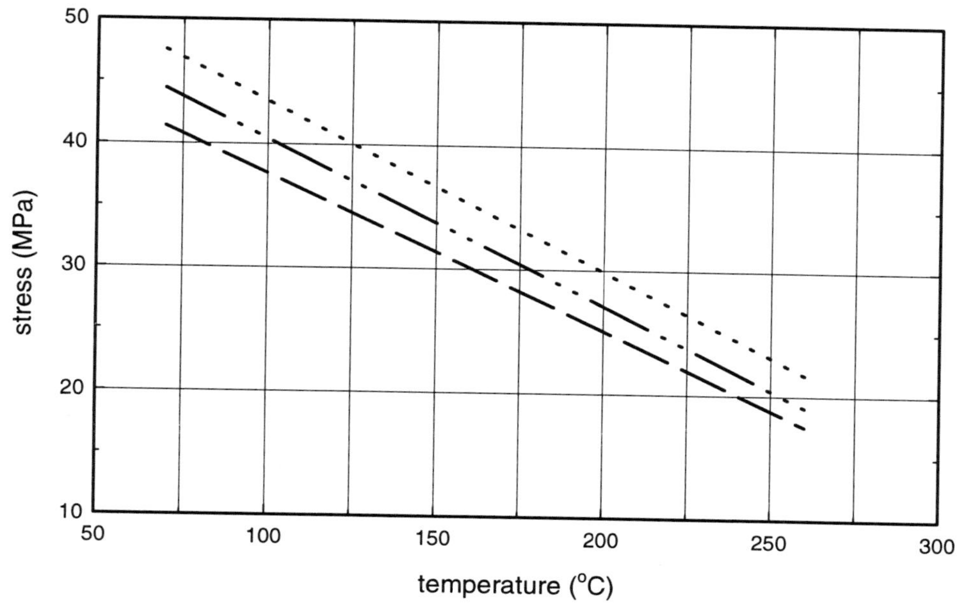

....	DuPont Vespel SP-1 Polyimide (45 - 60 Rockwell E, unfilled; machined spec.); tension-compression; 30 Hz; 10^5 cycles to failure [fig d]
—·· —··	DuPont Vespel SP-1 Polyimide (45 - 60 Rockwell E, unfilled; machined spec.); tension-compression; 30 Hz; 10^6 cycles to failure [fig d]
— — — —	DuPont Vespel SP-1 Polyimide (45 - 60 Rockwell E, unfilled; machined spec.); tension-compression; 30 Hz; 10^7 cycles to failure [fig d]
Reference No.	37

GRAPH 239: **Fatigue Stress vs. Temperature at Various Cycles to Failure for Machined DuPont Vespel SP-21 Polyimide.**

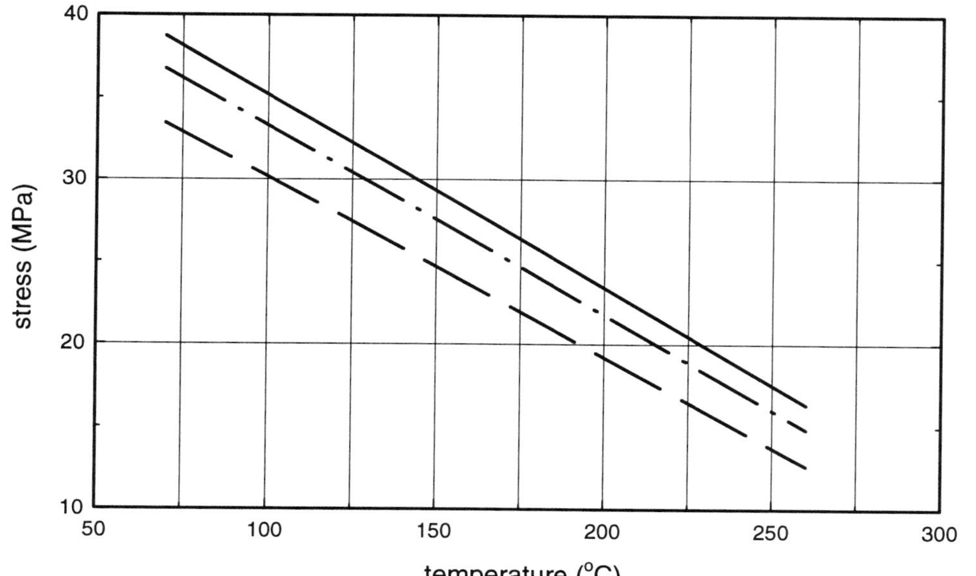

———	DuPont Vespel SP-21 Polyimide (25 - 45 Rockwell E; machined spec.; 15% graphite); tension- compression; 30 Hz; 10^5 cycles to failure [fig d]
—·—·—	DuPont Vespel SP-21 Polyimide (25 - 45 Rockwell E; machined spec.; 15% graphite); tension- compression; 30 Hz; 10^6 cycles to failure [fig d]
— — —	DuPont Vespel SP-21 Polyimide (25 - 45 Rockwell E; machined spec.; 15% graphite); tension- compression; 30 Hz; 10^7 cycles to failure [fig d]
Reference No.	37

Chapter 57

Vinyl Ester

Fatigue Crack Propagation

Dow Chemical: Derakane 411-45, Derakane 510A-40, Derakane 8084, Derakane 470-36, Derakane 510N

Vinyl ester resins (VER) exhibited superior fatigue behavior to polyester resins, both at room temperature and at 100°C (212°F). Specifically, the elastomer modified VER, Derakane 8084, has the highest resistance to crack propagation at room temperature and the brominated VER, Derakane 510A, has the highest resistance to crack propagation at 100°C (212°F).

Reference: Barron, D.L. (Dow), Ke lley, D.H. (Dow), Bl ankenship, L.T. (Dow), *Fracture Mechanics Approach Differentiates Fatigue Performance of Thermoset Resin Systems,* 44th Annual Conference, SPI Composites Institute, conference proceedings - Society of the Plastics Industry, 1989.

Effect of Modifiers on Fatigue Behavior

Dow Chemical: Derakane 470-36 (modifier: 5% carboxyl terminated butadiene-acrylonitrile (CTBN) rubber (1.5 micron diameter)); **Derakane 470-36** (modifier: 5% dispersed acrylate rubber (DAR))

The fatigue resistance of novolac vinyl ester resin was significantly improved by two types of rubber modifier, a CTBN rubber and an insoluble particulate rubber. The fracture mechanics approach to fatigue differentiated between the two types of rubber even though all other mechanical tests showed no differentiation.

Reference: Barron, D.L. (Dow), Kelley, D.H. (Dow), Blankenship, L.T. (Dow), *Fracture Mechanics Approach Differentiates Fatigue Performance of Thermoset Resin Systems,* 44th Annual Conference, SPI Composites Institute, conference proceedings - Society of the Plastics Industry, 1989.

GRAPH 240: Fatigue Crack Propagation (FCP) Rate vs. Stress Intensity Range at 100°C for Dow Chemical Dekrakane Vinyl Ester.

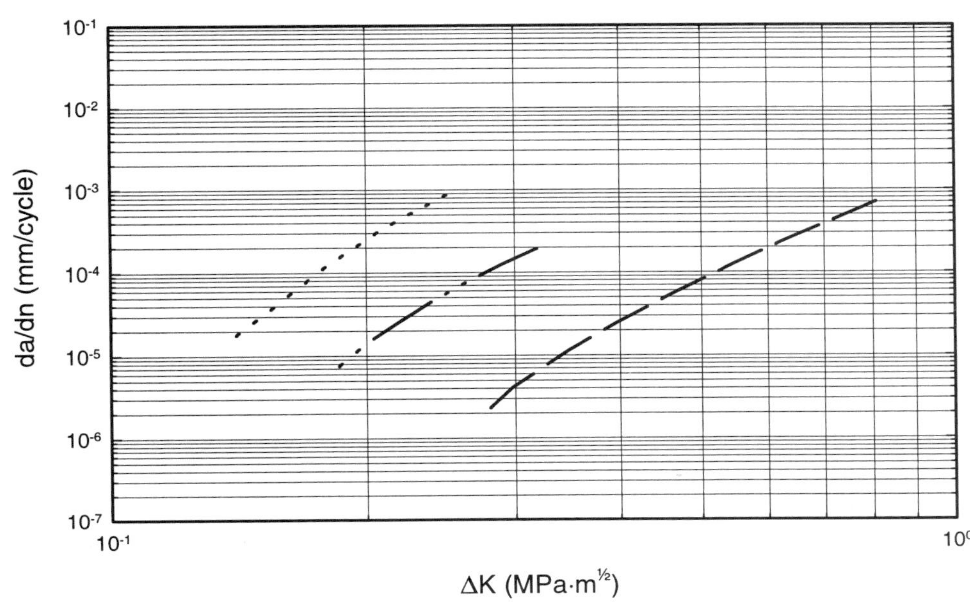

....	Dow Derakane 470-36 Epoxy Novalec Vinyl Ester; tension- tension; 20 Hz; stress ratio: 0.6; 100°C [fig s]
—··—··	Dow Derakane 510N Brominated Novolac Epoxy Vinyl Ester; tension-tension; 20 Hz; stress ratio: 0.6; 100°C [fig s]
– – – –	Dow Derakane 510A-40 Brominated Bisphenol A Epoxy Vinyl Ester; tension-tension; 20 Hz; stress ratio: 0.6; 100°C [fig s]
Reference No.	397

Chapter 58

Olefinic Thermoplastic Elastomer

Fatigue Properties

Advanced Elastomer Systems: Santoprene

Flex fatigue of Santoprene rubber is extremely high compared to many thermoset rubber compounds. The softer grades of Santoprene rubber are superior in fatigue resistance to natural rubber specially compounded for high fatigue resistance. Because of the great variety of flex fatigue applications, Santoprene rubber should be tested for fatigue resistance in accordance with each performance requirement. Tests using a Monsanto Fatigue-to-Failure test show failure at <3.4 megacycles in flex fatigue for 55, 66 and 73 Shore A hardness Santoprene grades.

Reference: *Santoprene Thermoplastic Rubber General Product Bulletin,* supplier technical report (TPE-02-11) - Advanced Elastomer Systems.

Advanced Elastomer Systems: Santoprene

The Ross flex test (ASTM D 1052) measures the formation and propagation of a crack through rubber during flex bending. Two sets of tests were run, using all hardnesseses of Santoprene, plus polychloroprene (CR), EPDM and chlorosulfonated polyethylene (CSM), flexing at 100 cycles/minute. The first test used specimens with no initial cut. After one million cycles of flexing, no failures were reported.

Further tests, using specimens with an initial cut of 0.10 inch (0.25 cm), produced substantially different results. After more than two million cycles of flexing, Santoprene rubber specimens showed no failures; whereas the three thermoset rubbers failed rapidly. This indicates outstanding fatigue resistance for Santoprene rubber. It must be kept in mind that the Ross flex test measures both the fatigue and tear of the rubber, rather than just tear.

Reference: *Santoprene Thermoplastic Elastomer Physical Properties,* supplier technical report (AES-1015) - Advanced Elastomer Systems, 1990.

Chapter 59

Polyester Thermoplastic Elastomer

Fatigue Properties

DuPont: Hytrel

The fatigue resistance of Hytrel is excellent. Data on the temperature rise (due to hysteresis) after twenty minutes for two of the softer grades of Hytrel (when tested in a Goodrich Flexometer) indicate the temperature rises fairly quickly and then remains roughly constant for the balance of the test. One of the outstanding properties of Hytrel polyester elastomer is its resistance to cut growth in flexure. Hytrel can endure more than million cycles without failure in the Ross and DeMattia pierced flex tests.

Sample size and shape, frequency of flexing, ambient temperature, and heat transfer all have significant effects on fatigue. For design purposes, tests simulating actual end-use conditions should be performed to determine the expected fatigue limit.

Reference: *Hytrel Polyester Elastomer Design Handbook,* supplier design guide (E-52083-1) - DuPont Company, 1988.

Chapter 60

Urethane Thermoplastic Elastomer

Fatigue Properties

Bayer: Texin

Resistance to breakdown under repeated stretching or deflections, for Texin elastomers, is largely a matter of heat buildup. Being thermoplastic, they fail when plastic flow occurs. In small section parts, such as the Zwick test specimen, heat dissipation is rapid enough to permit excellent fatigue resistance, the failure mechanism then being progressive crack growth. In the case of the compressive fatigue test, the thicker specimen heats up and fails relatively quickly. Endurance is thus largely controlled by rate of energy input and part geometry.

Reference: *Texin Urethane Elastomer - An Engineering Handbook,* supplier design guide - Bayer, 1993.

Chapter 61

Polybutadiene

Fatigue Properties

Enichem Polybutadiene: (chemical type: high cis 1,4-polybutadiene; features: tire sidewall compound; note: neodymium based catalyst); **Polybutadiene:** (features: tire sidewall compound; material compostion: 50 phr HAF carbon black; note: Ziegler-Natta catalyst based on cobalt); **Polybutadiene:** (features: tire sidewall compound; material compostion: 50 phr HAF carbon black; note: Ziegler-Natta catalyst based on titanium)

Fatigue and crack growth resistance of high 1,4 cis polybutadienes and their blends are strongly dependent on the steric purity of the polybutadiene. Fracture resistance is improved by strain-induced crystallization which acts as a self reinforcing mechanism in delaying crack propagation. In the case of tire applications where high fracture resistance is requested together with low hysteresis, polybutadienes having very high 1,4 cis content and high chain linearity should be used; neodymium polybutadiene fully meets these specifications so that it is a powerful tool to extend life.

Reference: Lauretti, E. (Enichem Elastomeri), Miani, B. (Enichem Elastomeri), Mistrali, F. (Enichem Elastomeri), *Improving Fatigue Resistance with Neodimium Polybutadiene (paper 62),* ACS Rubber Division 144th Meeting - Orlando, Florida, conference proceedings1993.

GRAPH 243: Fatigue Crack Propagation (FCP) Rate vs. Tearing Energy Under Simulated Normal Sidewall Service Conditions at 40 Hz Test Freuency for Polybutadiene Rubber With Different Catalysts.

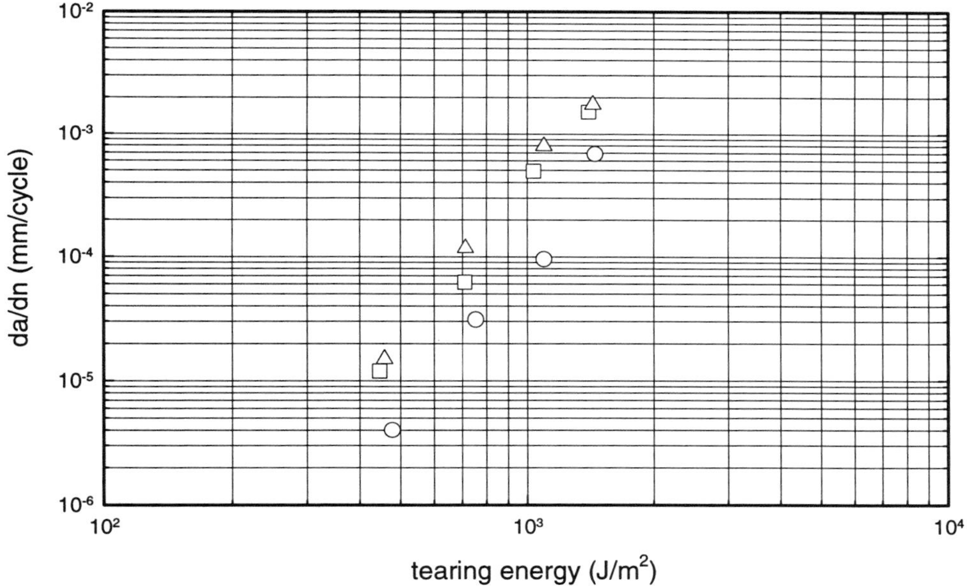

○	EniChem Polybutadiene (high cis 1,4-polybutadiene; tire sidewall compound; neodymium based catalyst); 40 Hz; 50°C; pure shear specimen; strain: 10-30% [fig ad]
□	Polybutadiene (tire sidewall compound; 50 phr HAF carb. bl.; cobalt based Ziegler-Natta catalyst); 40 Hz; 50°C; pure shear specimen; strain: 10-30% [fig ad]
△	Polybutadiene (tire sidewall compound; 50 phr HAF carb. bl.; Ziegler-Natta catalyst based on titanium); 40 Hz; 50°C; pure shear specimen; strain: 10-30% [fig ad]
Reference No.	409

Chapter 62

Butadiene/ Natural Rubber Compound

GRAPH 244: **Fatigue Crack Propagation (FCP) Rate vs. Tearing Energy Under Simulated Normal Sidewall Service Conditions at 20 Hz Test Frequency for Butadiene/ Natural Rubber Compound With Different Catalysts.**

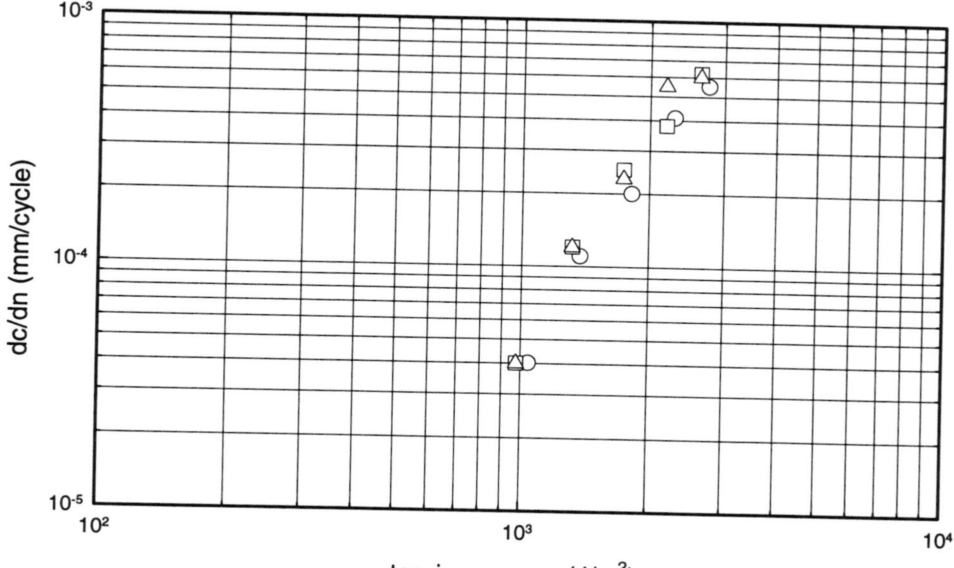

○	BR/ NR (tire sidewall compound; neodymium based catalyst); 20 Hz; 50°C; pure shear specimen; strain: 10-30% [fig ad]
□	BR/ NR (tire sidewall compound; cobalt based Ziegler-Natta catalyst); 20 Hz; 50°C; pure shear specimen; strain: 10-30% [fig ad]
△	BR/ NR (tire sidewall compound; Ziegler-Natta catalyst based on titanium); 20 Hz; 50°C; pure shear specimen; strain: 10-30% [fig ad]
Reference No.	409

Chapter 63

Bromoisobutylene-Isoprene Copolymer

Fatigue Crack Propagation

BIIR

Comparing the fatigue crack growth (FCG) rates at the same strain level and temperature, it can be seen that the BIIR compound had FCG rates that were typically two orders of magnitude lower than epichlorohydrin (ECO) and high styrene SBR (HS-SBR). Comparison at equal strain is appropriate in this case because the specific application intended is innerliners which experience strain-controlled deformation.

Reference: Young, D.G. (Exxon), Danik, J.A. (Exxon), *Fatigue Tests Can Provide Useful Information,* Rubber & Plastics News, trade journal1994.

Chapter 64

Natural Rubber

Fatigue Properties

Natural Rubber (note: vulcanizate)

The fatigue life of an unfilled NR vulcanizate varies with the maximum strain of the fatigue cycle. As expected, fatigue life decreases with increasing strain. When only mechanical stresses are involved, like in tests carried out in a vacuum, the life becomes virtually infinite as the maximum strain is reduced towards 100%. This latter strain is actually a mechanical fatigue limit, below which no failure occurs in the absence of chemical attack. The oxygen present in the laboratory atmosphere has two effects: it reduces the fatigue limit and reduces also the life at higher strains. At strains below the reduced fatigue limit, failure is due to the attack of ozone present in small quantities in the laboratory atmosphere; the ozone effect, however, is much smaller at strains over the fatigue limit.

Reference: Royo, J., *Fatigue Testing of Rubber Materials and Articles,* Polymer Testing 11 (1992), technical journal - Elsevier Science Publishers Ltd., 1992.

Effect of Filler Content on Fatigue Behavior

Natural Rubber (rubber stock: belt rubber; fiber conntent: 6 pph; fiber length: 6.4 mm; reinforcement: aramid fiber); **Natural Rubber** (note: rubber stock: belt rubber; fiber conntent: 6 pph; fiber length: 6.4 mm; reinforcement: nylon 6 (low modulus)); **Natural Rubber** (note: rubber stock: belt rubber; fiber conntent: 6 pph; fiber length: 6.4 mm; reinforcement: nylon 6 fiber); **Natural Rubber** (note: rubber stock: belt rubber; fiber conntent: 6 pph; fiber length: 6.4 mm; reinforcement: PET fiber)

Reinforcement of rubber with short fiber results in various property changes of rubber which are relevant for many applications. In pneumatic tires, however, the relatively poor fatigue resistance of the short fiber - reinforced rubber (SFRR) has so far hindered use of this interesting and promising material. Various methods of making the SFRR were explored to examine the effect of process variables and fiber characteristics on the fatigue resistance of SFRR and it was found that, when the fiber and rubber are combined properly and processed under certain conditions, the resulting SFRR has fatigue resistance which is comparable to or surpasses that of the unreinforced rubber. Both with regular nylon 6 and a modified nylon 6, it was possible to achieve improvements in fatigue life simultaneously with increasing the modulus of composite. Reinforcement with PET and aramid yields results which were in line with the prediction, namely higher stress concentration with increased fiber modulus.

The results of this study show the following:

1) The challenging goal of simultaneously increasing tensile modulus of elastomers and their fatigue resistance is achievable by short fiber reinforcement of rubber.

2) The range of combination of properties and composite structure variable permitting such simultaneous improvement is relatively narrow.

3) The failure mechanism of short fiber reinforced rubber is affected by interfacial adhesion and modulus of the fiber.

4) The simultaneous improvement in modulus and fatigue resistance requires sufficient fiber/rubber adhesion and sufficiently low fiber/rubber modulus ratio.

5) If fiber/rubber adhesion is sufficient to prevent pull out, the fatigue resistance increases with the increasing energy to break of the fiber.

Reference: Prevorsek, D.C. (Allied-Signal), Kwon, Y.D. (Allied-Signal), Beringer, C.W. (Allied-Signal), Feldste, *Mechanics of Short Fiber Reinforced Elastomers (paper 71),* ACS Rubber Division 1990 Fall Meeting - Washington D.C., conference proceedings - American Chemical Society, 1990.

GRAPH 245: **Fatigue Cycles to Failure vs. Strain in Tension and Different Environments for Unfilled Natural Latex Rubber Vulcanizate.**

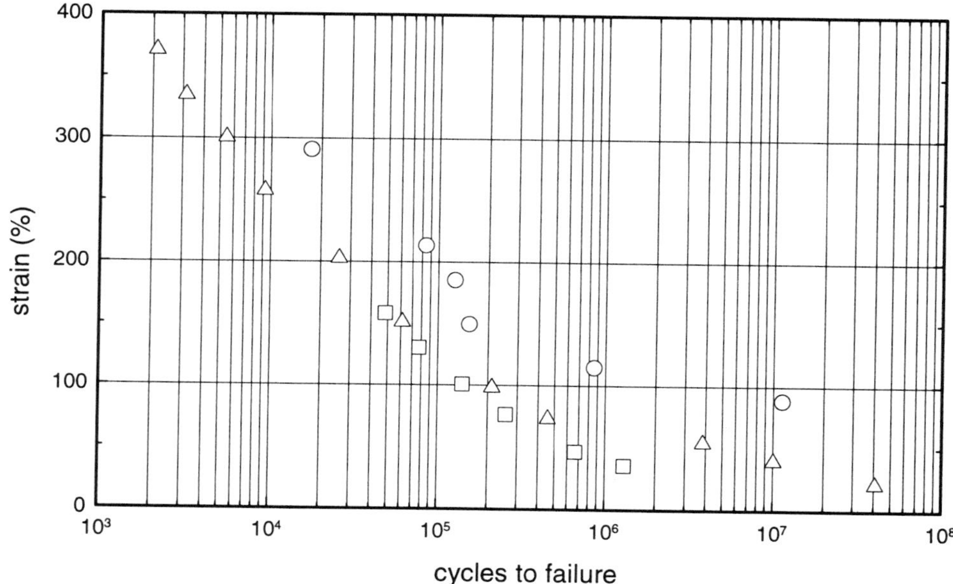

○	Natural Rubber (vulcanizate); tension-zero; vacuo (1E-6)
□	Natural Rubber (vulcanizate); tension-zero; laboratory atmosphere
△	Natural Rubber (vulcanizate); tension-zero; ozone chamber
Reference No.	399

GRAPH 246: **Fatigue Crack Propagation (FCP) Rate vs. Tearing Energy Under Simulated Normal Sidewall Service Conditions for Natural Latex Rubber.**

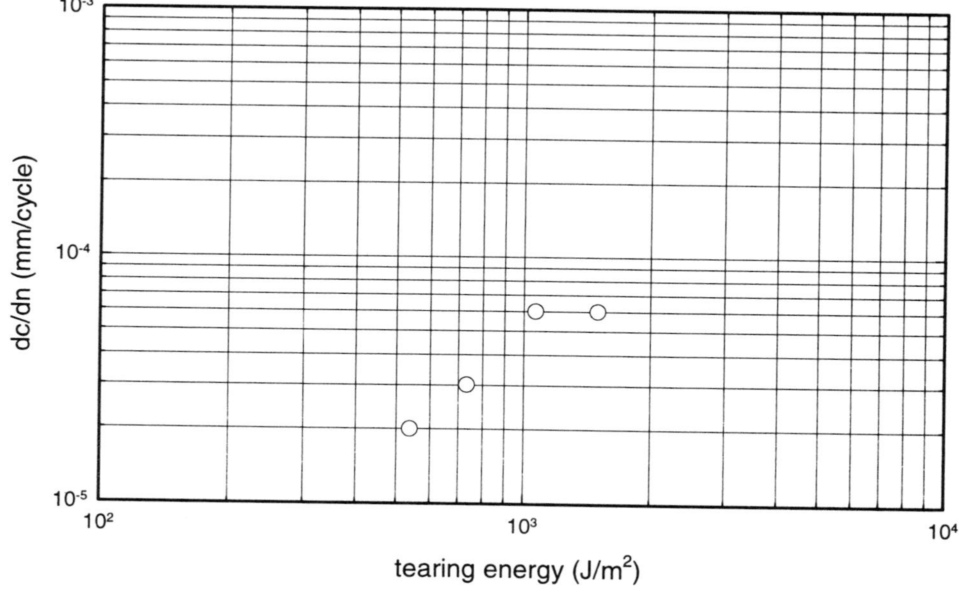

○	Natural Rubber; 20 Hz; 50°C; pure shear specimen; strain: 10-30% [fig ad]
Reference No.	409

GRAPH 247: **Fatigue Crack Propagation (FCP) Rate vs. Tearing Energy for Natural Latex Rubber Gum Vulcanizate.**

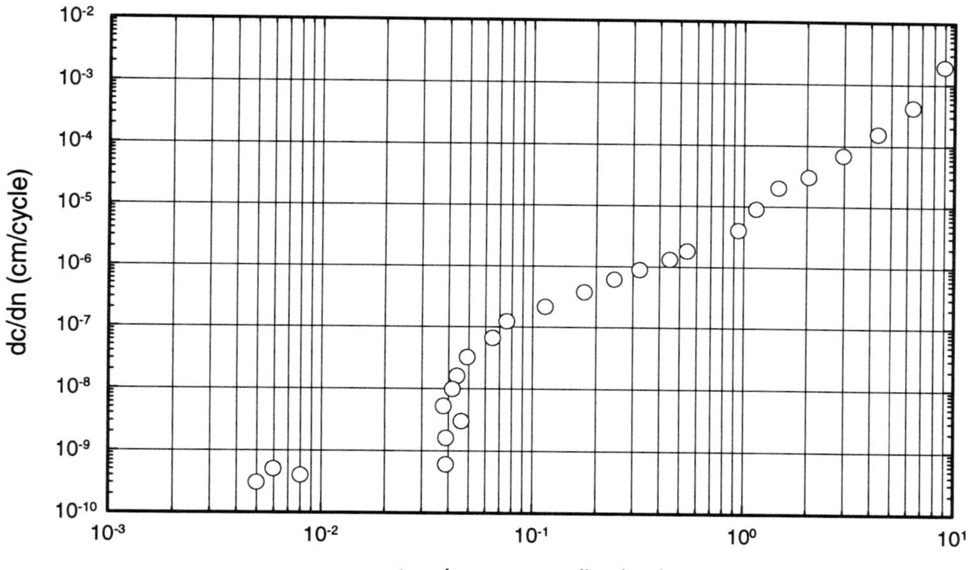

○	Natural Rubber; 1.67 Hz; laboratory atmosphere; non-relaxing conditions
Reference No.	399

Chapter 65

Polyisoprene Rubber

Fatigue Crack Propagation

Polyisoprene

Annealing of polyisoprene based elastomers can result in improvement or in deterioration of its failure properties, depending on the deformation of the rubber during the annealing. These alterations in performance, moreover, are readily induced at strains within the range of conventional fatigue and crack growth experimentation.

Reference: Roland, C.M. (Naval Research Laboratory), Sobieski, J.W. (Geo-Centers, Inc.), *Anomalous Fatigue Behavior in Polyisoprene (paper 35),* ACS Rubber Division 134th Meeting - Cincinnati, Ohio, conference proceedings - American Chemical Society, 1988.

Chapter 66

Polyurethane

Fatigue Properties

Gallagher Corporation:

Fatigue is an important consideration in the design of parts for cyclic dynamic applications. Fatigue and cut growth resulting from cyclic stress-strain are related. When testing for fatigue resistance for a specific application it's important to test at a strain energy experienced by the part in actual service. Strain energy is a function of both the modulus of the material and the strain cycle the material sees. Also important with polyurethanes is to test at various levels of stoichiometry (polymer to curative ratio) because we have seen dramatic improvements in flex fatigue resistance by this chemical adjustment.

Reference: *Design and Application Guide Cast & Molded Components of Polyurethane Elastomers,* supplier technical report - Gallagher Corporation, 1994.

Chapter 67

Silicone

Fatigue Properties

Silicone

When fatigue testing, take care to avoid affecting results by testing at too high a frequency. Silicones are strain rate dependent to some extent, and testing at 5 Hz has been shown to cause a noticeable rise in specimen temperature. Thus, testing for reasonable times and at reasonable frequencies will always leave us in the position of having to make extrapolations and apply judgment in developing design guidelines. An interesting problem with silicone fatigue testing is the fact that the silicone is likely to continue to cure during the test. If all specimens are not put under load at the same time after sample preparation, differences in the degree of cure may produce misleading results.

Reference: Sandberg, L.B. (Michicgan Technological University), Rinntala, A.E. (Rowe Engineering), *Resistance of Structural Silicones to Creep Rupture and Fatigue,* Building Sealants: Materials, Properties and Performance - ASTM STP 1069 - Symposium Proceedings, conference proceedings - American Society for Testing and Materials, 1990.

Effect of Modifiers on Fatigue Behavior

Silicone

It was shown that relative fatigue life increased in proportion to the silica's surface area, but after the silica's surface area reached 300 m^2/g, relative fatigue life decreased. It is inferred from this result that the more the silica's surface area increased, the more the agglomeration of silica was promoted, which caused the starting point of fracture.

Reference: Omura, N. (Shin-Etsu), Takahashi, M. (Shin-Etsu), Nakamura, T. (Shin-Etsu), *Silicone Rubber and its Fatigue Properties,* Rubber & Plastics News, trade journal, 1991.

GRAPH 248: Fatigue Cycles to Failure vs. Stress in Tension for Structural Adhesive Silicone.

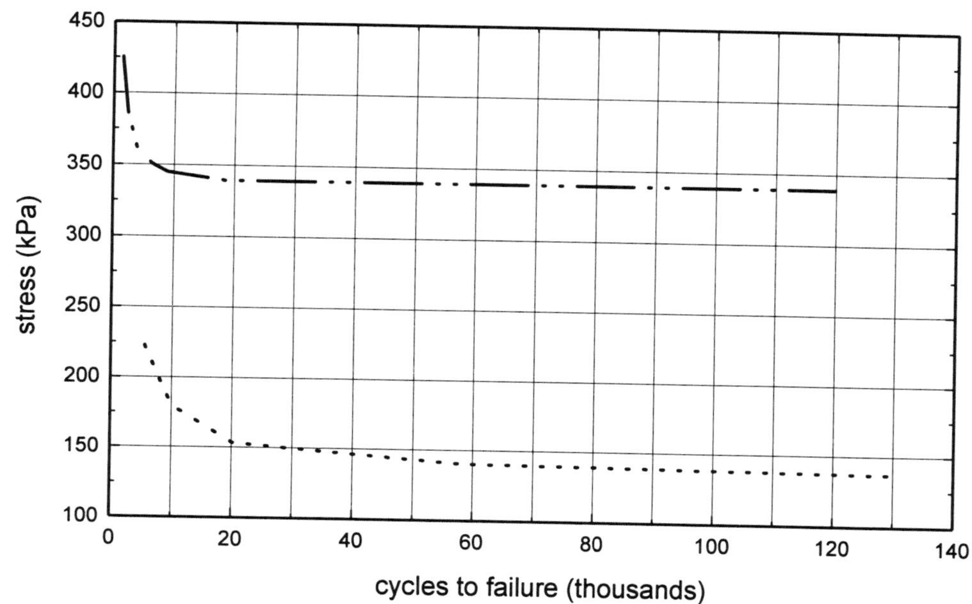

....	Silicone (neutral cure; structural sealant); tension-tension [fig d, l]
—··—··	Silicone (insulating glass; structural sealant; two part silicone); tension- tension [fig d, n]
Reference No.	398

GRAPH 249: Fatigue Cycles to Failure vs. Stress in Shear and Tension for Structural Adhesive Silicone.

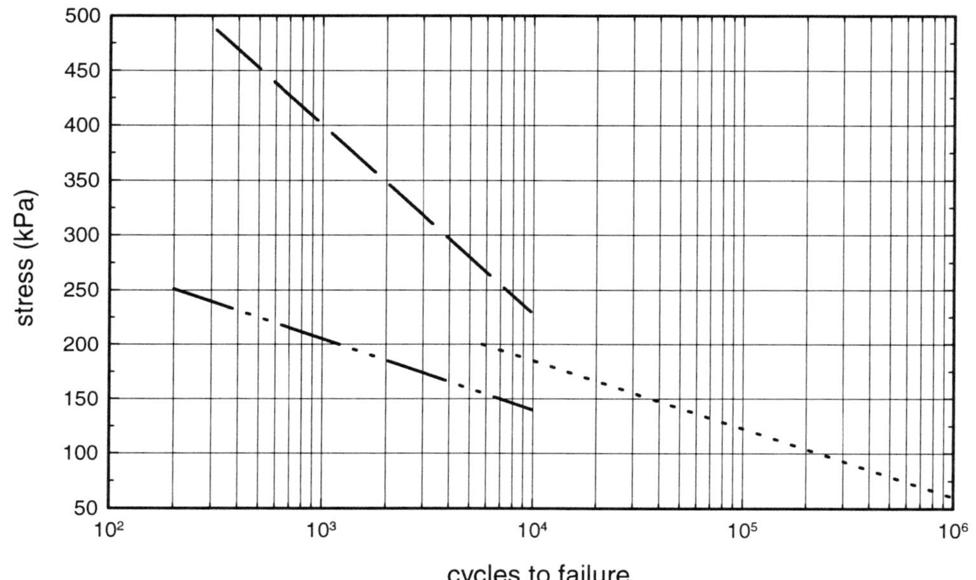

···············	Silicone (neutral cure; structural sealant); tension-tension [fig d, l]
—·-·—·-	Silicone (acetoxy silicone; structural sealant); tension-tension [fig d, p]
— — — —	Silicone (insulating glass; structural sealant; two part silicone); tension- tension [fig d, n]
Reference No.	398

GRAPH 250: Fatigue Cycles to Failure vs. Stress in Shear for Acetoxy Silicone.

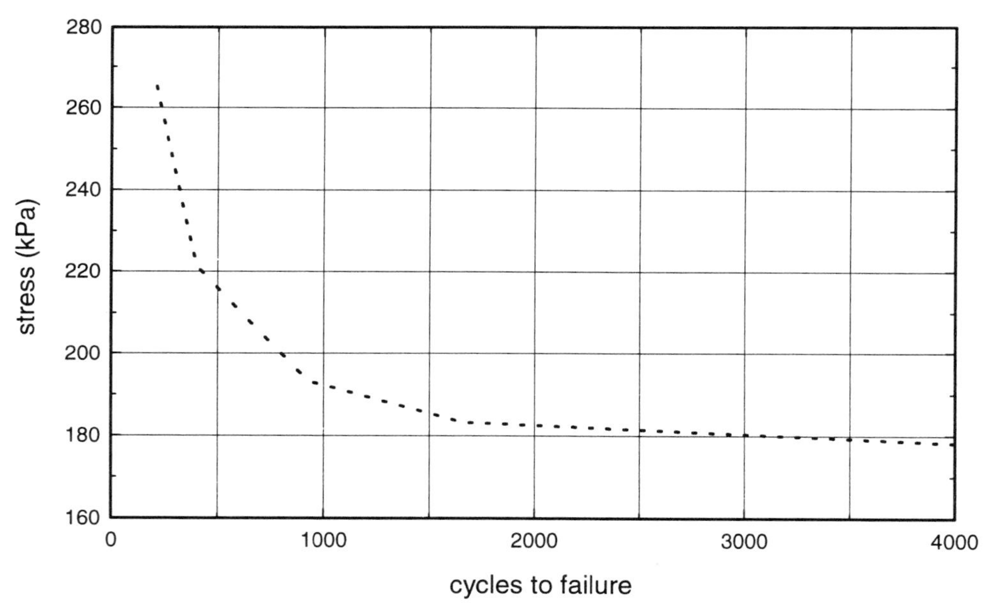

···············	Silicone (acetoxy silicone; structural sealant); shear [fig d, p]
Reference No.	398

Chapter 68

Styrene-Butadiene Copolymer

GRAPH 251: **Fatigue Crack Propagation (FCP) Rate vs. Tearing Energy for Styrene Butadiene Rubber.**

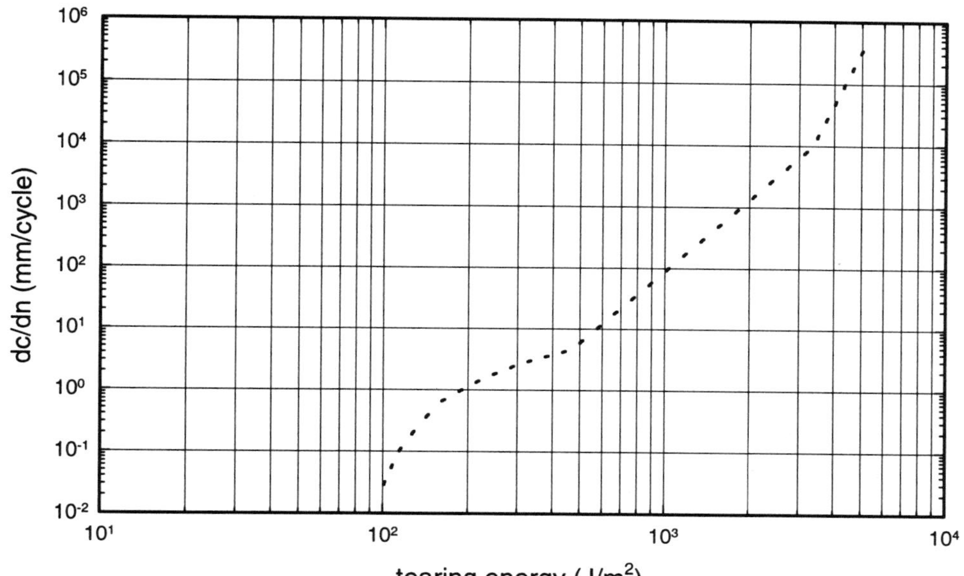

○	SBR (gum vulcanizate); 27°C; pure shear specimen; 40 Hz pulse, 5 Hz overall cycle [fig ad]
Reference No.	378

Chapter 69

Acetal Resin

Friction Properties

DuPont: Delrin

Delrin acetal resins have excellent frictional and wear characteristics, and can be used in applications where lubricants are not permitted or desirable. Continuous or initial lubrication of the surface extends the application range for Delrin acetal resin.

The coefficient of friction depends upon many variables, including: equipment, temperature, clearance, material, mating surface finish, pressure, velocity. Delrin acetal resin offers low stick slip design possibilities, as the static coefficient of friction is lower than the dynamic. Resins filled with low friction materials may be preferable for specific applications where a low coefficient of friction is essential. One example is highly loaded bearings with short service life, running for very short time periods. Delrin AF, filled with Teflon fibers, has the lowest coefficient of friction of all grades of Delrin.

Reference: *Delrin Design Handbook For Du Pont Engineering Plastics,* supplier design guide (E-62619) - Du Pont Company, 1987.

DuPont: Delrin 500

The static coefficient of friction diminishes when the normal load is rising. This shows that the adhesion component of friction is more significant during those tests. The surface roughness of the polymeric samples has a marked effect on static friction. The static coefficient of friction increases when the surface roughness of the polymeric samples diminishes. This is due to the appreciable adhesion when smooth surfaces are in contact.

Reference: Benabdallah, H. (Ecole Polytechnique de Montreal), Fisa, B. (Ecole Polytechnique de Montreal), *Static Friction Characteristics of Some Thermoplastics,* ANTEC 1989, conference proceedings - Society of Plastics Engineers, 1989.

Abrasion Resistance

DuPont: Delrin 100 (features: 120 Rockwell R hardness, 94 Rockwell M hardness, surface lubricity, low flow; melt flow index: 1.0 g/10 min.); **Delrin 500** (features: general purpose grade, 94 Rockwell M hardness, 120 Rockwell R hardness, surface lubricity; melt flow index: 6.0 g/10 min.); **Delrin 500CL** (features: 120 Rockwell R hardness, 90 Rockwell M hardness, low coefficient of friction, low wear grade, chemical lubricant; melt flow index: 6.0 g/10 min.); **Delrin 500F** (features: 92 Rockwell M hardness, 120 Rockwell R hardness, enhanced mold release, fast cycling); **Delrin 900F** (features: 120 Rockwell R hardness, fast cycling, 92 Rockwell M hardness, enhanced mold release); **Delrin** (unmodified)

A dramatic reduction in wear can be seen as material hardness increases. The most dramatic differences can be seen where Delrin acetal resin is matched with Zytel 101 nylon resin. Data on the wear performances of Delrin 500, 900 F and 500 CL against mild steel have been determined. Comparable data have also been obtained to show the suitability of Delrin acetal resin with aluminum and brass. Most other engineering resins do not perform well running against soft metals due to the tendency to pick up metal particles in the plastic surface. Because Delrin acetal resin is harder, this tendency is reduced.

The actual wear performance of specific resins will vary depending upon load, speed, mating surface, lubrication, and clearance. The wear performance of Delrin 500F and 900F are slightly better than those of Delrin 100 and 500. Delrin 500CL offers wear performance that is superior to all filled resins, even to those having a lower coefficient of friction. It is especially preferred when running against soft steel and non-ferrous metals.

Reference: *Delrin Design Handbook For Du Pont Engineering Plastics,* supplier design guide (E-62619) - Du Pont Company, 1987.

DuPont: Delrin

Delrin acetal resin has good abrasion resistance. Two tests were used to compare the properties of Delrin acetal resin with Zytel 101 nylon resin and other materials. Zytel nylon resins are well known for their outstanding abrasion resistance. Although not comparable to nylon, Delrin acetal resins are superior to most other engineering plastics.

Reference: *Delrin Design Handbook For Du Pont Engineering Plastics,* supplier design guide (E-62619) - Du Pont Company, 1987.

Hardness

DuPont: Delrin

Hardness of Delrin acetal resin is usually reported in terms of Rockwell Hardness (ASTM D785), which measures surface penetration with a steel ball under specified loading conditions. The Rockwell hardness scales which indicate indenter diameter and load are M and R. For the different grades of Delrin acetal resins, there are only slight variations in Rockwell hardness values.

Delrin acetal resin is unusually hard for a material with such good toughness. This combination has been a distinct advantage in many applications. The high hardness enhances friction and wear characteristics against a wide variety of mating materials; improves damage resistance and speeds up ejection from the mold, as well as being the key characteristic in applications such as printer wheels, etc.

Reference: *Delrin Design Handbook For Du Pont Engineering Plastics,* supplier design guide (E-62619) - Du Pont Company, 1987.

GRAPH 252: **Static Coefficient of Friction vs. Load Against Steel of Various Surface Roughnesses for DuPont Delrin 500 Acetal Resin.**

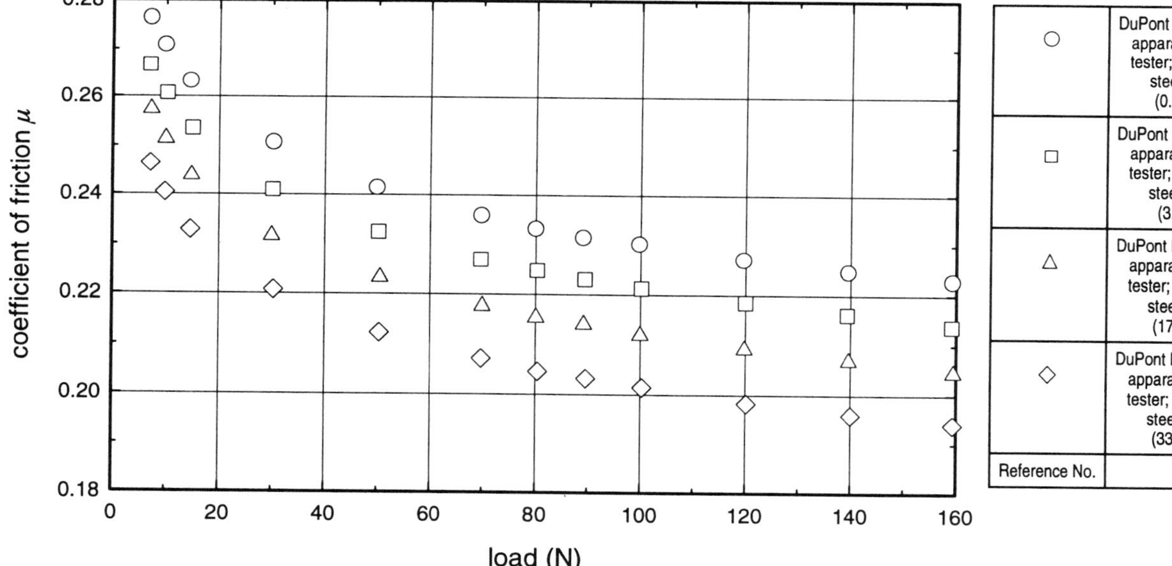

Chapter 70

Acetal Copolymer

Friction and Wear Properties

BASF AG: Ultraform N2310P (lubricated); **Ultraform N2320Y** (modifier: molybdenum disulfide); **Ultraform N2720XM210** (developmental material; filler: Wollastonite); **Ultraform N2770K** (filler: chalk); **Ultraform** (unmodified)

Ultraform's smooth and hard surface and its crystalline structure makes it well suited for the production of parts that are subjected to sliding friction. Even if the system is unlubricated, the wear will be very slight, and the coefficient of friction will be adequately low. In systems in which Ultraform is paired with other materials, an increase in the roughness of the other material will entail a reduction in the coefficient of sliding friction of the Ultraform. The wear, however, will increase.

Superior tribological characteristics are displayed by Ultraform grades N 2310P, N2320Y, N2770K, and N2720XM210, the optimum grade for any specific application depends on the conditions of use, e.g. speed, load, surface roughness, etc. The optimum tribological properties of Ultraform N 2310P and N 2770K occur at low roughness heights, i.e., when paired with a smooth material. In contrast, Ultraform N 2320Y and N 2720XM210 display optimum frictional properties at higher load and/or high surface roughnesses.

Reference: *Ultraform Polyacetal (POM) Product Line, Properties, Processing,* supplier design guide (B 563/1e - (888) 4.91) - BASF Aktiengesellschaft, 1991.

Hoechst AG: Hostaform C9021TF (Ball indentation hardness: 120 (358/30); features: low coefficient of friction; melt flow index: 8 g/10 min. (190/2.16); modifier: PTFE); **Hostaform** (unmodified)

As a result of its high hardness and smooth surface, Hostaform has good sliding properties (low friction). The following values for Hostaform against Hostaform will serve as a guide to the assessment of sliding properties: static friction coefficient of approximately 0.35 and dynamic friction coefficient of approximately 0.25. These values are considerably lower for the Hostaflon (PTFE) modified Hostaform C 9021TF. The phenomenon of stick-slip is due to the difference between the static and dynamic friction coefficients.

Reference: *Technical Plastics Calculations, Design, Applications - A.1.1 Grades and Properties Hostaform,* supplier design guide (HBKP 395/A.1.1 E-8127/022) - Hoechst AG, 1987.

Hoechst Celanese: Celcon LW90 (features: 80 Rockwell M hardness, low wear grade; melt flow index: 9.0 g/10 min.); **Celcon LW90S2** (features: low wear grade, 75 Rockwell M hardness; material compostion: 2% silicone; melt flow index: 9.0 g/10 min.); **Celcon LW90SC** (features: low wear grade; material compostion: 20% silicone; product form: concentrate); **Celcon LWGCS2** (features: 83 Rockwell M hardness; material compostion: 2% silicone; reinforcement: glass fiber)

The Celcon Low Wear series - LW90, LW90SC, LW90S2 and LWGCS2 - offer low wear performance for a broad range of applications. Celcon LW90 is formulated for material handling components that operate against metals at high speeds under low loads. LW90S2 is a 2% silicone modified grade for low wear performance in both plastic to plastic and plastic to metal applications operating under various combinations of speeds and loads. (For higher mechanical strength and stiffness Celcon LWGCS2 offers a preblended glass coupled grade containing 2% silicone polymer). LW90SC provides a 20% silicone polymer concentrate for custom letdown ratios. In addition to enhanced frictional properties, Celcon silicone modified grades also improve flow and mold release during injection molding of complex parts like gears and rollers.

Reference: *Celcon Acetal Copolymer Specialty Grades Tailored To Meet the Needs of a More Demanding Marketplace,* supplier technical report (10M/8-85, CS-29) - Hoechst Celanese Corporation, 1985.

GRAPH 253: **Dynamic Coefficient of Friction vs. Average Surface Roughness for BASF Ultraform Acetal Copolymer.**

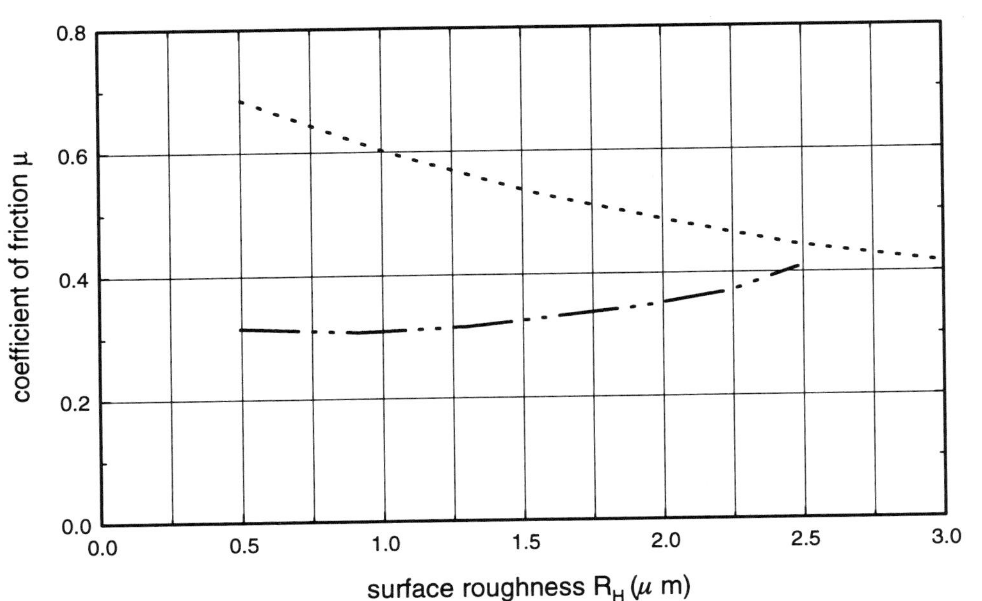

....	BASF AG Ultraform N2320 Acetal Copol. (moderate flow, gen. purp. grade); mating surface: steel (54-56 Rockwell C, 5-8 μin); 40°C; speed: 0.5 m/sec; pressure: 1 MPa
—··—··	BASF AG Ultraform N2310P Acetal Copol. (lubricated); mating surface: steel (54-56 Rockwell C, 5-8 μin); 40°C; speed: 0.5 m/sec; pressure: 1 MPa
Reference No.	181

GRAPH 254: **Wear Rate vs. Average Surface Roughness for BASF Ultraform Acetal Copolymer.**

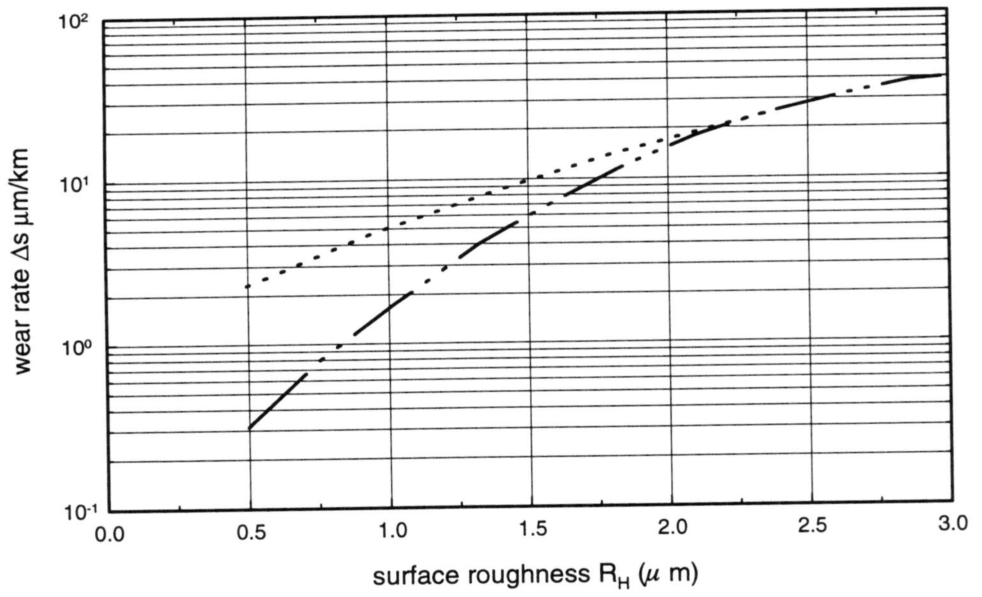

....	BASF AG Ultraform N2320 Acetal Copol. (moderate flow, gen. purp. grade); mating surface: steel (54-56 Rockwell C, 5-8 μin); 40°C; speed: 0.5 m/sec; pressure: 1 MPa
—··—··	BASF AG Ultraform N2310P Acetal Copol. (lubricated); mating surface: steel (54-56 Rockwell C, 5-8 μin); 40°C; speed: 0.5 m/sec; pressure: 1 MPa
Reference No.	181

Chapter 71

Ethylene-Tetrafluoroethylene Copolymer

Surface Properties

DuPont: Tefzel 200 (features: general purpose grade, 50 Rockwell R hardness, 75 Shore D hardness); **Tefzel HT-2004** (features: 74 Rockwell R hardness; material compostion: 25% glass fiber reinforcement)

Unlike many other polymers, the addition of glass reinforcement improves the frictional and wear properties of Tefzel. For example, the dynamic coefficient of friction (100 psi at >10 fpm) for Tefzel 200 is 0.4 but drops to 0.3 for Tefzel HT-2004 at these conditions. The wear factor also improves from 6000 x 10^{-10} to 16 x 10^{-10} $in^3 \cdot min/ft\ lb \cdot hr$. These improved frictional and wear characteristics, combined with outstanding creep resistance, suggest that the glass-reinforced resin be favored for bearing applications. Tefzel HT-2004 also appears to be less abrasive on mating surfaces than most glass-reinforced polymers. The static coefficient of friction for Tefzel HT-2004 is dependent on bearing pressure. Dynamic friction is dependent on pressure and rubbing velocity (PV). The generation of frictional heat is dependent on coefficient of friction and the PV factor. For "Tefzel" HT-2004, temperature buildup begins at about a PV of 20,000. High wear rates begin at a PV above 15,000.

The rate of wear depends on the type of metal rubbing surface and on other factors such as finish, lubrication, and clearances. Lubrication, harder shaft surfaces, and high finishes all improve wear rates. Minimum diametral clearances of 0.3 to 0.5% are suggested for sleeve bearings. The wear rate of both "Tefzel" and the metal is much higher for aluminum than for steel. Therefore, if aluminum is the mating metal, an anodized surface is suggested. The wear factor of Tefzel HT-2004 against steel is about one-tenth that of 33% glass-reinforced nylon.

Reference: *Tefzel Fluoropolymer Design Handbook,* supplier design guide (E-31301-1) - Du Pont Company, 1973.

GRAPH 255: Coefficient of Friction vs. PV Limit Against Steel for DuPont Tefzel HT2004 Ethylene Tetrafluoroethylene Copolymer.

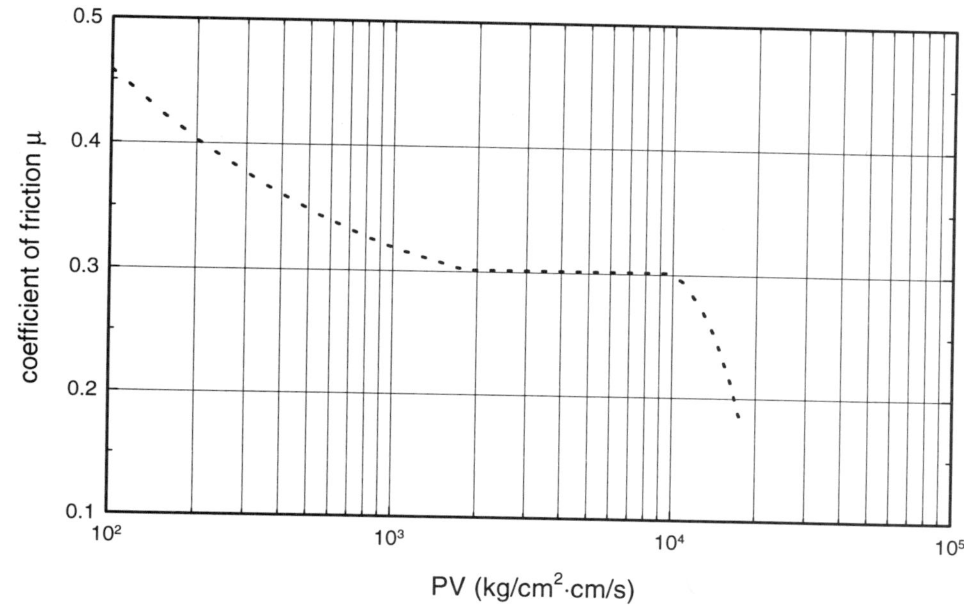

....	DuPont Tefzel HT-2004 ETFE (74 Rockwell R; 25% glass fiber); mating surface steel
Reference No.	205

Chapter 72

Fluorinated Ethylene-Propylene Copolymer

GRAPH 256: Coefficient of Friction vs. Load for DuPont Teflon Fluorinated Ethylene Propylene Copolymer.

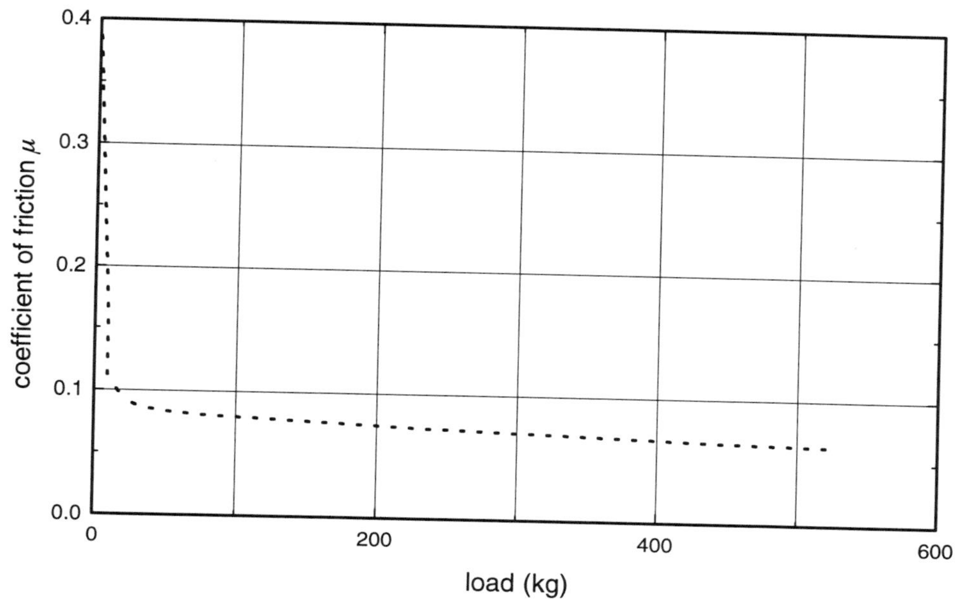

....	DuPont Teflon FEP (25 Rockwell R, 96 Shore A, 59 Shore D); pressure: 0.0014- 517 kPa; 23°C; speed: < 0.6m/min; mating surface: steel
Reference No.	339

GRAPH 257: Coefficient of Friction vs. Sliding Speed at Different Pressures for DuPont Teflon Fluorinated Ethylene Propylene Copolymer.

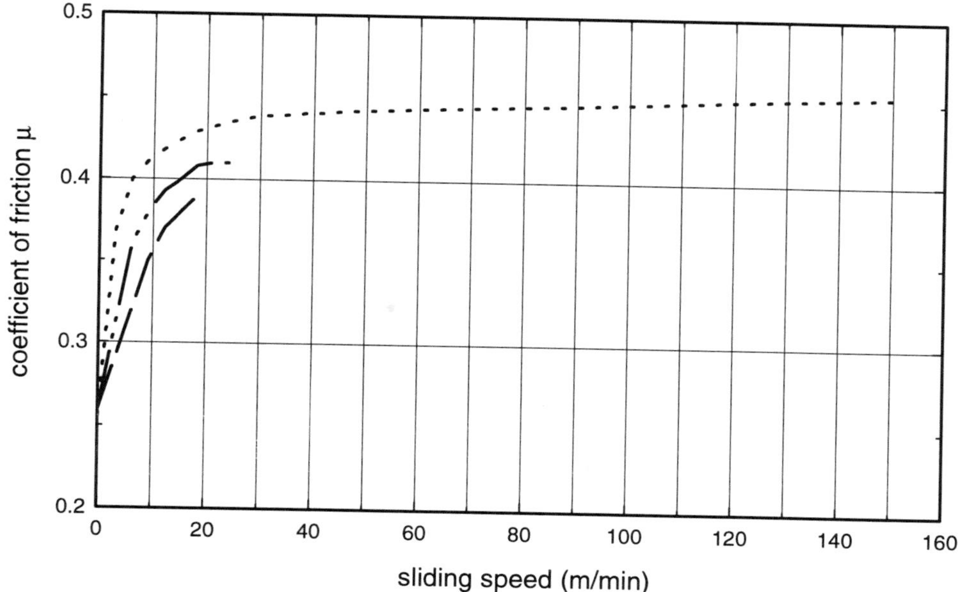

....	DuPont Teflon FEP (25 Rockwell R, 96 Shore A, 59 Shore D); pressure: 6.9 kPa; temperature: 24-66 °C; mating surface: steel
—··—··	DuPont Teflon FEP (25 Rockwell R, 96 Shore A, 59 Shore D); pressure: 69 kPa; 24-66 °C; mating surface: steel
— — — —	DuPont Teflon FEP (25 Rockwell R, 96 Shore A, 59 Shore D); pressure: 689 kPa; 24-66 °C; mating surface: steel
Reference No.	339

Chapter 73

Perfluoroalkoxy Resin

Surface Properties

DuPont: Teflon PFA 340 (features: general purpose grade, 55 Shore D hardness, moderate molecular weight)

Frictional and wear tests have been run on Teflon PFA to indicate its level of performance - unfilled - in mechanical applications, such as bearings, seals, etc. Tests were run on molded thrust bearings at 100 psi (0.7 MPa) against AISI 1018, Rc 20, 16 AA steel; tests were run at ambient conditions in air with no lubrication. Results indicate a limiting PV value of 5,000, but wear rate, rather than PV, will likely be the critical parameter. At PV = 1,000, for instance, Teflon PFA will wear 5 mm (3/16 in.) per 1,000 hours. Wear factors decreased over the PV range 1,000 to 5,000 from 1840 x 10^{-10} to 700 x 10^{-10}. Coefficient of friction ran 0.236.

Reference: *Handbook Of Properties For Teflon PFA,* supplier design guide (E-96679) - Du Pont Company, 1987.

Effect of Temperature on Friction and Wear Properties

LNP Engineering Plastics: Fluoromelt FP-PC-1003 (material compostion: 15% carbon fiber reinforcement)

Fluoromelt melt processible fluoropolymers offer many properties similar to those of PTFE materials, and some even greater. For example, the chemical and thermal properties of PFA approximate those of PTFE, and deformation resistance of PFA at elevated temperature is superior. Wear properties of the carbon fiber reinforced PFA, however, were not as good as those of the filled PTFE composites, but significantly better up to 204°C (400°F) than those of almost all of the other composites tested.

Reference: *Lubricomp Internally Lubricated Reinforced Thermoplastics and Fluoropolymer Composites,* supplier design guide (254-691) - LNP Engineering Plastics, 1991.

Chapter 74

Polytetrafluoroethylene

Surface Properties

DuPont: Teflon (features: 98 Shore A hardness, 58 Rockwell R hardness, 52 Shore D hardness)

Teflon has a smooth surface with a slippery feel. Because of Teflon's low coefficient of friction, there have been many practical nonlubricated and minimally lubricated mechanical systems developed. Teflon resins exhibit low friction in nonlubricated applications, especially at low surface velocities and pressures higher than 0.030 MPa (5 psi). The coefficient of friction increases rapidly with sliding speeds up to about 35 m/min (100 fpm), under all pressure conditions. This pattern of behavior prevents "stick-slip" tendencies. Moreover, no "squeaking" or noise occurs, even at the slowest speeds. Above 45.7 m/min (150 fpm), sliding velocity has relatively little effect at combinations of pressure and velocity below the composition's PV limit. Static friction of Teflon resins decreases with increases in pressure. The incorporation of fillers does not appreciably alter the coefficient of friction of Teflon.

The PV limits of all Teflon TFE resin matrix compositions approach zero (0) at 288°C to 316°C (550°F to 600°F) ambient temperature. In other words, the limiting surface temperature for operation of Teflon compositions is 288°C to 316°C (550°F to 600°F), regardless of the cause of the temperature. Reduced ambient temperatures, below 21°C (70°F), and/or cooling will provide increased PV limits. PV limit does not necessarily define useful combinations of pressure and velocity since wear is not considered in its determination. The useful PV limit of a material cannot exceed the PV limit and must take into account the composition's wear characteristics and the allowable wear for the application.

Reference: *Teflon Fluorocarbon Resin Mechanical Design Data,* supplier design guide (E-05561-3) - DuPont Company.

Friction and Wear Properties

ICI: Fluon (features: 60-65 Shore D hardness, unfilled)

Shooter and Thomas first published about PTFE's low coefficient of friction. They measured the coefficient of friction using a Bowden-Leben machine with loads of between 1 and 4 kg (2.2 and 8.8 lb) and sliding velocities from 0.1 to 10 mm/s (0.02 to 2 fpm). The reported coefficient was 0.04. Other workers report that while Amonton's law is fairly well obeyed at moderate loads the coefficient of friction rises steeply at very light loads, say below 100 g (3 1/2 oz.). Thompson et al. who studied the coefficient at high loads found the extremely low figure of 0.016 at a load of 1360 kg (3000 lb). The coefficient of friction is dependent also on the sliding velocity, a high speed resulting in a high coefficient. By combining a low load and a high sliding velocity of 1.89 m/s (370 ft/min), Flom and Porile found the high value of 0.36 for the coefficient.

Shooter and Thomas claimed that, at the very low speeds they used, the coefficient was independent of temperature over the range 20°C to 200°C (68°F to 392°F). However, later work has shown that temperature has some effect. King and Tabor report that the coefficient remains steady at about 0.1 over the range 100°C to -45°C (212°F to - 49°F). On further cooling the coefficient rises to about 0.2 but does not alter further even when the polymer is cooled to -80°C (-112°F). For the behavior at elevated temperatures the best guide is the work of McLaren and Tabor who demonstrated that the coefficient of friction fell with increase in temperature.

Makinson and Tabor have also examined the effect of sliding velocity and substantially agree with the variation in coefficient of friction with velocity given above. They have found that whereas at low velocities a thin continous film of PTFE is laid down on the other slide surface (in this case glass), at higher velocities the PTFE is torn off in discrete fragments.

Of less general importance than the dependence on load, velocity and temperature, but still of interest is the obsrevation of Tabor and Williams that the coefficient is influenced by the orientation of the polymer. The coefficient being about 30% higher when sliding was across the chains than when it was along them.

Reference: *Physical Properties Of Unfilled And Filled Fluon Polytetrafluoroethylene,* supplier design guide (Technical Service Note F12/13) - ICI PLC, 1981.

LNP Engineering Plastics: Lubricomp FC-103 (material compostion: 15% milled glass) **Lubricomp FC-182** (material compostion: 15% bronze, 5% molybdenum disulfide) **Lubricomp FC-191** (material compostion: 25% carbon fiber/ graphite); **Lubricomp PC-185** (reinforcement: polyphenylene sulfide)

Filled PTFE composites provide an excellent balance of bearing properties and chemical resistance, along with high elongation. Wear factors changed little if any at temperatures of 204°C (400°F) in all PTFE composites tested. The lowest wear factor measured at 260°C (500°F) was for the carbon/ graphite reinforced PTFE materials. The lowest wear factor at 204°C (400°F) was for the PPS filled composite, making it well suited for valve seats and sealing rings.

Bronze filled PTFE, with the highest modulus and thermal conductivity and the lowest coefficient of thermal expansion and deformation under load, also demonstrated the highest LPV values at all test temperatures. These composites are used in transmission seals and compressor rings. Friction values for the filled PTFE composites were the lowest of almost all materials tested, at all test temperatures.

Reference: *Lubricomp Internally Lubricated Reinforced Thermoplastics and Fluoropolymer Composites,* supplier design guide (254-691) - LNP Engineering Plastics, 1991.

ICI: Fluon (features: 60-65 Shore D hardness, unfilled)

The mechanisms responsible for the wear of PTFE are not fully understood, but it is generally thought that adhesion and the freeing of transferred wear fragments, either in terms of surface energy or by virtue of fatigue, are of major importance. It is known that when PTFE is rubbed against other materials a transfer takes place and it is believed that the wear process involves the laying down and subsequent removal of such transferred layers. An ideal situation is given as having a highly oriented mono-molecular layer of PTFE bonded to the metal surface which then rubs against as smooth a mating surface of PTFE as possible.

What is not clear is exactly how and why fillers and conditions affect both the initial laying down and subsequent removal of the PTFE particles. It is suggested that a minimum temperature at the interface is required to promote adequate bonding and that certain fillers function by causing frictional heat. It is also clear that surface finish will affect this transfer, and while there is wide agreement that too rough a mating surface will cause rapid wear, one school of thought suggests that too smooth a surface finish leads to high wear rates, while others suggest that it is not so. The answer may be that although too fine a finish may well inhibit good transfer, many filled compounds are sufficiently abrasive to roughen the mating surface adequately. However, if the filler or environmental conditions are too abrasive, rapid wear will occur through ploughing. The entrapment of wear debris can have a similar effect.

It has been suggested that chemical reactions at the interface may be important. Buckley and Johnson consider that wear is related to the decomposition mechanism and hence to the temperature at the interface, while Hargreaves and Tantam suggest lead oxide can be an oxygen carrier to other metals, giving selective oxidation of roughnesses on the mating surface. Mitchell and Pratt have noted the formation of copper fluoride at the interface of bronze-filled PTFE, presumably caused by local degradation of the PTFE and bronze. They do not, however, attribute the reduction in wear accompanying the formation of copper fluoride to the chemical action, but rather to the fact that the area of contact at the interface increases with time, which reduces the interface temperature. Vinogradov did, however, attribute a reduction in friction between copper and PTFE to the formation of the solid lubricant copper fluoride.

Reference: *Physical Properties Of Unfilled And Filled Fluon Polytetrafluoroethylene,* supplier design guide (Technical Service Note F12/13) - ICI PLC, 1981.

Effect of Fillers and Additives on Surface Properties

ICI: Fluon

Thompson et al. suggest that, using molybdenum disulphide (MoS_2), asbestos, carbon, graphite, and copper as fillers, as the volume of filler increases the coefficient of friction increases from 0.016 (no filler) to about 0.030 (30% of filler), but that there is little difference in this effect between the various fillers. For a similar range of fillers, Milz and Sargent showed the coefficient to be independent both of the type of filler and its volume addition. In particular, MoS_2 and graphite showed no advantage over glass fiber, asbestos and copper. Their results for all types ranged from 0.09 to 0.22 depending on velocity, load, etc. They concluded that the filler was effectively encapsulated and the friction was that of PTFE only. O'Rourke originally came to the same conclusion but later states that friction is dependent more upon the volume than the type of filler although cadmium oxide is claimed to be an exception. At the very low temperatures of liquid oxygen and nitrogen and under conditions of high vacuum there is considerable variation in the coefficient, but this does not appear to be correlated with either filler type or volume. In practical tests with the Wankel engine using various grades of PTFE as a seal, the coefficient was again found independently of the filler, while in a laboratory test, Ganz and Parkhomenko state that the type of filler is important; however, they appear to quote the filler content as % by weight so that filler type and filler volume are not separable. They again found MoS_2 and graphite fillers to give high coefficients of 0.26 to 0.34.

The evidence of Mitchell and Pratt is that filler type has a greater effect than filler volume, with MoS_2 giving a lower coefficient than unfilled PTFE. They found bronze had little effect and kieselguhr increased it by 25%. Work done by ICI suggests that volume of filler is not directly related to friction coefficient but fillers in general raise the coefficient under these particular test conditions by a factor of about two.

It has also been suggested that the addition of MoS_2 and carbon to glass fiber compounds reduces the coefficient of friction, although figures quoted show only a marginal decrease. Tests carried out by ICI have not confirmed this and Buckley et al. found no improvement when working under vacuum. Similarly, practical tests showed no advantages for adding MoS_2 to glass although this combination was suggested as a possible means of reducing the scoring of shafts, and for use in very dry gases. It is conceivable that after prolonged continuous running under dry conditions, the MoS_2 is not subject to the rise in friction reported for PTFE. There is therefore conflicting evidence as to the effect of filler type and volume upon the coefficient of friction of PTFE.

Filler particle size and shape

It is difficult to separate the effects of particle size and shape from those of filler type, since specific forms of particle tend to be used with specific types of filler (e.g., glass fiber, irregular particles of graphite and MoS_2, spherical bronze, etc.). Moreover, in much of the published work no details of filler particle are given.

The most explicit work in this field is that of Speerschneide and Li where, with the very hard particles of alumina (Al_2O_3), they found spherical particles gave coefficients of friction similar to that of unfilled PTFE (0.05 - 0.08) whereas irregular particles gave significantly higher results (0.14 - 0.15). They attributed this increase to cleavage of the irregular Al_2O_3 which saturates the surface until the coefficient is that of Al_2O_3 on steel. The abrasive nature of the filler also gives a 'rough' surface finish to the steel, thereby giving a coefficient approximately double that of a 'smooth' steel surface. This effect is less likely to occur with softer fillers, and this has been found true with bronze, where no difference in friction has been found between spherical and irregular particles, although Thompson et al. suggest that particle size can have an effect in extreme cases.

Environments

Work with filled PTFE at low temperature and in contact with liquid oxygen and nitrogen shows the coefficient to rise with the passage of time (e.g., 0.18 to 0.43 in 23 hours), which tends to confirm the work of Steijn with unfilled PTFE. This same effect at room temperature has been found by ICI and by Mitchell and Pratt although actual coefficients are lower (0.07 to 0.20 in 20 hours). There is some evidence therefore that the coefficient of friction increases in the presence of liquid oxygen or nitrogen. High coefficients (0.2 to 0.4) were also found by Buckley et al. for filled PTFE under high vacuum, but some of the fillers, notably copper, silver and powdered coke gave coefficients lower than for unfilled PTFE under the same conditions. The reasons for these effects are not known: the effects may be due to temperature of environment, or the mechanisms may be similar to that experienced with graphite where the low coefficient of friction is attributed to the presence of absorbed gases at the crystallite interfaces where cleavage occurs.

Reference: *Physical Properties Of Unfilled And Filled Fluon Polytetrafluoroethylene,* supplier design guide (Technical Service Note F12/13) - ICI PLC, 1981.

Effect of Operating Variables on Friction Properties

ICI: Fluon (features: 60-65 Shore D hardness, unfilled)

Most studies on the friction of PTFE have been carried out with unfilled PTFE. While the exact mechanisms involved are still not fully understood, a picture emerges in which the 'dry' coefficient of friction is dependent upon the pressure, the speed, the temperature, the mating surface, the orientation of the PTFE, the environment and the time of running. Coefficients from 0.016 to 0.36 have been quoted. The classical laws of dry friction state that the friction force is independent of the apparent area of contact, making the friction force proportional to load rather than pressure. Many investigators quote the coefficient of friction of PTFE as a function of load and show it to rise steeply at very light loads (below 5 lb) and decrease with increasing load. RB Lewis does not support this, but suggests the coefficient of friction (μ) is proportional to the applied pressure P (lbf/in^2) according to a given formula.

Speed

The coefficient of friction falls markedly at low speeds (below 50 mm/s; 10 fpm) and increases with increasing speed.

Temperature

The coefficient of friction appears to be stable over the range - 45°C to 100°C (-49°F to 212°F) but to rise at lower temperatures and fall at higher temperatures.

Mating Surface

Work by Steijn showed that sliding of PTFE against steel gave lower coefficients of friction than sliding bulk PTFE against bulk PTFE. He suggests that when mating areas are large, friction is primarily due to adhesion.

Orientation

It has been shown that the coefficient of friction can be affected by up to 30% depending upon the orientation of the PTFE molecules.

Environment

Steijn showed that prolonged and continous running under dry nitrogen (5-10 parts per million of water) gave rise to intermittently high coefficients of friction, but this was alleviated as soon as normally moist air (50% RH) was admitted. The short term tests at temperatures from - 1°C to + 60°C (30°F to 140°F) in helium, oxygen, nitrogen and air showed no such effect and neither did tests in air at room temperature with relative humidities in the range 12% to 54%. The friction of PTFE in vacuum (10^{-9} mm H_g) was studied by Buckley and Johnson who obtained coefficients of friction of 0.25 with a load of 1kg. They also report the coefficient to be constant over the speed range <50 mm/s-5 m/s (<10-1000 fpm). This high figure could well be attributed to the relatively small loads applied, but may be linked with Steijn's observations regarding very dry atmospheres.

Several investigators have shown that the coefficient of friction is decreased dramatically by the addition of lubricants. This is not surprising since, if a full film of oil is present, the friction is virtually independent of the mating surfaces.

Time of Running

The work of Steijn shows that the coefficient of friction for PTFE on PTFE is influenced by the number of traverses, the time lapse between runs, the nature (especially velocity) of the preceding sliding, and the thermal history of the sliding components. Mitchell and Pratt demonstrated a similar increase in friction with time for PTFE on steel, up to a steady level (from 0.05 to 0.20 in 4 hours), and showed this to be due to a change in the surface of the PTFE rather than a change in the surface of the steel (i.e., the transfer of PTFE to the steel).

Reference: *Physical Properties Of Unfilled And Filled Fluon Polytetrafluoroethylene,* supplier design guide (Technical Service Note F12/13) - ICI PLC, 1981.

Effect of Mating Surface on Wear Properties

ICI: Fluon (features: 60-65 Shore D hardness, unfilled)

At room temperatures and above it is generally agreed that a hard, approximately 900 VPN (Vickers Pyramid Number), mating surface is beneficial. Softer materials can be used providing the filler will not abrade them. The materials with good dry bearing properties of their own (e.g., bronze) are preferred to the softer, more easily damaged materials (e.g., aluminum). There is some divergence of opinion as to the suitability of chromium plating. Pratt shows chromium plating to be advantageous whereas O'Rourke et al. show it to give poor results. The answer might well be that the fillers used by Pratt were less abrasive than those use by O'Rourke.

The surface finish of a material is generally quoted as a mean of the 'peaks and valleys' of its surface as detected by traversing a diamond stylus across it. This does not fully specify a surface however, since a turned and a ground surface of the same value will be different. It is now generally accepted that a ground surface is superior to a turned surface and that above 0.75 µm the wear rate of the filled PTFE will increase. The existence of a lower limit is still in dispute and so the best compromise is to use a ground surface finish of 0.2-0.4 µm.

'Lubricant' is a very general term and it used to be stated that any liquid will act as a lubricant and be beneficial to PTFE. To some extent this is true in that, if hydrodynamic conditions are established, no wear will take place, but filled PTFE may run under conditions of boundary lubrication. Hydrocarbon oils are generally advantageous, with a significant reduction in wear rates. This is not so with water. O'Rourke confirmed that the wear factor increased for unfilled and various filled PTFE compounds when running against steel with water boundary lubrication. Tests by ICI have shown that boundary lubrication with water gave a reduction of wear life of 50% when filled PTFE ran against steel.

Reference: *Physical Properties Of Unfilled And Filled Fluon Polytetrafluoroethylene,* supplier design guide (Technical Service Note F12/13) - ICI PLC, 1981.

Effect of Temperature on Wear Properties

ICI: Fluon (features: 60-65 Shore D hardness, unfilled)

RB Lewis suggests that each material has two wear rates, mild wear and severe wear which is attributed to a rise in temperature at the interface. The actual temperature at which the transition occurs is reported to depend upon the load. He concludes that the PV value at which transition begins depends upon the application geometry, ambient temperature, and manner and amount of cooling, while the slope of the transition depends upon the application parameters and properties of the compositions. The mild wear is reported to be characterized by wear of the surface layers while severe wear is characterized by bulk removal of material. Similar conclusions were drawn by Summers and Smith who considered their composite to be hard granules in a cement of softer materials. He suggests the mild wear region corresponds to a gradual attrition of the hard granules, the change to severe wear occurring when the 'cement' becomes softened by heat and the granules are plucked bodily out of the matrix. From work carried out by Mitchell and Pratt and in ICI's laboratories, it is concluded that the entrapment of wear debris as well as surface temperature is a very important factor in determining whether severe wear occurs or not. For example, although the difference in running conditions between a thrust washer and a piston ring is mainly considered to be one of interface temperature, it is also true that wear debris is far less likely to become entrapped in the piston ring. It is also true that differences in wear rate can be attributed to differences in behavior (abrasive or otherwise) when trapped wear debris is present. Whatever the mechanisms, it is generally accepted that an increase in interface temperature increases the wear rate.

Reference: *Physical Properties Of Unfilled And Filled Fluon Polytetrafluoroethylene,* supplier design guide (Technical Service Note F12/13) - ICI PLC, 1981.

GRAPH 258: **Coefficient of Friction vs. Load for DuPont Teflon Polytetrafluoroethylene.**

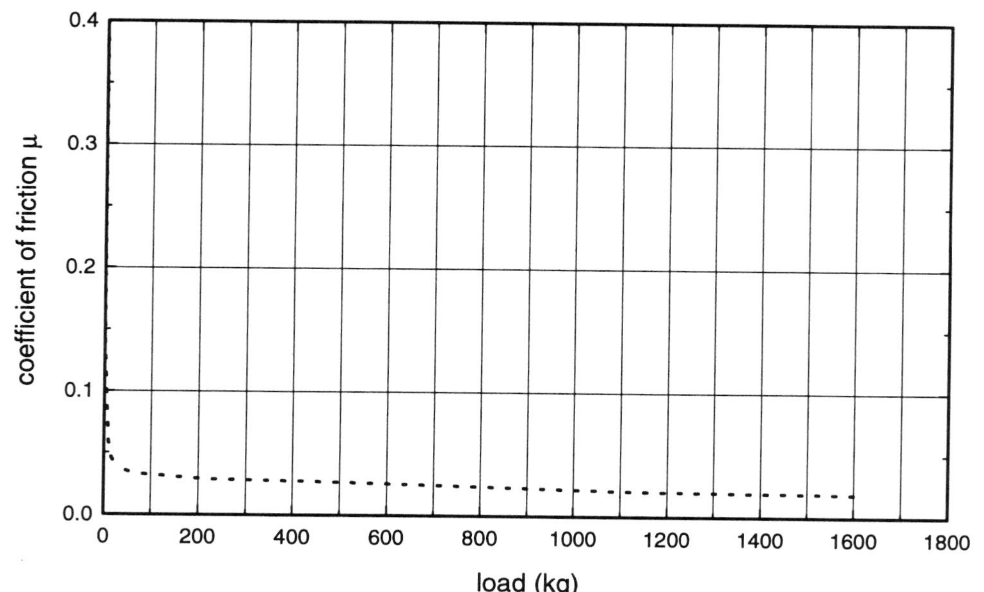

GRAPH 259: **Coefficient of Friction vs. Sliding Speed at Different Pressures for DuPont Teflon Polytetrafluoroethylene.**

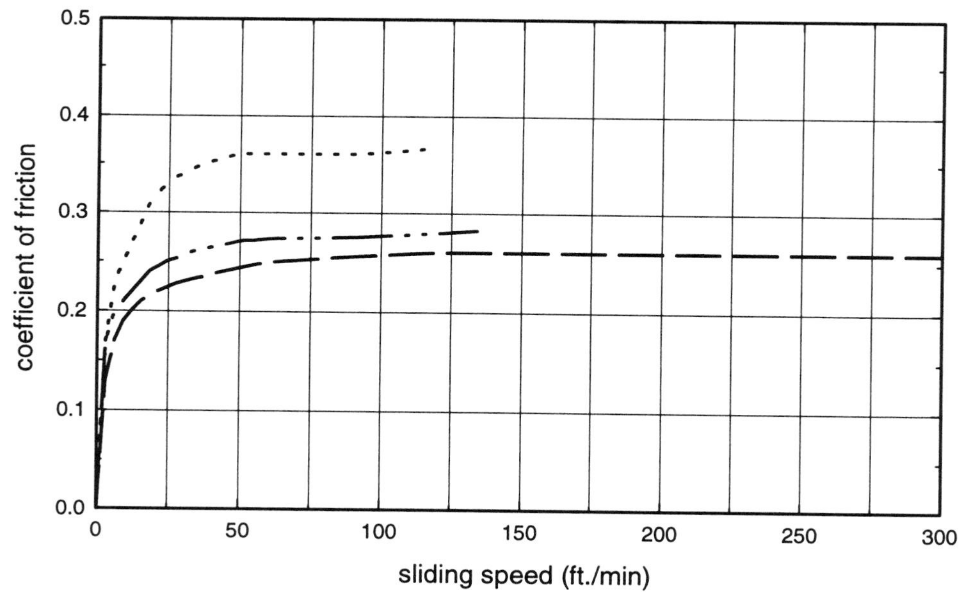

Chapter 75

Polyvinylidene Fluoride

Effect of Temperature on Friction and Wear Properties

LNP Engineering Plastics: Fluoromelt FP-PC-1003 (material compostion: 15% carbon fiber reinforcement)

The test specimens of carbon fiber reinforced PVDF were unable to sustain the conditions of the wear test at 149°C (300°F), probably because the PV limit of the composite at that temperature may have been exceeded. The PVDF composite also had the lowest LPV value of all materials investigated.

Reference: *Lubricomp Internally Lubricated Reinforced Thermoplastics and Fluoropolymer Composites,* supplier design guide (254-691) - LNP Engineering Plastics, 1991.

Chapter 76

Nylon 46

Friction and Wear Properties

DSM: Stanyl

Stanyl 46 nylon has excellent abrasion resistance and, at elevated temperatures, performs better than other plastics. Its smooth, tough surface and stiffness at elevated temperatures make Stanlyl 46 nylon an ideal material for sliding parts. Among the standard engineering thermoplastics, nylons serve traditionally as excellent bearing materials. Within the nylon family, those with a higher melting point and level of crystallinity have the best wear resistance. Therefore Stanyl 46 nylon (due to its high level of crystallinity) is an excellent material for bearing applications.

Reference: *Stanyl 46 Nylon General Information,* supplier design guide (MBC-PP-492-5M) - DSM, 1992.

Friction and Wear Properties

BASF: Ultramid B

By virtue of its smooth, tough and hard surface, its crystalline structure, its high resistance to heat, and its resistance to lubricants, fuels and solvents, Ultramid is an ideal material for parts subjected to sliding friction. A striking feature is its performance in the absence of lubrication. Unlubricated metal systems tend to seize, whereas Ultramid pairs can function satisfactorily even without lubrication. Friction and wear are properties of the system and depend on many parameters, e.g. on the materials paired, the roughness of the surfaces, the geometry of the parts in contact, the nature of any intermediate medium, e.g. lubricant, and the stresses brought about by external conditions, e.g., load, speed, and temperature.

The main factors that affect the amount of wear and the coefficient of sliding friction for Ultramid are the hardness and roughness of the material with which it is paired, the pressure applied, the length of the path of contact, the temperature of the rubbing surfaces, and the nature of the lubrication. Drop erosion and cavitation frequently occur in parts for water pumps. Ultramid has proved to be superior to aluminum in withstanding erosion of this nature. Examples of articles that may be eroded by granular solids entrained in streams of air or liquids are fans and spoilers. By virtue of its flexibility, Ultramid is very resistant to this form of erosion.

Reference: *Ultramid Nylon Resins Product Line, Properties, Processing,* supplier design guide (B 568/1e/4.91) - BASF Corporation, 1991.

GRAPH 260: Dynamic Coefficient of Friction vs. Sliding Time for DuPont Zytel 211 Nylon 6.

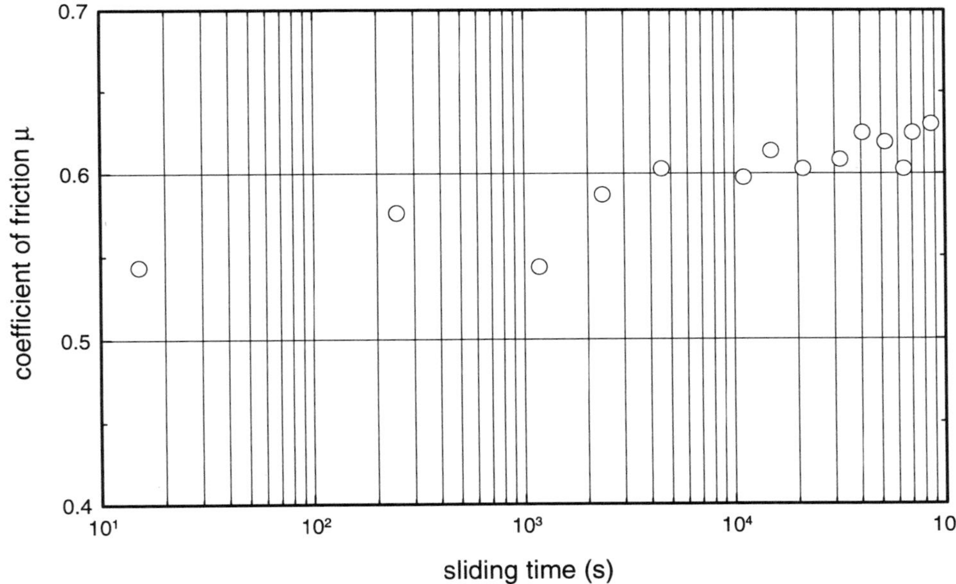

○	DuPont Zytel 211 Nylon 6; apparatus: pin on plane; mating surface: steel (highly polished); speed: 0.15 m/sec; load: 22.3 N
Reference No.	380

GRAPH 261: **Surface Velocity vs. Pressure for Ube Nylon 6.**

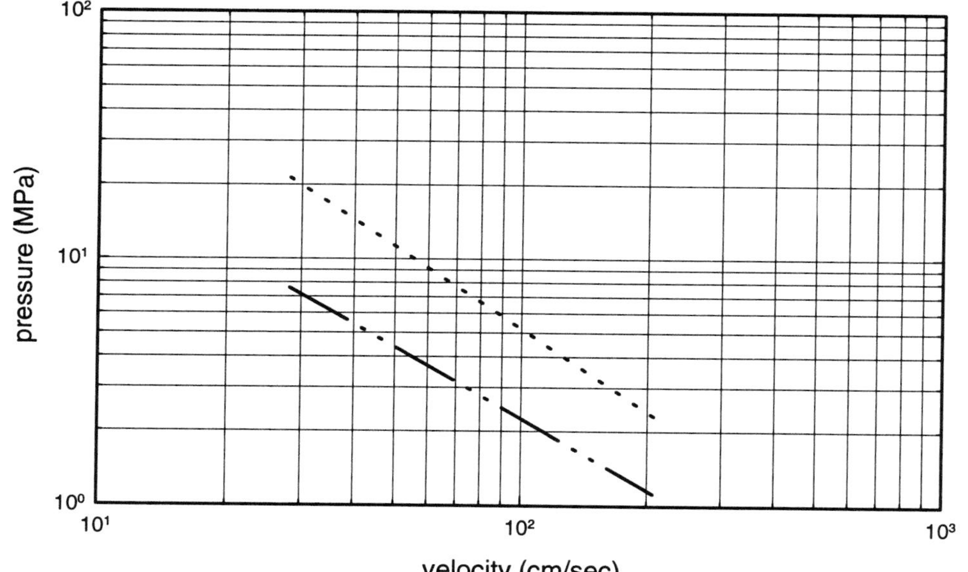

···············	Ube Nylon 6 (10% carbon fiber); mating surface: steel (S45C); 23°C; DAM
—··—··	Ube Nylon 6 (unfilled); mating surface: steel (S45C); 23°C; DAM
Reference No.	238

Chapter 78

Nylon 66

Friction and Wear Properties

BASF: Ultramid A3R (features: high flow; modifier: polyethylene; note: stabilized; : , noiseless friction bearings);

By virtue of its smooth, tough and hard surface, its crystalline structure, its high resistance to heat, and its resistance to lubricants, fuels and solvents, Ultramid is an ideal material for parts subjected to sliding friction. A striking feature is its performance in the absence of lubrication. Unlubricated metal systems tend to seize, whereas Ultramid pairs can function satisfactorily even without lubrication.

Friction and wear are properties of the system and depend on many parameters, e.g. on the materials paired, the roughness of the surfaces, the geometry of the parts in contact, the nature of any intermediate medium, e.g. lubricant, and the stresses brought about by external conditions, e.g., load, speed, and temperature.

The main factors that affect the amount of wear and the coefficient of sliding friction for Ultramid are the hardness and roughness of the material with which it is paired, the pressure applied, the length of the path of contact, the temperature of the rubbing surfaces, and the nature of the lubrication. The Ultramid A resins, particularly A3R, and the mineral-filled products feature low coefficients of friction and rates of wear, i.e. wear intensity S in µm/km. Drop erosion and cavitation frequently occur in parts for water pumps. Ultramid has proved to be superior to aluminum in withstanding erosion of this nature. Examples of articles that may be eroded by granular solids entrained in streams of air or liquids are fans and spoilers. By virtue of its flexibility, Ultramid is very resistant to this form of erosion.

Reference: *Ultramid Nylon Resins Product Line, Properties, Processing,* supplier design guide (B 568/1e/4.91) - BASF Corporation, 1991.

DuPont: Minlon (reinforcement: mineral); **Zytel ST (super tough), Zytel 70613L, Zytel 70633L** (reinforcement: mineral); **DuPont: Zytel 101** (features: general purpose grade, unlubricated); **Zytel 70G 13L** (features: lubricated, general purpose grade; material compostion: 13% glass fiber reinforcement); **Zytel 70G 33L** (features: lubricated, general purpose grade; material compostion: 33% glass fiber reinforcement); **Zytel** (reinforcement: glass fiber);

Zytel nylon resins have excellent friction and wear characteristics, and Zytel can be used without lubricant in many applications. However, continous or initial lubrication of the surface extends the range of applicability. The measured coefficient of friction depends upon many variables, including equipment, temperature, clearance, material, hardness and finish of the mating surface. The values are also dependent on pressure and velocity. Data on coefficients of friction indicate that there is little variation over a temperature range of 23°C to 121°C (73°F to 250°F) and rubbing velocities of 0.04 to 2.0 m/sec (8 to 400 fpm) In any application where friction is critical, it is recommended that measurements be made under simulated operating conditions.

The coefficient of friction for Zytel ST is 0.29 and is determined by using the Thrust Washer Test against carbon steel, at a speed of 51 mm/s (10 fpm) and 2.1 MPa (300 psi). The samples were conditioned to equilibrium moisture content at 50% R.H. GRZ nylon resins retain much of the natural lubricity and excellent wear resistance of unreinforced nylons. The static coefficients for Zytel 70G 13 L and 70G 33L range from 0.15 to 0.40. Coefficient of Friction of Minlon is in the same range as that of the Glass-Reinforced Zytel resins.

Reference: *Design Handbook For Du Pont Engineering Plastics - Module II,* supplier design guide (E-42267) - Du Pont Engineering Polymers.

DuPont: Zytel 101

The static coefficient of friction diminishes when the normal load is rising. This shows that the adhesion component of friction is more significant during those tests. The surface roughness of the polymeric samples has a marked effect on static friction. The static coefficient of friction increases when the surface roughness of the polymeric samples diminish. This is due to the appreciable adhesion when smooth surfaces are in contact.

Reference: Benabdallah, H. (Ecole Polytechnique de Montreal), Fisa, B. (Ecole Polytechnique de Montreal), *Static Friction Characteristics of Some Thermoplastics,* ANTEC 1989, conference proceedings - Society of Plastics Engineers, 1989.

Abrasion Resistance

DuPont: Minlon (reinforcement: mineral); **Zytel 101** (features: general purpose grade, unlubricated); **Zytel 42** (features: low flow; process type: extrusion); **Zytel** (reinforcement: glass fiber); **Zytel 158** (Nylon 612)

Abrasion resistance of plastics is measured by a variety of tests. For nylon, the usual test is the non-standard Taber abrasion adapted from ASTM D1044. Other tests have been used for measuring the resistance to abrasion of plastic materials. In all of these tests, the unreinforced Zytel resins are outstanding among plastics in resistance to abrasion. A resilient material like Zytel can deform under load and return to its original dimensions without wear. For example, worm gears have operated more than 18 months with little or no wear; whereas, metal gears in the same equipment had the teeth worn to a knife edge in three to six months. Zytel 42, an extrusion grade nylon, and Zytel 158 nylon resin are even more resistant to abrasion in the Taber test than Zytel 101. The glass-reinforced nylons, GRZ, and the mineral-reinforced nylons, Minlon, show greater wear than the unreinforced Zytel nylon resins.

The abrasion resistance of Zytel 101 nylon resin in both the Taber and the Ball Mill tests was compared with other types of plastics. Zytel 101 shows far less material loss than any other plastic. Weight loss of various materials relative to Zytel in Taber and Ball Mill abrasion tests are as follows:

Material	Taber	Ball Mill
Zytel	1	1
Polystyrene	9-20	15-20
Terpolymer of styrene, butadiene &	9	10-20
Cellulose acetate	9-10	-
Cellulose acetate butyrate	9-15	10-20
Methyl methacrylate	2-15	10-20
Melamine formaldehyde (molded)	-	15-20
Phenol formaldehyde (moldings)	4-12	-
Hard rubber	-	10
Die cast aluminum	-	11
Mild steel	-	15-20

Reference: *Design Handbook For Du Pont Engineering Plastics - Module II,* supplier design guide (E-42267) - Du Pont Engineering Polymers.

Effect of Fillers and Additives on Wear Properties

DuPont: Nylon 66 (reinforcement: glass fiber); **Nylon 66** (features: unmodified resin); **Nylon 66** (reinforcement: aramid fiber)

At PV = 2500 psi · fpm (P = 250 psi, V = 10 fpm), aramid fiber lowers the nylon 66 wear rate by a factor of 4 and also lowers the coefficient of friction. These are very important changes because nylon 66 is generally considered to be self lubricating. The very low weight loss of the steel washer indicates the nonabrasive nature of the aramid fiber. In contrast, glass fiber has adverse effects; it scratches the steel and causes severe abrasion. Surface profiles of these two washers, measured by a moving diamond stylus in order to show the surface valleys and hills, clearly show the erosion of the steel washer surface by glass fiber.

Preliminary study showed that aramid fiber increases the PV limit of nylon 66 from 9000 to 15,000 at V = 10 fpm, and from 2500 to about 10,000 at V = 100 fpm.

Reference: Wu, Y.T., *How Short Aramid Fiber Improves Wear Resistance,* Modern Plastics, trade journal - McGraw-Hill, 1988.

LNP Engineering Plastics: Lubricomp RAL-4022 (material compostion: 10% aramid fiber, 10% PTFE modifier);

Adding aramid fiber to a nylon 66 resin can reduce the wear factor by more than two thirds. The aramid reinforced nylon 66 composite produces extremely low wear on a carbon steel mating surface. The wear factor with the aramid reinforced nylon 66 composite is only one-fourth that of either glass or the carbon reinforced analog. The addition of PTFE lubricant lowers wear factors still further, while also lowering the coefficient of friction.

Reference: *Lubricomp Internally Lubricated Reinforced Thermoplastics and Fluoropolymer Composites,* supplier design guide (254-691) - LNP Engineering Plastics, 1991.

Effect of Operating Variables on Friction and Wear Properties

DuPont: Nylon 66 (features: unmodified resin)

To understand the effect of pressure (P) and velocity (V) on wear rate, tests were conducted on neat nylon 66 at two velocities (10 fpm and 100 fpm) and P ranges from 50 psi to 1500 psi. This gives PV values of 500 to 15,000. At V = 10 fpm and PV less than 2500, the wear rate is linearly dependent on PV. When PV is greater than 2500, the wear rate plateaus until it suddenly jumps up at high PV (15,000). this sudden increase in wear factor indicates that there has been a change in wear mode caused by surface softening or melting. For nylon 66 at V = 10 fpm, this surface softening occurs somewhere near PV = 10,000. Examination of the nylon 66 discs tested at PV = 10,000 using a microscope shows strings of material hanging from the outer edges and some scattered radial cracking near the edges. This indicates moderate softening of the material and possibly a mild thermal fatigue. Material losss due to craking leads to a higher weight loss and relatively low thickness loss. The PV limit of nylon 66 has been exceeded at 10,000.

A similar examination was made on a disc tested at PV = 5000 and V = 10 fpm. The surface was smooth and no cracking was observed. It was concluded that the PV limit for nylon 66 at V = 10 fpm is slightly lower than 10,000. When PV = 15,000 and V = 10 fpm, frictional heating was so excessive that the test disc was melted and squeezed back to a mushroom shape. Thickness change was high but the weight loss was small. Neat nylon 66 should not be used at this PV condition.

Test results for nylon 66 at V = 100 fpm show that the increase in velocity has dramatic effects on the PV limit and washer surface temperature. AT PV = 2500, the same mild stringing and edge cracking phenomena were observed, indicating the PV limit was reached. AT PV = 5000, edge cracking due to thermal fatigue was very severe. As a result, the wear data at PV > 2500 can only be used as references, not in design. It has been concluded that the PV limit for nylon 66 at V = 100 fpm is 2500 psi · fpm.

Reference: Wu, Y.T., *How Short Aramid Fiber Improves Wear Resistance,* Modern Plastics, trade journal - McGraw-Hill, 1988.

Effect of Temperature on Friction and Wear Properties

LNP Engineering Plastics: Lubricomp RFL-4036 (material compostion: 30% glass fiber reinforcement, 15% PTFE modifier);

The nylon 66 composite (15% PTFE, 30% glass fiber) retained a high percentage of room temperature LPV value throughout the range of test temperatures. Wear resistance at elevated temperatures is only fair, however, so the material can be considered only for intermittent use in service to about 204°C (400°F).

In general the addition of PTFE to the fiber reinforced composites improves tribological properties at all temperatures. Also, composites containing carbon fiber reinforcement have better bearing properties than those of their glass fiber containing analogs.

Reference: *Lubricomp Internally Lubricated Reinforced Thermoplastics and Fluoropolymer Composites,* supplier design guide (254-691) - LNP Engineering Plastics, 1991.

GRAPH 262: Static Coefficient of Friction vs. Load Against Steel of Various Surface Roughness for DuPont Zytel 101 Nylon 66.

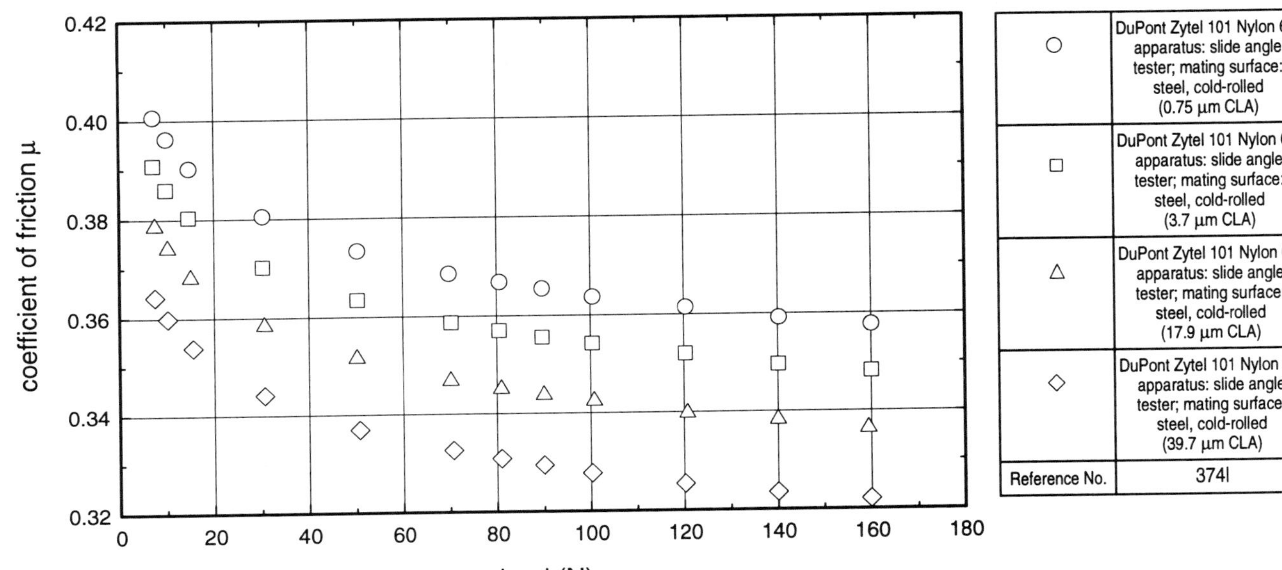

GRAPH 263: Surface Velocity vs. Pressure for Ube Nylon 66.

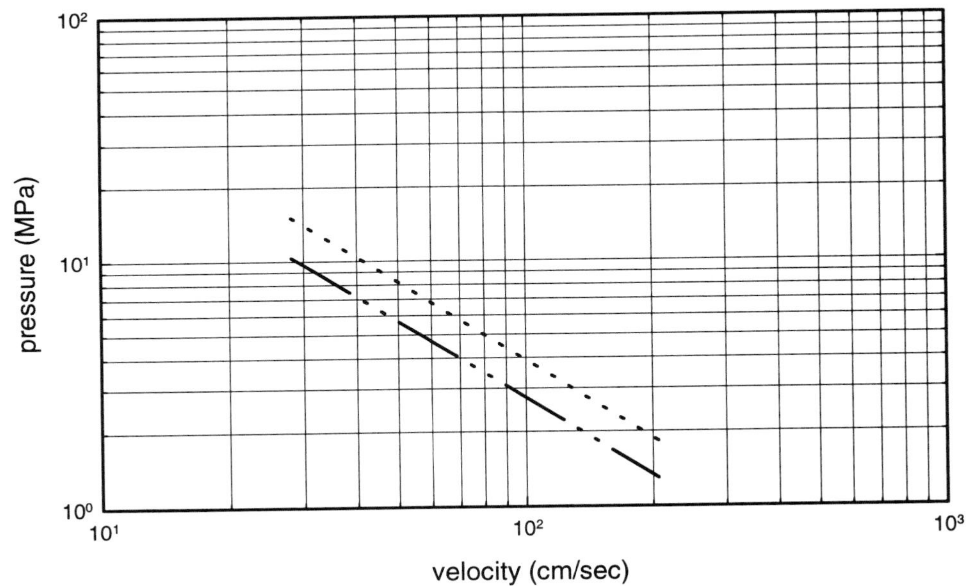

··············	Ube Nylon 66 (30% glass fiber); mating surface: steel (S45C); 23°C; DAM
—·—··	Ube Nylon 66 (unfilled); mating surface: steel (S45C); 23°C; DAM
Reference No.	238

Chapter 79

Polybenzimidazole

Friction and Wear Properties

Hoechst Celanese: Celazole

Unfilled melt-processable PBI compound was tested with virgin PBI, unfilled polyimide and polyamideimide as the controls, while melt-processable PBI lube grade was compared to virgin PBI, unfilled polyimide, PBI lube grade and lube grade polyimide. The comparative results indicate that the unfilled melt-processable PBI exhibits outstanding wear and friction properties among these high performance unfilled engineering plastics. This nature of the unfilled melt-processable PBI compounds, when combined with selected lubricants, gives the melt-processable PBI lube grade good wear and friction performance (low wear rate, low running temperature and low friction coefficient) among the high performance lube grade plastics. Further, all reinforced melt-processable compounds have improved impact resistance and tensile properties when compared to virgin PBI.

Reference: Willis, E.N., Wang, L.C., *Ultra-High Performance Melt Processable Polybenzimidazole Parts,* SAE Technical Paper Series - International Congress & Exposition, conference proceedings (920818) - SAE International, 1992.

Chapter 80

Polycarbonate

Friction Properties

Dow Chemical: Calibre (features: transparent)

The coefficients of friction of Calibre polycarbonate in contact with itself are 0.30 - 0.35 for static friction, and 0.35 - 0.40 for kinetic friction. Generally, these values are consistent for Calibre polycarbonate in contact with polished stainless steel.

Reference: *Calibre Engineering Thermoplastics Basic Design Manual,* supplier design guide (301-1040-1288) - Dow Chemical Company, 1988.

Wear Properties

Dow Chemical: Calibre (features: transparent)

Calibre polycarbonate can function well in an abrasive environment. Calibre polycarbonate loses about 0.003 cm^3 in volume and about 2.5% in light transmission (from 90% to 87.5%) after 100 cycles.

Reference: *Calibre Engineering Thermoplastics Basic Design Manual,* supplier design guide (301-1040-1288) - Dow Chemical Company, 1988.

Bisphenol A PC (note: neat resin)

The wear rate vs. sliding distance curve for neat polycarbonate showed its tribological behavior to be very poor. Polycarbonate could not form a coherent transfer film on the metal counterpart leading to a very high wear rate of 1.0 x 10^{-13} m^3/Nm.

Reference: Bolvari, Anne (LNP), *Analysis of Dry Sliding Wear and Friction Data of Internally Lubricated Composites,* ANTEC 1994, conference proceedings - Society of Plastics Engineers, 1994.

Effect of Weld Line and Notch on Fatigue Behavior

Bisphenol A PC (reinforcement: glass fiber); **Bisphenol A PC** (modifier: PTFE; reinforcement: glass fiber); **Bisphenol A PC** (material compostion: 15% PTFE modifier)

The addition of PTFE to polycarbonate resulted in a material with very good tribological properties due to the formation of a PTFE transfer film on the counterpart. The equilibrium wear rate of this compound was 1.2 x 10^{-15} m^3/Nm which was two orders of magnitude better than neat polycarbonate. The addition of glass fiber to PTFE/polycarbonate further improved the wear rates to an equilibrium value of 5.0 x 10^{-16} m^3/Nm. The shape of the wear rate vs sliding distance curve showed that an abrasive wear as well as an adhesive wear process was contributing to the overall wear rate. The wear rate vs. sliding distance curve for a glass fiber/polycarbonate composite without PTFE showed a rapid rise in wear rate. Real time friction data showed the same trend. Without a coherent transfer film, glass fiber abrasion became the major wear mechanism.

Reference: Bolvari, Anne (LNP), *Analysis of Dry Sliding Wear and Friction Data of Internally Lubricated Composites,* ANTEC 1994, conference proceedings - Society of Plastics Engineers, 1994.

GRAPH 264: Dynamic Coefficient of Friction vs. Time for Modified and Unmodified Bisphenol A Polycarbonate.

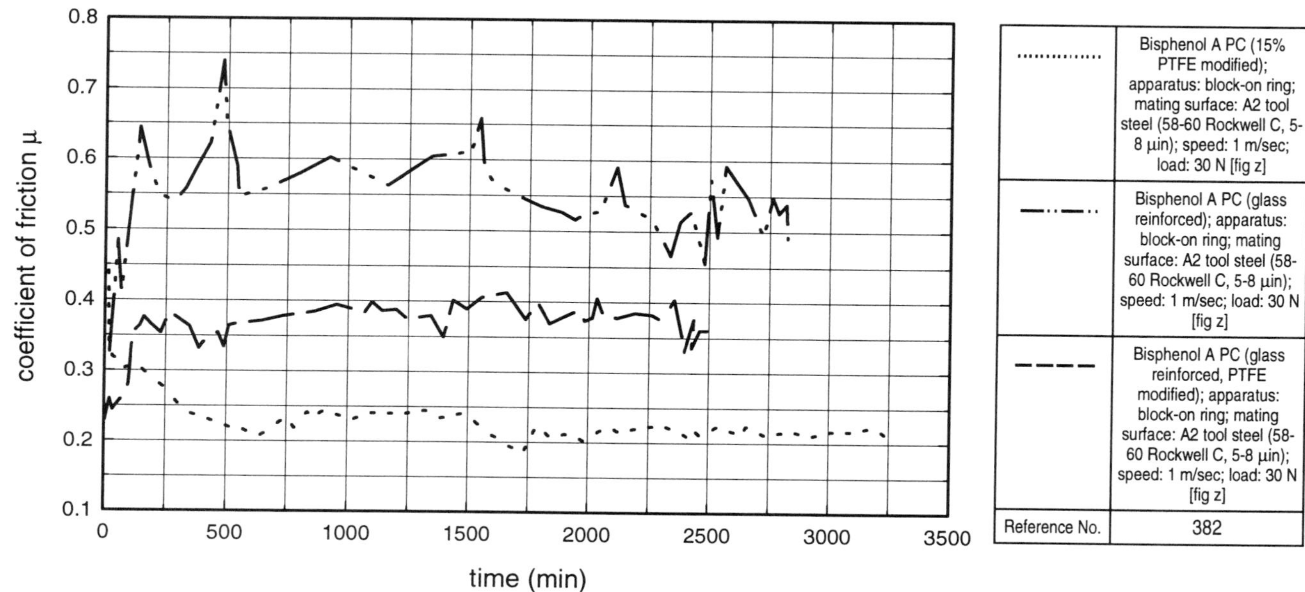

GRAPH 265: Wear Rate vs. Sliding Distance for Modified and Unmodified Bisphenol A Polycarbonate.

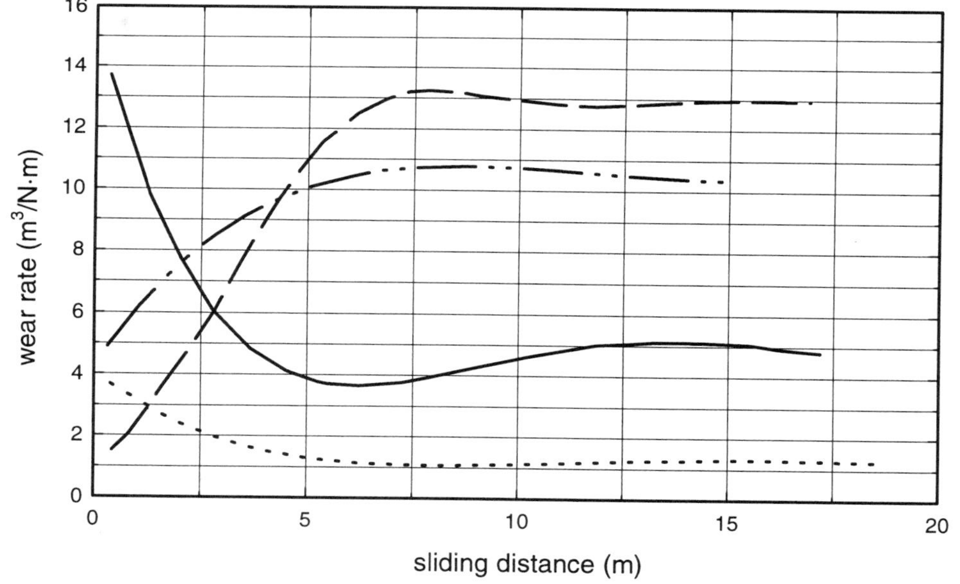

···············	Bisphenol A PC (15% PTFE modified); apparatus: block-on ring; mating surface: A2 tool steel (58-60 Rockwell C, 5-8 μin); speed: 1 m/sec; load: 30 N [fig z]
—·· —··	Bisphenol A PC (neat resin); apparatus: block-on ring; mating surface: A2 tool steel (58-60 Rockwell C, 5-8 μin); speed: 1 m/sec; load: 30 N [fig z]
– – – –	Bisphenol A PC (glass reinforced); apparatus: block-on ring; mating surface: A2 tool steel (58-60 Rockwell C, 5-8 μin); speed: 1 m/sec; load: 30 N [fig z]
———	Bisphenol A PC (glass reinforced, PTFE modified); apparatus: block-on ring; mating surface: A2 tool steel (58-60 Rockwell C, 5-8 μin); speed: 1 m/sec; load: 30 N [fig z]
Reference No.	382

Chapter 81

Polybutylene Terephthalate

Surface Properties

Hoechst AG: Celanex

Celanex has good surface properties, such as hardness, abrasion resistance and low friction. In sliding contact with metal or other thermoplastics, Celanex also displays good friction properties. The material's high hardness and low friction coefficient mean that components made from Celanex in many cases exhibit higher abrasion resistance than parts made from other polymers or even metal.

Reference: *Celanex Polybutylene Terephthalate (PBT),* supplier design guide (BYKR 123 E 9070/014) - Hoechst AG, 1990.

Friction and Wear Properties

BASF AG: Ultradur B

By virtue of its low coefficient of friction and high resistance to wear, Ultradur is an ideal material for sliding parts. The coefficient of friction and the wear depend on the pressure applied, the temperature of the rubbing surfaces, the length of the path of contact, and the roughness and hardness of the mating surface. The rubbing velocity is not a significant factor, unless it is responsible for heating the rubbing surfaces. If the average roughness height of the steel mating surface is reduced to R_v = 0.2 - 0.3 μm (0.008 - 0.012 mil) from R_v = 0.5 μm (0.02 mil), the rate of wear can be halved, but the coefficient of sliding friction will be increased to about 0.4. If the hardness of the mating metal surface is less than 50 HRC, higher rates of wear and coefficients of friction can be expected.

Reference: *Ultradur Polybutylene Terephthalate (PBT) Product Line, Properties, Processing,* supplier design guide (B 575/1e - (819) 4.91) - BASF Aktiengesellschaft, 1991.

GRAPH 266: **Dynamic Coefficient of Friction vs. Pressure for BASF Ultradur Polybutylene Terephthalate Polyester.**

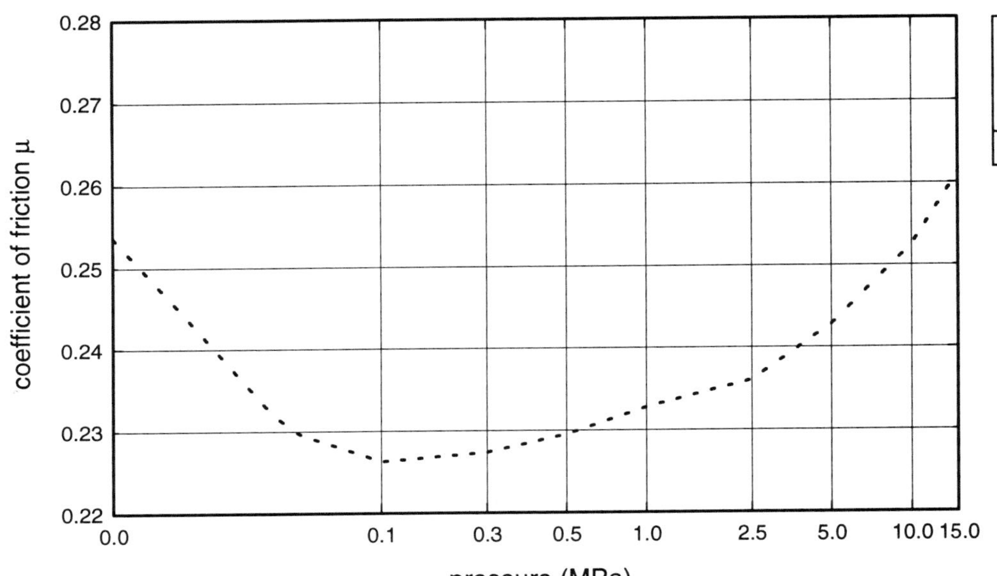

....	BASF AG Ultradur B PBT (unmodified resin, unlubricated); mating surface: steel (0.5 μm roughness); unlubricated
Reference No.	180

GRAPH 267: **Dynamic Coefficient of Friction vs. Pressure for Huls Vestodur 2000 Polybutylene Terephthalate Polyester.**

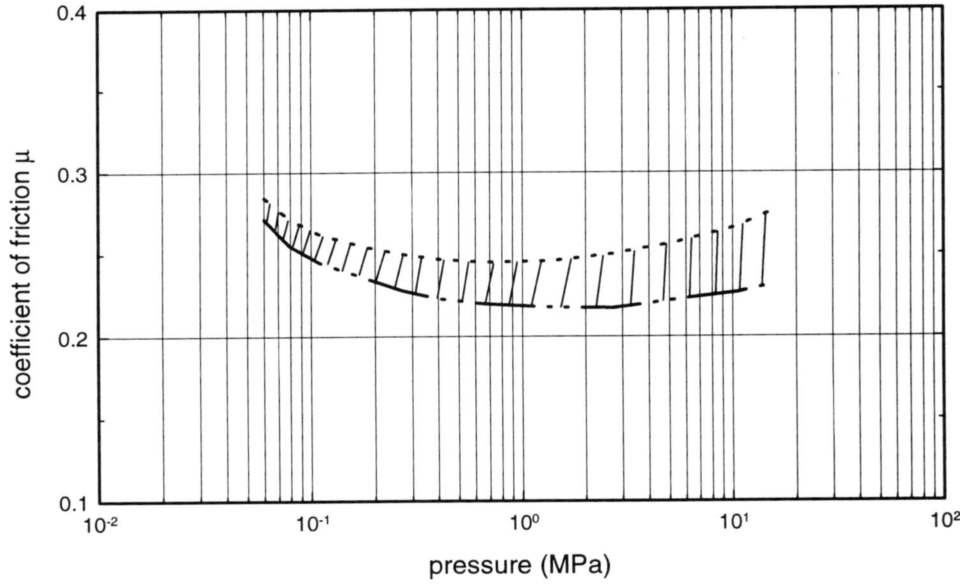

....	Huls Vestodur 2000 PBT (unmodified resin, gen. purp. grade, 77 Shore D; 150 (H30) Ball ind. hardness); velocity: 0.5 m/s; upper limit values; mating surface: steel
—·-—··	Huls Vestodur 2000 PBT (unmodified resin, gen. purp. grade, 77 Shore D; 150 (H30) Ball ind. hardness); velocity: 0.5 m/s; lower limit values; mating surface: steel
Reference No.	403

GRAPH 268: Wear Rate vs. Pressure for BASF Ultradur Polybutylene Terephthalate Polyester.

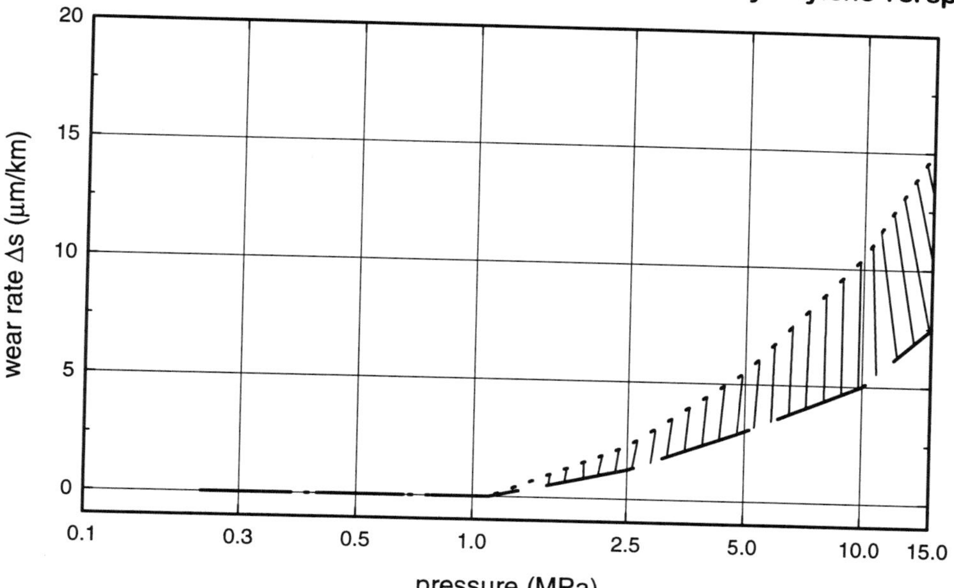

....	BASF AG Ultradur B PBT (unmodified resin, unlubricated); mating surface: steel (0.5 μm roughness); unlubricated; high range of values
– – – –	BASF AG Ultradur B PBT (unmodified resin, unlubricated); mating surface: steel (0.5 μm roughness); unlubricated; low range of values
Reference No.	180

Chapter 82

Liquid Crystal Polymer

Friction Properties

Hoechst AG: Vectra

Vectra has high surface hardness with Rockwell hardness scale M values of 60 to 98. Vectra moldings have good sliding properties; slide elements made from Vectra have coefficients of dynamic friction against steel in the range 0.1 to 0.4. The sliding properties of Vectra can be made even better by additives, such as Vectra grades A 430 and A 435, which have good sliding properties through the addition of PTFE. The grades in the 400 series also exhibit good wear behavior.

Reference: *Vectra Polymer Materials,* supplier design guide (B 121 BR E 9102/014) - Hoechst AG, 1991.

Chapter 83

Polyamideimide

Friction and Wear Properties

Amoco Performance Products: Torlon 4275 (features: wear resistant; material compostion: 3% fluorocarbon, 30% graphite powder); **Torlon 4301** (features: wear resistant; material compostion: 20% graphite powder, 3% fluorocarbon); **Torlon 4347** (features: wear resistant; material compostion: 12% graphite powder, 8% fluorocarbon)

New possibilities in the design of moving parts are opened by Torlon wear-resistant grades; 4301, 4275, and 4347. These grades offer high compressive strength and modulus, excellent creep resistance and outstanding retention of strength and modulus at elevated temperatures, as well as self-lubricity and low coefficients of thermal expansion, which make them prime candidates for wear surfaces in severe service. Torlon bearings are dependable in lubricated, unlubricated, and marginally lubricated service. Some typical applications which lend themselves to this unique set of properties are ball bearings, thrust washers, piston rings, vanes, valve seats, bushings and wear pads.

Low wear factors (K) are characteristic of wear resistant materials. Fluorocarbons, which have low coefficients of friction, have very low wear factors but limited mechanical properties and poor creep resistance. At low PV's, Torlon wear resistant grades have wear factors comparable to filled polytetrafluorethylene (PTFE), a fluorocarbon, but Torlon polymers offer superior creep resistance and strength. Torlon polymers have wear factors similar to those of more expensive polymide resins. In addition, Torlon resins are injection moldable, polyimides are not.

Amoco has developed wear data from tests run on an unlubricated thrust washer. They have found this procedure to be more reproducible than tests using journal bearings, and the results can predict journal bearing test results with a fair degree of confidence. All three wear resistant grades are useful in excess of 50,000 PV. Most other engineering resins fail far below this level. Under the test conditions, Torlon wear resistant grades are similar in wear rate to the more expensive polyimide Vespel SP21. At low PVs, Torlon wear rates are similar to filled PTFE.

The impressive performance of Torlon bearing grades in nonlubricated environments is insurance against catastrophic part failure or seizure upon lube loss in a normally lubricated environment. In a transmission lubricated with hydrocarbon fluid, Torlon thrust washers are performing well at PVs of 1,300,000. In a water lubricated hydraulic motor vane, excellent performance has been attained at over 2,000,000 PV. The wear resistance of Torlon parts depends on proper post-cure. A thorough and complete post-cure is necessary to achieve maximum wear resistance. To illustrate the dependence of wear resistance on post-cure, a sample of Torlon 4301 was post-cured through a specified cycle (cure cycle consisted of one day at each of the following temperatures: 149°C, 216°C, 243°C followed by post-cure at 260°C (300°F, 420°F, 470°F and 500°F respectively)) and tested for wear resistance at various points in time. In this case, the Wear Factor, K reached a minimum after eleven days, indicating achievement of maximum wear resistance.

The length of post-cure will depend on part configuration, thickness, and to some extent on conditions of molding. Very long exposure to 260°F (500°C) is not detrimental to Torlon parts. The suitability of shorter cycles must be verified experimentally.

Reference: *Torlon Engineering Polymers / Design Manual,* supplier design guide (F-49893) - Amoco Performance Products.

Chapter 84

Polyaryletherketone

Friction and Wear Properties

BASF AG: Ultrapek KR4190 (features: lubricated, developmental material, tribological properties); **Ultrapek**

By virtue of its low coefficient of friction and its resistance to wear, Ultrapek attracts interest for applications that involve sliding friction. Ultrapek's inherently good resistance to wear can be further improved by reinforcement with fibers. The incorporation of lubricants and fiber reinforcement further improves this property profile in Ultrapek KR 4190. These modifications make Ultrapek KR 4190 ideally suited for bearing applications, where it can frequently replace traditional metals.

Reference: *Ultrapek Product Line, Properties, Processing,* supplier design guide (B 607 e/10.92) - BASF Aktiengesellschaft, 1992.

Chapter 85

Polyetheretherketone

Surface Properties

Victrex USA: Victrex PEEK 450FC30 (features: bearing grade, formerly Victrex PEEK D450HF30, tribological properties; note: 30% graphite, carbon fiber, PTFE filled); **Victrex PEEK D450HT15** (features: no longer commercially available, tribological properties, bearing grade); **Victrex PEEK**

Comparison of the tribological data of various bearing materials must be treated with caution since limiting PV, coefficient of friction and wear factor all depend on the method of measurement, temperature, surface speed, and load for the latter two. To enable a meaningful comparison of materials, samples were tested under the same conditions using the same equipment. Metal bearings were lubricated at the beginning of the test. In addition to PEEK, materials tested included lubricated nylon grades, polyimide, acetal, filled PTFE, resin impregnated carbon, lubricated porous bronze and white metal.

Limiting PV

PEEK D450HF30 exhibits an extremely high value of LPV over a wide temperature range at a high surface velocity of 183 m/min (7200 in/min) which is greater than or comparable with that of other bearing materials tested. At the highest temperature and surface velocity used, PEEK D450HT15 exhibits the maximum value of LPV of materials tested.

Coefficient of Friction

PEEK D450HF30 and PEEK D450HT15 exhibit the lowest coefficient of friction of the materials tested at 200°C, 183 m/min (390°F, 7200 in/min) under a 20 kg (44 lbs) load. At high temperatures and/or surface velocities, PEEK D450HT15 exhibits approximately half the coefficient of friction of PEEK D450HF30.

Wear Factor

PEEK D450HF30 and PEEK D450HT15 exhibit extremely low wear factors at 200°C, 183 m/min (390°F, 7200 in/min) under a 20 kg (44 lbs) load, which are lower or comparable to that of the other bearing materials tested.

Reference: *Victrex PEEK For Bearing Applications,* supplier design guide (VKT1/0986) - ICI Advanced Materials, 1986.

Effect of Fillers, Additives and Chemical Modification

Victrex USA: Victrex PEEK 450FC30 (features: bearing grade, formerly Victrex PEEK D450HF30, tribological properties; note: 30% graphite, carbon fiber, PTFE filled); **Victrex PEEK D450HT15** (features: no longer commercially available, tribological properties, bearing grade); **Victrex PEEK**

Two types of additives can be used to improve the tribological behavior of 'Victrex' PEEK. The addition of solid lubricants such as PTFE and graphite can decrease the coefficient of friction over a wide temperature range. Reinforcement with fibers such as carbon fiber can increase modulus, heat distortion temperature, creep and fatigue resistance and thermal conductivity. This generally results in decreased wear and increased LPV values and may also reduce the coefficient of friction. The LPV of a thermoplastic is directly related to its thermal conductivity and creep resistance.

A combination of solid lubricant(s) and reinforcing fiber(s) can be used to yield versatile, ultimate material(s). One such composite is development grade PEEK D450HF30 which has been specially formulated to optimize both tribological and mechanical properties. Another development grade PEEK D450HT15 has been specially formulated to optimize tribological properties at high temperatures and speeds and to minimize the anisotropic nature of thermal expansion.

Reference: *Victrex PEEK For Bearing Applications,* supplier design guide (VKT1/0986) - ICI Advanced Materials, 1986.

Effect of Operating Variables on Tribological Properties

Victrex USA: Victrex PEEK 450FC30 (features: bearing grade, formerly Victrex PEEK D450HF30, tribological properties; note: 30% graphite, carbon fiber, PTFE filled); **Victrex PEEK**

In general, an increase in temperature will result in a decrease in LPV and an increase in both the wear rate and coefficient of friction. For PEEK D450HF30, increasing temperature causes a decrease in LPV, a small change in coefficient of friction and can cause an increase in wear rate.

Increasing surface velocity also generally causes a decrease in LPV and an increase in wear rate and coefficient of friction. PEEK D450HF30 exhibits a lower LPV at increased velocity. Both coefficient of friction and wear rate tend to increase slightly with surface velocity.

Reference: *Victrex PEEK For Bearing Applications,* supplier design guide (VKT1/0986) - ICI Advanced Materials, 1986.

Effect of Temperature on Friction and Wear Properties

LNP Engineering Plastics: Lubricomp LCL-4033 EM (material compostion: 15% PTFE modifier, 15% carbon fiber reinforcement); **Victrex PEEK 150FC30** (filler: proprietary); **Victrex PEEK 450CA30** (material compostion: 30% carbon fiber reinforcement)

The Victrex PEEK composites with a deflection temperature of 316°C (600°F), are particularly suited for use at elevated temperatures. These materials have excellent wear factors, and coefficients of friction are comparable to those of PTFE composites at temperatures to 260°C (500°F). The carbon fiber reinforced Victrex PEEK composite had the highest LPV values at 204°C (400°F) and 260°C (500°F) of all materials investigated and was the only injection moldable composite that had a measurable wear factor at 260°C (500°F); all others failed at or below that temperature.

In general the addition of PTFE to the fiber reinforced composites improves tribological properties at all temperatures. Also, composites containing carbon fiber reinforcement have better bearing properties than those of their glass fiber containing analogs.

Reference: *Lubricomp Internally Lubricated Reinforced Thermoplastics and Fluoropolymer Composites,* supplier design guide (254-691) - LNP Engineering Plastics, 1991.

GRAPH 269: **Coefficient of Friction vs. Temperature for Victrex PEEK D450FC30 Polyetheretherketone.**

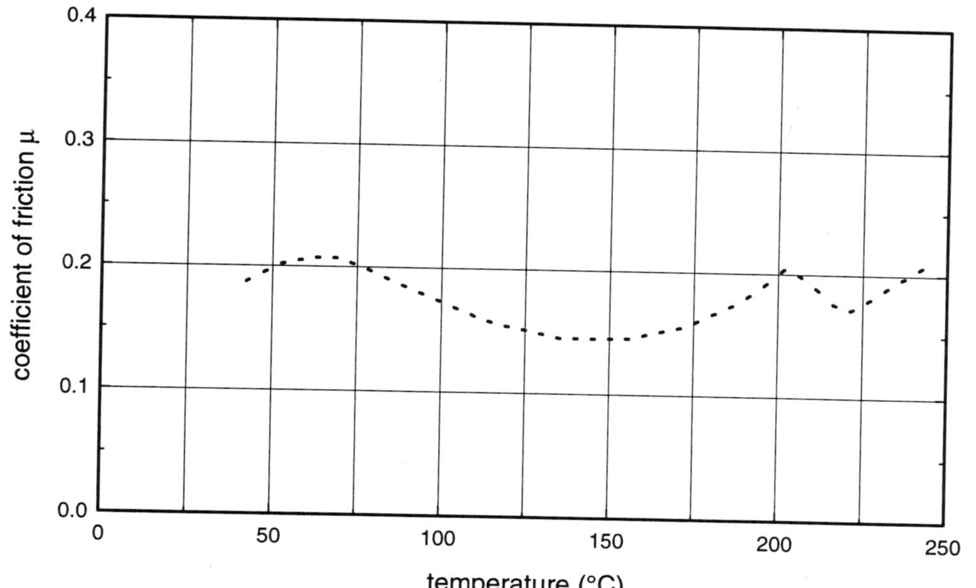

....	Victrex 450FC30 PEEK (tribological properties, was Victrex D450HF30, bearing grade; 30% graphite/ carbon/ PTFE); load: 18.8 kg; speed: 10.2 m/min; apparatus: Ansler wear tester, pad on ring: mating surface: carbon steel (EN8, 0.3-0.4 μm R_a
Reference No.	338

GRAPH 270: **Surface Velocity vs. Limiting PV at Two Temperatures for Victrex PEEK D450HF30 Polyetheretherketone.**

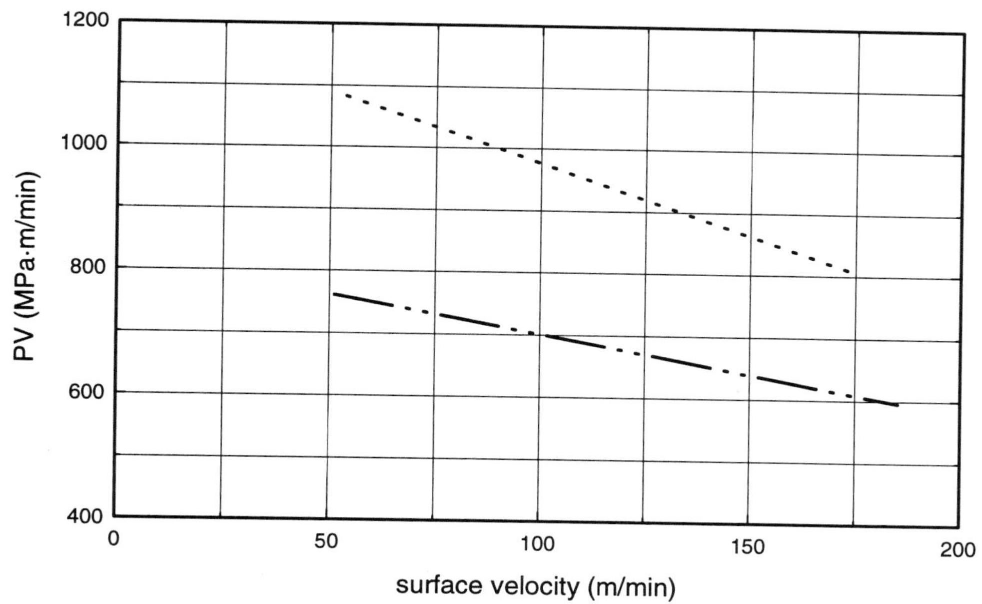

....	Victrex 450FC30 PEEK (tribological properties, was Victrex D450HF30, bearing grade; 30% graphite/ carbon/ PTFE); 20°C
—··—	Victrex 450FC30 PEEK (tribological properties, was Victrex D450HF30, bearing grade; 30% graphite/ carbon/ PTFE); 200°C; apparatus: Ansler wear tester, pad on ring: mating surface: carbon steel (EN8, 0.3-0.4 μm R_a
Reference No.	338

Chapter 86

Polyethylene

Abrasion Resistance

Marlex PE

Abrasion Resistance depends to a great extent on the hardness of the material and the texture of the abrasive. For instance, a coarse abrasive may not abrade a soft material (such as low density polyethylene) as much as it will a hard material (such as styrene) because the soft material will deflect rather than abrade. The converse is often found to be true as well. That is, a fine abrasive will attack a soft material to a greater extent than it will a hard material. Obviously, abrasion tests are difficult to interpret and in order to obtain definitive data for a given application it is desirable to run an actual use test in the final service of the item. Typical results for Marlex polyethylenes and polypropylenes usually run between 1-5 mg/1000 revolutions depending on molecular weight, density and possibly other factors.

Reference: *Engineering Properties Of Marlex Resins,* supplier design guide (TSM-243) - Phillips 66 Company, 1983.

Chapter 87

High Density Polyethylene

GRAPH 271: **Dynamic Coefficient of Friction vs. Pressure for BASF Lupolen 5261ZS High Density Polyethylene.**

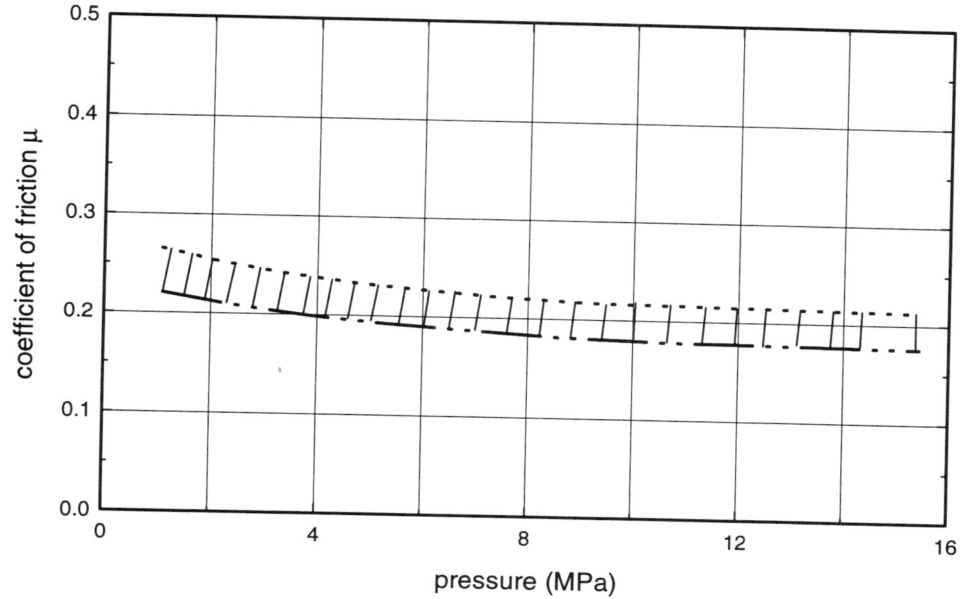

....	BASF AG Lupolen 5261ZS HDPE (0.951 g/cm³ density; 2 g/10 min. MFI); upper limit vales; appartus: peg and disk; mating surface: steel (100 Cr6 800 Hv, 0.15 μm R_Z); 25°C
—··—··	BASF AG Lupolen 5261ZS HDPE (0.951 g/cm³ density; 2 g/10 min. MFI); lower limit values; appartus: peg and disk; mating surface: steel (100 Cr6 800 Hv, 0.15 μm R_Z); 25°C
Reference No.	402

GRAPH 272: **Dynamic Coefficient of Friction vs. Sliding Time for DuPont Canada Sclair 2908-UV8A High Density Polyethylene.**

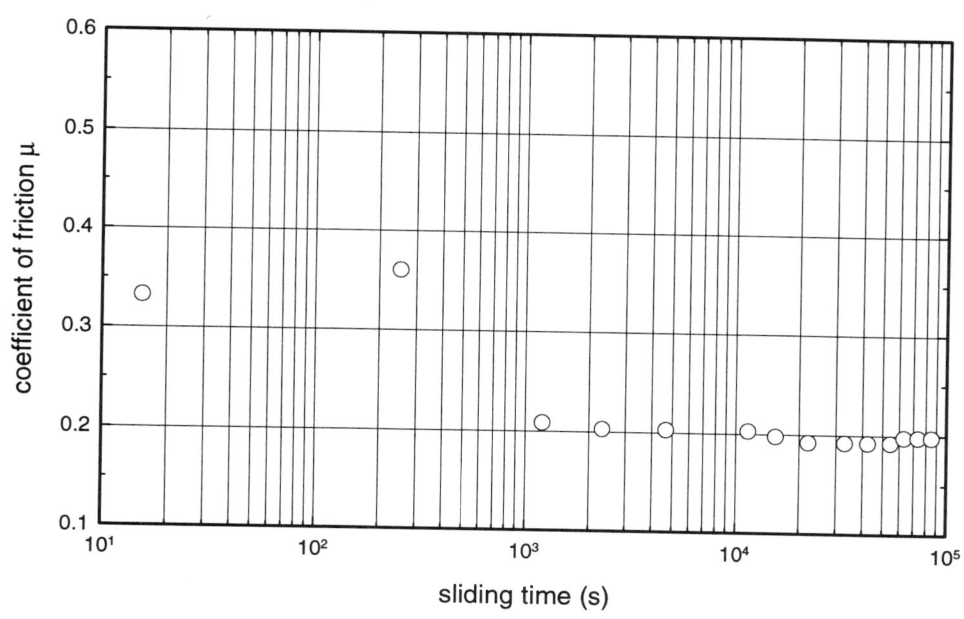

○	DuPont Sclair 2908-UV8A HDPE (extrusion grade); apparatus: pin on plane; mating surface: steel (highly polished); speed: 0.15 m/sec; load: 22.3 N
Reference No.	380

Chapter 88

Ultrahigh Molecular Weight Polyethylene

Friction and Wear Properties

Hoechst Celanese: Hostalen GUR

Hostalen GUR UHMW Polymer has self-lubricating and high-slip properties, making it ideal for areas where sliding contact is encountered. While it is advisable to perform actual operating tests, which are somewhat influenced by environment, the following data are offered for comparison:

Coefficient of Friction (ASTM D 1894)

Material	Static	Kinetic
Mild steel vs. Mild steel	0.30-0.40	0.25-0.35
Mild steel vs. UHMW-PE	0.15-0.20	0.12-0.20
UHMW-PE vs. UHMW-PE	0.20-0.30	0.20-0.30

Hostalen GUR is ideally suited for lining applications in the coal industry because of its resistance to wear, its impact and flexural strength, and its water repellent properties. In addition, it helps eliminate operational problems such as abrasive wear and the caking and freezing of moist coal dust on container wall surfaces.

For a number of years, Hostalen GUR has been used for lining silos, coal chutes, conveyor troughs and railcars. It has also been used successfully on rotary drum filters, as a stripping blade material for filter cake, for bearing shells and bushings, and the lining of coal dust air separators. During this period, there has been minimal evidence of moist coal caking and only minimal abrasive wear from coal.

Reference: *Hostalen GUR Ultra High Molecular Weight Polymer Product Guide,* supplier technical report - Hoechst Celanese, 1983.

Hoechst AG: Hostalen GUR

Hostalen GUR is an excellent sliding material. Comparative tests made with other plastic materials have shown that Hostalen GUR possesses self-lubricating properties, particularly in dry sliding movement against metal surfaces such as steel, brass or copper. These properties are invaluable in emergency conditions; thus for example bushes made from Hostalen GUR, in which metal shafts rotate, can tolerate extraneous materials (such as dust, sand etc.) or misalignment. These conditions do not cause seizing of the shaft. It is, however, important to make adequate provision for the dissipation of frictional heat in order to ensure trouble-free performance of the bushes.

The recommended permissible LPV values for bearing bushes made from Hostalen GUR are about 4 MPa · m/min for dry running and about 6-7 MPa · m/min for lubricated operation. These limiting values are however governed by a number of other factors including bush dimensions, bush clearance, the material and surface finish of the rotating partner and the method of frictional heat removal. The maximum permissible loading pressure for Hostalen GUR bearings is in the region of 10 MPa. It has been found in practice that the limiting PV values do not remain constant over the whole loading range. At higher sliding speeds they should be set somewhat lower and at low speeds somewhat higher. It is difficult to give an estimate for speeds in excess of 120 m/min, which until now has been regarded as the upper limit for lubricated GUR bearings and for creep rates. Under such conditions bearing performance is determined by other criteria.

The paraffinic tactile quality of molded parts made from Hostalen GUR prevents caking or freezing on the surface by many, especially moist, bulk materials.

Reference: *Hoechst Plastics Hostalen GUR,* supplier design guide (HKR112E8102/14) - Hoechst AG.

Solidur 1900

The static coefficient of friction diminishes when the normal load is rising. This shows that the adhesion component of friction is more significant during those tests. The surface roughness of the polymeric samples has a marked effect on static friction. The static coefficient of friction increases when the surface roughness of the polymeric samples diminishes. This is due to the appreciable adhesion when smooth surfaces are in contact.

Reference: Benabdallah, H. (Ecole Polytechnique de Montreal)% Fisa, B. (Ecole Polytechnique de Montreal), *Static Friction Characteristics of Some Thermoplastics,* ANTEC 1989, conference proceedings - Society of Plastics Engineers, 1989.

Solidur Plastics: Solidur 10 100 (features: virgin resin); **Solidur 10 DS** (features: crosslinked); **Solidur Ceram P** (note: UHMWPE with strenghthening additive); **Solidur Marble** (note: blend of virgin and reclaimed UHMWPE);

Solidur UHMWPE is naturally self lubricating and super slick, making it ideal for wear strips, slide plates, bearings, bushings, as well as for lining chutes and bunkers - all areas where sliding on contact is desireable. While it is impossible to duplicate field conditions, laboratory data are offered for comparison.

		Coefficient of Friction	
Sliding Surface	**Sliding**	**Static**	**Kinetic**
Solidur on itself	2 fpm	0.20	0.20
Solidur on chrome-plated steel	2 fpm	0.14	0.12
Solidur on stainless steel	2 fpm	0.28	0.13
Solidur on cold-rolled steel	2 fpm	0.27	0.15
Solidur on brass	2 fpm	0.34	0.20
Industrial Belting			
on Solidur	2 fpm	0.40	0.27
on chrome-plated steel	2 fpm	0.75	0.81

Comparison of Dynamic Coefficient of Friction on Polished Steel

	Dry	Water	Oil
Solidur 10 100	0.10-0.22	0.05-0.10	0.05-0.08
Solidur DS	0.08-0.10	0	0
Ceram P	0.08-0.10	0	0
Solidur 10 802 AST	0.08-0.10	0	0
Solidur Marble	0.15-0.18	0	0
Nylon 6	.15-.40	.14-.19	.02-.11
Nylon 66	.15-.40	.14-.19	.02-.11
Nylon / MoS_2	.12-.20	.10-.12	.08-.10
PTFE	.04-.25	.04-.08	.04-.05
Acetal copolymer	.15-.35	.10-.20	.05-.10

Reference: *Solidur UHMW-PE Product Selection Guide,* supplier marketing literature - Solidur Plastics Company, 1988.

Abrasion Resistance

Hoechst Celanese: Hostalen GUR

The abrasion resistance of Hostalen GUR UHMW Polymer is due to its ultra high molecular weight. Abrasive wear is measured in the laboratory by a "sand/water-slurry test," in which a sample is rotated at high velocity in a sand/water slurry for 24 hours. The following table gives an indication of the abrasion resistance of Hostalen GUR UHMW Polymer. A value of 100 was designated for the amount of volumetric abrasion loss to the GUR specimen. The values shown for other materials are their relative loss compared to Hostalen GUR. The higher the figure, the greater the abrasion loss.

Material	Relative Abrasion
Hostalen GUR (UHMW-PE)	100
Stainless steel	160
Polytetrafluoroethylene (PTFE)	530
PTFE (25% glass fiber)	570
Low density polyethylene (LDPE)	600
Polypropylene (PP)	660
Acetal copolymer (POM)	700
Polyvinyl chloride (PVC)	920
Polymethylmethacrylate (PMMA)	1800
Phenolic resin (PF)	2500
Beechwood	2700
Epoxy resin (EP)	3400

Abrasive wear is a complex phenomenon and is not readily duplicated in laboratory tests. However, years of field experience with Hostalen GUR sheet and profiles have shown it to be a good choice as a protective lining material.

Reference: *Hostalen GUR Ultra High Molecular Weight Polymer Product Guide,* supplier technical report - Hoechst Celanese, 1983.

Solidur Plastics: Solidur 10 100 (features: virgin resin); **Solidur 10 DS** (features: crosslinked); **Solidur Ceram P** (note: UHMWPE with strenghthening additive); **Solidur Marble** (note: blend of virgin and reclaimed UHMWPE);

The table gives an indication of the abrasion resistance of Solidur UHMW PE. The value of 100 designates the amount of volumetric abrasion loss of the Solidur 10 100 specimen. The values for other materials represent volumetric loss in comparison with Solidur. The higher the figure, the more abrasion loss.

Material	Specific Gravity (g/cm^3)	Relative Volumetric Abrasion
Solidur Ceram P	0.98	65
Solidur DS	0.96	85
Solidur Marble	0.94	90
Solidur 10 100	0.94	100
Carbon steel	7.45	160
Nylon	1.15	210
PTFE	2.26	530
Stainless steel	7.85	550
PTFE (25% glass filled)	2.55	570
LDPE (low density PE)	0.92	600
PP (polypropylene)	0.90	660
POM (acetal copolymer)	1.42	700
PVC (polyvinyl chloride)	1.33	920
PMMA (polymethylacrylate)	1.31	1800
PF (phenolic resin)	1.40	2500
Beechwood	0.83	2700
EP (epoxy resin)	1.53	3400

Reference: *Solidur UHMW-PE Product Selection Guide,* supplier marketing literature - Solidur Plastics Company, 1988.

Effect of Fillers, Additives and Chemical Modification on Surface Properties

Hoechst Celanese: Hostalen GUR

The addition of 5% microglass spheres increases wear resistance and is commonly used for suction box covers in the pulp and paper industry. Chemical cross-linking with 0.3 to 0.5% (active ingredient) organic peroxides has been found to improve wear resistance by as much as 30% over non-modified resins, while reducing deformation under load. Thin-film transparency improves, and density is lowered because of a reduction in crystallinity. Silicone oil, waxes, greases, and molybdenum disulfide (normally, 2 to 5 wt%) can be added to UHMWPE to reduce by a slight amount the already low coefficient of friction properties.

Reference: Stein, Harvey L., *Ultrahigh Molecular Weight Polyethylenes (UHMWPE),* Engineered Materials Handbook, Vol. 2, Engineering Plastics, reference book - ASM International, 1988.

GRAPH 274: Static Coefficient of Friction vs. Load Against Steel of Various Surface Roughnesses for Solidur Hercules 1900 Ultrahigh Molecular Weight Polyethylene.

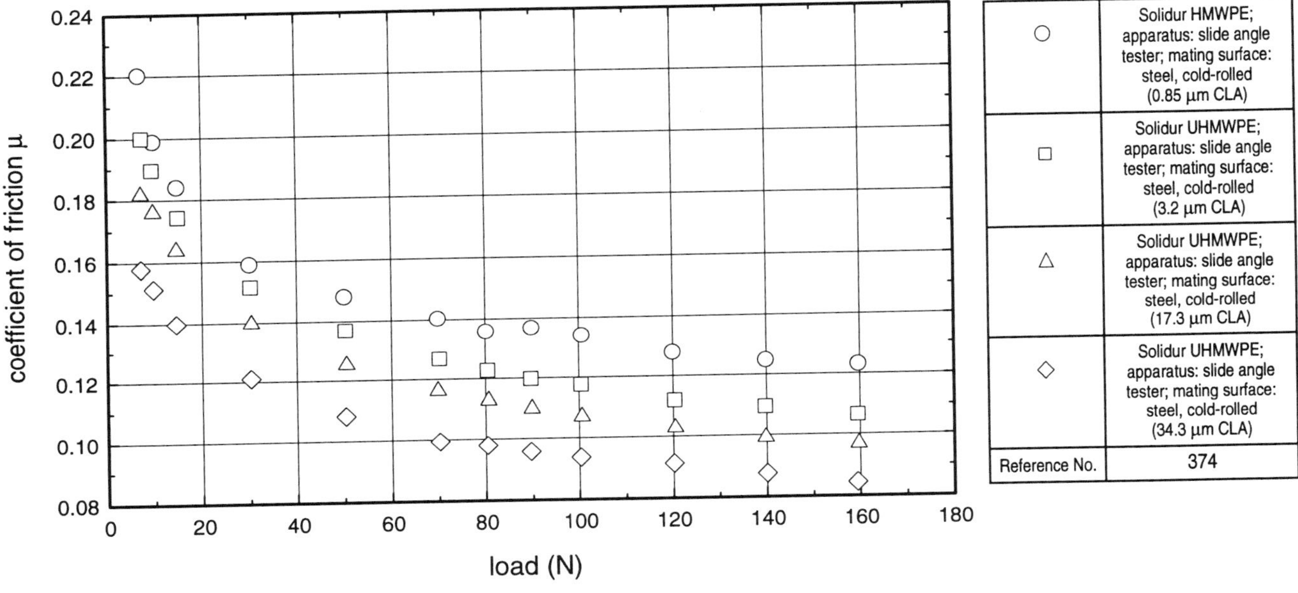

GRAPH 275: **Dynamic Coefficient of Friction vs. Pressure for Hoechst Celanese Hostalen GUR Ultrahigh Molecular Weight Polyethylene.**

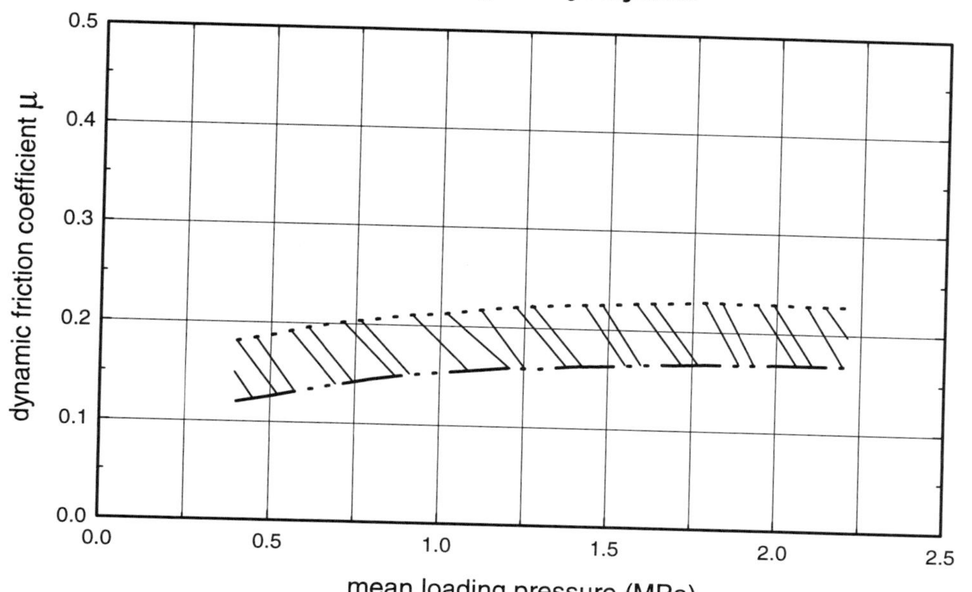

....	Hoechst AG Hostalen GUR UHMWPE; velocity: 10 m/min; upper limit values; mating surface: steel (hardened, polished, 2-2.5 μm R_Z)
—··—··	Hoechst AG Hostalen GUR UHMWPE; velocity: 10 m/min; lower limit values; mating surface: steel (hardened, polished, 2-2.5 μm R_Z)
Reference No.	327

GRAPH 276: **Dynamic Coefficient of Friction vs. Sliding Speed for Hoechst Celanese Hostalen GUR Ultrahigh Molecular Weight Polyethylene.**

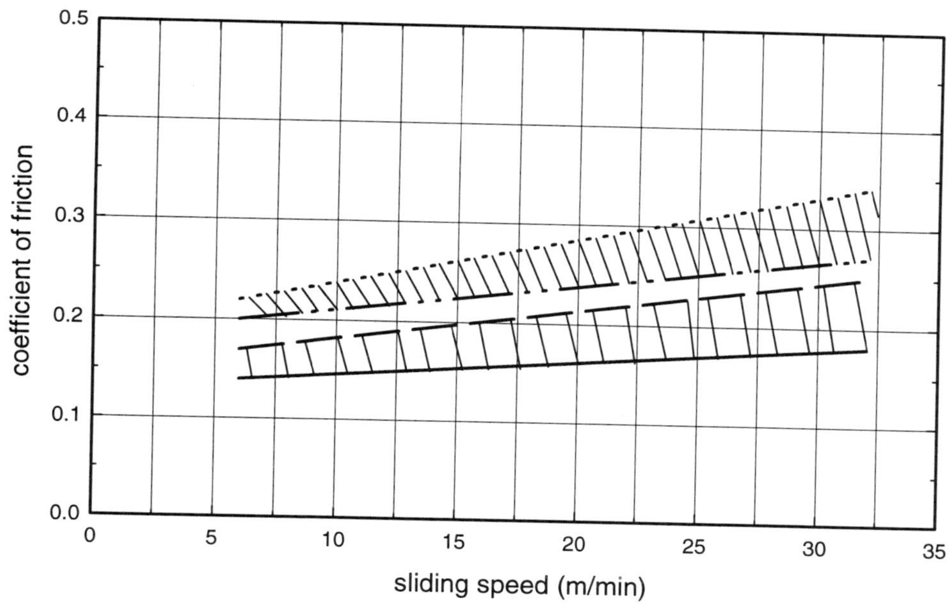

....	Hoechst AG Hostalen GUR UHMWPE; pressure: 1.24 MPa; upper limit values; mating surface: steel (hardened, polished, 2-2.5 μm R_Z)
—··—··	Hoechst AG Hostalen GUR UHMWPE; pressure: 1.24 MPa; lower limit values; mating surface: steel (hardened, polished, 2-2.5 μm R_Z)
– – – –	Hoechst AG Hostalen GUR UHMWPE; pressure: 0.26 MPa; upper limit values; mating surface: steel (hardened, polished, 2-2.5 μm R_Z)
———	Hoechst AG Hostalen GUR UHMWPE; pressure: 0.26 MPa; lower limit values; mating surface: steel (hardened, polished, 2-2.5 μm R_Z)
Reference No.	327

GRAPH 277: PV Value vs. Sliding Speed for unlubricated Hoechst Celanese Hostalen GUR Ultrahigh Molecular Weight Polyethylene Bearings.

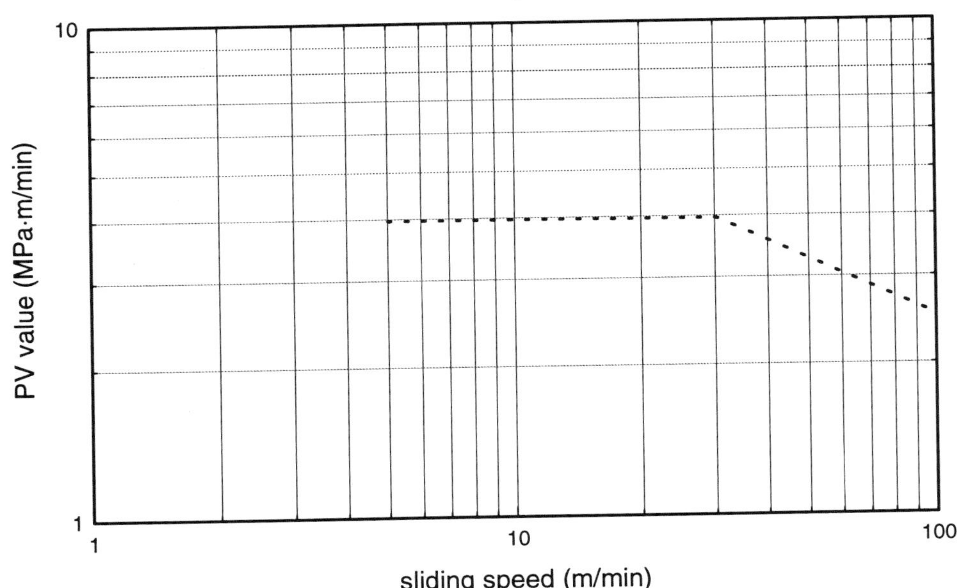

....	Hoechst AG Hostalen GUR UHMWPE; unlubricated; mating surface: steel (hardened, polished, <= 2.5 μm); bearing ratio: 1.0
Reference No.	327

GRAPH 278: Bearing Load vs. Shaft Speed for unlubricated Hoechst Celanese Hostalen GUR Ultrahigh Molecular Weight Polyethylene Bearings.

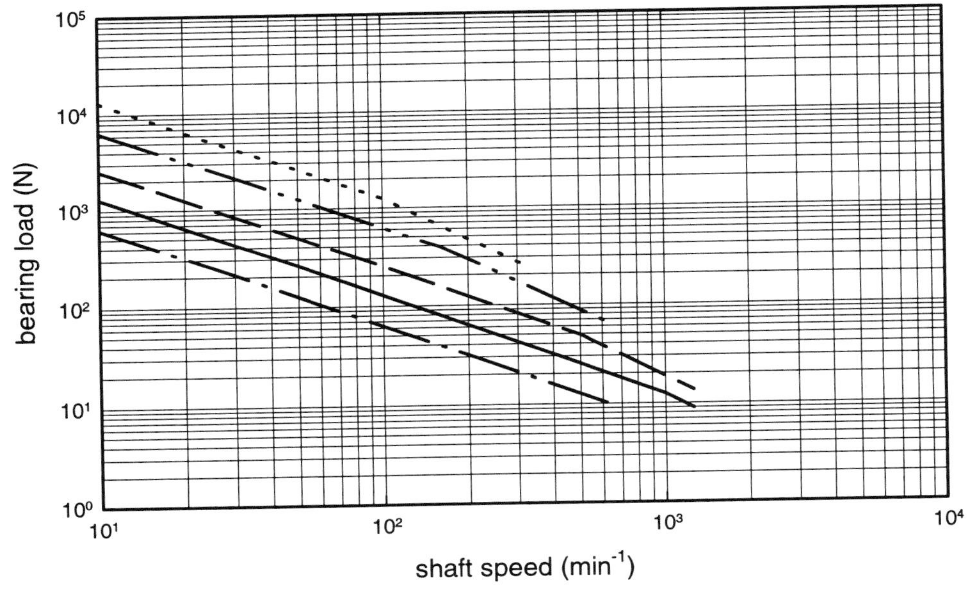

....	Hoechst AG Hostalen GUR UHMWPE; diameter: 100 mm; bearing ratio: 1.0; mating surface: steel (hardened, polished, <= 2.5 μm); unlubricated
—··—··	Hoechst AG Hostalen GUR UHMWPE; diameter: 50 mm; bearing ratio: 1.0; mating surface: steel (hardened, polished, <= 2.5 μm); unlubricated
— — — —	Hoechst AG Hostalen GUR UHMWPE; diameter: 20 mm; bearing ratio: 1.0; mating surface: steel (hardened, polished, <= 2.5 μm); unlubricated
————	Hoechst AG Hostalen GUR UHMWPE; diameter: 10 mm; bearing ratio: 1.0; mating surface: steel (hardened, polished, <= 2.5 μm); unlubricated
—·—·—	Hoechst AG Hostalen GUR UHMWPE; diameter: 5 mm; bearing ratio: 1.0; mating surface: steel (hardened, polished, <= 2.5 μm); unlubricated
Reference No.	327

GRAPH 279: Wear Rate vs. Pressure for BASF Lupolen UHM Ultrahigh Molecular Weight Polyethylene.

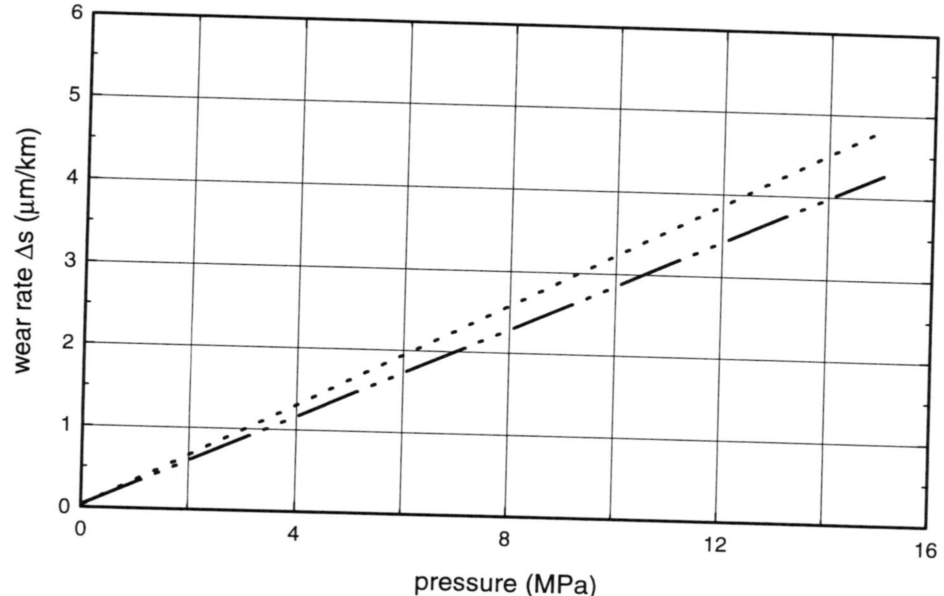

....	BASF AG Lupolen UHM200 UHMWPE (0.934 g/cm³ density; <0.1 g/10 min. MFI; 3x10⁶ g/mol mol. wgt.); mating surface: 100 CR 6 800 HV steel (roughness: 0.15 μm); 23°C ambient temp.; <40°C surface temp.; speed: 0.5 m/sec; 20 km between measured points
—··—··	BASF AG Lupolen UHM300 UHMWPE (0.929 g/cm³ density; <0.1 g/10 min. MFI; 6 x 10⁶ g/mol mol. wgt.); mating surface: 100 CR 6 800 HV steel (roughness: 0.15 μm); 23°C ambient temp.; <40°C surface temp.; speed: 0.5 m/sec; 20 km between measured points
Reference No.	402

Chapter 89

Polypropylene

Friction and Wear Properties

Hoechst AG: Hostacom (reinforcement: glass fiber)

Hostacom can be used as a material for sliding elements, though only in selected applications. An example is conveyor belt guide bars in the beverage industry. Experience has shown that reinforcing materials such as glass fibers have varying effects on wear properties. Hostacom grades are not employed as special bearing materials.

Reference: *Hostacom Reinforced Polypropylene,* supplier design guide (B115BRE9072/046) - Hoechst AG, 1992.

Abrasion Resistance

Marlex (product form: film)

Typical results for Marlex polyethylenes and polypropylenes usually run between 1-5 mg/1000 revolutions depending on molecular weight, density and possibly other factors.

Abrasion Resistance depends to a great extent on the hardness of the material and the texture of the abrasive. For instance, a coarse abrasive may not abrade a soft material (such as low density polyethylene) as much as it will a hard material (such as styrene) because the soft material will deflect rather than abrade. The converse is often found to be true as well. That is, a fine abrasive will attack a soft material to a greater extent than it will a hard material. Obviously, abrasion tests are difficult to interpret and in order to obtain definitive data for a given application it is desirable to run an actual use test in the final service of the item.

Reference: *Engineering Properties Of Marlex Resins,* supplier design guide (TSM-243) - Phillips 66 Company, 1983.

Chapter 90

Polyphenylene Sulfide

Friction and Wear Properties

PPS (reinforcement: aramid fiber); **PPS** (reinforcement: glass fiber)

Polyphenylene sulfide (PPS) has excellent temperature capability and chemical resistance. When properly formulated (e.g., with PTFE) it has the potential to be a high performance wear material replacing bronze. PPS is also brittle. It is mostly used with glass fiber reinforcement, even in wear applications. Glass fiber has a detrimental abrasion effect on the counter surface and on the PPS matrix itself.

Wear data show that aramid fiber can upgrade PPS to a super wear resistant material. The wear rate (K factor) for PPS/ aramid is in the same range as that of nylon 66/ aramid and is only 1/18 of that for PPS/ glass. The abrasion on steel washer for PPS/ aramid is a more dramatic 1/278 of the PPS/ glass material. The washer surface is only slightly marked by the PPS/ aramid material after 202 hours, whereas it is severely grooved by the PPS/ glass into a 2800 μin channel after only 42 hours of wear.

The friction coefficient between PPS/ aramid and the steel washer is 0.349, lower than that for PPS/ glass. This implies some lubricity effect by the aramid fiber. With the addition of a lubricating material such as PTFE, the PPS/ aramid/ PTFE may be the material of choice for such high temperature wear parts as bearings, piston rings, and compressor vanes.

Reference: Wu, Y.T., *How Short Aramid Fiber Improves Wear Resistance,* Modern Plastics, trade journal - McGraw-Hill, 1988.

Phillips 66: Ryton R4 (features: 89 Shore D hardness)

There is little difference shown in the static or dynamic coefficient of friction for R-4. A considerable reduction in the coefficient of friction can be expected with the incorporation of selected antifriction materials.

Reference: *Ryton Polyphenylene Sulfide Resins Engineering Properties Guide*, supplier design guide - Phillips 66 Company, 1989.

Effect of Temperature on Friction and Wear Properties

LNP Engineering Plastics: Lubricomp O-BG (filler: proprietary); **Lubricomp OC-1006** (material compostion: 30% carbon fiber reinforcement); **Lubricomp OCL-4036** (material compostion: 30% carbon fiber reinforcement, 15% PTFE modifier); **Lubricomp OFL-4036** (material compostion: 30% glass fiber reinforcement, 30% glass fiber reinforcement, 15% PTFE modifier)

The PPS composites, with an excellent combination of mechanical and thermal properties, are good candidates for gear and bearing applications at elevated temperatures. Wear factors are low at temperatures to 204°C (400°F) - less than those of unmodified nylon 66 at room temperature. LPV values of all PPS composites were acceptable at all test temperatures to 204°C (400°F).

In general the addition of PTFE to the fiber reinforced composites improves tribological properties at all temperatures. Also, composites containing carbon fiber reinforcement have better bearing properties than those of their glass fiber containing analogs.

Reference: *Lubricomp Internally Lubricated Reinforced Thermoplastics and Fluoropolymer Composites,* supplier design guide (254-691) - LNP Engineering Plastics, 1991.

Chapter 91

Polysulfone

Friction and Wear Properties

Amoco Performance Products: Udel (features: transparent, amber tint)

A decrease in coefficient of friction with increasing velocity (at about 150 fpm (45.7 m/min)) occurs when the polymer starts to transfer to steel counter surface. The results of Taber abrasion testing show that polysulfone loses 20 mg by weight after 1,000 cycles under a 1,000 g load on a CS-17 wheel.

Reference: *Udel Polysulfone Design Engineering Handbook,* supplier design guide (F-47178) - Amoco Performance Products, Inc., 1988.

Chapter 92

Polyethersulfone

Surface Properties

ICI: Victrex PES 4800HF30 (features: tribological properties, bearing grade; note: formulated to optimize tribological properties and mechanical properties); **Victrex PES 4800HL15** (features: tribological properties, bearing grade; note: formulated to optimize tribological properties and price)

Comparison of the tribological data of various bearing materials must be treated with caution since limiting PV, coefficient of friction and wear factor all depend on the method of measurement, temperature, surface speed, and load for the latter two. To enable a meaningful comparison of materials, samples were tested under the same conditions using the same equipment. Metal bearings were lubricated at the beginning of the test. Other materials tested included lubricated nylon grades, PEEK, polysulfone, polyetherimide, polyphenylene sulfide, acetal, polyimide, PTEE, bronze and white metal.

PES D4800HL15 and PES D4800HF30 exhibit high values of LPV over a range of conditions. At 20°C and 30.5 m/min (68°F, 1200 in/min), PES D4800HL15 and PES D4800HF30 exhibit extremely high values of LPV, greater than or comparable with that of the other bearing materials tested. At higher temperatures and surface velocities, the 'Victrex' PES and 'Victrex' PEEK range of bearing grades exhibit from good (PES D4800HL15) to excellent (PEEK D450HT15) LPV values.

The 'Victrex' PES and 'Victrex' PEEK bearing grades exhibit the lowest coefficients of friction of the materials tested at 200°C, 183 m/min (390°F, 7200 in/min). At a surface velocity of 30.5 m/min (1200 in/min) PES D4800HF30 exhibits a lower comparable mean coefficient of friction to the bearing grades PEEK D450HF30 and PEEK D450HT15, over the temperature range studied. PES D4800HL15 and PES D4800HF30 exhibit low wear rates at 20°C and 183 m/min (68°F, 7200 in/min) which are lower or comparable to that of the other materials tested.

Reference: *Victrex PES Bearing Applications,* supplier design guide (VST5/0286) - ICI Advanced Materials, 1986.

Effect of Fillers, Additives and Chemical Modification on Surface Properties

ICI: Victrex PES 4800HF30 (features: tribological properties, bearing grade; note: formulated to optimize tribological properties and mechanical properties); **Victrex PES 4800HL15** (features: tribological properties, bearing grade; note: formulated to optimize tribological properties and price)

Two types of additive can be used to improve the tribological behavior of 'Victrex' PES. The addition of solid lubricants such as PTFE and graphite can decrease the coefficient of friction over a wide temperature range. Reinforcement with fibers such as carbon-fiber can increase modulus, heat distortion temperature, creep and fatigue resistance and thermal conductivity. This generally results in decreased wear and increased LPV values and may also reduce the coefficient of friction. The LPV of a thermoplastic is directly related to its thermal conductivity and creep resistance.

A combination of solid lubricant(s) and reinforcing fiber(s) can be used to yield versatile, ultimate materials. One such composite is development grade PES D4800HF30 which has been specially formulated to optimize both tribological and mechanical properties. Another development grade PES D4800HL15 has been specially formulated to optimize tribological properties and price, and to minimize the directional dependence of thermal expansion. For both of these 'Victrex' PES bearing grades, the thermal expansion is approximately linear up to 200°C (390°F), enabling a single coefficient of thermal expansion to be quoted over this temperature range.

Reference: *Victrex PES Bearing Applications,* supplier design guide (VST5/0286) - ICI Advanced Materials, 1986.

Effect of Operating Variables on Tribological Properties

ICI: Victrex PES 4800HF30 (features: tribological properties, bearing grade; note: formulated to optimize tribological properties and mechanical properties); **Victrex PES 4800HL15** (features: tribological properties, bearing grade; note: formulated to optimize tribological properties and price)

In general, an increase in temperature will result in a decrease in LPV and an increase in both the wear rate and coefficient of friction. For PES D4800HF30 and PES D4800HL15, increasing temperature causes a decrease in LPV, a small change in coefficient of friction, and can cause an increase in wear rate. Increasing surface velocity also generally causes a decrease in LPV and an increase in wear rate and coefficient of friction. PES D4800HF30 and PES D4800HL15 both exhibit a lower LPV at increased velocity. Coefficient of friction tends to increase slightly with surface velocity, but wear rate varies little and can even decrease.

Reference: *Victrex PES Bearing Applications,* supplier design guide (VST5/0286) - ICI Advanced Materials, 1986.

GRAPH 280: **Surface Velocity vs. Limiting PV at Two Temperatures for Victrex PES D4800HF30 Polyethersulfone.**

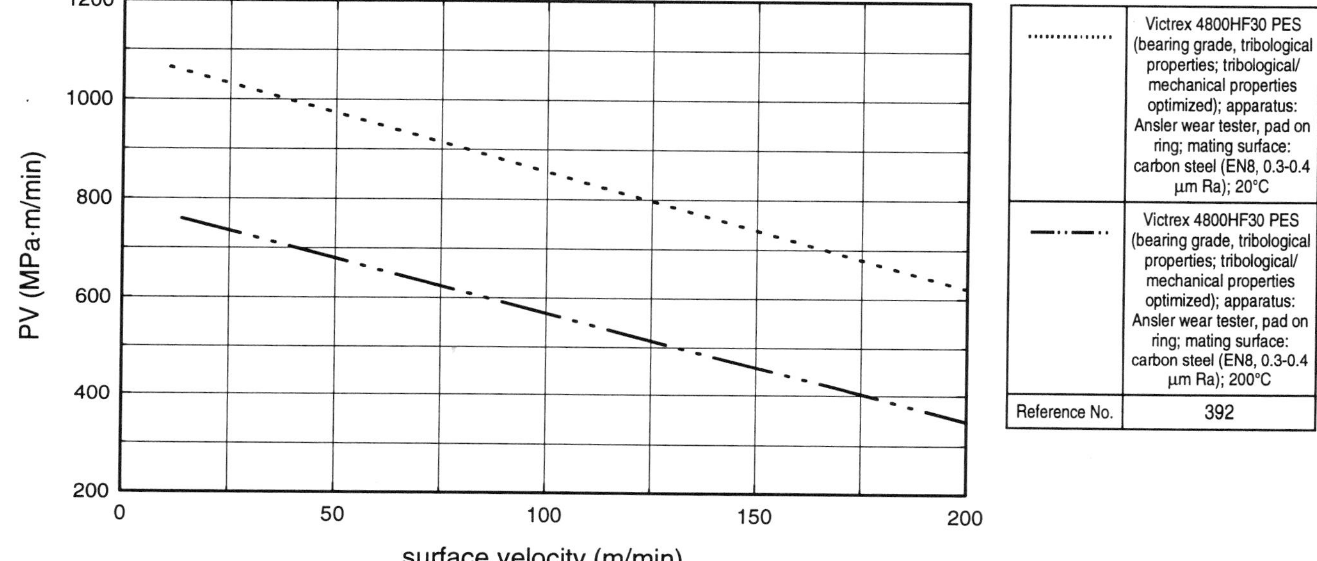

GRAPH 281: Surface Velocity vs. Limiting PV at Two Temperatures for ICI Victrex PES D4800HL15 Polyethersulfone.

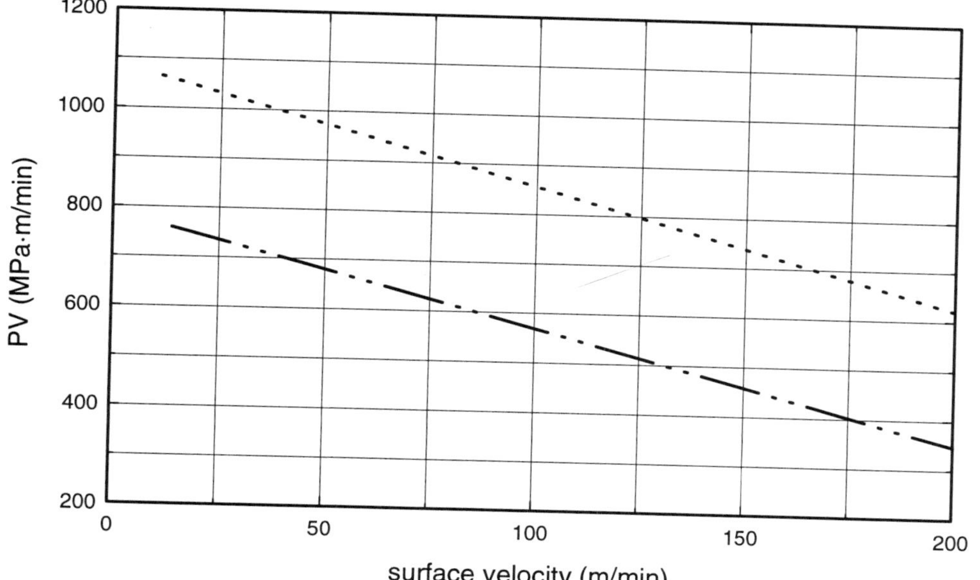

....	Victrex 4800HL15 PES (tribological properties, bearing grade; tribological properties/ price optimized); apparatus: Ansler wear tester, pad on ring; mating surface: carbon steel (EN8, 0.3-0.4 µm Ra); 20°C
—··—··	Victrex 4800HL15 PES (tribological properties, bearing grade; tribological properties/ price optimized); apparatus: Ansler wear tester, pad on ring; mating surface: carbon steel (EN8, 0.3-0.4 µm Ra); 200°C
Reference No.	392

Chapter 93

Impact Resistant Polystyrene

Surface Properties

Dow Chemical: Styron XL-8035MFD (features: 45 Rockwell M hardness, 85 Rockwell L hardness; : , floppy disk grade)

Styron XL 8035 MFD (microfloppy diskette) resin provides improved abrasion resistance over standard polystyrene resins - an especially important advantage in MFD applications. The lower coefficient of friction for Styron XL-8035 MFD resin results in a more lubricated surface. Therefore, MFDs made with Styron XL-8035 MFD resin eject from disk drives more easily and are more resistant to dusting, scratching, and overall wear.

Reference: *Styron XL-8035 MFD High Impact Polystyrene Resin for Microfloppy Diskettes,* supplier marketing literature (301-1607-791X SMG) - Dow Chemical Company, 1991.

Chapter 94

High Density Polyethylene/ Nylon 6 Alloy

Friction and Wear Properties

HDPE/PA (note: high density polyethylene (Sclair 2908-UV8A) containing 5 wt.% of ionomer (Surlyn 9020) as a compatiblizer blended with 20 wt.% Nylon 6 (Zytel 211))

A blend containing 20% of PA 6 showed optimum combined properties; it showed a 21% increase in modulus and a 50% increase of yield stress without deterioration of the tribological properties, HDPE taken as reference. Further investigations of this blend have found a limiting PV factor of 0.35 MPa · m · s^{-1} which is higher than the common virgin plastics. Friction and wear tests conducted on this material showed that it behaves in the same manner as the HDPE which is beneficial. It shows an almost constant coefficient of friction for the first 200 m of sliding then a decrease to a stable value which may represent a steady state of friction. This is due to the deposit and renewal of a transfer film of HDPE on the track, acting as a lubricant.

Reference: Benabdallah, S.H. (Royal Military College of Canada), *Mechanical and Tribological Properties of PEHD/PA 6 Blends,* ANTEC 1992, conference proceedings - Society of Plastics Engineers, 1992.

GRAPH 282: **Dynamic Coefficient of Friction vs. Sliding Time for High Density Polyethylene/ Nylon 6 Alloy.**

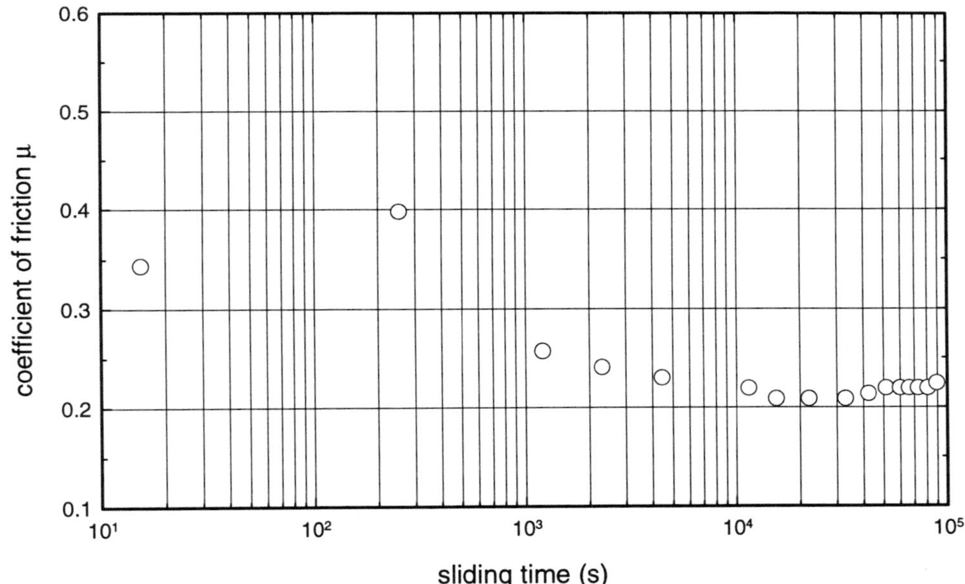

○	HDPE/PA (HDPE w/ 5% ionomer/ 20% Nylon 6); apparatus: pin on plane; mating surface: steel (highly polished); speed: 0.15 m/sec; load: 22.3 N
Reference No.	380

Chapter 95

Polyimide

Friction and Wear Properties

Rhone Poulenc: Kinel 5504 (features: 120 Rockwell M hardness; material compostion: 65% glass fiber reinforcement; note: 6 mm long glass fibers; process type: compression); **Kinel**

Kinel, intended to withstand friction, contains graphite, molybdenum disulfide, or PTFE, and exhibits a very low degree of wear during friction with a metal at very high PV values particularly for high pressures applied at low velocity. In contrast to findings with other self-lubricating materials, the frictional properties are maintained, or even improved at high temperature.

Kinel moving parts are usually used in association with metal counterparts, but Kinel parts can be used together. In a lubricated assembly, fiberglass reinforced Kinel can be used. Kinel 5504, with a steel counterpart demonstrates a low coefficient of friction (0.075 to 0.095), less than that observed with an aluminum alloy rubbing against steel.

Reference: *Kinel Polyimide Compounds, Applications, Properties,* supplier design guide - Rhone-Poulenc.

Friction Properties

DuPont: Vespel SP-211 (material compostion: 15 wt% graphite, 10 wt.% PTFE modifier); **Vespel**

The coefficients of friction for filled SP compositions undergo a transition at about 149°C (300°F). Below this temperature the frictional behavior is similar to that of 66 nylon, but above 149°C (300°F) the frictional forces drop sharply, and in the range of 204°C to 538°C, (400°F to 1000°F) the friction characteristics of SP compositions remain independent of temperature. The friction transition is not associated with wear transition.. The magnitude of the transition, and the wear rate below 149°C (300°F), are greatly reduced in SP-211.

The designer must allow for the higher frictional forces, resulting from two separate phenomena, which may be present during start-up. One is the transfer of a layer of SP polyimide resin/filler composition to the mating surface and the second is the temperature transition for SP polyimide resins. During restart, it may not be necessary under service conditions to break in a new layer, but the temperature effect is reversible and will continue to operate at each restart.

Reference: *VESPEL - Using Vespel Bearings Design and Technical Data,* supplier design guide (E-61500 5-88) - DuPont Company, 1988.

Wear Properties

DuPont: Vespel

Like many other plastics, Vespel parts provide corrosion-resistant, gall-resistant, low-wear and low-friction surfaces - even in non-lubricated environments. Unlike thermoplastics, Vespel parts won't melt or soften even at 482°C (900°F). They have excellent oxidative stability and creep resistance. This helps Vespel parts deliver consistently superior wear performance over a broad temperature range - even when the temperature at the bearing interface is substantially higher than ambient, due to the frictional heat generated by operating at high pressures and/or velocities. Vespel bearings have performed at PV values up to 300,000 psi · fpm, unlubricated, and PV values up to 1,000,000 psi · fpm, lubricated.

Unlike most thermoset plastics, Vespel parts are naturally tough and don't require the addition of abrasive fillers or fibers to enhance part performance. This reduces wear of the mating surface. The superior oxidative ability of Vespel is also an advantage when compared with thermosets.

Reference: *VESPEL - A Publication on Du Pont's high-performance Vespel Polyimide Parts and Shapes,* supplier marketing literature (H-36046) - DuPont Company, 1991.

Effect of Mating Surface on Wear Properties

DuPont: Vespel

The wear performance of Vespel parts can be substantially affected by the hardness of the mating material and its surface finish. Unlubricated bearing wear rates can be reduced by increasing the hardness and decreasing the roughness of the mating surface. In general, a ground surface finish on the mating material is preferable to a turned surface. A fine polishing operation is often beneficial. The finishing operation should be in the same direction as the bearing motion relative to the mating surface.

Aluminum and zinc are not good mating surfaces for plastic bearings because the softness of these materials can lead to rapid wear. If used, aluminum should be hardened or, preferably, anodized. Die-cast aluminum with high silica content is very abrasive to Vespel. Chrome plating is not necessary and may cause greater wear than a polished steel surface. The porosity of the chrome tends to micro-machine the Vespel surface. In general, the metal surface should be as hard and smooth as is practical.

Plastic is not a good mating material for Vespel bearings and, if used, should be limited to low PV conditions. The softness of a plastic mating surface can lead to high wear. In addition, since plastics are relatively poor thermal conductors, plastic-to-plastic bearing interfaces run hotter than plastic-to-metal interfaces, so metal-plastic bearing systems have higher PV limits than plastic-plastic bearing systems.

Reference: *VESPEL - Using Vespel Bearings Design and Technical Data,* supplier design guide (E-61500 5-88) - DuPont Company, 1988.

Effect of Temperature on Wear Properties

DuPont: Vespel SP-21 (material compostion: 15 wt% graphite); Vespel

Wear characteristics of Vespel bearings will be moderate even at high PVs if sufficient cooling is provided. Wear can be severe at any PV if the ambient temperature is too high. The wear resistance of a Vespel bearing operating at a temperature below its limit can be predicted from an experimentally determined Wear Factor. The wear factor is derived from an equation relating the volume of material removed by wear in a given time per unit of load and surface velocity.

The wear rate of a plastic material operating in air is proportional to the product of pressure and velocity (PV) if the surface temperature does not exceed a critical value called Wear Transition Temperature. Above the wear transition temperature, wear increases dramatically. For SP resins, the wear transition temperature is in the range of 482°C to 538°C (900°F to 1000°F) in vacuum or inert gases, and 371°C to 399°C (700°F to 750°F) in air. The wear factor of Vespel bearings made with SP-21 resin is essentially constant over a wide range of operating conditions, as long as surface temperature does not exceed the wear transition temperature.

Reference: *VESPEL - Using Vespel Bearings Design and Technical Data,* supplier design guide (E-61500 5-88) - DuPont Company, 1988.

GRAPH 283: **Dynamic Coefficient of Friction vs. Pressure for Kinel Polyimide**

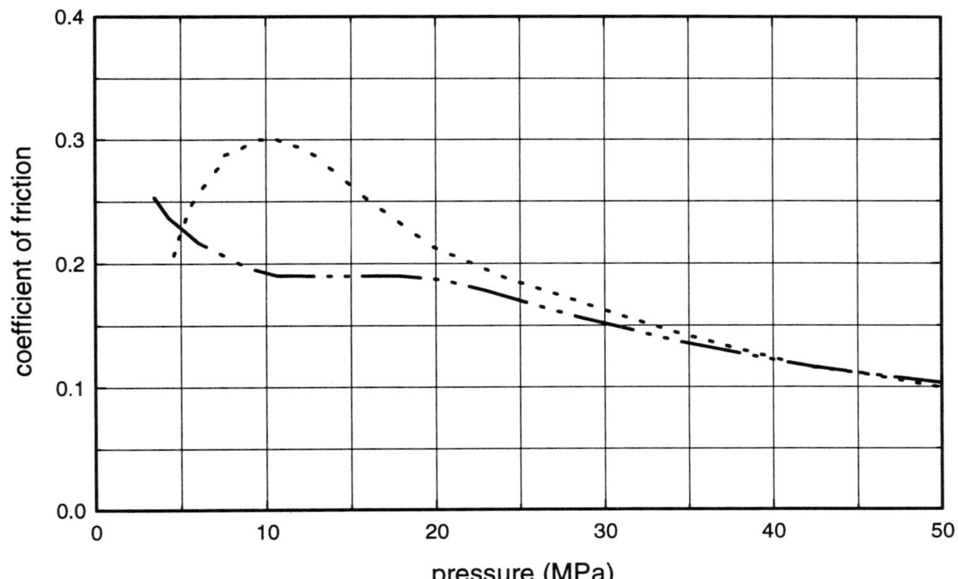

....	ICI Kinel 5517; (110 Rockwell M; compression, w/ graphite); w/ graphite, w/ MoS_2); mating surface: steel (XC38, 27 Rockwell C); speed: 30 m/min
—··—	ICI Kinel 5520 (compression; mating surface: steel (XC38, 27 Rockwell C); speed: 30 m/min
Reference No.	360

GRAPH 284: **Coefficient of Friction vs. Temperature Against Mild Steel for DuPont Vespel Polyimide.**

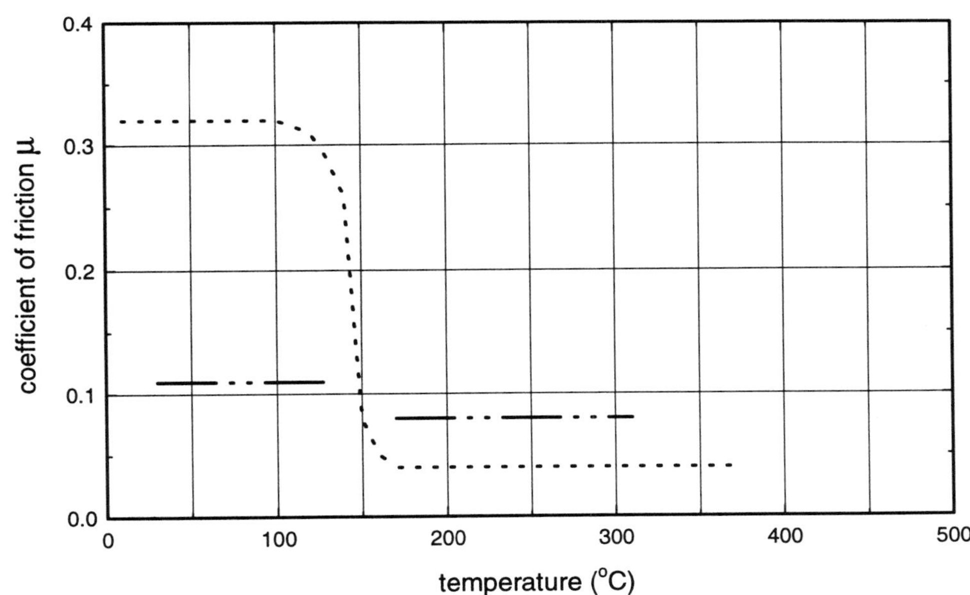

....	DuPont Vespel SP-21 Polyimide (15% graphite); mating surface: steel (mild carbon); unlubricated
—··—	DuPont Vespel SP-211 Polyimide (15% graphite, 10% PTFE modified); mating surface: steel (mild carbon); unlubricated
Reference No.	344

GRAPH 285: Dynamic Coefficient of Friction vs. Temperature Against Stainless Steel for Kinel Polyimide.

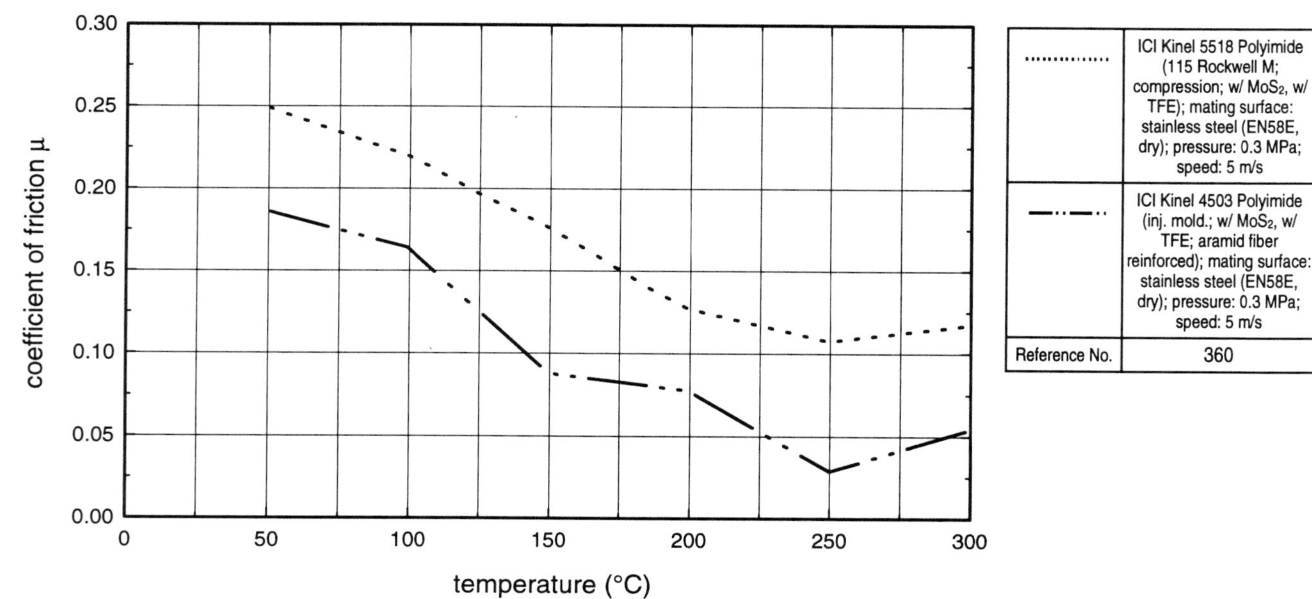

GRAPH 286: Dynamic Coefficient of Friction vs. Temperature for Kinel Polyimide

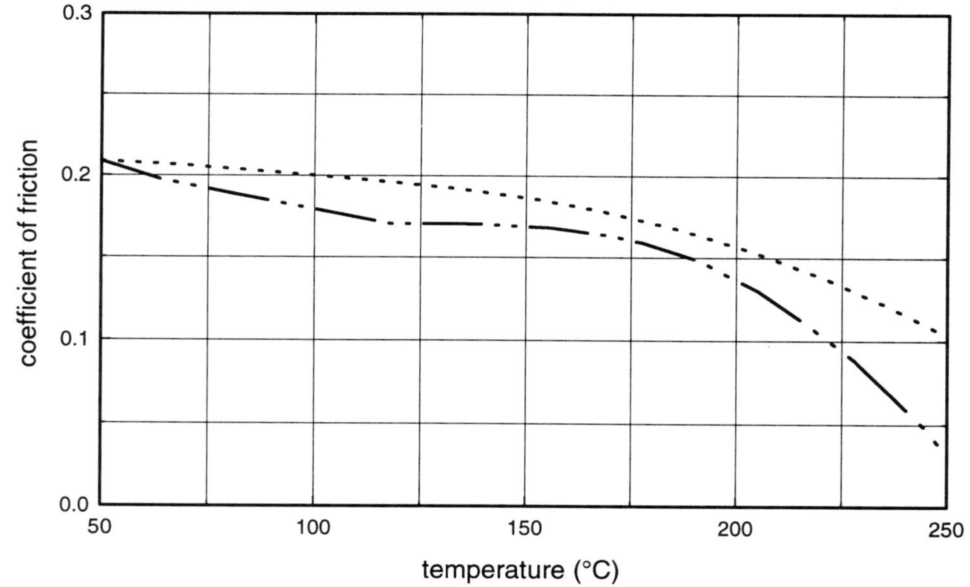

....	ICI Kinel 5517; (110 Rockwell M; compression; w/ graphite, w/ MoS_2); mating surface: steel (XC38, 27 Rockwell C); pressure: 13.8 MPa; speed: 1.5 m/min
—··—··	ICI Kinel 5520 (compression; w/ graphite); mating surface: steel (XC38, 27 Rockwell C); pressure: 13.8 MPa; speed: 1.5 m/min
Reference No.	360

GRAPH 287: Wear Rate vs. Temperature Against Stainless Steel for Kinel Polyimide.

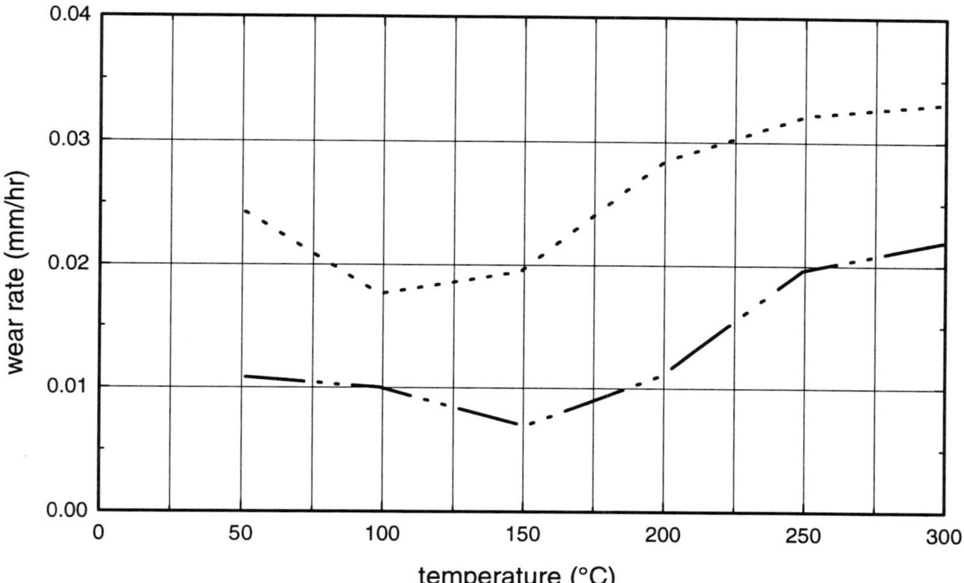

....	ICI Kinel 5518 Polyimide (115 Rockwell M; compression; w/ MoS_2, w/ TFE); mating surface: stainless steel (EN58E, dry); pressure: 0.3 MPa; speed: 5 m/s
—··—··	ICI Kinel 4503 Polyimide (inj. mold.; w/ MoS_2, w/ TFE; , aramid fiber reinforced); mating surface: stainless steel (EN58E, dry); pressure: 0.3 MPa; speed: 5 m/s
Reference No.	360

GRAPH 288: Wear Factor vs. Temperature Against Mild Steel for DuPont Vespel Polyimide.

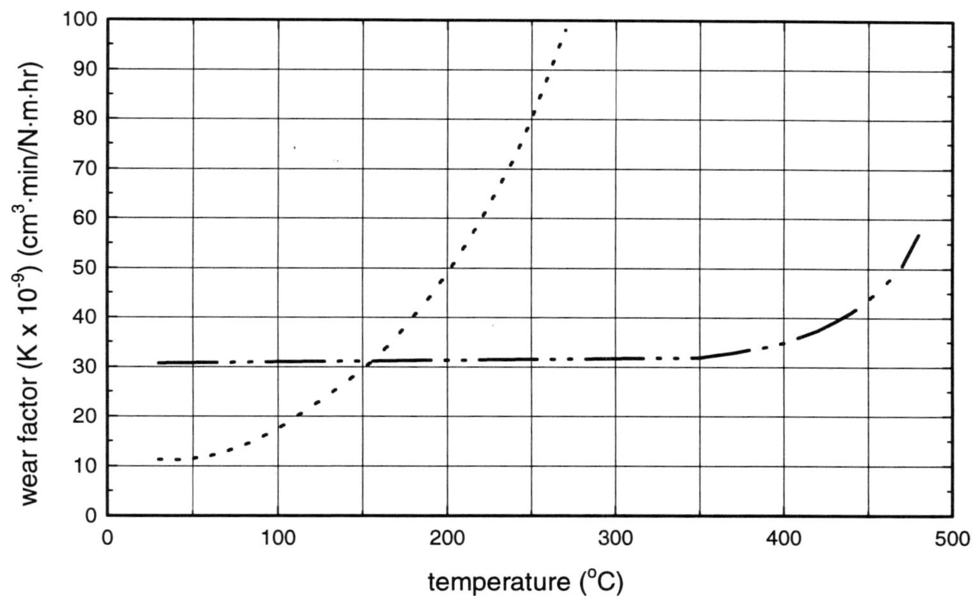

....	DuPont Vespel SP-211 Polyimide (15% graphite, 10% PTFE modified); mating surface: steel (mild carbon); unlubricated
—··—··	DuPont Vespel SP-21 Polyimide (15% graphite); mating surface: steel (mild carbon); unlubricated
Reference No.	344

GRAPH 289: **Wear Factor vs. Surface Temperature Against Carbon Steel for DuPont Vespel SP-21 Polyimide.**

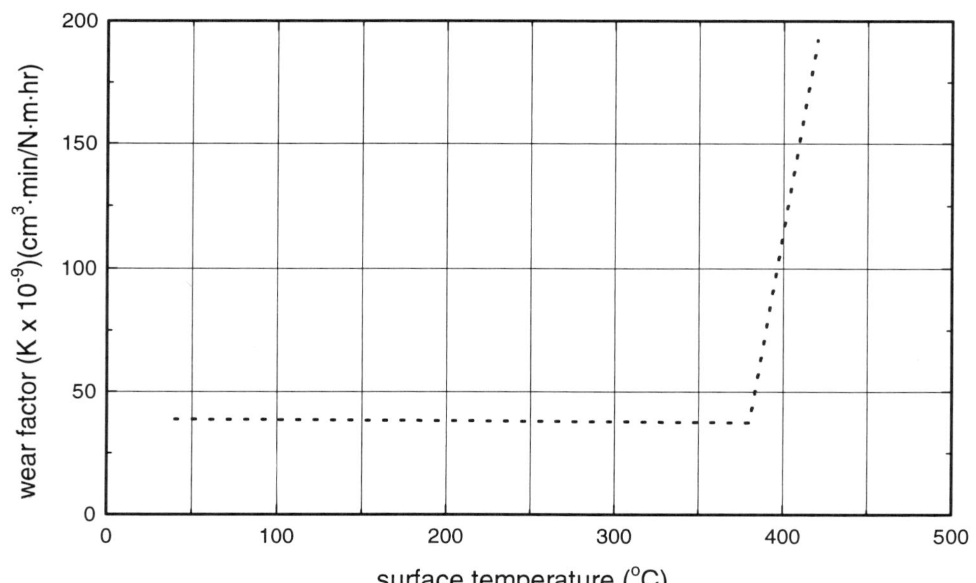

....	DuPont Vespel SP-21 Polyimide (15% graphite); mating surface: carbon steel; thrust bearing tester; PV: 310- 155,000 N/cm^2
Reference No.	344

GRAPH 290: **Wear Factor Multiplier vs. Mating Material Surface Finish for DuPont Vespel Polyimide.**

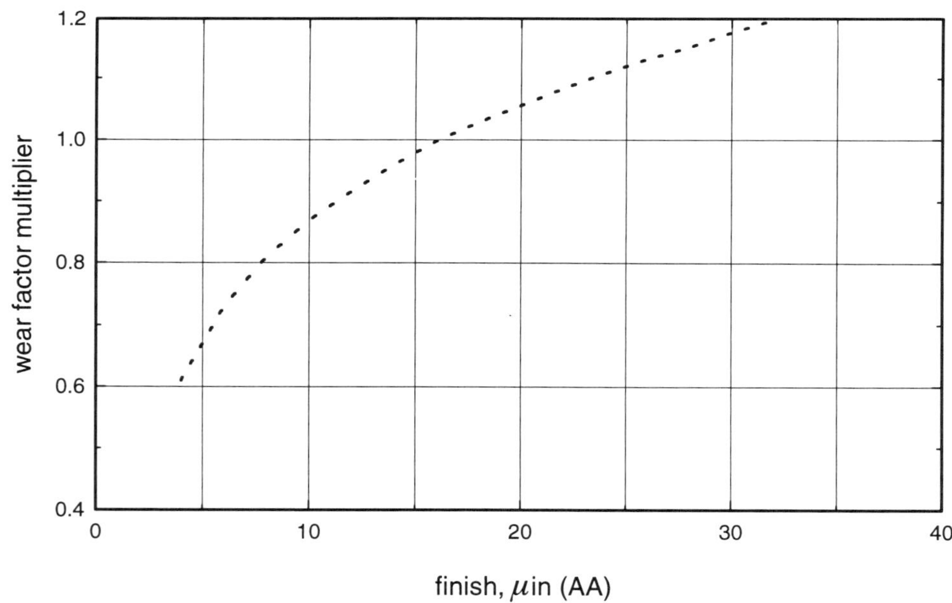

....	DuPont Vespel Polyimide; thrust bearing tester; unlubricated
Reference No.	344

GRAPH 291: Wear Factor Multiplier vs. Mating Material Hardness for DuPont Vespel Polyimide.

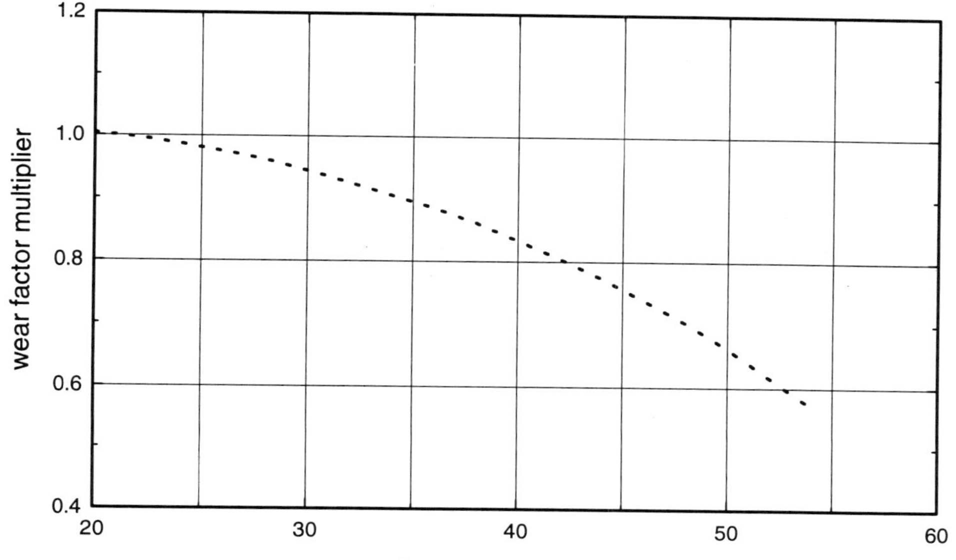

....	DuPont Vespel Polyimide; thrust bearing tester; unlubricated
Reference No.	344

Chapter 96

Polyester Thermoplastic Elastomer

Friction and Wear Properties

DuPont: Hytrel

Hytrel polyester elastomer has excellent wear properties in many applications. As can be seen from test data, test conditions have a great influence on the coefficient of friction. Therefore, it is difficult to predict frictional forces unless testing is performed under conditions that simulate the end use.

Reference: *Hytrel Polyester Elastomer Design Handbook,* supplier design guide (E-52083-1) - DuPont Company, 1988.

Chapter 97

Urethane Thermoplastic Elastomer

Abrasion Resistance

Dow Chemical: Pellethane (features: general purpose grade)

Pellethane resins provide good abrasion resistance, which is necessary for coated fabrics used in applications such as belting, life rafts and airplane escape slides.

Reference: *Typical Physical Properties of Pellethane Thermoplastic Polyurethane Elastomers,* supplier data sheets (306-183-1293X SMG) - Dow Chemical Company, 1993.

Bayer: Texin

As in the case of fatigue, abrasion resistance depends to some extent on heat dissipation. For instance, in intermittent sandblasting, Texin materials outwear steel, but in continuous sandblasting, will soon soften. Where temperature is not excessive, Texin provides the excellent abrasion resistance for which urethane elastomers are well known.

Reference: *Texin Urethane Elastomer - An Engineering Handbook,* supplier design guide - Bayer, 1993.

Chapter 98

Styrenic Thermoplastic Elastomer

Effect of Weld Line and Notch on Fatigue Behavior

Shell Chemical: Kraton K1102 (chemical type: styrene butadiene styrene block copolymer (SBS)); **Kraton K1102** (filler: 10% ground rubber from scrap tires); **Kraton K1102** (filler: 20% ground rubber from scrap tires); **Kraton K1102** (filler: 30% ground rubber from scrap tires)

SBS - Ground Rubber (SBS-GR) composite with 10% GR content has minimum mass loss compared with other SBS-GR composites and virgin SBS. With the exception of the SBS - 10% GR composite mass loss increased with increasing GR content. This trend continued for different loads. The coefficient of friction increases with increasing GR content when applied pressure is 80 KPa. When applied pressure is 100 KPa or above, the coefficient of friction changed little. The reason that the SBS-10% GR composite has the best wear properties is unclear at this time. It may be related to the size distribution of the GR filler particles and their relative size as compared to carbon black.

Adding 10% ground rubber to styrene budatdiene block copolymer leads to an improvement in the wear behavior of SBS. Increasing the abraded time leads to increasing mass loss for both SBS and SBS-GR composites. The coefficient of friction decreases and the mass loss increases with increasing applied pressure for SBS and SBS-GR composites. It is possible to use ground rubber from scrap tires in some SBS products for improving their wear resistance. It may be possible to replace the costly carbon black filler in some SBS-filler composites with GR without sacrificing much on abrasion properties.

Reference: Li, Z. (University of Florida),El-Rahman, M. (University of Florida),Beatty, C. (University of Flori, *Wear/ Abrasion Studies of Styrene-Butadiene Block Copolymer Containing Recycled Rubber,* ANTEC 1994, conference proceedings - Society of Plastics Engineers, 1994.

GRAPH 292: Abrasion Time vs. Coefficient of Friction at 80 kPa pressure for Recycled Rubber Modified Shell Kraton Styrenic Thermoplastic Elastomer.

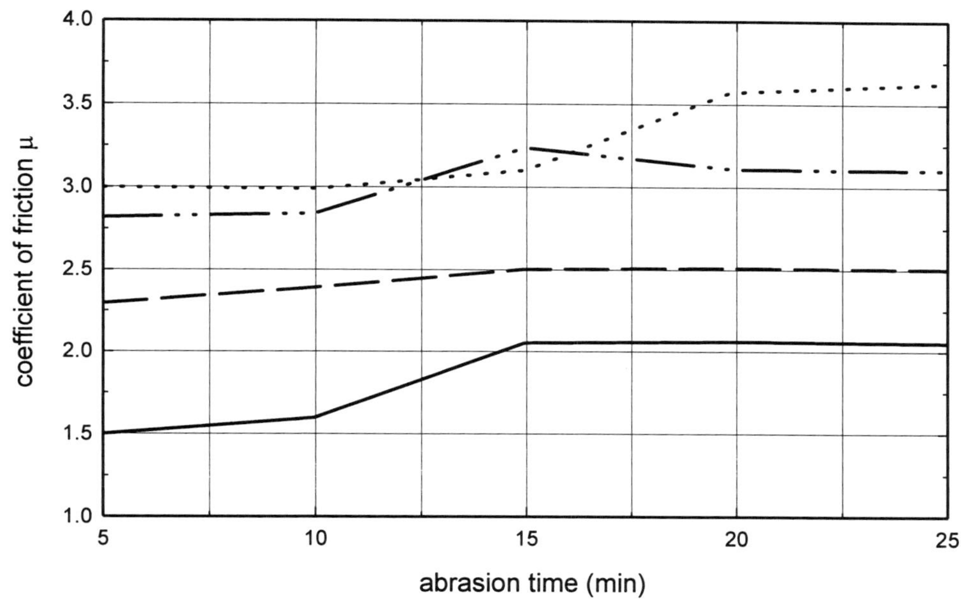

....	Shell Kraton K1102 Styrenic TPE (SBS; 30% scrap tire rubber); apparatus: pin-on-disk; pressure: 80 kPa; speed: 5 m/min; steel surface roughness: 0.05 μm
—··—··	Shell Kraton K1102 Styrenic TPE (SBS; 20% scrap tire rubber); apparatus: pin-on-disk; pressure: 80 kPa; speed: 5 m/min; steel surface roughness: 0.05 μm
----	Shell Kraton K1102 Styrenic TPE (SBS; 10% scrap tire rubber); apparatus: pin-on-disk; pressure: 80 kPa; speed: 5 m/min; steel surface roughness: 0.05 μm
——	Shell Kraton K1102 Styrenic TPE (SBS); apparatus: pin-on-disk; pressure: 80 kPa; speed: 5 m/min; steel surface roughness: 0.05 μm
Reference No.	385

GRAPH 293: Abrasion Time vs. Coefficient of Friction at 140 kPa pressure for Recycled Rubber Modified Shell Kraton Styrenic Thermoplastic Elastomer.

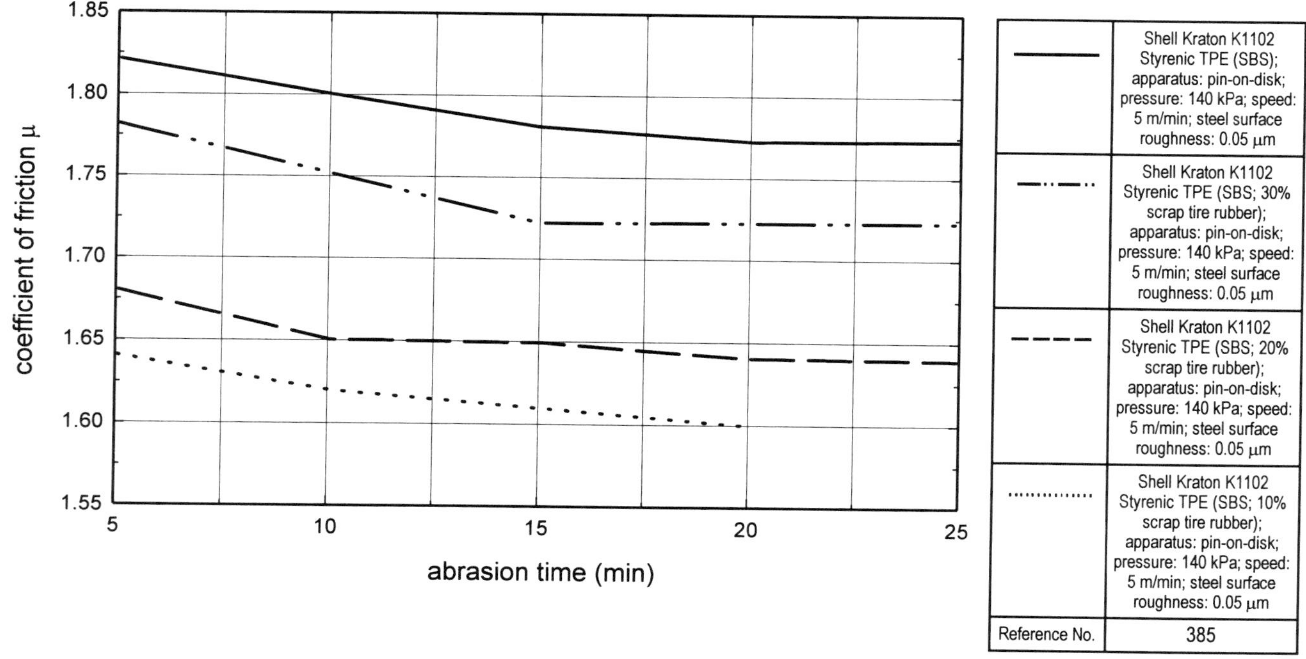

GRAPH 294: Abrasion Time vs. Mass Loss at 80 kPa pressure for Recycled Rubber Modified Shell Kraton Styrenic Thermoplastic Elastomer.

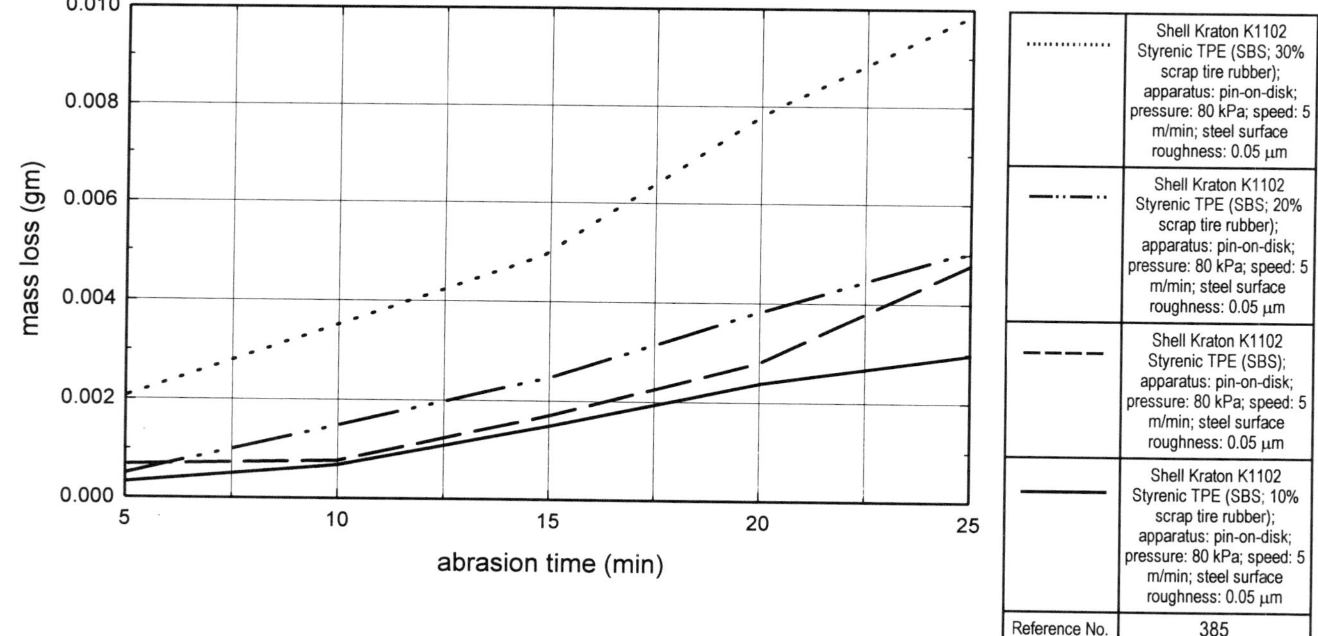

GRAPH 295: **Abrasion Time vs. Mass Loss at 140 kPa pressure for Recycled Rubber Modified Shell Kraton Styrenic Thermoplastic Elastomer.**

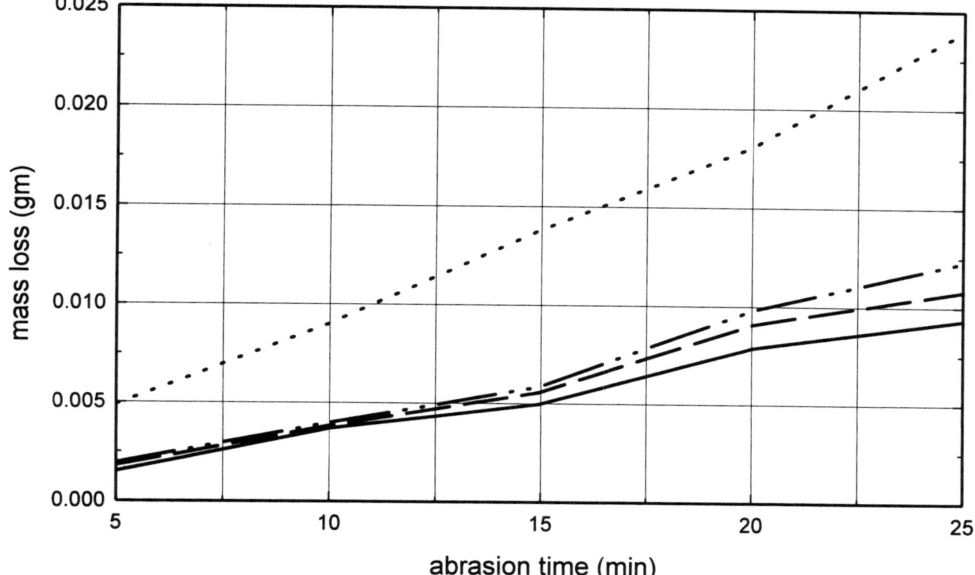

dotted	Shell Kraton K1102 Styrenic TPE (SBS; 30% scrap tire rubber); apparatus: pin-on-disk; pressure: 140 kPa; speed: 5 m/min; steel surface roughness: 0.05 μm
dash-dot-dot	Shell Kraton K1102 Styrenic TPE (SBS; 20% scrap tire rubber); apparatus: pin-on-disk; pressure: 140 kPa; speed: 5 m/min; steel surface roughness: 0.05 μm
dashed	Shell Kraton K1102 Styrenic TPE (SBS); apparatus: pin-on-disk; pressure: 140 kPa; speed: 5 m/min; steel surface roughness: 0.05 μm
solid	Shell Kraton K1102 Styrenic TPE (SBS; 10% scrap tire rubber); apparatus: pin-on-disk; pressure: 140 kPa; speed: 5 m/min; steel surface roughness: 0.05 μm
Reference No.	385

Chapter 99

Polyurethane

Friction Properties

Mearthane Products Corp.: Mearthane

Frictional properties of urethane elastomers, like conventional rubbers, depend upon the hardness of the elastomers. The harder urethanes exhibit a surprisingly low coefficient of friction. This factor makes Mearthane an outstanding candidate for bearings, seals, liners, and similar applications. This may be lowered further by the addition of certain additives such as graphite, molybdenum disulphide, teflon, etc. during the production of the elastomer. These additives give urethane elastomers self-lubricating properties and further improve the already high-abrasion resistance. This material is excellent for oscillating or sliding motion but thorough testing is recommended before use for rotating motion to ensure that heat buildup does not reach destructive levels.

The coefficient of friction of urethane parts varies directly with the relative velocity of the contact surfaces and inversely with the normal force to the contact area. Contact surface is an important consideration on low-friction applications of urethanes. In comparing the kinetic coefficient of friction of a urethane elastomer running against steel machined to rms 10 and rms 125 finish, the least drag is experienced with the highly polished rms 10 steel surface. It has been shown that as the temperature of the urethane part increases its coefficient of friction increases. Thus, in applications where frictional drag is to be minimal, designs involving urethanes should provide contact surfaces which will aid in dissipating the heat of friction.

An excellent example of the high efficiency of urethane elastomers in stressed parts is a sprocket gear, cast of 93-96 durometer elastomer, for the chain drive mechanism of a motorized bicycle. This sprocket was cast, with 0.008 inch tolerance permissible, and shows no appreciable wear after 60,000 miles of test operation. Shoes of the slide block type for leaf springs, also of 93-96 durometer, are giving comparable service. The urethane elastomer shoes have an added advantage of noiseless operation.

Reference: *Mearthane Products Corp. - The Polyurethane Leader Technical Guide,* supplier technical report - Mearthane Products Corp., 1993.

Abrasion Resistance

TSE Industries: Millathane 76

Millathane 76 requires very little reinforcement to have good abrasion resistance. However, its performance is improved by the addition of fine particular size blacks such as N-330. The best abrasion resistance is obtained between 60 and 70 durometer A. The abrasion drops off at higher hardness because of the dilution of the polyurethane polymer due to large amounts of filler.

Fine particular size N-220 and N-330 carbon blacks have better abrasion resistance than N-774 and N-990 black, as tested by Taber Abrasion. Comparing each carbon black at 20 phr, 40 phr and 60 phr, wear resistance is best for each carbon black at a loading of 40 phr.

Reference: *Millathane 76 - A Millable Urethane Elastomer,* supplier design guide - TSE Industries, 1993.

Mearthane Products Corp.: Mearthane

Mearthane parts can provide twenty times the resistance to abrasion obtained with the best conventional elastomers. The design of Mearthane parts for abrasive service must be carefully considered for each individual application. Hysteresis (caused by repeated flexing or rotary motion under load), will lower the abrasion resistance (and the service life) of the part. Because of this factor, laboratory abrasion tests can be misleading: The compound with the best abrasion index based on standard laboratory tests may not provide the best performance in service. Hard (75 Shore D) Mearthane is replacing conventional plastics and even metals where severe abrasion is a factor.

Reference: *Mearthane Products Corp. - The Polyurethane Leader Technical Guide,* supplier technical report - Mearthane Products Corp., 1993.

GRAPH 296: **Dynamic Coefficient of Friction vs. Load Against Various Contact Surfaces for Gallagher Corporation GC 1575 Urethane.**

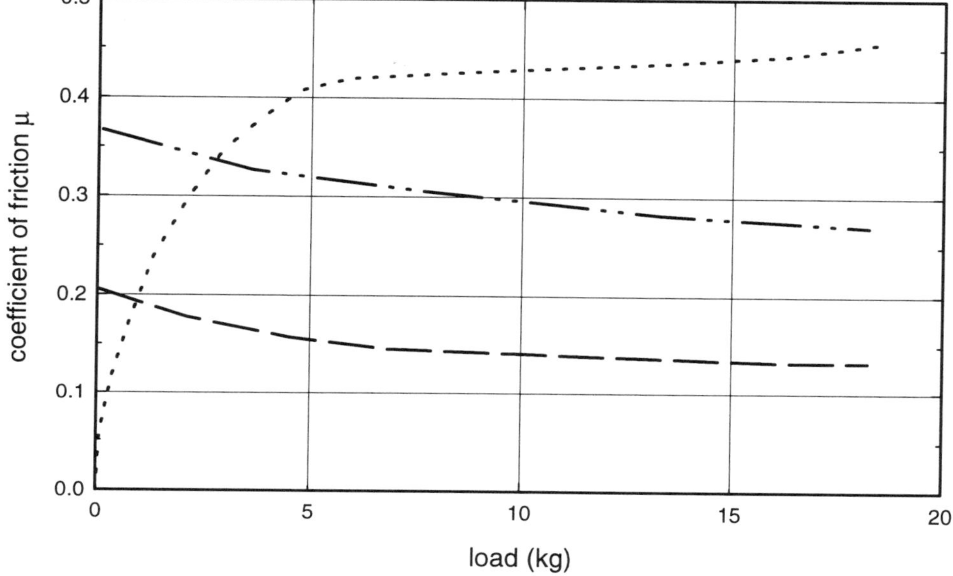

....	Gallagher GC1575 (73 Shore D); mating surface: smooth plastic; 23°C; speed: 0.38 m/sec; unlubricated
—··—··	Gallagher GC1575 (73 Shore D); mating surface: steel (125 RMS); 23°C; speed: 0.38 m/sec; unlubricated
– – – –	Gallagher GC1575 (73 Shore D); mating surface: steel (10 RMS); 23°C; speed: 0.38 m/sec; unlubricated
Reference No.	406

GRAPH 297: **Dynamic Coefficient of Friction vs. Load At Various Surface Velocities for Gallagher Corporation GC 1575 Urethane.**

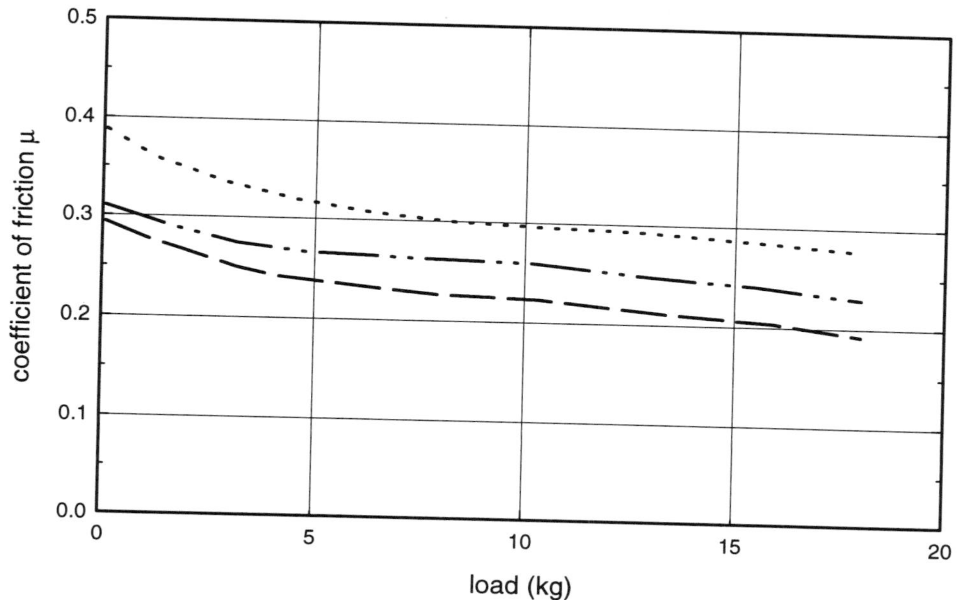

....	Gallagher GC1575 (73 Shore D); mating surface: steel (125 RMS); 23°C; speed: 0.38 m/sec; unlubricated
—··—··	Gallagher GC1575 (73 Shore D); mating surface: steel (125 RMS); 23°C; speed: 0.254 m/sec; unlubricated
— — — —	Gallagher GC1575 (73 Shore D); mating surface: steel (125 RMS); 23°C; speed: 0.127 m/sec; unlubricated
Reference No.	406

GRAPH 298: **Dynamic Coefficient of Friction vs. Temperature for Gallagher Corporation GC 1575 Urethane.**

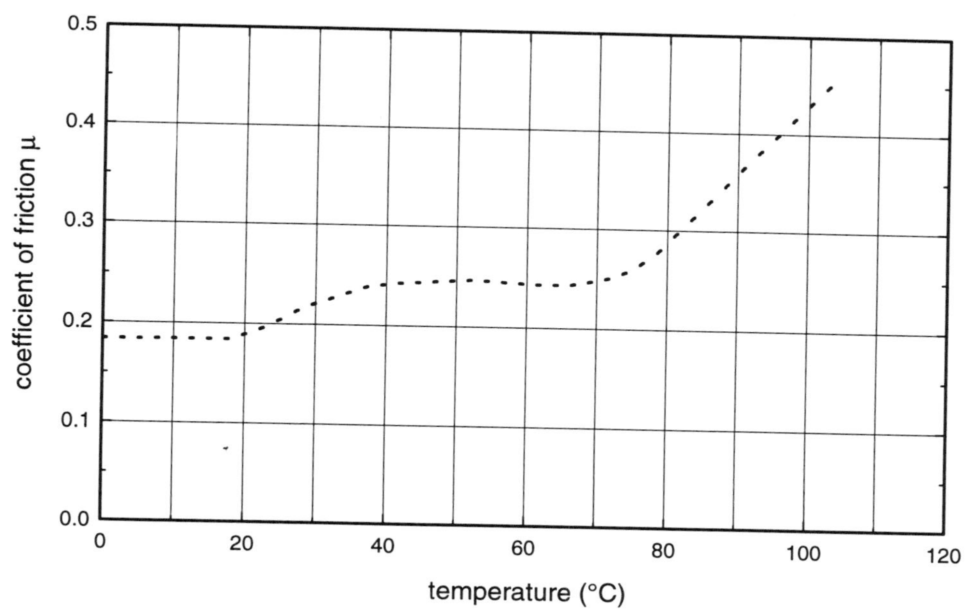

....	Gallagher GC1575 (73 Shore D); mating surface: steel (125 RMS); load: 9.1 kg; speed: 0.26 m/sec; unlubricated
Reference No.	406

Appendix I

Coefficient of Friction Data

ACETAL RESIN against mating surface

Material		Mating Surface		Test Conditions						Coefficient Of Friction		Source
Supplier / Grade	Material Note	Mating Surface Material	Mating Surface Note	Test Method	Test Conditions Note	Temp. (°C)	Pressure (psi) (MPa)	Speed (fpm) (m/min)	PV (psi x fpm) (MPa x m/min)	Static	Dynamic	
DuPont Delrin 500	gen. purp. grade, surface lubricity, 120 Rockwell R, 94 Rockwell M; 6.0 g/10 min. MFI	acetal	Delrin 500, unlubricated	thrust washer test	apparatus: Faville Le Valley rotating disk tester	23	305 (2.1)	10 (3)	3050 (6.4)	0.3	0.4	201
LNP K-1000	unmodified		K-1000 (unmodified)	thrust washer		23	40 (0.28)	50 (15.2)	2000 (4.3)	0.19	0.15	453
LNP KL-4040	20% PTFE		KL-4040 (20% PTFE)	"		23	40 (0.28)	50 (15.2)	2000 (4.3)	0.1	0.09	453
DuPont Delrin 500	gen. purp. grade, surface lubricity, 120 Rockwell R, 94 Rockwell M; 6.0 g/10 min. MFI	nylon 66	Zytel 101, unlubricated	thrust washer test	apparatus: Faville Le Valley rotating disk tester	23	305 (2.1)	10 (3)	3050 (6.4)	0.1	0.2	201
LNP K-1000	unmodified		R-1000 (unmodified)	thrust washer		23	40 (0.28)	50 (15.2)	2000 (4.3)	0.04	0.05	453
LNP K-1000	unmodified		RFL-4036 (30% glass fiber, 15% PTFE)	"		23	40 (0.28)	50 (15.2)	2000 (4.3)	0.07	0.09	453
LNP K-1000	unmodified		RF-1006 (30% glass fiber)	"		23	40 (0.28)	50 (15.2)	2000 (4.3)	0.07	0.09	453
LNP KFL-4036	30% glass fiber, 15% PTFE		RL-4040 (20% PTFE)	"		23	40 (0.28)	50 (15.2)	2000 (4.3)	0.06	0.07	453
LNP KFL-4036	30% glass fiber, 15% PTFE		RFL-4036 (30% glass fiber, 15% PTFE)	"		23	40 (0.28)	50 (15.2)	2000 (4.3)	0.08	0.08	453
LNP KFL-4036	30% glass fiber, 15% PTFE		RCL-4036 (30% carbon fiber, 15% PTFE)	"		23	40 (0.28)	50 (15.2)	2000 (4.3)	0.1	0.13	453
LNP KL-4040	20% PTFE		RL-4040 (20% PTFE)	"		23	40 (0.28)	50 (15.2)	2000 (4.3)	0.03	0.04	453
LNP KL-4040	20% PTFE		R-1000 (unmodified)	"		23	40 (0.28)	50 (15.2)	2000 (4.3)	0.05	0.06	453
LNP KL-4040	20% PTFE		RF-1006 (30% glass fiber)	"		23	40 (0.28)	50 (15.2)	2000 (4.3)	0.05	0.06	453
LNP KL-4040	20% PTFE		RFL-4036 (30% glass fiber, 15% PTFE)	"		23	40 (0.28)	50 (15.2)	2000 (4.3)	0.05	0.07	453
LNP KL-4040	20% PTFE		RCL-4036 (30% carbon fiber, 15% PTFE)	"		23	40 (0.28)	50 (15.2)	2000 (4.3)	0.06	0.09	453
LNP KFL-4036	30% glass fiber, 15% PTFE	polycarbonate	DFL-4036 (30% glass fiber, 15% PTFE)	"		23	40 (0.28)	50 (15.2)	2000 (4.3)	0.08	0.1	453
LNP KL-4040	20% PTFE		D-1000; unmodified	"		23	40 (0.28)	50 (15.2)	2000 (4.3)	0.06	0.04	453
LNP KL-4040	20% PTFE		DFL-4036 (30% glass fiber, 15% PTFE)	"		23	40 (0.28)	50 (15.2)	2000 (4.3)	0.09	0.11	453
Bay Resins Lubriplas POM-1100	unmodified resin	steel		"			40 (0.28)	50 (15.2)	2000 (4.3)	0.14	0.21	435
Bay Resins Lubriplas POM-1100G25	25% glass fiber			"			40 (0.28)	50 (15.2)	2000 (4.3)	0.23	0.3	435
Bay Resins Lubriplas POM-1100TF20	20% PTFE			"			40 (0.28)	50 (15.2)	2000 (4.3)	0.07	0.15	435
Bay Resins Lubriplas POM-1100TF20	20% PTFE			"			630 (4.34)	20 (6.1)	12600 (26.5)		0.14	435
Bay Resins Lubriplas POM-1100TF20	20% PTFE			"			250 (1.72)	20 (6.1)	5000 (10.5)		0.15	435
Bay Resins Lubriplas POM-1100TF20	20% PTFE			"			200 (1.38)	63 (19)	12600 (26.5)		0.15	435

Material		Mating Surface		Test Conditions						Coefficient Of Friction		Source
Supplier / Grade	Material Note	Mating Surface Material	Mating Surface Note	Test Method	Test Conditions Note	Temp. (°C)	Pressure (psi) (MPa)	Speed (fpm) (m/min)	PV (psi x fpm) (MPa x m/min)	Static	Dynamic	
ACETAL RESIN against mating surface →												
Bay Resins Lubriplas POM-1100TF20	20% PTFE	**steel**		"			32 (0.22)	63 (19)	2000 (4.2)		0.2	435
Bay Resins Lubriplas POM-1100TF20	20% PTFE			"			80 (0.55)	63 (19)	5000 (10.5)		0.22	435
Bay Resins Lubriplas POM-1100TF20	20% PTFE			"			63 (0.43)	200 (61)	12600 (26.2)		0.24	435
Bay Resins Lubriplas POM-1100TF20	20% PTFE			"			100 (0.69)	20 (6.1)	2000 (4.2)		0.24	435
Bay Resins Lubriplas POM-1100TF20	20% PTFE			"			25 (0.17)	200 (61)	5000 (10.5)		0.27	435
Bay Resins Lubriplas POM-1100TF20	20% PTFE			"			10 (0.07)	200 (61)	2000 (4.3)		0.41	435
Bay Resins Lubriplas POM-1100UH20	20% UHMWPE			"			40 (0.28)	50 (15.2)	2000 (4.3)	0.1	0.18	435
DuPont Delrin	standard grade						40 (0.28)	50 (15.2)	2000 (4.3)		0.2	449
DuPont Delrin 100	120 Rockwell R, surface lubricity, low flow, 94 Rockwell M; 1.0 g/10 min. MFI		surface finish: 16 µm; unlubricated; 20 Rockwell C	thrust washer test	apparatus: Faville Le Valley rotating disk tester	23	305 (2.1)	10 (3)	3050 (6.4)	0.2	0.35	201
DuPont Delrin 100AF	low flow, NSF Standard 14, low COF, low wear, 78 Rockwell M, 118 Rockwell R; PTFE fiber modified; 1.0 g/10 min. MFI		"	"	"		290 (2)	10 (3)	2900 (6.1)	0.08	0.14	350
DuPont Delrin 100ST	impact modified, high impact, low flow, surface lubricity, 58 Rockwell M, 105 Rockwell R; 1.0 g/10 min. MFI		"	"	"		290 (2)	10 (3)	2900 (6.1)		0.14	350
DuPont Delrin 107	weatherable, UV stabilized, low flow, 120 Rockwell R, 94 Rockwell M; 1.0 g/10 min. MFI		"	"	"		290 (2)	10 (3)	2900 (6.1)	0.2	0.35	350
DuPont Delrin 150E	120 Rockwell R, low flow, low die deposit, 94 Rockwell M; 1.0 g/10 min. MFI; extrus.		"	"	"		290 (2)	10 (3)	2900 (6.1)	0.2	0.35	350
DuPont Delrin 150SA	FDA grade, low die deposit, NSF Standard 14, 94 Rockwell M, low flow, 120 Rockwell R; 1.0 g/10 min. MFI; extrus.		"	"	"		290 (2)	10 (3)	2900 (6.1)	0.2	0.35	350
DuPont Delrin 500	gen. purp. grade, surface lubricity, 120 Rockwell R, 94 Rockwell M; 6.0 g/10 min. MFI		"	"	"	23	305 (2.1)	10 (3)	3050 (6.4)	0.2	0.35	201
DuPont Delrin 500AF	low wear, low COF, NSF Standard 14, 118 Rockwell R, 78 Rockwell M, surface lubricity; 20% PTFE, PTFE fiber modified; 6.0 g/10 min. MFI		"	"	"	23	305 (2.1)	10 (3)	3050 (6.4)	0.08	0.14	201
DuPont Delrin 500AF	surface lubricity, 78 Rockwell M, low COF, 118 Rockwell R, low wear, NSF Standard 14; PTFE fiber modified, 20% PTFE; 6.0 g/10 min. MFI		"	"	"		290 (2)	10 (3)	2900 (6.1)	0.08	0.14	350
DuPont Delrin 500CL	120 Rockwell R, low COF, low wear, 90 Rockwell M, chemical lubricant; 6.0 g/10 min. MFI		"	"	"		290 (2)	10 (3)	2900 (6.1)	0.1	0.2	350
DuPont Delrin 500CL	chemical lubricant, low wear, low COF, 90 Rockwell M, 120 Rockwell R; 6.0 g/10 min. MFI		"	"	"	23	305 (2.1)	10 (3)	3050 (6.4)	0.1	0.2	201
DuPont Delrin 500CL	low wear, chemical lubricant		unlubricated		load: 22.7 kg					0.12-0.14		441
DuPont Delrin 500CL	low wear, chemical lubricant		"				250 (1.72)	20 (6.1)	5000 (10.5)		0.20-0.21	441

Material		Mating Surface		Test Conditions						Coefficient Of Friction		Source
Supplier / Grade	Material Note	Mating Surface Material	Mating Surface Note	Test Method	Test Conditions Note	Temp. (°C)	Pressure (psi) (MPa)	Speed (fpm) (m/min)	PV (psi x fpm) (MPa x m/min)	Static	Dynamic	
ACETAL RESIN against mating surface												
DuPont Delrin 500F	enhanced mold release, fast cycling, 120 Rockwell R, 92 Rockwell M	**steel**	surface finish: 16 µm; unlubricated; 20 Rockwell C	thrust washer test	apparatus: Faville Le Valley rotating disk tester	23	305 (2.1)	10 (3)	3050 (6.4)		0.2	201
DuPont Delrin 500T	impact modified, surface lubricity, 79 Rockwell M, 117 Rockwell R, toughness; 6.0 g/10 min. MFI		"	"	"		290 (2)	10 (3)	2900 (6.1)		0.17	350
DuPont Delrin 500TL	FDA grade, low wear, surface lubricity, low COF, 93 Rockwell M, 123 Rockwell R; PTFE powder modified, 1.5% PTFE; 6.0 g/10 min. MFI		"	"	"		290 (2)	10 (3)	2900 (6.1)	0.13	0.13	350
DuPont Delrin 507	UV stabilized, gen. purp. grade, 120 Rockwell R, surface lubricity, 94 Rockwell M		"	"	"		290 (2)	10 (3)	2900 (6.1)	0.2	0.35	350
DuPont Delrin 550SA	gen. purp. grade, 120 Rockwell R, fast cycling, 94 Rockwell M; 6.0 g/10 min. MFI; extrus.		"	"	"		290 (2)	10 (3)	2900 (6.1)	0.2	0.35	350
DuPont Delrin 570	20% glass fiber						40 (0.28)	50 (15.2)	2000 (4.3)		0.34	449
DuPont Delrin 570	90 Rockwell M, high modulus, surface lubricity, 118 Rockwell R; 20% glass fiber; 6.0 g/10 min. MFI		surface finish: 16 µm; unlubricated; 20 Rockwell C	thrust washer test	apparatus: Faville Le Valley rotating disk tester		290 (2)	10 (3)	2900 (6.1)		0.35	350
DuPont Delrin 577	weatherable, UV stabilized, 90 Rockwell M, 118 Rockwell R, high modulus, surface lubricity; 20% glass fiber; 6.0 g/10 min. MFI		"	"	"		290 (2)	10 (3)	2900 (6.1)		0.35	350
DuPont Delrin 900	high flow, 94 Rockwell M, 120 Rockwell R, surface lubricity; 11.0 g/10 min. MFI		"	"	"	23	305 (2.1)	10 (3)	3050 (6.4)	0.2	0.35	201
DuPont Delrin 900F	fast cycling, 92 Rockwell M, enhanced mold release, 120 Rockwell R		"	"	"	23	305 (2.1)	10 (3)	3050 (6.4)		0.2	201
DuPont Delrin AF	low wear, PTFE modified		unlubricated		load: 22.7 kg					0.13-0.14		441
DuPont Delrin AF	low wear, PTFE modified		"				250 (1.72)	20 (6.1)	5000 (10.5)		0.15-0.16	441
Ensinger Ensital	standard grade						40 (0.28)	50 (15.2)	2000 (4.3)		0.21	449
Ensinger Ensital HPV 13	13% PTFE						40 (0.28)	50 (15.2)	2000 (4.3)		0.16	449
LNP Fulton 404	20% PTFE		carbon steel; surface finish: 12-16 µin; 18-20 Rockwell C	thrust washer		23	40 (0.28)	50 (15.2)	2000 (4.3)	0.07	0.15	453
LNP Fulton 404D	20% PTFE		"	"		23	40 (0.28)	50 (15.2)	2000 (4.3)	0.07	0.12	453
LNP Fulton 441	2% silicone		"	"		23	40 (0.28)	50 (15.2)	2000 (4.3)	0.09	0.12	453
LNP Fulton 441D	2% silicone		"	"		23	40 (0.28)	50 (15.2)	2000 (4.3)	0.08	0.11	453
LNP K-1000	unmodified		"	"		23	40 (0.28)	50 (15.2)	2000 (4.3)	0.14	0.21	453
LNP KC-1004	20% carbon fiber		"	"		23	40 (0.28)	50 (15.2)	2000 (4.3)	0.11	0.14	453
LNP KF-1006	30% glass fiber		"	"		23	40 (0.28)	50 (15.2)	2000 (4.3)	0.25	0.34	453
LNP KFL-4036	30% glass fiber, 15% PTFE		"	"		23	40 (0.28)	50 (15.2)	2000 (4.3)	0.2	0.28	453
LNP KFL-4536	30% glass fiber, 13% PTFE, 2% silicone		"	"		23	40 (0.28)	50 (15.2)	2000 (4.3)	0.18	0.25	453

Material		Mating Surface		Test Conditions						Coefficient Of Friction		Source
Supplier / Grade	Material Note	Mating Surface Material	Mating Surface Note	Test Method	Test Conditions Note	Temp. (°C)	Pressure (psi) (MPa)	Speed (fpm) (m/min)	PV (psi x fpm) (MPa x m/min)	Static	Dynamic	
ACETAL RESIN against mating surface →												
LNP KL-4010	5% PTFE	**steel**	carbon steel; surface finish: 12-16 μin; 18-20 Rockwell C	thrust washer		23	40 (0.28)	50 (15.2)	2000 (4.3)	0.12	0.18	453
LNP KL-4020	10% PTFE		"	"		23	40 (0.28)	50 (15.2)	2000 (4.3)	0.1	0.17	453
LNP KL-4030	15% PTFE		"	"		23	40 (0.28)	50 (15.2)	2000 (4.3)	0.08	0.16	453
LNP KL-4050	25% PTFE		"	"		23	40 (0.28)	50 (15.2)	2000 (4.3)	0.06	0.13	453
LNP KL-4320	10% graphite		"	"		23	40 (0.28)	50 (15.2)	2000 (4.3)	0.16	0.22	453
LNP KL-4540	18% PTFE, 2% silicone		"	"		23	40 (0.28)	50 (15.2)	2000 (4.3)	0.06	0.11	453
LNP KL-4540D	18% PTFE, 2% silicone		"	"		23	40 (0.28)	50 (15.2)	2000 (4.3)	0.06	0.1	453
Polymer Acetron GP	general purpose, porosity free		unlubricated		load: 22.7 kg					0.12-0.28		441
Polymer Acetron GP	general purpose, porosity free		"				250 (1.72)	20 (6.1)	5000 (10.5)		0.15-0.35	441
Polymer Acetron NS	w/ composite solid lubricants to improve wear		"		load: 22.7 kg					0.13-0.19		441
Polymer Acetron NS	w/ composite solid lubricants to improve wear		"				250 (1.72)	20 (6.1)	5000 (10.5)		0.15-0.19	441
RTP 800			C1018 steel; surface finish: 14-17 μin; 15-25 Rockwell C	thrust washer	apparatus: Falex Model No. 6	23	40 (0.28)	50 (15.2)	2000 (4.3)	0.14	0.21	457
RTP 800 AR 5 TFE 10	10% PTFE, 5% aramid fiber		"	"	"	23	40 (0.28)	50 (15.2)	2000 (4.3)	0.07	0.07	457
RTP 800 AR 5 TFE 10 SI 3	10% PTFE, 3% silicone, 5% aramid fiber		"	"	"	23	40 (0.28)	50 (15.2)	2000 (4.3)	0.05	0.05	457
RTP 800 SI 2	2% silicone		"	"	"	23	40 (0.28)	50 (15.2)	2000 (4.3)	0.09	0.12	457
RTP 800 TFE 10	10% PTFE		"	"	"	23	40 (0.28)	50 (15.2)	2000 (4.3)	0.11	0.18	457
RTP 800 TFE 15	15% PTFE		"	"	"	23	40 (0.28)	50 (15.2)	2000 (4.3)	0.08	0.15	457
RTP 800 TFE 18 SI 2	18% PTFE, 2% silicone		"	"	"	23	40 (0.28)	50 (15.2)	2000 (4.3)	0.06	0.1	457
RTP 800 TFE 20	20% PTFE		"	"	"	23	40 (0.28)	50 (15.2)	2000 (4.3)	0.07	0.12	457
RTP 805	30% glass fiber		"	"	"	23	40 (0.28)	50 (15.2)	2000 (4.3)	0.26	0.34	457
RTP 805 TFE 15	15% PTFE, 30% glass fiber		"	"	"	23	40 (0.28)	50 (15.2)	2000 (4.3)	0.2	0.27	457
Thermofil G-20FG-0214	20% glass fiber, PTFE lubricated			ASTM D1894		23				0.18	0.25	459
Thermofil G-9900-0215	20% PTFE			"		23				0.05	0.1	459
Thermofil G-9900-0257	PTFE/ oil lubricated			"		23				0.06	0.1	459

Material		Mating Surface		Test Conditions						Coefficient Of Friction		Source
Supplier / Grade	Material Note	Mating Surface Material	Mating Surface Note	Test Method	Test Conditions Note	Temp. (°C)	Pressure (psi) (MPa)	Speed (fpm) (m/min)	PV (psi x fpm) (MPa x m/min)	Static	Dynamic	
ACETAL RESIN against mating surface												
Unspecified grade	general purpose	**steel**	unlubricated		load: 22.7 kg					0.12-0.28		441
Unspecified grade	PTFE modified		polished CR steel, dry	ASTM D1894							0.1-0.15	330
Unspecified grade	transverse to flow		ball (diameter: 13 mm)	sliding against steel	load: 6 N			2 (0.6)			0.12	70
Unspecified grade	parallel to flow		"	"	"			2 (0.6)			0.15	70
Unspecified grade	general purpose		unlubricated				250 (1.72)	20 (6.1)	5000 (10.5)		0.15-0.35	441
Unspecified grade			polished CR steel, dry	ASTM D1894							0.16-0.42	330
Unspecified grade			carbon EN8, dry	pad on ring (Amsler wear tester)		20		100 (30.5)			0.24 (average of COF @ LPV and 50% of LPV load)	338
Unspecified grade			"	"		20		600 (183)			0.34 (average of COF @ LPV and 50% of LPV load)	338
ACETAL COPOLYMER against mating surface												
Hoechst Cel. Celcon M25	high molecular weight, 78 Rockwell M; 2.5 g/10 min. MFI	**acetal**	Celcon	ASTM D1894							0.35	210
Hoechst Cel. Celcon M270	low molecular weight, high flow, 80 Rockwell M; 27.0 g/10 min. MFI		"	"							0.35	210
Hoechst Cel. Celcon M450	high flow, 80 Rockwell M; 45 g/10 min. MFI		"	"							0.35	210
Hoechst Cel. Celcon M90	gen. purp. grade, 80 Rockwell M; 9.0 g/10 min. MFI		"	"							0.35	210
Hoechst Cel. Celcon LW90	low wear, 80 Rockwell M; 9.0 g/10 min. MFI	**aluminum**		thrust washer test							<0.06	321
Hoechst Cel. Celcon LW90S2	low wear, 75 Rockwell M, low wear; 2% silicone; 9.0 g/10 min. MFI			"							0.13	321
Hoechst Cel. Celcon LWGCS2	83 Rockwell M; glass reinforced, 2% silicone			"							0.13	321
Hoechst Cel. Celcon M25	high molecular weight, 78 Rockwell M; 2.5 g/10 min. MFI			ASTM D1894							0.15	210
Hoechst Cel. Celcon M270	low molecular weight, high flow, 80 Rockwell M; 27.0 g/10 min. MFI			"							0.15	210
Hoechst Cel. Celcon M450	high flow, 80 Rockwell M; 45 g/10 min. MFI			"							0.15	210
Hoechst Cel. Celcon M90	80 Rockwell M; 9.0 g/10 min. MFI			thrust washer test							0.15	321
Hoechst Cel. Celcon M90	gen. purp. grade, 80 Rockwell M; 9.0 g/10 min. MFI			ASTM D1894							0.15	210

Material		Mating Surface		Test Conditions						Coefficient Of Friction		Source
Supplier / Grade	Material Note	Mating Surface Material	Mating Surface Note	Test Method	Test Conditions Note	Temp. (°C)	Pressure (psi) (MPa)	Speed (fpm) (m/min)	PV (psi x fpm) (MPa x m/min)	Static	Dynamic	
ACETAL COPOLYMER against mating surface →												
Hoechst Cel. Celcon M25	high molecular weight, 78 Rockwell M; 2.5 g/10 min. MFI	**brass**		ASTM D1894							0.15	210
Hoechst Cel. Celcon M270	low molecular weight, high flow, 80 Rockwell M; 27.0 g/10 min. MFI			"							0.15	210
Hoechst Cel. Celcon M450	high flow, 80 Rockwell M; 45 g/10 min. MFI			"							0.15	210
Hoechst Cel. Celcon M90	gen. purp. grade, 80 Rockwell M; 9.0 g/10 min. MFI			"							0.15	210
ACETAL COPOLYMER against mating surface →												
Akzo AC-80/TF/10	10% PTFE	**steel**	carbon steel; surface finish: 16 µin; 18-22 Rockwell C	thrust washer	apparatus: Faville-LeVally	23	40 (0.28)	50 (15.2)	2000 (4.3)	0.1	0.17	458
Akzo AC-80/TF/30	30% PTFE		"	"	"	23	40 (0.28)	50 (15.2)	2000 (4.3)	0.06	0.13	458
Akzo AC-8O/TF/15	15% PTFE		"	"	"	23	40 (0.28)	50 (15.2)	2000 (4.3)	0.08	0.16	458
Akzo AC-8O/TF/20	20% PTFE		"	"	"	23	40 (0.28)	50 (15.2)	2000 (4.3)	0.07	0.15	458
Akzo J-80/20/TF/15	20% glass fiber, 15% PTFE		"	"	"	23	40 (0.28)	50 (15.2)	2000 (4.3)	0.2	0.28	458
Akzo J-80/30/TF/15	30% glass fiber, 15% PTFE		"	"	"	23	40 (0.28)	50 (15.2)	2000 (4.3)	0.2	0.28	458
Ferro Star-L POM CD11	1% silicon oil, lubricated			ASTM D3028						0.16		452
Ferro Star-L POM CD11	1% silicon oil, lubricated			"	load: 12 N			39 (12)			0.1	452
Ferro Star-L POM CD11	1% silicon oil, lubricated			"	load: 43 N			39 (12)			0.11	452
Ferro Star-L POM CD52	20% PTFE, lubricated			"						0.13		452
Ferro Star-L POM CD52	20% PTFE, lubricated			"	load: 12 N			39 (12)			0.07	452
Ferro Star-L POM CD52	20% PTFE, lubricated			"	load: 43 N			39 (12)			0.09	452
Ferro Star-L POM CD61	13% PTFE, 2% silicon, lubricated			"						0.1		452
Ferro Star-L POM CD61	13% PTFE, 2% silicon, lubricated			"	load: 12 N			39 (12)			0.08	452
Ferro Star-L POM CD61	13% PTFE, 2% silicon, lubricated			"	load: 43 N			39 (12)			0.09	452
Hoechst Cel. Celcon LW90	low wear, 80 Rockwell M; 9.0 g/10 min. MFI			thrust washer test							<0.06	321
Hoechst Cel. Celcon LW90S2	low wear, 75 Rockwell M, low wear; 2% silicone; 9.0 g/10 min. MFI			"							0.14	321
Hoechst Cel. Celcon LWGCS2	83 Rockwell M; glass reinforced, 2% silicone			thrust washer test							0.15	321
Hoechst Cel. Celcon M25	high molecular weight, 78 Rockwell M; 2.5 g/10 min. MFI			ASTM D1894							0.15	210

Material		Mating Surface		Test Conditions						Coefficient Of Friction		Source
Supplier / Grade	Material Note	Mating Surface Material	Mating Surface Note	Test Method	Test Conditions Note	Temp. (°C)	Pressure (psi) (MPa)	Speed (fpm) (m/min)	PV (psi x fpm) (MPa x m/min)	Static	Dynamic	
ACETAL COPOLYMER against mating surface												
Hoechst Cel. Celcon M270	low molecular weight, high flow, 80 Rockwell M; 27.0 g/10 min. MFI	**steel**		ASTM D1894							0.15	210
Hoechst Cel. Celcon M450	high flow, 80 Rockwell M; 45 g/10 min. MFI			"							0.15	210
Hoechst Cel. Celcon M90	80 Rockwell M; 9.0 g/10 min. MFI			thrust washer test							0.15	321
Hoechst Cel. Celcon M90	gen. purp. grade, 80 Rockwell M; 9.0 g/10 min. MFI			ASTM D1894							0.15	210
Unspecified grade	unmodified		carbon steel; surface finish: 16 µin; 18-22 Rockwell C	thrust washer	apparatus: Faville-LeVally	23	40 (0.28)	50 (15.2)	2000 (4.3)	0.18	0.21	458
ACRYLIC (PMMA) against mating surface												
Unspecified grade		**steel**		ASTM D1894							0.4	78
FLUOROPOLYMER, ECTFE against mating surface												
Ausimont	93 Rockwell R hardness, 75 Shore D	**steel**			Custom Scientifics Friction-O-Meter			98 (30)		0.15	0.65	440
LNP FP C-1000	unmodified		carbon steel; surface finish: 12-16 µin; 18-20 Rockwell C	thrust washer		23	40 (0.28)	50 (15.2)	2000 (4.3)	0.27	0.29	453
LNP FP-CC-1003	15% carbon fiber		"	"		23	40 (0.28)	50 (15.2)	2000 (4.3)	0.15	0.17	453
LNP FP-CL 4020	10% PTFE		"	"		23	40 (0.28)	50 (15.2)	2000 (4.3)	0.06	0.11	453
FLUOROPOLYMER, ETFE against mating surface												
DuPont Tefzel 200	gen. purp. grade, 50 Rockwell R, 75 Shore D	**steel**	AISI 1018; 16 AA; unlubricated; 20 Rockwell C	thrust bearing test		23	100 (0.69)	10.2 (3.1)	1020 (2.1)		0.4	205
DuPont Tefzel HT-2004	74 Rockwell R; 25% glass fiber		"	"		23	100 (0.69)	10.2 (3.1)	1020 (2.1)	0.3	0.3	205
DuPont Tefzel HT-2004	74 Rockwell R; 25% glass fiber						100 (0.69)			0.31		205
DuPont Tefzel HT-2004	74 Rockwell R; 25% glass fiber						493 (3.4)			0.34		205
DuPont Tefzel HT-2004	74 Rockwell R; 25% glass fiber						49 (0.34)			0.38		205
DuPont Tefzel HT-2004	74 Rockwell R; 25% glass fiber						10 (0.07)			0.51		205
DuPont	63-72 Shore D			ASTM D1894				<10 (<3)			0.23	340
LNP FP EEL 4024	20% glass fiber, 10% PTFE		carbon steel; surface finish: 12-16 µin; 18-20 Rockwell C	thrust washer		23	40 (0.28)	50 (15.2)	2000 (4.3)	0.14	0.16	453
LNP FP EF 1006	30% glass fiber		"	"		23	40 (0.28)	50 (15.2)	2000 (4.3)	0.17	0.18	453

Material		Mating Surface		Test Conditions						Coefficient Of Friction		Source
Supplier / Grade	Material Note	Mating Surface Material	Mating Surface Note	Test Method	Test Conditions Note	Temp. (°C)	Pressure (psi) (MPa)	Speed (fpm) (m/min)	PV (psi x fpm) (MPa x m/min)	Static	Dynamic	
FLUOROPOLYMER, ETFE against mating surface												
LNP FP-E-1000	unmodified	**steel**	carbon steel; surface finish: 12-16 µin; 18-20 Rockwell C	thrust washer		23	40 (0.28)	50 (15.2)	2000 (4.3)	0.5	0.4	453
LNP FP-EC-1003	15% carbon fiber		"	"		23	40 (0.28)	50 (15.2)	2000 (4.3)	0.11	0.18	453
LNP FP-EF 1003	15% glass fiber		"	"		23	40 (0.28)	50 (15.2)	2000 (4.3)	0.2	0.2	453
LNP FP-EF-1002	10% glass fiber		"	"		23	40 (0.28)	50 (15.2)	2000 (4.3)	0.46	0.33	453
LNP FP-EF-1004	20% glass fiber		"	"		23	40 (0.28)	50 (15.2)	2000 (4.3)	0.19	0.2	453
LNP FP-EF-1005	25% glass fiber		"	"		23	40 (0.28)	50 (15.2)	2000 (4.3)	0.18	0.19	453
LNP FP-EL-4060	30% PTFE		AISI 1141; surface finish: 12-16 µin	"		23	40 (0.28)	50 (15.2)	2000 (4.3)	0.09	0.15	453
LNP FP-EL-4060	30% PTFE		AISI 1141; surface finish: 8-12 µin	"		23	40 (0.28)	50 (15.2)	2000 (4.3)	0.1	0.12	453
LNP FP-EL-4060	30% PTFE		AISI 1141; surface finish: 50-70 µin	"		23	40 (0.28)	50 (15.2)	2000 (4.3)	0.1	0.12	453
LNP FP-EL-4060	30% PTFE		carbon steel; surface finish: 12-16 µin; 18-20 Rockwell C	"		23	40 (0.28)	50 (15.2)	2000 (4.3)	0.1	0.12	453
LNP FP-EL-4320	10% graphite		"	"		23	40 (0.28)	50 (15.2)	2000 (4.3)	0.13	0.23	453
FLUOROPOLYMER, FEP against mating surface												
DuPont	56 Shore D	**steel**		ASTM D1894				<10 (<3)			0.2	340
LNP FP-FC-1002	10% carbon fiber		carbon steel; surface finish: 12-16 µin; 18-20 Rockwell C	thrust washer		23	40 (0.28)	50 (15.2)	2000 (4.3)	0.1	0.11	453
LNP FP-FF 1004M	20% milled glass		"	"		23	40 (0.28)	50 (15.2)	2000 (4.3)	0.11	0.12	453
LNP FP-FF-1003M	15% milled glass		"	"		23	40 (0.28)	50 (15.2)	2000 (4.3)	0.11	0.12	453
LNP PE F-1000	unmodified		"	"		23	40 (0.28)	50 (15.2)	2000 (4.3)	0.11	0.16	453
FLUOROPOLYMER, PFA against mating surface												
DuPont Teflon PFA 340	gen. purp. grade, 55 Shore D, moderate mol. wgt.	**steel**	AISI 1018; 16AA; 20 Rockwell C	thrust bearing wear test	test duration: 103 hours; conditions: room temperature, ambient air, unlubricated		100 (0.69)	3 (0.9)	300 (0.6)		0.21	39
DuPont Teflon PFA 340	gen. purp. grade, 55 Shore D, moderate mol. wgt.		"	"	"		100 (0.69)	10 (3)	1000 (2.1)		0.214	39
DuPont Teflon PFA 340	gen. purp. grade, 55 Shore D, moderate mol. wgt.		AISI 1018; 16AA; 20 Rockwell C	thrust bearing wear test	test duration: 103 hours; conditions: room temperature, ambient air, unlubricated		100 (0.69)	30 (9.1)	3000 (6.3)		0.229	39

Material		Mating Surface		Test Conditions						Coefficient Of Friction		Source
Supplier / Grade	Material Note	Mating Surface Material	Mating Surface Note	Test Method	Test Conditions Note	Temp. (°C)	Pressure (psi) (MPa)	Speed (fpm) (m/min)	PV (psi x fpm) (MPa x m/min)	Static	Dynamic	
FLUOROPOLYMER, PFA against mating surface												
DuPont Teflon PFA 340	gen. purp. grade, 55 Shore D, moderate mol. wgt.	**steel**	AISI 1018; 16AA; 20 Rockwell C	thrust bearing wear test	room temperature, ambient air, unlubricated				5000 (10.5)		0.236	39
DuPont Teflon PFA 340	gen. purp. grade, 55 Shore D, moderate mol. wgt.		"	"	test duration: 103 hours; conditions: room temperature, ambient air, unlubricated		100 (0.69)	50 (15.2)	5000 (10.5)		0.289	39
DuPont	15% glass fiber		"	"	room temperature, ambient air, unlubricated				5000 (10.5)		0.16	39
DuPont	15% graphite		"	"	"				5000 (10.5)		0.2	39
DuPont	5% glass fiber, 5% MoS_2		"	"	"				5000 (10.5)		0.2	39
DuPont	60 Shore D			ASTM D1894				<10 (<3)			0.2	340
DuPont	25% glass fiber		AISI 1018; 16AA; 20 Rockwell C	thrust bearing wear test	room temperature, ambient air, unlubricated				5000 (10.5)		0.325	39
LNP FP PE 1002	10% glass fiber		carbon steel; surface finish: 12-16 μin; 18-20 Rockwell C	thrust washer		23	40 (0.28)	50 (15.2)	2000 (4.3)	0.12	0.14	453
LNP FP PML-3312	10% mineral, 5% PTFE		"	"		23	40 (0.28)	50 (15.2)	2000 (4.3)	0.13	0.15	453
LNP FP-P-1000	unmodified		"	"		23	40 (0.28)	50 (15.2)	2000 (4.3)	0.12	0.15	453
LNP FP-PC 1003M	15% milled glass		"	"		23	40 (0.28)	50 (15.2)	2000 (4.3)	0.11	0.18	453
LNP FP-PC-1003	15% carbon fiber		cold rolled steel; surface finish: 12-16 μin; 22 Rockwell C	"		23	40 (0.28)	50 (15.2)	2000 (4.3)	0.11	0.18	453
LNP FP-PC-1003	15% carbon fiber		"	"		93	40 (0.28)	50 (15.2)	2000 (4.3)	0.17	0.21	453
LNP FP-PC-1003	15% carbon fiber		"	"		149	40 (0.28)	50 (15.2)	2000 (4.3)	0.28	0.33	453
LNP FP-PC-1003	15% carbon fiber		"	"		204	40 (0.28)	50 (15.2)	2000 (4.3)	0.36	0.43	453
LNP FP-PF 1004M	20% milled glass		carbon steel; surface finish: 12-16 μin; 18-20 Rockwell C	"		23	40 (0.28)	50 (15.2)	2000 (4.3)	0.13	0.15	453
LNP FP-PL-4020	10% PTFE		"	"		23	40 (0.28)	50 (15.2)	2000 (4.3)	0.06	0.11	453
FLUOROPOLYMER, CTFE against mating surface												
3M Kel-F 81	amorphous Kel-F 81	**aluminum**								0.23	0.15	96
3M Kel-F 81	crystalline Kel-F 81									0.23	0.18	96
3M Kel-F 81	amorphous Kel-F 81	**polyester PET**	Mylar							0.27	0.25	96
3M Kel-F 81	crystalline Kel-F 81		"							0.3	0.24	96

Material		Mating Surface		Test Conditions						Coefficient Of Friction		Source
Supplier / Grade	Material Note	Mating Surface Material	Mating Surface Note	Test Method	Test Conditions Note	Temp. (°C)	Pressure (psi) (MPa)	Speed (fpm) (m/min)	PV (psi x fpm) (MPa x m/min)	Static	Dynamic	
FLUOROPOLYMER, CTFE against mating surface												
3M Kel-F 81	crystalline Kel-F 81	**PCTFE**	Kel-F 81 (amorphous)							0.33	0.27	96
3M Kel-F 81	amorphous Kel-F 81		"							0.36	0.34	96
FLUOROPOLYMER, TFE against mating surface												
LNP FC-103	15% milled glass	**aluminum**	surface finish: 12-16 µin	thrust washer		23	33.3 (0.23)	150 (45.7)	5000 (10.5)	0.24	0.19	453
LNP FC-113/SM	15% synergistic MoS_2		"	"		23	33.3 (0.23)	150 (45.7)	5000 (10.5)	0.1	0.2	453
LNP FC-122	10% graphite powder		"	"		23	33.3 (0.23)	150 (45.7)	5000 (10.5)	0.07	0.13	453
LNP FC-132	10% coke flour		"	"		23	33.3 (0.23)	150 (45.7)	5000 (10.5)	0.12	0.21	453
LNP FC-146	60% bronze		"	"		23	33.3 (0.23)	150 (45.7)	5000 (10.5)	0.1	0.14	453
LNP PC-142	PPS filler, lubricant		"	"		23	33.3 (0.23)	150 (45.7)	5000 (10.5)	0.14	0.25	453
LNP PC-149	PPS filler, lubricant		"	"		23	33.3 (0.23)	150 (45.7)	5000 (10.5)	0.13	0.28	453
LNP PC-158	PPS filler, lubricant		"	"		23	33.3 (0.23)	150 (45.7)	5000 (10.5)	0.21	0.32	453
LNP PC-161	Polyoxybenzoate filler		"	"		23	33.3 (0.23)	150 (45.7)	5000 (10.5)	0.12	0.21	453
LNP PC-184	PPS filler, lubricant		"	"		23	33.3 (0.23)	150 (45.7)	5000 (10.5)	0.12	0.19	453
LNP PDX-81199	mineral filler		"	"		23	33.3 (0.23)	150 (45.7)	5000 (10.5)	0.13	0.18	453
Ausimont Halon 1005	5% glass fiber	**steel**		ASTM D1894						0.3	0.2	439
Ausimont Halon 1015	15% glass fiber			"						0.45	0.25	439
Ausimont Halon 1018	18% glass fiber			"						0.5	0.27	439
Ausimont Halon 1020	20% glass fiber			"						0.55	0.3	439
Ausimont Halon 1025	25% glass fiber			"						0.6	0.35	439
Ausimont Halon 1030	30% glass fiber			"						0.7	0.4	439
Ausimont Halon 1035	35% glass fiber			"						0.8	0.45	439
Ausimont Halon 1205	20% glass fiber, 5% graphite			"						0.6	0.35	439
Ausimont Halon 1206	5% glass fiber, 5% MoS_2			"						0.26	0.21	439
Ausimont Halon 1211	15% glass fiber, 5% MoS_2			"						0.3	0.24	439

Material		Mating Surface		Test Conditions						Coefficient Of Friction		Source
Supplier / Grade	Material Note	Mating Surface Material	Mating Surface Note	Test Method	Test Conditions Note	Temp. (°C)	Pressure (psi) (MPa)	Speed (fpm) (m/min)	PV (psi x fpm) (MPa x m/min)	Static	Dynamic	
FLUOROPOLYMER, TFE against mating surface		→										
Ausimont Halon 1223	23% glass fiber, 2% MoS_2	**steel**		ASTM D1894						0.5	0.25	439
Ausimont Halon 1230	20% glass fiber, 5% MoS_2, 5% graphite			"						0.6	0.4	439
Ausimont Halon 1240	20% glass fiber, 20% MoS_2			"						0.75	0.45	439
Ausimont Halon 1410	10% glass fiber, 10% carbon fiber			"						0.4	0.3	439
Ausimont Halon 1416	5% glass fiber, 10% carbon fiber			"						0.55	0.3	439
Ausimont Halon 2010	10% graphite			"						0.35	0.25	439
Ausimont Halon 2015	15% graphite			"						0.35	0.25	439
Ausimont Halon 2021	5% MoS_2			"						0.23	0.17	439
Ausimont Halon 3040	40% bronze			"						0.65	0.25	439
Ausimont Halon 3050	50% bronze			"						0.65	0.3	439
Ausimont Halon 3060	60% bronze			"						0.65	0.35	439
Ausimont Halon 3205	55% bronze, 5% MoS_2			"						0.55	0.3	439
Ausimont Halon 4010	10% carbon fiber			"						0.4	0.35	439
Ausimont Halon 4015	15% carbon fiber			"						0.45	0.4	439
Ausimont Halon 4022	25% carbon/ graphite			"						0.42	0.38	439
Ausimont Halon 4025	25% carbon fiber			"						0.55	0.45	439
Ausimont Halon 4026	25% carbon/ graphite			"						0.45	0.4	439
Ausimont Halon 8105	5% mineral filler			"						0.25	0.19	439
Ausimont Halon 8115	15% mineral filler			"						0.4	0.23	439
Ausimont Halon 9350	50% stainless steel			"						0.6	0.25	439
Ausimont Halon 9360	60% stainless steel			"						0.65	0.25	439
Ausimont Halon 9370	70% stainless steel			"						0.65	0.18	439
Ausimont Halon G80	virgin resin			"						0.2	0.15	439
DuPont	98 Shore A, 58 Rockwell R, 52 Shore D		carbon steel; unlubricated; 20-25 Rocwell C	thrust washer		23	800-1000 (5.5-6.9)	10 (3)	8000-10000 (16.8-21)		0.1	339

Material		Mating Surface		Test Conditions						Coefficient Of Friction		Source
Supplier / Grade	Material Note	Mating Surface Material	Mating Surface Note	Test Method	Test Conditions Note	Temp. (°C)	Pressure (psi) (MPa)	Speed (fpm) (m/min)	PV (psi x fpm) (MPa x m/min)	Static	Dynamic	
FLUOROPOLYMER, TFE against mating surface →												
DuPont	50-65 Shore D	**steel**		ASTM D1894				<10 (<3)			0.1	340
DuPont	58 Shore D; 15% glass fiber, 5% MoS_2		carbon steel; unlubricated; 20-25 Rocwell C	thrust washer		23	800-1000 (5.5-6.9)	10 (3)	8000-10000 (16.8-21)		0.12-0.13	339
DuPont	20% glass fiber, 5% graphite		"	"		23	800-1000 (5.5-6.9)	10 (3)	8000-10000 (16.8-21)		0.12-0.15	339
DuPont	55 Shore D; 15% graphite		"	"		23	800-1000 (5.5-6.9)	10 (3)	8000-10000 (16.8-21)		0.12-0.16	339
DuPont	98 Shore A, 58 Rockwell R, 52 Shore D		"	"		23	80-100 (0.55-0.69)	100 (30.5)	8000-10000 (16.8-21)		0.13	339
DuPont	65 Shore D; 60% bronze		"	"		23	800-1000 (5.5-6.9)	10 (3)	8000-10000 (16.8-21)		0.14-0.22	339
DuPont	65 Shore D; 60% bronze		"	"		23	8-10 (0.05-0.07)	1000 (304.8)	8000-10000 (16.8-21)		0.16-0.24	339
DuPont	56 Shore D; 25% glass fiber		"	"		23	800-1000 (5.5-6.9)	10 (3)	8000-10000 (16.8-21)		0.17-0.21	339
DuPont	58 Shore D; 15% glass fiber, 5% MoS_2		"	"		23	8-10 (0.05-0.07)	1000 (304.8)	8000-10000 (16.8-21)		0.19-0.24	339
DuPont	54 Shore D; 15% glass fiber		"	"		23	800-1000 (5.5-6.9)	10 (3)	8000-10000 (16.8-21)		0.20-0.22	339
DuPont	55 Shore D; 15% graphite		"	"		23	80-100 (0.55-0.69)	100 (30.5)	8000-10000 (16.8-21)		0.20-0.26	339
DuPont	20% glass fiber, 5% graphite		"	"		23	8-10 (0.05-0.07)	1000 (304.8)	8000-10000 (16.8-21)		0.24-0.37	339
DuPont	20% glass fiber, 5% graphite		"	"		23	80-100 (0.55-0.69)	100 (30.5)	8000-10000 (16.8-21)		0.24-0.50	339
DuPont	56 Shore D; 25% glass fiber		"	"		23	80-100 (0.55-0.69)	100 (30.5)	8000-10000 (16.8-21)		0.26-0.29	339
DuPont	54 Shore D; 15% glass fiber		"	"		23	80-100 (0.55-0.69)	100 (30.5)	8000-10000 (16.8-21)		0.27-0.40	339
DuPont	55 Shore D; 15% graphite		"	"		23	8-10 (0.05-0.07)	1000 (304.8)	8000-10000 (16.8-21)		0.30-0.31	339
DuPont	56 Shore D; 25% glass fiber		"	"		23	8-10 (0.05-0.07)	1000 (304.8)	8000-10000 (16.8-21)		0.30-0.45	339
DuPont	58 Shore D; 15% glass fiber, 5% MoS_2		"	"		23	80-100 (0.55-0.69)	100 (30.5)	8000-10000 (16.8-21)		0.32-0.35	339
DuPont	65 Shore D; 60% bronze		"	"		23	80-100 (0.55-0.69)	100 (30.5)	8000-10000 (16.8-21)		0.35-0.50	339
DuPont	54 Shore D; 15% glass fiber		"	"		23	8-10 (0.05-0.07)	1000 (304.8)	8000-10000 (16.8-21)		0.37-0.50	339
DuPont	98 Shore A, 58 Rockwell R, 52 Shore D		"	"		23	8-10 (0.05-0.07)	1000 (304.8)	8000-10000 (16.8-21)		unstable	339
DuPont	98 Shore A, 58 Rockwell R, 52 Shore D		"	"			500 (3.45)			0.05-0.08		339
DuPont	20% glass fiber, 5% graphite		"	"			500 (3.45)			0.08-0.10		339
DuPont	55 Shore D; 15% graphite		"	"			500 (3.45)			0.08-0.10		339

Material		Mating Surface		Test Conditions						Coefficient Of Friction		Source
Supplier / Grade	Material Note	Mating Surface Material	Mating Surface Note	Test Method	Test Conditions Note	Temp. (°C)	Pressure (psi) (MPa)	Speed (fpm) (m/min)	PV (psi x fpm) (MPa x m/min)	Static	Dynamic	
FLUOROPOLYMER, TFE against mating surface →												
DuPont	58 Shore D; 15% glass fiber, 5% MoS_2	**steel**	carbon steel; unlubricated; 20-25 Rocwell C	thrust washer			500 (3.45)			0.08-0.10		339
DuPont	65 Shore D; 60% bronze		"	"			500 (3.45)			0.08-0.10		339
DuPont	54 Shore D; 15% glass fiber		"	"			500 (3.45)			0.1-0.13		339
DuPont	56 Shore D; 25% glass fiber		"	"			500 (3.45)			0.1-0.13		339
ICI Fluon VB60	70-75 Shore D; 30% glass fiber, 60% bronze		420 S 37, T condition, surface finish: 0.3 µm				290 (2)	3.9 (1.2)	1131 (2.4)		0.16	71
ICI Fluon VG15	65-70 Shore D; 15% glass fiber		"				290 (2)	3.9 (1.2)	1131 (2.4)		0.1	71
ICI Fluon VG25	70-75 Shore D; 25% glass fiber, 30% carbon fiber		"				290 (2)	3.9 (1.2)	1131 (2.4)		0.11	71
ICI Fluon VP25	70-75 Shore D; 25% powdered coke		"				290 (2)	3.9 (1.2)	1131 (2.4)		0.17	71
ICI Fluon VR15	65-70 Shore D; 15% graphite		"				290 (2)	3.9 (1.2)	1131 (2.4)		0.22	71
ICI Fluon VX1	70-75 Shore D; 30% glass/ complex salts		"				290 (2)	3.9 (1.2)	1131 (2.4)		0.12	71
ICI Fluon VX2	70-75 Shore D; 63% bronze, graphite		"				290 (2)	3.9 (1.2)	1131 (2.4)		0.2	71
ICI	15% glass fiber, 5% MoS_2		"				290 (2)	3.9 (1.2)	1131 (2.4)		0.19	71
LNP FC-103	15% milled glass		surface finish: 12-16 µin	thrust washer		23	33.3 (0.23)	150 (45.7)	5000 (10.5)	0.05	0.09	453
LNP FC-103	15% milled glass		cold rolled steel; surface finish: 12-16 µin; 22 Rockwell C	"		23	33.3 (0.23)	150 (45.7)	5000 (10.5)	0.05	0.09	453
LNP FC-103	15% milled glass		"	"		93	33.3 (0.23)	150 (45.7)	5000 (10.5)	0.06	0.09	453
LNP FC-103	15% milled glass		"	"		149	33.3 (0.23)	150 (45.7)	5000 (10.5)	0.06	0.09	453
LNP FC-103	15% milled glass		"	"		204	33.3 (0.23)	150 (45.7)	5000 (10.5)	0.08	0.12	453
LNP FC-103	15% milled glass		"	"		260	33.3 (0.23)	150 (45.7)	5000 (10.5)	0.09	0.13	453
LNP FC-113/SM	15% synergistic MoS&&2@l		surface finish: 12-16 µin	"		23	33.3 (0.23)	150 (45.7)	5000 (10.5)	0.08	0.13	453
LNP FC-122	10% graphite powder		"	"		23	33.3 (0.23)	150 (45.7)	5000 (10.5)	0.05	0.07	453
LNP FC-132	10% coke flour		"	"		23	33.3 (0.23)	150 (45.7)	5000 (10.5)	0.06	0.07	453
LNP FC-146	60% bronze		"	"		23	33.3 (0.23)	150 (45.7)	5000 (10.5)	0.07	0.13	453
LNP FC-182	55% bronze, 5% MoS_2		cold rolled steel; surface finish: 12-16 µin; 22 Rockwell C	"		23	33.3 (0.23)	150 (45.7)	5000 (10.5)	0.07	0.13	453
LNP FC-182	55% bronze, 5% MoS_2		"	"		93	33.3 (0.23)	150 (45.7)	5000 (10.5)	0.08	0.13	453

Material		Mating Surface		Test Conditions						Coefficient Of Friction		Source
Supplier / Grade	Material Note	Mating Surface Material	Mating Surface Note	Test Method	Test Conditions Note	Temp. (°C)	Pressure (psi) (MPa)	Speed (fpm) (m/min)	PV (psi x fpm) (MPa x m/min)	Static	Dynamic	
FLUOROPOLYMER, TFE against mating surface →												
LNP FC-182	55% bronze, 5% MoS_2	**steel**	cold rolled steel; surface finish: 12-16 μin; 22 Rockwell C	thrust washer		149	33.3 (0.23)	150 (45.7)	5000 (10.5)	0.09	0.14	453
LNP FC-182	55% bronze, 5% MoS_2		"	"		204	33.3 (0.23)	150 (45.7)	5000 (10.5)	0.1	0.15	453
LNP FC-191	25% carbon/ graphite		"	"		23	33.3 (0.23)	150 (45.7)	5000 (10.5)	0.08	0.09	453
LNP FC-191	25% carbon/ graphite		"	"		93	33.3 (0.23)	150 (45.7)	5000 (10.5)	0.11	0.13	453
LNP FC-191	25% carbon/ graphite		"	"		149	33.3 (0.23)	150 (45.7)	5000 (10.5)	0.12	0.13	453
LNP FC-191	25% carbon/ graphite		"	"		204	33.3 (0.23)	150 (45.7)	5000 (10.5)	0.14	0.15	453
LNP FC-191	25% carbon/ graphite		"	"		260	33.3 (0.23)	150 (45.7)	5000 (10.5)	0.16	0.18	453
LNP PC-142	PPS filler, lubricant		"	"		23	33.3 (0.23)	150 (45.7)	5000 (10.5)	0.08	0.14	453
LNP PC-149	PPS filler, lubricant		"	"		23	33.3 (0.23)	150 (45.7)	5000 (10.5)	0.08	0.12	453
LNP PC-158	PPS filler, lubricant		"	"		23	33.3 (0.23)	150 (45.7)	5000 (10.5)	0.07	0.11	453
LNP PC-161	Polyoxybenzoate filler		surface finish: 12-16 μin	"		23	33.3 (0.23)	150 (45.7)	5000 (10.5)	0.05	0.13	453
LNP PC-184	PPS filler, lubricant		"	"		23	33.3 (0.23)	150 (45.7)	5000 (10.5)	0.09	0.11	453
LNP PC-185	PPS filler		cold rolled steel; surface finish: 12-16 μin; 22 Rockwell C	"		23	33.3 (0.23)	150 (45.7)	5000 (10.5)	0.07	0.13	453
LNP PC-185	PPS filler		"	"		93	33.3 (0.23)	150 (45.7)	5000 (10.5)	0.09	0.14	453
LNP PC-185	PPS filler		"	"		149	33.3 (0.23)	150 (45.7)	5000 (10.5)	0.11	0.15	453
LNP PC-185	PPS filler		"	"		204	33.3 (0.23)	150 (45.7)	5000 (10.5)	0.11	0.16	453
LNP PC-185	PPS filler		"	"		260	33.3 (0.23)	150 (45.7)	5000 (10.5)	0.15	0.19	453
LNP PDX-81199	mineral filler		surface finish: 12-16 μin	"		23	33.3 (0.23)	150 (45.7)	5000 (10.5)	0.11	0.14	453
Unspecified grade	carbon filled		carbon EN8, dry	pad on ring (Amsler wear tester)		20		600 (183)			0.25 (average of COF @ LPV and 50% of LPV load)	338

Material		Mating Surface		Test Conditions						Coefficient Of Friction		Source
Supplier / Grade	Material Note	Mating Surface Material	Mating Surface Note	Test Method	Test Conditions Note	Temp. (°C)	Pressure (psi) (MPa)	Speed (fpm) (m/min)	PV (psi x fpm) (MPa x m/min)	Static	Dynamic	
FLUOROPOLYMER, PVDF against mating surface												
Atochem		**steel**	gound steel plate								0.48	437
Ensinger Ensikem	standard grade						40 (0.28)	50 (15.2)	2000 (4.3)		0.24	449
LNP FP V-1000	unmodified		carbon steel; surface finish: 12-16 μin; 18-20 Rockwell C	thrust washer		23	40 (0.28)	50 (15.2)	2000 (4.3)	0.21	0.24	453
LNP FP VM 3850	25% mineral fiber		"	"		23	40 (0.28)	50 (15.2)	2000 (4.3)	0.11	0.12	453
LNP FP-VC-1003	15% carbon fiber		"	"		23	40 (0.28)	50 (15.2)	2000 (4.3)	0.25	0.25	453
LNP FP-VC-1003	15% carbon fiber		cold rolled steel; surface finish: 12-16 μin; 22 Rockwell C	"		23	40 (0.28)	50 (15.2)	2000 (4.3)	0.25	0.25	453
LNP FP-VC-1003	15% carbon fiber		"	"		93	40 (0.28)	50 (15.2)	2000 (4.3)	0.34	0.39	453
LNP FP-VCL-4024	20% carbon fiber, 10% PTFE		carbon steel; surface finish: 12-16 μin; 18-20 Rockwell C	"		23	40 (0.28)	50 (15.2)	2000 (4.3)	0.09	0.2	453
LNP FP-VM-2550	25% mineral fiber		"	"		23	40 (0.28)	50 (15.2)	2000 (4.3)	0.11	0.12	453
Solvay Solef 1008	79 Shore D, injection molding			"						0.45	0.34	444
Solvay Solef 1010	77 Shore D, extrusion			"						0.45	0.34	444
Solvay Solef 1012	79 Shore D, semi-finished products			"						0.45	0.34	444
Solvay Solef 3108	82 Shore D, anti-static			ASTM D1894						0.33	0.23	444
Solvay Solef 3208	78 Shore D, low friction			"						0.2	0.15	444
Solvay Solef 5708	79 Shore D, rotational molding			"						0.45	0.34	444
Solvay Solef 6010	77 Shore D, compression molding			"						0.45	0.34	444
Solvay Solef 8808	82 Shore D, carbon fiber reinforced			"						0.33	0.23	444
Solvay Solef 8908	81 Shore D, mica reinforced			"						0.28	0.25	444
NYLON against mating surface												
LNP V-1000	unmodified, high impact	**steel**	carbon steel; surface finish: 12-16 μin; 18-20 Rockwell C	thrust washer		23	40 (0.28)	50 (15.2)	2000 (4.3)	0.22	0.28	453
LNP VFL-4036	high impact, 30% glass fiber, 15% PTFE		"	"		23	40 (0.28)	50 (15.2)	2000 (4.3)	0.23	0.18	453
LNP VL-4040	high impact, 20% PTFE		"	"		23	40 (0.28)	50 (15.2)	2000 (4.3)	0.18	0.24	453
LNP VL-4410	high impact, 2% silicone		"	"		23	40 (0.28)	50 (15.2)	2000 (4.3)	0.12	0.21	453

Material: Supplier / Grade	Material: Material Note	Mating Surface: Mating Surface Material	Mating Surface: Mating Surface Note	Test Conditions: Test Method	Test Conditions: Test Conditions Note	Test Conditions: Temp. (°C)	Test Conditions: Pressure (psi) (MPa)	Test Conditions: Speed (fpm) (m/min)	Test Conditions: PV (psi x fpm) (MPa x m/min)	Coefficient Of Friction: Static	Coefficient Of Friction: Dynamic	Source
NYLON against mating surface												
LNP VL-4530	high impact, 13% PTFE, 2% silicone	steel	carbon steel; surface finish: 12-16 µin; 18-20 Rockwell C	thrust washer		23	40 (0.28)	50 (15.2)	2000 (4.3)	0.15	0.2	453
RTP 200H	high impact		C1018 steel; surface finish: 14-17 µin; 15-25 Rockwell C	"	apparatus: Falex Model No. 6	23	40 (0.28)	50 (15.2)	2000 (4.3)	0.23	0.3	457
RTP 200H TFE 20	20% PTFE, high impact		"	"	"	23	40 (0.28)	50 (15.2)	2000 (4.3)	0.11	0.12	457
RTP 205H	30% glass fiber, high impact		"	"	"	23	40 (0.28)	50 (15.2)	2000 (4.3)	0.19	0.35	457
RTP 205H TFE 15	15% PTFE, 30% glass fiber, high impact		"	"	"	23	40 (0.28)	50 (15.2)	2000 (4.3)	0.18	0.26	457
				ASTM D1894							0.4	78
NYLON, AMORPHOUS against mating surface												
LNP X-1000	unmodified	steel	carbon steel; surface finish: 12-16 µin; 18-20 Rockwell C	thrust washer		23	40 (0.28)	50 (15.2)	2000 (4.3)	0.23	0.32	453
LNP XC-1006	30% carbon fiber		"	"		23	40 (0.28)	50 (15.2)	2000 (4.3)	0.19	0.24	453
LNP XF-1006	30% glass fiber		"	"		23	40 (0.28)	50 (15.2)	2000 (4.3)	0.28	0.34	453
LNP XFL-4036	30% glass fiber, 15% PTFE		"	"		23	40 (0.28)	50 (15.2)	2000 (4.3)	0.2	0.26	453
LNP XL-4040	20% PTFE		"	"		23	40 (0.28)	50 (15.2)	2000 (4.3)	0.13	0.22	453
NYLON 11 against mating surface												
LNP HFL-4325	25% glass fiber, 10% graphite	steel	carbon steel; surface finish: 12-16 µin; 18-20 Rockwell C	thrust washer		23	40 (0.28)	50 (15.2)	2000 (4.3)	0.18	0.22	453
LNP PDX-4208	85% bronze		"	"		23	40 (0.28)	50 (15.2)	2000 (4.3)	0.15	0.15	453
LNP PDX-5156	83% bronze, 3% MoS_2		"	"		23	40 (0.28)	50 (15.2)	2000 (4.3)	0.12	0.12	453
RTP 200C TFE 20	20% PTFE		C1018 steel; surface finish: 14-17 µin; 15-25 Rockwell C	"	apparatus: Falex Model No. 6	23	40 (0.28)	50 (15.2)	2000 (4.3)	0.11	0.15	457
RTP 203C TFE 20	20% PTFE, 20% glass fiber		"	"	"	23	40 (0.28)	50 (15.2)	2000 (4.3)	0.16	0.2	457
NYLON 12 against mating surface												
LNP S-1000	unmodified	steel	carbon steel; surface finish: 12-16 µin; 18-20 Rockwell C	thrust washer		23	40 (0.28)	50 (15.2)	2000 (4.3)	0.21	0.27	453
LNP SFL-4036	30% glass fiber, 15% PTFE		"	"		23	40 (0.28)	50 (15.2)	2000 (4.3)	0.15	0.19	453
LNP SL-4040	15% PTFE		"	"		23	40 (0.28)	50 (15.2)	2000 (4.3)	0.09	0.16	453

Material: Supplier / Grade	Material: Material Note	Mating Surface: Mating Surface Material	Mating Surface: Mating Surface Note	Test Conditions: Test Method	Test Conditions: Test Conditions Note	Test Conditions: Temp. (°C)	Test Conditions: Pressure (psi) (MPa)	Test Conditions: Speed (fpm) (m/min)	Test Conditions: PV (psi x fpm) (MPa x m/min)	Coefficient Of Friction: Static	Coefficient Of Friction: Dynamic	Source
NYLON 12 against mating surface												
LNP SL-4610	2% silicone	steel	carbon steel; surface finish: 12-16 μin; 18-20 Rockwell C	thrust washer		23	40 (0.28)	50 (15.2)	2000 (4.3)	0.18	0.17	453
RTP 205F TFE 20	20% PTFE, 30% glass fiber		C1018 steel; surface finish: 14-17 μin; 15-25 Rockwell C	"	apparatus: Falex Model No. 6	23	40 (0.28)	50 (15.2)	2000 (4.3)	0.16	0.2	457
RTP 20OF TFE 20	20% PTFE		"	"	"	23	40 (0.28)	50 (15.2)	2000 (4.3)	0.11	0.15	457
NYLON 6 against mating surface												
Akzo J-3/30/MS/5	30% glass fiber, 5% MoS_2	steel	C1018 steel; surface finish: 14-17 μin; 15-25 Rockwell C	thrust washer	apparatus: Falex Model No. 6	23	40 (0.28)	50 (15.2)	2000 (4.3)	0.24	0.31	458
Bay Resins Lubriplas PA-211	unmodified resin			"			40 (0.28)	50 (15.2)	2000 (4.3)	0.22	0.26	435
Bay Resins Lubriplas PA-211G13	13% glass fiber			"			40 (0.28)	50 (15.2)	2000 (4.3)	0.24	0.28	435
Bay Resins Lubriplas PA-211G30TF15	30% glass fiber, 15% PTFE			"			40 (0.28)	50 (15.2)	2000 (4.3)	0.2	0.24	435
Bay Resins Lubriplas PA-211G33	33% glass fiber			"			40 (0.28)	50 (15.2)	2000 (4.3)	0.26	0.32	435
Bay Resins Lubriplas PA-211TF20	20% PTFE			"			40 (0.28)	50 (15.2)	2000 (4.3)	0.1	0.19	435
Ensinger Vekton	standard grade, cast						40 (0.28)	50 (15.2)	2000 (4.3)		0.26	449
Ensinger Vekton 6PAL	oil impregnated, cast						40 (0.28)	50 (15.2)	2000 (4.3)	0.16	0.18	449
Ensinger	standard grade						40 (0.28)	50 (15.2)	2000 (4.3)		0.26	449
Ferro Star-C PA6 20Y	20% carbon fiber			ASTM D3028						0.22		452
Ferro Star-C PA6 30Y	30% carbon fiber			ASTM D3028						0.2		452
Ferro Star-L PA6 30VD31	30% glass, 5% MoS_2, lubricated			"						0.3		452
Ferro Star-L PA6 30VD31	30% glass, 5% MoS_2, lubricated			"	load: 12 N			39 (12)			0.3	452
Ferro Star-L PA6 30VD31	30% glass, 5% MoS_2, lubricated			"	load: 43 N			39 (12)			0.32	452
Ferro Star-L PA6 30VD51	30% glass, 15% PTFE, lubricated			"						0.26		452
Ferro Star-L PA6 30VD51	30% glass, 15% PTFE, lubricated			"	load: 12 N			39 (12)			0.09	452
Ferro Star-L PA6 30VD51	30% glass, 15% PTFE, lubricated			"	load: 43 N			39 (12)			0.1	452
Ferro Star-L PA6 30VD61	30% glass, 13% PTFE, 2% silicon, lubricated			"						0.19		452
Ferro Star-L PA6 30VD61	30% glass, 13% PTFE, 2% silicon, lubricated			"	load: 12 N			39 (12)			0.12	452
Ferro Star-L PA6 30VD61	30% glass, 13% PTFE, 2% silicon, lubricated			"	load: 43 N			39 (12)			0.14	452

Material		Mating Surface		Test Conditions						Coefficient Of Friction		Source
Supplier / Grade	Material Note	Mating Surface Material	Mating Surface Note	Test Method	Test Conditions Note	Temp. (°C)	Pressure (psi) (MPa)	Speed (fpm) (m/min)	PV (psi x fpm) (MPa x m/min)	Static	Dynamic	
NYLON 6 against mating surface												
Ferro Star-L PA6 D51	15% PTFE, lubricated	**steel**		ASTM D3028						0.18		452
Ferro Star-L PA6 D51	15% PTFE, lubricated			"	load: 12 N			39 (12)			0.11	452
Ferro Star-L PA6 D51	15% PTFE, lubricated			"	load: 43 N			39 (12)			0.11	452
LNP Ny-Kon P	<5% MoS_2		carbon steel; surface finish: 12-16 µin; 18-20 Rockwell C	thrust washer		23	40 (0.28)	50 (15.2)	2000 (4.3)	0.28	0.3	453
LNP P-1000	unmodified		"	"		23	40 (0.28)	50 (15.2)	2000 (4.3)	0.22	0.26	453
LNP PC-1006	30% carbon fiber		"	"		23	40 (0.28)	50 (15.2)	2000 (4.3)	0.18	0.21	453
LNP PF-1006	30% glass fiber		"	"		23	40 (0.28)	50 (15.2)	2000 (4.3)	0.26	0.32	453
LNP PFL-4036	30% glass fiber, 15% PTFE		"	"		23	40 (0.28)	50 (15.2)	2000 (4.3)	0.2	0.25	453
LNP PFL-4216	30% glass fiber, <5% MoS_2		"	"		23	40 (0.28)	50 (15.2)	2000 (4.3)	0.26	0.32	453
LNP PFL-4218	40% glass fiber, <5% MoS_2		"	"		23	40 (0.28)	50 (15.2)	2000 (4.3)	0.28	0.34	453
LNP PFL-4536	30% glass fiber, 13% PTFE, 2% silicone		"	"		23	40 (0.28)	50 (15.2)	2000 (4.3)	0.17	0.2	453
LNP PL-4030	15% PTFE		"	"		23	40 (0.28)	50 (15.2)	2000 (4.3)	0.13	0.15	453
LNP PL-4040	20% PTFE		"	"		23	40 (0.28)	50 (15.2)	2000 (4.3)	0.1	0.19	453
LNP PL-431 0	5% graphite		"	"		23	40 (0.28)	50 (15.2)	2000 (4.3)	0.16	0.19	453
LNP PL-4410	2% silicone		"	"		23	40 (0.28)	50 (15.2)	2000 (4.3)	0.1	0.12	453
LNP PL-4540	18% PTFE, 2% silicone		"	"		23	40 (0.28)	50 (15.2)	2000 (4.3)	0.09	0.11	453
Polymer Nylatron GSM	cast, MoS_2 modified		unlubricated	"	load: 22.7 kg					0.17-0.27		441
Polymer Nylatron GSM	cast, MoS_2 modified		"	"			250 (1.72)	20 (6.1)	5000 (10.5)		0.17-0.25	441
Polymer Nylatron MC901	cast, general purpose, high heat		"	"	load: 22.7 kg					0.18-0.25		441
Polymer Nylatron MC901	cast, general purpose, high heat		"	"			250 (1.72)	20 (6.1)	5000 (10.5)		0.17-0.28	441
Polymer Nylatron NSM	w/ additives to improve bearing properties		"	"	load: 22.7 kg					0.17-0.26		441
Polymer Nylatron NSM	w/ additives to improve bearing properties		"	"			250 (1.72)	20 (6.1)	5000 (10.5)		0.19-0.28	441
RTP 200A			C1018 steel; surface finish: 14-17 µin; 15-25 Rockwell C	"	apparatus: Falex Model No. 6	23	40 (0.28)	50 (15.2)	2000 (4.3)	0.2	0.28	457

Material		Mating Surface		Test Conditions						Coefficient Of Friction		Source
Supplier / Grade	Material Note	Mating Surface Material	Mating Surface Note	Test Method	Test Conditions Note	Temp. (°C)	Pressure (psi) (MPa)	Speed (fpm) (m/min)	PV (psi x fpm) (MPa x m/min)	Static	Dynamic	
NYLON 6 against mating surface												
RTP 200A TFF 20	20% PTFE	**steel**	C1018 steel; surface finish: 14-17 µin; 15-25 Rockwell C	thrust washer	apparatus: Falex Model No. 6	23	40 (0.28)	50 (15.2)	2000 (4.3)	0.12	0.2	457
RTP 205A	30% glass fiber		"	"	"	23	40 (0.28)	50 (15.2)	2000 (4.3)	0.25	0.32	457
RTP 205A TFE 15	15% PTFE, 30% glass fiber		"	"	"	23	40 (0.28)	50 (15.2)	2000 (4.3)	0.17	0.23	457
Thermofil N-15FG-0100	15% glass fiber			ASTM D1894		23				0.24	0.28	459
Thermofil N-30FG-0100	30% glass fiber			"		23				0.27	0.3	459
Thermofil N-30FG-0214	30% glass fiber, PTFE lubricated			"		23				0.21	0.25	459
Thermofil N-30NF-0100	30% graphite fiber			"		23				0.15	0.2	459
Thermofil N-40-MF-0100	40% mineral			"		23				0.2	0.25	459
Thermofil N-40BG-0100	40% glass bead			"		23				0.22	0.24	459
Thermofil N-40FM-0100	40% glass/ mineral			"		23				0.25	0.3	459
Unspecified grade	unmodified		carbon steel; surface finish: 16 µin; 18-22 Rockwell C	thrust washer	apparatus: Faville-LeVally	23	40 (0.28)	50 (15.2)	2000 (4.3)	0.22	0.26	458
NYLON 610 against mating surface												
LNP QFL-4036	30% glass fiber, 15% PTFE	**polycarbonate**	DFL-4034 (20% glass fiber, 15% PTFE)	thrust washer		23	40 (0.28)	50 (15.2)	2000 (4.3)	0.05	0.07	453
LNP Ny-Kon Q	<5% MoS_2	**steel**	carbon steel; surface finish: 12-16 µin; 18-20 Rockwell C	"		23	40 (0.28)	50 (15.2)	2000 (4.3)	0.3	0.31	453
LNP OL-4540	18% PTFE, 2% silicone		"	"		23	40 (0.28)	50 (15.2)	2000 (4.3)	0.09	0.11	453
LNP Q-1000	unmodified		"	"		23	40 (0.28)	50 (15.2)	2000 (4.3)	0.23	0.31	453
LNP QC-1006	30% carbon fiber		"	thrust washer		23	40 (0.28)	50 (15.2)	2000 (4.3)	0.2	0.25	453
LNP QF-1006	30% glass fiber		"	"		23	40 (0.28)	50 (15.2)	2000 (4.3)	0.26	0.34	453
LNP QFL-4036	30% glass fiber, 15% PTFE		"	"		23	40 (0.28)	50 (15.2)	2000 (4.3)	0.23	0.31	453
LNP QFL-4536	30% glass fiber, 13% PTFE, 2% silicone		"	"		23	40 (0.28)	50 (15.2)	2000 (4.3)	0.19	0.24	453
LNP QL-4040	20% PTFE		"	"		23	40 (0.28)	50 (15.2)	2000 (4.3)	0.12	0.2	453
LNP QL-4410	2% silicone		"	"		23	40 (0.28)	50 (15.2)	2000 (4.3)	0.1	0.12	453
RTP 200B			C1018 steel; surface finish: 14-17 µin; 15-25 Rockwell C	"	apparatus: Falex Model No. 6	23	40 (0.28)	50 (15.2)	2000 (4.3)	0.2	0.3	457
RTP 205B	30% glass fiber		"	"	"	23	40 (0.28)	50 (15.2)	2000 (4.3)	0.26	0.34	457

Supplier / Grade	Material Note	Mating Surface Material	Mating Surface Note	Test Method	Test Conditions Note	Temp. (°C)	Pressure (psi) (MPa)	Speed (fpm) (m/min)	PV (psi x fpm) (MPa x m/min)	Coefficient Of Friction Static	Coefficient Of Friction Dynamic	Source
NYLON 610 against mating surface												
RTP 205B TFE 15	15% PTFE, 30% glass fiber	steel	C1018 steel; surface finish: 14-17 µin; 15-25 Rockwell C	thrust washer	apparatus: Falex Model No. 6	23	40 (0.28)	50 (15.2)	2000 (4.3)	0.24	0.31	457
NYLON 612 against mating surface												
Akzo J-4/30/TF/15	30% glass fiber, 15% PTFE	steel	carbon steel; surface finish: 16 µin; 18-22 Rockwell C	thrust washer	apparatus: Faville-LeVally	23	40 (0.28)	50 (15.2)	2000 (4.3)	0.2	0.25	458
Akzo J-4/CF/30/TF/10	30% carbon fiber, 10% PTFE		"	"	"	23	40 (0.28)	50 (15.2)	2000 (4.3)	0.14	0.18	458
Akzo J-4/CF/30/TF/13/SI/2	30% carbon fiber, 13% PTFE, 2% silicone		"	"	"	23	40 (0.28)	50 (15.2)	2000 (4.3)	0.1	0.11	458
Akzo NY-4/TF/10	10% PTFE		"	"	"	23	40 (0.28)	50 (15.2)	2000 (4.3)	0.14	0.22	458
LNP I-1000	unmodified		carbon steel; surface finish: 12-16 µin; 18-20 Rockwell C	"		23	40 (0.28)	50 (15.2)	2000 (4.3)	0.24	0.31	453
LNP IC-1006	30% carbon fiber		"	"		23	40 (0.28)	50 (15.2)	2000 (4.3)	0.19	0.23	453
LNP IF-1006	30% glass fiber		"	"		23	40 (0.28)	50 (15.2)	2000 (4.3)	0.27	0.33	453
LNP IFL-4036	30% glass fiber, 15% PTFE		"	"		23	40 (0.28)	50 (15.2)	2000 (4.3)	0.24	0.3	453
LNP IFL-4536	30% glass fiber, 13% PTFE, 2% silicone		"	"		23	40 (0.28)	50 (15.2)	2000 (4.3)	0.19	0.22	453
LNP IL-4040	20% PTFE		"	"		23	40 (0.28)	50 (15.2)	2000 (4.3)	0.12	0.19	453
LNP IL-4410	2% silicone		"	"		23	40 (0.28)	50 (15.2)	2000 (4.3)	0.1	0.12	453
LNP IL-4540	18% PTFE, 2% silicone		"	"		23	40 (0.28)	50 (15.2)	2000 (4.3)	0.08	0.1	453
LNP Ny-Kon I	<5% MoS_2		"	"		23	40 (0.28)	50 (15.2)	2000 (4.3)	0.33	0.33	453
RTP 200D			C1018 steel; surface finish: 14-17 µin; 15-25 Rockwell C	"	apparatus: Falex Model No. 6	23	40 (0.28)	50 (15.2)	2000 (4.3)	0.2	0.3	457
RTP 200D TFE 20	20% PTFE		"	thrust washer	apparatus: Falex Model No. 6	23	40 (0.28)	50 (15.2)	2000 (4.3)	0.12	0.2	457
RTP 205D	30% glass fiber		"	"	"	23	40 (0.28)	50 (15.2)	2000 (4.3)	0.26	0.34	457
RTP 205D TFE 15	15% PTFE, 30% glass fiber		"	"	"	23	40 (0.28)	50 (15.2)	2000 (4.3)	0.24	0.31	457
Thermofil N6-30FG-0100	30% glass fiber			ASTM D1894		23				0.25	0.31	459
Thermofil N6-30FG-0214	30% glass fiber, PTFE lubricated			"		23				0.24	0.3	459
Thermofil N6-30FG-0282	30% glass fiber, lubricated			"		23				0.2	0.24	459
Thermofil N6-30FG-0500	30% glass fiber, flame retardant			"		23				0.3	0.34	459
Thermofil N6-30NF-0100	30% graphite fiber			"		23				0.15	0.21	459

Material		Mating Surface		Test Conditions						Coefficient Of Friction		Source
Supplier / Grade	Material Note	Mating Surface Material	Mating Surface Note	Test Method	Test Conditions Note	Temp. (°C)	Pressure (psi) (MPa)	Speed (fpm) (m/min)	PV (psi x fpm) (MPa x m/min)	Static	Dynamic	
NYLON 612 against mating surface →												
Thermofil N6-9900-0500	flame retardant	**steel**		ASTM D1894		23				0.24	0.3	459
Thermofil R-20NF-0214	20% graphite fiber, PTFE lubricated			"		23				0.18	0.2	459
Thermofil R-30FG-0100	30% glass fiber			"		23				0.21	0.2	459
Thermofil R-30FG-0214	30% glass fiber, PTFE lubricated			"		23				0.21	0.2	459
Thermofil R-40NF-0100	40% graphite fiber			"		23				0.15	0.14	459
Thermofil R-9900-0200	lubricated			"		23				0.1	0.16	459
Thermofil R-9900-0214	PTFE lubricated			"		23				0.35	0.18	459
Unspecified grade	unmodified		carbon steel; surface finish: 16 µin; 18-22 Rockwell C	thrust washer	apparatus: Faville-LeVally	23	40 (0.28)	50 (15.2)	2000 (4.3)	0.27	0.31	458
NYLON 66 against mating surface →												
DuPont Zytel 101	gen. purp. grade, unlubricated	**acetal**	Delrin, unlubricated	thrust washer		23	20 (0.14)	94 (28.8)	1880 (4)	0.13-0.20	0.08-0.11	68
LNP R-1000	unmodified		K-1000 (unmodified)	"		23	40 (0.28)	50 (15.2)	2000 (4.3)	0.06	0.07	453
LNP RCL-4036	30% carbon fiber, 15% PTFE		KFL-4036 (30% glass fiber, 15% PTFE)	"		23	40 (0.28)	50 (15.2)	2000 (4.3)	0.05	0.07	453
LNP RF-1006	30% glass fiber		KL-4040 (20% PTFE)	"		23	40 (0.28)	50 (15.2)	2000 (4.3)	0.05	0.06	453
LNP RFL-4036	30% glass fiber, 15% PTFE		"	"		23	40 (0.28)	50 (15.2)	2000 (4.3)	0.05	0.06	453
LNP RL-4040	20% PTFE		"	"		23	40 (0.28)	50 (15.2)	2000 (4.3)	0.03	0.04	453
LNP RL-4410	2% silicone		KFL-4036 (30% glass fiber, 15% PTFE)	"		23	40 (0.28)	50 (15.2)	2000 (4.3)	0.1	0.11	453
Hoechst Cel. Nylon 1500	33% glass fiber	**aluminum**		ASTM D1894							0.12	317
Hoechst Cel. Nylon 1500	33% glass fiber			"							0.13	317
Hoechst Cel. Nylon 1600	43% glass fiber			"							0.12	317
Hoechst Cel. Nylon 1600	43% glass fiber			"							0.13	317
LNP RAL-4022	10% aramid fiber, 10% PTFE		2024 aluminum; surface finish: 50-70 µin	thrust washer		23	40 (0.28)	50 (15.2)	2000 (4.3)	0.09	0.17	453
LNP RAL-4022	10% aramid fiber, 10% PTFE		2024 aluminum; surface finish: 8-12 µin	"		23	40 (0.28)	50 (15.2)	2000 (4.3)	0.1	0.17	453
LNP RAL-4022	10% aramid fiber, 10% PTFE		2024 aluminum; surface finish: 12-16 µin	"		23	40 (0.28)	50 (15.2)	2000 (4.3)	0.11	0.16	453
LNP RCL-4036	30% carbon fiber, 15% PTFE		"	"		23	40 (0.28)	50 (15.2)	2000 (4.3)	0.12	0.12	453

Material		Mating Surface		Test Conditions						Coefficient Of Friction		Source
Supplier / Grade	Material Note	Mating Surface Material	Mating Surface Note	Test Method	Test Conditions Note	Temp. (°C)	Pressure (psi) (MPa)	Speed (fpm) (m/min)	PV (psi x fpm) (MPa x m/min)	Static	Dynamic	
NYLON 66 against mating surface →												
LNP RCL-4036	30% carbon fiber, 15% PTFE	**aluminum**	2024 aluminum; surface finish: 50-70 µin	thrust washer		23	40 (0.28)	50 (15.2)	2000 (4.3)	0.12	0.13	453
LNP RCL-4036	30% carbon fiber, 15% PTFE		2024 aluminum; surface finish: 8-12 µin	"		23	40 (0.28)	50 (15.2)	2000 (4.3)	0.13	0.14	453
LNP RF-1006	30% glass fiber		2024 aluminum; surface finish: 12-16 µin	"		23	40 (0.28)	50 (15.2)	2000 (4.3)	0.15	0.2	453
LNP RF-1006	30% glass fiber		2024 aluminum; surface finish: 50-70 µin	"		23	40 (0.28)	50 (15.2)	2000 (4.3)	0.16	0.21	453
LNP RF-1006	30% glass fiber		2024 aluminum; surface finish: 8-12 µin	"		23	40 (0.28)	50 (15.2)	2000 (4.3)	0.18	0.2	453
LNP RFL-4036	30% glass fiber, 15% PTFE		2024 aluminum; surface finish: 50-70 µin	"		23	40 (0.28)	50 (15.2)	2000 (4.3)	0.14	0.19	453
LNP RFL-4036	30% glass fiber, 15% PTFE		2024 aluminum; surface finish: 8-12 µin	"		23	40 (0.28)	50 (15.2)	2000 (4.3)	0.15	0.18	453
LNP RFL-4036	30% glass fiber, 15% PTFE		2024 aluminum; surface finish: 12-16 µin	"		23	40 (0.28)	50 (15.2)	2000 (4.3)	0.15	0.18	453
LNP RL-4040	20% PTFE		"	"		23	40 (0.28)	50 (15.2)	2000 (4.3)	0.06	0.09	453
LNP RL-4040	20% PTFE		2024 aluminum; surface finish: 8-12 µin	"		23	40 (0.28)	50 (15.2)	2000 (4.3)	0.07	0.09	453
LNP RL-4040	20% PTFE		2024 aluminum; surface finish: 50-70 µin	"		23	40 (0.28)	50 (15.2)	2000 (4.3)	0.08	0.1	453
DuPont Zytel 101	gen. purp. grade, unlubricated	**brass**	oil lubricated	Neely (or boundary film) testing machine			1552 (10.7)	157 (48)	243664 (512.2)		0.08-0.14	68
DuPont Zytel 101	gen. purp. grade, unlubricated		water lubricated	"			1044 (7.2)	157 (48)	163908 (344.5)		0.3-0.5	68
Hoechst Cel. Nylon 1500	33% glass fiber			ASTM D1894							0.24	317
Hoechst Cel. Nylon 1500	33% glass fiber			"							0.11	317
Hoechst Cel. Nylon 1600	43% glass fiber			"							0.11	317
Hoechst Cel. Nylon 1600	43% glass fiber			ASTM D1894							0.24	317
LNP RAL-4022	10% aramid fiber, 10% PTFE		70/30 brass; surface finish: 50-70 µin	thrust washer		23	40 (0.28)	50 (15.2)	2000 (4.3)	0.1	0.14	453
LNP RAL-4022	10% aramid fiber, 10% PTFE		70/30 brass; surface finish: 8-16 µin	"		23	40 (0.28)	50 (15.2)	2000 (4.3)	0.12	0.15	453
LNP RC-1006	30% carbon fiber		70/30 brass; surface finish: 50-70 µin	"		23	40 (0.28)	50 (15.2)	2000 (4.3)	0.18	0.18	453
LNP RC-1006	30% carbon fiber		70/30 brass; surface finish: 8-16 µin	"		23	40 (0.28)	50 (15.2)	2000 (4.3)	0.21	0.21	453
LNP RCL-4036	30% carbon fiber, 15% PTFE		70/30 brass; surface finish: 50-70 µin	"		23	40 (0.28)	50 (15.2)	2000 (4.3)	0.13	0.14	453
LNP RCL-4036	30% carbon fiber, 15% PTFE		70/30 brass; surface finish: 8-16 µin	"		23	40 (0.28)	50 (15.2)	2000 (4.3)	0.15	0.15	453
LNP RF-1006	30% glass fiber		70/30 brass; surface finish: 50-70 µin	"		23	40 (0.28)	50 (15.2)	2000 (4.3)	0.16	0.19	453

Material		Mating Surface		Test Conditions						Coefficient Of Friction		Source
Supplier / Grade	Material Note	Mating Surface Material	Mating Surface Note	Test Method	Test Conditions Note	Temp. (°C)	Pressure (psi) (MPa)	Speed (fpm) (m/min)	PV (psi x fpm) (MPa x m/min)	Static	Dynamic	
NYLON 66 against mating surface												
LNP RF-1006	30% glass fiber	**brass**	70/30 brass; surface finish: 8-16 μin	thrust washer		23	40 (0.28)	50 (15.2)	2000 (4.3)	0.17	0.22	453
LNP RFL-4036	30% glass fiber, 15% PTFE		70/30 brass; surface finish: 50-70 μin	"		23	40 (0.28)	50 (15.2)	2000 (4.3)	0.16	0.15	453
LNP RFL-4036	30% glass fiber, 15% PTFE		70/30 brass; surface finish: 8-16 μin	"		23	40 (0.28)	50 (15.2)	2000 (4.3)	0.18	0.15	453
LNP RFL-4536	30% glass fiber, 13% PTFE, 2% silicone		"	"		23	40 (0.28)	50 (15.2)	2000 (4.3)	0.18	0.17	453
LNP RFL-4536	30% glass fiber, 13% PTFE, 2% silicone		70/30 brass; surface finish: 50-70 μin	"		23	40 (0.28)	50 (15.2)	2000 (4.3)	0.19	0.18	453
LNP RL-4040	20% PTFE		70/30 brass; surface finish: 8-16 μin	"		23	40 (0.28)	50 (15.2)	2000 (4.3)	0.06	0.09	453
LNP RL-4040	20% PTFE		70/30 brass; surface finish: 50-70 μin	"		23	40 (0.28)	50 (15.2)	2000 (4.3)	0.05	0.09	453
DuPont Zytel 101	gen. purp. grade, unlubricated	**nylon 66**	Zytel 101, unlubricated	"		23	20 (0.14)	94 (28.8)	1880 (4)	0.36-0.46	0.11-0.19	68
DuPont Zytel 101	gen. purp. grade, unlubricated		"	Neely (or boundary film) testing machine			1044 (7.2)	157 (48)	163908 (344.5)		0.04-0.13	68
DuPont Zytel 101	gen. purp. grade, unlubricated		Zytel 101, oil lubricated	"			1044 (7.2)	157 (48)	163908 (344.5)		0.07-0.08	68
DuPont Zytel 101	gen. purp. grade, unlubricated		Zytel 101, water lubricated	"			1044 (7.2)	157 (48)	163908 (344.5)		0.08-0.14	68
Hoechst Cel. Nylon 1500	33% glass fiber		Hoechst Celanese Nylon 1500	ASTM D1894							0.23	317
Hoechst Cel. Nylon 1600	43% glass fiber		"	"							0.23	317
Hoechst Cel. Nylon 1600	43% glass fiber		"	"							0.23	317
LNP R-1000	unmodified		R-1000 (unmodified)	thrust washer		23	40 (0.28)	50 (15.2)	2000 (4.3)	0.06	0.07	453
LNP R-1000	unmodified		RF-1006 (30% glass fiber)	"		23	40 (0.28)	50 (15.2)	2000 (4.3)	0.07	0.08	453
LNP RAL-4022	10% aramid fiber, 10% PTFE		RCL-4036 (30% carbon fiber, 15% PTFE)	"		23	40 (0.28)	50 (15.2)	2000 (4.3)	0.1	0.13	453
LNP RAL-4022	10% aramid fiber, 10% PTFE		"	"		23	40 (0.28)	50 (15.2)	2000 (4.3)	0.1	0.13	453
LNP RC-1006	30% carbon fiber		"	"		23	40 (0.28)	50 (15.2)	2000 (4.3)	0.06	0.11	453
LNP RC-1006	30% carbon fiber		RF-1006 (30% glass fiber)	"		23	40 (0.28)	50 (15.2)	2000 (4.3)	0.09	0.18	453
LNP RC-1006	30% carbon fiber		RFL-4036 (30% glass fiber, 15% PTFE)	"		23	40 (0.28)	50 (15.2)	2000 (4.3)	0.11	0.12	453
LNP RC-1006	30% carbon fiber		R-1000 (unmodified)	"		23	40 (0.28)	50 (15.2)	2000 (4.3)	0.26	0.16	453
LNP RC-1008	40% carbon fiber		RCL-4536 (30% carbon fiber, 13% PTFE, 2% silicone)	"		23	40 (0.28)	50 (15.2)	2000 (4.3)	0.12	0.14	453
LNP RCL-4036	30% carbon fiber, 15% PTFE		RFL-4036 (30% glass fiber, 15% PTFE)	"		23	40 (0.28)	50 (15.2)	2000 (4.3)	0.08	0.08	453

Supplier / Grade	Material Note	Mating Surface Material	Mating Surface Note	Test Method	Test Conditions Note	Temp. (°C)	Pressure (psi) (MPa)	Speed (fpm) (m/min)	PV (psi x fpm) (MPa x m/min)	Coefficient Of Friction Static	Coefficient Of Friction Dynamic	Source
NYLON 66 against mating surface		→										
LNP RCL-4036	30% carbon fiber, 15% PTFE	**nylon 66**	RF-1006 (30% glass fiber)	thrust washer		23	40 (0.28)	50 (15.2)	2000 (4.3)	0.08	0.11	453
LNP RCL-4036	30% carbon fiber, 15% PTFE		RCL-4036 (30% carbon fiber, 15% PTFE)	"		23	40 (0.28)	50 (15.2)	2000 (4.3)	0.1	0.11	453
LNP RCL-4536	30% carbon fiber, 13% PTFE, 2% silicone		RCL-4536 (30% carbon fiber, 13% PTFE, 2% silicone)	"		23	40 (0.28)	50 (15.2)	2000 (4.3)	0.11	0.15	453
LNP RF-1006	30% glass fiber		RFL-4036 (30% glass fiber, 15% PTFE)	"		23	40 (0.28)	50 (15.2)	2000 (4.3)	0.07	0.09	453
LNP RF-1006	30% glass fiber		RF-1006 (30% glass fiber)	"		23	40 (0.28)	50 (15.2)	2000 (4.3)	0.12	0.12	453
LNP RF-1006	30% glass fiber		RL-4040 (20% PTFE)	"		23	40 (0.28)	50 (15.2)	2000 (4.3)	0.05	0.07	453
LNP RFL-4036	30% glass fiber, 15% PTFE		"	"		23	40 (0.28)	50 (15.2)	2000 (4.3)	0.05	0.06	453
LNP RFL-4036	30% glass fiber, 15% PTFE		RF-1006 (30% glass fiber)	"		23	40 (0.28)	50 (15.2)	2000 (4.3)	0.07	0.1	453
LNP RFL-4036	30% glass fiber, 15% PTFE		RCL-4036 (30% carbon fiber, 15% PTFE)	"		23	40 (0.28)	50 (15.2)	2000 (4.3)	0.1	0.15	453
LNP RFL-4036	30% glass fiber, 15% PTFE		RFL-4036 (30% glass fiber, 15% PTFE)	"		23	40 (0.28)	50 (15.2)	2000 (4.3)	0.11	0.12	453
LNP RL-4040	20% PTFE		RF-1006 (30% glass fiber)	"		23	40 (0.28)	50 (15.2)	2000 (4.3)	0.09	0.09	453
LNP RL-4040	20% PTFE		RCL-4036 (30% carbon fiber, 15% PTFE)	"		23	40 (0.28)	50 (15.2)	2000 (4.3)	0.05	0.06	453
LNP RL-4040	20% PTFE		RL-4040 (20% PTFE)	"		23	40 (0.28)	50 (15.2)	2000 (4.3)	0.05	0.08	453
LNP RL-4040	20% PTFE		RFL-4036 (30% glass fiber, 15% PTFE)	"		23	40 (0.28)	50 (15.2)	2000 (4.3)	0.06	0.06	453
LNP RL-4410	2% silicone		RF-1006 (30% glass fiber)	"		23	40 (0.28)	50 (15.2)	2000 (4.3)	0.09	0.14	453
LNP RL-4410	2% silicone		RCL-4036 (30% carbon fiber, 15% PTFE)	"		23	40 (0.28)	50 (15.2)	2000 (4.3)	0.1	0.13	453
LNP RL-4530	13% PTFE, 2% silicone		RF-1006 (30% glass fiber)	"		23	40 (0.28)	50 (15.2)	2000 (4.3)	0.06	0.07	453
LNP RL-4530	13% PTFE, 2% silicone	**nylon 66**	RCL-4036 (30% carbon fiber, 15% PTFE)	thrust washer		23	40 (0.28)	50 (15.2)	2000 (4.3)	0.1	0.1	453
LNP RL-4530	13% PTFE, 2% silicone		RFL-4036 (30% glass fiber, 15% PTFE)	"		23	40 (0.28)	50 (15.2)	2000 (4.3)	0.06	0.06	453
LNP R-1000	unmodified	**polycarbonate**	D-1000; unmodified	"		23	40 (0.28)	50 (15.2)	2000 (4.3)	0.06	0.05	453
LNP R-1000	unmodified		DFL-4036 (30% glass fiber, 15% PTFE)	"		23	40 (0.28)	50 (15.2)	2000 (4.3)	0.15	0.18	453
LNP RC-1006	30% carbon fiber		"	"		23	40 (0.28)	50 (15.2)	2000 (4.3)	0.09	0.1	453
LNP RF-1006	30% glass fiber		DF-1006 (30% glass fiber)	"		23	40 (0.28)	50 (15.2)	2000 (4.3)	0.16	0.27	453
LNP RL-4040	20% PTFE		"	"		23	40 (0.28)	50 (15.2)	2000 (4.3)	0.08	0.12	453

Material		Mating Surface		Test Conditions						Coefficient Of Friction		Source
Supplier / Grade	Material Note	Mating Surface Material	Mating Surface Note	Test Method	Test Conditions Note	Temp. (°C)	Pressure (psi) (MPa)	Speed (fpm) (m/min)	PV (psi x fpm) (MPa x m/min)	Static	Dynamic	
NYLON 66 against mating surface												
LNP RL-4040	20% PTFE	**polycarbonate**	DFL-4036 (30% glass fiber, 15% PTFE)	thrust washer		23	40 (0.28)	50 (15.2)	2000 (4.3)	0.06	0.07	453
LNP RL-4410	2% silicone		D-1000; unmodified	"		23	40 (0.28)	50 (15.2)	2000 (4.3)	0.08	0.08	453
LNP RL-4410	2% silicone		DFL-4036 (30% glass fiber, 15% PTFE)	"		23	40 (0.28)	50 (15.2)	2000 (4.3)	0.1	0.14	453
LNP RL-4530	13% PTFE, 2% silicone		"	"		23	40 (0.28)	50 (15.2)	2000 (4.3)	0.06	0.06	453
LNP RL-4530	13% PTFE, 2% silicone		D-1000; unmodified	"		23	40 (0.28)	50 (15.2)	2000 (4.3)	0.06	0.06	453
Akzo G-1/30/MS/5	30% long glass fiber, 5% MoS_2	**steel**	carbon steel; surface finish: 16 µin; 18-22 Rockwell C	"	apparatus: Faville-LeVally	23	40 (0.28)	50 (15.2)	2000 (4.3)	0.24	0.31	458
Akzo G-1/30/SI/2	30% long glass fiber, 2% silicone		"	"	"	23	40 (0.28)	50 (15.2)	2000 (4.3)	14	0.15	458
Akzo G-1/30/TF/15	30% long glass fiber,15% PTFE		"	"	"	23	40 (0.28)	50 (15.2)	2000 (4.3)		0.25	458
Akzo J-1/30/MS/5	30% glass fiber, 5% MoS_2		"	"	"	23	40 (0.28)	50 (15.2)	2000 (4.3)	0.24	0.31	458
Akzo J-1/30/SI/3	30% glass fiber, 3% silicone		"	"	"	23	40 (0.28)	50 (15.2)	2000 (4.3)	0.14	0.15	458
Akzo J-1/30/TF/15	30% glass fiber,15% PTFE		"	"	"	23	40 (0.28)	50 (15.2)	2000 (4.3)	0.2	0.25	458
Akzo J-1/33/TF/13/SI/2	33% glass fiber,13% PTFE, 2% silicone		"	"	"	23	40 (0.28)	50 (15.2)	2000 (4.3)	0.12	0.14	458
Akzo J-1/CF/15/TF/20	15% carbon fiber, 20% PTFE		"	"	"	23	40 (0.28)	50 (15.2)	2000 (4.3)	0.11	0.15	458
Akzo J-1/CF/30/TF/13/SI/2	30% carbon fiber, 13% PTFE, 2% silicone		"	"	"	23	40 (0.28)	50 (15.2)	2000 (4.3)	0.1	0.11	458
Akzo J-1/CF/30/TF/15	30% carbon fiber, 15% PTFE		"	"	"	23	40 (0.28)	50 (15.2)	2000 (4.3)	0.11	0.15	458
Akzo NY-1/MS/5	5% MoS_2		"	"	"	23	40 (0.28)	50 (15.2)	2000 (4.3)	0.28	0.3	458
Akzo NY-1/MS/5/TF/30	5% MoS_2, 30% PTFE		"	"	"	23	40 (0.28)	50 (15.2)	2000 (4.3)	0.12	0.15	458
Akzo NY-1/SI/5	5% silicone		"	thrust washer	apparatus: Faville-LeVally	23	40 (0.28)	50 (15.2)	2000 (4.3)	0.08	0.08	458
Akzo NY-1/TF/10	10% PTFE		"	"	"	23	40 (0.28)	50 (15.2)	2000 (4.3)	0.12	0.2	458
Akzo NY-1/TF/15	15% PTFE		"	"	"	23	40 (0.28)	50 (15.2)	2000 (4.3)	0.11	0.18	458
Akzo NY-1/TF/30	30% PTFE		"	"	"	23	40 (0.28)	50 (15.2)	2000 (4.3)	0.1	0.18	458
Bay Resins Lubriplas PA-111	unmodified resin			"			40 (0.28)	50 (15.2)	2000 (4.3)	0.2	0.28	435
Bay Resins Lubriplas PA-111	unmodified resin			"			63 (0.43)	20 (6.1)	1260 (2.6)		0.43	435
Bay Resins Lubriplas PA-111	unmodified resin			"			100 (0.69)	20 (6.1)	2000 (4.2)		0.44	435

Material		Mating Surface		Test Conditions						Coefficient Of Friction		Source
Supplier / Grade	Material Note	Mating Surface Material	Mating Surface Note	Test Method	Test Conditions Note	Temp. (°C)	Pressure (psi) (MPa)	Speed (fpm) (m/min)	PV (psi x fpm) (MPa x m/min)	Static	Dynamic	
NYLON 66 against mating surface												
Bay Resins Lubriplas PA-111	unmodified resin	**steel**	carbon steel; surface finish: 16 µin; 18-22 Rockwell C	thrust washer			20 (0.14)	63 (19)	1260 (2.6)		0.5	435
Bay Resins Lubriplas PA-111	unmodified resin			"			158 (1.09)	20 (6.1)	3170 (6.7)		0.5	435
Bay Resins Lubriplas PA-111	unmodified resin			"			32 (0.22)	63 (19)	2000 (4.2)		0.58	435
Bay Resins Lubriplas PA-111	unmodified resin			"			10 (0.07)	200 (61)	2000 (4.3)		0.61	435
Bay Resins Lubriplas PA-111	unmodified resin			"			6.3 (0.04)	200 (61)	1260 (2.4)		0.61	435
Bay Resins Lubriplas PA-111	unmodified resin			"			250 (1.72)	20 (6.1)	5000 (10.5)		0.67	435
Bay Resins Lubriplas PA-111	unmodified resin			"			25 (0.17)	200 (61)	5000 (10.5)		0.71	435
Bay Resins Lubriplas PA-111	unmodified resin			"			50 (0.34)	63 (19)	3170 (6.7)		0.71	435
Bay Resins Lubriplas PA-111	unmodified resin			"			16 (0.11)	200 (61)	3170 (6.7)		0.82	435
Bay Resins Lubriplas PA-111	unmodified resin			"			80 (0.55)	63 (19)	5000 (10.5)		0.88	435
Bay Resins Lubriplas PA-111C	<5% MoS_2			"			40 (0.28)	50 (15.2)	2000 (4.3)	0.28	0.3	435
Bay Resins Lubriplas PA-111CF30	30% carbon fiber			"			40 (0.28)	50 (15.2)	2000 (4.3)	0.16	0.2	435
Bay Resins Lubriplas PA-111G13	13% glass fiber			"			40 (0.28)	50 (15.2)	2000 (4.3)	0.21	0.28	435
Bay Resins Lubriplas PA-111G30TF15	30% glass fiber, 15% PTFE			"			40 (0.28)	50 (15.2)	2000 (4.3)	0.19	0.26	435
Bay Resins Lubriplas PA-111G33	33% glass fiber			"			40 (0.28)	50 (15.2)	2000 (4.3)	0.25	0.32	435
Bay Resins Lubriplas PA-111TF20	20% PTFE			"			40 (0.28)	50 (15.2)	2000 (4.3)	0.1	0.18	435
BASF Ultramid A3K	high flow, heat stabilized		Cr 6/800/HV; surface finish: 2.0-2.6 µm	peg-and-disc apparatus			145 (1)	98 (30)	14210 (29.9)		0.4-0.53	93
BASF Ultramid A3K	high flow, heat stabilized		Cr 6/800/HV; surface finish: 0.15-0.2 µm	"			145 (1)	98 (30)	14210 (29.9)		0.45-0.6	93
BASF Ultramid A3R	noiseless bearings; high flow; PE modified; stabilized		"	peg-and-disc apparatus			145 (1)	98 (30)	14210 (29.9)		0.32-0.42	93
BASF Ultramid A3R	noiseless bearings; high flow; PE modified; stabilized		Cr 6/800/HV; surface finish: 2.0-2.6 µm	"			145 (1)	98 (30)	14210 (29.9)		0.4-0.5	93
BASF Ultramid A3WC6	high flow, heat stabilized; 30% carbon fiber		Cr 6/800/HV; surface finish: 0.15-0.2 µm	"			145 (1)	98 (30)	14210 (29.9)		0.4-0.5	93
BASF Ultramid A3WC6	high flow, heat stabilized; 30% carbon fiber		Cr 6/800/HV; surface finish: 2.0-2.6 µm	"			145 (1)	98 (30)	14210 (29.9)		0.4-0.5	93
BASF Ultramid A3WG6	high flow, heat stabilized; 30% glass fiber		"	"			145 (1)	98 (30)	14210 (29.9)		0.55-0.65	93
BASF Ultramid A3WG6	high flow, heat stabilized; 30% glass fiber		Cr 6/800/HV; surface finish: 0.15-0.2 µm	"			145 (1)	98 (30)	14210 (29.9)		0.6-0.7	93

Material		Mating Surface		Test Conditions						Coefficient Of Friction		Source
Supplier / Grade	Material Note	Mating Surface Material	Mating Surface Note	Test Method	Test Conditions Note	Temp. (°C)	Pressure (psi) (MPa)	Speed (fpm) (m/min)	PV (psi x fpm) (MPa x m/min)	Static	Dynamic	
NYLON 66 against mating surface		→										
BASF Ultramid A4	moderate flow	**steel**	Cr 6/800/HV; surface finish: 2.0-2.6 µm	peg-and-disc apparatus			145 (1)	98 (30)	14210 (29.9)		0.4-0.53	93
BASF Ultramid A4	moderate flow		Cr 6/800/HV; surface finish: 0.15-0.2 µm	"			145 (1)	98 (30)	14210 (29.9)		0.45-0.6	93
DuPont Maranyl A108	tribological properties, formerly by LNP; w/ MoS_2, w/ graphite		carbon EN8, dry	pad on ring (Amsler wear tester)		20		100 (30.5)			0.23 (average of COF @ LPV and 50% of LPV load)	338
DuPont Maranyl A108	tribological properties, formerly by LNP; w/ MoS_2, w/ graphite		"	"		20		600 (183)			0.76 (average of COF @ LPV and 50% of LPV load)	338
DuPont Maranyl A198	formerly by LNP, tribological properties; glass reinforced; w/ graphite		"	"		20		100 (30.5)			0.15 (average of COF @ LPV and 50% of LPV load)	338
DuPont Maranyl A198	formerly by LNP, tribological properties; glass reinforced; w/ graphite		"	"		200		600 (183)			0.33 (average of COF @ LPV and 50% of LPV load)	338
DuPont Maranyl A198	formerly by LNP, tribological properties; glass reinforced; w/ graphite		"	"		20		600 (183)			0.34 (average of COF @ LPV and 50% of LPV load)	338
DuPont Zytel 101	gen. purp. grade, unlubricated		water lubricated	Neely (or boundary film) testing machine			1044 (7.2)	157 (48)	163908 (344.5)		0.3-0.5	68
DuPont Zytel 101	gen. purp. grade, unlubricated		unlubricated	thrust washer		23	20 (0.14)	94 (28.8)	1880 (4)	0.31-0.74	0.17-0.43	68
DuPont Zytel 101	gen. purp. grade, unlubricated		oil lubricated	Neely (or boundary film) testing machine			1552 (10.7)	157 (48)	243664 (512.2)		0.02-0.11	68
DuPont	20% aramid fiber			thrust washer	ASTM D3702	23	250 (1.72)	10 (3)	2500 (5.3)		0.39	454
DuPont	33% glass fiber			"	"	23	250 (1.72)	10 (3)	2500 (5.3)		0.42	454
DuPont				"	"	23	250 (1.72)	10 (3)	2500 (5.3)		0.435	454
DuPont	33% glass fiber			"	"	23	40 (0.28)	50 (15.2)	2000 (4.3)		0.476	454
DuPont				"	"	23	40 (0.28)	50 (15.2)	2000 (4.3)		0.574	454
Ensinger Ensilon	standard grade						40 (0.28)	50 (15.2)	2000 (4.3)		0.28	449

Material		Mating Surface		Test Conditions						Coefficient Of Friction		Source
Supplier / Grade	Material Note	Mating Surface Material	Mating Surface Note	Test Method	Test Conditions Note	Temp. (°C)	Pressure (psi) (MPa)	Speed (fpm) (m/min)	PV (psi x fpm) (MPa x m/min)	Static	Dynamic	
NYLON 66 against mating surface		→										
Ensinger Ensilon CF20	20% carbon fiber	**steel**					40 (0.28)	50 (15.2)	2000 (4.3)		0.2	449
Ensinger Ensilon GF30	30% glass fiber						40 (0.28)	50 (15.2)	2000 (4.3)		0.31	449
Ferro Star-C PA66 10Y	10% carbon fiber			ASTM D3028						0.22		452
Ferro Star-C PA66 20Y	20% carbon fiber			"						0.2		452
Ferro Star-C PA66 30Y	30% carbon fiber			"						0.19		452
Ferro Star-L PA66 15YD52	15% carbon fiber, 20% PTFE, lubricated			"						0.11		452
Ferro Star-L PA66 15YD52	15% carbon fiber, 20% PTFE, lubricated			"	load: 12 N			39 (12)			0.07	452
Ferro Star-L PA66 15YD52	15% carbon fiber, 20% PTFE, lubricated			"	load: 43 N			39 (12)			0.08	452
Ferro Star-L PA66 15YD61	15% carbon fiber, 13% PTFE, 2% silicon, lubricated			"						0.12		452
Ferro Star-L PA66 15YD61	15% carbon fiber, 13% PTFE, 2% silicon, lubricated			"	load: 12 N			39 (12)			0.08	452
Ferro Star-L PA66 15YD61	15% carbon fiber, 13% PTFE, 2% silicon, lubricated			"	load: 43 N			39 (12)			0.08	452
Ferro Star-L PA66 30VD31	30% glass, 5% MoS_2, lubricated			"						0.3		452
Ferro Star-L PA66 30VD31	30% glass, 5% MoS_2, lubricated			"	load: 43 N			39 (12)			0.28	452
Ferro Star-L PA66 30VD31	30% glass, 5% MoS_2, lubricated			"	load: 12 N			39 (12)			0.31	452
Ferro Star-L PA66 30VD51	30% glass, 15% PTFE, lubricated			"						0.22		452
Ferro Star-L PA66 30VD51	30% glass, 15% PTFE, lubricated			"	load: 12 N			39 (12)			0.08	452
Ferro Star-L PA66 30VD51	30% glass, 15% PTFE, lubricated			"	load: 43 N			39 (12)			0.08	452
Ferro Star-L PA66 30VD61	30% glass, 13% PTFE, 2% silicon, lubricated			"						0.17		452
Ferro Star-L PA66 30VD61	30% glass, 13% PTFE, 2% silicon, lubricated			"	load: 12 N			39 (12)			0.1	452
Ferro Star-L PA66 30VD61	30% glass, 13% PTFE, 2% silicon, lubricated			ASTM D3028	load: 43 N			39 (12)			0.11	452
Ferro Star-L PA66 30YD51	30% carbon fiber, 15% PTFE, lubricated			"						0.16		452
Ferro Star-L PA66 30YD51	30% carbon fiber, 15% PTFE, lubricated			"	load: 12 N			39 (12)			0.09	452
Ferro Star-L PA66 30YD51	30% carbon fiber, 15% PTFE, lubricated			"	load: 43 N			39 (12)			0.11	452
Ferro Star-L PA66 30YD61	30% carbon fiber, 13% PTFE, 2% silicon, lubricated			"						0.13		452

Material		Mating Surface		Test Conditions						Coefficient Of Friction		Source
Supplier / Grade	Material Note	Mating Surface Material	Mating Surface Note	Test Method	Test Conditions Note	Temp. (°C)	Pressure (psi) (MPa)	Speed (fpm) (m/min)	PV (psi x fpm) (MPa x m/min)	Static	Dynamic	
NYLON 66 against mating surface												
Ferro Star-L PA66 30YD61	30% carbon fiber, 13% PTFE, 2% silicon, lubricated	**steel**		ASTM D3028	load: 12 N			39 (12)			0.08	452
Ferro Star-L PA66 30YD61	30% carbon fiber, 13% PTFE, 2% silicon, lubricated			"	load: 43 N			39 (12)			0.1	452
Ferro Star-L PA66 D51	15% PTFE, lubricated			"						0.17		452
Ferro Star-L PA66 D51	15% PTFE, lubricated			"	load: 12 N			39 (12)			0.1	452
Ferro Star-L PA66 D51	15% PTFE, lubricated			"	load: 43 N			39 (12)			0.12	452
Ferro Star-L PA66 D61	13% PTFE, 2% silicon, lubricated			"	"			39 (12)			0.13	452
Ferro Star-L PA66 D61	13% PTFE, 2% silicon, lubricated			"						0.1		452
Ferro Star-L PA66 D61	13% PTFE, 2% silicon, lubricated			"	load: 12 N			39 (12)			0.1	452
Hoechst Cel. Nylon 1500	33% glass fiber			ASTM D1894							0.15	317
Hoechst Cel. Nylon 1500	33% glass fiber			"							0.17	317
Hoechst Cel. Nylon 1600	43% glass fiber			"							0.15	317
Hoechst Cel. Nylon 1600	43% glass fiber			"							0.17	317
LNP Ny Kon R	<5% MoS_2		carbon steel; surface finish: 12-16 µin; 18-20 Rockwell C	thrust washer		23	40 (0.28)	50 (15.2)	2000 (4.3)	0.28	0.3	453
LNP R-1000	unmodified		"	"		23	40 (0.28)	50 (15.2)	2000 (4.3)	0.2	0.28	453
LNP RAL-4022	10% aramid fiber, 10% PTFE		AISI 304 stainless steel; surface finish: 50-70 µin	"		23	40 (0.28)	50 (15.2)	2000 (4.3)	0.07	0.11	453
LNP RAL-4022	10% aramid fiber, 10% PTFE		AISI 304 stainless steel; surface finish: 8-16 µin	"		23	40 (0.28)	50 (15.2)	2000 (4.3)	0.08	0.11	453
LNP RAL-4022	10% aramid fiber, 10% PTFE		AISI 440 stainless steel; surface finish: 50-70 µin	"		23	40 (0.28)	50 (15.2)	2000 (4.3)	0.08	0.12	453
LNP RAL-4022	10% aramid fiber, 10% PTFE		AISI 440 stainless steel; surface finish: 8-16 µin	"		23	40 (0.28)	50 (15.2)	2000 (4.3)	0.1	0.13	453
LNP RAL-4022	10% aramid fiber, 10% PTFE		AISI 1141; surface finish: 50-70 µin	"		23	40 (0.28)	50 (15.2)	2000 (4.3)	0.11	0.12	453
LNP RAL-4022	10% aramid fiber, 10% PTFE		AISI 1141; surface finish: 12-16 µin	"		23	40 (0.28)	50 (15.2)	2000 (4.3)	0.12	0.13	453
LNP RAL-4022	10% aramid fiber, 10% PTFE		carbon steel; surface finish: 12-16 µin; 18-20 Rockwell C	thrust washer		23	40 (0.28)	50 (15.2)	2000 (4.3)	0.12	0.13	453
LNP RAL-4022	10% aramid fiber, 10% PTFE		AISI 1141; surface finish: 8-12 µin	"		23	40 (0.28)	50 (15.2)	2000 (4.3)	0.18	0.21	453
LNP RC-1004	20% carbon fiber		carbon steel; surface finish: 12-16 µin; 18-20 Rockwell C	"		23	40 (0.28)	50 (15.2)	2000 (4.3)	0.16	0.2	453
LNP RC-1006	30% carbon fiber		AISI 440 stainless steel; surface finish: 8-16 µin	"		23	40 (0.28)	50 (15.2)	2000 (4.3)	0.08	0.28	453

Material		Mating Surface		Test Conditions						Coefficient Of Friction		Source
Supplier / Grade	Material Note	Mating Surface Material	Mating Surface Note	Test Method	Test Conditions Note	Temp. (°C)	Pressure (psi) (MPa)	Speed (fpm) (m/min)	PV (psi x fpm) (MPa x m/min)	Static	Dynamic	
NYLON 66 against mating surface		→										
LNP RC-1006	30% carbon fiber	**steel**	AISI 304 stainless steel; surface finish: 50-70 µin	thrust washer		23	40 (0.28)	50 (15.2)	2000 (4.3)	0.11	0.17	453
LNP RC-1006	30% carbon fiber		AISI 304 stainless steel; surface finish: 8-16 µin	"		23	40 (0.28)	50 (15.2)	2000 (4.3)	0.11	0.21	453
LNP RC-1006	30% carbon fiber		AISI 1141; surface finish: 8-12 µin	"		23	40 (0.28)	50 (15.2)	2000 (4.3)	0.13	0.14	453
LNP RC-1006	30% carbon fiber		AISI 440 stainless steel; surface finish: 50-70 µin	"		23	40 (0.28)	50 (15.2)	2000 (4.3)	0.13	0.16	453
LNP RC-1006	30% carbon fiber		AISI 1141; surface finish: 12-16 µin	"		23	40 (0.28)	50 (15.2)	2000 (4.3)	0.16	0.2	453
LNP RC-1006	30% carbon fiber		carbon steel; surface finish: 12-16 µin; 18-20 Rockwell C	"		23	40 (0.28)	50 (15.2)	2000 (4.3)	0.16	0.2	453
LNP RC-1006	30% carbon fiber		AISI 1141; surface finish: 50-70 µin	"		23	40 (0.28)	50 (15.2)	2000 (4.3)	0.17	0.21	453
LNP RC-1008	40% carbon fiber		carbon steel; surface finish: 12-16 µin; 18-20 Rockwell C	"		23	40 (0.28)	50 (15.2)	2000 (4.3)	0.13	0.18	453
LNP RCL-4036	30% carbon fiber, 15% PTFE		AISI 440 stainless steel; surface finish: 8-16 µin	"		23	40 (0.28)	50 (15.2)	2000 (4.3)	0.1	0.23	453
LNP RCL-4036	30% carbon fiber, 15% PTFE		carbon steel; surface finish: 12-16 µin; 18-20 Rockwell C	"		23	40 (0.28)	50 (15.2)	2000 (4.3)	0.11	0.15	453
LNP RCL-4036	30% carbon fiber, 15% PTFE		AISI 1141; surface finish: 12-16 µin	"		23	40 (0.28)	50 (15.2)	2000 (4.3)	0.11	0.15	453
LNP RCL-4036	30% carbon fiber, 15% PTFE		AISI 1141; surface finish: 8-12 µin	"		23	40 (0.28)	50 (15.2)	2000 (4.3)	0.12	0.15	453
LNP RCL-4036	30% carbon fiber, 15% PTFE		AISI 1141; surface finish: 50-70 µin	"		23	40 (0.28)	50 (15.2)	2000 (4.3)	0.13	0.16	453
LNP RCL-4036	30% carbon fiber, 15% PTFE		AISI 440 stainless steel; surface finish: 50-70 µin	"		23	40 (0.28)	50 (15.2)	2000 (4.3)	0.15	0.21	453
LNP RCL-4036	30% carbon fiber, 15% PTFE		AISI 304 stainless steel; surface finish: 50-70 µin	"		23	40 (0.28)	50 (15.2)	2000 (4.3)	0.16	0.2	453
LNP RCL-4036	30% carbon fiber, 15% PTFE		AISI 304 stainless steel; surface finish: 8-16 µin	"		23	40 (0.28)	50 (15.2)	2000 (4.3)	0.17	0.22	453
LNP RCL-4536	30% carbon fiber, 13% PTFE, 2% silicone		carbon steel; surface finish: 12-16 µin; 18-20 Rockwell C	"		23	40 (0.28)	50 (15.2)	2000 (4.3)	0.1	0.11	453
LNP RF 100-10	50% glass fiber		"	"		23	40 (0.28)	50 (15.2)	2000 (4.3)	0.28	0.35	453
LNP RF-1002	10% glass fiber		"	"		23	40 (0.28)	50 (15.2)	2000 (4.3)	0.21	0.28	453
LNP RF-1004	20% glass fiber		"	"		23	40 (0.28)	50 (15.2)	2000 (4.3)	0.23	0.3	453
LNP RF-1006	30% glass fiber		AISI 304 stainless steel; surface finish: 50-70 µin	"		23	40 (0.28)	50 (15.2)	2000 (4.3)	0.1	0.18	453
LNP RF-1006	30% glass fiber		AISI 304 stainless steel; surface finish: 8-16 µin	thrust washer		23	40 (0.28)	50 (15.2)	2000 (4.3)	0.12	0.22	453
LNP RF-1006	30% glass fiber		AISI 440 stainless steel; surface finish: 50-70 µin	"		23	40 (0.28)	50 (15.2)	2000 (4.3)	0.15	0.18	453
LNP RF-1006	30% glass fiber		AISI 1141; surface finish: 8-12 µin	"		23	40 (0.28)	50 (15.2)	2000 (4.3)	0.16	0.21	453

Material		Mating Surface		Test Conditions						Coefficient Of Friction		Source
Supplier / Grade	Material Note	Mating Surface Material	Mating Surface Note	Test Method	Test Conditions Note	Temp. (°C)	Pressure (psi) (MPa)	Speed (fpm) (m/min)	PV (psi x fpm) (MPa x m/min)	Static	Dynamic	
NYLON 66 against mating surface												
LNP RF-1006	30% glass fiber	steel	AISI 440 stainless steel; surface finish: 8-16 µin	thrust washer		23	40 (0.28)	50 (15.2)	2000 (4.3)	0.2	0.22	453
LNP RF-1006	30% glass fiber		AISI 1141; surface finish: 50-70 µin	"		23	40 (0.28)	50 (15.2)	2000 (4.3)	0.22	0.28	453
LNP RF-1006	30% glass fiber		AISI 1141; surface finish: 12-16 µin	"		23	40 (0.28)	50 (15.2)	2000 (4.3)	0.25	0.31	453
LNP RF-1006 HS	30% glass fiber		carbon steel; surface finish: 12-16 µin; 18-20 Rockwell C	"		23	40 (0.28)	50 (15.2)	2000 (4.3)	0.25	0.31	453
LNP RF-1008	40% glass fiber		"	"		23	40 (0.28)	50 (15.2)	2000 (4.3)	0.26	0.33	453
LNP RFL-4036	30% glass fiber, 15% PTFE		AISI 304 stainless steel; surface finish: 50-70 µin	"		23	40 (0.28)	50 (15.2)	2000 (4.3)	0.13	0.15	453
LNP RFL-4036	30% glass fiber, 15% PTFE		AISI 440 stainless steel; surface finish: 50-70 µin	"		23	40 (0.28)	50 (15.2)	2000 (4.3)	0.13	0.15	453
LNP RFL-4036	30% glass fiber, 15% PTFE		AISI 440 stainless steel; surface finish: 8-16 µin	"		23	40 (0.28)	50 (15.2)	2000 (4.3)	0.14	0.21	453
LNP RFL-4036	30% glass fiber, 15% PTFE		AISI 304 stainless steel; surface finish: 8-16 µin	"		23	40 (0.28)	50 (15.2)	2000 (4.3)	0.17	0.18	453
LNP RFL-4036	30% glass fiber, 15% PTFE		AISI 1141; surface finish: 50-70 µin	"		23	40 (0.28)	50 (15.2)	2000 (4.3)	0.17	0.2	453
LNP RFL-4036	30% glass fiber, 15% PTFE		AISI 1141; surface finish: 12-16 µin	"		23	40 (0.28)	50 (15.2)	2000 (4.3)	0.19	0.26	453
LNP RFL-4036	30% glass fiber, 15% PTFE		carbon steel; surface finish: 12-16 µin; 18-20 Rockwell C	"		23	40 (0.28)	50 (15.2)	2000 (4.3)	0.19	0.26	453
LNP RFL-4036	30% glass fiber, 15% PTFE		AISI 1141; surface finish: 8-12 µin	"		23	40 (0.28)	50 (15.2)	2000 (4.3)	0.2	0.26	453
LNP RFL-4036	30% glass fiber, 15% PTFE		cold rolled steel; surface finish: 12-16 µin; 22 Rockwell C	"		93	40 (0.28)	50 (15.2)	2000 (4.3)	0.29	0.24	453
LNP RFL-4036	30% glass fiber, 15% PTFE		"	"		149	40 (0.28)	50 (15.2)	2000 (4.3)	0.36	0.32	453
LNP RFL-4036	30% glass fiber, 15% PTFE		"	"		204	40 (0.28)	50 (15.2)	2000 (4.3)	0.37	0.4	453
LNP RFL-4216	30% glass fiber, <5% MoS_2		carbon steel; surface finish: 12-16 µin; 18-20 Rockwell C	"		23	40 (0.28)	50 (15.2)	2000 (4.3)	0.24	0.31	453
LNP RFL-4218	40% glass fiber, <5% MoS_2		"	"		23	40 (0.28)	50 (15.2)	2000 (4.3)	0.26	0.33	453
LNP RFL-4416	30% glass fiber, 2% silicone		"	"		23	40 (0.28)	50 (15.2)	2000 (4.3)	0.19	0.26	453
LNP RFL-4536	30% glass fiber, 13% PTFE, 2% silicone		AISI 440 stainless steel; surface finish: 50-70 µin	"		23	40 (0.28)	50 (15.2)	2000 (4.3)	0.1	0.16	453
LNP RFL-4536	30% glass fiber, 13% PTFE, 2% silicone		AISI 440 stainless steel; surface finish: 8-16 µin	"		23	40 (0.28)	50 (15.2)	2000 (4.3)	0.1	0.4	453
LNP RFL-4536	30% glass fiber, 13% PTFE, 2% silicone		carbon steel; surface finish: 12-16 µin; 18-20 Rockwell C	"		23	40 (0.28)	50 (15.2)	2000 (4.3)	0.12	0.14	453
LNP RFL-4536	30% glass fiber, 13% PTFE, 2% silicone		AISI 304 stainless steel; surface finish: 50-70 µin	"		23	40 (0.28)	50 (15.2)	2000 (4.3)	0.15	0.18	453
LNP RFL-4536	30% glass fiber, 13% PTFE, 2% silicone		AISI 1141; surface finish: 50-70 µin	"		23	40 (0.28)	50 (15.2)	2000 (4.3)	0.16	0.19	453

Material		Mating Surface		Test Conditions						Coefficient Of Friction		Source
Supplier / Grade	Material Note	Mating Surface Material	Mating Surface Note	Test Method	Test Conditions Note	Temp. (°C)	Pressure (psi) (MPa)	Speed (fpm) (m/min)	PV (psi x fpm) (MPa x m/min)	Static	Dynamic	
NYLON 66 against mating surface		→										
LNP RFL-4536	30% glass fiber, 13% PTFE, 2% silicone	**steel**	AISI 304 stainless steel; surface finish: 8-16 µin	thrust washer		23	40 (0.28)	50 (15.2)	2000 (4.3)	0.17	0.2	453
LNP RFL-4536	30% glass fiber, 13% PTFE, 2% silicone		AISI 1141; surface finish: 12-16 µin	"		23	40 (0.28)	50 (15.2)	2000 (4.3)	0.18	0.2	453
LNP RFL-4536	30% glass fiber, 13% PTFE, 2% silicone		AISI 1141; surface finish: 8-12 µin	"		23	40 (0.28)	50 (15.2)	2000 (4.3)	0.2	0.26	453
LNP RFL-4616	30% glass fiber, 2% silicone		carbon steel; surface finish: 12-16 µin; 18-20 Rockwell C	"		23	40 (0.28)	50 (15.2)	2000 (4.3)	0.14	0.15	453
LNP RL-4010	5% PTFE		"	"		23	40 (0.28)	50 (15.2)	2000 (4.3)	0.13	0.2	453
LNP RL-4040	20% PTFE		AISI 440 stainless steel; surface finish: 50-70 µin	"		23	40 (0.28)	50 (15.2)	2000 (4.3)	0.08	0.11	453
LNP RL-4040	20% PTFE		AISI 440 stainless steel; surface finish: 8-16 µin	"		23	40 (0.28)	50 (15.2)	2000 (4.3)	0.1	0.12	453
LNP RL-4040	20% PTFE		AISI 1141; surface finish: 12-16 µin	"		23	40 (0.28)	50 (15.2)	2000 (4.3)	0.1	0.14	453
LNP RL-4040	20% PTFE		carbon steel; surface finish: 12-16 µin; 18-20 Rockwell C	"		23	40 (0.28)	50 (15.2)	2000 (4.3)	0.1	0.18	453
LNP RL-4040	20% PTFE		AISI 1141; surface finish: 50-70 µin	"		23	40 (0.28)	50 (15.2)	2000 (4.3)	0.11	0.13	453
LNP RL-4040	20% PTFE		AISI 304 stainless steel; surface finish: 50-70 µin	"		23	40 (0.28)	50 (15.2)	2000 (4.3)	0.04	0.09	453
LNP RL-4040	20% PTFE		AISI 304 stainless steel; surface finish: 8-16 µin	"		23	40 (0.28)	50 (15.2)	2000 (4.3)	0.05	0.09	453
LNP RL-4040	20% PTFE		AISI 1141; surface finish: 8-12 µin	"		23	40 (0.28)	50 (15.2)	2000 (4.3)	0.05	0.1	453
LNP RL-4040FR(94VO)	flame retardant, 20% PTFE		carbon steel; surface finish: 12-16 µin; 18-20 Rockwell C	"		23	40 (0.28)	50 (15.2)	2000 (4.3)	0.12	0.19	453
LNP RL-4310	5% graphite		"	"		23	40 (0.28)	50 (15.2)	2000 (4.3)	0.15	0.2	453
LNP RL-4410	2% silicone		"	"		23	40 (0.28)	50 (15.2)	2000 (4.3)	0.09	0.09	453
LNP RL-4540	18% PTFE, 2% silicone		"	"		23	40 (0.28)	50 (15.2)	2000 (4.3)	0.06	0.08	453
LNP RL-4610	2% silicone		"	"		23	40 (0.28)	50 (15.2)	2000 (4.3)	0.19	0.19	453
LNP RL-4730	13% PTFE, 2% silicone		"	"		23	40 (0.28)	50 (15.2)	2000 (4.3)	0.11	0.18	453
LNP Verton RF-700-1OHS	50% glass fiber		"	"		23	40 (0.28)	50 (15.2)	2000 (4.3)	0.26	0.32	453
LNP Verton RF-7007HS	35% glass fiber		"	"		23	40 (0.28)	50 (15.2)	2000 (4.3)	0.24	0.3	453
Polymer Nylatron NS	w/ additives to improve bearing properties		unlubricated	"	load: 22.7 kg					0.13-0.22		441
Polymer Nylatron NS	w/ additives to improve bearing properties		"	"			250 (1.72)	20 (6.1)	5000 (10.5)		0.13-0.26	441
RTP 200			C1018 steel; surface finish: 14-17 µin; 15-25 Rockwell C	"	apparatus: Falex Model No. 6	23	40 (0.28)	50 (15.2)	2000 (4.3)	0.2	0.28	457

Material		Mating Surface		Test Conditions						Coefficient Of Friction		Source
Supplier / Grade	Material Note	Mating Surface Material	Mating Surface Note	Test Method	Test Conditions Note	Temp. (°C)	Pressure (psi) (MPa)	Speed (fpm) (m/min)	PV (psi x fpm) (MPa x m/min)	Static	Dynamic	
NYLON 66 against mating surface												
RTP 200 AR 15	15% aramid fiber	**steel**	C1018 steel; surface finish: 14-17 μin; 15-25 Rockwell C	thrust washer	apparatus: Falex Model No. 6	23	40 (0.28)	50 (15.2)	2000 (4.3)	0.15	0.2	457
RTP 200 AR 15 TFE 15	15% PTFE, 15% aramid fiber		"	"	"	23	40 (0.28)	50 (15.2)	2000 (4.3)	0.1	0.15	457
RTP 200 TFE 10	10% PTFE		"	"	"	23	40 (0.28)	50 (15.2)	2000 (4.3)	0.12	0.18	457
RTP 200 TFE 20	20% PTFE		"	"	"	23	40 (0.28)	50 (15.2)	2000 (4.3)	0.1	0.16	457
RTP 200 TFE 20 SI	20% PTFE, 0.5% silicone		"	"	"	23	40 (0.28)	50 (15.2)	2000 (4.3)	0.05	0.06	457
RTP 201 TFE 5 SI	5% PTFE, 0.5% silicone, 10% glass fiber		"	"	"	23	40 (0.28)	50 (15.2)	2000 (4.3)	0.12	0.15	457
RTP 203 TFE 10	10% PTFE, 20% glass fiber		"	"	"	23	40 (0.28)	50 (15.2)	2000 (4.3)	0.18	0.25	457
RTP 203 TFE 15	15% PTFE, 20% glass fiber		"	"	"	23	40 (0.28)	50 (15.2)	2000 (4.3)	0.17	0.23	457
RTP 203 TFE 20	20% PTFE, 20% glass fiber		"	"	"	23	40 (0.28)	50 (15.2)	2000 (4.3)	0.15	0.21	457
RTP 205	30% glass fiber		"	"	"	23	40 (0.28)	50 (15.2)	2000 (4.3)	0.25	0.31	457
RTP 205 SI 2	2% silicone, 30% glass fiber		"	"	"	23	40 (0.28)	50 (15.2)	2000 (4.3)	0.13	0.15	457
RTP 205 TFE 13 SI 2	13% PTFE, 2% silicone, 30% glass fiber		"	"	"	23	40 (0.28)	50 (15.2)	2000 (4.3)	0.12	0.14	457
RTP 205 TFE 15	30% glass fiber		"	"	"	23	40 (0.28)	50 (15.2)	2000 (4.3)	0.18	0.25	457
RTP 205 TFE 20	20% PTFE, 30% glass fiber		"	"	"	23	40 (0.28)	50 (15.2)	2000 (4.3)	0.16	0.23	457
RTP 205 TFE 5	5% PTFE, 30% glass fiber		"	"	"	23	40 (0.28)	50 (15.2)	2000 (4.3)	0.23	0.3	457
RTP 281 TFE 20	20% PTFE, 10% carbon fiber		"	"	"	23	40 (0.28)	50 (15.2)	2000 (4.3)	0.1	0.13	457
RTP 283 TFE 1 0	10% PTFE, 20% carbon fiber		"	"	"	23	40 (0.28)	50 (15.2)	2000 (4.3)	0.12	0.15	457
RTP 285	30% carbon fiber		"	"	"	23	40 (0.28)	50 (15.2)	2000 (4.3)	0.15	0.2	457
RTP 285 TFE 13 SI 2	13% PTFE, 2% silicone, 30% carbon fiber		"	"	"	23	40 (0.28)	50 (15.2)	2000 (4.3)	0.1	0.12	457
RTP 285 TFE 15	15% PTFE, 30% carbon fiber		"	"	"	23	40 (0.28)	50 (15.2)	2000 (4.3)	0.11	0.16	457
RTP 287 TFE 10	10% PTFE, 40% carbon fiber		"	"	"	23	40 (0.28)	50 (15.2)	2000 (4.3)	0.13	0.16	457
Thermofil N3-13FG-0100	13% glass fiber			ASTM D1894		23				0.2	0.25	459
Thermofil N3-13FG-0700	13% glass fiber, impact modified			"		23				0.22	0.27	459
Thermofil N3-15G-0560	15% glass fiber, flame retardant			"		23				0.2	0.31	459

Material		Mating Surface		Test Conditions						Coefficient Of Friction		Source
Supplier / Grade	Material Note	Mating Surface Material	Mating Surface Note	Test Method	Test Conditions Note	Temp. (°C)	Pressure (psi) (MPa)	Speed (fpm) (m/min)	PV (psi x fpm) (MPa x m/min)	Static	Dynamic	
NYLON 66 against mating surface												
Thermofil N3-20NF-0100	20% graphite fiber	**steel**		thrust washer		23				0.16	0.2	459
Thermofil N3-30FG-0214	30% glass fiber, PTFE lubricated			"		23				0.25	0.07	459
Thermofil N3-30FG-0231	30% glass fiber, MoS_2 lubricated			"		23				0.29	0.15	459
Thermofil N3-30FG-0282	30% glass fiber, lubricated			"		23				0.12	0.14	459
Thermofil N3-30FG-0560	30% glass fiber, flame retardant			"		23				0.23	0.3	459
Thermofil N3-30NF-0214	30% graphite fiber, PTFE lubricated			"		23				0.1	0.11	459
Thermofil N3-33FG-0100	33% glass fiber			"		23				0.23	0.3	459
Thermofil N3-40NF-0100	40% graphite fiber			"		23				0.13	0.18	459
Thermofil N3-43FG-0100	43% glass fiber			"		23				0.25	0.31	459
Thermofil N3-9900-0200	lubricated			"		23				0.05	0.06	459
Thermofil N3-9900-0214	PTFE lubricated			"		23				0.2	0.06	459
Thermofil N3-9900-0231	MoS_2 lubricated			"		23				0.17	0.09	459
Thermofil N3-9900-0279	40% carbon fiber			"		23				0.15	0.21	459
Thermofil N3-9900-0560	flame retardant			"		23				0.18	0.22	459
Unspecified grade	general purpose		unlubricated	thrust washer	load: 22.7 kg					0.17-0.27		441
Unspecified grade	unmodified		carbon steel; surface finish: 16 µin; 18-22 Rockwell C	"	apparatus: Faville-LeVally	23	40 (0.28)	50 (15.2)	2000 (4.3)	0.2	0.28	458
Unspecified grade	impact modified		polished CR steel, dry	ASTM D1894							0.17-0.34	330
Unspecified grade	general purpose		unlubricated	thrust washer			250 (1.72)	20 (6.1)	5000 (10.5)		0.19-0.43	441
POLYARYLAMIDE against mating surface												
Solvay Ixef 1022	50% glass fiber	**steel**	XC 45				167 (1.15)	33 (10)	5511 (11.6)		0.4	135
POLYPHTHALAMIDE against mating surface												
RTP 4005 TFE 15	15% PTFE, 30% glass fiber	**steel**	C1018 steel; surface finish: 14-17 µin; 15-25 Rockwell C	thrust washer	apparatus: Falex Model No. 6	23	40 (0.28)	50 (15.2)	2000 (4.3)	0.2	0.28	457
RTP 4080 TFE 15	15% PTFE, 30% carbon fiber		"	"	"	23	40 (0.28)	50 (15.2)	2000 (4.3)		0.2	460

Material		Mating Surface		Test Conditions						Coefficient Of Friction		Source
Supplier / Grade	Material Note	Mating Surface Material	Mating Surface Note	Test Method	Test Conditions Note	Temp. (°C)	Pressure (psi) (MPa)	Speed (fpm) (m/min)	PV (psi x fpm) (MPa x m/min)	Static	Dynamic	
POLYPHTHALAMIDE against mating surface												
RTP 4083 TFE 15 S12	15% PTFE, 2% silicone, 20% carbon fiber	**steel**	C1018 steel; surface finish: 14-17 µin; 15-25 Rockwell C	thrust washer	apparatus: Falex Model No. 6	23	40 (0.28)	50 (15.2)	2000 (4.3)	0.13	0.15	457
POLYBENZIMIDAZOLE against mating surface												
Hoechst Cel. Celazole DL80	lubricated, heterocyclic, 50 Rockwell A	**steel**	AISI 1018, surface finish: 16 RMS	thrust bearing	apparatus: LRI-1 Wear and Friction Tester		100 (0.69)	984 (300)	98400 (207)	0.29	0.07	318
Hoechst Cel. Celazole DTL60	lubricated, melt processable, heterocyclic, 25 Rockwell K		"	"	"		100 (0.69)	984 (300)	98400 (207)	0.3	0.08	318
Hoechst Cel. Celazole DTU60	melt processable, heterocyclic, 50 Rockwell K		"	"	"		2000 (13.8)	12.5 (3.8)	25000 (52.6)		0.21	318
Hoechst Cel. Celazole U60			"	"	"		50 (0.34)	20 (6.1)	10000 (2.1)	0.22	0.28	320
Hoechst Cel. Celazole U60			"	"	"		1000 (6.89)	50 (15.2)	50000 (105.1)	0.27	0.19	320
Hoechst Cel. Celazole U60	heterocyclic, virgin resin, 50 Rockwell A		"	"	"		100 (0.69)	984 (300)	98400 (207)	0.63	0.42	318
Hoechst Cel. Celazole U60			"	"	"		40 (0.28)	50 (15.2)	2000 (4.3)		0.21	320
Hoechst Cel. Celazole U60			"	"	"		50 (0.34)	800 (244)	40000 (84.1)		0.21	320
Hoechst Cel. Celazole U60			"	"	"		20 (0.14)	100 (30.5)	2000 (4.3)		0.21	320
Hoechst Cel. Celazole U60			"	"	"		100 (0.69)	100 (30.5)	10000 (21)		0.22	320
Hoechst Cel. Celazole U60			"	"	"		100 (0.69)	500 (152)	50000 (105.1)		0.23	320
Hoechst Cel. Celazole U60	heterocyclic, virgin resin, 50 Rockwell A		"	"	"		2000 (13.8)	12.5 (3.8)	25000 (52.6)		0.3	318
Hoechst Cel. Celazole U60			"	"	"		50 (0.34)	900 (274)	45000 (94.6)		0.35	320
POLYCARBONATE against mating surface												
LNP DFL-4044	20% glass fiber, 20% PTFE	**ABS**	A-1000; unmodified	thrust washer		23	40 (0.28)	50 (15.2)	2000 (4.3)	0.04	0.06	453
LNP DFL-4036	30% glass fiber, 15% PTFE	**brass**	70/30 brass; surface finish: 50-70 µin	"		23	40 (0.28)	50 (15.2)	2000 (4.3)	0.1	0.15	453
LNP DFL-4036	30% glass fiber, 15% PTFE		70/30 brass; surface finish: 8-16 µin	"		23	40 (0.28)	50 (15.2)	2000 (4.3)	0.1	0.17	453
LNP DFL-4034	20% glass fiber, 15% PTFE	**nylon 610**	QFL-4036 (30% glass fiber, 15% PTFE)	"		23	40 (0.28)	50 (15.2)	2000 (4.3)	0.05	0.07	453
LNP DL-4020	10% PTFE		"	"		23	40 (0.28)	50 (15.2)	2000 (4.3)	0.04	0.06	453
LNP D-1000	unmodified	**nylon 66**	RL-4530 (13% PTFE, 2% silicone)	"		23	40 (0.28)	50 (15.2)	2000 (4.3)	0.04	0.04	453
LNP D-1000	unmodified		RF-1006 (30% glass fiber)	"		23	40 (0.28)	50 (15.2)	2000 (4.3)	0.11	0.17	453

Material		Mating Surface		Test Conditions						Coefficient Of Friction		Source
Supplier / Grade	Material Note	Mating Surface Material	Mating Surface Note	Test Method	Test Conditions Note	Temp. (°C)	Pressure (psi) (MPa)	Speed (fpm) (m/min)	PV (psi x fpm) (MPa x m/min)	Static	Dynamic	
POLYCARBONATE against mating surface		→										
LNP D-1000	unmodified	**nylon 66**	R-1000 (unmodified)	thrust washer		23	40 (0.28)	50 (15.2)	2000 (4.3)	0.25	0.04	453
LNP DCL-4042	10% carbon fiber, 20% PTFE		RCL-4036 (30% carbon fiber, 15% PTFE)	"		23	40 (0.28)	50 (15.2)	2000 (4.3)	0.11	0.07	453
LNP DCL-4042	10% carbon fiber, 20% PTFE		RCL-4036 (30% carbon fiber, 15% PTFE)	thrust washer		23	40 (0.28)	50 (15.2)	2000 (4.3)	0.11	0.07	453
LNP DCL-4536	30% carbon fiber, 13% PTFE, 2% silicone		RCL-4536 (30% carbon fiber, 13% PTFE, 2% silicone)	"		23	40 (0.28)	50 (15.2)	2000 (4.3)	0.13	0.18	453
LNP DF-1006	30% glass fiber		RFL-4036 (30% glass fiber, 15% PTFE)	"		23	40 (0.28)	50 (15.2)	2000 (4.3)	0.07	0.08	453
LNP DFL-4036	30% glass fiber, 15% PTFE		"	"		23	40 (0.28)	50 (15.2)	2000 (4.3)	0.05	0.06	453
LNP DFL-4036	30% glass fiber, 15% PTFE		RF-1006 (30% glass fiber)	"		23	40 (0.28)	50 (15.2)	2000 (4.3)	0.05	0.07	453
LNP DFL-4036	30% glass fiber, 15% PTFE		RCL-4036 (30% carbon fiber, 15% PTFE)	"		23	40 (0.28)	50 (15.2)	2000 (4.3)	0.11	0.24	453
Dow Calibre 300-15	transparent, gen. purp. grade, 72 Rockwell M; 15 g/10 min. MFI	**polycarbonate**		ASTM D1894						0.31	0.38	78
Dow Calibre 300-4	transparent, gen. purp. grade, 74 Rockwell M; 4 g/10 min. MFI			"						0.31	0.37	78
Dow Calibre 800-6	transparent, flame retardant, 72 Rockwell M; 6 g/10 min. MFI			"						0.32	0.37	78
LNP DF-1004	20% glass fiber		DF-1004 (20% glass fiber)	thrust washer		23	40 (0.28)	50 (15.2)	2000 (4.3)	0.3	0.32	453
LNP DFL-4036	30% glass fiber, 15% PTFE		DFL-4036 (30% glass fiber, 15% PTFE)	"		23	40 (0.28)	50 (15.2)	2000 (4.3)	0.06	0.08	453
LNP DFL-4044	20% glass fiber, 20% PTFE		D-1000; unmodified	"		23	40 (0.28)	50 (15.2)	2000 (4.3)	0.06	0.08	453
LNP DFL-4036	30% glass fiber, 15% PTFE	**polyphenylene sulfide**	OFL-4036 (30% glass fiber, 15% PTFE)	"		23	40 (0.28)	50 (15.2)	2000 (4.3)	0.1	0.11	453
Akzo G-50/20/TF/10	20% long glass fiber, 10% PTFE	**steel**	carbon steel; surface finish: 16 µin; 18-22 Rockwell C	"	apparatus: Faville-LeVally	23	40 (0.28)	50 (15.2)	2000 (4.3)	0.2	0.24	458
Akzo G-50/20/TF/15	20% long glass fiber, 15% PTFE		"	"	"	23	40 (0.28)	50 (15.2)	2000 (4.3)	0.19	0.22	458
Akzo J-50/20/TF/15	20% glass fiber, 15% PTFE		"	"	"	23	40 (0.28)	50 (15.2)	2000 (4.3)	0.19	0.22	458
Akzo J-50/20/TF/I 0	20% glass fiber, 10% PTFE		"	"	"	23	40 (0.28)	50 (15.2)	2000 (4.3)	0.2	0.24	458
Akzo J-50/30/TF/10	30% glass fiber, 10% PTFE		"	"	"	23	40 (0.28)	50 (15.2)	2000 (4.3)	0.21	0.25	458
Akzo J-50/3O/TF/15	30% glass fiber, 15% PTFE		"	"	"	23	40 (0.28)	50 (15.2)	2000 (4.3)	0.18	0.2	458
Akzo PC-50/TF/10	10% PTFE		"	"	"	23	40 (0.28)	50 (15.2)	2000 (4.3)	0.11	0.17	458
Akzo PC-50/TF/15	15% PTFE		"	"	"	23	40 (0.28)	50 (15.2)	2000 (4.3)	0.09	0.15	458
Akzo PC-50/TF/30	30% PTFE		"	"	"	23	40 (0.28)	50 (15.2)	2000 (4.3)	6.08	0.14	458

Material		Mating Surface		Test Conditions						Coefficient Of Friction		Source
Supplier / Grade	Material Note	Mating Surface Material	Mating Surface Note	Test Method	Test Conditions Note	Temp. (°C)	Pressure (psi) (MPa)	Speed (fpm) (m/min)	PV (psi x fpm) (MPa x m/min)	Static	Dynamic	
POLYCARBONATE against mating surface												
Bay Resins Lubriplas PC-1100	unmodified resin	**steel**		thrust washer			40 (0.28)	50 (15.2)	2000 (4.3)	0.3	0.38	435
Bay Resins Lubriplas PC-1100CF30	30% carbon fiber			"			40 (0.28)	50 (15.2)	2000 (4.3)	0.18	0.16	435
Bay Resins Lubriplas PC-1100G10	10% glass fiber			thrust washer			40 (0.28)	50 (15.2)	2000 (4.3)	0.22	0.2	435
Bay Resins Lubriplas PC-1100G30	30% glass fiber			"			40 (0.28)	50 (15.2)	2000 (4.3)	0.24	0.21	435
Bay Resins Lubriplas PC-1100G30TF15	30% glass fiber, 15% PTFE			"			40 (0.28)	50 (15.2)	2000 (4.3)	0.18	0.2	435
Bay Resins Lubriplas PC-1100G40	40% glass fiber			"			40 (0.28)	50 (15.2)	2000 (4.3)	0.25	0.22	435
Bay Resins Lubriplas PC-1100TF15	15% PTFE			"			40 (0.28)	50 (15.2)	2000 (4.3)	0.09	0.15	435
Bay Resins Lubriplas PC-1100TF20	20% PTFE			"			40 (0.28)	50 (15.2)	2000 (4.3)	0.08	0.14	435
Dow Calibre 300-15	transparent, gen. purp. grade, 72 Rockwell M; 15 g/10 min. MFI		stainless steel	ASTM D1894						0.32	0.37	78
Dow Calibre 300-4	transparent, gen. purp. grade, 74 Rockwell M; 4 g/10 min. MFI		"	"						0.33	0.38	78
Dow Calibre 800-6	transparent, flame retardant, 72 Rockwell M; 6 g/10 min. MFI		"	"						0.32	0.35	78
Dow	transparent			"							0.55	78
Ensinger Ensicar	standard grade						40 (0.28)	50 (15.2)	2000 (4.3)		0.38	449
Ferro Star-C PC B20Y	20% carbon fiber			ASTM D3028						0.19		452
Ferro Star-C PC B30Y	. 30% carbon fiber			"						0.18		452
Ferro Star-L PC B20YD51	20% carbon fiber, 15% PTFE, lubricated			"						0.16		452
Ferro Star-L PC B20YD51	20% carbon fiber, 15% PTFE, lubricated			"	load: 12 N			39 (12)			0.09	452
Ferro Star-L PC B20YD51	20% carbon fiber, 15% PTFE, lubricated			"	load: 43 N			39 (12)			0.1	452
Ferro Star-L PC B30VD51	30% glass, 15% PTFE, lubricated			"						0.21		452
Ferro Star-L PC B30VD51	30% glass, 15% PTFE, lubricated			"	load: 12 N			39 (12)			0.11	452
Ferro Star-L PC B30VD51	30% glass, 15% PTFE, lubricated			"	load: 43 N			39 (12)			0.12	452
Ferro Star-L PC B30YD52	30% carbon fiber, 20% PTFE, lubricated			"						0.15		452
Ferro Star-L PC B30YD52	30% carbon fiber, 20% PTFE, lubricated			"	load: 12 N			39 (12)			0.1	452
Ferro Star-L PC B30YD52	30% carbon fiber, 20% PTFE, lubricated			"	load: 43 N			39 (12)			0.1	452

Material		Mating Surface		Test Conditions						Coefficient Of Friction		Source
Supplier / Grade	Material Note	Mating Surface Material	Mating Surface Note	Test Method	Test Conditions Note	Temp. (°C)	Pressure (psi) (MPa)	Speed (fpm) (m/min)	PV (psi x fpm) (MPa x m/min)	Static	Dynamic	
POLYCARBONATE against mating surface		→										
Ferro Star-L PC BD52	20% PTFE, lubricated	**steel**		ASTM D3028						0.1		452
Ferro Star-L PC BD52	20% PTFE, lubricated			"	load: 12 N			39 (12)			0.18	452
Ferro Star-L PC BD52	20% PTFE, lubricated			"	load: 43 N			39 (12)			0.22	452
GE Lexan 3412	20% glass fiber						40 (0.28)	50 (15.2)	2000 (4.3)		0.24	449
LNP D-1000	unmodified		carbon steel; surface finish: 12-16 µin; 18-20 Rockwell C	thrust washer		23	40 (0.28)	50 (15.2)	2000 (4.3)	0.31	0.38	453
LNP DC-1006	30% carbon fiber		"	"		23	40 (0.28)	50 (15.2)	2000 (4.3)	0.18	0.17	453
LNP DCL-4532	10% carbon fiber, 13% PTFE, 2% silicone		"	"		23	40 (0.28)	50 (15.2)	2000 (4.3)	0.2	0.19	453
LNP DF-1006	30% glass fiber		"	"		23	40 (0.28)	50 (15.2)	2000 (4.3)	0.23	0.22	453
LNP DFL-4034	20% glass fiber, 15% PTFE		"	"		23	40 (0.28)	50 (15.2)	2000 (4.3)	0.19	0.22	453
LNP DFL-4036	30% glass fiber, 15% PTFE		AISI 304 stainless steel; surface finish: 50-70 µin	"		23	40 (0.28)	50 (15.2)	2000 (4.3)	0.09	0.15	453
LNP DFL-4036	30% glass fiber, 15% PTFE		AISI 304 stainless steel; surface finish: 8-16 µin	"		23	40 (0.28)	50 (15.2)	2000 (4.3)	0.1	0.16	453
LNP DFL-4036	30% glass fiber, 15% PTFE		AISI 1141; surface finish: 50-70 µin	"		23	40 (0.28)	50 (15.2)	2000 (4.3)	0.17	0.2	453
LNP DFL-4036	30% glass fiber, 15% PTFE		AISI 1141; surface finish: 12-16 µin	"		23	40 (0.28)	50 (15.2)	2000 (4.3)	0.18	0.2	453
LNP DFL-4036	30% glass fiber, 15% PTFE		carbon steel; surface finish: 12-16 µin; 18-20 Rockwell C	"		23	40 (0.28)	50 (15.2)	2000 (4.3)	0.18	0.2	453
LNP DFL-4036	30% glass fiber, 15% PTFE		AISI 440 stainless steel; surface finish: 50-70 µin	"		23	40 (0.28)	50 (15.2)	2000 (4.3)	0.19	0.23	453
LNP DFL-4036	30% glass fiber, 15% PTFE		AISI 440 stainless steel; surface finish: 8-16 µin	"		23	40 (0.28)	50 (15.2)	2000 (4.3)	0.2	0.2	453
LNP DFL-4036	30% glass fiber, 15% PTFE		AISI 1141; surface finish: 8-12 µin	"		23	40 (0.28)	50 (15.2)	2000 (4.3)	0.22	0.25	453
LNP DFL-4536	30% glass fiber, 13% PTFE, 2% silicone		carbon steel; surface finish: 12-16 µin; 18-20 Rockwell C	"		23	40 (0.28)	50 (15.2)	2000 (4.3)	0.17	0.19	453
LNP DL 4040	20% PTFE		"	"		23	40 (0.28)	50 (15.2)	2000 (4.3)	0.08	0.14	453
LNP DL-4010	5% PTFE		"	"		23	40 (0.28)	50 (15.2)	2000 (4.3)	0.14	0.2	453
LNP DL-4020	10% PTFE		"	"		23	40 (0.28)	50 (15.2)	2000 (4.3)	0.11	0.17	453
LNP DL-4030	15% PTFE		"	"		23	40 (0.28)	50 (15.2)	2000 (4.3)	0.09	0.15	453
LNP DL-4530	13% PTFE, 2% silicone		"	"		23	40 (0.28)	50 (15.2)	2000 (4.3)	0.06	0.09	453
RTP 300			C1018 steel; surface finish: 14-17 µin; 15-25 Rockwell C	"	apparatus: Falex Model No. 6	23	40 (0.28)	50 (15.2)	2000 (4.3)	0.31	0.38	457

Material		Mating Surface		Test Conditions						Coefficient Of Friction		Source
Supplier / Grade	Material Note	Mating Surface Material	Mating Surface Note	Test Method	Test Conditions Note	Temp. (°C)	Pressure (psi) (MPa)	Speed (fpm) (m/min)	PV (psi x fpm) (MPa x m/min)	Static	Dynamic	
POLYCARBONATE against mating surface												
RTP 300 AR 15	15% aramid fiber	**steel**	C1018 steel; surface finish: 14-17 µin; 15-25 Rockwell C	thrust washer	apparatus: Falex Model No. 6	23	40 (0.28)	50 (15.2)	2000 (4.3)	0.1	0.15	457
RTP 300 AR 15 TFE 15	15% PTFE, 15% aramid fiber		"	"	"	23	40 (0.28)	50 (15.2)	2000 (4.3)	0.08	0.1	457
RTP 300 TFE 10	10% PTFE		C1018 steel; surface finish: 14-17 µin; 15-25 Rockwell C	thrust washer	apparatus: Falex Model No. 6	23	40 (0.28)	50 (15.2)	2000 (4.3)	0.11	0.16	457
RTP 300 TFE 15	15% PTFE		"	"	"	23	40 (0.28)	50 (15.2)	2000 (4.3)	0.09	0.15	457
RTP 300 TFE 15 SI	15% PTFE, 0.5% silicone		"	"	"	23	40 (0.28)	50 (15.2)	2000 (4.3)	0.07	0.1	457
RTP 300 TFE 20	20% PTFE		"	"	"	23	40 (0.28)	50 (15.2)	2000 (4.3)	0.08	0.14	457
RTP 300 TFE 5	5% PTFE		"	"	"	23	40 (0.28)	50 (15.2)	2000 (4.3)	0.13	0.16	457
RTP 301 TFE 5	5% PTFE, 10% glass fiber		"	"	"	23	40 (0.28)	50 (15.2)	2000 (4.3)	0.14	0.16	457
RTP 302 TFE 10	10% PTFE, 15% glass fiber		"	"	"	23	40 (0.28)	50 (15.2)	2000 (4.3)	0.2	0.24	457
RTP 302 TFE 15	15% PTFE, 15% glass fiber		"	"	"	23	40 (0.28)	50 (15.2)	2000 (4.3)	0.17	0.2	457
RTP 303 TFE 1 0	10% PTFE, 20% glass fiber		"	"	"	23	40 (0.28)	50 (15.2)	2000 (4.3)	0.2	0.23	457
RTP 303 TFE 15	15% PTFE, 20% glass fiber		"	"	"	23	40 (0.28)	50 (15.2)	2000 (4.3)	0.18	0.21	457
RTP 303 TFE 20	20% PTFE, 20% glass fiber		"	"	"	23	40 (0.28)	50 (15.2)	2000 (4.3)	0.17	0.2	457
RTP 303 TFE 20 SI 2	20% PTFE, 2% silicone, 20% glass fiber		"	"	"	23	40 (0.28)	50 (15.2)	2000 (4.3)	0.09	0.1	457
RTP 305	30% glass fiber		"	"	"	23	40 (0.28)	50 (15.2)	2000 (4.3)	0.22	0.22	457
RTP 305 TFE 13 SI 2	13% PTFE, 2% silicone, 30% glass fiber		"	"	"	23	40 (0.28)	50 (15.2)	2000 (4.3)	0.16	0.2	457
RTP 305 TFE 15	15% PTFE, 30% glass fiber		"	"	"	23	40 (0.28)	50 (15.2)	2000 (4.3)	0.17	0.2	457
RTP 306 TFE 20	35% glass fiber		"	"	"	23	40 (0.28)	50 (15.2)	2000 (4.3)	0.16	0.18	457
RTP 385	30% carbon fiber		"	"	"	23	40 (0.28)	50 (15.2)	2000 (4.3)	0.2	0.2	457
RTP 385 TFE 13 SI 2	13% PTFE, 2% silicone, 30% carbon fiber		"	"	"	23	40 (0.28)	50 (15.2)	2000 (4.3)	0.16	0.2	457
Unspecified grade	unmodified		carbon steel; surface finish: 16 µin; 18-22 Rockwell C	"	apparatus: Faville-LeVally	23	40 (0.28)	50 (15.2)	2000 (4.3)	0.31	0.38	458
POLYESTER, PBT against mating surface												
DuPont Rynite 6125	heat stabilized, 120 Rockwell R	**polyester PBT**	Rynite 6125	ASTM D1894	ISO 8295					0.35		356
DuPont Rynite 6129	extrusion grade, 120 Rockwell R, low flow		Rynite 6129	"	"					0.35		356

Material		Mating Surface		Test Conditions						Coefficient Of Friction		Source
Supplier / Grade	Material Note	Mating Surface Material	Mating Surface Note	Test Method	Test Conditions Note	Temp. (°C)	Pressure (psi) (MPa)	Speed (fpm) (m/min)	PV (psi x fpm) (MPa x m/min)	Static	Dynamic	
POLYESTER, PBT against mating surface												
DuPont Rynite 6130	moderate flow, inj. mold. grade, gen. purp. grade, 120 Rockwell R	**polyester PBT**	Rynite 6130	ASTM D1894	ISO 8295					0.35		356
DuPont Rynite 6131	high flow, 120 Rockwell R		Rynite 6131	"	"					0.35		356
DuPont Rynite FR7915	flame retardant, 122 Rockwell R; 15% glass fiber		Rynite FR7915	ASTM D1894	ISO 8295					0.3		356
DuPont Rynite FR7930	flame retardant, 117 Rockwell R; 30% glass fiber		Rynite FR7930	"	"					0.3		356
DuPont Rynite FR7930F	high flow, flame retardant, 117 Rockwell R; 30% glass fiber		Rynite FR7930F	"	"					0.3		356
LNP W-1000	unmodified		W-1000 (unmodified)	thrust washer		23	40 (0.28)	50 (15.2)	2000 (4.3)	0.17	0.24	453
LNP WL-4040	20% PTFE		WL-4040 (20% PTFE)	"		23	40 (0.28)	50 (15.2)	2000 (4.3)	0.1	0.08	453
Bay Resins Lubriplas PBT-1100	unmodified resin	**steel**		"			40 (0.28)	50 (15.2)	2000 (4.3)	0.19	0.24	435
Bay Resins Lubriplas PBT-1100CF30	30% carbon fiber			"			40 (0.28)	50 (15.2)	2000 (4.3)	0.11	0.14	435
Bay Resins Lubriplas PBT-1100G15	15% glass fiber			"			40 (0.28)	50 (15.2)	2000 (4.3)	0.2	0.25	435
Bay Resins Lubriplas PBT-1100G20	20% glass fiber			"			40 (0.28)	50 (15.2)	2000 (4.3)	0.21	0.26	435
Bay Resins Lubriplas PBT-1100G30	30% glass fiber			"			40 (0.28)	50 (15.2)	2000 (4.3)	0.22	0.27	435
Bay Resins Lubriplas PBT-1100G30TF15	30% glass fiber, 15% PTFE			"			40 (0.28)	50 (15.2)	2000 (4.3)	0.16	0.2	435
Bay Resins Lubriplas PBT-1100TF20	20% PTFE			"			40 (0.28)	50 (15.2)	2000 (4.3)	0.09	0.16	435
Bayer Pocan B1505	gen. purp. grade, unfilled		lapped plate	ASTM D1894		120				0.073	0.047	227
Bayer Pocan B1505	gen. purp. grade, unfilled		"	"		80				0.078	0.061	227
Bayer Pocan B1505	gen. purp. grade, unfilled		"	"		140				0.089	0.046	227
Bayer Pocan B1505	gen. purp. grade, unfilled		"	"		23				0.125	0.101	227
Bayer Pocan B3235	natural resin; 30% glass fiber		"	"		120				0.06	0.045	227
Bayer Pocan B3235	natural resin; 30% glass fiber		"	"		80				0.08	0.061	227
Bayer Pocan B3235	natural resin; 30% glass fiber		"	"		140				0.09	0.053	227
Bayer Pocan B3235	natural resin; 30% glass fiber		"	"		23				0.119	0.113	227
Bayer Pocan B4235	flame retardant; 30% glass fiber		"	"		120				0.066	0.046	227
Bayer Pocan B4235	flame retardant; 30% glass fiber		"	"		140				0.067	0.043	227

Material		Mating Surface		Test Conditions						Coefficient Of Friction		Source
Supplier / Grade	Material Note	Mating Surface Material	Mating Surface Note	Test Method	Test Conditions Note	Temp. (°C)	Pressure (psi) (MPa)	Speed (fpm) (m/min)	PV (psi x fpm) (MPa x m/min)	Static	Dynamic	
POLYESTER, PBT against mating surface												
Bayer Pocan B4235	flame retardant; 30% glass fiber	**steel**	lapped plate	ASTM D1894		80				0.08	0.066	227
Bayer Pocan B4235	flame retardant; 30% glass fiber		"	"		23				0.122	0.113	227
DuPont Rynite 6125	heat stabilized, 120 Rockwell R			ASTM D1894	ISO 8295					0.19		356
DuPont Rynite 6129	extrusion grade, 120 Rockwell R, low flow			"	"					0.19		356
DuPont Rynite 6130	moderate flow, inj. mold. grade, gen. purp. grade, 120 Rockwell R			"	"					0.19		356
DuPont Rynite 6131	high flow, 120 Rockwell R			"	"					0.19		356
DuPont Rynite FR7915	flame retardant, 122 Rockwell R; 15% glass fiber			"	"					0.2		356
DuPont Rynite FR7930	flame retardant, 117 Rockwell R; 30% glass fiber			"	"					0.2		356
DuPont Rynite FR7930F	high flow, flame retardant, 117 Rockwell R; 30% glass fiber			"	"					0.2		356
Ensinger	standard grade						40 (0.28)	50 (15.2)	2000 (4.3)		0.25	449
Ferro Star-C PBT 20Y	20% carbon fiber			ASTM D3028						0.19		452
Ferro Star-C PBT 30Y	30% carbon fiber			"						0.18		452
Hoechst Cel. Celanex 2000	high flow, gen. purp. grade, 75 Rockwell M			ASTM D1894							0.13	315
Hoechst Cel. Celanex 2000K	moderate flow, 75 Rockwell M, keycap grade			"							0.13	315
Hoechst Cel. Celanex 2002	moderate flow, gen. purp. grade, 78 Rockwell M			"							0.13	315
Hoechst Cel. Celanex 2003	gen. purp. grade, mod.-high flow			"							0.13	315
Hoechst Cel. Celanex 2003K	keycap grade, mod.-high flow			"							0.13	315
Hoechst Cel. Celanex 3112	flame retardant, 88 Rockwell M; 13% glass fiber			"							0.12	315
Hoechst Cel. Celanex 3112	flame retardant, 88 Rockwell M; 13% glass fiber			"						0.15		315
Hoechst Cel. Celanex 3200	gen. purp. grade, 90 Rockwell M; 15% glass fiber			"						0.17		315
Hoechst Cel. Celanex 3200	gen. purp. grade, 90 Rockwell M; 15% glass fiber			"							0.15	315
Hoechst Cel. Celanex 3210	flame retardant, 90 Rockwell M; 18% glass fiber			"							0.12	315
Hoechst Cel. Celanex 3211	copolymer; flame retardant, 93 Rockwell M; 19% glass fiber			"							0.14	315
Hoechst Cel. Celanex 3211	copolymer; flame retardant, 93 Rockwell M; 19% glass fiber			"						0.21		315

POLYESTER, PBT against mating surface

Material: Supplier / Grade	Material: Material Note	Mating Surface: Mating Surface Material	Mating Surface: Mating Surface Note	Test Conditions: Test Method	Test Conditions: Test Conditions Note	Test Conditions: Temp. (°C)	Test Conditions: Pressure (psi) (MPa)	Test Conditions: Speed (fpm) (m/min)	Test Conditions: PV (psi x fpm) (MPa x m/min)	Coefficient Of Friction: Static	Coefficient Of Friction: Dynamic	Source
Hoechst Cel. Celanex 3300	gen. purp. grade, 90 Rockwell M; 30% glass fiber	**steel**		ASTM D1894						0.25		315
Hoechst Cel. Celanex 3300	gen. purp. grade, 90 Rockwell M; 30% glass fiber			"							0.12	315
Hoechst Cel. Celanex 3310	flame retardant, 90 Rockwell M; 30% glass fiber			ASTM D1894							0.12	315
Hoechst Cel. Celanex 3311	copolymer; flame retardant, 94 Rockwell M; 30% glass fiber			"							0.14	315
Hoechst Cel. Celanex 3311	copolymer; flame retardant, 94 Rockwell M; 30% glass fiber			"						0.22		315
Hoechst Cel. Celanex 3400	gen. purp. grade, 93 Rockwell M; 40% glass fiber			"						0.18		315
Hoechst Cel. Celanex 3400	gen. purp. grade, 93 Rockwell M; 40% glass fiber			"							0.14	315
Hoechst Cel. Celanex 5300	good surface, 93 Rockwell M; 30% glass fiber			"							0.13	315
Hoechst Cel. Celanex 6400	low warp grade, good surface, 86 Rockwell M; 40% mineral/ glass			"							0.14	315
Hoechst Cel. Celanex 6400	low warp grade, good surface, 86 Rockwell M; 40% mineral/ glass			"						0.2		315
Hoechst Cel. Celanex 7700	low warp grade, flame retardant, 76 Rockwell M; 35% mineral/ glass			"						0.19		315
Hoechst Cel. Celanex 7700	low warp grade, flame retardant, 76 Rockwell M; 35% mineral/ glass			"							0.15	315
LNP W-1000	unmodified		carbon steel; surface finish: 12-16 µin; 18-20 Rockwell C	thrust washer		23	40 (0.28)	50 (15.2)	2000 (4.3)	0.19	0.25	453
LNP WC-1006	30% carbon fiber		"	"		23	40 (0.28)	50 (15.2)	2000 (4.3)	0.12	0.15	453
LNP WF-1006	30% glass fiber		"	"		23	40 (0.28)	50 (15.2)	2000 (4.3)	0.23	0.27	453
LNP WFL-4036	30% glass fiber, 15% PTFE		"	"		23	40 (0.28)	50 (15.2)	2000 (4.3)	0.16	0.21	453
LNP WFL-4536	30% glass fiber, 13% PTFE, 2% silicone		"	"		23	40 (0.28)	50 (15.2)	2000 (4.3)	0.11	0.12	453
LNP WL-4040	20% PTFE		"	"		23	40 (0.28)	50 (15.2)	2000 (4.3)	0.09	0.17	453
LNP WL-4410	2% silicone		"	"		23	40 (0.28)	50 (15.2)	2000 (4.3)	0.09	0.16	453
LNP WL-4540	18% PTFE, 2% silicone		"	"		23	40 (0.28)	50 (15.2)	2000 (4.3)	0.08	0.13	453
RTP 1000			C1018 steel; surface finish: 14-17 µin; 15-25 Rockwell C	"	apparatus: Falex Model No. 6	23	40 (0.28)	50 (15.2)	2000 (4.3)	0.2	0.25	457
RTP 1002 TFE 15	15% PTFE, 15% glass fiber		"	"	"	23	40 (0.28)	50 (15.2)	2000 (4.3)	0.15	0.2	457
RTP 1005	30% glass fiber		"	"	"	23	40 (0.28)	50 (15.2)	2000 (4.3)	0.23	0.27	457
RTP 1005 TFE 15	15% PTFE, 30% glass fiber		"	"	"	23	40 (0.28)	50 (15.2)	2000 (4.3)	0.15	0.2	457

Material		Mating Surface		Test Conditions						Coefficient Of Friction		Source
Supplier / Grade	Material Note	Mating Surface Material	Mating Surface Note	Test Method	Test Conditions Note	Temp. (°C)	Pressure (psi) (MPa)	Speed (fpm) (m/min)	PV (psi x fpm) (MPa x m/min)	Static	Dynamic	
POLYESTER, PBT against mating surface												
RTP 1085	30% carbon fiber	**steel**	C1018 steel; surface finish: 14-17 µin; 15-25 Rockwell C	thrust washer	apparatus: Falex Model No. 6	23	40 (0.28)	50 (15.2)	2000 (4.3)	0.12	0.15	457
RTP 1085 TFE 15	15% PTFE, 30% carbon fiber		"	"	"	23	40 (0.28)	50 (15.2)	2000 (4.3)	0.1	0.13	457
Thermofil E-30FG-0287	30% glass fiber, lubricated			ASTM D1894		23				0.11	0.14	459
Thermofil E-30NF-0100	30% graphite fiber			"		23				0.11	0.13	459
POLYESTER, PET against mating surface												
DuPont Rynite 415HP	impact modified, 58 Rockwell M, 111 Rockwell R; 15% glass fiber	**polyester PET**	Rynite 530	ASTM D1894	ISO 8295					0.42		356
DuPont Rynite 530	100 Rockwell M, 120 Rockwell R; 30% glass fiber		"	"	"					0.18		356
DuPont Rynite 545	100 Rockwell M, 120 Rockwell R; 45% glass fiber		"	"	"					0.17		356
DuPont Rynite 555	105 Rockwell M, 121 Rockwell R; 55% glass fiber		"	"	"					0.27		356
DuPont Rynite 935	low warp grade, 85 Rockwell M, 119 Rockwell R; 35% mica/ glass		"	"	"					0.21		356
DuPont Rynite FR515	flame retardant, 88 Rockwell M, 120 Rockwell R; 15% glass fiber		"	"	"					0.21		356
DuPont Rynite FR530	flame retardant, 120 Rockwell R, 100 Rockwell M; 30% glass fiber		"	"	"					0.18		356
DuPont Rynite FR543	flame retardant, 122 Rockwell R, 102 Rockwell M; 43% glass fiber		"	"	"					0.18		356
DuPont Rynite FR945	low warp grade, flame retardant, 120 Rockwell R, 92 Rockwell M; 45% mica/ glass		"	"	"					0.2		356
DuPont Rynite 415HP	impact modified, 58 Rockwell M, 111 Rockwell R; 15% glass fiber	**steel**		"	"					0.27		356
DuPont Rynite 530	100 Rockwell M, 120 Rockwell R; 30% glass fiber			"	"					0.17		356
DuPont Rynite 545	100 Rockwell M, 120 Rockwell R; 45% glass fiber			"	"					0.2		356
DuPont Rynite 555	105 Rockwell M, 121 Rockwell R; 55% glass fiber			"	"					0.18		356
DuPont Rynite 935	low warp grade, 85 Rockwell M, 119 Rockwell R; 35% mica/ glass			"	"					0.19		356
DuPont Rynite FR515	flame retardant, 88 Rockwell M, 120 Rockwell R; 15% glass fiber			"	"					0.18		356
DuPont Rynite FR530	flame retardant, 120 Rockwell R, 100 Rockwell M; 30% glass fiber			"	"					0.19		356
DuPont Rynite FR543	flame retardant, 122 Rockwell R, 102 Rockwell M; 43% glass fiber			"	"					0.16		356
DuPont Rynite FR945	low warp grade, flame retardant, 120 Rockwell R, 92 Rockwell M; 45% mica/ glass			"	"					0.2		356
Ensinger Ensitep	standard grade						40 (0.28)	50 (15.2)	2000 (4.3)		0.25	449

Material		Mating Surface		Test Conditions						Coefficient Of Friction		Source
Supplier / Grade	Material Note	Mating Surface Material	Mating Surface Note	Test Method	Test Conditions Note	Temp. (°C)	Pressure (psi) (MPa)	Speed (fpm) (m/min)	PV (psi x fpm) (MPa x m/min)	Static	Dynamic	
POLYESTER, PET against mating surface												
RTP 1107 TFE 10	10% PTFE, 40% glass fiber	**steel**	C1018 steel; surface finish: 14-17 µin; 15-25 Rockwell C	thrust washer	apparatus: Falex Model No. 6	23	40 (0.28)	50 (15.2)	2000 (4.3)	0.13	0.15	457
Thermofil E2-30FG-7100	30% glass fiber, lubricated			ASTM D1894		23				0.16	0.21	459
POLYESTER COPOLYMER against mating surface												
Eastman PCTA 6761	amorphous, 105 Rockwell R, 54 Rockwell L	**glass**		ASTM D1894							0.29	448
Eastman PCTA 6761	amorphous, 105 Rockwell R, 54 Rockwell L	**polyester copolymer**	PCTA 6761	"							1.48	448
Eastman PCTA 6761	amorphous, 105 Rockwell R, 54 Rockwell L	**steel**	stainless steel	"							0.32	448
LIQUID CRYSTAL POLYMER against mating surface												
DuPont Zenite 6130	high heat grade, 3.2 mm thick, 63 Rockwell M, 110 Rockwell R; 30% glass fiber	**liquid crystal polymer**	Zenite	ASTM D1894						0.29	0.23	354
DuPont Zenite 6130	3.2 mm thick, 61 Rockwell M, 108 Rockwell R; 30% glass fiber	**steel**		"						0.09	0.12	354
DuPont Zenite 6130	high heat grade, 3.2 mm thick, 63 Rockwell M, 110 Rockwell R; 30% glass fiber			"						0.1	0.1	354
Hoechst AG Vectra A130	80 Rockwell M; 30% glass fiber, Vectra A950 matrix; parallel to flow		ball (diameter: 13 mm)	sliding against steel	load: 6 N			2 (0.6)			0.38	70
Hoechst AG Vectra A130	80 Rockwell M; 30% glass fiber, Vectra A950 matrix; transverse to flow		"	"	"			2 (0.6)			0.38	70
Hoechst AG Vectra A230	80 Rockwell M; Vectra A950 matrix, 30% carbon fiber; transverse to flow		"	"	"			2 (0.6)			0.33	70
Hoechst AG Vectra A230	80 Rockwell M; Vectra A950 matrix, 30% carbon fiber; parallel to flow		"	"	"			2 (0.6)			0.37	70
Hoechst AG Vectra A422	75 Rockwell M; 40% glass fiber/ graphite, Vectra A950 matrix; transverse to flow		"	"	"			2 (0.6)			0.35	70
Hoechst AG Vectra A422	75 Rockwell M; 40% glass fiber/ graphite, Vectra A950 matrix; parallel to flow		"	"	"			2 (0.6)			0.43	70
Hoechst AG Vectra A430	32 Rockwell M; 25% PTFE modified, Vectra A950 matrix; transverse to flow		"	"	"			2 (0.6)			0.1	70
Hoechst AG Vectra A430	32 Rockwell M; 25% PTFE modified, Vectra A950 matrix; parallel to flow		"	"	"			2 (0.6)			0.15	70
Hoechst AG Vectra A435	49 Rockwell M; Vectra A950 matrix, 35% glass fiber/ graphite; transverse to flow		"	"	"			2 (0.6)			0.18	70
Hoechst AG Vectra A435	49 Rockwell M; Vectra A950 matrix, 35% glass fiber/ graphite; parallel to flow		"	"	"			2 (0.6)			0.28	70
Hoechst AG Vectra A530	65 Rockwell M; Vectra A950 matrix, 30% mineral filler; parallel to flow		"	"	"			2 (0.6)			0.31	70
Hoechst AG Vectra A530	65 Rockwell M; Vectra A950 matrix, 30% mineral filler; transverse to flow		"	"	"			2 (0.6)			0.38	70
Hoechst AG Vectra A625	60 Rockwell M; 25% graphite, Vectra A950 matrix; transverse to flow		"	"	"			2 (0.6)			0.28	70

Material		Mating Surface		Test Conditions						Coefficient Of Friction		Source
Supplier / Grade	Material Note	Mating Surface Material	Mating Surface Note	Test Method	Test Conditions Note	Temp. (°C)	Pressure (psi) (MPa)	Speed (fpm) (m/min)	PV (psi x fpm) (MPa x m/min)	Static	Dynamic	
LIQUID CRYSTAL POLYMER against mating surface →												
Hoechst AG Vectra A625	60 Rockwell M; 25% graphite, Vectra A950 matrix; parallel to flow	**steel**	ball (diameter: 13 mm)	sliding against steel	load: 6 N			2 (0.6)			0.3	70
Hoechst AG Vectra B130	98 Rockwell M; 30% glass fiber, Vectra B950 matrix; transverse to flow		"	"	"			2 (0.6)			0.36	70
Hoechst AG Vectra B130	98 Rockwell M; 30% glass fiber, Vectra B950 matrix; parallel to flow		"	"	"			2 (0.6)			0.38	70
Hoechst AG Vectra B230	100 Rockwell M; Vectra B950 matrix, 30% carbon fiber; transverse to flow		"	"	"			2 (0.6)			0.3	70
Hoechst AG Vectra B230	100 Rockwell M; Vectra B950 matrix, 30% carbon fiber; parallel to flow		"	"	"			2 (0.6)			0.32	70
Hoechst AG Vectra C130	75 Rockwell M; 30% glass fiber, Vectra C950 matrix; parallel to flow		"	"	"			2 (0.6)			0.4	70
Hoechst AG Vectra C130	75 Rockwell M; 30% glass fiber, Vectra C950 matrix; transverse to flow		"	"	"			2 (0.6)			0.4	70
POLYAMIDEIMIDE against mating surface →												
Amoco Torlon 4301	wear resistant; 20% graphite powd., 3% fluorocarbon	**aluminum**	A380 casting alloy; -28 Rockwell C	thrust washer			1000 (6.89)	50 (15.2)	50000 (105.1)	0.04	0.13	20
Amoco Torlon 4301	wear resistant; 20% graphite powd., 3% fluorocarbon		A360 casting alloy; -24 Rockwell C	"			1000 (6.89)	50 (15.2)	50000 (105.1)	0.05	0.13	20
Amoco Torlon 4301	wear resistant; 20% graphite powd., 3% fluorocarbon		A380 casting alloy; -28 Rockwell C	"			50 (0.34)	900 (274)	45000 (94.6)	0.1	0.18	20
Amoco Torlon 4301	wear resistant; 20% graphite powd., 3% fluorocarbon		A360 casting alloy; -24 Rockwell C	"			50 (0.34)	900 (274)	45000 (94.6)	0.11	0.2	20
Amoco Torlon 4301	wear resistant; 20% graphite powd., 3% fluorocarbon	**brass**	free cutting; -15 Rockwell C	"			1000 (6.89)	50 (15.2)	50000 (105.1)	0.04	0.23	20
Amoco Torlon 4301	wear resistant; 20% graphite powd., 3% fluorocarbon		"	"			50 (0.34)	900 (274)	45000 (94.6)	0.11	0.26	20
Amoco Torlon 4301	wear resistant; 20% graphite powd., 3% fluorocarbon	**stainless steel**	316 stainless steel; 17 Rockwell C	"			1000 (6.89)	50 (15.2)	50000 (105.1)	0.07	0.12	20
Amoco Torlon 4301	wear resistant; 20% graphite powd., 3% fluorocarbon		"	"			50 (0.34)	900 (274)	45000 (94.6)	0.13	0.2	20
3M Torlon 4203	unfilled	**steel**	AISI 1018, surface finish: 16 RMS	thrust bearing	apparatus: LRI-1 Wear and Friction Tester		2000 (13.8)	12.5 (3.8)	25000 (52.6)		0.38	318
Amoco Torlon 4275	wear resistant; 3% fluorocarbon, 30% graphite powd.		C1018 steel; 24 Rockwell C	thrust washer			50 (0.34)	200 (61)	10000 (20.7)	0.02	0.19	20
Amoco Torlon 4275	wear resistant; 3% fluorocarbon, 30% graphite powd.		"	"			50 (0.34)	900 (274)	45000 (94.6)	0.07	0.15	20
Amoco Torlon 4275	wear resistant; 3% fluorocarbon, 30% graphite powd.		"	"			1000 (6.89)	50 (15.2)	50000 (105.1)	0.14	0.11	20
Amoco Torlon 4301	wear resistant; 20% graphite powd., 3% fluorocarbon		"	"			50 (0.34)	200 (61)	10000 (20.7)	0.06	0.27	20
Amoco Torlon 4301	wear resistant; 20% graphite powd., 3% fluorocarbon		C1018; lubricated; 24 Rockwell C	"			50 (0.34)	900 (274)	45000 (94.6)	0.08	0.1	20
Amoco Torlon 4301	wear resistant; 20% graphite powd., 3% fluorocarbon		C1018 steel; soft; 6 Rockwell C	"			1000 (6.89)	50 (15.2)	50000 (105.1)	0.08	0.13	20
Amoco Torlon 4301	wear resistant; 20% graphite powd., 3% fluorocarbon		C1018 steel; 24 Rockwell C	"			1000 (6.89)	50 (15.2)	50000 (105.1)	0.11	0.12	20

Material		Mating Surface		Test Conditions						Coefficient Of Friction		Source
Supplier / Grade	Material Note	Mating Surface Material	Mating Surface Note	Test Method	Test Conditions Note	Temp. (°C)	Pressure (psi) (MPa)	Speed (fpm) (m/min)	PV (psi x fpm) (MPa x m/min)	Static	Dynamic	
POLYAMIDEIMIDE against mating surface												
Amoco Torlon 4301	wear resistant; 20% graphite powd., 3% fluorocarbon	**steel**	C1018 steel; 24 Rockwell C	thrust washer			50 (0.34)	900 (274)	45000 (94.6)	0.13	0.14	20
Amoco Torlon 4301	wear resistant; 20% graphite powd., 3% fluorocarbon		C1018 steel; soft; 6 Rockwell C	"			50 (0.34)	900 (274)	45000 (94.6)	0.15	0.17	20
Amoco	wear resistant; 12% graphite powd., 8% fluorocarbon		C1018 steel; 24 Rockwell C	"			50 (0.34)	200 (61)	10000 (20.7)	0.02	0.19	20
Amoco	wear resistant; 12% graphite powd., 8% fluorocarbon		"	"			1000 (6.89)	50 (15.2)	50000 (105.1)	0.08	0.11	20
Amoco	wear resistant; 12% graphite powd., 8% fluorocarbon		"	"			50 (0.34)	900 (274)	45000 (94.6)	0.08	0.13	20
POLYETHERIMIDE against mating surface												
GE Ultem	standard grade	**steel**					40 (0.28)	50 (15.2)	2000 (4.3)		0.17	449
GE Ultem 2300	30% glass fiber						40 (0.28)	50 (15.2)	2000 (4.3)		0.24	449
GE Ultem 4000	reinforced, internal lubricant, 85 Rockwell M			ASTM D1894						0.25	0.24	51
GE Ultem 4001	internal lubricant, 110 Rockwell M			"							0.25	51
LNP E-1000	unmodified		carbon steel; surface finish: 12-16 µin; 18-20 Rockwell C	thrust washer		23	40 (0.28)	50 (15.2)	2000 (4.3)	0.18	0.17	453
LNP EC-1006	30% carbon fiber		"	"		23	40 (0.28)	50 (15.2)	2000 (4.3)	0.2	0.22	453
LNP EC-1006	30% carbon fiber		cold rolled steel; surface finish: 12-16 µin; 22 Rockwell C	"		93	40 (0.28)	50 (15.2)	2000 (4.3)	0.27	0.33	453
LNP EC-1006	30% carbon fiber		"	"		149	40 (0.28)	50 (15.2)	2000 (4.3)	0.3	0.36	453
LNP EF-1006	30% glass fiber		carbon steel; surface finish: 12-16 µin; 18-20 Rockwell C	"		23	40 (0.28)	50 (15.2)	2000 (4.3)	0.22	0.24	453
LNP EFL-4036	30% glass fiber, 15% PTFE		"	"		23	40 (0.28)	50 (15.2)	2000 (4.3)	0.19	0.2	453
LNP EFL-4036	30% glass fiber, 15% PTFE		cold rolled steel; surface finish: 12-16 µin; 22 Rockwell C	"		93	40 (0.28)	50 (15.2)	2000 (4.3)	0.24	0.29	453
LNP EFL-4036	30% glass fiber, 15% PTFE		"	"		149	40 (0.28)	50 (15.2)	2000 (4.3)	0.32	0.38	453
RTP 2100 AR 15	15% aramid fiber		C1018 steel; surface finish: 14-17 µin; 15-25 Rockwell C	"	apparatus: Falex Model No. 6	23	40 (0.28)	50 (15.2)	2000 (4.3)	0.08	0.1	457
RTP 2100 AR 15 TFE 15	15% PTFE, 15% aramid fiber		"	"	"	23	40 (0.28)	50 (15.2)	2000 (4.3)	0.07	0.08	457
RTP 2105 TFE 15	15% PTFE, 30% glass fiber		"	"	"	23	40 (0.28)	50 (15.2)	2000 (4.3)	0.14	0.16	457
RTP 2185 TFE 13 SI2	13% PTFE, 2% silicone, 30% carbon fiber		"	"	"	23	40 (0.28)	50 (15.2)	2000 (4.3)	0.11	0.14	457
RTP 2185 TFE 15	15% PTFE, 30% carbon fiber		"	"	"	23	40 (0.28)	50 (15.2)	2000 (4.3)	0.12	0.14	457

Material		Mating Surface		Test Conditions						Coefficient Of Friction		Source
Supplier / Grade	Material Note	Mating Surface Material	Mating Surface Note	Test Method	Test Conditions Note	Temp. (°C)	Pressure (psi) (MPa)	Speed (fpm) (m/min)	PV (psi x fpm) (MPa x m/min)	Static	Dynamic	
POLYETHERIMIDE against mating surface												
Thermofil W-10FG-0100	10% glass fiber	**steel**		ASTM D1894		23				0.26	0.3	459
Thermofil W-30NF-0100	30% graphite fiber			"		23				0.16	0.13	459
Thermofil W-30NF-0214	30% graphite fiber, PTFE lubricated			"		23				0.15	0.22	459
POLYARYLETHERKETONE against mating surface												
BASF AG Ultrapek A2000C6	moderate flow; 30% carbon fiber	**steel**	Cr 6/800/HV				145 (1)	98 (30)	14210 (29.9)		0.3	27
BASF AG Ultrapek A2000C6	moderate flow; 30% carbon fiber		"				145 (1)	98 (30)	14210 (29.9)		0.45	27
BASF AG Ultrapek A3000	low flow		"				145 (1)	98 (30)	14210 (29.9)		0.51	27
BASF AG Ultrapek A3000	low flow		"				145 (1)	98 (30)	14210 (29.9)		0.56	27
BASF AG Ultrapek KR4190	lubricated, developmental material, tribological properties; modified grade		"				145 (1)	98 (30)	14210 (29.9)		0.25	27
BASF AG Ultrapek KR4190	lubricated, developmental material, tribological properties; modified grade		"				145 (1)	98 (30)	14210 (29.9)		0.26	27
POLYETHERETHERKETONE against mating surface												
Ensinger	standard grade	**steel**					40 (0.28)	50 (15.2)	2000 (4.3)		0.25	449
LNP LCL-4033EM	15% carbon fiber, 15% PTFE		cold rolled steel; surface finish: 12-16 µin; 22 Rockwell C	thrust washer		149	40 (0.28)	50 (15.2)	2000 (4.3)	0.07	0.15	453
LNP LCL-4033EM	15% carbon fiber, 15% PTFE		"	"		260	40 (0.28)	50 (15.2)	2000 (4.3)	0.17	0.23	453
LNP LCL-4033EM	15% carbon fiber, 15% PTFE		carbon steel; surface finish: 12-16 µin; 18-20 Rockwell C	"		23	40 (0.28)	50 (15.2)	2000 (4.3)	0.18	0.2	453
LNP LFL-4036	30% glass fiber, 15% PTFE		"	"		23	40 (0.28)	50 (15.2)	2000 (4.3)	0.27	0.28	453
LNP LL-4040	20% PTFE		"	"		23	40 (0.28)	50 (15.2)	2000 (4.3)	0.19	0.23	453
RTP 2205 TFE	15% PTFE, 15% silicone, 30% glass fiber		C1018 steel; surface finish: 14-17 µin; 15-25 Rockwell C	"	apparatus: Falex Model No. 6	23	40 (0.28)	50 (15.2)	2000 (4.3)	0.15	0.2	457
RTP 2285TFE	15% PTFE, 15% silicone, 30% carbon fiber		"	"	"	23	40 (0.28)	50 (15.2)	2000 (4.3)	0.14	0.18	457
Thermofil K2-30NF-0100	30% graphite fiber			ASTM D1894		23				0.3	0.25	459
Victrex 450CA30	30% carbon fiber						40 (0.28)	50 (15.2)	2000 (4.3)		0.13	449
Victrex 450GL30	30% glass fiber						40 (0.28)	50 (15.2)	2000 (4.3)		0.3	449

Material		Mating Surface		Test Conditions						Coefficient Of Friction		Source
Supplier / Grade	Material Note	Mating Surface Material	Mating Surface Note	Test Method	Test Conditions Note	Temp. (°C)	Pressure (psi) (MPa)	Speed (fpm) (m/min)	PV (psi x fpm) (MPa x m/min)	Static	Dynamic	
POLYETHERETHERKETONE against mating surface												
Victrex PEEK 15OFC30	proprietary filler	**steel**	cold rolled steel; surface finish: 12-16 μin; 22 Rockwell C	thrust washer		149	40 (0.28)	50 (15.2)	2000 (4.3)	0.1	0.21	453
Victrex PEEK 15OFC30	proprietary filler		"	"		260	40 (0.28)	50 (15.2)	2000 (4.3)	0.13	0.22	453
Victrex PEEK 15OFC30	proprietary filler		carbon steel; surface finish: 12-16 μin; 18-20 Rockwell C	"		23	40 (0.28)	50 (15.2)	2000 (4.3)	0.18	0.2	453
Victrex PEEK 450CA30	30% carbon fiber		cold rolled steel; surface finish: 12-16 μin; 22 Rockwell C	"		93	40 (0.28)	50 (15.2)	2000 (4.3)	0.11	0.06	453
Victrex PEEK 450CA30	30% carbon fiber		"	"		204	40 (0.28)	50 (15.2)	2000 (4.3)	0.12	0.09	453
Victrex PEEK 450CA30	30% carbon fiber		"	"		260	40 (0.28)	50 (15.2)	2000 (4.3)	0.15	0.14	453
Victrex PEEK 450CA30	30% carbon fiber		"	"		149	40 (0.28)	50 (15.2)	2000 (4.3)	0.15	0.14	453
Victrex PEEK 450CA30	30% carbon fiber		carbon steel; surface finish: 12-16 μin; 18-20 Rockwell C	"		23	40 (0.28)	50 (15.2)	2000 (4.3)	0.19	0.13	453
Victrex PEEK 450CA30	124 Rockwell R, 107 Rockwell M; 30% carbon fiber		carbon EN8, dry	pad on ring (Amsler wear tester)		20		100 (30.5)			0.16 (average of COF @ LPV and 50% of LPV load)	338
Victrex PEEK 450CA30	124 Rockwell R, 107 Rockwell M; 30% carbon fiber		"	"		200		100 (30.5)			0.23 (average of COF @ LPV and 50% of LPV load)	338
Victrex PEEK 450CA30	124 Rockwell R, 107 Rockwell M; 30% carbon fiber		"	"		200		600 (183)			0.25 (average of COF @ LPV and 50% of LPV load)	338
Victrex PEEK 450CA30	124 Rockwell R, 107 Rockwell M; 30% carbon fiber		"	"		20		600 (183)			0.28 (average of COF @ LPV and 50% of LPV load)	338
Victrex PEEK 450FC30	was Victrex D450HF30, tribological properties, bearing grade; 30% graphite/ carbon/ PTFE		"	"		20		100 (30.5)			0.11 (average of COF @ LPV and 50% of LPV load)	338
Victrex PEEK 450FC30	was Victrex D450HF30, tribological properties, bearing grade; 30% graphite/ carbon/ PTFE		"	"		200		100 (30.5)			0.12 (average of COF @ LPV and 50% of LPV load)	338

Material		Mating Surface		Test Conditions						Coefficient Of Friction		Source
Supplier / Grade	Material Note	Mating Surface Material	Mating Surface Note	Test Method	Test Conditions Note	Temp. (°C)	Pressure (psi) (MPa)	Speed (fpm) (m/min)	PV (psi x fpm) (MPa x m/min)	Static	Dynamic	
POLYETHERETHERKETONE against mating surface →												
Victrex PEEK 450FC30	was Victrex D450HF30, tribological properties, bearing grade; 30% graphite/ carbon/ PTFE	**steel**	carbon EN8, dry	pad on ring (Amsler wear tester)		200		600 (183)			0.14 (average of COF @ LPV and 50% of LPV load)	338
Victrex PEEK 450FC30	was Victrex D450HF30, tribological properties, bearing grade; 30% graphite/ carbon/ PTFE		"	"		20		600 (183)			0.17 (average of COF @ LPV and 50% of LPV load)	338
Victrex PEEK 450G	gen. purp. grade, 126 Rockwell R, 99 Rockwell M		"	"		200		100 (30.5)			0.15 (average of COF @ LPV and 50% of LPV load)	338
Victrex PEEK 450G	gen. purp. grade, 126 Rockwell R, 99 Rockwell M		"	"		20		100 (30.5)			0.34 (average of COF @ LPV and 50% of LPV load)	338
Victrex PEEK 450G	gen. purp. grade, 126 Rockwell R, 99 Rockwell M		"	"		200		600 (183)			0.51 (average of COF @ LPV and 50% of LPV load)	338
Victrex PEEK 450G	gen. purp. grade, 126 Rockwell R, 99 Rockwell M		"	"		20		600 (183)			0.58 (average of COF @ LPV and 50% of LPV load)	338
Victrex PEEK 450GL30	30% glass fiber		carbon steel; surface finish: 12-16 µin; 18-20 Rockwell C	thrust washer		23	40 (0.28)	50 (15.2)	2000 (4.3)	0.28	0.3	453
Victrex PEEK 450G	unmodified		"	"		23	40 (0.28)	50 (15.2)	2000 (4.3)	0.2	0.25	453
Victrex PEEK D450HT15	no longer available, tribological properties, bearing grade		carbon EN8, dry	pad on ring (Amsler wear tester)		200		100 (30.5)			0.05 (average of COF @ LPV and 50% of LPV load)	338
Victrex PEEK D450HT15	no longer available, tribological properties, bearing grade		"	"		200		600 (183)			0.06 (average of COF @ LPV and 50% of LPV load)	338

Material		Mating Surface		Test Conditions						Coefficient Of Friction		Source
Supplier / Grade	Material Note	Mating Surface Material	Mating Surface Note	Test Method	Test Conditions Note	Temp. (°C)	Pressure (psi) (MPa)	Speed (fpm) (m/min)	PV (psi x fpm) (MPa x m/min)	Static	Dynamic	
POLYETHERETHERKETONE against mating surface												
Victrex PEEK D450HT15	no longer available, tribological properties, bearing grade	**steel**	carbon EN8, dry	pad on ring (Amsler wear tester)		20		600 (183)			0.09 (average of COF @ LPV and 50% of LPV load)	338
Victrex PEEK D450HT15	no longer available, tribological properties, bearing grade		"	"		20		100 (30.5)			0.18 (average of COF @ LPV and 50% of LPV load)	338
POLYETHYLENE against mating surface												
Unspecified grade	film	**steel**	64 mm square x 6.4 mm thick block	sliding block				0.42 (0.13)		0.31	0.22	101
POLYETHYLENE, HDPE against mating surface												
LNP FL-4030	15% PTFE	**steel**	carbon steel; surface finish: 12-16 µin; 18-20 Rockwell C	thrust washer		23	40 (0.28)	50 (15.2)	2000 (4.3)	0.11	0.15	453
Unspecified grade				ASTM D1894							0.26	78
POLYETHYLENE, UHMWPE against mating surface												
Solidur Solidur 10 100	virgin resin	**brass**						0.17 (0.051)		0.34	0.2	436
Unspecified grade		**nylon 610**	QFL-4036 (30% glass fiber, 15% PTFE)	thrust washer		23	40 (0.28)	50 (15.2)	2000 (4.3)	0.06	0.08	453
Unspecified grade		**polycarbonate**	DL-4010 (5% PTFE)	"		23	40 (0.28)	50 (15.2)	2000 (4.3)	0.08	0.09	453
Solidur Solidur 10 100	virgin resin	**steel**	chrome plated steel					0.17 (0.051)		0.14	0.12	436
Solidur Solidur 10 100	virgin resin		cold rolled steel					0.17 (0.051)		0.27	0.15	436
Solidur Solidur 10 100	virgin resin		stainless steel					0.17 (0.051)		0.28	0.13	436
Solidur Solidur 10 100	virgin resin	**UHMWPE**	Solidur					0.17 (0.051)		0.2	0.2	436
POLYPROPYLENE against mating surface												
LNP M-1000	unmodified	**steel**	carbon steel; surface finish: 12-16 µin; 18-20 Rockwell C	thrust washer		23	40 (0.28)	50 (15.2)	2000 (4.3)	0.19	0.23	394
LNP MFL-4034HS	15% PTFE, 20% silicone		"	"		23	40 (0.28)	50 (15.2)	2000 (4.3)	0.09	0.09	453
LNP MFX-700-10	50% long glass fiber, chemically coupled		"	"		23	40 (0.28)	50 (15.2)	2000 (4.3)	0.13	0.15	394

Material		Mating Surface		Test Conditions						Coefficient Of Friction		Source
Supplier / Grade	Material Note	Mating Surface Material	Mating Surface Note	Test Method	Test Conditions Note	Temp. (°C)	Pressure (psi) (MPa)	Speed (fpm) (m/min)	PV (psi x fpm) (MPa x m/min)	Static	Dynamic	
POLYPROPYLENE against mating surface												
LNP ML-404OHS	20% PTFE	steel	carbon steel; surface finish: 12-16 µin; 18-20 Rockwell C	thrust washer		23	40 (0.28)	50 (15.2)	2000 (4.3)	0.08	0.11	453
Unspecified grade	film		64 mm square x 6.4 mm thick block	sliding block				0.42 (0.13)		0.37	0.24	101
Unspecified grade				ASTM D1894							0.33	78
MODIFIED PPE against mating surface												
GE Noryl 731	gen. purp. grade, 119 Rockwell R	brass	Noryl 731								0.17-0.20	430
GE Noryl GFN2	gen. purp. grade, 106 Rockwell L; 20% glass fiber		Noryl GFN2								0.30-0.35	430
GE Noryl GFN3	gen. purp. grade, 108 Rockwell L; 30% glass fiber		Noryl GFN3								0.39-0.42	430
GE Noryl SE1	gen. purp. grade, flame retardant, 119 Rockwell R, electronics grade		Noryl SE1								0.55-0.66	430
GE Noryl SE1-GFN2	flame retardant, 106 Rockwell L; 20% glass fiber		Noryl SE1 GFN2								0.23-0.25	430
GE Noryl SE1-GFN3	flame retardant, 108 Rockwell L; 30% glass fiber		Noryl SE1 GFN3								0.29-0.35	430
GE Noryl SE100	gen. purp. grade, flame retardant, 115 Rockwell R		Noryl SE100								0.27-0.31	430
GE Noryl 731	gen. purp. grade, 119 Rockwell R	copper	Noryl 731								0.39-0.42	430
GE Noryl GFN2	gen. purp. grade, 106 Rockwell L; 20% glass fiber		Noryl GFN2								0.19-0.21	430
GE Noryl GFN3	gen. purp. grade, 108 Rockwell L; 30% glass fiber		Noryl GFN3								0.38-0.41	430
GE Noryl SE1	gen. purp. grade, flame retardant, 119 Rockwell R, electronics grade		Noryl SE1								0.27-0.31	430
GE Noryl SE1-GFN2	flame retardant, 106 Rockwell L; 20% glass fiber		Noryl SE1 GFN2								0.36-0.39	430
GE Noryl SE1-GFN3	flame retardant, 108 Rockwell L; 30% glass fiber		Noryl SE1 GFN3								0.36-0.42	430
GE Noryl SE100	gen. purp. grade, flame retardant, 115 Rockwell R		Noryl SE100								0.30-0.41	430
GE Noryl 731	gen. purp. grade, 119 Rockwell R	modified PPO	Noryl 731								0.26-.028	430
GE Noryl GFN2	gen. purp. grade, 106 Rockwell L; 20% glass fiber		Noryl GFN2								0.25-0.44	430
GE Noryl GFN3	gen. purp. grade, 108 Rockwell L; 30% glass fiber		Noryl GFN3								0.36-0.49	430
GE Noryl HS2000	121 Rockwell R; 10% mineral filler		Noryl HS2000								0.19-0.40	430
GE Noryl SE1	gen. purp. grade, flame retardant, 119 Rockwell R, electronics grade		Noryl SE1								0.20-0.33	430
GE Noryl SE1-GFN2	flame retardant, 106 Rockwell L; 20% glass fiber		Noryl SE1 GFN2								0.25-0.33	430

Material		Mating Surface		Test Conditions						Coefficient Of Friction		Source
Supplier / Grade	Material Note	Mating Surface Material	Mating Surface Note	Test Method	Test Conditions Note	Temp. (°C)	Pressure (psi) (MPa)	Speed (fpm) (m/min)	PV (psi x fpm) (MPa x m/min)	Static	Dynamic	
MODIFIED PPE against mating surface												
GE Noryl SE1-GFN3	flame retardant, 108 Rockwell L; 30% glass fiber	**modified PPO**	Noryl SE1 GFN3								0.39-0.50	430
GE Noryl SE100	gen. purp. grade, flame retardant, 115 Rockwell R		Noryl SE100								0.36-0.49	430
Ferro Star-L PPO 25VD23	25% glass, 10% graphite, lubricated	**steel**		ASTM D3028						0.29		452
Ferro Star-L PPO 25VD23	25% glass, 10% graphite, lubricated			"	load: 12 N			39 (12)			0.16	452
Ferro Star-L PPO 25VD23	25% glass, 10% graphite, lubricated			"	load: 43 N			39 (12)			0.23	452
GE Noryl	standard grade						40 (0.28)	50 (15.2)	2000 (4.3)		0.39	449
GE Noryl 731	gen. purp. grade, 119 Rockwell R		stainless steel								0.30-0.36	430
GE Noryl GFN2	gen. purp. grade, 106 Rockwell L; 20% glass fiber		"								0.21-0.23	430
GE Noryl GFN3	gen. purp. grade, 108 Rockwell L; 30% glass fiber		"								0.26-0.30	430
GE Noryl SE1	gen. purp. grade, flame retardant, 119 Rockwell R, electronics grade		"								0.13-0.20	430
GE Noryl SE1-GFN2	flame retardant, 106 Rockwell L; 20% glass fiber		"								0.22-0.25	430
GE Noryl SE1-GFN3	30% glass fiber, flame retardant						40 (0.28)	50 (15.2)	2000 (4.3)		0.27	449
GE Noryl SE1-GFN3	flame retardant, 108 Rockwell L; 30% glass fiber		stainless steel								0.28-0.36	430
GE Noryl SE100	gen. purp. grade, flame retardant, 115 Rockwell R		"								0.21-0.27	430
LNP Z-1000	unmodified		carbon steel; surface finish: 12-16 µin; 18-20 Rockwell C	thrust washer		23	40 (0.28)	50 (15.2)	2000 (4.3)	0.32	0.39	453
LNP ZBL-4326	30% glass beads, 10% graphite		"	"	test conducted in water	23	40 (0.28)	50 (15.2)	2000 (4.3)	0.21	0.18	453
LNP ZBL-4326	30% glass beads, 10% graphite		"	"		23	40 (0.28)	50 (15.2)	2000 (4.3)	0.29	0.22	453
LNP ZF-1006	30% glass fiber		"	"		23	40 (0.28)	50 (15.2)	2000 (4.3)	0.26	0.27	453
LNP ZFL-4036	30% glass fiber, 15% PTFE		"	"		23	40 (0.28)	50 (15.2)	2000 (4.3)	0.2	0.22	453
LNP ZFL-4323	15% glass fiber, 10% graphite		"	"		23	40 (0.28)	50 (15.2)	2000 (4.3)	0.09	0.11	453
LNP ZL-4030	15% PTFE		"	"		23	40 (0.28)	50 (15.2)	2000 (4.3)	0.1	0.16	453
LNP ZML-4334	20% mineral, 15% graphite		"	"		23	40 (0.28)	50 (15.2)	2000 (4.3)	0.27	0.21	453
Mitsubishi Iupiace AHF 6010	low COF		carbon	"	test duration: 0.5 hours		73 (0.5)	59 (18)	4307 (9.1)		0.17	362
Mitsubishi Iupiace AHF 6010	low COF		"	"	test duration: 1 hour		73 (0.5)	59 (18)	4307 (9.1)		0.18	362

Material		Mating Surface		Test Conditions						Coefficient Of Friction		Source
Supplier / Grade	Material Note	Mating Surface Material	Mating Surface Note	Test Method	Test Conditions Note	Temp. (°C)	Pressure (psi) (MPa)	Speed (fpm) (m/min)	PV (psi x fpm) (MPa x m/min)	Static	Dynamic	
MODIFIED PPE against mating surface												
Mitsubishi Iupiace AHF 6020	low COF	**steel**	carbon	thrust washer	test duration: 0.5 hours		73 (0.5)	59 (18)	4307 (9.1)		0.15	362
Mitsubishi Iupiace AHF 6020	low COF		"	"	test duration: 1 hour		73 (0.5)	59 (18)	4307 (9.1)		0.16	362
Mitsubishi Iupiace AVF 6010	flame ret., low COF		"	"	test duration: 0.5 hours		73 (0.5)	59 (18)	4307 (9.1)		0.17	362
Mitsubishi Iupiace AVF 6010	flame ret., low COF		"	"	test duration: 1 hour		73 (0.5)	59 (18)	4307 (9.1)		0.18	362
Mitsubishi Iupiace AVF 6020	flame ret., low COF		"	"	test duration: 0.5 hours		73 (0.5)	59 (18)	4307 (9.1)		0.13	362
Mitsubishi Iupiace AVF 6020	flame ret., low COF		"	"	test duration: 1 hour		73 (0.5)	59 (18)	4307 (9.1)		0.14	362
Mitsubishi Iupiace GHF 3010	10% glass fiber		"	"	test duration: 0.5 hours		73 (0.5)	59 (18)	4307 (9.1)		0.23	362
Mitsubishi Iupiace GHF 3010	10% glass fiber		"	"	test duration: 1 hour		73 (0.5)	59 (18)	4307 (9.1)		0.24	362
Mitsubishi Iupiace GHF 3015	20% glass fiber		"	"	test duration: 0.5 hours		73 (0.5)	59 (18)	4307 (9.1)		0.25	362
Mitsubishi Iupiace GHF 3015	20% glass fiber		"	"	test duration: 1 hour		73 (0.5)	59 (18)	4307 (9.1)		0.28	362
Mitsubishi Iupiace GHF 3020	30% glass fiber		"	"	test duration: 0.5 hours		73 (0.5)	59 (18)	4307 (9.1)		0.25	362
Mitsubishi Iupiace GHF 3020	30% glass fiber		"	"	test duration: 1 hour		73 (0.5)	59 (18)	4307 (9.1)		0.27	362
Unspecified grade				ASTM D1894							0.35	78
POLYPHENYLENE SULFIDE against mating surface												
LNP OFL-4036	30% glass fiber, 15% PTFE	**brass**	70/30 brass; surface finish: 8-16 μin	thrust washer		23	40 (0.28)	50 (15.2)	2000 (4.3)	0.1	0.14	453
LNP OFL-4036	30% glass fiber, 15% PTFE		70/30 brass; surface finish: 50-70 μin	"		23	40 (0.28)	50 (15.2)	2000 (4.3)	0.11	0.14	453
LNP OFL-4036	30% glass fiber, 15% PTFE	**polycarbonate**	DL-4020 (10% PTFE)	"		23	40 (0.28)	50 (15.2)	2000 (4.3)	0.04	0.06	453
Bayer Tedur KU1-9500	no longer available	**polyphenylene sulfide**	Tedur							0.33	0.33	428
Bayer Tedur KU1-9510-1	no longer available; 40% glass fiber		"							0.28	0.26	428
Bayer Tedur KU1-9520-1	no longer available; 50% mineral/ glass		"							0.45	0.44	428
Bayer Tedur KU1-9523	no longer available; 60% mineral/ glass		"							0.46	0.43	428
LNP OFL-4036	30% glass fiber, 15% PTFE		OFL-4326 (30% glass fiber, 10% graphite powder)	thrust washer		23	40 (0.28)	50 (15.2)	2000 (4.3)	0.16	0.13	453
LNP OFL-4036	30% glass fiber, 15% PTFE		OFL-4036 (30% glass fiber, 15% PTFE)	"		23	40 (0.28)	50 (15.2)	2000 (4.3)	0.08	0.07	453
LNP OFL-4326	30% glass fiber, 10% graphite powder		"	"		23	40 (0.28)	50 (15.2)	2000 (4.3)	0.13	0.12	453

Material		Mating Surface		Test Conditions						Coefficient Of Friction		Source
Supplier / Grade	Material Note	Mating Surface Material	Mating Surface Note	Test Method	Test Conditions Note	Temp. (°C)	Pressure (psi) (MPa)	Speed (fpm) (m/min)	PV (psi x fpm) (MPa x m/min)	Static	Dynamic	
POLYPHENYLENE SULFIDE against mating surface →												
Akzo J-1300/30/TF/15	30% glass fiber, 15% PTFE	steel	carbon steel; surface finish: 16 µin; 18-22 Rockwell C	thrust washer	apparatus: Faville-LeVally	23	40 (0.28)	50 (15.2)	2000 (4.3)	0.15	0.17	458
Akzo J-1300/CF/20/MS/10/TF/15	20% carbon fiber, 10% MoS_2, 15% PTFE		"	"	"	23	40 (0.28)	50 (15.2)	2000 (4.3)	0.53		458
Akzo J-1300/CF/30/TF/15	30% carbon fiber, 15% PTFE		"	"	"	23	40 (0.28)	50 (15.2)	2000 (4.3)	0.15	0.16	458
Bayer Tedur KU1-9500	no longer available		lapped							0.18	0.14	428
Bayer Tedur KU1-9510-1	no longer available; 40% glass fiber		"							0.18	0.16	428
Bayer Tedur KU1-9520-1	no longer available; 50% mineral/ glass		"							0.23	0.19	428
Bayer Tedur KU1-9523	no longer available; 60% mineral/ glass		"							0.25	0.22	428
Ensinger Ensifide	standard grade						40 (0.28)	50 (15.2)	2000 (4.3)		0.24	449
Ferro Star-C PPS 20Y	20% carbon fiber			ASTM D3028						0.24		452
Ferro Star-C PPS 30Y	30% carbon fiber			"						0.22		452
GE Supec W331	116 Rockwell R; 30% glass fiber, 15% PTFE			ASTM D3702			40 (0.28)	50 (15.2)	2000 (4.3)	0.2	0.37	429
Hoechst AG Fortron 1140 A1	high heat grade, moderate flow, 100 Rockwell M; 40% glass fiber, 30% glass fiber									0.4		426
Hoechst AG Fortron 1140-AO	40% glass fiber						40 (0.28)	50 (15.2)	2000 (4.3)		0.29	449
LNP Lubricomp O-BG	proprietary filler		carbon steel; surface finish: 12-16 µin; 18-20 Rockwell C	thrust washer		23	40 (0.28)	50 (15.2)	2000 (4.3)	0.07	0.18	453
LNP Lubricomp O-BG	proprietary filler		cold rolled steel; surface finish: 12-16 µin; 22 Rockwell C	"		204	40 (0.28)	50 (15.2)	2000 (4.3)	0.09	0.2	453
LNP Lubricomp O-BG	proprietary filler		"	"		149	40 (0.28)	50 (15.2)	2000 (4.3)	0.11	0.17	453
LNP O-1000	unmodified		carbon steel; surface finish: 12-16 µin; 18-20 Rockwell C	"		23	40 (0.28)	50 (15.2)	2000 (4.3)	0.3	0.24	453
LNP OC-1006	30% carbon fiber		"	"		23	40 (0.28)	50 (15.2)	2000 (4.3)	0.23	0.2	453
LNP OC-1006	30% carbon fiber		cold rolled steel; surface finish: 12-16 µin; 22 Rockwell C	"		93	40 (0.28)	50 (15.2)	2000 (4.3)	0.34	0.38	453
LNP OC-1006	30% carbon fiber		"	"		149	40 (0.28)	50 (15.2)	2000 (4.3)	0.34	0.38	453
LNP OC-1006	30% carbon fiber		"	"		204	40 (0.28)	50 (15.2)	2000 (4.3)	0.34	0.38	453
LNP OCL-4036	30% carbon fiber, 15% PTFE		carbon steel; surface finish: 12-16 µin; 18-20 Rockwell C	"		23	40 (0.28)	50 (15.2)	2000 (4.3)	0.13	0.15	453
LNP OCL-4036	30% carbon fiber, 15% PTFE		cold rolled steel; surface finish: 12-16 µin; 22 Rockwell C	"		93	40 (0.28)	50 (15.2)	2000 (4.3)	0.28	0.31	453

Material		Mating Surface		Test Conditions						Coefficient Of Friction		Source
Supplier / Grade	Material Note	Mating Surface Material	Mating Surface Note	Test Method	Test Conditions Note	Temp. (°C)	Pressure (psi) (MPa)	Speed (fpm) (m/min)	PV (psi x fpm) (MPa x m/min)	Static	Dynamic	
POLYPHENYLENE SULFIDE against mating surface												
LNP OCL-4036	30% carbon fiber, 15% PTFE	**steel**	cold rolled steel; surface finish: 12-16 µin; 22 Rockwell C	thrust washer		149	40 (0.28)	50 (15.2)	2000 (4.3)	0.3	0.32	453
LNP OCL-4036	30% carbon fiber, 15% PTFE		"	"		204	40 (0.28)	50 (15.2)	2000 (4.3)	0.3	0.32	453
LNP OF-1008	40% glass fiber		carbon steel; surface finish: 12-16 µin; 18-20 Rockwell C	"		23	40 (0.28)	50 (15.2)	2000 (4.3)	0.38	0.29	453
LNP OFL-4036	30% glass fiber, 15% PTFE		AISI 440 stainless steel; surface finish: 8-16 µin	"		23	40 (0.28)	50 (15.2)	2000 (4.3)	0.13	0.15	453
LNP OFL-4036	30% glass fiber, 15% PTFE		AISI 440 stainless steel; surface finish: 50-70 µin	"		23	40 (0.28)	50 (15.2)	2000 (4.3)	0.15	0.16	453
LNP OFL-4036	30% glass fiber, 15% PTFE		AISI 1141; surface finish: 12-16 µin	"		23	40 (0.28)	50 (15.2)	2000 (4.3)	0.15	0.17	453
LNP OFL-4036	30% glass fiber, 15% PTFE		carbon steel; surface finish: 12-16 µin; 18-20 Rockwell C	"		23	40 (0.28)	50 (15.2)	2000 (4.3)	0.15	0.17	453
LNP OFL-4036	30% glass fiber, 15% PTFE		AISI 1141; surface finish: 50-70 µin	"		23	40 (0.28)	50 (15.2)	2000 (4.3)	0.16	0.17	453
LNP OFL-4036	30% glass fiber, 15% PTFE		AISI 1141; surface finish: 8-12 µin	"		23	40 (0.28)	50 (15.2)	2000 (4.3)	0.2	0.18	453
LNP OFL-4036	30% glass fiber, 15% PTFE		cold rolled steel; surface finish: 12-16 µin; 22 Rockwell C	"		93	40 (0.28)	50 (15.2)	2000 (4.3)	0.33	0.38	453
LNP OFL-4036	30% glass fiber, 15% PTFE		"	"		149	40 (0.28)	50 (15.2)	2000 (4.3)	0.33	0.38	453
LNP OFL-4036	30% glass fiber, 15% PTFE		"	"		204	40 (0.28)	50 (15.2)	2000 (4.3)	0.34	0.38	453
LNP OFL-4036	30% glass fiber, 15% PTFE		AISI 304 stainless steel; surface finish: 50-70 µin	"		23	40 (0.28)	50 (15.2)	2000 (4.3)	0.1	0.15	453
LNP OFL-4036	30% glass fiber, 15% PTFE		AISI 304 stainless steel; surface finish: 8-16 µin	"		23	40 (0.28)	50 (15.2)	2000 (4.3)	0.1	0.17	453
LNP OFL-4536	30% glass fiber, 13% PTFE, 2% silicone		carbon steel; surface finish: 12-16 µin; 18-20 Rockwell C	"		23	40 (0.28)	50 (15.2)	2000 (4.3)	0.2	0.22	453
LNP OL-4040	20% PTFE		"	"		23	40 (0.28)	50 (15.2)	2000 (4.3)	0.08	0.1	453
Phillips Ryton R4	123 Rockwell R; 40% glass fiber			flat block against steel ring (Alpha Molykote LFW-1 apparatus)	load: 6.8 kg					0.5		102
Phillips Ryton R4	123 Rockwell R; 40% glass fiber			"	"			55 (16.8)			0.53	102
Phillips Ryton R4	123 Rockwell R; 40% glass fiber			"	"			29 (8.8)			0.55	102
RTP 1300			C1018 steel; surface finish: 14-17 µin; 15-25 Rockwell C	thrust washer	apparatus: Falex Model No. 6	23	40 (0.28)	50 (15.2)	2000 (4.3)	0.28	0.24	457
RTP 1300 AR 15	15% aramid fiber		"	"	"	23	40 (0.28)	50 (15.2)	2000 (4.3)	0.12	0.15	457
RTP 1300 AR 15 TFE 15	15% PTFE, 15% aramid fiber		"	"	"	23	40 (0.28)	50 (15.2)	2000 (4.3)	0.11	0.14	457
RTP 1302 TFE 10	10% PTFE, 15% glass fiber		"	"	"	23	40 (0.28)	50 (15.2)	2000 (4.3)	0.13	0.15	457

Material		Mating Surface		Test Conditions						Coefficient Of Friction		Source
Supplier / Grade	Material Note	Mating Surface Material	Mating Surface Note	Test Method	Test Conditions Note	Temp. (°C)	Pressure (psi) (MPa)	Speed (fpm) (m/min)	PV (psi x fpm) (MPa x m/min)	Static	Dynamic	
POLYPHENYLENE SULFIDE against mating surface												
RTP 1303 TFE 20	20% PTFE, 20% glass fiber	**steel**	C1018 steel; surface finish: 14-17 µin; 15-25 Rockwell C	thrust washer	apparatus: Falex Model No. 6	23	40 (0.28)	50 (15.2)	2000 (4.3)	0.12	0.12	457
RTP 1305	30% glass fiber		"	"	"	23	40 (0.28)	50 (15.2)	2000 (4.3)	0.35	0.28	457
RTP 1307 TFE 10	10% PTFE, 40% glass fiber		"	"	"	23	40 (0.28)	50 (15.2)	2000 (4.3)	0.19	0.2	457
RTP 1378	15% PTFE, 30% glass fiber		"	"	"	23	40 (0.28)	50 (15.2)	2000 (4.3)	0.13	0.14	457
RTP 1385 TFE 15	15% PTFE, 30% carbon fiber		"	"	"	23	40 (0.28)	50 (15.2)	2000 (4.3)	0.18	0.18	457
RTP 1387 TFE 10	10% PTFE, 40% carbon fiber		"	"	"	23	40 (0.28)	50 (15.2)	2000 (4.3)	0.19	0.2	457
Thermofil T-20NF-0100	20% graphite fiber			ASTM D1894		23				0.25	0.23	459
Thermofil T-30FG-0214	30% glass fiber, PTFE lubricated			"		23				0.14	0.15	459
Thermofil T-30NF-0214	30% graphite fiber, PTFE lubricated			"		23				0.16	0.18	459
Thermofil T-40FG-0100	40% glass fiber			"		23				0.35	0.25	459
Thermofil T-40NF-0100	40% graphite fiber			"		23				0.15	0.13	459
Unspecified grade	unmodified		carbon steel; surface finish: 16 µin; 18-22 Rockwell C	thrust washer	apparatus: Faville-LeVally	23	40 (0.28)	50 (15.2)	2000 (4.3)	0.22	0.27	458
Unspecified grade	20% aramid fiber			"	ASTM D3702	23	250 (1.72)	10 (3)	2500 (5.3)		0.349	454
Unspecified grade	40% glass fiber			"	"	23	250 (1.72)	10 (3)	2500 (5.3)		0.425	454

Supplier / Grade	Material Note	Mating Surface Material	Mating Surface Note	Test Method	Test Conditions Note	Temp. (°C)	Pressure (psi) (MPa)	Speed (fpm) (m/min)	PV (psi x fpm) (MPa x m/min)	Static	Dynamic	Source
POLYSULFONE against mating surface												
Amoco	transparent, amber tint	**polysulfone**								0.67		15
Amoco	transparent, amber tint	**steel**					191 (1.32)			0.43		15
Amoco	transparent, amber tint						86 (0.59)			0.45		15
Amoco	transparent, amber tint						191 (1.32)	900 (274)	171900 (361)		0.2	15
Amoco	transparent, amber tint						191 (1.32)	700 (213)	133700 (281)		0.23	15
Amoco	transparent, amber tint						86 (0.59)	900 (274)	77400 (163)		0.23	15
Amoco	transparent, amber tint						86 (0.59)	700 (213)	60200 (126)		0.27	15
Amoco	transparent, amber tint						191 (1.32)	500 (152)	95500 (201)		0.29	15
Amoco	transparent, amber tint						86 (0.59)	500 (152)	43000 (90.4)		0.35	15

Material		Mating Surface		Test Conditions						Coefficient Of Friction		Source
Supplier / Grade	Material Note	Mating Surface Material	Mating Surface Note	Test Method	Test Conditions Note	Temp. (°C)	Pressure (psi) (MPa)	Speed (fpm) (m/min)	PV (psi x fpm) (MPa x m/min)	Static	Dynamic	
POLYSULFONE against mating surface												
Amoco	transparent, amber tint	**steel**					191 (1.32)	300 (91.4)	57300 (120)		0.4	15
Amoco	transparent, amber tint						86 (0.59)	300 (91.4)	25800 (54.2)		0.45	15
Amoco	transparent, amber tint						191 (1.32)	200 (61)	38200 (80.3)		0.465	15
Amoco	transparent, amber tint						191 (1.32)	100 (30.5)	19100 (40.1)		0.48	15
Amoco	transparent, amber tint						191 (1.32)	150 (45.7)	28650 (60.2)		0.48	15
Amoco	transparent, amber tint						86 (0.59)	200 (61)	17200 (36.2)		0.485	15
Amoco	transparent, amber tint						86 (0.59)	100 (30.5)	8600 (18.1)		0.5	15
Amoco	transparent, amber tint						86 (0.59)	150 (45.7)	12900 (27.1)		0.5	15
BASF AG Ultrason S 2010	moderate flow, gen. purp. grade		Cr 6/800/HV; surface finish: 0.15 µm	peg-and-disc apparatus			145 (1)	98 (30)	14210 (29.9)		0.55	28
BASF AG Ultrason S 2010	moderate flow, gen. purp. grade		Cr 6/800/HV; surface finish: 2.5 µm	"			145 (1)	98 (30)	14210 (29.9)		0.6	28
BASF AG Ultrason S 2010G4	moderate flow; 20% glass fiber		Cr 6/800/HV; surface finish: 0.15 µm	"			145 (1)	98 (30)	14210 (29.9)		0.49	28
BASF AG Ultrason S 2010G4	moderate flow; 20% glass fiber		Cr 6/800/HV; surface finish: 2.5 µm	"			145 (1)	98 (30)	14210 (29.9)		0.55	28
BASF AG Ultrason S 2010G4	moderate flow; 20% glass fiber		"	"			145 (1)	98 (30)	14210 (29.9)		0.56	28
BASF AG Ultrason S 2010G4	moderate flow; 20% glass fiber		Cr 6/800/HV; surface finish: 0.15 µm	"			145 (1)	98 (30)	14210 (29.9)		0.56	28
Ensinger Ensifone	standard grade						40 (0.28)	50 (15.2)	2000 (4.3)		0.37	449
LNP G-1000	unmodified		carbon steel; surface finish: 12-16 µin; 18-20 Rockwell C	thrust washer		23	40 (0.28)	50 (15.2)	2000 (4.3)	0.29	0.37	453
LNP GC-1006	30% carbon fiber		"	"		23	40 (0.28)	50 (15.2)	2000 (4.3)	0.17	0.14	453
LNP GF-1006	30% glass fiber		"	"		23	40 (0.28)	50 (15.2)	2000 (4.3)	0.24	0.22	453
LNP GFL 4022	10% glass fiber, 10% PTFE		"	"		23	40 (0.28)	50 (15.2)	2000 (4.3)	0.15	0.21	453
LNP GFL-4036	30% glass fiber, 15% PTFE		"	"		23	40 (0.28)	50 (15.2)	2000 (4.3)	0.16	0.19	453
LNP GL-4030	15% PTFE		"	"		23	40 (0.28)	50 (15.2)	2000 (4.3)	0.09	0.14	453
LNP GL-4520	8% PTFE, 2% silicone		"	"		23	40 (0.28)	50 (15.2)	2000 (4.3)	0.17	0.15	453
LNP GML-4334	20% mineral, 15% graphite		"	"	test conducted in water	23	40 (0.28)	50 (15.2)	2000 (4.3)	0.14	0.1	453
LNP GML-4334	20% mineral, 15% graphite		"	"		23	40 (0.28)	50 (15.2)	2000 (4.3)	0.18	0.18	453

Material		Mating Surface		Test Conditions						Coefficient Of Friction		Source
Supplier / Grade	Material Note	Mating Surface Material	Mating Surface Note	Test Method	Test Conditions Note	Temp. (°C)	Pressure (psi) (MPa)	Speed (fpm) (m/min)	PV (psi x fpm) (MPa x m/min)	Static	Dynamic	
POLYSULFONE against mating surface												
RTP 900 TFE 15	15% PTFE	**steel**	C1018 steel; surface finish: 14-17 µin; 15-25 Rockwell C	thrust washer	apparatus: Falex Model No. 6	23	40 (0.28)	50 (15.2)	2000 (4.3)	0.1	0.14	457
RTP 900 TFE 20	20% PTFE		“	“	“	23	40 (0.28)	50 (15.2)	2000 (4.3)	0.08	0.12	457
RTP 905 TFE 15	15% PTFE, 30% glass fiber		“	“	“	23	40 (0.28)	50 (15.2)	2000 (4.3)	0.16	0.2	457
Thermofil S-30FG-0214	30% glass fiber, PTFE lubricated			ASTM D1894		23				0.24	0.22	459
Thermofil S-30NF-0100	30% graphite fiber			“		23				0.16	0.13	459
POLYETHERSULFONE against mating surface												
BASF AG Ultrason E 2010	moderate flow, gen. purp. grade	**steel**	Cr 6/800/HV; surface finish: 0.15 µm	peg-and-disc apparatus			145 (1)	98 (30)	14210 (29.9)		0.63	28
BASF AG Ultrason E 2010	moderate flow, gen. purp. grade		Cr 6/800/HV; surface finish: 2.5 µm	“			145 (1)	98 (30)	14210 (29.9)		0.68	28
BASF AG Ultrason E 2010G4	moderate flow; 20% glass fiber		“	“			145 (1)	98 (30)	14210 (29.9)		0.42	28
BASF AG Ultrason E 2010G4	moderate flow; 20% glass fiber		“	“			145 (1)	98 (30)	14210 (29.9)		0.48	28
BASF AG Ultrason E 2010G4	moderate flow; 20% glass fiber		Cr 6/800/HV; surface finish: 0.15 µm	“			145 (1)	98 (30)	14210 (29.9)		0.51	28
BASF AG Ultrason E 2010G4	moderate flow; 20% glass fiber		“	“			145 (1)	98 (30)	14210 (29.9)		0.55	28
Ensinger	standard grade						40 (0.28)	50 (15.2)	2000 (4.3)		0.32	449
LNP JFL-4036	30% glass fiber, 15% PTFE		carbon steel; surface finish: 12-16 µin; 18-20 Rockwell C	thrust washer		23	40 (0.28)	50 (15.2)	2000 (4.3)	0.16	0.2	453
LNP JFL-4036	30% glass fiber, 15% PTFE		cold rolled steel; surface finish: 12-16 µin; 22 Rockwell C	“		93	40 (0.28)	50 (15.2)	2000 (4.3)	0.27	0.3	453
LNP JL-4030	15% PTFE		carbon steel; surface finish: 12-16 µin; 18-20 Rockwell C	“		23	40 (0.28)	50 (15.2)	2000 (4.3)	0.09	0.12	453
RTP 1405 TFE 15	15% PTFE, 30% glass fiber		C1018 steel; surface finish: 14-17 µin; 15-25 Rockwell C	“	apparatus: Falex Model No. 6	23	40 (0.28)	50 (15.2)	2000 (4.3)	0.14	0.16	457
RTP 1485 TFE 15	15% PTFE, 30% carbon fiber		“	“	“	23	40 (0.28)	50 (15.2)	2000 (4.3)	0.12	0.14	457
Thermofil K-20FG-0100	20% glass fiber			ASTM D1894		23				0.24	0.27	459
Thermofil K-30FG-0100	30% glass fiber			“		23				0.22	0.25	459
Thermofil K-30FG-0214	30% glass fiber, PTFE lubricated			“		23				0.15	0.19	459
Thermofil K-30NF-0100	30% graphite fiber			“		23				0.17	0.16	459
Victrex PES 4101GL30	30% glass fiber		carbon steel; surface finish: 12-16 µin; 18-20 Rockwell C	thrust washer		23	40 (0.28)	50 (15.2)	2000 (4.3)	0.23	0.21	453

Material		Mating Surface		Test Conditions						Coefficient Of Friction		Source
Supplier / Grade	Material Note	Mating Surface Material	Mating Surface Note	Test Method	Test Conditions Note	Temp. (°C)	Pressure (psi) (MPa)	Speed (fpm) (m/min)	PV (psi x fpm) (MPa x m/min)	Static	Dynamic	
POLYETHERSULFONE against mating surface												
Victrex PES 4800G	unmodified	**steel**	carbon steel; surface finish: 12-16 µin; 18-20 Rockwell C	thrust washer		23	40 (0.28)	50 (15.2)	2000 (4.3)	0.27	0.32	453
Victrex PES-PDX-86174	15% PTFE, 30% carbon fiber		cold rolled steel; surface finish: 12-16 µin; 22 Rockwell C	"		93	40 (0.28)	50 (15.2)	2000 (4.3)	0.24	0.29	453
POLYURETHANE, RIGID against mating surface												
LNP T-1000	unmodified	**steel**	carbon steel; surface finish: 12-16 µin; 18-20 Rockwell C	thrust washer		23	40 (0.28)	50 (15.2)	2000 (4.3)	0.32	0.37	453
LNP TF-1006	30% glass fiber		"	"		23	40 (0.28)	50 (15.2)	2000 (4.3)	0.3	0.34	453
LNP TFL-4036	30% glass fiber, 15% PTFE		"	"		23	40 (0.28)	50 (15.2)	2000 (4.3)	0.2	0.25	453
LNP TFL-4536	30% glass fiber, 13% PTFE, 2% silicone		"	"		23	40 (0.28)	50 (15.2)	2000 (4.3)	0.18	0.24	453
LNP TL-4030	15% PTFE		"	"		23	40 (0.28)	50 (15.2)	2000 (4.3)	0.27	0.32	453
LNP TL-4410	2% silicone		"	"		23	40 (0.28)	50 (15.2)	2000 (4.3)	0.25	0.31	453
ABS against mating surface												
Dow	floppy disks	**ABS**	Magnum ABS MFD grade							0.36	0.24	341
LNP AFL-4044	20% glass fiber, 20% PTFE		A-1000; unmodified	thrust washer		23	40 (0.28)	50 (15.2)	2000 (4.3)	0.08	0.1	453
LNP AFL 4044	20% glass fiber, 20% PTFE	**polycarbonate**	D-1000; unmodified	"		23	40 (0.28)	50 (15.2)	2000 (4.3)	0.1	0.12	453
Ensinger Ensidur	standard grade	**steel**					40 (0.28)	50 (15.2)	2000 (4.3)		0.35	449
LNP A-1000	unmodified		carbon steel; surface finish: 12-16 µin; 18-20 Rockwell C	thrust washer		23	40 (0.28)	50 (15.2)	2000 (4.3)	0.3	0.35	453
LNP AFL-4036	30% glass fiber, 15% PTFE		"	"		23	40 (0.28)	50 (15.2)	2000 (4.3)	0.16	0.2	453
LNP AL-4030	15% PTFE		"	"		23	40 (0.28)	50 (15.2)	2000 (4.3)	0.13	0.16	453
LNP AL-4410	2% silicone		"	"		23	40 (0.28)	50 (15.2)	2000 (4.3)	0.11	0.14	453
RTP 600			C1018 steel; surface finish: 14-17 µin; 15-25 Rockwell C	"	apparatus: Falex Model No. 6	23	40 (0.28)	50 (15.2)	2000 (4.3)	0.3	0.35	457
RTP 600 SI 2	2% silicone		"	"	"	23	40 (0.28)	50 (15.2)	2000 (4.3)	0.11	0.15	457
Thermofil G-20NF-0100	20% graphite fiber			ASTM D1894		23				0.1	0.13	459
Thermofil G-30FG-0100	30% glass fiber			"		23				0.23	0.32	459

Material		Mating Surface		Test Conditions						Coefficient Of Friction		Source
Supplier / Grade	Material Note	Mating Surface Material	Mating Surface Note	Test Method	Test Conditions Note	Temp. (°C)	Pressure (psi) (MPa)	Speed (fpm) (m/min)	PV (psi x fpm) (MPa x m/min)	Static	Dynamic	
POLYSTYRENE against mating surface												
Ferro Star-L PS D28	28% graphite, lubricated	**steel**		ASTM D3028						0.24		452
Ferro Star-L PS D28	28% graphite, lubricated			"	load: 12 N			39 (12)			0.28	452
Ferro Star-L PS D28	28% graphite, lubricated			"	load: 43 N			39 (12)			0.33	452
LNP C-1000	unmodified		carbon steel; surface finish: 12-16 µin; 18-20 Rockwell C	thrust washer		23	40 (0.28)	50 (15.2)	2000 (4.3)	0.28	0.32	453
LNP CL-4030	15% PTFE		"	"		23	40 (0.28)	50 (15.2)	2000 (4.3)	0.12	0.14	453
LNP CL-4410	2% silicone		"	"		23	40 (0.28)	50 (15.2)	2000 (4.3)	0.06	0.08	453
Unspecified grade				ASTM D1894							0.4	78
Thermofil A-20FG-0100	20% glass fiber			"		23				0.21	0.27	459
POLYSTYRENE, IPS against mating surface												
Dow Styron XL-8035MFD	floppy disks; 45 Rockwell M, 85 Rockwell L	**impact polystyrene**	Styron XL-8035 MFD							0.27	0.21	341
Dow	electronics grade		electronics grade							0.54	0.46	341
SAN against mating surface												
LNP B-1000	unmodified	**steel**	carbon steel; surface finish: 12-16 µin; 18-20 Rockwell C	thrust washer		23	40 (0.28)	50 (15.2)	2000 (4.3)	0.28	0.33	453
LNP BBL-4326	30% glass beads, 10% graphite		"	"		23	40 (0.28)	50 (15.2)	2000 (4.3)	0.17	0.21	453
LNP BFL-4036	30% glass fiber, 15% PTFE		"	"		23	40 (0.28)	50 (15.2)	2000 (4.3)	0.13	0.18	453
LNP BFL-4536	30% glass fiber, 13% PTFE, 2% silicone		"	"		23	40 (0.28)	50 (15.2)	2000 (4.3)	0.12	0.18	453
LNP BL-4030	15% PTFE		"	"		23	40 (0.28)	50 (15.2)	2000 (4.3)	0.11	0.14	453
LNP BL-4410	2% silicone		"	"		23	40 (0.28)	50 (15.2)	2000 (4.3)	0.11	0.13	453
LNP BL-4530	13% PTFE, 2% silicone		"	"		23	40 (0.28)	50 (15.2)	2000 (4.3)	0.1	0.13	453
Unspecified grade				ASTM D1894							0.5	78
POLYVINYL CHLORIDE against mating surface												
Geon Co.		**polyvinyl chloride**		ASTM D1894						0.25	0.17	189
Geon Co.		**steel**		"						0.21	0.25	189

Material		Mating Surface		Test Conditions						Coefficient Of Friction		Source
Supplier / Grade	Material Note	Mating Surface Material	Mating Surface Note	Test Method	Test Conditions Note	Temp. (°C)	Pressure (psi) (MPa)	Speed (fpm) (m/min)	PV (psi x fpm) (MPa x m/min)	Static	Dynamic	
POLYOLEFIN ALLOY against mating surface												
Hoechst Cel. Hostalloy 731	0.95 g/cm³ density; 68 Shore D	**steel**	polished CR steel, oil	ASTM D1894							0.05-0.08	330
Hoechst Cel. Hostalloy 731	0.95 g/cm³ density; 68 Shore D		polished CR steel, water	"							0.05-0.1	330
Hoechst Cel. Hostalloy 731	0.95 g/cm³ density; 68 Shore D		polished CR steel, dry	"							0.10-0.22	330
TPE, POLYESTER against mating surface												
DuPont Hytrel 4056	40 Shore D	**steel**		ASTM D1894	apparatus: moving sled					0.32	0.29	347
DuPont Hytrel 5556	55 Shore D			"	"					0.22	0.18	347
DuPont Hytrel 6346	63 Shore D			"	"					0.3	0.21	347
DuPont Hytrel 7246	72 Shore D			"	"					0.23	0.16	347
LNP Y-1000	unmodified		carbon steel; surface finish: 12-16 µin; 18-20 Rockwell C	thrust washer		23	40 (0.28)	50 (15.2)	2000 (4.3)	0.27	0.59	453
LNP YF-1006	30% glass fiber		"	"		23	40 (0.28)	50 (15.2)	2000 (4.3)	0.25	0.4	453
LNP YFL-4536	30% glass fiber, 13% PTFE, 2% silicone		"	"		23	40 (0.28)	50 (15.2)	2000 (4.3)	0.19	0.2	453
LNP YL-4030	15% PTFE		"	"		23	40 (0.28)	50 (15.2)	2000 (4.3)	0.22	0.25	453
LNP YL-4410	2% silicone		"	"		23	40 (0.28)	50 (15.2)	2000 (4.3)	0.21	0.22	453
LNP YL-4530	13% PTFE, 2% silicone		"	"		23	40 (0.28)	50 (15.2)	2000 (4.3)	0.2	0.21	453
DuPont Hytrel 4056	40 Shore D			"	ASTM D3702						0.85	347
DuPont Hytrel 5556	55 Shore D			"	"						0.94	347
DuPont Hytrel 6346	63 Shore D			"	"						0.9	347
DuPont Hytrel 7246	72 Shore D			"	"						0.9	347
TPE, URETHANE (TPAU) against mating surface												
Bayer Texin 355D	55 Shore D	**aluminum**		ASTM D1894							0.4	427
Bayer Texin 480A	high heat grade, flame retardant, 123 Rockwell R, 83 Shore A, 100 Rockwell M; 30% glass fiber			"							0.61	427
Bayer Texin 591A	91 Shore A			"							0.55	427
Bayer Texin 355D	55 Shore D	**brass**		"							0.46	427

Material		Mating Surface		Test Conditions						Coefficient Of Friction		Source
Supplier / Grade	Material Note	Mating Surface Material	Mating Surface Note	Test Method	Test Conditions Note	Temp. (°C)	Pressure (psi) (MPa)	Speed (fpm) (m/min)	PV (psi x fpm) (MPa x m/min)	Static	Dynamic	
TPE, URETHANE (TPAU) against mating surface												
Bayer Texin 480A	high heat grade, flame retardant, 123 Rockwell R, 83 Shore A, 100 Rockwell M; 30% glass fiber	**brass**		ASTM D1894							0.7	427
Bayer Texin 591A	91 Shore A			"							0.64	427
Bayer Texin 355D	55 Shore D	**steel**		"							0.31	427
Bayer Texin 480A	high heat grade, flame retardant, 123 Rockwell R, 83 Shore A, 100 Rockwell M; 30% glass fiber			"							0.53	427
Bayer Texin 591A	91 Shore A			"							0.38	427
POLYIMIDE against mating surface												
DuPont Vespel SP-1	unfilled, 45 - 60 Rockwell E; machined spec.	**steel**	unlubricated		in air					0.35		345
DuPont Vespel SP-1	unfilled, 45 - 60 Rockwell E; machined spec.		"		"				25000 (52.5)		0.29	345
DuPont Vespel SP-1	unfilled; direct formed		"		"				25000 (52.5)		0.29	345
DuPont Vespel SP-1	unfilled		AISI 1018, surface finish: 16 RMS	thrust bearing	apparatus: LRI-1 Wear and Friction Tester		2000 (13.8)	12.5 (3.8)	25000 (52.6)		0.36	318
DuPont Vespel SP-21	lubricated		"	"	"		100 (0.69)	984 (300)	98400 (207)	0.26	0.12	318
DuPont Vespel SP-21	15% graphite		unlubricated	thrust bearing test						0.3		344
DuPont Vespel SP-21	25 - 45 Rockwell E; machined spec.; 15% graphite		"		in air					0.3		345
DuPont Vespel SP-21	15% graphite		carbon EN8, dry	pad on ring (Amsler wear tester)		200		100 (30.5)			0.08 (average of COF @ LPV and 50% of LPV load)	338
DuPont Vespel SP-21	15% graphite		"	"		20		100 (30.5)			0.08 (average of COF @ LPV and 50% of LPV load)	338
DuPont Vespel SP-21	15% graphite		unlubricated	thrust bearing test			100 (0.69)	1000 (305)	100000 (210)		0.12	344
DuPont Vespel SP-21	25 - 45 Rockwell E; machined spec.; 15% graphite		"		in air				100000 (210)		0.12	345
DuPont Vespel SP-21	direct formed; 15% graphite		"		"				100000 (210)		0.12	345

Material		Mating Surface		Test Conditions						Coefficient Of Friction		Source
Supplier / Grade	Material Note	Mating Surface Material	Mating Surface Note	Test Method	Test Conditions Note	Temp. (°C)	Pressure (psi) (MPa)	Speed (fpm) (m/min)	PV (psi x fpm) (MPa x m/min)	Static	Dynamic	
POLYIMIDE against mating surface												
DuPont Vespel SP-21	15% graphite	**steel**	carbon EN8, dry	pad on ring (Amsler wear tester)		200		600 (183)			0.21 (average of COF @ LPV and 50% of LPV load)	338
DuPont Vespel SP-21	15% graphite		unlubricated	thrust bearing test			49 (0.34)	500 (152)	24500 (51.5)		0.24	344
DuPont Vespel SP-21	25 - 45 Rockwell E; machined spec.; 15% graphite		"		in air				25000 (52.5)		0.24	345
DuPont Vespel SP-21	direct formed; 15% graphite		"		"				25000 (52.5)		0.24	345
DuPont Vespel SP-21	15% graphite		carbon EN8, dry	pad on ring (Amsler wear tester)		20		600 (183)			0.24 (average of COF @ LPV and 50% of LPV load)	338
DuPont Vespel SP-21	15% graphite		unlubricated	thrust bearing test			100 (0.69)	299 (91.2)	29900 (62.8)		0.28	344
DuPont Vespel SP-21	15% graphite		"	"			100 (0.69)	100.4 (30.6)	10040 (21.1)		0.3	344
DuPont Vespel SP-211	1 - 20 Rockwell E; machined spec.; 15% graphite, 10% PTFE		"		in air					0.2		345
DuPont Vespel SP-211	15% graphite, 10% PTFE modified		"	thrust bearing test						0.2		344
DuPont Vespel SP-211	1 - 20 Rockwell E; machined spec.; 15% graphite, 10% PTFE		"		in air				100000 (210)		0.08	345
DuPont Vespel SP-211	15% graphite, 10% PTFE modified		"	thrust bearing test			100 (0.69)	1000 (305)	100000 (210)		0.08	344
DuPont Vespel SP-211	direct formed; 15% graphite, 10% PTFE		"		in air				100000 (210)		0.08	345
DuPont Vespel SP-211	1 - 20 Rockwell E; machined spec.; 15% graphite, 10% PTFE		"		"				25000 (52.5)		0.12	345
DuPont Vespel SP-211	15% graphite, 10% PTFE modified		"	thrust bearing test			49 (0.34)	500 (152)	24500 (51.5)		0.12	344
DuPont Vespel SP-211	direct formed; 15% graphite, 10% PTFE		"		in air				25000 (52.5)		0.12	345
DuPont Vespel SP-211	15% graphite, 10% PTFE modified		"	thrust bearing test			100 (0.69)	299 (91.2)	29900 (62.8)		0.2	344
DuPont Vespel SP-211	15% graphite, 10% PTFE modified		"	"			100 (0.69)	100.4 (30.6)	10040 (21.1)		0.24	344
DuPont Vespel SP-22	40 wt% graphite		"	"						0.27		344

Material		Mating Surface		Test Conditions						Coefficient Of Friction		Source
Supplier / Grade	Material Note	Mating Surface Material	Mating Surface Note	Test Method	Test Conditions Note	Temp. (°C)	Pressure (psi) (MPa)	Speed (fpm) (m/min)	PV (psi x fpm) (MPa x m/min)	Static	Dynamic	
POLYIMIDE against mating surface												
DuPont Vespel SP-22	5 - 25 Rockwell E; machined spec.; 40 wt% graphite	**steel**	unlubricated		in air					0.27		345
DuPont Vespel SP-22	40 wt% graphite		"	thrust bearing test			100 (0.69)	1000 (305)	100000 (210)		0.09	344
DuPont Vespel SP-22	5 - 25 Rockwell E; machined spec.; 40 wt% graphite		"		in air				100000 (210)		0.09	345
DuPont Vespel SP-22	direct formed; 40 wt% graphite		"		"				100000 (210)		0.09	345
DuPont Vespel SP-22	40 wt% graphite		"	thrust bearing test			49 (0.34)	500 (152)	24500 (51.5)		0.2	344
DuPont Vespel SP-22	40 wt% graphite		"	"			100 (0.69)	299 (91.2)	29900 (62.8)		0.21	344
DuPont Vespel SP-22	40 wt% graphite		"	"			100 (0.69)	100.4 (30.6)	10040 (21.1)		0.24	344
DuPont Vespel SP-22	5 - 25 Rockwell E; machined spec.; 40 wt% graphite		"		in air				25000 (52.5)		0.3	345
DuPont Vespel SP-22	direct formed; 40 wt% graphite		"		"				25000 (52.5)		0.3	345
DuPont Vespel SP-3	40 - 55 Rockwell E; machined spec., machined spec.; 15% MoS_2		"		in vacuum					0.03		345
DuPont Vespel SP-3	40 - 55 Rockwell E; machined spec., machined spec.; 15% MoS_2		"		in air				100000 (210)		0.17	345
DuPont Vespel SP-3	40 - 55 Rockwell E; machined spec., machined spec.; 15% MoS_2		"		"				25000 (52.5)		0.25	345
Rhone Pou. Kinel 4503	inj. mold.; w/ TFE, w/ MoS_2; , aramid fiber reinforced		carbon steel; surface finish: 12-16 µin; unlubricated; 18-22 Rockwell C	thrust washer			51 (0.35)	50 (15.2)	2522 (5.3)		0.24	360
Rhone Pou. Kinel 4503	inj. mold.; w/ TFE, w/ MoS_2; , aramid fiber reinforced		"	"			127 (0.875)	200 (61)	25410 (53.4)		0.3	360
Rhone Pou. Kinel 4512	94 Rockwell M; inj. mold.; w/ graphite; , aramid fiber reinforced						57 (0.39)	118 (36)	6726 (14)		0.15	92
Rhone Pou. Kinel 5505	110 Rockwell M; 25% graphite						57 (0.39)	118 (36)	6726 (14)		0.10-0.25	92
Rhone Pou. Kinel 5508	100 Rockwell M; 40% graphite; compression						57 (0.39)	118 (36)	6726 (14)		0.10-0.2	92
Rhone Pou. Kinel 5517	110 Rockwell M; compression; w/ graphite, w/ MoS_2						57 (0.39)	118 (36)	6726 (14)		0.10-0.25	92
Rhone Pou. Kinel 5517	95 Rockwell M; sintering; w/ graphite, w/ MoS_2						57 (0.39)	118 (36)	6726 (14)		0.10-0.25	92
Rhone Pou. Kinel 5517	110 Rockwell M; compression; w/ graphite, w/ MoS_2		carbon steel; surface finish: 12-16 µin; unlubricated; 18-22 Rockwell C	thrust washer			51 (0.35)	50 (15.2)	2522 (5.3)		0.23	360
Rhone Pou. Kinel 5517	110 Rockwell M; compression; w/ graphite, w/ MoS_2		"	"			127 (0.875)	200 (61)	25410 (53.4)		0.29	360
Rhone Pou. Kinel 5518	115 Rockwell M; compression; w/ MoS_2, w/ TFE						57 (0.39)	118 (36)	6726 (14)		0.10-0.25	92
Rhone Pou. Kinel 5520	compression		carbon steel; surface finish: 12-16 µin; unlubricated; 18-22 Rockwell C	thrust washer			127 (0.875)	200 (61)	25410 (53.4)		0.17	360

Material		Mating Surface			Test Conditions					Coefficient Of Friction		Source
Supplier / Grade	Material Note	Mating Surface Material	Mating Surface Note	Test Method	Test Conditions Note	Temp. (°C)	Pressure (psi) (MPa)	Speed (fpm) (m/min)	PV (psi x fpm) (MPa x m/min)	Static	Dynamic	
POLYIMIDE against mating surface		→										
Rhone Pou. Kinel 5520	compression	**steel**	carbon steel; surface finish: 12-16 µin; unlubricated; 18-22 Rockwell C	thrust washer			51 (0.35)	50 (15.2)	2522 (5.3)		0.23	360
Ube Upimol R	gen. purp. grade			S45C (ANSI 1045)	load: 50 N			98 (30)			0.25	123
Ube Upimol S	high heat grade			"	"			98 (30)			0.31	123

Appendix II

Wear Factor (K) and Wear Rate

Material		Mating Surface		Test Conditions						K Factor 10^{-10} x (in³-min/ft-lb-hr)		Results Note	Source
Supplier / Grade	Material Note	Mating Surface Material	Mating Surface Note	Test Method	Test Conditions Note	Temp. (°C)	Pressure (psi) (MPa)	Speed (fpm) (m/min)	PV (psi x fpm) (MPa x m/min)	Material	Mating Surface		
ACETAL RESIN against mating surface		→											
LNP K-1000	unmodified	**acetal**	K-1000 (unmodified)	thrust washer		23	40 (0.28)	50 (15.2)	2000 (4.3)	14000	10200		453
LNP KL-4040	20% PTFE		KL-4040 (20% PTFE)	"		23	40 (0.28)	50 (15.2)	2000 (4.3)	40	35		453
LNP K-1000	unmodified	**nylon 66**	RFL-4036 (30% glass fiber, 15% PTFE)	"		23	40 (0.28)	50 (15.2)	2000 (4.3)	35000	115		453
LNP K-1000	unmodified		RF-1006 (30% glass fiber)	"		23	40 (0.28)	50 (15.2)	2000 (4.3)	5500	25		453
LNP K-1000	unmodified		R-1000 (unmodified)	"		23	40 (0.28)	50 (15.2)	2000 (4.3)	60	50		453
LNP KFL-4036	30% glass fiber, 15% PTFE		RL-4040 (20% PTFE)	"		23	40 (0.28)	50 (15.2)	2000 (4.3)	10	40		453
LNP KFL-4036	30% glass fiber, 15% PTFE		RFL-4036 (30% glass fiber, 15% PTFE)	"		23	40 (0.28)	50 (15.2)	2000 (4.3)	200	70		453
LNP KFL-4036	30% glass fiber, 15% PTFE		RCL-4036 (30% carbon fiber, 15% PTFE)	"		23	40 (0.28)	50 (15.2)	2000 (4.3)	30	80		453
LNP KL-4040	20% PTFE		RFL-4036 (30% glass fiber, 15% PTFE)	"		23	40 (0.28)	50 (15.2)	2000 (4.3)	20	12		453
LNP KL-4040	20% PTFE		R-1000 (unmodified)	"		23	40 (0.28)	50 (15.2)	2000 (4.3)	20	35		453
LNP KL-4040	20% PTFE		RL-4040 (20% PTFE)	"		23	40 (0.28)	50 (15.2)	2000 (4.3)	25	12		453
LNP KL-4040	20% PTFE		RF-1006 (30% glass fiber)	"		23	40 (0.28)	50 (15.2)	2000 (4.3)	25	20		453
LNP KL-4040	20% PTFE		RCL-4036 (30% carbon fiber, 15% PTFE)	"		23	40 (0.28)	50 (15.2)	2000 (4.3)	10	20		453
LNP KFL-4036	30% glass fiber, 15% PTFE	**polycarbonate**	DFL-4036 (30% glass fiber, 15% PTFE)	"		23	40 (0.28)	50 (15.2)	2000 (4.3)	1460	210		453
LNP KL-4040	20% PTFE		"	"		23	40 (0.28)	50 (15.2)	2000 (4.3)	11700	10		453
LNP KL-4040	20% PTFE		D-1000; unmodified	"		23	40 (0.28)	50 (15.2)	2000 (4.3)	380	26		453
Bay Resins Lubriplas POM-1100	unmodified resin	**steel**		"			40 (0.28)	50 (15.2)	2000 (4.3)	65			435
Bay Resins Lubriplas POM-1100G25	25% glass fiber			"			40 (0.28)	50 (15.2)	2000 (4.3)	200			435
Bay Resins Lubriplas POM-1100TF20	20% PTFE			"			630 (4.34)	20 (6.1)	12600 (26.5)	24			435
Bay Resins Lubriplas POM-1100TF20	20% PTFE			"			25 (0.17)	200 (61)	5000 (10.5)	92			435
Bay Resins Lubriplas POM-1100TF20	20% PTFE			"			250 (1.72)	20 (6.1)	5000 (10.5)	40			435
Bay Resins Lubriplas POM-1100TF20	20% PTFE			"			80 (0.55)	63 (19)	5000 (10.5)	42			435

Material		Mating Surface		Test Conditions						K Factor 10^{-10} x (in³-min/ft-lb-hr)		Results Note	Source
Supplier / Grade	Material Note	Mating Surface Material	Mating Surface Note	Test Method	Test Conditions Note	Temp. (°C)	Pressure (psi) (MPa)	Speed (fpm) (m/min)	PV (psi x fpm) (MPa x m/min)	Material	Mating Surface		
ACETAL RESIN against mating surface →													
Bay Resins Lubriplas POM-1100TF20	20% PTFE	steel		thrust washer			32 (0.22)	63 (19)	2000 (4.2)	43			435
Bay Resins Lubriplas POM-1100TF20	20% PTFE			"			200 (1.38)	63 (19)	12600 (26.5)	45			435
Bay Resins Lubriplas POM-1100TF20	20% PTFE			"			100 (0.69)	20 (6.1)	2000 (4.2)	53			435
Bay Resins Lubriplas POM-1100TF20	20% PTFE			"			10 (0.07)	200 (61)	2000 (4.3)	110			435
Bay Resins Lubriplas POM-1100TF20	20% PTFE			"			40 (0.28)	50 (15.2)	2000 (4.3)	14			435
Bay Resins Lubriplas POM-1100TF20	20% PTFE			"			63 (0.43)	200 (61)	12600 (26.2)	143			435
Bay Resins Lubriplas POM-1100UH20	20% UHMWPE			"			40 (0.28)	50 (15.2)	2000 (4.3)	10			435
DuPont Delrin	standard grade						40 (0.28)	50 (15.2)	2000 (4.3)	55			449
DuPont Delrin 500CL	low wear, chemical lubricant						42.2 (0.29)	118 (36)	5000 (10.5)	176			441
DuPont Delrin 570	20% glass fiber						40 (0.28)	50 (15.2)	2000 (4.3)	245			449
DuPont Delrin AF	low wear, PTFE modified						42.2 (0.29)	118 (36)	5000 (10.5)	65			441
Ensinger Ensital	standard grade						40 (0.28)	50 (15.2)	2000 (4.3)	65			449
LNP Fulton 404	20% PTFE		carbon steel; surface finish: 12-16 μin; 18-20 Rockwell C	thrust washer		23	40 (0.28)	50 (15.2)	2000 (4.3)	14			453
LNP Fulton 404D	20% PTFE		"	"		23	40 (0.28)	50 (15.2)	2000 (4.3)	13			453
LNP Fulton 441	2% silicone		"	"		23	40 (0.28)	50 (15.2)	2000 (4.3)	27			453
LNP Fulton 441D	2% silicone		"	"		23	40 (0.28)	50 (15.2)	2000 (4.3)	20			453
LNP K-1000	unmodified		"	"		23	40 (0.28)	50 (15.2)	2000 (4.3)	65			453
LNP KC-1004	20% carbon fiber		"	"		23	40 (0.28)	50 (15.2)	2000 (4.3)	40			453
LNP KF-1006	30% glass fiber		"	"		23	40 (0.28)	50 (15.2)	2000 (4.3)	245			453
LNP KFL-4036	30% glass fiber, 15% PTFE		"	"		23	40 (0.28)	50 (15.2)	2000 (4.3)	200			453
LNP KFL-4536	30% glass fiber, 13% PTFE, 2% silicone		"	"		23	40 (0.28)	50 (15.2)	2000 (4.3)	180			453
LNP KL-4010	5% PTFE		"	"		23	40 (0.28)	50 (15.2)	2000 (4.3)	40			453

Material		Mating Surface		Test Conditions						K Factor 10^{-10} x (in³-min/ft-lb-hr)		Results Note	Source
Supplier / Grade	Material Note	Mating Surface Material	Mating Surface Note	Test Method	Test Conditions Note	Temp. (°C)	Pressure (psi) (MPa)	Speed (fpm) (m/min)	PV (psi x fpm) (MPa x m/min)	Material	Mating Surface		
ACETAL RESIN against mating surface													
LNP KL-4020	10% PTFE	**steel**	carbon steel; surface finish: 12-16 µin; 18-20 Rockwell C	thrust washer		23	40 (0.28)	50 (15.2)	2000 (4.3)	30			453
LNP KL-4030	15% PTFE		"	"		23	40 (0.28)	50 (15.2)	2000 (4.3)	20			453
LNP KL-4050	25% PTFE		"	"		23	40 (0.28)	50 (15.2)	2000 (4.3)	21			453
LNP KL-4320	10% graphite		"	"		23	40 (0.28)	50 (15.2)	2000 (4.3)	60			453
LNP KL-4540	18% PTFE, 2% silicone		"	"		23	40 (0.28)	50 (15.2)	2000 (4.3)	9			453
LNP KL-4540D	18% PTFE, 2% silicone		"	"		23	40 (0.28)	50 (15.2)	2000 (4.3)	7			453
Polymer Acetron GP	general purpose, porosity free						42.2 (0.29)	118 (36)	5000 (10.5)	200			441
Polymer Acetron NS	w/ composite solid lubricants to improve wear						42.2 (0.29)	118 (36)	5000 (10.5)	23			441
RTP 800			C1018 steel; surface finish: 14-17 µin; 15-25 Rockwell C	thrust washer	apparatus: Falex Model No. 6	23	40 (0.28)	50 (15.2)	2000 (4.3)	65			457
RTP 800 AR 5 TFE 10	10% PTFE, 5% aramid fiber		"	"	"	23	40 (0.28)	50 (15.2)	2000 (4.3)	30			457
RTP 800 AR 5 TFE 10 SI 3	10% PTFE, 3% silicone, 5% aramid fiber		"	"	"	23	40 (0.28)	50 (15.2)	2000 (4.3)	23			457
RTP 800 SI 2	2% silicone		"	"	"	23	40 (0.28)	50 (15.2)	2000 (4.3)	30			457
RTP 800 TFE 10	10% PTFE		"	"	"	23	40 (0.28)	50 (15.2)	2000 (4.3)	35			457
RTP 800 TFE 15	15% PTFE		"	"	"	23	40 (0.28)	50 (15.2)	2000 (4.3)	25			457
RTP 800 TFE 18 SI 2	18% PTFE, 2% silicone		"	"	"	23	40 (0.28)	50 (15.2)	2000 (4.3)	10			457
RTP 800 TFE 20	20% PTFE		"	"	"	23	40 (0.28)	50 (15.2)	2000 (4.3)	15			457
RTP 805	30% glass fiber		"	"	"	23	40 (0.28)	50 (15.2)	2000 (4.3)	245			457
RTP 805 TFE 15	15% PTFE, 30% glass fiber		"	"	"	23	40 (0.28)	50 (15.2)	2000 (4.3)	210			457
Thermofil G-20FG-0214	20% glass fiber, PTFE lubricated			ASTM D1894		23				200			459
Thermofil G-9900-0215	20% PTFE			"		23				10			459
Thermofil G-9900-0257	PTFE/ oil lubricated			"		23				7			459
Unspecified grade	general purpose						42.2 (0.29)	118 (36)	5000 (10.5)	216			441
Unspecified grade			carbon EN8, dry	pad on ring (Amsler wear tester)		20		100 (30.5)				wear rate: 40 µm/hour; maximum load: 50 kg	338

Material		Mating Surface		Test Conditions						K Factor 10^{-10} x (in³-min/ft-lb-hr)		Results Note	Source
Supplier / Grade	Material Note	Mating Surface Material	Mating Surface Note	Test Method	Test Conditions Note	Temp. (°C)	Pressure (psi) (MPa)	Speed (fpm) (m/min)	PV (psi x fpm) (MPa x m/min)	Material	Mating Surface		
ACETAL RESIN against mating surface													
Unspecified grade	transverse to flow	**steel**	shaft; surface finish: 0.1 µm	sliding on a ground shaft	test duration: 60 hours			446 (136)				wear: 6 mm³	70
Unspecified grade	parallel to flow		"	"	"			446 (136)				wear: 8 mm³	70
ACETAL COPOLYMER against mating surface													
Akzo AC-80/TF/10	10% PTFE	**steel**	carbon steel; surface finish: 16 µin; 18-22 Rockwell C	thrust washer	apparatus: Faville-LeVally	23	40 (0.28)	50 (15.2)	2000 (4.3)	30			458
Akzo AC-80/TF/30	30% PTFE		"	"	"	23	40 (0.28)	50 (15.2)	2000 (4.3)	12			458
Akzo AC-8O/TF/15	15% PTFE		"	"	"	23	40 (0.28)	50 (15.2)	2000 (4.3)	20			458
Akzo AC-8O/TF/20	20% PTFE		"	"	"	23	40 (0.28)	50 (15.2)	2000 (4.3)	14			458
Akzo J-80/20/TF/15	20% glass fiber, 15% PTFE		"	"	"	23	40 (0.28)	50 (15.2)	2000 (4.3)	200			458
Akzo J-80/30/TF/15	30% glass fiber, 15% PTFE		"	"	"	23	40 (0.28)	50 (15.2)	2000 (4.3)	200			458
Unspecified grade	unmodified		"	"	"	23	40 (0.28)	50 (15.2)	2000 (4.3)	65			458
FLUOROPOLYMER, ECTFE against mating surface													
LNP FP C-1000	unmodified	**steel**	carbon steel; surface finish: 12-16 µin; 18-20 Rockwell C	thrust washer		23	40 (0.28)	50 (15.2)	2000 (4.3)	1000			453
LNP FP-CC-1003	15% carbon fiber		"	"		23	40 (0.28)	50 (15.2)	2000 (4.3)	18			453
LNP FP-CL 4020	10% PTFE		"	"		23	40 (0.28)	50 (15.2)	2000 (4.3)	27			453
FLUOROPOLYMER, ETFE against mating surface													
DuPont Tefzel HT-2004	74 Rockwell R; 25% glass fiber	**aluminum**	LM 24M; surface finish: 406 nm; unlubricated	thrust bearing test		23	300 (2.1)	10 (3.1)	3000 (6.5)	1220	1200		205
DuPont Tefzel HT-2004	74 Rockwell R; 25% glass fiber		"	"		23	100 (0.69)	50 (15.2)	5000 (10.5)	480	390		205
DuPont Tefzel HT-2004	74 Rockwell R; 25% glass fiber	**steel**	AISI 1018; surface finish: 406 nm, unlubricated	"		23	1000 (6.9)	10 (3.1)	10000 (21.4)	14	6		205
DuPont Tefzel HT-2004	74 Rockwell R; 25% glass fiber		"	"		23	1000 (6.9)	5 (1.5)	5000 (10.3)	16	4		205
DuPont Tefzel HT-2004	74 Rockwell R; 25% glass fiber		"	"		23	1000 (6.9)	15 (4.6)	15000 (31.7)	19	13		205

Material		Mating Surface		Test Conditions						K Factor 10^{-10} x (in^3-min/ft-lb-hr)		Results Note	Source
Supplier / Grade	Material Note	Mating Surface Material	Mating Surface Note	Test Method	Test Conditions Note	Temp. (°C)	Pressure (psi) (MPa)	Speed (fpm) (m/min)	PV (psi x fpm) (MPa x m/min)	Material	Mating Surface		
FLUOROPOLYMER, ETFE against mating surface													
DuPont Tefzel HT-2004	74 Rockwell R; 25% glass fiber	steel	AISI 1018; surface finish: 406 nm, unlubricated	thrust bearing test		23	1000 (6.9)	17.5 (5.3)	17500 (36.6)	30	16		205
DuPont Tefzel HT-2004	74 Rockwell R; 25% glass fiber		"	"		23	1000 (6.9)	20 (6.1)	20000 (42)	fail			205
LNP FP EEL 4024	20% glass fiber, 10% PTFE		carbon steel; surface finish: 12-16 µin; 18-20 Rockwell C	thrust washer		23	40 (0.28)	50 (15.2)	2000 (4.3)	11			453
LNP FP EF 1006	30% glass fiber		"	"		23	40 (0.28)	50 (15.2)	2000 (4.3)	10			453
LNP FP-E-1000	unmodified		"	"		23	40 (0.28)	50 (15.2)	2000 (4.3)	5000			453
LNP FP-EC-1003	15% carbon fiber		"	"		23	40 (0.28)	50 (15.2)	2000 (4.3)	10			453
LNP FP-EF 1003	15% glass fiber		"	"		23	40 (0.28)	50 (15.2)	2000 (4.3)	25			453
LNP FP-EF-1002	10% glass fiber		"	"		23	40 (0.28)	50 (15.2)	2000 (4.3)	28			453
LNP FP-EF-1004	20% glass fiber		"	"		23	40 (0.28)	50 (15.2)	2000 (4.3)	22			453
LNP FP-EF-1005	25% glass fiber		"	"		23	40 (0.28)	50 (15.2)	2000 (4.3)	15			453
LNP FP-EL-4060	30% PTFE		AISI 1141; surface finish: 8-12 µin	"		23	40 (0.28)	50 (15.2)	2000 (4.3)	10	2.1		453
LNP FP-EL-4060	30% PTFE		AISI 1141; surface finish: 50-70 µin	"		23	40 (0.28)	50 (15.2)	2000 (4.3)	12	2.4		453
LNP FP-EL-4060	30% PTFE		AISI 1141; surface finish: 12-16 µin	"		23	40 (0.28)	50 (15.2)	2000 (4.3)	7	1.3		453
LNP FP-EL-4060	30% PTFE		carbon steel; surface finish: 12-16 µin; 18-20 Rockwell C	"		23	40 (0.28)	50 (15.2)	2000 (4.3)	9			453
LNP FP-EL-4320	10% graphite		"	"		23	40 (0.28)	50 (15.2)	2000 (4.3)	165			453
LNP FP-FF 1004M	20% milled glass		"	"		23	40 (0.28)	50 (15.2)	2000 (4.3)	50			453
FLUOROPOLYMER, FEP against mating surface													
DuPont	96 Shore A, 59 Shore D, 25 Rockwell R	steel	carbon steel; surface finish: 12-20 µin; unlubricated; 20-25 Rockwell C	thrust washer		23			at PV values below PV limit	>5000			339
DuPont	10 vol.% bronze		"	"		23			at PV values below PV limit	10			339
DuPont	15% glass fiber		"	"		23			at PV values below PV limit	30			339
LNP FP-FC-1002	10% carbon fiber		carbon steel; surface finish: 12-16 µin; 18-20 Rockwell C	"		23	40 (0.28)	50 (15.2)	2000 (4.3)	15			453

Material		Mating Surface		Test Conditions						K Factor 10^{-10} x (in^3-min/ft-lb-hr)		Results Note	Source
Supplier / Grade	Material Note	Mating Surface Material	Mating Surface Note	Test Method	Test Conditions Note	Temp. (°C)	Pressure (psi) (MPa)	Speed (fpm) (m/min)	PV (psi x fpm) (MPa x m/min)	Material	Mating Surface		
FLUOROPOLYMER, FEP against mating surface													
LNP FP-FF 1004M	20% milled glass	**steel**	carbon steel; surface finish: 12-16 µin; 18-20 Rockwell C	thrust washer		23	40 (0.28)	50 (15.2)	2000 (4.3)	28			453
LNP FP-FF-1003M	15% milled glass		"	"		23	40 (0.28)	50 (15.2)	2000 (4.3)	25			453
LNP PE F-1000	unmodified		"	"		23	40 (0.28)	50 (15.2)	2000 (4.3)	1100			453
FLUOROPOLYMER, PFA against mating surface													
DuPont Teflon PFA 340	gen. purp. grade, 55 Shore D, moderate mol. wgt.	**steel**	AISI 1018; 16AA; 20 Rockwell C	thrust bearing wear test	test duration: 103 hours; conditions: room temperature, ambient air, unlubricated		100 (0.69)	3 (0.9)	300 (0.6)	1591			39
DuPont Teflon PFA 340	gen. purp. grade, 55 Shore D, moderate mol. wgt.		"	"	"		100 (0.69)	10 (3)	1000 (2.1)	1837			39
DuPont Teflon PFA 340	gen. purp. grade, 55 Shore D, moderate mol. wgt.		"	"	"		100 (0.69)	50 (15.2)	5000 (10.5)	694			39
DuPont Teflon PFA 340	gen. purp. grade, 55 Shore D, moderate mol. wgt.		"	"	room temperature, ambient air, unlubricated				5000 (10.5)	700			39
DuPont Teflon PFA 340	gen. purp. grade, 55 Shore D, moderate mol. wgt.		"	"	test duration: 103 hours; conditions: room temperature, ambient air, unlubricated		100 (0.69)	30 (9.1)	3000 (6.3)	983			39
DuPont	25% glass fiber		"	"	room temperature, ambient air, unlubricated				5000 (10.5)	13			39
DuPont	15% graphite		"	"	"				5000 (10.5)	147			39
DuPont	5% glass fiber, 5% MoS_2		"	"	"				5000 (10.5)	15			39
DuPont	15% glass fiber		"	"	"				5000 (10.5)	16			39
LNP FP PE 1002	10% glass fiber		carbon steel; surface finish: 12-16 µin; 18-20 Rockwell C	thrust washer		23	40 (0.28)	50 (15.2)	2000 (4.3)	20			453
LNP FP PML-3312	10% mineral, 5% PTFE		"	"		23	40 (0.28)	50 (15.2)	2000 (4.3)	5			453
LNP FP-P-1000	unmodified		"	"		23	40 (0.28)	50 (15.2)	2000 (4.3)	1250			453
LNP FP-PC 1003M	15% milled glass		"	"		23	40 (0.28)	50 (15.2)	2000 (4.3)	16			453
LNP FP-PC-1003	15% carbon fiber		cold rolled steel; surface finish: 12-16 µin; 22 Rockwell C	"		204	40 (0.28)	50 (15.2)	2000 (4.3)	105			453
LNP FP-PC-1003	15% carbon fiber		"	"		23	40 (0.28)	50 (15.2)	2000 (4.3)	13			453
LNP FP-PC-1003	15% carbon fiber		"	"		149	40 (0.28)	50 (15.2)	2000 (4.3)	38			453

Material		Mating Surface		Test Conditions						K Factor 10^{-10} x (in^3-min/ft-lb-hr)		Results Note	Source
Supplier / Grade	Material Note	Mating Surface Material	Mating Surface Note	Test Method	Test Conditions Note	Temp. (°C)	Pressure (psi) (MPa)	Speed (fpm) (m/min)	PV (psi x fpm) (MPa x m/min)	Material	Mating Surface		
FLUOROPOLYMER, PFA against mating surface													
LNP FP-PC-1003	15% carbon fiber	**steel**	cold rolled steel; surface finish: 12-16 µin; 22 Rockwell C	thrust washer		93	40 (0.28)	50 (15.2)	2000 (4.3)	18			453
LNP FP-PF 1004M	20% milled glass		carbon steel; surface finish: 12-16 µin; 18-20 Rockwell C	"		23	40 (0.28)	50 (15.2)	2000 (4.3)	15			453
LNP FP-PL-4020	10% PTFE		"	"		23	40 (0.28)	50 (15.2)	2000 (4.3)	5			453
FLUOROPOLYMER, TFE against mating surface													
LNP FC-103	15% milled glass	**aluminum**	surface finish: 12-16 µin	thrust washer		23	33.3 (0.23)	150 (45.7)	5000 (10.5)	188	38		453
LNP FC-113/SM	15% synergistic MoS_2		"	"		23	33.3 (0.23)	150 (45.7)	5000 (10.5)	1	<0.1		453
LNP FC-122	10% graphite powder		"	"		23	33.3 (0.23)	150 (45.7)	5000 (10.5)	187	0.8		453
LNP FC-132	10% coke flour		"	"		23	33.3 (0.23)	150 (45.7)	5000 (10.5)	47	0.9		453
LNP FC-146	60% bronze		"	"		23	33.3 (0.23)	150 (45.7)	5000 (10.5)	1.2	0.3		453
LNP PC-142	PPS filler, lubricant		"	"		23	33.3 (0.23)	150 (45.7)	5000 (10.5)	23	0.4		453
LNP PC-149	PPS filler, lubricant		"	"		23	33.3 (0.23)	150 (45.7)	5000 (10.5)	1.2	0.2		453
LNP PC-158	PPS filler, lubricant		"	"		23	33.3 (0.23)	150 (45.7)	5000 (10.5)	2.7	0.2		453
LNP PC-161	Polyoxybenzoate filler		"	"		23	33.3 (0.23)	150 (45.7)	5000 (10.5)	4.1	<0.1		453
LNP PC-184	PPS filler, lubricant		"	"		23	33.3 (0.23)	150 (45.7)	5000 (10.5)	1.5	<0.1		453
LNP PDX-81199	mineral filler		"	"		23	33.3 (0.23)	150 (45.7)	5000 (10.5)	2.8	0.6		453
Ausimont Halon 1005	5% glass fiber	**steel**			Ausimont test method			100 (30.5)		8			439
Ausimont Halon 1015	15% glass fiber				"			100 (30.5)		7			439
Ausimont Halon 1018	18% glass fiber				"			100 (30.5)		7			439
Ausimont Halon 1020	20% glass fiber				"			100 (30.5)		6			439
Ausimont Halon 1025	25% glass fiber				"			100 (30.5)		6			439
Ausimont Halon 1030	30% glass fiber				"			100 (30.5)		6			439

Material		Mating Surface		Test Conditions						K Factor 10^{-10} x (in³-min/ft-lb-hr)		Results Note	Source
Supplier / Grade	Material Note	Mating Surface Material	Mating Surface Note	Test Method	Test Conditions Note	Temp. (°C)	Pressure (psi) (MPa)	Speed (fpm) (m/min)	PV (psi x fpm) (MPa x m/min)	Material	Mating Surface		
FLUOROPOLYMER, TFE against mating surface													
Ausimont Halon 1035	35% glass fiber	steel			Ausimont test method			100 (30.5)		6			439
Ausimont Halon 1205	20% glass fiber, 5% graphite				"			100 (30.5)		6			439
Ausimont Halon 1206	5% glass fiber, 5% MoS_2				"			100 (30.5)		6			439
Ausimont Halon 1211	15% glass fiber, 5% MoS_2				"			100 (30.5)		6			439
Ausimont Halon 1223	23% glass fiber, 2% MoS_2				"			100 (30.5)		6			439
Ausimont Halon 1230	20% glass fiber, 5% MoS_2, 5% graphite				"			100 (30.5)		9			439
Ausimont Halon 1240	20% glass fiber, 20% MoS_2				"			100 (30.5)		8			439
Ausimont Halon 1410	10% glass fiber, 10% carbon fiber				"			100 (30.5)		8			439
Ausimont Halon 1416	5% glass fiber, 10% carbon fiber				"			100 (30.5)		8			439
Ausimont Halon 2010	10% graphite				"			100 (30.5)		40			439
Ausimont Halon 2015	15% graphite				"			100 (30.5)		30			439
Ausimont Halon 2021	5% MoS_2				"			100 (30.5)		8			439
Ausimont Halon 3040	40% bronze				"			100 (30.5)		5			439
Ausimont Halon 3050	50% bronze				"			100 (30.5)		5			439
Ausimont Halon 3060	60% bronze				"			100 (30.5)		5			439
Ausimont Halon 3205	55% bronze, 5% MoS_2				"			100 (30.5)		4			439
Ausimont Halon 4010	10% carbon fiber				"			100 (30.5)		15			439
Ausimont Halon 4015	15% carbon fiber				"			100 (30.5)		15			439
Ausimont Halon 4022	25% carbon/ graphite				"			100 (30.5)		8			439
Ausimont Halon 4025	25% carbon fiber				"			100 (30.5)		8			439
Ausimont Halon 4026	25% carbon/ graphite				"			100 (30.5)		8			439

Material		Mating Surface		Test Conditions						K Factor 10^{-10} x (in^3-min/ft-lb-hr)		Results Note	Source
Supplier / Grade	Material Note	Mating Surface Material	Mating Surface Note	Test Method	Test Conditions Note	Temp. (°C)	Pressure (psi) (MPa)	Speed (fpm) (m/min)	PV (psi x fpm) (MPa x m/min)	Material	Mating Surface		
FLUOROPOLYMER, TFE against mating surface													
Ausimont Halon 8105	5% mineral filler	**steel**			Ausimont test method			100 (30.5)		7			439
Ausimont Halon 8115	15% mineral filler				"			100 (30.5)		6			439
Ausimont Halon 9350	50% stainless steel				"			100 (30.5)		5			439
Ausimont Halon 9360	60% stainless steel				"			100 (30.5)		5			439
Ausimont Halon 9370	70% stainless steel				"			100 (30.5)		5			439
DuPont	55 Shore D; 15% graphite		carbon steel; surface finish: 12-20 µin; unlubricated; 20-25 Rockwell C	thrust washer		23			at PV values below PV limit	34			339
DuPont	65 Shore D; 60% bronze		"	"		23			at PV values below PV limit	6			339
DuPont	58 Shore D; 15% glass fiber, 5% MoS_2		"	"		23			at PV values below PV limit	9			339
DuPont	56 Shore D; 25% glass fiber		"	"		23			at PV values below PV limit	10			339
DuPont	amorphous; 25% carbon		"	"		23			at PV values below PV limit	11.5			339
DuPont	20% glass fiber, 5% graphite		"	"		23			at PV values below PV limit	15			339
DuPont	54 Shore D; 15% glass fiber		"	"		23			at PV values below PV limit	16			339
DuPont	98 Shore A, 58 Rockwell R, 52 Shore D		"	"		23			at PV values below PV limit	2500			339
LNP FC-103	15% milled glass		surface finish: 12-16 µin	"		23	33.3 (0.23)	150 (45.7)	5000 (10.5)	7	0.4		453
LNP FC-103	15% milled glass		cold rolled steel; surface finish: 12-16 µin; 22 Rockwell C	"		23	33.3 (0.23)	150 (45.7)	5000 (10.5)	7			453
LNP FC-103	15% milled glass		"	"		93	33.3 (0.23)	150 (45.7)	5000 (10.5)	7			453
LNP FC-103	15% milled glass		"	"		149	33.3 (0.23)	150 (45.7)	5000 (10.5)	7			453
LNP FC-103	15% milled glass		"	"		204	33.3 (0.23)	150 (45.7)	5000 (10.5)	7			453
LNP FC-103	15% milled glass		"	"		260	33.3 (0.23)	150 (45.7)	5000 (10.5)	70			453

Material		Mating Surface		Test Conditions						K Factor 10^{-10} x (in³-min/ft-lb-hr)		Results Note	Source
Supplier / Grade	Material Note	Mating Surface Material	Mating Surface Note	Test Method	Test Conditions Note	Temp. (°C)	Pressure (psi) (MPa)	Speed (fpm) (m/min)	PV (psi x fpm) (MPa x m/min)	Material	Mating Surface		
FLUOROPOLYMER, TFE against mating surface		→											
LNP FC-113/SM	15% synergistic MoS_2	**steel**	surface finish: 12-16 µin	thrust washer		23	33.3 (0.23)	150 (45.7)	5000 (10.5)	3	<0.1		453
LNP FC-122	10% graphite powder		"	"		23	33.3 (0.23)	150 (45.7)	5000 (10.5)	6	0.2		453
LNP FC-132	10% coke flour		"	"		23	33.3 (0.23)	150 (45.7)	5000 (10.5)	12	0.7		453
LNP FC-146	60% bronze		"	"		23	33.3 (0.23)	150 (45.7)	5000 (10.5)	5	0.3		453
LNP FC-182	55% bronze, 5% MoS_2		cold rolled steel; surface finish: 12-16 µin; 22 Rockwell C	"		204	33.3 (0.23)	150 (45.7)	5000 (10.5)	31			453
LNP FC-182	55% bronze, 5% MoS_2		"	"		23	33.3 (0.23)	150 (45.7)	5000 (10.5)	5			453
LNP FC-182	55% bronze, 5% MoS_2		"	"		93	33.3 (0.23)	150 (45.7)	5000 (10.5)	5			453
LNP FC-182	55% bronze, 5% MoS_2		"	"		149	33.3 (0.23)	150 (45.7)	5000 (10.5)	5			453
LNP FC-191	25% carbon/ graphite		"	"		260	33.3 (0.23)	150 (45.7)	5000 (10.5)	45			453
LNP FC-191	25% carbon/ graphite		"	"		23	33.3 (0.23)	150 (45.7)	5000 (10.5)	6			453
LNP FC-191	25% carbon/ graphite		"	"		93	33.3 (0.23)	150 (45.7)	5000 (10.5)	6			453
LNP FC-191	25% carbon/ graphite		"	"		204	33.3 (0.23)	150 (45.7)	5000 (10.5)	10			453
LNP FC-191	25% carbon/ graphite		"	"		149	33.3 (0.23)	150 (45.7)	5000 (10.5)	8			453
LNP PC-142	PPS filler, lubricant		surface finish: 12-16 µin	"		23	33.3 (0.23)	150 (45.7)	5000 (10.5)	1	0.1		453
LNP PC-149	PPS filler, lubricant		"	"		23	33.3 (0.23)	150 (45.7)	5000 (10.5)	2	<0.1		453
LNP PC-158	PPS filler, lubricant		"	"		23	33.3 (0.23)	150 (45.7)	5000 (10.5)	4	<0.1		453
LNP PC-161	Polyoxybenzoate filler		"	"		23	33.3 (0.23)	150 (45.7)	5000 (10.5)	5	<0.1		453
LNP PC-184	PPS filler, lubricant		"	"		23	33.3 (0.23)	150 (45.7)	5000 (10.5)	1	<0.1		453
LNP PC-185	PPS filler		cold rolled steel; surface finish: 12-16 µin; 22 Rockwell C	"		23	33.3 (0.23)	150 (45.7)	5000 (10.5)	1			453
LNP PC-185	PPS filler		"	"		93	33.3 (0.23)	150 (45.7)	5000 (10.5)	1			453

Material		Mating Surface		Test Conditions						K Factor 10^{-10} x (in^3-min/ft-lb-hr)		Results Note	Source
Supplier / Grade	Material Note	Mating Surface Material	Mating Surface Note	Test Method	Test Conditions Note	Temp. (°C)	Pressure (psi) (MPa)	Speed (fpm) (m/min)	PV (psi x fpm) (MPa x m/min)	Material	Mating Surface		
FLUOROPOLYMER, TFE against mating surface													
LNP PC-185	PPS filler	**steel**	cold rolled steel; surface finish: 12-16 μin; 22 Rockwell C	thrust washer		149	33.3 (0.23)	150 (45.7)	5000 (10.5)	1			453
LNP PC-185	PPS filler		"	"		204	33.3 (0.23)	150 (45.7)	5000 (10.5)	5			453
LNP PC-185	PPS filler		"	"		260	33.3 (0.23)	150 (45.7)	5000 (10.5)	80			453
LNP PDX-81199	mineral filler		surface finish: 12-16 μin	"		23	33.3 (0.23)	150 (45.7)	5000 (10.5)	1.4	0.6		453
Unspecified grade	carbon		carbon EN8, dry	pad on ring (Amsler wear tester)		20		600 (183)				wear rate: 250 μm/hour; maximum load: 25 kg	338
Unspecified grade	carbon		C1018 steel; 24 Rockwell C	thrust washer			50 (0.34)	200 (61)	10000 (20.7)	5			20
FLUOROPOLYMER, PVDF against mating surface													
Ensinger Ensikem	standard grade	**steel**					40 (0.28)	50 (15.2)	2000 (4.3)	1000			449
LNP FP V-1000	unmodified		carbon steel; surface finish: 12-16 μin; 18-20 Rockwell C	thrust washer		23	40 (0.28)	50 (15.2)	2000 (4.3)	1000			453
LNP FP VM 3850	25% mineral fiber		"	"		23	40 (0.28)	50 (15.2)	2000 (4.3)	>100			453
LNP FP-VC-1003	15% carbon fiber		"	"		23	40 (0.28)	50 (15.2)	2000 (4.3)	14			453
LNP FP-VC-1003	15% carbon fiber		cold rolled steel; surface finish: 12-16 μin; 22 Rockwell C	"		23	40 (0.28)	50 (15.2)	2000 (4.3)	14			453
LNP FP-VC-1003	15% carbon fiber		"	"		93	40 (0.28)	50 (15.2)	2000 (4.3)	75			453
LNP FP-VCL-4024	20% carbon fiber, 10% PTFE		carbon steel; surface finish: 12-16 μin; 18-20 Rockwell C	"		23	40 (0.28)	50 (15.2)	2000 (4.3)	11			453
LNP FP-VM-2550	25% mineral fiber		"	"		23	40 (0.28)	50 (15.2)	2000 (4.3)	100			453
NYLON against mating surface													
LNP V-1000	unmodified, high impact	**steel**	carbon steel; surface finish: 12-16 μin; 18-20 Rockwell C	thrust washer		23	40 (0.28)	50 (15.2)	2000 (4.3)	200			453
LNP VFL-4036	high impact, 30% glass fiber, 15% PTFE		"	"		23	40 (0.28)	50 (15.2)	2000 (4.3)	10			453
LNP VL-4040	high impact, 20% PTFE		"	"		23	40 (0.28)	50 (15.2)	2000 (4.3)	30			453
LNP VL-4410	high impact, 2% silicone		"	"		23	40 (0.28)	50 (15.2)	2000 (4.3)	38			453

Material		Mating Surface		Test Conditions						K Factor 10^-10^ x (in³-min/ft·lb·hr)		Results Note	Source
Supplier / Grade	Material Note	Mating Surface Material	Mating Surface Note	Test Method	Test Conditions Note	Temp. (°C)	Pressure (psi) (MPa)	Speed (fpm) (m/min)	PV (psi x fpm) (MPa x m/min)	Material	Mating Surface		
NYLON against mating surface													
LNP VL-4530	high impact, 13% PTFE, 2% silicone	steel	carbon steel; surface finish: 12-16 µin; 18-20 Rockwell C	thrust washer		23	40 (0.28)	50 (15.2)	2000 (4.3)	45			453
RTP 200H	high impact		C1018 steel; surface finish: 14-17 µin; 15-25 Rockwell C	"	apparatus: Falex Model No. 6	23	40 (0.28)	50 (15.2)	2000 (4.3)	225			457
RTP 200H TFE 20	20% PTFE, high impact		"	"	"	23	40 (0.28)	50 (15.2)	2000 (4.3)	14			457
RTP 205H	30% glass fiber, high impact		"	"	"	23	40 (0.28)	50 (15.2)	2000 (4.3)	100			457
RTP 205H TFE 15	15% PTFE, 30% glass fiber, high impact		"	"	"	23	40 (0.28)	50 (15.2)	2000 (4.3)	20			457
NYLON, AMORPHOUS against mating surface													
LNP X-1000	unmodified	steel	carbon steel; surface finish: 12-16 µin; 18-20 Rockwell C	thrust washer		23	40 (0.28)	50 (15.2)	2000 (4.3)	600			453
LNP XC-1006	30% carbon fiber		"	"		23	40 (0.28)	50 (15.2)	2000 (4.3)	90			453
LNP XF-1006	30% glass fiber		"	"		23	40 (0.28)	50 (15.2)	2000 (4.3)	350			453
LNP XFL-4036	30% glass fiber, 15% PTFE		"	"		23	40 (0.28)	50 (15.2)	2000 (4.3)	22			453
LNP XL-4040	20% PTFE		"	"		23	40 (0.28)	50 (15.2)	2000 (4.3)	20			453
NYLON 11 against mating surface													
LNP HFL-4325	25% glass fiber, 10% graphite	steel	carbon steel; surface finish: 12-16 µin; 18-20 Rockwell C	thrust washer		23	40 (0.28)	50 (15.2)	2000 (4.3)	30			453
LNP PDX-4208	85% bronze		"	"		23	40 (0.28)	50 (15.2)	2000 (4.3)	82			453
LNP PDX-5156	83% bronze, 3% MoS_2		"	"		23	40 (0.28)	50 (15.2)	2000 (4.3)	65			453
RTP 200C TFE 20	20% PTFE		C1018 steel; surface finish: 14-17 µin; 15-25 Rockwell C	"	apparatus: Falex Model No. 6	23	40 (0.28)	50 (15.2)	2000 (4.3)	25			457
RTP 203C TFE 20	20% PTFE, 20% glass fiber		"	"	"	23	40 (0.28)	50 (15.2)	2000 (4.3)	40			457
NYLON 12 against mating surface													
LNP S-1000	unmodified	steel	carbon steel; surface finish: 12-16 µin; 18-20 Rockwell C	thrust washer		23	40 (0.28)	50 (15.2)	2000 (4.3)	180			453
LNP SFL-4036	30% glass fiber, 15% PTFE		"	"		23	40 (0.28)	50 (15.2)	2000 (4.3)	35			453

Material		Mating Surface		Test Conditions						K Factor 10^{-10} x (in^3-min/ft-lb-hr)		Results Note	Source
Supplier / Grade	Material Note	Mating Surface Material	Mating Surface Note	Test Method	Test Conditions Note	Temp. (°C)	Pressure (psi) (MPa)	Speed (fpm) (m/min)	PV (psi x fpm) (MPa x m/min)	Material	Mating Surface		
NYLON 12 against mating surface													
LNP SL-4040	15% PTFE	**steel**	carbon steel; surface finish: 12-16 µin; 18-20 Rockwell C	thrust washer		23	40 (0.28)	50 (15.2)	2000 (4.3)	30			453
LNP SL-4610	2% silicone		"	"		23	40 (0.28)	50 (15.2)	2000 (4.3)	155			453
RTP 205F TFE 20	20% PTFE, 30% glass fiber		C1018 steel; surface finish: 14-17 µin; 15-25 Rockwell C	"	apparatus: Falex Model No. 6	23	40 (0.28)	50 (15.2)	2000 (4.3)	35			457
RTP 20OF TFE 20	20% PTFE		"	"	"	23	40 (0.28)	50 (15.2)	2000 (4.3)	25			457
NYLON 6 against mating surface													
Akzo G-3/40/MS/5	40% long glass fiber, 5% MoS_2	**steel**	carbon steel; surface finish: 16 µin; 18-22 Rockwell C	thrust washer	apparatus: Faville-LeVally	23	40 (0.28)	50 (15.2)	2000 (4.3)	70			458
Akzo J-3/30/MS/5	30% glass fiber, 5% MoS_2		"	"	"	23	40 (0.28)	50 (15.2)	2000 (4.3)	75			458
Bay Resins Lubriplas PA-211	unmodified resin			"			40 (0.28)	50 (15.2)	2000 (4.3)	200			435
Bay Resins Lubriplas PA-211G13	13% glass fiber			"			40 (0.28)	50 (15.2)	2000 (4.3)	120			435
Bay Resins Lubriplas PA-211G30TF15	30% glass fiber, 15% PTFE			"			40 (0.28)	50 (15.2)	2000 (4.3)	18			435
Bay Resins Lubriplas PA-211G33	33% glass fiber			"			40 (0.28)	50 (15.2)	2000 (4.3)	90			435
Bay Resins Lubriplas PA-211TF20	20% PTFE			"			40 (0.28)	50 (15.2)	2000 (4.3)	15			435
Ensinger Vekton	standard grade, cast						40 (0.28)	50 (15.2)	2000 (4.3)	200			449
Ensinger Vekton 6PAL	oil impregnated, cast						40 (0.28)	50 (15.2)	2000 (4.3)	26			449
Ensinger	standard grade						40 (0.28)	50 (15.2)	2000 (4.3)	200			449
LNP Ny-Kon P	<5% MoS_2		carbon steel; surface finish: 12-16 µin; 18-20 Rockwell C	thrust washer		23	40 (0.28)	50 (15.2)	2000 (4.3)	160			453
LNP P-1000	unmodified		"	"		23	40 (0.28)	50 (15.2)	2000 (4.3)	200			453
LNP PC-1006	30% carbon fiber		"	"		23	40 (0.28)	50 (15.2)	2000 (4.3)	30			453
LNP PF-1006	30% glass fiber		"	"		23	40 (0.28)	50 (15.2)	2000 (4.3)	90			453
LNP PFL-4036	30% glass fiber, 15% PTFE		"	"		23	40 (0.28)	50 (15.2)	2000 (4.3)	17			453
LNP PFL-4216	30% glass fiber, <5% MoS_2		"	"		23	40 (0.28)	50 (15.2)	2000 (4.3)	80			453
LNP PFL-4218	40% glass fiber, <5% MoS_2		"	"		23	40 (0.28)	50 (15.2)	2000 (4.3)	75			453

Material		Mating Surface		Test Conditions						K Factor 10^{-10} x (in³-min/ft-lb-hr)		Results Note	Source
Supplier / Grade	Material Note	Mating Surface Material	Mating Surface Note	Test Method	Test Conditions Note	Temp. (°C)	Pressure (psi) (MPa)	Speed (fpm) (m/min)	PV (psi x fpm) (MPa x m/min)	Material	Mating Surface		
NYLON 6 against mating surface													
LNP PFL-4536	30% glass fiber, 13% PTFE, 2% silicone	**steel**	carbon steel; surface finish: 12-16 µin; 18-20 Rockwell C	thrust washer		23	40 (0.28)	50 (15.2)	2000 (4.3)	10			453
LNP PL-4030	15% PTFE		"	"		23	40 (0.28)	50 (15.2)	2000 (4.3)	30			453
LNP PL-4040	20% PTFE		"	"		23	40 (0.28)	50 (15.2)	2000 (4.3)	15			453
LNP PL-431 0	5% graphite		"	"		23	40 (0.28)	50 (15.2)	2000 (4.3)	60			453
LNP PL-4410	2% silicone		"	"		23	40 (0.28)	50 (15.2)	2000 (4.3)	50			453
LNP PL-4540	18% PTFE, 2% silicone		"	"		23	40 (0.28)	50 (15.2)	2000 (4.3)	11			453
Polymer Nylatron GSM	cast, MoS_2 modified			journal bearing			42.2 (0.29)	118 (36)	5000 (10.5)	83			441
Polymer Nylatron MC901	cast, general purpose, high heat			"			42.2 (0.29)	118 (36)	5000 (10.5)	87			441
Polymer Nylatron NSM	w/ additives to improve bearing properties			"			42.2 (0.29)	118 (36)	5000 (10.5)	9			441
RTP 200A			C1018 steel; surface finish: 14-17 µin; 15-25 Rockwell C	thrust washer	apparatus: Falex Model No. 6	23	40 (0.28)	50 (15.2)	2000 (4.3)	200			457
RTP 200A TFF 20	20% PTFE		"	"	"	23	40 (0.28)	50 (15.2)	2000 (4.3)	14			457
RTP 205A	30% glass fiber		"	"	"	23	40 (0.23)	50 (15.2)	2000 (4.3)	100			457
RTP 205A TFE 15	15% PTFE, 30% glass fiber		"	"	"	23	40 (0.28)	50 (15.2)	2000 (4.3)	18			457
Thermofil N-15FG-0100	15% glass fiber			ASTM D1894		23				150			459
Thermofil N-30FG-0100	30% glass fiber			"		23				85			459
Thermofil N-30FG-0214	30% glass fiber, PTFE lubricated			"		23				17			459
Thermofil N-30NF-0100	30% graphite fiber			"		23				25			459
Thermofil N-40-MF-0100	40% mineral			"		23				150			459
Thermofil N-40BG-0100	40% glass bead			"		23				140			459
Thermofil N-40FM-0100	40% glass/ mineral			"		23				100			459
Unspecified grade	unmodified		carbon steel; surface finish: 16 µin; 18-22 Rockwell C	thrust washer	apparatus: Faville-LeVally	23	40 (0.28)	50 (15.2)	2000 (4.3)	200			458
NYLON 610 against mating surface													
LNP QFL-4036	30% glass fiber, 15% PTFE	**polycarbonate**	DFL-4034 (20% glass fiber, 15% PTFE)	"		23	40 (0.28)	50 (15.2)	2000 (4.3)	12	15		453

Material		Mating Surface		Test Conditions						K Factor 10^{-10} x (in³-min/ft-lb-hr)		Results Note	Source
Supplier / Grade	Material Note	Mating Surface Material	Mating Surface Note	Test Method	Test Conditions Note	Temp. (°C)	Pressure (psi) (MPa)	Speed (fpm) (m/min)	PV (psi x fpm) (MPa x m/min)	Material	Mating Surface		
NYLON 610 against mating surface													
LNP Ny-Kon Q	<5% MoS_2	steel	carbon steel; surface finish: 12-16 µin; 18-20 Rockwell C	thrust washer		23	40 (0.28)	50 (15.2)	2000 (4.3)	145			453
LNP OL-4540	18% PTFE, 2% silicone		"	"		23	40 (0.28)	50 (15.2)	2000 (4.3)	10			453
LNP Q-1000	unmodified		"	"		23	40 (0.28)	50 (15.2)	2000 (4.3)	180			453
LNP QC-1006	30% carbon fiber		"	"		23	40 (0.28)	50 (15.2)	2000 (4.3)	25			453
LNP QF-1006	30% glass fiber		"	"		23	40 (0.28)	50 (15.2)	2000 (4.3)	80			453
LNP QFL-4036	30% glass fiber, 15% PTFE		"	"		23	40 (0.28)	50 (15.2)	2000 (4.3)	15			453
LNP QFL-4536	30% glass fiber, 13% PTFE, 2% silicone		"	"		23	40 (0.28)	50 (15.2)	2000 (4.3)	9			453
LNP QL-4040	20% PTFE		"	"		23	40 (0.28)	50 (15.2)	2000 (4.3)	15			453
LNP QL-4410	2% silicone		"	"		23	40 (0.28)	50 (15.2)	2000 (4.3)	46			453
RTP 200B			C1018 steel; surface finish: 14-17 µin; 15-25 Rockwell C	"	apparatus: Falex Model No. 6	23	40 (0.28)	50 (15.2)	2000 (4.3)	200			457
RTP 205B	30% glass fiber		"	"	"	23	40 (0.28)	50 (15.2)	2000 (4.3)	80			457
RTP 205B TFE 15	15% PTFE, 30% glass fiber		"	"	"	23	40 (0.28)	50 (15.2)	2000 (4.3)	18			457
NYLON 612 against mating surface													
Akzo J-4/30/TF/15	30% glass fiber, 15% PTFE	steel	carbon steel; surface finish: 16 µin; 18-22 Rockwell C	thrust washer	apparatus: Faville-LeVally	23	40 (0.28)	50 (15.2)	2000 (4.3)	17			458
Akzo J-4/CF/30/TF/10	30% carbon fiber, 10% PTFE		"	"	"	23	40 (0.28)	50 (15.2)	2000 (4.3)	12			458
Akzo NY-4/TF/10	10% PTFE		"	"	"	23	40 (0.28)	50 (15.2)	2000 (4.3)	20			458
LNP I-1000	unmodified		carbon steel; surface finish: 12-16 µin; 18-20 Rockwell C	"		23	40 (0.28)	50 (15.2)	2000 (4.3)	190			453
LNP IC-1006	30% carbon fiber		"	"		23	40 (0.28)	50 (15.2)	2000 (4.3)	25			453
LNP IF-1006	30% glass fiber		"	"		23	40 (0.28)	50 (15.2)	2000 (4.3)	85			453
LNP IFL-4036	30% glass fiber, 15% PTFE		"	"		23	40 (0.28)	50 (15.2)	2000 (4.3)	16			453
LNP IFL-4536	30% glass fiber, 13% PTFE, 2% silicone		"	"		23	40 (0.28)	50 (15.2)	2000 (4.3)	9			453
LNP IL-4040	20% PTFE		"	"		23	40 (0.28)	50 (15.2)	2000 (4.3)	16			453

Material		Mating Surface		Test Conditions						K Factor 10^{-10} x (in³-min/ft-lb-hr)		Results Note	Source
Supplier / Grade	Material Note	Mating Surface Material	Mating Surface Note	Test Method	Test Conditions Note	Temp. (°C)	Pressure (psi) (MPa)	Speed (fpm) (m/min)	PV (psi x fpm) (MPa x m/min)	Material	Mating Surface		
NYLON 612 against mating surface													
LNP IL-4410	2% silicone	**steel**	carbon steel; surface finish: 12-16 µin; 18-20 Rockwell C	thrust washer		23	40 (0.28)	50 (15.2)	2000 (4.3)	48			453
LNP IL-4540	18% PTFE, 2% silicone		"	"		23	40 (0.28)	50 (15.2)	2000 (4.3)	10			453
LNP Ny-Kon I	<5% MoS_2		"	"		23	40 (0.28)	50 (15.2)	2000 (4.3)	145			453
RTP 200D			C1018 steel; surface finish: 14-17 µin; 15-25 Rockwell C	"	apparatus: Falex Model No. 6	23	40 (0.28)	50 (15.2)	2000 (4.3)	200			457
RTP 200D TFE 20	20% PTFE		"	"	"	23	40 (0.28)	50 (15.2)	2000 (4.3)	17			457
RTP 205D	30% glass fiber		"	"	"	23	40 (0.28)	50 (15.2)	2000 (4.3)	80			457
RTP 205D TFE 15	15% PTFE, 30% glass fiber		"	"	"	23	40 (0.28)	50 (15.2)	2000 (4.3)	18			457
Thermofil N6-30FG-0100	30% glass fiber			ASTM D1894		23				80			459
Thermofil N6-30FG-0214	30% glass fiber, PTFE lubricated			"		23				16			459
Thermofil N6-30FG-0282	30% glass fiber, lubricated			"		23				9			459
Thermofil N6-30FG-0500	30% glass fiber, flame retardant			"		23				90			459
Thermofil N6-30NF-0100	30% graphite fiber			"		23				20			459
Thermofil N6-9900-0500	flame retardant			"		23				200			459
Thermofil R-20NF-0214	20% graphite fiber, PTFE lubricated			"		23				12			459
Thermofil R-30FG-0100	30% glass fiber			"		23				180			459
Thermofil R-30FG-0214	30% glass fiber, PTFE lubricated			"		23				16			459
Thermofil R-40NF-0100	40% graphite fiber			"		23				80			459
Thermofil R-9900-0200	lubricated			"		23				35			459
Thermofil R-9900-0214	PTFE lubricated			"		23				12			459
Unspecified grade	unmodified		carbon steel; surface finish: 16 µin; 18-22 Rockwell C	thrust washer	apparatus: Faville-LeVally	23	40 (0.28)	50 (15.2)	2000 (4.3)	190			458
NYLON 66 against mating surface													
LNP R-1000	unmodified	**acetal**	K-1000 (unmodified)	thrust washer		23	40 (0.28)	50 (15.2)	2000 (4.3)	75	55		453
LNP RCL-4036	30% carbon fiber, 15% PTFE		KFL-4036 (30% glass fiber, 15% PTFE)	"		23	40 (0.28)	50 (15.2)	2000 (4.3)	60	28		453
LNP RF-1006	30% glass fiber		KL-4040 (20% PTFE)	"		23	40 (0.28)	50 (15.2)	2000 (4.3)	30	38		453

Material		Mating Surface		Test Conditions						K Factor 10^{-10} x (in³-min/ft-lb-hr)		Results Note	Source
Supplier / Grade	Material Note	Mating Surface Material	Mating Surface Note	Test Method	Test Conditions Note	Temp. (°C)	Pressure (psi) (MPa)	Speed (fpm) (m/min)	PV (psi x fpm) (MPa x m/min)	Material	Mating Surface		
NYLON 66 against mating surface													
LNP RFL-4036	30% glass fiber, 15% PTFE	**acetal**	KL-4040 (20% PTFE)	thrust washer		23	40 (0.28)	50 (15.2)	2000 (4.3)	25	32		453
LNP RL-4040	20% PTFE		"	"		23	40 (0.28)	50 (15.2)	2000 (4.3)	45	30		453
LNP RL-4410	2% silicone		KFL-4036 (30% glass fiber, 15% PTFE)	"		23	40 (0.28)	50 (15.2)	2000 (4.3)	50	940		453
LNP RAL-4022	10% aramid fiber, 10% PTFE	**aluminum**	2024 aluminum; surface finish: 12-16 µin	"		23	40 (0.28)	50 (15.2)	2000 (4.3)	48	4		453
LNP RAL-4022	10% aramid fiber, 10% PTFE		2024 aluminum; surface finish: 8-12 µin	"		23	40 (0.28)	50 (15.2)	2000 (4.3)	500	8		453
LNP RAL-4022	10% aramid fiber, 10% PTFE		2024 aluminum; surface finish: 50-70 µin	"		23	40 (0.28)	50 (15.2)	2000 (4.3)	128	5		453
LNP RCL-4036	30% carbon fiber, 15% PTFE		2024 aluminum; surface finish: 12-16 µin	"		23	40 (0.28)	50 (15.2)	2000 (4.3)	175	105		453
LNP RCL-4036	30% carbon fiber, 15% PTFE		2024 aluminum; surface finish: 50-70 µin	"		23	40 (0.28)	50 (15.2)	2000 (4.3)	247	151		453
LNP RCL-4036	30% carbon fiber, 15% PTFE		2024 aluminum; surface finish: 8-12 µin	"		23	40 (0.28)	50 (15.2)	2000 (4.3)	2500	600		453
LNP RF-1006	30% glass fiber		2024 aluminum; surface finish: 50-70 µin	"		23	40 (0.28)	50 (15.2)	2000 (4.3)	2000	100		453
LNP RF-1006	30% glass fiber		2024 aluminum; surface finish: 8-12 µin	"		23	40 (0.28)	50 (15.2)	2000 (4.3)	>4,000	>4,000		453
LNP RF-1006	30% glass fiber		2024 aluminum; surface finish: 12-16 µin	"		23	40 (0.28)	50 (15.2)	2000 (4.3)	400	265		453
LNP RFL-4036	30% glass fiber, 15% PTFE		"	"		23	40 (0.28)	50 (15.2)	2000 (4.3)	320	175		453
LNP RFL-4036	30% glass fiber, 15% PTFE		2024 aluminum; surface finish: 50-70 µin	"		23	40 (0.28)	50 (15.2)	2000 (4.3)	1800	75		453
LNP RFL-4036	30% glass fiber, 15% PTFE		2024 aluminum; surface finish: 8-12 µin	"		23	40 (0.28)	50 (15.2)	2000 (4.3)	2250	1500		453
LNP RL-4040	20% PTFE		"	"		23	40 (0.28)	50 (15.2)	2000 (4.3)	250	10		453
LNP RL-4040	20% PTFE		2024 aluminum; surface finish: 12-16 µin	"		23	40 (0.28)	50 (15.2)	2000 (4.3)	26	6		453
LNP RL-4040	20% PTFE		2024 aluminum; surface finish: 50-70 µin	"		23	40 (0.28)	50 (15.2)	2000 (4.3)	105	6		453
LNP RAL-4022	10% aramid fiber, 10% PTFE	**brass**	70/30 brass; surface finish: 8-16 µin	"		23	40 (0.28)	50 (15.2)	2000 (4.3)	16	0.2		453
LNP RAL-4022	10% aramid fiber, 10% PTFE		70/30 brass; surface finish: 50-70 µin	"		23	40 (0.28)	50 (15.2)	2000 (4.3)	19	0.3		453
LNP RC-1006	30% carbon fiber		"	"		23	40 (0.28)	50 (15.2)	2000 (4.3)	34	19		453
LNP RC-1006	30% carbon fiber		70/30 brass; surface finish: 8-16 µin	"		23	40 (0.28)	50 (15.2)	2000 (4.3)	40	34		453

Material		Mating Surface		Test Conditions						K Factor 10^-10 x (in³-min/ft-lb-hr)		Results Note	Source
Supplier / Grade	Material Note	Mating Surface Material	Mating Surface Note	Test Method	Test Conditions Note	Temp. (°C)	Pressure (psi) (MPa)	Speed (fpm) (m/min)	PV (psi x fpm) (MPa x m/min)	Material	Mating Surface		
NYLON 66 against mating surface →													
LNP RCL-4036	30% carbon fiber, 15% PTFE	**brass**	70/30 brass; surface finish: 8-16 µin	thrust washer		23	40 (0.28)	50 (15.2)	2000 (4.3)	18	5		453
LNP RCL-4036	30% carbon fiber, 15% PTFE		70/30 brass; surface finish: 50-70 µin	"		23	40 (0.28)	50 (15.2)	2000 (4.3)	13	6		453
LNP RF-1006	30% glass fiber		"	"		23	40 (0.28)	50 (15.2)	2000 (4.3)	52	17		453
LNP RF-1006	30% glass fiber		70/30 brass; surface finish: 8-16 µin	"		23	40 (0.28)	50 (15.2)	2000 (4.3)	53	42		453
LNP RFL-4036	30% glass fiber, 15% PTFE		70/30 brass; surface finish: 50-70 µin	"		23	40 (0.28)	50 (15.2)	2000 (4.3)	18	12		453
LNP RFL-4036	30% glass fiber, 15% PTFE		70/30 brass; surface finish: 8-16 µin	"		23	40 (0.28)	50 (15.2)	2000 (4.3)	21	12		453
LNP RFL-4536	30% glass fiber, 13% PTFE, 2% silicone		70/30 brass; surface finish: 50-70 µin	"		23	40 (0.28)	50 (15.2)	2000 (4.3)	18	12		453
LNP RFL-4536	30% glass fiber, 13% PTFE, 2% silicone		70/30 brass; surface finish: 8-16 µin	"		23	40 (0.28)	50 (15.2)	2000 (4.3)	20	12		453
LNP RL-4040	20% PTFE		70/30 brass; surface finish: 50-70 µin	"		23	40 (0.28)	50 (15.2)	2000 (4.3)	21	0.3		453
LNP RL-4040	20% PTFE		70/30 brass; surface finish: 8-16 µin	"		23	40 (0.28)	50 (15.2)	2000 (4.3)	8	0.2		453
LNP R-1000	unmodified	**nylon 66**	RF-1006 (30% glass fiber)	"		23	40 (0.28)	50 (15.2)	2000 (4.3)	22000	350		453
LNP R-1000	unmodified		R-1000 (unmodified)	"		23	40 (0.28)	50 (15.2)	2000 (4.3)	2500	1100		453
LNP RAL-4022	10% aramid fiber, 10% PTFE		RCL-4036 (30% carbon fiber, 15% PTFE)	"		23	40 (0.28)	50 (15.2)	2000 (4.3)	300	40		453
LNP RAL-4022	10% aramid fiber, 10% PTFE		"	"		23	40 (0.28)	50 (15.2)	2000 (4.3)	300	40		453
LNP RC-1006	30% carbon fiber		RF-1006 (30% glass fiber)	"		23	40 (0.28)	50 (15.2)	2000 (4.3)	300	45		453
LNP RC-1006	30% carbon fiber		RCL-4036 (30% carbon fiber, 15% PTFE)	"		23	40 (0.28)	50 (15.2)	2000 (4.3)	90	100		453
LNP RC-1006	30% carbon fiber		RFL-4036 (30% glass fiber, 15% PTFE)	"		23	40 (0.28)	50 (15.2)	2000 (4.3)	130	30		453
LNP RC-1006	30% carbon fiber		R-1000 (unmodified)	"		23	40 (0.28)	50 (15.2)	2000 (4.3)	1400	1700		453
LNP RC-1008	40% carbon fiber		RCL-4536 (30% carbon fiber, 13% PTFE, 2% silicone)	"		23	40 (0.28)	50 (15.2)	2000 (4.3)	45	50		453
LNP RCL-4036	30% carbon fiber, 15% PTFE		RFL-4036 (30% glass fiber, 15% PTFE)	"		23	40 (0.28)	50 (15.2)	2000 (4.3)	75	20		453
LNP RCL-4036	30% carbon fiber, 15% PTFE		RCL-4036 (30% carbon fiber, 15% PTFE)	"		23	40 (0.28)	50 (15.2)	2000 (4.3)	80	90		453
LNP RCL-4036	30% carbon fiber, 15% PTFE		RF-1006 (30% glass fiber)	"		23	40 (0.28)	50 (15.2)	2000 (4.3)	800	80		453

Material		Mating Surface		Test Conditions						K Factor 10^{-10} x (in³-min/ft-lb-hr)		Results Note	Source
Supplier / Grade	Material Note	Mating Surface Material	Mating Surface Note	Test Method	Test Conditions Note	Temp. (°C)	Pressure (psi) (MPa)	Speed (fpm) (m/min)	PV (psi x fpm) (MPa x m/min)	Material	Mating Surface		
NYLON 66 against mating surface →													
LNP RCL-4536	30% carbon fiber, 13% PTFE, 2% silicone	**nylon 66**	RCL-4536 (30% carbon fiber, 13% PTFE, 2% silicone)	thrust washer		23	40 (0.28)	50 (15.2)	2000 (4.3)	25	30		453
LNP RF-1006	30% glass fiber		RL-4040 (20% PTFE)	"		23	40 (0.28)	50 (15.2)	2000 (4.3)	25	40		453
LNP RF-1006	30% glass fiber		RF-1006 (30% glass fiber)	"		23	40 (0.28)	50 (15.2)	2000 (4.3)	600	600		453
LNP RF-1006	30% glass fiber		RFL-4036 (30% glass fiber, 15% PTFE)	"		23	40 (0.28)	50 (15.2)	2000 (4.3)	350	400		453
LNP RFL-4036	30% glass fiber, 15% PTFE		RF-1006 (30% glass fiber)	"		23	40 (0.28)	50 (15.2)	2000 (4.3)	350	500		453
LNP RFL-4036	30% glass fiber, 15% PTFE		RCL-4036 (30% carbon fiber, 15% PTFE)	"		23	40 (0.28)	50 (15.2)	2000 (4.3)	30	140		453
LNP RFL-4036	30% glass fiber, 15% PTFE		RFL-4036 (30% glass fiber, 15% PTFE)	"		23	40 (0.28)	50 (15.2)	2000 (4.3)	100	115		453
LNP RFL-4036	30% glass fiber, 15% PTFE		RL-4040 (20% PTFE)	"		23	40 (0.28)	50 (15.2)	2000 (4.3)	15	25		453
LNP RL-4040	20% PTFE		RF-1006 (30% glass fiber)	"		23	40 (0.28)	50 (15.2)	2000 (4.3)	15	5		453
LNP RL-4040	20% PTFE		RCL-4036 (30% carbon fiber, 15% PTFE)	"		23	40 (0.28)	50 (15.2)	2000 (4.3)	20	15		453
LNP RL-4040	20% PTFE		RFL-4036 (30% glass fiber, 15% PTFE)	"		23	40 (0.28)	50 (15.2)	2000 (4.3)	30	15		453
LNP RL-4040	20% PTFE		RL-4040 (20% PTFE)	"		23	40 (0.28)	50 (15.2)	2000 (4.3)	35	30		453
LNP RL-4410	2% silicone		RF-1006 (30% glass fiber)	"		23	40 (0.28)	50 (15.2)	2000 (4.3)	350	35		453
LNP RL-4410	2% silicone		RCL-4036 (30% carbon fiber, 15% PTFE)	"		23	40 (0.28)	50 (15.2)	2000 (4.3)	200	30		453
LNP RL-4530	13% PTFE, 2% silicone		"	"		23	40 (0.28)	50 (15.2)	2000 (4.3)	40	20		453
LNP RL-4530	13% PTFE, 2% silicone		RF-1006 (30% glass fiber)	"		23	40 (0.28)	50 (15.2)	2000 (4.3)	425	30		453
LNP RL-4530	13% PTFE, 2% silicone		RFL-4036 (30% glass fiber, 15% PTFE)	"		23	40 (0.28)	50 (15.2)	2000 (4.3)	60	20		453
LNP R-1000	unmodified	**polycarbonate**	D-1000; unmodified	"		23	40 (0.28)	50 (15.2)	2000 (4.3)	54000	975000		453
LNP R-1000	unmodified		DFL-4036 (30% glass fiber, 15% PTFE)	"		23	40 (0.28)	50 (15.2)	2000 (4.3)	1400	10		453
LNP RC-1006	30% carbon fiber		"	"		23	40 (0.28)	50 (15.2)	2000 (4.3)	130	110		453
LNP RF-1006	30% glass fiber		DF-1006 (30% glass fiber)	"		23	40 (0.28)	50 (15.2)	2000 (4.3)	13000	18500		453
LNP RL-4040	20% PTFE		"	"		23	40 (0.28)	50 (15.2)	2000 (4.3)	150	150		453

Material		Mating Surface		Test Conditions						K Factor 10^{-10} x (in^3-min/ft-lb-hr)		Results Note	Source
Supplier / Grade	Material Note	Mating Surface Material	Mating Surface Note	Test Method	Test Conditions Note	Temp. (°C)	Pressure (psi) (MPa)	Speed (fpm) (m/min)	PV (psi x fpm) (MPa x m/min)	Material	Mating Surface		
NYLON 66 against mating surface													
LNP RL-4040	20% PTFE	**polycarbonate**	DFL-4036 (30% glass fiber, 15% PTFE)	thrust washer		23	40 (0.28)	50 (15.2)	2000 (4.3)	23	13		453
LNP RL-4410	2% silicone		"	"		23	40 (0.28)	50 (15.2)	2000 (4.3)	34000	20		453
LNP RL-4410	2% silicone		D-1000; unmodified	"		23	40 (0.28)	50 (15.2)	2000 (4.3)	408	30		453
LNP RL-4530	13% PTFE, 2% silicone		"	"		23	40 (0.28)	50 (15.2)	2000 (4.3)	31	32		453
LNP RL-4530	13% PTFE, 2% silicone		DFL-4036 (30% glass fiber, 15% PTFE)	"		23	40 (0.28)	50 (15.2)	2000 (4.3)	140	20		453
Akzo G-1/30/MS/5	30% long glass fiber, 5% MoS_2	**steel**	carbon steel; surface finish: 16 µin; 18-22 Rockwell C	"	apparatus: Faville-LeVally	23	40 (0.28)	50 (15.2)	2000 (4.3)	75			458
Akzo G-1/30/SI/2	30% long glass fiber, 2% silicone		"	"	"	23	40 (0.28)	50 (15.2)	2000 (4.3)	100			458
Akzo G-1/30/TF/15	30% long glass fiber,15% PTFE		"	"	"	23	40 (0.28)	50 (15.2)	2000 (4.3)	17			458
Akzo J-1/30/MS/5	30% glass fiber, 5% MoS_2		"	"	"	23	40 (0.28)	50 (15.2)	2000 (4.3)	75			458
Akzo J-1/30/SI/3	30% glass fiber, 3% silicone		"	"	"	23	40 (0.28)	50 (15.2)	2000 (4.3)	65			458
Akzo J-1/30/TF/15	30% glass fiber,15% PTFE		"	"	"	23	40 (0.28)	50 (15.2)	2000 (4.3)	17			458
Akzo J-1/33/TF/13/SI/2	33% glass fiber,13% PTFE, 2% silicone		"	"	"	23	40 (0.28)	50 (15.2)	2000 (4.3)	9			458
Akzo J-1/CF/30/TF/13/SI/2	30% carbon fiber, 13% PTFE, 2% silicone		"	"	"	23	40 (0.28)	50 (15.2)	2000 (4.3)	6			458
Akzo J-1/CF/30/TF/15	30% carbon fiber, 15% PTFE		"	"	"	23	40 (0.28)	50 (15.2)	2000 (4.3)	10			458
Akzo NY-1/MS/5	5% MoS_2		"	"	"	23	40 (0.28)	50 (15.2)	2000 (4.3)	160			458
Akzo NY-1/MS/5/TF/30	5% MoS_2, 30% PTFE		"	"	"	23	40 (0.28)	50 (15.2)	2000 (4.3)	12			458
Akzo NY-1/SI/5	5% silicone		"	"	"	23	40 (0.28)	50 (15.2)	2000 (4.3)	40			458
Akzo NY-1/TF/10	10% PTFE		"	"	"	23	40 (0.28)	50 (15.2)	2000 (4.3)	20			458
Akzo NY-1/TF/15	15% PTFE		"	"	"	23	40 (0.28)	50 (15.2)	2000 (4.3)	12			458
Akzo NY-1/TF/30	30% PTFE		"	"	"	23	40 (0.28)	50 (15.2)	2000 (4.3)	10			458
Bay Resins Lubriplas PA-111	unmodified resin			"			25 (0.17)	200 (61)	5000 (10.5)	10			435
Bay Resins Lubriplas PA-111	unmodified resin			"			80 (0.55)	63 (19)	5000 (10.5)	10			435

Material		Mating Surface		Test Conditions						K Factor 10^{-10} x (in³-min/ft-lb-hr)		Results Note	Source
Supplier / Grade	Material Note	Mating Surface Material	Mating Surface Note	Test Method	Test Conditions Note	Temp. (°C)	Pressure (psi) (MPa)	Speed (fpm) (m/min)	PV (psi x fpm) (MPa x m/min)	Material	Mating Surface		
NYLON 66 against mating surface													
Bay Resins Lubriplas PA-111	unmodified resin	steel		thrust washer			20 (0.14)	63 (19)	1260 (2.6)	1150			435
Bay Resins Lubriplas PA-111	unmodified resin			"			10 (0.07)	200 (61)	2000 (4.3)	125			435
Bay Resins Lubriplas PA-111	unmodified resin			"			16 (0.11)	200 (61)	3170 (6.7)	13			435
Bay Resins Lubriplas PA-111	unmodified resin			"			50 (0.34)	63 (19)	3170 (6.7)	173			435
Bay Resins Lubriplas PA-111	unmodified resin			"			40 (0.28)	50 (15.2)	2000 (4.3)	200			435
Bay Resins Lubriplas PA-111	unmodified resin			"			6.3 (0.04)	200 (61)	1260 (2.4)	208			435
Bay Resins Lubriplas PA-111	unmodified resin			"			250 (1.72)	20 (6.1)	5000 (10.5)	24			435
Bay Resins Lubriplas PA-111	unmodified resin			"			158 (1.09)	20 (6.1)	3170 (6.7)	380			435
Bay Resins Lubriplas PA-111	unmodified resin			"			32 (0.22)	63 (19)	2000 (4.2)	480			435
Bay Resins Lubriplas PA-111	unmodified resin			"			100 (0.69)	20 (6.1)	2000 (4.2)	500			435
Bay Resins Lubriplas PA-111	unmodified resin			"			63 (0.43)	20 (6.1)	1260 (2.6)	585			435
Bay Resins Lubriplas PA-111C	<5% MoS_2			"			40 (0.28)	50 (15.2)	2000 (4.3)	150			435
Bay Resins Lubriplas PA-111CF30	30% carbon fiber			"			40 (0.28)	50 (15.2)	2000 (4.3)	20			435
Bay Resins Lubriplas PA-111G13	13% glass fiber			"			40 (0.28)	50 (15.2)	2000 (4.3)	80			435
Bay Resins Lubriplas PA-111G30TF15	30% glass fiber, 15% PTFE			"			40 (0.28)	50 (15.2)	2000 (4.3)	15			435
Bay Resins Lubriplas PA-111G33	33% glass fiber			"			40 (0.28)	50 (15.2)	2000 (4.3)	75			435
Bay Resins Lubriplas PA-111TF20	20% PTFE			"			40 (0.28)	50 (15.2)	2000 (4.3)	12			435
BASF Ultramid A3K	high flow, heat stabilized		Cr 6/800/HV; surface finish: 0.15-0.2 µm	peg-and-disc apparatus			145 (1)	98 (30)	14210 (29.9)			wear rate: 2.8-3.3 µm/km	93
BASF Ultramid A3K	high flow, heat stabilized		Cr 6/800/HV; surface finish: 2.0-2.6 µm	"			145 (1)	98 (30)	14210 (29.9)			wear rate: 7.2-8.6 µm/km	93
BASF Ultramid A3R	noiseless bearings; high flow; PE modified; stabilized		Cr 6/800/HV; surface finish: 0.15-0.2 µm	"			145 (1)	98 (30)	14210 (29.9)			wear rate: 2.5-3.2 µm/km	93
BASF Ultramid A3R	noiseless bearings; high flow; PE modified; stabilized		Cr 6/800/HV; surface finish: 2.0-2.6 µm	"			145 (1)	98 (30)	14210 (29.9)			wear rate: 3.8-5.1 µm/km	93
BASF Ultramid A3WC6	high flow, heat stabilized; 30% carbon fiber		Cr 6/800/HV; surface finish: 0.15-0.2 µm	"			145 (1)	98 (30)	14210 (29.9)			wear rate: 0.4-0.7 µm/km	93

Material		Mating Surface		Test Conditions						K Factor 10^{-10} x (in³-min/ft-lb-hr)		Results Note	Source
Supplier / Grade	Material Note	Mating Surface Material	Mating Surface Note	Test Method	Test Conditions Note	Temp. (°C)	Pressure (psi) (MPa)	Speed (fpm) (m/min)	PV (psi x fpm) (MPa x m/min)	Material	Mating Surface		
NYLON 66 against mating surface													
BASF Ultramid A3WC6	high flow, heat stabilized; 30% carbon fiber	**steel**	Cr 6/800/HV; surface finish: 2.0-2.6 µm	peg-and-disc apparatus			145 (1)	98 (30)	14210 (29.9)			wear rate: 2-3 µm/km	93
BASF Ultramid A3WG6	high flow, heat stabilized; 30% glass fiber		Cr 6/800/HV; surface finish: 0.15-0.2 µm	"			145 (1)	98 (30)	14210 (29.9)			wear rate: 2-3 µm/km	93
BASF Ultramid A3WG6	high flow, heat stabilized; 30% glass fiber		Cr 6/800/HV; surface finish: 2.0-2.6 µm	"			145 (1)	98 (30)	14210 (29.9)			wear rate: 3.5-5 µm/km	93
BASF Ultramid A4	moderate flow		Cr 6/800/HV; surface finish: 0.15-0.2 µm	"			145 (1)	98 (30)	14210 (29.9)			wear rate: 2.8-3.3 µm/km	93
BASF Ultramid A4	moderate flow		Cr 6/800/HV; surface finish: 2.0-2.6 µm	"			145 (1)	98 (30)	14210 (29.9)			wear rate: 7.2-8.6 µm/km	93
DuPont Maranyl A198	formerly by LNP, tribological properties; glass reinforced; w/ graphite		carbon EN8, dry	pad on ring (Amsler wear tester)		20		100 (30.5)				wear rate: 250 µm/hour; maximum load: 110 kg	338
DuPont				thrust washer	ASTM D3702	23	40 (0.28)	50 (15.2)	2000 (4.3)	727			454
DuPont	20% aramid fiber			"	"	23	40 (0.28)	50 (15.2)	2000 (4.3)	88			454
DuPont				"	"	23	250 (1.72)	10 (3)	2500 (5.3)	917			454
DuPont	33% glass fiber			"	"	23	40 (0.28)	50 (15.2)	2000 (4.3)	137			454
DuPont	20% aramid fiber			"	"	23	250 (1.72)	10 (3)	2500 (5.3)	239			454
DuPont	33% glass fiber			"	"	23	250 (1.72)	10 (3)	2500 (5.3)	424			454
Ensinger Ensilon	standard grade						40 (0.28)	50 (15.2)	2000 (4.3)	200			449
Ensinger Ensilon CF20	20% carbon fiber						40 (0.28)	50 (15.2)	2000 (4.3)	4			449
Ensinger Ensilon GF30	30% glass fiber						40 (0.28)	50 (15.2)	2000 (4.3)	75			449
LNP Ny Kon R	<5% MoS_2		carbon steel; surface finish: 12-16 µin; 18-20 Rockwell C	thrust washer		23	40 (0.28)	50 (15.2)	2000 (4.3)	150			453
LNP R-1000	unmodified		"	"		23	40 (0.28)	50 (15.2)	2000 (4.3)	200			453
LNP RAL-4022	10% aramid fiber, 10% PTFE		AISI 304 stainless steel; surface finish: 8-16 µin	"		23	40 (0.28)	50 (15.2)	2000 (4.3)	18	0.2		453
LNP RAL-4022	10% aramid fiber, 10% PTFE		AISI 440 stainless steel; surface finish: 8-16 µin	"		23	40 (0.28)	50 (15.2)	2000 (4.3)	18	0.3		453
LNP RAL-4022	10% aramid fiber, 10% PTFE		AISI 1141; surface finish: 8-12 µin	"		23	40 (0.28)	50 (15.2)	2000 (4.3)	18	0.3		453
LNP RAL-4022	10% aramid fiber, 10% PTFE		AISI 440 stainless steel; surface finish: 50-70 µin	"		23	40 (0.28)	50 (15.2)	2000 (4.3)	23	0.3		453
LNP RAL-4022	10% aramid fiber, 10% PTFE		AISI 1141; surface finish: 12-16 µin	"		23	40 (0.28)	50 (15.2)	2000 (4.3)	13	0.1		453

Material		Mating Surface		Test Conditions						K Factor 10^-10 x (in³-min/ft-lb-hr)		Results Note	Source
Supplier / Grade	Material Note	Mating Surface Material	Mating Surface Note	Test Method	Test Conditions Note	Temp. (°C)	Pressure (psi) (MPa)	Speed (fpm) (m/min)	PV (psi x fpm) (MPa x m/min)	Material	Mating Surface		
NYLON 66 against mating surface													
LNP RAL-4022	10% aramid fiber, 10% PTFE	steel	carbon steel; surface finish: 12-16 µin; 18-20 Rockwell C	thrust washer		23	40 (0.28)	50 (15.2)	2000 (4.3)	13			453
LNP RAL-4022	10% aramid fiber, 10% PTFE		AISI 1141; surface finish: 50-70 µin	"		23	40 (0.28)	50 (15.2)	2000 (4.3)	14	0.2		453
LNP RAL-4022	10% aramid fiber, 10% PTFE		AISI 304 stainless steel; surface finish: 50-70 µin	"		23	40 (0.28)	50 (15.2)	2000 (4.3)	39	0.4		453
LNP RC-1004	20% carbon fiber		carbon steel; surface finish: 12-16 µin; 18-20 Rockwell C	"		23	40 (0.28)	50 (15.2)	2000 (4.3)	40			453
LNP RC-1006	30% carbon fiber		AISI 304 stainless steel; surface finish: 50-70 µin	"		23	40 (0.28)	50 (15.2)	2000 (4.3)	34	0.1		453
LNP RC-1006	30% carbon fiber		AISI 1141; surface finish: 8-12 µin	"		23	40 (0.28)	50 (15.2)	2000 (4.3)	36	1.3		453
LNP RC-1006	30% carbon fiber		AISI 440 stainless steel; surface finish: 8-16 µin	"		23	40 (0.28)	50 (15.2)	2000 (4.3)	41	0.5		453
LNP RC-1006	30% carbon fiber		AISI 440 stainless steel; surface finish: 50-70 µin	"		23	40 (0.28)	50 (15.2)	2000 (4.3)	50	0.1		453
LNP RC-1006	30% carbon fiber		AISI 1141; surface finish: 12-16 µin	"		23	40 (0.28)	50 (15.2)	2000 (4.3)	20	0.6		453
LNP RC-1006	30% carbon fiber		carbon steel; surface finish: 12-16 µin; 18-20 Rockwell C	"		23	40 (0.28)	50 (15.2)	2000 (4.3)	20			453
LNP RC-1006	30% carbon fiber		AISI 304 stainless steel; surface finish: 8-16 µin	"		23	40 (0.28)	50 (15.2)	2000 (4.3)	24	0.1		453
LNP RC-1006	30% carbon fiber		AISI 1141; surface finish: 50-70 µin	"		23	40 (0.28)	50 (15.2)	2000 (4.3)	30	0.5		453
LNP RC-1008	40% carbon fiber		carbon steel; surface finish: 12-16 µin; 18-20 Rockwell C	"		23	40 (0.28)	50 (15.2)	2000 (4.3)	14			453
LNP RCL-4036	30% carbon fiber, 15% PTFE		"	"		23	40 (0.28)	50 (15.2)	2000 (4.3)	10			453
LNP RCL-4036	30% carbon fiber, 15% PTFE		AISI 1141; surface finish: 12-16 µin	"		23	40 (0.28)	50 (15.2)	2000 (4.3)	13	0.4		453
LNP RCL-4036	30% carbon fiber, 15% PTFE		AISI 304 stainless steel; surface finish: 8-16 µin	"		23	40 (0.28)	50 (15.2)	2000 (4.3)	15	0.2		453
LNP RCL-4036	30% carbon fiber, 15% PTFE		AISI 1141; surface finish: 50-70 µin	"		23	40 (0.28)	50 (15.2)	2000 (4.3)	15	0.5		453
LNP RCL-4036	30% carbon fiber, 15% PTFE		AISI 440 stainless steel; surface finish: 8-16 µin	"		23	40 (0.28)	50 (15.2)	2000 (4.3)	16	0.2		453
LNP RCL-4036	30% carbon fiber, 15% PTFE		AISI 304 stainless steel; surface finish: 50-70 µin	"		23	40 (0.28)	50 (15.2)	2000 (4.3)	17	0.3		453
LNP RCL-4036	30% carbon fiber, 15% PTFE		AISI 1141; surface finish: 8-12 µin	"		23	40 (0.28)	50 (15.2)	2000 (4.3)	27	1		453
LNP RCL-4036	30% carbon fiber, 15% PTFE		AISI 440 stainless steel; surface finish: 50-70 µin	"		23	40 (0.28)	50 (15.2)	2000 (4.3)	45	0.4		453
LNP RCL-4536	30% carbon fiber, 13% PTFE, 2% silicone		carbon steel; surface finish: 12-16 µin; 18-20 Rockwell C	"		23	40 (0.28)	50 (15.2)	2000 (4.3)	6			453

Material		Mating Surface		Test Conditions						K Factor 10⁻¹⁰ x (in³-min/ft-lb-hr)		Results Note	Source
Supplier / Grade	Material Note	Mating Surface Material	Mating Surface Note	Test Method	Test Conditions Note	Temp. (°C)	Pressure (psi) (MPa)	Speed (fpm) (m/min)	PV (psi x fpm) (MPa x m/min)	Material	Mating Surface		
NYLON 66 against mating surface		→											
LNP RF 100-10	50% glass fiber	**steel**	carbon steel; surface finish: 12-16 µin; 18-20 Rockwell C	thrust washer		23	40 (0.28)	50 (15.2)	2000 (4.3)	60			453
LNP RF-1002	10% glass fiber		"	"		23	40 (0.28)	50 (15.2)	2000 (4.3)	80			453
LNP RF-1004	20% glass fiber		"	"		23	40 (0.28)	50 (15.2)	2000 (4.3)	80			453
LNP RF-1006	30% glass fiber		AISI 1141; surface finish: 12-16 µin	"		23	40 (0.28)	50 (15.2)	2000 (4.3)	75	1.1		453
LNP RF-1006	30% glass fiber		AISI 304 stainless steel; surface finish: 8-16 µin	"		23	40 (0.28)	50 (15.2)	2000 (4.3)	33	0.5		453
LNP RF-1006	30% glass fiber		AISI 304 stainless steel; surface finish: 50-70 µin	"		23	40 (0.28)	50 (15.2)	2000 (4.3)	45	0.1		453
LNP RF-1006	30% glass fiber		AISI 440 stainless steel; surface finish: 8-16 µin	"		23	40 (0.28)	50 (15.2)	2000 (4.3)	48	0.5		453
LNP RF-1006	30% glass fiber		AISI 440 stainless steel; surface finish: 50-70 µin	"		23	40 (0.28)	50 (15.2)	2000 (4.3)	49	0.2		453
LNP RF-1006	30% glass fiber		AISI 1141; surface finish: 50-70 µin	"		23	40 (0.28)	50 (15.2)	2000 (4.3)	100	0.8		453
LNP RF-1006	30% glass fiber		AISI 1141; surface finish: 8-12 µin	"		23	40 (0.28)	50 (15.2)	2000 (4.3)	142	2.2		453
LNP RF-1006 HS	30% glass fiber		carbon steel; surface finish: 12-16 µin; 18-20 Rockwell C	"		23	40 (0.28)	50 (15.2)	2000 (4.3)	75			453
LNP RF-1008	40% glass fiber		"	"		23	40 (0.28)	50 (15.2)	2000 (4.3)	70			453
LNP RFL-4036	30% glass fiber, 15% PTFE		cold rolled steel; surface finish: 12-16 µin; 22 Rockwell C	"		93	40 (0.28)	50 (15.2)	2000 (4.3)	60			453
LNP RFL-4036	30% glass fiber, 15% PTFE		"	"		204	40 (0.28)	50 (15.2)	2000 (4.3)	700			453
LNP RFL-4036	30% glass fiber, 15% PTFE		AISI 304 stainless steel; surface finish: 8-16 µin	"		23	40 (0.28)	50 (15.2)	2000 (4.3)	12	0.2		453
LNP RFL-4036	30% glass fiber, 15% PTFE		AISI 304 stainless steel; surface finish: 50-70 µin	"		23	40 (0.28)	50 (15.2)	2000 (4.3)	13	0.2		453
LNP RFL-4036	30% glass fiber, 15% PTFE		AISI 440 stainless steel; surface finish: 50-70 µin	"		23	40 (0.28)	50 (15.2)	2000 (4.3)	16	0.2		453
LNP RFL-4036	30% glass fiber, 15% PTFE		AISI 1141; surface finish: 50-70 µin	"		23	40 (0.28)	50 (15.2)	2000 (4.3)	16	0.6		453
LNP RFL-4036	30% glass fiber, 15% PTFE		AISI 1141; surface finish: 12-16 µin	"		23	40 (0.28)	50 (15.2)	2000 (4.3)	16	0.6		453
LNP RFL-4036	30% glass fiber, 15% PTFE		carbon steel; surface finish: 12-16 µin; 18-20 Rockwell C	"		23	40 (0.28)	50 (15.2)	2000 (4.3)	16			453
LNP RFL-4036	30% glass fiber, 15% PTFE		cold rolled steel; surface finish: 12-16 µin; 22 Rockwell C	"		149	40 (0.28)	50 (15.2)	2000 (4.3)	300			453
LNP RFL-4036	30% glass fiber, 15% PTFE		AISI 440 stainless steel; surface finish: 8-16 µin	"		23	40 (0.28)	50 (15.2)	2000 (4.3)	22	0.2		453

Material		Mating Surface		Test Conditions						K Factor 10^{-10} x (in^3·min/ft·lb·hr)		Results Note	Source
Supplier / Grade	Material Note	Mating Surface Material	Mating Surface Note	Test Method	Test Conditions Note	Temp. (°C)	Pressure (psi) (MPa)	Speed (fpm) (m/min)	PV (psi x fpm) (MPa x m/min)	Material	Mating Surface		
NYLON 66 against mating surface													
LNP RFL-4036	30% glass fiber, 15% PTFE	**steel**	AISI 1141; surface finish: 8-12 µin	thrust washer		23	40 (0.28)	50 (15.2)	2000 (4.3)	30	1.1		453
LNP RFL-4216	30% glass fiber, <5% MoS_2		carbon steel; surface finish: 12-16 µin; 18-20 Rockwell C	“		23	40 (0.28)	50 (15.2)	2000 (4.3)	75			453
LNP RFL-4218	40% glass fiber, <5% MoS_2		“	“		23	40 (0.28)	50 (15.2)	2000 (4.3)	70			453
LNP RFL-4416	30% glass fiber, 2% silicone		“	“		23	40 (0.28)	50 (15.2)	2000 (4.3)	65			453
LNP RFL-4536	30% glass fiber, 13% PTFE, 2% silicone		AISI 1141; surface finish: 12-16 µin	“		23	40 (0.28)	50 (15.2)	2000 (4.3)	9	0.6		453
LNP RFL-4536	30% glass fiber, 13% PTFE, 2% silicone		carbon steel; surface finish: 12-16 µin; 18-20 Rockwell C	“		23	40 (0.28)	50 (15.2)	2000 (4.3)	9			453
LNP RFL-4536	30% glass fiber, 13% PTFE, 2% silicone		AISI 440 stainless steel; surface finish: 50-70 µin	“		23	40 (0.28)	50 (15.2)	2000 (4.3)	18	0.3		453
LNP RFL-4536	30% glass fiber, 13% PTFE, 2% silicone		AISI 1141; surface finish: 8-12 µin	“		23	40 (0.28)	50 (15.2)	2000 (4.3)	20	0.9		453
LNP RFL-4536	30% glass fiber, 13% PTFE, 2% silicone		AISI 1141; surface finish: 50-70 µin	“		23	40 (0.28)	50 (15.2)	2000 (4.3)	20	1		453
LNP RFL-4536	30% glass fiber, 13% PTFE, 2% silicone		AISI 304 stainless steel; surface finish: 50-70 µin	“		23	40 (0.28)	50 (15.2)	2000 (4.3)	26	0.7		453
LNP RFL-4536	30% glass fiber, 13% PTFE, 2% silicone		AISI 440 stainless steel; surface finish: 8-16 µin	“		23	40 (0.28)	50 (15.2)	2000 (4.3)	59	2		453
LNP RFL-4536	30% glass fiber, 13% PTFE, 2% silicone		AISI 304 stainless steel; surface finish: 8-16 µin	“		23	40 (0.28)	50 (15.2)	2000 (4.3)	15	0.2		453
LNP RFL-4616	30% glass fiber, 2% silicone		carbon steel; surface finish: 12-16 µin; 18-20 Rockwell C	“		23	40 (0.28)	50 (15.2)	2000 (4.3)	100			453
LNP RL-4010	5% PTFE		“	“		23	40 (0.28)	50 (15.2)	2000 (4.3)	80			453
LNP RL-4040	20% PTFE		AISI 304 stainless steel; surface finish: 8-16 µin	“		23	40 (0.28)	50 (15.2)	2000 (4.3)	7	0.1		453
LNP RL-4040	20% PTFE		AISI 440 stainless steel; surface finish: 8-16 µin	“		23	40 (0.28)	50 (15.2)	2000 (4.3)	7	0.1		453
LNP RL-4040	20% PTFE		AISI 1141; surface finish: 12-16 µin	“		23	40 (0.28)	50 (15.2)	2000 (4.3)	12	0.1		453
LNP RL-4040	20% PTFE		AISI 440 stainless steel; surface finish: 50-70 µin	“		23	40 (0.28)	50 (15.2)	2000 (4.3)	12	0.2		453
LNP RL-4040	20% PTFE		carbon steel; surface finish: 12-16 µin; 18-20 Rockwell C	“		23	40 (0.28)	50 (15.2)	2000 (4.3)	12			453
LNP RL-4040	20% PTFE		AISI 304 stainless steel; surface finish: 50-70 µin	“		23	40 (0.28)	50 (15.2)	2000 (4.3)	13	0.3		453
LNP RL-4040	20% PTFE		AISI 1141; surface finish: 8-12 µin	“		23	40 (0.28)	50 (15.2)	2000 (4.3)	16	0.2		453
LNP RL-4040	20% PTFE		AISI 1141; surface finish: 50-70 µin	“		23	40 (0.28)	50 (15.2)	2000 (4.3)	24	0.2		453

Material		Mating Surface		Test Conditions						K Factor 10^{-10} x (in^3-min/ft-lb-hr)		Results Note	Source
Supplier / Grade	Material Note	Mating Surface Material	Mating Surface Note	Test Method	Test Conditions Note	Temp. (°C)	Pressure (psi) (MPa)	Speed (fpm) (m/min)	PV (psi x fpm) (MPa x m/min)	Material	Mating Surface		
NYLON 66 against mating surface													
LNP RL-4040FR(94VO)	flame retardant, 20% PTFE	**steel**	carbon steel; surface finish: 12-16 μin; 18-20 Rockwell C	thrust washer		23	40 (0.28)	50 (15.2)	2000 (4.3)	25			453
LNP RL-4310	5% graphite		"	"		23	40 (0.28)	50 (15.2)	2000 (4.3)	55			453
LNP RL-4410	2% silicone		"	"		23	40 (0.28)	50 (15.2)	2000 (4.3)	40			453
LNP RL-4540	18% PTFE, 2% silicone		"	"		23	40 (0.28)	50 (15.2)	2000 (4.3)	6			453
LNP RL-4610	2% silicone		"	"		23	40 (0.28)	50 (15.2)	2000 (4.3)	155			453
LNP RL-4730	13% PTFE, 2% silicone		"	"		23	40 (0.28)	50 (15.2)	2000 (4.3)	25			453
LNP Verton RF-700-1OHS	50% glass fiber		"	"		23	40 (0.28)	50 (15.2)	2000 (4.3)	30			453
LNP Verton RF-7007HS	35% glass fiber		"	"		23	40 (0.28)	50 (15.2)	2000 (4.3)	40			453
Polymer Nylatron NS	w/ additives to improve bearing properties			journal bearing			42.2 (0.29)	118 (36)	5000 (10.5)	30			441
RTP 200			C1018 steel; surface finish: 14-17 μin; 15-25 Rockwell C	thrust washer	apparatus: Falex Model No. 6	23	40 (0.28)	50 (15.2)	2000 (4.3)	200			457
RTP 200 AR 15	15% aramid fiber		"	"	"	23	40 (0.28)	50 (15.2)	2000 (4.3)	50			457
RTP 200 AR 15 TFE 15	15% PTFE, 15% aramid fiber		"	"	"	23	40 (0.28)	50 (15.2)	2000 (4.3)	10			457
RTP 200 TFE 10	10% PTFE		"	"	"	23	40 (0.28)	50 (15.2)	2000 (4.3)	35			457
RTP 200 TFE 20	20% PTFE		"	"	"	23	40 (0.28)	50 (15.2)	2000 (4.3)	13			457
RTP 200 TFE 20 SI	20% PTFE, 0.5% silicone		"	"	"	23	40 (0.28)	50 (15.2)	2000 (4.3)	10			457
RTP 201 TFE 5 SI	5% PTFE, 0.5% silicone, 10% glass fiber		"	"	"	23	40 (0.28)	50 (15.2)	2000 (4.3)	50			457
RTP 203 TFE 10	10% PTFE, 20% glass fiber		"	"	"	23	40 (0.28)	50 (15.2)	2000 (4.3)	40			457
RTP 203 TFE 15	15% PTFE, 20% glass fiber		"	"	"	23	40 (0.28)	50 (15.2)	2000 (4.3)	20			457
RTP 203 TFE 20	20% PTFE, 20% glass fiber		"	"	"	23	40 (0.28)	50 (15.2)	2000 (4.3)	15			457
RTP 205	30% glass fiber		"	"	"	23	40 (0.28)	50 (15.2)	2000 (4.3)	75			457
RTP 205 SI 2	2% silicone, 30% glass fiber		"	"	"	23	40 (0.28)	50 (15.2)	2000 (4.3)	50			457
RTP 205 TFE 13 SI 2	13% PTFE, 2% silicone, 30% glass fiber		"	"	"	23	40 (0.28)	50 (15.2)	2000 (4.3)	10			457

Material		Mating Surface		Test Conditions						K Factor 10^{-10} x (in³-min/ft-lb-hr)		Results Note	Source
Supplier / Grade	Material Note	Mating Surface Material	Mating Surface Note	Test Method	Test Conditions Note	Temp. (°C)	Pressure (psi) (MPa)	Speed (fpm) (m/min)	PV (psi x fpm) (MPa x m/min)	Material	Mating Surface		
NYLON 66 against mating surface →													
RTP 205 TFE 15	30% glass fiber	**steel**	C1018 steel; surface finish: 14-17 µin; 15-25 Rockwell C	thrust washer	apparatus: Falex Model No. 6	23	40 (0.28)	50 (15.2)	2000 (4.3)	20			457
RTP 205 TFE 20	20% PTFE, 30% glass fiber		"	"	"	23	40 (0.28)	50 (15.2)	2000 (4.3)	15			457
RTP 205 TFE 5	5% PTFE, 30% glass fiber		"	"	"	23	40 (0.28)	50 (15.2)	2000 (4.3)	50			457
RTP 281 TFE 20	20% PTFE, 10% carbon fiber		"	"	"	23	40 (0.28)	50 (15.2)	2000 (4.3)	25			457
RTP 283 TFE 1 0	10% PTFE, 20% carbon fiber		"	"	"	23	40 (0.28)	50 (15.2)	2000 (4.3)	30			457
RTP 285	30% carbon fiber		"	"	"	23	40 (0.28)	50 (15.2)	2000 (4.3)	25			457
RTP 285 TFE 13 SI 2	13% PTFE, 2% silicone, 30% carbon fiber		"	"	"	23	40 (0.28)	50 (15.2)	2000 (4.3)	8			457
RTP 285 TFE 15	15% PTFE, 30% carbon fiber		"	"	"	23	40 (0.28)	50 (15.2)	2000 (4.3)	12			457
RTP 287 TFE 10	10% PTFE, 40% carbon fiber		"	"	"	23	40 (0.28)	50 (15.2)	2000 (4.3)	14			457
Thermofil N3-13FG-0100	13% glass fiber			ASTM D1894		23				85			459
Thermofil N3-13FG-0700	13% glass fiber, impact modified			"		23				90			459
Thermofil N3-15G-0560	15% glass fiber, flame retardant			"		23				90			459
Thermofil N3-20NF-0100	20% graphite fiber			"		23				40			459
Thermofil N3-30FG-0214	30% glass fiber, PTFE lubricated			"		23				16			459
Thermofil N3-30FG-0231	30% glass fiber, MoS_2 lubricated			"		23				40			459
Thermofil N3-30FG-0282	30% glass fiber, lubricated			"		23				9			459
Thermofil N3-30FG-0560	30% glass fiber, flame retardant			"		23				80			459
Thermofil N3-30NF-0214	30% graphite fiber, PTFE lubricated			"		23				6			459
Thermofil N3-33FG-0100	33% glass fiber			"		23				75			459
Thermofil N3-40NF-0100	40% graphite fiber			"		23				14			459
Thermofil N3-43FG-0100	43% glass fiber			"		23				70			459
Thermofil N3-9900-0200	lubricated			"		23				35			459
Thermofil N3-9900-0214	PTFE lubricated			"		23				15			459
Thermofil N3-9900-0231	MoS_2 lubricated			"		23				30			459
Thermofil N3-9900-0279	40% carbon fiber			"		23				30			459
Thermofil N3-9900-0560	flame retardant			"		23				100			459

Material		Mating Surface		Test Conditions						K Factor 10^{-10} x (in³-min/ft-lb-hr)		Results Note	Source
Supplier / Grade	Material Note	Mating Surface Material	Mating Surface Note	Test Method	Test Conditions Note	Temp. (°C)	Pressure (psi) (MPa)	Speed (fpm) (m/min)	PV (psi x fpm) (MPa x m/min)	Material	Mating Surface		
NYLON 66 against mating surface													
Unspecified grade	unmodified	**steel**	carbon steel; surface finish: 16 µin; 18-22 Rockwell C	thrust washer	apparatus: Faville-LeVally	23	40 (0.28)	50 (15.2)	2000 (4.3)	200			458
Unspecified grade	unmodified		unlubricated	"	ASTM D3702		150 (1.03)	100 (30.5)	15000 (31.4)	21		surface cracks occured	454
Unspecified grade	unmodified		"	"	"		1000 (6.89)	10 (3)	10000 (20.7)	211		surface cracks occured	454
Unspecified grade	unmodified		"	"	"		500 (3.45)	10 (3)	5000 (10.5)	419			454
Unspecified grade	unmodified		"	"	"		25 (0.17)	100 (30.5)	2500 (5.3)	424			454
Unspecified grade	unmodified		"	"	"		100 (0.69)	100 (30.5)	10000 (21)	550		surface cracks occured	454
Unspecified grade	unmodified		"	"	"		50 (0.34)	10 (3)	500 (1.1)	566			454
Unspecified grade	unmodified		"	"	"		50 (0.34)	100 (30.5)	5000 (10.5)	574		surface cracks occured	454
Unspecified grade	unmodified		"	"	"		1500 (10.34)	10 (3)	15000 (31)	637		surface cracks occured	454
Unspecified grade	unmodified		"	"	"		100 (0.69)	10 (3)	1000 (2.1)	695			454
Unspecified grade	general purpose			journal bearing			42.2 (0.29)	118 (36)	5000 (10.5)	72			441
Unspecified grade	unmodified		unlubricated	thrust washer	ASTM D3702		250 (1.72)	10 (3)	2500 (5.3)	911			454
POLYPHTHALAMIDE against mating surface													
RTP 4005 TFE 15	15% PTFE, 30% glass fiber	**steel**	C1018 steel; surface finish: 14-17 µin; 15-25 Rockwell C	thrust washer	apparatus: Falex Model No. 6	23	40 (0.28)	50 (15.2)	2000 (4.3)	25			457
RTP 4080 TFE 15	15% PTFE, 30% carbon fiber		"	"	"	23	40 (0.28)	50 (15.2)	2000 (4.3)	20			460
RTP 4083 TFE 15 S12	15% PTFE, 2% silicone, 20% carbon fiber		"	"	"	23	40 (0.28)	50 (15.2)	2000 (4.3)	15			457
POLYBENZIMIDAZOLE against mating surface													
Hoechst Cel. Celazole DL80	lubricated, heterocyclic, 50 Rockwell A	**steel**	AISI 1018, surface finish: 16 RMS	thrust bearing	apparatus: LRI-1 Wear and Friction Tester		100 (0.69)	984 (300)	98400 (207)			wear rate: 10 µm/hour; sliding surface temp.: 156°C	318
Hoechst Cel. Celazole DL80	lubricated, 50 Rockwell A		"	"	test duration: 4.5 hours; apparatus: LRI-1 Wear and Friction Tester		100 (0.69)	1000 (304.8)	100000 (210)	39			320

Material		Mating Surface		Test Conditions						K Factor 10^{-10} x (in³-min/ft-lb-hr)		Results Note	Source
Supplier / Grade	Material Note	Mating Surface Material	Mating Surface Note	Test Method	Test Conditions Note	Temp. (°C)	Pressure (psi) (MPa)	Speed (fpm) (m/min)	PV (psi x fpm) (MPa x m/min)	Material	Mating Surface		
POLYBENZIMIDAZOLE against mating surface													
Hoechst Cel. Celazole DTL60	lubricated, melt processable, heterocyclic, 25 Rockwell K	**steel**	AISI 1018, surface finish: 16 RMS	thrust bearing	apparatus: LRI-1 Wear and Friction Tester		100 (0.69)	984 (300)	98400 (207)			wear rate: 11 µm/hour; sliding surface temp.: 145°C	318
Hoechst Cel. Celazole DTU60	melt processable, heterocyclic, 50 Rockwell K		"	"	"		2000 (13.8)	12.5 (3.8)	25000 (52.6)			wear rate: 3 µm/hour; sliding surface temp.: 148°C	318
Hoechst Cel. Celazole U60			"	"	"		100 (0.69)	500 (152)	50000 (105.1)	7.5			320
Hoechst Cel. Celazole U60			"	"	"		20 (0.14)	100 (30.5)	2000 (4.3)	9.66			320
Hoechst Cel. Celazole U60	heterocyclic, virgin resin, 50 Rockwell A		"	"	"		2000 (13.8)	12.5 (3.8)	25000 (52.6)			wear rate: 12 µm/hour; sliding surface temp.: 153°C	318
Hoechst Cel. Celazole U60			"	"	"		100 (0.69)	100 (30.5)	10000 (21)	6.9			320
Hoechst Cel. Celazole U60			"	"	"		40 (0.28)	50 (15.2)	2000 (4.3)	1.17			320
Hoechst Cel. Celazole U60			"	"	"		50 (0.34)	20 (6.1)	10000 (2.1)	22.5			320
Hoechst Cel. Celazole U60			"	"	"		1000 (6.890000000000001)	50 (15.2)	50000 (105.1)	28.9			320
Hoechst Cel. Celazole U60			"	"	"		50 (0.34)	800 (244)	40000 (84.09999999999999)	3.2			320
Hoechst Cel. Celazole U60	heterocyclic, virgin resin, 50 Rockwell A		"	"	"		100 (0.69)	984 (300)	98400 (207)			wear rate: 363 µm/hour; sliding surface temp.: 393°C	318
POLYCARBONATE against mating surface													
LNP DFL-4044	20% glass fiber, 20% PTFE	**ABS**	A-1000; unmodified	thrust washer		23	40 (0.28)	50 (15.2)	2000 (4.3)	65	1900		453
LNP DFL-4036	30% glass fiber, 15% PTFE	**brass**	70/30 brass; surface finish: 8-16 µin	"		23	40 (0.28)	50 (15.2)	2000 (4.3)	22	8		453
LNP DFL-4036	30% glass fiber, 15% PTFE		70/30 brass; surface finish: 50-70 µin	"		23	40 (0.28)	50 (15.2)	2000 (4.3)	12	4		453
LNP DFL-4034	20% glass fiber, 15% PTFE	**nylon 610**	QFL-4036 (30% glass fiber, 15% PTFE)	"		23	40 (0.28)	50 (15.2)	2000 (4.3)	14	18		453
LNP DL-4020	10% PTFE		"	"		23	40 (0.28)	50 (15.2)	2000 (4.3)	30	15		453
LNP D-1000	unmodified	**nylon 66**	RF-1006 (30% glass fiber)	"		23	40 (0.28)	50 (15.2)	2000 (4.3)	275000	3000		453
LNP D-1000	unmodified		RL-4530 (13% PTFE, 2% silicone)	"		23	40 (0.28)	50 (15.2)	2000 (4.3)	4500	100		453

Material		Mating Surface		Test Conditions						K Factor 10⁻¹⁰ x (in³-min/ft-lb-hr)		Results Note	Source
Supplier / Grade	Material Note	Mating Surface Material	Mating Surface Note	Test Method	Test Conditions Note	Temp. (°C)	Pressure (psi) (MPa)	Speed (fpm) (m/min)	PV (psi x fpm) (MPa x m/min)	Material	Mating Surface		
POLYCARBONATE against mating surface →													
LNP D-1000	unmodified	**nylon 66**	R-1000 (unmodified)	thrust washer		23	40 (0.28)	50 (15.2)	2000 (4.3)	49000	1200		453
LNP DCL-4042	10% carbon fiber, 20% PTFE		RCL-4036 (30% carbon fiber, 15% PTFE)	"		23	40 (0.28)	50 (15.2)	2000 (4.3)	650	80		453
LNP DCL-4042	10% carbon fiber, 20% PTFE		"	"		23	40 (0.28)	50 (15.2)	2000 (4.3)	650	80		453
LNP DCL-4536	30% carbon fiber, 13% PTFE, 2% silicone		RCL-4536 (30% carbon fiber, 13% PTFE, 2% silicone)	"		23	40 (0.28)	50 (15.2)	2000 (4.3)	200	150		453
LNP DF-1006	30% glass fiber		RFL-4036 (30% glass fiber, 15% PTFE)	"		23	40 (0.28)	50 (15.2)	2000 (4.3)	7000	460		453
LNP DFL-4036	30% glass fiber, 15% PTFE		RF-1006 (30% glass fiber)	"		23	40 (0.28)	50 (15.2)	2000 (4.3)	190	165		453
LNP DFL-4036	30% glass fiber, 15% PTFE		RCL-4036 (30% carbon fiber, 15% PTFE)	"		23	40 (0.28)	50 (15.2)	2000 (4.3)	100	120		453
LNP DFL-4036	30% glass fiber, 15% PTFE		RFL-4036 (30% glass fiber, 15% PTFE)	"		23	40 (0.28)	50 (15.2)	2000 (4.3)	17	34		453
LNP DF-1004	20% glass fiber	**polycarbonate**	DF-1004 (20% glass fiber)	"		23	40 (0.28)	50 (15.2)	2000 (4.3)	3500	4000		453
LNP DFL-4036	30% glass fiber, 15% PTFE		DFL-4036 (30% glass fiber, 15% PTFE)	"		23	40 (0.28)	50 (15.2)	2000 (4.3)	310	350		453
LNP DFL-4044	20% glass fiber, 20% PTFE		D-1000; unmodified	"		23	40 (0.28)	50 (15.2)	2000 (4.3)	210	750		453
LNP DFL-4036	30% glass fiber, 15% PTFE	**polyphenylene sulfide**	OFL-4036 (30% glass fiber, 15% PTFE)	"		23	40 (0.28)	50 (15.2)	2000 (4.3)	700	550		453
Akzo G-50/20/TF/10	20% long glass fiber, 10% PTFE	**steel**	carbon steel; surface finish: 16 µin; 18-22 Rockwell C	"	apparatus: Faville-LeVally	23	40 (0.28)	50 (15.2)	2000 (4.3)	50			458
Akzo G-50/20/TF/15	20% long glass fiber, 15% PTFE		"	"	"	23	40 (0.28)	50 (15.2)	2000 (4.3)	40			458
Akzo J-50/20/TF/15	20% glass fiber, 15% PTFE		"	"	"	23	40 (0.28)	50 (15.2)	2000 (4.3)	40			458
Akzo J-50/20/TF/l 0	20% glass fiber, 10% PTFE		"	"	"	23	40 (0.28)	50 (15.2)	2000 (4.3)	40			458
Akzo J-50/30/TF/10	30% glass fiber, 10% PTFE		"	"	"	23	40 (0.28)	50 (15.2)	2000 (4.3)	40			458
Akzo J-50/3O/TF/15	30% glass fiber, 15% PTFE		"	"	"	23	40 (0.28)	50 (15.2)	2000 (4.3)	30			458
Akzo PC-50/TF/10	10% PTFE		"	"	"	23	40 (0.28)	50 (15.2)	2000 (4.3)	85			458
Akzo PC-50/TF/15	15% PTFE		"	"	"	23	40 (0.28)	50 (15.2)	2000 (4.3)	75			458
Akzo PC-50/TF/30	30% PTFE		"	"	"	23	40 (0.28)	50 (15.2)	2000 (4.3)	65			458
Bay Resins Lubriplas PC-1100	unmodified resin			"			40 (0.28)	50 (15.2)	2000 (4.3)	2500			435

Material		Mating Surface		Test Conditions						K Factor 10^{-10} x (in^3-min/ft-lb-hr)		Results Note	Source
Supplier / Grade	Material Note	Mating Surface Material	Mating Surface Note	Test Method	Test Conditions Note	Temp. (°C)	Pressure (psi) (MPa)	Speed (fpm) (m/min)	PV (psi x fpm) (MPa x m/min)	Material	Mating Surface		
POLYCARBONATE against mating surface													
Bay Resins Lubriplas PC-1100CF30	30% carbon fiber	steel		thrust washer			40 (0.28)	50 (15.2)	2000 (4.3)	80			435
Bay Resins Lubriplas PC-1100G10	10% glass fiber			"			40 (0.28)	50 (15.2)	2000 (4.3)	200			435
Bay Resins Lubriplas PC-1100G30	30% glass fiber			"			40 (0.28)	50 (15.2)	2000 (4.3)	175			435
Bay Resins Lubriplas PC-1100G30TF15	30% glass fiber, 15% PTFE			"			40 (0.28)	50 (15.2)	2000 (4.3)	30			435
Bay Resins Lubriplas PC-1100G40	40% glass fiber			"			40 (0.28)	50 (15.2)	2000 (4.3)	165			435
Bay Resins Lubriplas PC-1100TF15	15% PTFE			"			40 (0.28)	50 (15.2)	2000 (4.3)	75			435
Bay Resins Lubriplas PC-1100TF20	20% PTFE			"			40 (0.28)	50 (15.2)	2000 (4.3)	70			435
Ensinger Ensicar	standard grade						40 (0.28)	50 (15.2)	2000 (4.3)	2500			449
GE Lexan 3412	20% glass fiber						40 (0.28)	50 (15.2)	2000 (4.3)	200			449
LNP D-1000	unmodified		carbon steel; surface finish: 12-16 µin; 18-20 Rockwell C	thrust washer		23	40 (0.28)	50 (15.2)	2000 (4.3)	2500			453
LNP DC-1006	30% carbon fiber		"	"		23	40 (0.28)	50 (15.2)	2000 (4.3)	85			453
LNP DCL-4532	10% carbon fiber, 13% PTFE, 2% silicone		"	"		23	40 (0.28)	50 (15.2)	2000 (4.3)	10			453
LNP DF-1006	30% glass fiber		"	"		23	40 (0.28)	50 (15.2)	2000 (4.3)	180			453
LNP DFL-4034	20% glass fiber, 15% PTFE		"	"		23	40 (0.28)	50 (15.2)	2000 (4.3)	40			453
LNP DFL-4036	30% glass fiber, 15% PTFE		AISI 440 stainless steel; surface finish: 50-70 µin	"		23	40 (0.28)	50 (15.2)	2000 (4.3)	50	1.3		453
LNP DFL-4036	30% glass fiber, 15% PTFE		AISI 1141; surface finish: 50-70 µin	"		23	40 (0.28)	50 (15.2)	2000 (4.3)	52	4.2		453
LNP DFL-4036	30% glass fiber, 15% PTFE		AISI 304 stainless steel; surface finish: 8-16 µin	"		23	40 (0.28)	50 (15.2)	2000 (4.3)	20	0.4		453
LNP DFL-4036	30% glass fiber, 15% PTFE		AISI 1141; surface finish: 12-16 µin	"		23	40 (0.28)	50 (15.2)	2000 (4.3)	22	3.5		453
LNP DFL-4036	30% glass fiber, 15% PTFE		AISI 440 stainless steel; surface finish: 8-16 µin	"		23	40 (0.28)	50 (15.2)	2000 (4.3)	24	0.5		453
LNP DFL-4036	30% glass fiber, 15% PTFE		AISI 304 stainless steel; surface finish: 50-70 µin	"		23	40 (0.28)	50 (15.2)	2000 (4.3)	29	0.5		453
LNP DFL-4036	30% glass fiber, 15% PTFE		carbon steel; surface finish: 12-16 µin; 18-20 Rockwell C	"		23	40 (0.28)	50 (15.2)	2000 (4.3)	30			453
LNP DFL-4036	30% glass fiber, 15% PTFE		AISI 1141; surface finish: 8-12 µin	"		23	40 (0.28)	50 (15.2)	2000 (4.3)	75	3.9		453

Material		Mating Surface		Test Conditions						K Factor 10^-10 x (in³-min/ft-lb-hr)		Results Note	Source
Supplier / Grade	Material Note	Mating Surface Material	Mating Surface Note	Test Method	Test Conditions Note	Temp. (°C)	Pressure (psi) (MPa)	Speed (fpm) (m/min)	PV (psi x fpm) (MPa x m/min)	Material	Mating Surface		
POLYCARBONATE against mating surface →													
LNP DFL-4536	30% glass fiber, 13% PTFE, 2% silicone	**steel**	carbon steel; surface finish: 12-16 μin; 18-20 Rockwell C	thrust washer		23	40 (0.28)	50 (15.2)	2000 (4.3)	27			453
LNP DL 4040	20% PTFE		"	"		23	40 (0.28)	50 (15.2)	2000 (4.3)	70			453
LNP DL-4010	5% PTFE		"	"		23	40 (0.28)	50 (15.2)	2000 (4.3)	125			453
LNP DL-4020	10% PTFE		"	"		23	40 (0.28)	50 (15.2)	2000 (4.3)	85			453
LNP DL-4030	15% PTFE		"	"		23	40 (0.28)	50 (15.2)	2000 (4.3)	75			453
LNP DL-4530	13% PTFE, 2% silicone		"	"		23	40 (0.28)	50 (15.2)	2000 (4.3)	40			453
RTP 300			C1018 steel; surface finish: 14-17 μin; 15-25 Rockwell C	"	apparatus: Falex Model No. 6	23	40 (0.28)	50 (15.2)	2000 (4.3)	2500			457
RTP 300 AR 15	15% aramid fiber		"	"	"	23	40 (0.28)	50 (15.2)	2000 (4.3)	130			457
RTP 300 AR 15 TFE 15	15% PTFE, 15% aramid fiber		"	"	"	23	40 (0.28)	50 (15.2)	2000 (4.3)	40			457
RTP 300 TFE 10	10% PTFE		"	"	"	23	40 (0.28)	50 (15.2)	2000 (4.3)	80			457
RTP 300 TFE 15	15% PTFE		"	"	"	23	40 (0.28)	50 (15.2)	2000 (4.3)	65			457
RTP 300 TFE 15 SI	15% PTFE, 0.5% silicone		"	"	"	23	40 (0.28)	50 (15.2)	2000 (4.3)	50			457
RTP 300 TFE 20	20% PTFE		"	"	"	23	40 (0.28)	50 (15.2)	2000 (4.3)	60			457
RTP 300 TFE 5	5% PTFE		"	"	"	23	40 (0.28)	50 (15.2)	2000 (4.3)	80			457
RTP 301 TFE 5	5% PTFE, 10% glass fiber		"	"	"	23	40 (0.28)	50 (15.2)	2000 (4.3)	80			457
RTP 302 TFE 10	10% PTFE, 15% glass fiber		"	"	"	23	40 (0.28)	50 (15.2)	2000 (4.3)	50			457
RTP 302 TFE 15	15% PTFE, 15% glass fiber		"	"	"	23	40 (0.28)	50 (15.2)	2000 (4.3)	40			457
RTP 303 TFE 1 0	10% PTFE, 20% glass fiber		"	"	"	23	40 (0.28)	50 (15.2)	2000 (4.3)	45			457
RTP 303 TFE 15	15% PTFE, 20% glass fiber		"	"	"	23	40 (0.28)	50 (15.2)	2000 (4.3)	42			457
RTP 303 TFE 20	20% PTFE, 20% glass fiber		"	"	"	23	40 (0.28)	50 (15.2)	2000 (4.3)	40			457
RTP 303 TFE 20 SI 2	20% PTFE, 2% silicone, 20% glass fiber		"	"	"	23	40 (0.28)	50 (15.2)	2000 (4.3)	35			457
RTP 305	30% glass fiber		"	"	"	23	40 (0.28)	50 (15.2)	2000 (4.3)	180			457

Material		Mating Surface		Test Conditions						K Factor 10^{-10} x (in^3-min/ft-lb-hr)		Results Note	Source
Supplier / Grade	Material Note	Mating Surface Material	Mating Surface Note	Test Method	Test Conditions Note	Temp. (°C)	Pressure (psi) (MPa)	Speed (fpm) (m/min)	PV (psi x fpm) (MPa x m/min)	Material	Mating Surface		
POLYCARBONATE against mating surface													
RTP 305 TFE 13 SI 2	13% PTFE, 2% silicone, 30% glass fiber	**steel**	C1018 steel; surface finish: 14-17 µin; 15-25 Rockwell C	thrust washer	apparatus: Falex Model No. 6	23	40 (0.28)	50 (15.2)	2000 (4.3)	30			457
RTP 305 TFE 15	15% PTFE, 30% glass fiber		"	"	"	23	40 (0.28)	50 (15.2)	2000 (4.3)	35			457
RTP 306 TFE 20	35% glass fiber		"	"	"	23	40 (0.28)	50 (15.2)	2000 (4.3)	27			457
RTP 385	30% carbon fiber		"	"	"	23	40 (0.28)	50 (15.2)	2000 (4.3)	90			457
RTP 385 TFE 13 SI 2	13% PTFE, 2% silicone, 30% carbon fiber		"	"	"	23	40 (0.28)	50 (15.2)	2000 (4.3)	25			457
Unspecified grade	unmodified		carbon steel; surface finish: 16 µin; 18-22 Rockwell C	"	apparatus: Faville-LeVally	23	40 (0.28)	50 (15.2)	2000 (4.3)	2500			458
POLYESTER, PBT against mating surface													
LNP W-1000	unmodified	**polyester PBT**	W-1000 (unmodified)	thrust washer		23	40 (0.28)	50 (15.2)	2000 (4.3)	3000	2500		453
LNP WL-4040	20% PTFE		WL-4040 (20% PTFE)	"		23	40 (0.28)	50 (15.2)	2000 (4.3)	45	40		453
Bay Resins Lubriplas PBT-1100	unmodified resin	**steel**		"			40 (0.28)	50 (15.2)	2000 (4.3)	200			435
Bay Resins Lubriplas PBT-1100CF30	30% carbon fiber			"			40 (0.28)	50 (15.2)	2000 (4.3)	25			435
Bay Resins Lubriplas PBT-1100G15	15% glass fiber			"			40 (0.28)	50 (15.2)	2000 (4.3)	105			435
Bay Resins Lubriplas PBT-1100G20	20% glass fiber			"			40 (0.28)	50 (15.2)	2000 (4.3)	100			435
Bay Resins Lubriplas PBT-1100G30	30% glass fiber			"			40 (0.28)	50 (15.2)	2000 (4.3)	85			435
Bay Resins Lubriplas PBT-1100G30TF15	30% glass fiber, 15% PTFE			"			40 (0.28)	50 (15.2)	2000 (4.3)	20			435
Bay Resins Lubriplas PBT-1100TF20	20% PTFE			"			40 (0.28)	50 (15.2)	2000 (4.3)	16			435
Ensinger	standard grade						40 (0.28)	50 (15.2)	2000 (4.3)	210			449
LNP W-1000	unmodified		carbon steel; surface finish: 12-16 µin; 18-20 Rockwell C	thrust washer		23	40 (0.28)	50 (15.2)	2000 (4.3)	210			453
LNP WC-1006	30% carbon fiber		"	"		23	40 (0.28)	50 (15.2)	2000 (4.3)	24			453
LNP WF-1006	30% glass fiber		"	"		23	40 (0.28)	50 (15.2)	2000 (4.3)	90			453
LNP WFL-4036	30% glass fiber, 15% PTFE		"	"		23	40 (0.28)	50 (15.2)	2000 (4.3)	20			453
LNP WFL-4536	30% glass fiber, 13% PTFE, 2% silicone		"	"		23	40 (0.28)	50 (15.2)	2000 (4.3)	12			453

Material		Mating Surface		Test Conditions						K Factor 10^{-10} x (in³-min/ft-lb-hr)		Results Note	Source
Supplier / Grade	Material Note	Mating Surface Material	Mating Surface Note	Test Method	Test Conditions Note	Temp. (°C)	Pressure (psi) (MPa)	Speed (fpm) (m/min)	PV (psi x fpm) (MPa x m/min)	Material	Mating Surface		
POLYESTER, PBT against mating surface													
LNP WL-4040	20% PTFE	**steel**	carbon steel; surface finish: 12-16 µin; 18-20 Rockwell C	thrust washer		23	40 (0.28)	50 (15.2)	2000 (4.3)	15			453
LNP WL-4410	2% silicone		"	"		23	40 (0.28)	50 (15.2)	2000 (4.3)	50			453
LNP WL-4540	18% PTFE, 2% silicone		"	"		23	40 (0.28)	50 (15.2)	2000 (4.3)	9			453
RTP 1000			C1018 steel; surface finish: 14-17 µin; 15-25 Rockwell C	"	apparatus: Falex Model No. 6	23	40 (0.28)	50 (15.2)	2000 (4.3)	210			457
RTP 1002 TFE 15	15% PTFE, 15% glass fiber		"	"	"	23	40 (0.28)	50 (15.2)	2000 (4.3)	16			457
RTP 1005	30% glass fiber		"	"	"	23	40 (0.28)	50 (15.2)	2000 (4.3)	90			457
RTP 1005 TFE 15	15% PTFE, 30% glass fiber		"	"	"	23	40 (0.28)	50 (15.2)	2000 (4.3)	20			457
RTP 1085	30% carbon fiber		"	"	"	23	40 (0.28)	50 (15.2)	2000 (4.3)	25			457
RTP 1085 TFE 15	15% PTFE, 30% carbon fiber		"	"	"	23	40 (0.28)	50 (15.2)	2000 (4.3)	10			457
Thermofil E-30FG-0287	30% glass fiber, lubricated			ASTM D1894		23				12			459
Thermofil E-30NF-0100	30% graphite fiber			"		23				22			459
POLYESTER, PET against mating surface													
Ensinger Ensitep	standard grade	**steel**					40 (0.28)	50 (15.2)	2000 (4.3)	210			449
RTP 1107 TFE 10	10% PTFE, 40% glass fiber		C1018 steel; surface finish: 14-17 µin; 15-25 Rockwell C	thrust washer	apparatus: Falex Model No. 6	23	40 (0.28)	50 (15.2)	2000 (4.3)	25			457
Thermofil E2-30FG-7100	30% glass fiber, lubricated			ASTM D1894		23				30			459
LIQUID CRYSTAL POLYMER against mating surface													
Hoechst AG Vectra A130	80 Rockwell M; 30% glass fiber, Vectra A950 matrix; parallel to flow	**steel**	shaft; surface finish: 0.1 µm	sliding on a ground shaft	test duration: 60 hours			446 (136)				wear: 4 mm³	70
Hoechst AG Vectra A130	80 Rockwell M; 30% glass fiber, Vectra A950 matrix; transverse to flow		"	"	"			446 (136)				wear: 7 mm³	70
Hoechst AG Vectra A230	80 Rockwell M; Vectra A950 matrix, 30% carbon fiber; parallel to flow		"	"	"			446 (136)				wear: 1 mm³	70
Hoechst AG Vectra A230	80 Rockwell M; Vectra A950 matrix, 30% carbon fiber; transverse to flow		"	"	"			446 (136)				wear: 1.5 mm³	70
Hoechst AG Vectra A422	75 Rockwell M; 40% glass fiber/ graphite, Vectra A950 matrix; parallel to flow		"	"	"			446 (136)				wear: 1.5 mm³	70
Hoechst AG Vectra A422	75 Rockwell M; 40% glass fiber/ graphite, Vectra A950 matrix; transverse to flow		"	"	"			446 (136)				wear: 2 mm³	70

Material		Mating Surface		Test Conditions						K Factor 10⁻¹⁰ x (in³-min/ft-lb-hr)		Results Note	Source
Supplier / Grade	Material Note	Mating Surface Material	Mating Surface Note	Test Method	Test Conditions Note	Temp. (°C)	Pressure (psi) (MPa)	Speed (fpm) (m/min)	PV (psi x fpm) (MPa x m/min)	Material	Mating Surface		
LIQUID CRYSTAL POLYMER against mating surface →													
Hoechst AG Vectra A430	32 Rockwell M; 25% PTFE modified, Vectra A950 matrix; parallel to flow	**steel**	shaft; surface finish: 0.1 µm	sliding on a ground shaft	test duration: 60 hours			446 (136)				wear: 0.5 mm^3	70
Hoechst AG Vectra A430	32 Rockwell M; 25% PTFE modified, Vectra A950 matrix; transverse to flow		"	"	"			446 (136)				wear: 1 mm^3	70
Hoechst AG Vectra A435	49 Rockwell M; Vectra A950 matrix, 35% glass fiber/ graphite; parallel to flow		"	"	"			446 (136)				wear: 1.5 mm^3	70
Hoechst AG Vectra A435	49 Rockwell M; Vectra A950 matrix, 35% glass fiber/ graphite; transverse to flow		"	"	"			446 (136)				wear: 2 mm^3	70
Hoechst AG Vectra A530	65 Rockwell M; Vectra A950 matrix, 30% mineral filler; parallel to flow		"	"	"			446 (136)				wear: 4.25 mm^3	70
Hoechst AG Vectra A530	65 Rockwell M; Vectra A950 matrix, 30% mineral filler; transverse to flow		"	"	"			446 (136)				wear: 7 mm^3	70
Hoechst AG Vectra A625	60 Rockwell M; 25% graphite, Vectra A950 matrix; transverse to flow		"	"	"			446 (136)				wear: 17 mm^3	70
Hoechst AG Vectra A625	60 Rockwell M; 25% graphite, Vectra A950 matrix; parallel to flow		"	"	"			446 (136)				wear: 17.5 mm^3	70
Hoechst AG Vectra B130	98 Rockwell M; 30% glass fiber, Vectra B950 matrix; parallel to flow		"	"	"			446 (136)				wear: 2.5 mm^3	70
Hoechst AG Vectra B130	98 Rockwell M; 30% glass fiber, Vectra B950 matrix; transverse to flow		"	"	"			446 (136)				wear: 3.5 mm^3	70
Hoechst AG Vectra B230	100 Rockwell M; Vectra B950 matrix, 30% carbon fiber; parallel to flow		"	"	"			446 (136)				wear: 1 mm^3	70
Hoechst AG Vectra B230	100 Rockwell M; Vectra B950 matrix, 30% carbon fiber; transverse to flow		"	"	"			446 (136)				wear: 1.5 mm^3	70
Hoechst AG Vectra C130	75 Rockwell M; 30% glass fiber, Vectra C950 matrix; parallel to flow		"	"	"			446 (136)				wear: 3.5 mm^3	70
Hoechst AG Vectra C130	75 Rockwell M; 30% glass fiber, Vectra C950 matrix; transverse to flow		"	"	"			446 (136)				wear: 6 mm^3	70
POLYAMIDEIMIDE against mating surface →													
Amoco Torlon 4301	wear resistant; 20% graphite powd., 3% fluorocarbon	**aluminum**	A360 casting alloy; -24 Rockwell C	thrust washer			1000 (6.9)	50 (15.2)	50000 (105.1)	62			20
Amoco Torlon 4301	wear resistant; 20% graphite powd., 3% fluorocarbon		A380 casting alloy; -28 Rockwell C	"			50 (0.34)	900 (274)	45000 (93.2)	64			20
Amoco Torlon 4301	wear resistant; 20% graphite powd., 3% fluorocarbon		A360 casting alloy; -24 Rockwell C	"			50 (0.34)	900 (274)	45000 ((93.2)	69			20
Amoco Torlon 4301	wear resistant; 20% graphite powd., 3% fluorocarbon		A380 casting alloy; -28 Rockwell C	"			1000 (6.9)	50 (15.2)	50000 (105.1)	37			20
Amoco Torlon 4301	wear resistant; 20% graphite powd., 3% fluorocarbon	**brass**	free cutting; -15 Rockwell C	"			1000 (6.9)	50 (15.2)	50000 (105.1)	62			20

Material		Mating Surface		Test Conditions						K Factor 10^{-10} x (in³-min/ft-lb-hr)		Results Note	Source
Supplier / Grade	Material Note	Mating Surface Material	Mating Surface Note	Test Method	Test Conditions Note	Temp. (°C)	Pressure (psi) (MPa)	Speed (fpm) (m/min)	PV (psi x fpm) (MPa x m/min)	Material	Mating Surface		
POLYAMIDEIMIDE against mating surface													
Amoco Torlon 4301	wear resistant; 20% graphite powd., 3% fluorocarbon	**brass**	free cutting; -15 Rockwell C	thrust washer			50 (0.34)	900 (274)	45000 (93.2)	111			20
Amoco Torlon 4301	wear resistant; 20% graphite powd., 3% fluorocarbon	**stainless steel**	316 stainless steel; 17 Rockwell C	"			50 (0.34)	900 (274)	45000 (93.2)	398			20
Amoco Torlon 4301	wear resistant; 20% graphite powd., 3% fluorocarbon		"	"			1000 (6.9)	50 (15.2)	50000 (105.1)	49			20
3M Torlon 4203	unfilled	**steel**	AISI 1018, surface finish: 16 RMS	thrust bearing	apparatus: LRI-1 Wear and Friction Tester		2000 (13.8)	12.5 (3.8)	25000 (52.6)			wear rate: 432 µm/hour; sliding surface temp.: 164°C	318
Amoco Torlon 4275	wear resistant; 3% fluorocarbon, 30% graphite powd.		C1018 steel; 24 Rockwell C	thrust washer			1000 (6.9)	50 (15.2)	50000 (105.1)	31			20
Amoco Torlon 4275	wear resistant; 3% fluorocarbon, 30% graphite powd.		"	"			50 (0.34)	900 (274)	45000 (93.2)	40			20
Amoco Torlon 4275	wear resistant; 3% fluorocarbon, 30% graphite powd.		"	"			50 (0.34)	200 (61)	10000 (20.7)	8			20
Amoco Torlon 4301	wear resistant; 20% graphite powd., 3% fluorocarbon		C1018 steel; soft; 6 Rockwell C	"			50 (0.34)	900 (274)	45000 (93.2)	74			20
Amoco Torlon 4301	wear resistant; 20% graphite powd., 3% fluorocarbon		C1018 steel; 24 Rockwell C	"			1000 (6.9)	50 (15.2)	50000 (105.1)	41			20
Amoco Torlon 4301	wear resistant; 20% graphite powd., 3% fluorocarbon		C1018 steel; soft; 6 Rockwell C	"			1000 (6.9)	50 (15.2)	50000 (105.1)	49			20
Amoco Torlon 4301	wear resistant; 20% graphite powd., 3% fluorocarbon		C1018 steel; 24 Rockwell C	"			50 (0.34)	900 (274)	45000 (93.2)	53			20
Amoco Torlon 4301	wear resistant; 20% graphite powd., 3% fluorocarbon		C1018; lubricated; 24 Rockwell C	"			50 (0.34)	900 (274)	45000 (93.2)	1			20
Amoco Torlon 4301	wear resistant; 20% graphite powd., 3% fluorocarbon		C1018 steel; 24 Rockwell C	"			50 (0.34)	200 (61)	10000 (20.7)	17			20
Amoco	wear resistant; 12% graphite powd., 8% fluorocarbon		"	"			50 (0.34)	900 (274)	45000 (93.2)	42			20
Amoco	wear resistant; 12% graphite powd., 8% fluorocarbon		"	"			50 (0.34)	200 (61)	10000 (20.7)	6			20
Amoco	wear resistant; 12% graphite powd., 8% fluorocarbon		"	"			1000 (6.9)	50 (15.2)	50000 (105.1)	24			20
POLYETHERIMIDE against mating surface													
GE Ultem 4000	reinforced, internal lubricant, 85 Rockwell M	**polyetherimide**	Ultem 4000						2000 (4.2)	1900			51
GE Ultem 4001	internal lubricant, 110 Rockwell M		Ultem 4001						2000 (4.2)	27			51

Material		Mating Surface		Test Conditions						K Factor 10^{-10} x (in³-min/ft-lb-hr)		Results Note	Source
Supplier / Grade	Material Note	Mating Surface Material	Mating Surface Note	Test Method	Test Conditions Note	Temp. (°C)	Pressure (psi) (MPa)	Speed (fpm) (m/min)	PV (psi x fpm) (MPa x m/min)	Material	Mating Surface		
POLYETHERIMIDE against mating surface													
GE Ultem	standard grade	steel					40 (0.28)	50 (15.2)	2000 (4.3)	4000			449
GE Ultem 2300	30% glass fiber						40 (0.28)	50 (15.2)	2000 (4.3)	130			449
GE Ultem 4000	reinforced, internal lubricant, 85 Rockwell M								2000 (4.2)	62			51
GE Ultem 4001	internal lubricant, 110 Rockwell M								2000 (4.2)	72			51
LNP E-1000	unmodified		carbon steel; surface finish: 12-16 µin; 18-20 Rockwell C	thrust washer		23	40 (0.28)	50 (15.2)	2000 (4.3)	4000			453
LNP EC-1006	30% carbon fiber		"	"		23	40 (0.28)	50 (15.2)	2000 (4.3)	70			453
LNP EC-1006	30% carbon fiber		cold rolled steel; surface finish: 12-16 µin; 22 Rockwell C	"		93	40 (0.28)	50 (15.2)	2000 (4.3)	77			453
LNP EC-1006	30% carbon fiber		"	"		149	40 (0.28)	50 (15.2)	2000 (4.3)	100			453
LNP EF-1006	30% glass fiber		carbon steel; surface finish: 12-16 µin; 18-20 Rockwell C	"		23	40 (0.28)	50 (15.2)	2000 (4.3)	130			453
LNP EFL-4036	30% glass fiber, 15% PTFE		cold rolled steel; surface finish: 12-16 µin; 22 Rockwell C	"		149	40 (0.28)	50 (15.2)	2000 (4.3)	100			453
LNP EFL-4036	30% glass fiber, 15% PTFE		carbon steel; surface finish: 12-16 µin; 18-20 Rockwell C	"		23	40 (0.28)	50 (15.2)	2000 (4.3)	35			453
LNP EFL-4036	30% glass fiber, 15% PTFE		cold rolled steel; surface finish: 12-16 µin; 22 Rockwell C	"		93	40 (0.28)	50 (15.2)	2000 (4.3)	40			453
RTP 2100 AR 15	15% aramid fiber		C1018 steel; surface finish: 14-17 µin; 15-25 Rockwell C	"	apparatus: Falex Model No. 6	23	40 (0.28)	50 (15.2)	2000 (4.3)	100			457
RTP 2100 AR 15 TFE 15	15% PTFE, 15% aramid fiber		"	"	"	23	40 (0.28)	50 (15.2)	2000 (4.3)	40			457
RTP 2105 TFE 15	15% PTFE, 30% glass fiber		"	"	"	23	40 (0.28)	50 (15.2)	2000 (4.3)	50			457
RTP 2185 TFE 13 SI2	13% PTFE, 2% silicone, 30% carbon fiber		"	"	"	23	40 (0.28)	50 (15.2)	2000 (4.3)	35			457
RTP 2185 TFE 15	15% PTFE, 30% carbon fiber		"	"	"	23	40 (0.28)	50 (15.2)	2000 (4.3)	35			457
Thermofil W-10FG-0100	10% glass fiber			ASTM D1894		23				500			459
Thermofil W-30NF-0100	30% graphite fiber			"		23				70			459
Thermofil W-30NF-0214	30% graphite fiber, PTFE lubricated			"		23				45			459

Material		Mating Surface		Test Conditions						K Factor 10^{-10} x (in^3-min/ft-lb-hr)		Results Note	Source
Supplier / Grade	Material Note	Mating Surface Material	Mating Surface Note	Test Method	Test Conditions Note	Temp. (°C)	Pressure (psi) (MPa)	Speed (fpm) (m/min)	PV (psi x fpm) (MPa x m/min)	Material	Mating Surface		
POLYARYLETHERKETONE against mating surface →													
BASF AG Ultrapek A2000C6	moderate flow; 30% carbon fiber	steel	Cr 6/800/HV				145 (1)	98 (30)	14210 (29.9)			wear rate: 0.3 µm/km; sliding surface temp.: 40°C	27
BASF AG Ultrapek A2000C6	moderate flow; 30% carbon fiber		"				145 (1)	98 (30)	14210 (29.9)			wear rate: 2 µm/km; sliding surface temp.: 120°C	27
BASF AG Ultrapek A3000	low flow		"				145 (1)	98 (30)	14210 (29.9)			wear rate: 2.4 µm/km; sliding surface temp.: 40°C	27
BASF AG Ultrapek A3000	low flow		"				145 (1)	98 (30)	14210 (29.9)			wear rate: 7 µm/km; sliding surface temp.: 120°C	27
BASF AG Ultrapek KR4190	lubricated, developmental material, tribological properties; modified grade		"				145 (1)	98 (30)	14210 (29.9)			wear rate: 0.46 µm/km; sliding surface temp.: 40°C	27
BASF AG Ultrapek KR4190	lubricated, developmental material, tribological properties; modified grade		"				145 (1)	98 (30)	14210 (29.9)			wear rate: 3.6 µm/km; sliding surface temp.: 120°C	27
POLYETHERETHERKETONE against mating surface →													
Ensinger	standard grade	steel					40 (0.28)	50 (15.2)	2000 (4.3)	200			449
LNP LCL-4033EM	15% carbon fiber, 15% PTFE		carbon steel; surface finish: 12-16 µin; 18-20 Rockwell C	thrust washer		23	40 (0.28)	50 (15.2)	2000 (4.3)	20			453
LNP LCL-4033EM	15% carbon fiber, 15% PTFE		cold rolled steel; surface finish: 12-16 µin; 22 Rockwell C	"		149	40 (0.28)	50 (15.2)	2000 (4.3)	35			453
LNP LCL-4033EM	15% carbon fiber, 15% PTFE		"	"		260	40 (0.28)	50 (15.2)	2000 (4.3)	50			453
LNP LFL-4036	30% glass fiber, 15% PTFE		carbon steel; surface finish: 12-16 µin; 18-20 Rockwell C	"		23	40 (0.28)	50 (15.2)	2000 (4.3)	110			453
LNP LL-4040	20% PTFE		"	"		23	40 (0.28)	50 (15.2)	2000 (4.3)	130			453
RTP 2205 TFE	15% PTFE, 15% silicone, 30% glass fiber		C1018 steel; surface finish: 14-17 µin; 15-25 Rockwell C	"	apparatus: Falex Model No. 6	23	40 (0.28)	50 (15.2)	2000 (4.3)	75			457
RTP 2285TFE	15% PTFE, 15% silicone, 30% carbon fiber		"	"	"	23	40 (0.28)	50 (15.2)	2000 (4.3)	55			457
Thermofil K2-30NF-0100	30% graphite fiber			ASTM D1894		23				100			459
Victrex 450CA30	30% carbon fiber						40 (0.28)	50 (15.2)	2000 (4.3)	60			449
Victrex 450GL30	30% glass fiber						40 (0.28)	50 (15.2)	2000 (4.3)	90			449
Victrex PEEK 15OFC30	proprietary filler		carbon steel; surface finish: 12-16 µin; 18-20 Rockwell C	thrust washer		23	40 (0.28)	50 (15.2)	2000 (4.3)	10			453

Material		Mating Surface		Test Conditions						K Factor 10⁻¹⁰ x (in³·min/ft·lb·hr)		Results Note	Source
Supplier / Grade	Material Note	Mating Surface Material	Mating Surface Note	Test Method	Test Conditions Note	Temp. (°C)	Pressure (psi) (MPa)	Speed (fpm) (m/min)	PV (psi x fpm) (MPa x m/min)	Material	Mating Surface		
POLYETHERETHERKETONE against mating surface													
Victrex PEEK 15OFC30	proprietary filler	**steel**	cold rolled steel; surface finish: 12-16 µin; 22 Rockwell C	thrust washer		149	40 (0.28)	50 (15.2)	2000 (4.3)	30			453
Victrex PEEK 15OFC30	proprietary filler		"	"		260	40 (0.28)	50 (15.2)	2000 (4.3)	40			453
Victrex PEEK 450CA30	30% carbon fiber		"	"		204	40 (0.28)	50 (15.2)	2000 (4.3)	120			453
Victrex PEEK 450CA30	30% carbon fiber		"	"		93	40 (0.28)	50 (15.2)	2000 (4.3)	60			453
Victrex PEEK 450CA30	30% carbon fiber		"	"		149	40 (0.28)	50 (15.2)	2000 (4.3)	60			453
Victrex PEEK 450CA30	30% carbon fiber		carbon steel; surface finish: 12-16 µin; 18-20 Rockwell C	"		23	40 (0.28)	50 (15.2)	2000 (4.3)	60			453
Victrex PEEK 450CA30	124 Rockwell R, 107 Rockwell M; 30% carbon fiber		carbon EN8, dry	pad on ring (Amsler wear tester)		200		100 (30.5)				wear rate: 200 µm/hour; maximum load: 120 kg	338
Victrex PEEK 450CA30	124 Rockwell R, 107 Rockwell M; 30% carbon fiber		"	"		20		600 (183)				wear rate: 225 µm/hour; maximum load: 22 kg	338
Victrex PEEK 450CA30	30% carbon fiber		cold rolled steel; surface finish: 12-16 µin; 22 Rockwell C	thrust washer		260	40 (0.28)	50 (15.2)	2000 (4.3)	170			453
Victrex PEEK 450CA30	124 Rockwell R, 107 Rockwell M; 30% carbon fiber		carbon EN8, dry	pad on ring (Amsler wear tester)		20		100 (30.5)				wear rate: 88 µm/hour; maximum load: 160 kg	338
Victrex PEEK 450FC30	was Victrex D450HF30, tribological properties, bearing grade; 30% graphite/ carbon/ PTFE		"	"		200		600 (183)				wear rate: 132 µm/hour; maximum load: 40 kg	338
Victrex PEEK 450FC30	was Victrex D450HF30, tribological properties, bearing grade; 30% graphite/ carbon/ PTFE		"	"		200		100 (30.5)				wear rate: 152 µm/hour; maximum load: 170 kg	338
Victrex PEEK 450FC30	was Victrex D450HF30, tribological properties, bearing grade; 30% graphite/ carbon/ PTFE		"	"		20		100 (30.5)				wear rate: 152 µm/hour; maximum load: 210 kg	338
Victrex PEEK 450FC30	was Victrex D450HF30, tribological properties, bearing grade; 30% graphite/ carbon/ PTFE		"	"		20		600 (183)				wear rate: 190 µm/hour; maximum load: 40 kg	338
Victrex PEEK 450G	gen. purp. grade, 126 Rockwell R, 99 Rockwell M		"	"		200		600 (183)				wear rate: 150 µm/hour; maximum load: 8 kg	338
Victrex PEEK 450G	gen. purp. grade, 126 Rockwell R, 99 Rockwell M		"	"		20		100 (30.5)				wear rate: 303 µm/hour; maximum load: 70 kg	338
Victrex PEEK 450G	gen. purp. grade, 126 Rockwell R, 99 Rockwell M		"	"		200		100 (30.5)				wear rate: 350 µm/hour; maximum load: 70 kg	338

Material		Mating Surface		Test Conditions						K Factor 10⁻¹⁰ x (in³-min/ft-lb-hr)		Results Note	Source
Supplier / Grade	Material Note	Mating Surface Material	Mating Surface Note	Test Method	Test Conditions Note	Temp. (°C)	Pressure (psi) (MPa)	Speed (fpm) (m/min)	PV (psi x fpm) (MPa x m/min)	Material	Mating Surface		
POLYETHERETHERKETONE against mating surface													
Victrex PEEK 450G	gen. purp. grade, 126 Rockwell R, 99 Rockwell M	**steel**	carbon EN8, dry	pad on ring (Amsler wear tester)		20		600 (183)				wear rate: 450 µm/hour; maximum load: 8 kg	338
Victrex PEEK 450GL30	30% glass fiber		carbon steel; surface finish: 12-16 µin; 18-20 Rockwell C	thrust washer		23	40 (0.28)	50 (15.2)	2000 (4.3)	90			453
Victrex PEEK 45OG	unmodified		"	"		23	40 (0.28)	50 (15.2)	2000 (4.3)	200			453
Victrex PEEK D450HT15	no longer available, tribological properties, bearing grade		carbon EN8, dry	pad on ring (Amsler wear tester)		200		100 (30.5)				wear rate: 107 µm/hour; maximum load: 210 kg	338
Victrex PEEK D450HT15	no longer available, tribological properties, bearing grade		"	"		20		600 (183)				wear rate: 150 µm/hour; maximum load: 40 kg	338
Victrex PEEK D450HT15	no longer available, tribological properties, bearing grade		"	"		20		100 (30.5)				wear rate: 300 µm/hour; maximum load: 110 kg	338
Victrex PEEK D450HT15	no longer available, tribological properties, bearing grade		"	"		200		600 (183)				wear rate: 50 µm/hour; maximum load: 40 kg	338
POLYETHYLENE, HDPE against mating surface													
LNP FL-4030	15% PTFE	**steel**	carbon steel; surface finish: 12-16 µin; 18-20 Rockwell C	thrust washer		23	40 (0.28)	50 (15.2)	2000 (4.3)	60			453
POLYETHYLENE, UHMWPE against mating surface													
Unspecified grade		**nylon 610**	QFL-4036 (30% glass fiber, 15% PTFE)	thrust washer		23	40 (0.28)	50 (15.2)	2000 (4.3)	18	15		453
Unspecified grade		**polycarbonate**	DL-4010 (5% PTFE)	"		23	40 (0.28)	50 (15.2)	2000 (4.3)	18	23		453
POLYETHYLENE against mating surface													
LNP M-1000	unmodified	**steel**	carbon steel; surface finish: 12-16 µin; 18-20 Rockwell C	thrust washer		23	40 (0.28)	50 (15.2)	2000 (4.3)	176			394
LNP MFL-4034HS	15% PTFE, 20% silicone		"	"		23	40 (0.28)	50 (15.2)	2000 (4.3)	45			453
LNP MFX-700-10	50% long glass fiber, chemically coupled		"	"		23	40 (0.28)	50 (15.2)	2000 (4.3)	73			394
LNP ML-404OHS	20% PTFE		"	"		23	40 (0.28)	50 (15.2)	2000 (4.3)	33			453

Material		Mating Surface		Test Conditions						K Factor 10^{-10} x (in³-min/ft-lb-hr)		Results Note	Source
Supplier / Grade	Material Note	Mating Surface Material	Mating Surface Note	Test Method	Test Conditions Note	Temp. (°C)	Pressure (psi) (MPa)	Speed (fpm) (m/min)	PV (psi x fpm) (MPa x m/min)	Material	Mating Surface		
MODIFIED, PPE against mating surface													
GE Noryl	standard grade	**steel**					40 (0.28)	50 (15.2)	2000 (4.3)	3000			449
GE Noryl SE1-GFN3	30% glass fiber, flame retardant						40 (0.28)	50 (15.2)	2000 (4.3)	230			449
LNP Z-1000	unmodified		carbon steel; surface finish: 12-16 µin; 18-20 Rockwell C	thrust washer		23	40 (0.28)	50 (15.2)	2000 (4.3)	3000			453
LNP ZBL-4326	30% glass beads, 10% graphite		"	"	test conducted in water	23	40 (0.28)	50 (15.2)	2000 (4.3)	50			453
LNP ZBL-4326	30% glass beads, 10% graphite		"	"		23	40 (0.28)	50 (15.2)	2000 (4.3)	145			453
LNP ZF-1006	30% glass fiber		"	"		23	40 (0.28)	50 (15.2)	2000 (4.3)	230			453
LNP ZFL-4036	30% glass fiber, 15% PTFE		"	"		23	40 (0.28)	50 (15.2)	2000 (4.3)	45			453
LNP ZFL-4323	15% glass fiber, 10% graphite		"	"		23	40 (0.28)	50 (15.2)	2000 (4.3)	11			453
LNP ZL-4030	15% PTFE		"	"		23	40 (0.28)	50 (15.2)	2000 (4.3)	100			453
LNP ZML-4334	20% mineral, 15% graphite		"	"		23	40 (0.28)	50 (15.2)	2000 (4.3)	850			453
POLYPHENYLENE SULFIDE against mating surface													
LNP OFL-4036	30% glass fiber, 15% PTFE	**brass**	70/30 brass; surface finish: 8-16 µin	"		23	40 (0.28)	50 (15.2)	2000 (4.3)	9	4		453
LNP OFL-4036	30% glass fiber, 15% PTFE		70/30 brass; surface finish: 50-70 µin	"		23	40 (0.28)	50 (15.2)	2000 (4.3)	10	4		453
LNP OFL-4036	30% glass fiber, 15% PTFE	**polycarbonate**	DL-4020 (10% PTFE)	"		23	40 (0.28)	50 (15.2)	2000 (4.3)	17	35		453
LNP OFL-4036	30% glass fiber, 15% PTFE	**polyphenylene sulfide**	OFL-4036 (30% glass fiber, 15% PTFE)	"		23	40 (0.28)	50 (15.2)	2000 (4.3)	140	160		453
LNP OFL-4036	30% glass fiber, 15% PTFE		OFL-4326 (30% glass fiber, 10% graphite powder)	"		23	40 (0.28)	50 (15.2)	2000 (4.3)	760	800		453
LNP OFL-4326	30% glass fiber, 10% graphite powder		OFL-4036 (30% glass fiber, 15% PTFE)	"		23	40 (0.28)	50 (15.2)	2000 (4.3)	710	750		453
Akzo J-1300/CF/30/TF/15	30% carbon fiber, 15% PTFE	**steel**	carbon steel; surface finish: 16 µin; 18-22 Rockwell C	"	apparatus: Faville-LeVally	23	40 (0.28)	50 (15.2)	2000 (4.3)	75			458
Ensinger Ensifide	standard grade						40 (0.28)	50 (15.2)	2000 (4.3)	540			449
GE Supec W331	116 Rockwell R; 30% glass fiber, 15% PTFE			ASTM D3702			40 (0.28)	50 (15.2)	2000 (4.3)	108			429
Hoechst AG Fortron 1140-AO	40% glass fiber						40 (0.28)	50 (15.2)	2000 (4.3)	240			449

Material		Mating Surface		Test Conditions						K Factor 10^{-10} x (in³-min/ft-lb-hr)		Results Note	Source
Supplier / Grade	Material Note	Mating Surface Material	Mating Surface Note	Test Method	Test Conditions Note	Temp. (°C)	Pressure (psi) (MPa)	Speed (fpm) (m/min)	PV (psi x fpm) (MPa x m/min)	Material	Mating Surface		
POLYPHENYLENE SULFIDE against mating surface →													
LNP Lubricomp O-BG	proprietary filler	**steel**	cold rolled steel; surface finish: 12-16 µin; 22 Rockwell C	thrust washer		149	40 (0.28)	50 (15.2)	2000 (4.3)	31			453
LNP Lubricomp O-BG	proprietary filler		carbon steel; surface finish: 12-16 µin; 18-20 Rockwell C	"		23	40 (0.28)	50 (15.2)	2000 (4.3)	35			453
LNP Lubricomp O-BG	proprietary filler		cold rolled steel; surface finish: 12-16 µin; 22 Rockwell C	"		204	40 (0.28)	50 (15.2)	2000 (4.3)	36			453
LNP O-1000	unmodified		carbon steel; surface finish: 12-16 µin; 18-20 Rockwell C	"		23	40 (0.28)	50 (15.2)	2000 (4.3)	540			453
LNP OC-1006	30% carbon fiber		cold rolled steel; surface finish: 12-16 µin; 22 Rockwell C	"		204	40 (0.28)	50 (15.2)	2000 (4.3)	210			453
LNP OC-1006	30% carbon fiber		carbon steel; surface finish: 12-16 µin; 18-20 Rockwell C	"		23	40 (0.28)	50 (15.2)	2000 (4.3)	160			453
LNP OC-1006	30% carbon fiber		cold rolled steel; surface finish: 12-16 µin; 22 Rockwell C	"		93	40 (0.28)	50 (15.2)	2000 (4.3)	160			453
LNP OC-1006	30% carbon fiber		"	"		149	40 (0.28)	50 (15.2)	2000 (4.3)	160			453
LNP OCL-4036	30% carbon fiber, 15% PTFE		carbon steel; surface finish: 12-16 µin; 18-20 Rockwell C	"		23	40 (0.28)	50 (15.2)	2000 (4.3)	75			453
LNP OCL-4036	30% carbon fiber, 15% PTFE		cold rolled steel; surface finish: 12-16 µin; 22 Rockwell C	"		93	40 (0.28)	50 (15.2)	2000 (4.3)	75			453
LNP OCL-4036	30% carbon fiber, 15% PTFE		"	"		149	40 (0.28)	50 (15.2)	2000 (4.3)	75			453
LNP OCL-4036	30% carbon fiber, 15% PTFE		"	"		204	40 (0.28)	50 (15.2)	2000 (4.3)	75			453
LNP OF-1008	40% glass fiber		carbon steel; surface finish: 12-16 µin; 18-20 Rockwell C	"		23	40 (0.28)	50 (15.2)	2000 (4.3)	240			453
LNP OFL-4036	30% glass fiber, 15% PTFE		AISI 1141; surface finish: 50-70 µin	"		23	40 (0.28)	50 (15.2)	2000 (4.3)	18	17		453
LNP OFL-4036	30% glass fiber, 15% PTFE		cold rolled steel; surface finish: 12-16 µin; 22 Rockwell C	"		204	40 (0.28)	50 (15.2)	2000 (4.3)	200			453
LNP OFL-4036	30% glass fiber, 15% PTFE		AISI 1141; surface finish: 8-12 µin	"		23	40 (0.28)	50 (15.2)	2000 (4.3)	28	27		453
LNP OFL-4036	30% glass fiber, 15% PTFE		AISI 304 stainless steel; surface finish: 50-70 µin	"		23	40 (0.28)	50 (15.2)	2000 (4.3)	8	0.5		453
LNP OFL-4036	30% glass fiber, 15% PTFE		AISI 440 stainless steel; surface finish: 8-16 µin	"		23	40 (0.28)	50 (15.2)	2000 (4.3)	11	0.3		453
LNP OFL-4036	30% glass fiber, 15% PTFE		AISI 1141; surface finish: 12-16 µin	"		23	40 (0.28)	50 (15.2)	2000 (4.3)	11	15		453
LNP OFL-4036	30% glass fiber, 15% PTFE		carbon steel; surface finish: 12-16 µin; 18-20 Rockwell C	"		23	40 (0.28)	50 (15.2)	2000 (4.3)	110			453

Material		Mating Surface		Test Conditions						K Factor 10^{-10} x (in³-min/ft·lb·hr)		Results Note	Source
Supplier / Grade	Material Note	Mating Surface Material	Mating Surface Note	Test Method	Test Conditions Note	Temp. (°C)	Pressure (psi) (MPa)	Speed (fpm) (m/min)	PV (psi x fpm) (MPa x m/min)	Material	Mating Surface		
POLYPHENYLENE SULFIDE against mating surface →													
LNP OFL-4036	30% glass fiber, 15% PTFE	**steel**	AISI 440 stainless steel; surface finish: 50-70 µin	thrust washer		23	40 (0.28)	50 (15.2)	2000 (4.3)	12	0.3		453
LNP OFL-4036	30% glass fiber, 15% PTFE		cold rolled steel; surface finish: 12-16 µin; 22 Rockwell C	"		93	40 (0.28)	50 (15.2)	2000 (4.3)	135			453
LNP OFL-4036	30% glass fiber, 15% PTFE		"	"		149	40 (0.28)	50 (15.2)	2000 (4.3)	150			453
LNP OFL-4036	30% glass fiber, 15% PTFE		AISI 304 stainless steel; surface finish: 8-16 µin	"		23	40 (0.28)	50 (15.2)	2000 (4.3)	4	0.2		453
LNP OFL-4536	30% glass fiber, 13% PTFE, 2% silicone		carbon steel; surface finish: 12-16 µin; 18-20 Rockwell C	"		23	40 (0.28)	50 (15.2)	2000 (4.3)	50			453
LNP OL-4040	20% PTFE		"	"		23	40 (0.28)	50 (15.2)	2000 (4.3)	55			453
RTP 1300			C1018 steel; surface finish: 14-17 µin; 15-25 Rockwell C	"	apparatus: Falex Model No. 6	23	40 (0.28)	50 (15.2)	2000 (4.3)	540			457
RTP 1300 AR 15	15% aramid fiber		"	"	"	23	40 (0.28)	50 (15.2)	2000 (4.3)	200			457
RTP 1300 AR 15 TFE 15	15% PTFE, 15% aramid fiber		"	"	"	23	40 (0.28)	50 (15.2)	2000 (4.3)	80			457
RTP 1302 TFE 10	10% PTFE, 15% glass fiber		"	"	"	23	40 (0.28)	50 (15.2)	2000 (4.3)	120			457
RTP 1303 TFE 20	20% PTFE, 20% glass fiber		"	"	"	23	40 (0.28)	50 (15.2)	2000 (4.3)	90			457
RTP 1305	30% glass fiber		"	"	"	23	40 (0.28)	50 (15.2)	2000 (4.3)	250			457
RTP 1307 TFE 10	10% PTFE, 40% glass fiber		"	"	"	23	40 (0.28)	50 (15.2)	2000 (4.3)	190			457
RTP 1378	15% PTFE, 30% glass fiber		"	"	"	23	40 (0.28)	50 (15.2)	2000 (4.3)	105			457
RTP 1385 TFE 15	15% PTFE, 30% carbon fiber		"	"	"	23	40 (0.28)	50 (15.2)	2000 (4.3)	80			457
RTP 1387 TFE 10	10% PTFE, 40% carbon fiber		"	"	"	23	40 (0.28)	50 (15.2)	2000 (4.3)	90			457
Thermofil T-20NF-0100	20% graphite fiber			ASTM D1894		23				190			459
Thermofil T-30FG-0214	30% glass fiber, PTFE lubricated			"		23				105			459
Thermofil T-30NF-0214	30% graphite fiber, PTFE lubricated			"		23				70			459
Thermofil T-40FG-0100	40% glass fiber			"		23				220			459
Thermofil T-40NF-0100	40% graphite fiber			"		23				75			459
Unspecified grade	20% aramid fiber			thrust washer	ASTM D3702	23	250 (1.72)	10 (3)	2500 (5.3)	270			454
Unspecified grade	40% glass fiber			"	"	23	250 (1.72)	10 (3)	2500 (5.3)	4867			454

Material		Mating Surface		Test Conditions						K Factor 10⁻¹⁰ x (in³-min/ft-lb-hr)		Results Note	Source
Supplier / Grade	Material Note	Mating Surface Material	Mating Surface Note	Test Method	Test Conditions Note	Temp. (°C)	Pressure (psi) (MPa)	Speed (fpm) (m/min)	PV (psi x fpm) (MPa x m/min)	Material	Mating Surface		
POLYSULFONE against mating surface													
BASF AG Ultrason S 2010	moderate flow, gen. purp. grade	**steel**	Cr 6/800/HV; surface finish: 2.5 µm	peg-and-disc apparatus			145 (1)	98 (30)	14210 (29.9)			wear rate: >1000 µm/km; sliding surface temp.: 40°C	28
BASF AG Ultrason S 2010	moderate flow, gen. purp. grade		Cr 6/800/HV; surface finish: 0.15 µm	"			145 (1)	98 (30)	14210 (29.9)			wear rate: 200 µm/km; sliding surface temp.: 40°C	28
BASF AG Ultrason S 2010G4	moderate flow; 20% glass fiber		Cr 6/800/HV; surface finish: 2.5 µm	"			145 (1)	98 (30)	14210 (29.9)			wear rate: 10 µm/km; sliding surface temp.: 100°C	28
BASF AG Ultrason S 2010G4	moderate flow; 20% glass fiber		Cr 6/800/HV; surface finish: 0.15 µm	"			145 (1)	98 (30)	14210 (29.9)			wear rate: 15 µm/km; sliding surface temp.: 100°C	28
BASF AG Ultrason S 2010G4	moderate flow; 20% glass fiber		"	"			145 (1)	98 (30)	14210 (29.9)			wear rate: 2 µm/km; sliding surface temp.: 40°C	28
BASF AG Ultrason S 2010G4	moderate flow; 20% glass fiber		Cr 6/800/HV; surface finish: 2.5 µm	"			145 (1)	98 (30)	14210 (29.9)			wear rate: 5 µm/km; sliding surface temp.: 40°C	28
Ensinger Ensifone	standard grade						40 (0.28)	50 (15.2)	2000 (4.3)	1500			449
LNP G-1000	unmodified		carbon steel; surface finish: 12-16 µin; 18-20 Rockwell C	thrust washer		23	40 (0.28)	50 (15.2)	2000 (4.3)	1500			453
LNP GC-1006	30% carbon fiber		"	"		23	40 (0.28)	50 (15.2)	2000 (4.3)	75			453
LNP GF-1006	30% glass fiber		"	"		23	40 (0.28)	50 (15.2)	2000 (4.3)	160			453
LNP GFL 4022	10% glass fiber, 10% PTFE		"	"		23	40 (0.28)	50 (15.2)	2000 (4.3)	60			453
LNP GFL-4036	30% glass fiber, 15% PTFE		"	"		23	40 (0.28)	50 (15.2)	2000 (4.3)	55			453
LNP GL-4030	15% PTFE		"	"		23	40 (0.28)	50 (15.2)	2000 (4.3)	46			453
LNP GL-4520	8% PTFE, 2% silicone		"	"		23	40 (0.28)	50 (15.2)	2000 (4.3)	250			453
LNP GML-4334	20% mineral, 15% graphite		"	"	test conducted in water	23	40 (0.28)	50 (15.2)	2000 (4.3)	30			453
LNP GML-4334	20% mineral, 15% graphite		"	"		23	40 (0.28)	50 (15.2)	2000 (4.3)	530			453
RTP 900 TFE 15	15% PTFE		C1018 steel; surface finish: 14-17 µin; 15-25 Rockwell C	"	apparatus: Falex Model No. 6	23	40 (0.28)	50 (15.2)	2000 (4.3)	46			457
RTP 900 TFE 20	20% PTFE		"	"	"	23	40 (0.28)	50 (15.2)	2000 (4.3)	40			457
RTP 905 TFE 15	15% PTFE, 30% glass fiber		"	"	"	23	40 (0.28)	50 (15.2)	2000 (4.3)	58			457
Thermofil S-30FG-0214	30% glass fiber, PTFE lubricated			ASTM D1894		23				50			459
Thermofil S-30NF-0100	30% graphite fiber			"		23				70			459

Material		Mating Surface		Test Conditions						K Factor 10^{-10} x (in³-min/ft-lb-hr)		Results Note	Source
Supplier / Grade	Material Note	Mating Surface Material	Mating Surface Note	Test Method	Test Conditions Note	Temp. (°C)	Pressure (psi) (MPa)	Speed (fpm) (m/min)	PV (psi x fpm) (MPa x m/min)	Material	Mating Surface		
POLYETHERSULFONE against mating surface													
BASF AG Ultrason E 2010	moderate flow, gen. purp. grade	steel	Cr 6/800/HV; surface finish: 0.15 µm	peg-and-disc apparatus			145 (1)	98 (30)	14210 (29.9)			wear rate: 280 µm/km; sliding surface temp.: 40°C	28
BASF AG Ultrason E 2010	moderate flow, gen. purp. grade		Cr 6/800/HV; surface finish: 2.5 µm	"			145 (1)	98 (30)	14210 (29.9)			wear rate: 880 µm/km; sliding surface temp.: 40°C	28
BASF AG Ultrason E 2010G4	moderate flow; 20% glass fiber		"	"			145 (1)	98 (30)	14210 (29.9)			wear rate: 60 µm/km; sliding surface temp.: 40°C	28
BASF AG Ultrason E 2010G4	moderate flow; 20% glass fiber		"	"			145 (1)	98 (30)	14210 (29.9)			wear rate: 1000 µm/km; sliding surface temp.: 100°C	28
BASF AG Ultrason E 2010G4	moderate flow; 20% glass fiber		Cr 6/800/HV; surface finish: 0.15 µm	"			145 (1)	98 (30)	14210 (29.9)			wear rate: 15 µm/km; sliding surface temp.: 100°C	28
BASF AG Ultrason E 2010G4	moderate flow; 20% glass fiber		"	"			145 (1)	98 (30)	14210 (29.9)			wear rate: 2.6 µm/km; sliding surface temp.: 40°C	28
Ensinger	standard grade						40 (0.28)	50 (15.2)	2000 (4.3)	1500			449
LNP JFL-4036	30% glass fiber, 15% PTFE		cold rolled steel; surface finish: 12-16 µin; 22 Rockwell C	thrust washer		93	40 (0.28)	50 (15.2)	2000 (4.3)	120			453
LNP JFL-4036	30% glass fiber, 15% PTFE		carbon steel; surface finish: 12-16 µin; 18-20 Rockwell C	"		23	40 (0.28)	50 (15.2)	2000 (4.3)	60			453
LNP JL-4030	15% PTFE		"	"		23	40 (0.28)	50 (15.2)	2000 (4.3)	39			453
RTP 1405 TFE 15	15% PTFE, 30% glass fiber		C1018 steel; surface finish: 14-17 µin; 15-25 Rockwell C	"	apparatus: Falex Model No. 6	23	40 (0.28)	50 (15.2)	2000 (4.3)	55			457
RTP 1485 TFE 15	15% PTFE, 30% carbon fiber		"	"	"	23	40 (0.28)	50 (15.2)	2000 (4.3)	35			457
Thermofil K-20FG-0100	20% glass fiber			ASTM D1894		23				170			459
Thermofil K-30FG-0100	30% glass fiber			"		23				160			459
Thermofil K-30FG-0214	30% glass fiber, PTFE lubricated			"		23				55			459
Thermofil K-30NF-0100	30% graphite fiber			"		23				80			459
Victrex Victrex PES 4101GL30	30% glass fiber		carbon steel; surface finish: 12-16 µin; 18-20 Rockwell C	thrust washer		23	40 (0.28)	50 (15.2)	2000 (4.3)	150			453
Victrex Victrex PES 4800G	unmodified		"	"		23	40 (0.28)	50 (15.2)	2000 (4.3)	1500			453
Victrex Victrex PES-PDX-86174	15% PTFE, 30% carbon fiber		cold rolled steel; surface finish: 12-16 µin; 22 Rockwell C	"		93	40 (0.28)	50 (15.2)	2000 (4.3)	97			453

Material		Mating Surface		Test Conditions						K Factor 10^{-10} x (in³-min/ft-lb-hr)		Results Note	Source
Supplier / Grade	Material Note	Mating Surface Material	Mating Surface Note	Test Method	Test Conditions Note	Temp. (°C)	Pressure (psi) (MPa)	Speed (fpm) (m/min)	PV (psi x fpm) (MPa x m/min)	Material	Mating Surface		
POLYURETHANE, RIGID against mating surface													
LNP T-1000	unmodified	**steel**	carbon steel; surface finish: 12-16 µin; 18-20 Rockwell C	thrust washer		23	40 (0.28)	50 (15.2)	2000 (4.3)	340			453
LNP TF-1006	30% glass fiber		“	“		23	40 (0.28)	50 (15.2)	2000 (4.3)	180			453
LNP TFL-4036	30% glass fiber, 15% PTFE		“	“		23	40 (0.28)	50 (15.2)	2000 (4.3)	35			453
LNP TFL-4536	30% glass fiber, 13% PTFE, 2% silicone		“	“		23	40 (0.28)	50 (15.2)	2000 (4.3)	30			453
LNP TL-4030	15% PTFE		“	“		23	40 (0.28)	50 (15.2)	2000 (4.3)	60			453
LNP TL-4410	2% silicone		“	“		23	40 (0.28)	50 (15.2)	2000 (4.3)	55			453
ABS against mating surface													
LNP AFL-4044	20% glass fiber, 20% PTFE	**ABS**	A-1000; unmodified	thrust washer		23	40 (0.28)	50 (15.2)	2000 (4.3)	800	1300		453
LNP AFL 4044	20% glass fiber, 20% PTFE	**polycarbonate**	D-1000; unmodified	“		23	40 (0.28)	50 (15.2)	2000 (4.3)	350	1200		453
Ensinger Ensidur	standard grade	**steel**					40 (0.28)	50 (15.2)	2000 (4.3)	3500			449
LNP A-1000	unmodified		carbon steel; surface finish: 12-16 µin; 18-20 Rockwell C	thrust washer		23	40 (0.28)	50 (15.2)	2000 (4.3)	3500			453
LNP AFL-4036	30% glass fiber, 15% PTFE		“	“		23	40 (0.28)	50 (15.2)	2000 (4.3)	75			453
LNP AL-4030	15% PTFE		“	“		23	40 (0.28)	50 (15.2)	2000 (4.3)	300			453
LNP AL-4410	2% silicone		“	“		23	40 (0.28)	50 (15.2)	2000 (4.3)	80			453
RTP 600			C1018 steel; surface finish: 14-17 µin; 15-25 Rockwell C	“	apparatus: Falex Model No. 6	23	40 (0.28)	50 (15.2)	2000 (4.3)	3500			457
RTP 600 SI 2	2% silicone		“	“	“	23	40 (0.28)	50 (15.2)	2000 (4.3)	85			457
Thermofil G-20NF-0100	20% graphite fiber			ASTM D1894		23				35			459
Thermofil G-30FG-0100	30% glass fiber			“		23				240			459
POLYSTYRENE against mating surface													
LNP C-1000	unmodified	**steel**	carbon steel; surface finish: 12-16 µin; 18-20 Rockwell C	thrust washer		23	40 (0.28)	50 (15.2)	2000 (4.3)	3000			453
LNP CL-4030	15% PTFE		“	“		23	40 (0.28)	50 (15.2)	2000 (4.3)	175			453

Material		Mating Surface		Test Conditions						K Factor 10^{-10} x (in³-min/ft-lb-hr)		Results Note	Source
Supplier / Grade	Material Note	Mating Surface Material	Mating Surface Note	Test Method	Test Conditions Note	Temp. (°C)	Pressure (psi) (MPa)	Speed (fpm) (m/min)	PV (psi x fpm) (MPa x m/min)	Material	Mating Surface		
POLYSTYRENE against mating surface													
LNP CL-4410	2% silicone	**steel**	carbon steel; surface finish: 12-16 µin; 18-20 Rockwell C	"		23	40 (0.28)	50 (15.2)	2000 (4.3)	37			453
POLYSTYRENE, GP against mating surface													
Thermofil A-20FG-0100	20% glass fiber	**steel**		ASTM D1894		23				500			459
SAN against mating surface													
LNP B-1000	unmodified	**steel**	carbon steel; surface finish: 12-16 µin; 18-20 Rockwell C	thrust washer		23	40 (0.28)	50 (15.2)	2000 (4.3)	3000			453
LNP BBL-4326	30% glass beads, 10% graphite		"	"		23	40 (0.28)	50 (15.2)	2000 (4.3)	105			453
LNP BFL-4036	30% glass fiber, 15% PTFE		"	"		23	40 (0.28)	50 (15.2)	2000 (4.3)	65			453
LNP BFL-4536	30% glass fiber, 13% PTFE, 2% silicone		"	"		23	40 (0.28)	50 (15.2)	2000 (4.3)	55			453
LNP BL-4030	15% PTFE		"	"		23	40 (0.28)	50 (15.2)	2000 (4.3)	200			453
LNP BL-4410	2% silicone		"	"		23	40 (0.28)	50 (15.2)	2000 (4.3)	70			453
LNP BL-4530	13% PTFE, 2% silicone		"	"		23	40 (0.28)	50 (15.2)	2000 (4.3)	60			453
TPE, POLYESTER against mating surface													
LNP Y-1000	unmodified	**steel**	carbon steel; surface finish: 12-16 µin; 18-20 Rockwell C	thrust washer		23	40 (0.28)	50 (15.2)	2000 (4.3)	1000			453
LNP YF-1006	30% glass fiber		"	"		23	40 (0.28)	50 (15.2)	2000 (4.3)	400			453
LNP YFL-4536	30% glass fiber, 13% PTFE, 2% silicone		"	"		23	40 (0.28)	50 (15.2)	2000 (4.3)	25			453
LNP YL-4030	15% PTFE		"	"		23	40 (0.28)	50 (15.2)	2000 (4.3)	40			453
LNP YL-4410	2% silicone		"	"		23	40 (0.28)	50 (15.2)	2000 (4.3)	30			453
LNP YL-4530	13% PTFE, 2% silicone		"	"		23	40 (0.28)	50 (15.2)	2000 (4.3)	5			453
POLYIMIDE against mating surface													
DuPont Vespel SP-1	unfilled	**steel**	AISI 1018, surface finish: 16 RMS	thrust bearing	apparatus: LRI-1 Wear and Friction Tester		100 (0.69)	984 (300)	98400 (207)			sample failed	318

Material		Mating Surface		Test Conditions						K Factor 10^{-10} x (in³-min/ft·lb·hr)		Results Note	Source
Supplier / Grade	Material Note	Mating Surface Material	Mating Surface Note	Test Method	Test Conditions Note	Temp. (°C)	Pressure (psi) (MPa)	Speed (fpm) (m/min)	PV (psi x fpm) (MPa x m/min)	Material	Mating Surface		
POLYIMIDE against mating surface													
DuPont Vespel SP-1	unfilled	**steel**	AISI 1018, surface finish: 16 RMS	thrust bearing	apparatus: LRI-1 Wear and Friction Tester		2000 (13.8)	12.5 (3.8)	25000 (52.6)			wear rate: 69 µm/hour; sliding surface temp.: 184°C	318
DuPont Vespel SP-21	15% graphite		carbon EN8, dry	pad on ring (Amsler wear tester)		20		600 (183)				wear rate: 50 µm/hour; maximum load: 30 kg	338
DuPont Vespel SP-21	15% graphite		"	"		200		100 (30.5)				wear rate: 64 µm/hour; maximum load: 210 kg	338
DuPont Vespel SP-21	15% graphite		"	"		20		100 (30.5)				wear rate: 80 µm/hour; maximum load: 210 kg	338
DuPont Vespel SP-21	15% graphite		"	"		200		600 (183)				wear rate: 125 µm/hour; maximum load: 20 kg	338
DuPont Vespel SP-21	lubricated		AISI 1018, surface finish: 16 RMS	thrust bearing	apparatus: LRI-1 Wear and Friction Tester		100 (0.69)	984 (300)	98400 (207)			wear rate: 20 µm/hour; sliding surface temp.: 203°C	318
DuPont Vespel SP-21	15% graphite		C1018 steel; 24 Rockwell C	thrust washer			50 (0.34)	900 (274)	45000 (94.5999999 9999999)	33			20
DuPont Vespel SP-21	15% graphite		"	"			1000 (6.89000000 0000001)	50 (15.2)	50000 (105.1)	35			20
DuPont Vespel SP-21	15% graphite		"	"			50 (0.34)	200 (61)	10000 (20.7)	6			20
Rhone Pou. Kinel 4503	inj. mold.; w/ TFE, w/ MoS_2; , aramid fiber reinforced		carbon steel; surface finish: 12-16 µin; unlubricated; 18-22 Rockwell C	"			51 (0.35)	50 (15.2)	2522 (5.3)	1.3			360
Rhone Pou. Kinel 4503	inj. mold.; w/ TFE, w/ MoS_2; , aramid fiber reinforced		"	"			127 (0.875)	200 (61)	25410 (53.4)	10.8			360
Rhone Pou. Kinel 4512	94 Rockwell M; inj. mold.; w/ graphite; , aramid fiber reinforced						95 (0.65)	98 (30)	9310 (19.6)			wear rate: 82.6 µm/hour	92
Rhone Pou. Kinel 4518	113 Rockwell M; inj. mold.; w/ TFE, w/ MoS_2						95 (0.65)	98 (30)	9310 (19.6)			wear rate: 30.7 µm/hour	92
Rhone Pou. Kinel 5505	110 Rockwell M; 25% graphite						43 (0.3)	98 (30)	4220 (8.9)			wear rate: 1.6 µm/hour	92
Rhone Pou. Kinel 5505	110 Rockwell M; 25% graphite						95 (0.65)	98 (30)	9310 (19.6)			wear rate: 29.5 µm/hour	92
Rhone Pou. Kinel 5508	100 Rockwell M; 40% graphite; compression						95 (0.65)	98 (30)	9310 (19.6)			wear rate: 47.2 µm/hour	92
Rhone Pou. Kinel 5508	100 Rockwell M; 40% graphite; compression						43 (0.3)	98 (30)	4220 (8.9)			wear rate: 5.1 µm/hour	92
Rhone Pou. Kinel 5517	110 Rockwell M; compression; w/ graphite, w/ MoS_2						43 (0.3)	98 (30)	4220 (8.9)			wear rate: 16 µm/hour	92
Rhone Pou. Kinel 5517	95 Rockwell M; sintering; w/ graphite, w/ MoS_2						43 (0.3)	98 (30)	4220 (8.9)			wear rate: 16 µm/hour	92

Material		Mating Surface		Test Conditions						K Factor 10^{-10} x (in^3-min/ft-lb-hr)		Results Note	Source
Supplier / Grade	Material Note	Mating Surface Material	Mating Surface Note	Test Method	Test Conditions Note	Temp. (°C)	Pressure (psi) (MPa)	Speed (fpm) (m/min)	PV (psi x fpm) (MPa x m/min)	Material	Mating Surface		
POLYIMIDE against mating surface		→											
Rhone Pou. Kinel 5517	110 Rockwell M; compression; w/ graphite, w/ MoS_2	**steel**					95 (0.65)	98 (30)	9310 (19.6)			wear rate: 19.6 µm/hour	92
Rhone Pou. Kinel 5517	95 Rockwell M; sintering; w/ graphite, w/ MoS_2						95 (0.65)	98 (30)	9310 (19.6)			wear rate: 19.6 µm/hour	92
Rhone Pou. Kinel 5517	110 Rockwell M; compression; w/ graphite, w/ MoS_2		carbon steel; surface finish: 12-16 µin; unlubricated; 18-22 Rockwell C	thrust washer			51 (0.35)	50 (15.2)	2522 (5.3)	0			360
Rhone Pou. Kinel 5517	110 Rockwell M; compression; w/ graphite, w/ MoS_2		"	"			127 (0.875)	200 (61)	25410 (53.4)	185.7			360
Rhone Pou. Kinel 5518	115 Rockwell M; compression; w/ MoS_2, w/ TFE						43 (0.3)	98 (30)	4220 (8.9)			wear rate: 5.9 µm/hour	92
Rhone Pou. Kinel 5518	115 Rockwell M; compression; w/ MoS_2, w/ TFE						95 (0.65)	98 (30)	9310 (19.6)			wear rate: 5.9 µm/hour	92
Rhone Pou. Kinel 5520	compression		carbon steel; surface finish: 12-16 µin; unlubricated; 18-22 Rockwell C	thrust washer			127 (0.875)	200 (61)	25410 (53.4)	79.6			360
Rhone Pou. Kinel 5520	compression		"	"			51 (0.35)	50 (15.2)	2522 (5.3)	4			360

Appendix III

Limiting PV Values

Material		Mating Surface		Test Method	Test Conditions Note	Temp.	Speed		Limiting PV		Note	Source
Supplier / Grade	Material Note	Mating Surface Material	Mating Surface Note			(°C)	(fpm)	(m/min)	(psi x fpm)	(MPa x m/min)		
ACETAL RESIN rubbing against		↓										
Bay Resins Lubriplas POM-1100	unmodified resin	steel		half journal bearing			10	3	4000	8.4		435
Bay Resins Lubriplas POM-1100	unmodified resin			"			100	30.5	3500	7.4		435
Bay Resins Lubriplas POM-1100	unmodified resin			"			1000	304.8	2500	5.3		435
Bay Resins Lubriplas POM-1100TF20	20% PTFE			"			1000	304.8	6000	12.6		435
Bay Resins Lubriplas POM-1100TF20	20% PTFE			"			10	3	10000	21		435
Bay Resins Lubriplas POM-1100TF20	20% PTFE			"			100	30.5	13000	27.3		435
DuPont Delrin 500CL	low wear, chemical lubricant		unlubricated				100	30.5	3500	7.4		441
DuPont Delrin AF	low wear, PTFE modified		"				100	30.5	11500	24.2		441
LNP Fulton 404	20% PTFE		carbon steel; surface finish: 12-16 µin; 18-20 Rockwell C	half cylindrical bearing		23	10	3	10000	21		453
LNP Fulton 404	20% PTFE		"	"		23	100	30.5	12500	26.3		453
LNP Fulton 404	20% PTFE		"	"		23	1000	304.8	5500	11.6		453
LNP Fulton 404D	20% PTFE		"	"		23	100	30.5	16000	33.6		453
LNP Fulton 404D	20% PTFE		"	"		23	1000	304.8	7000	14.7		453
LNP Fulton 404D	20% PTFE		"	"		23	10	3	12000	25.2		453
LNP Fulton 441	2% silicone		"	"		23	10	3	6000	12.6		453
LNP Fulton 441	2% silicone		"	"		23	100	30.5	9000	18.9		453
LNP Fulton 441	2% silicone		"	"		23	1000	304.8	4000	8.4		453
LNP Fulton 441D	2% silicone		"	"		23	1000	304.8	5000	10.5		453
LNP Fulton 441D	2% silicone		"	"		23	10	3	7000	14.7		453
LNP Fulton 441D	2% silicone		"	"		23	100	30.5	12000	25.2		453
LNP K-1000	unmodified		"	"		23	10	3	4000	8.4		453
LNP K-1000	unmodified		"	"		23	100	30.5	3500	7.4		453
LNP K-1000	unmodified		"	"		23	1000	304.8	2500	5.3		453
LNP KC-1004	20% carbon fiber		"	"		23	100	30.5	20000	42		453
LNP KC-1004	20% carbon fiber		"	"		23	1000	304.8	15000	31.5		453
LNP KC-1004	20% carbon fiber		"	"		23	10	3	13000	27.3		453
LNP KFL-4036	30% glass fiber, 15% PTFE		"	"		23	10	3	15000	31.5		453
LNP KFL-4036	30% glass fiber, 15% PTFE		"	"		23	100	30.5	17500	36.8		453
LNP KFL-4036	30% glass fiber, 15% PTFE		"	"		23	1000	304.8	12500	26.3		453

Material		Mating Surface		Test Method	Test Conditions Note	Temp.	Speed		Limiting PV		Note	Source
Supplier / Grade	Material Note	Mating Surface Material	Mating Surface Note			(°C)	(fpm)	(m/min)	(psi x fpm)	(MPa x m/min)		
ACETAL RESIN rubbing against		↴										
LNP KFL-4536	30% glass fiber, 13% PTFE, 2% silicone	**steel**	carbon steel; surface finish: 12-16 µin; 18-20 Rockwell C	half cylindrical bearing		23	100	30.5	19000	39.9		453
LNP KFL-4536	30% glass fiber, 13% PTFE, 2% silicone		"	"		23	1000	304.8	15000	31.5		453
LNP KFL-4536	30% glass fiber, 13% PTFE, 2% silicone		"	"		23	10	3	12000	25.2		453
LNP KL-4540	18% PTFE, 2% silicone		"	"		23	10	3	8000	16.8		453
LNP KL-4540	18% PTFE, 2% silicone		"	"		23	100	30.5	15000	31.5		453
LNP KL-4540	18% PTFE, 2% silicone		"	"		23	1000	304.8	12000	25.2		453
LNP KL-4540D	18% PTFE, 2% silicone		"	"		23	100	30.5	18000	37.8		453
LNP KL-4540D	18% PTFE, 2% silicone		"	"		23	1000	304.8	14000	29.4		453
LNP KL-4540D	18% PTFE, 2% silicone		"	"		23	10	3	9000	18.9		453
Polymer Acetron GP	general purpose, porosity free		unlubricated				100	30.5	2700	5.7		441
Polymer Acetron NS	w/ composite solid lubricants to improve wear		"				100	30.5	10000	21		441
RTP 800	unmodified		C1018 steel; surface finish: 14-17 µin; 15-25 Rockwell C	thrust washer	apparatus: Falex Model No. 6	23	100	30.5	4000	8.4		457
RTP 800 AR 5 TFE 10	10% PTFE, 5% aramid fiber		"	"	"	23	100	30.5	15000	31.5		457
RTP 800 AR 5 TFE 10 SI 3	10% PTFE, 3% silicone, 5% aramid fiber		"	"	"	23	100	30.5	17000	35.7		457
RTP 800 SI 2	2% silicone		"	"	"	23	100	30.5	8000	16.8		457
RTP 800 TFE 10	10% PTFE		"	"	"	23	100	30.5	10000	21		457
RTP 800 TFE 15	15% PTFE		"	"	"	23	100	30.5	13000	27.3		457
RTP 800 TFE 18 SI 2	18% PTFE, 2% silicone		"	"	"	23	100	30.5	15000	31.5		457
RTP 800 TFE 20	20% PTFE		"	"	"	23	100	30.5	14000	29.4		457
RTP 805	30% glass fiber		"	"	"	23	100	30.5	10000	21		457
RTP 805 TFE 15	15% PTFE, 30% glass fiber		"	"	"	23	100	30.5	17000	35.7		457
Thermofil G-20FG-0214	20% glass fiber, PTFE lubricated					23	100	30.5	17000	35.7		459
Thermofil G-9900-0215	20% PTFE					23	100	30.5	18000	37.8		459
Thermofil G-9900-0257	PTFE/ oil lubricated					23	100	30.5	18000	37.8		459
Unspecified grade	general purpose		unlubricated				100	30.5	2500	5.3		441
Unspecified grade			carbon EN8, dry	pad on ring (Amsler wear tester)		20	100	30.5	128996	271	maximum load: 50 kg	338
Unspecified grade			"	"		20	600	183	33796	71	maximum load: 5 kg	338

Material		Mating Surface		Test Method	Test Conditions Note	Temp.	Speed		Limiting PV		Note	Source
Supplier / Grade	Material Note	Mating Surface Material	Mating Surface Note			(°C)	(fpm)	(m/min)	(psi x fpm)	(MPa x m/min)		
ACETAL RESIN rubbing against												
Unspecified grade	unfilled	**steel**							3500	7.4	maximum contact temperature: 201°C; guideline values for reference only; actual values depend on test conditions	344
Unspecified grade	PTFE modified								7500	15.8	maximum contact temperature: 201°C; guideline values for reference only; actual values depend on test conditions	344
ACETAL COPOLYMER rubbing against												
Akzo AC-8O/TF/20	20% PTFE	**steel**	carbon steel; surface finish: 16 µin; 18-22 Rockwell C	thrust washer	apparatus: Faville-LeVally	23	100	30.5	1250	2.6		458
Akzo J-80/20/TF/15	20% glass fiber, 15% PTFE		"	"	"	23	100	30.5	17500	36.8		458
Akzo J-80/30/TF/15	30% glass fiber, 15% PTFE		"	"	"	23	100	30.5	17500	36.8		458
Hoechst Cel. Celcon LW90	low wear, 80 Rockwell M; 9.0 g/10 min. MFI			thrust washer test	apparatus: Faville-LeVally LFW-6		200	61	30000	63		321
Hoechst Cel. Celcon LW90	low wear, 80 Rockwell M; 9.0 g/10 min. MFI			"	"		50	15.2	>50000	>105		321
Hoechst Cel. Celcon LW90	low wear, 80 Rockwell M; 9.0 g/10 min. MFI			"	"		100	30.5	30000	63		321
Hoechst Cel. Celcon LW90S2	low wear, 75 Rockwell M, low wear; 2% silicone; 9.0 g/10 min. MFI			"	"		100	30.5	60000	126		321
Hoechst Cel. Celcon LW90S2	low wear, 75 Rockwell M, low wear; 2% silicone; 9.0 g/10 min. MFI			"	"		200	61	40000	84		321
Hoechst Cel. Celcon LW90S2	low wear, 75 Rockwell M, low wear; 2% silicone; 9.0 g/10 min. MFI			"	"		50	15.2	>50000	>105		321
Hoechst Cel. Celcon LWGCS2	83 Rockwell M; glass reinforced, 2% silicone			"	"		200	61	30000	63		321
Hoechst Cel. Celcon LWGCS2	83 Rockwell M; glass reinforced, 2% silicone			"	"		50	15.2	20000	42		321
Hoechst Cel. Celcon LWGCS2	83 Rockwell M; glass reinforced, 2% silicone			"	"		100	30.5	25000	52.5		321
Hoechst Cel. Celcon M90	80 Rockwell M; 9.0 g/10 min. MFI			"	"		100	30.5	30000	63		321
Hoechst Cel. Celcon M90	80 Rockwell M; 9.0 g/10 min. MFI			"	"		200	61	20000	42		321
Hoechst Cel. Celcon M90	80 Rockwell M; 9.0 g/10 min. MFI			"	"		50	15.2	40000	84		321
Unspecified grade	unmodified		carbon steel; surface finish: 16 µin; 18-22 Rockwell C	thrust washer	apparatus: Faville-LeVally	23	100	30.5	3500	7.4		458

Material		Mating Surface		Test Method	Test Conditions Note	Temp.	Speed		Limiting PV		Note	Source
Supplier / Grade	Material Note	Mating Surface Material	Mating Surface Note			(°C)	(fpm)	(m/min)	(psi x fpm)	(MPa x m/min)		
FLUOROPOLYMER, FEP rubbing against												
DuPont	15% glass fiber	**steel**	carbon steel; surface finish: 12-20 µin; unlubricated; 20-25 Rockwell C	thrust washer		23	10	3	4500	9.5		339
DuPont	96 Shore A, 59 Shore D, 25 Rockwell R		"	"		23	10	3	600	1.3		339
DuPont	10 vol.% bronze		"	"		23	10	3	9000	18.9		339
DuPont	15% glass fiber		"	"		23	100	30.5	10000	21		339
DuPont	10 vol.% bronze		"	"		23	100	30.5	12000	25.2		339
DuPont	96 Shore A, 59 Shore D, 25 Rockwell R		"	"		23	100	30.5	800	1.7		339
DuPont	96 Shore A, 59 Shore D, 25 Rockwell R		"	"		23	1000	304.8	1000	2.1		339
DuPont	10 vol.% bronze		"	"		23	1000	304.8	10000	21		339
DuPont	15% glass fiber		"	"		23	1000	304.8	8000	16.8		339
FLUOROPOLYMER, PFA rubbing against												
LNP FP-PC 1003M	15% milled glass	**steel**	carbon steel; surface finish: 12-16 µin; 18-20 Rockwell C	half cylindrical bearing		23	100	30.5	27000	56.7		453
LNP FP-PC 1003M	15% milled glass		"	"		23	1000	304.8	25000	52.5		453
LNP FP-PC 1003M	15% milled glass		"	"		23	10	3	30000	63		453
LNP FP-PC-1003	15% carbon fiber		cold rolled steel; surface finish: 12-16 µin; 22 Rockwell C	journal bearing	apparatus: Faville-LeVally LFW5	260	100	30.5	15000	31.5		453
LNP FP-PC-1003	15% carbon fiber		"	"	"	204	100	30.5	18000	37.8		453
LNP FP-PC-1003	15% carbon fiber		"	"	"	149	100	30.5	23000	48.3		453
LNP FP-PC-1003	15% carbon fiber		"	"	"	23	100	30.5	27000	56.7		453
LNP FP-PC-1003	15% carbon fiber		"	"	"	93	100	30.5	27000	56.7		453
LNP FP-PC-1003	15% carbon fiber		"	"	"	260	800	244	13000	27.3		453
LNP FP-PC-1003	15% carbon fiber		"	"	"	204	800	244	17000	35.7		453
LNP FP-PC-1003	15% carbon fiber		"	"	"	149	800	244	23000	48.3		453
LNP FP-PC-1003	15% carbon fiber		"	"	"	23	800	244	25000	52.5		453
LNP FP-PC-1003	15% carbon fiber		"	"	"	93	800	244	25000	52.5		453

Material		Mating Surface		Test Method	Test Conditions Note	Temp.	Speed		Limiting PV		Note	Source
Supplier / Grade	Material Note	Mating Surface Material	Mating Surface Note			(°C)	(fpm)	(m/min)	(psi x fpm)	(MPa x m/min)		
FLUOROPOLYMER, TFE rubbing against												
Ausimont Halon 1005	5% glass fiber	steel			Ausimont test method		100	30.5	10000	21		439
Ausimont Halon 1015	15% glass fiber				"		100	30.5	10000	21		439
Ausimont Halon 1018	18% glass fiber				"		100	30.5	10000	21		439
Ausimont Halon 1020	20% glass fiber				"		100	30.5	11000	23.1		439
Ausimont Halon 1025	25% glass fiber				"		100	30.5	11000	23.1		439
Ausimont Halon 1030	30% glass fiber				"		100	30.5	11000	23.1		439
Ausimont Halon 1035	35% glass fiber				"		100	30.5	11000	23.1		439
Ausimont Halon 1205	20% glass fiber, 5% graphite				"		100	30.5	12500	26.3		439
Ausimont Halon 1206	5% glass fiber, 5% MoS_2				"		100	30.5	11000	23.1		439
Ausimont Halon 1211	15% glass fiber, 5% MoS_2				"		100	30.5	12000	25.2		439
Ausimont Halon 1223	23% glass fiber, 2% MoS_2				"		100	30.5	12000	25.2		439
Ausimont Halon 1230	20% glass fiber, 5% MoS_2, 5% graphite				"		100	30.5	13000	27.3		439
Ausimont Halon 1240	20% glass fiber, 20% MoS_2				"		100	30.5	15000	31.5		439
Ausimont Halon 1410	10% glass fiber, 10% carbon fiber				"		100	30.5	17000	35.7		439
Ausimont Halon 1416	5% glass fiber, 10% carbon fiber				"		100	30.5	16000	33.6		439
Ausimont Halon 2010	10% graphite				"		100	30.5	15000	31.5		439
Ausimont Halon 2015	15% graphite				"		100	30.5	15000	31.5		439
Ausimont Halon 2021	5% MoS_2				"		100	30.5	11000	23.1		439
Ausimont Halon 3040	40% bronze				"		100	30.5	12000	25.2		439
Ausimont Halon 3050	50% bronze				"		100	30.5	12500	26.3		439
Ausimont Halon 3060	60% bronze				"		100	30.5	13000	27.3		439
Ausimont Halon 3205	55% bronze, 5% MoS_2				"		100	30.5	15000	31.5		439
Ausimont Halon 4010	10% carbon fiber				"		100	30.5	15000	31.5		439
Ausimont Halon 4015	15% carbon fiber				"		100	30.5	15000	31.5		439
Ausimont Halon 4022	25% carbon/ graphite				"		100	30.5	17000	35.7		439
Ausimont Halon 4025	25% carbon fiber				"		100	30.5	17000	35.7		439
Ausimont Halon 4026	25% carbon/ graphite				"		100	30.5	17000	35.7		439
Ausimont Halon 8105	5% mineral filler				"		100	30.5	10000	21		439
Ausimont Halon 8115	15% mineral filler				"		100	30.5	13000	27.3		439
Ausimont Halon 9350	50% stainless steel				"		100	30.5	13000	27.3		439

Material		Mating Surface		Test Method	Test Conditions Note	Temp.	Speed		Limiting PV		Note	Source
Supplier / Grade	Material Note	Mating Surface Material	Mating Surface Note			(°C)	(fpm)	(m/min)	(psi x fpm)	(MPa x m/min)		
FLUOROPOLYMER, TFE rubbing against												
Ausimont Halon 9360	60% stainless steel	steel			Ausimont test method		100	30.5	14000	29.4		439
Ausimont Halon 9370	70% stainless steel				"		100	30.5	14500	30.5		439
Ausimont Halon G80	virgin resin				"		100	30.5	2200	4.6		439
DuPont	54 Shore D; 15% glass fiber		carbon steel; surface finish: 12-20 µin; unlubricated; 20-25 Rockwell C	thrust washer		23	10	3	10000	21		339
DuPont	56 Shore D; 25% glass fiber		"	"		23	10	3	10000	21		339
DuPont	55 Shore D; 15% graphite		"	"		23	10	3	10000	21		339
DuPont	20% glass fiber, 5% graphite		"	"		23	10	3	11000	23.1		339
DuPont	58 Shore D; 15% glass fiber, 5% MoS_2		"	"		23	10	3	11000	23.1		339
DuPont	98 Shore A, 58 Rockwell R, 52 Shore D		"	"		23	10	3	1200	2.5		339
DuPont	amorphous; 25% carbon		"	"		23	10	3	14000	29.4		339
DuPont	65 Shore D; 60% bronze		"	"		23	10	3	15000	31.5		339
DuPont	54 Shore D; 15% glass fiber		"	"		23	100	30.5	12500	26.3		339
DuPont	56 Shore D; 25% glass fiber		"	"		23	100	30.5	13000	27.3		339
DuPont	58 Shore D; 15% glass fiber, 5% MoS_2		"	"		23	100	30.5	14000	29.4		339
DuPont	20% glass fiber, 5% graphite		"	"		23	100	30.5	15000	31.5		339
DuPont	55 Shore D; 15% graphite		"	"		23	100	30.5	17000	35.7		339
DuPont	98 Shore A, 58 Rockwell R, 52 Shore D		"	"		23	100	30.5	1800	3.8		339
DuPont	65 Shore D; 60% bronze		"	"		23	100	30.5	18500	38.9		339
DuPont	amorphous; 25% carbon		"	"		23	100	30.5	20000	42		339
DuPont	54 Shore D; 15% glass fiber		"	"		23	1000	304.8	15000	31.5		339
DuPont	56 Shore D; 25% glass fiber		"	"		23	1000	304.8	16000	33.6		339
DuPont	58 Shore D; 15% glass fiber, 5% MoS_2		"	"		23	1000	304.8	17500	36.8		339
DuPont	65 Shore D; 60% bronze		"	"		23	1000	304.8	22000	46.2		339
DuPont	20% glass fiber, 5% graphite		"	"		23	1000	304.8	22000	46.2		339
DuPont	98 Shore A, 58 Rockwell R, 52 Shore D		"	"		23	1000	304.8	2500	5.3		339
DuPont	55 Shore D; 15% graphite		"	"		23	1000	304.8	28000	58.8		339
DuPont	amorphous; 25% carbon		"	"		23	1000	304.8	30000	63		339
LNP FC-103	15% milled glass		cold rolled steel; surface finish: 12-16 µin; 22 Rockwell C	journal bearing	apparatus: Faville-LeVally LFW5	260	100	30.5	16000	33.6		453
LNP FC-103	15% milled glass		"	"	"	204	100	30.5	17000	35.7		453

Material		Mating Surface		Test Method	Test Conditions Note	Temp.	Speed		Limiting PV		Note	Source
Supplier / Grade	Material Note	Mating Surface Material	Mating Surface Note			(°C)	(fpm)	(m/min)	(psi x fpm)	(MPa x m/min)		
FLUOROPOLYMER, TFE rubbing against												
LNP FC-103	15% milled glass	**steel**	cold rolled steel; surface finish: 12-16 µin; 22 Rockwell C	journal bearing	apparatus: Faville-LeVally LFW5	23	100	30.5	18000	37.8		453
LNP FC-103	15% milled glass		"	"	"	93	100	30.5	18000	37.8		453
LNP FC-103	15% milled glass		"	"	"	149	100	30.5	18000	37.8		453
LNP FC-103	15% milled glass		"	"	"	260	800	244	12000	25.2		453
LNP FC-103	15% milled glass		"	"	"	204	800	244	14000	29.4		453
LNP FC-103	15% milled glass		"	"	"	23	800	244	16000	33.6		453
LNP FC-103	15% milled glass		"	"	"	93	800	244	16000	33.6		453
LNP FC-103	15% milled glass		"	"	"	149	800	244	16000	33.6		453
LNP FC-182	55% bronze, 5% MoS_2		"	"	"	260	100	30.5	17000	35.7		453
LNP FC-182	55% bronze, 5% MoS_2		"	"	"	204	100	30.5	18000	37.8		453
LNP FC-182	55% bronze, 5% MoS_2		"	"	"	23	100	30.5	21000	44.1		453
LNP FC-182	55% bronze, 5% MoS_2		"	"	"	93	100	30.5	21000	44.1		453
LNP FC-182	55% bronze, 5% MoS_2		"	"	"	149	100	30.5	21000	44.1		453
LNP FC-182	55% bronze, 5% MoS_2		"	"	"	23	800	244	17000	35.7		453
LNP FC-182	55% bronze, 5% MoS_2		"	"	"	93	800	244	17000	35.7		453
LNP FC-182	55% bronze, 5% MoS_2		"	"	"	149	800	244	17000	35.7		453
LNP FC-182	55% bronze, 5% MoS_2		"	"	"	204	800	244	17000	35.7		453
LNP FC-182	55% bronze, 5% MoS_2		"	"	"	260	800	244	17000	35.7		453
LNP FC-191	25% carbon/ graphite		"	"	"	260	100	30.5	16000	33.6		453
LNP FC-191	25% carbon/ graphite		"	"	"	204	100	30.5	17000	35.7		453
LNP FC-191	25% carbon/ graphite		"	"	"	23	100	30.5	20000	42		453
LNP FC-191	25% carbon/ graphite		"	"	"	93	100	30.5	20000	42		453
LNP FC-191	25% carbon/ graphite		"	"	"	149	100	30.5	20000	42		453
LNP FC-191	25% carbon/ graphite		"	"	"	260	800	244	10000	21		453
LNP FC-191	25% carbon/ graphite		"	"	"	204	800	244	13000	27.3		453
LNP FC-191	25% carbon/ graphite		"	"	"	23	800	244	15000	31.5		453
LNP FC-191	25% carbon/ graphite		"	"	"	93	800	244	15000	31.5		453
LNP FC-191	25% carbon/ graphite		"	"	"	149	800	244	15000	31.5		453
LNP PC-185	PPS filler		"	"	"	23	100	30.5	15000	31.5		453

Material		Mating Surface		Test Method	Test Conditions Note	Temp.	Speed		Limiting PV		Note	Source
Supplier / Grade	Material Note	Mating Surface Material	Mating Surface Note			(°C)	(fpm)	(m/min)	(psi x fpm)	(MPa x m/min)		
FLUOROPOLYMER, TFE rubbing against												
LNP PC-185	PPS filler	**steel**	cold rolled steel; surface finish: 12-16 µin; 22 Rockwell C	journal bearing	apparatus: Faville-LeVally LFW5	93	100	30.5	15000	31.5		453
LNP PC-185	PPS filler		"	"	"	149	100	30.5	15000	31.5		453
LNP PC-185	PPS filler		"	"	"	23	800	244	10000	21		453
LNP PC-185	PPS filler		"	"	"	93	800	244	10000	21		453
LNP PC-185	PPS filler		"	"	"	149	800	244	10000	21		453
LNP PC-185	PPS filler		"	"	"	260	800	244	8000	16.8		453
LNP PC-185	PPS filler		"	"	"	204	800	244	9000	18.9		453
LNP PC-185	PPS filler		"	"	"	260	100	30.5	13000	27.3		453
LNP PC-185	PPS filler		"	"	"	204	100	30.5	14000	29.4		453
Unspecified grade	carbon		carbon EN8, dry	pad on ring (Amsler wear tester)		20	600	183	212772	447	maximum load: 25 kg	338
Unspecified grade	15 - 25% glass fiber						100	30.5	12500	26.3	maximum contact temperature: 260°C; guideline values for reference only; actual values depend on test conditions	344
Unspecified grade	unfilled						100	30.5	1800	3.8	maximum contact temperature: 260°C; guideline values for reference only; actual values depend on test conditions	344
Unspecified grade	60% bronze						100	30.5	18500	38.9	maximum contact temperature: 260°C; guideline values for reference only; actual values depend on test conditions	344
Unspecified grade	45 - 60 Rockwell E						100	30.5	20000	42	maximum contact temperature: 260°C; guideline values for reference only; actual values depend on test conditions	344

Material		Mating Surface		Test Method	Test Conditions Note	Temp.	Speed		Limiting PV		Note	Source
Supplier / Grade	Material Note	Mating Surface Material	Mating Surface Note			(°C)	(fpm)	(m/min)	(psi x fpm)	(MPa x m/min)		
FLUOROPOLYMER, PVDF rubbing against												
LNP FP-VC-1003	15% carbon fiber	**steel**	cold rolled steel; surface finish: 12-16 µin; 22 Rockwell C	journal bearing	apparatus: Faville-LeVally LFW5	93	100	30.5	7000	14.7		453
LNP FP-VC-1003	15% carbon fiber		"	"	"	149	100	30.5	7000	14.7		453
LNP FP-VC-1003	15% carbon fiber		"	"	"	23	800	244	<5,000	<10.5		453
LNP FP-VC-1003	15% carbon fiber		"	"	"	93	800	244	<5,000	<10.5		453
LNP FP-VC-1003	15% carbon fiber		"	"	"	149	800	244	<5,000	<10.5		453
LNP FP-VC-1003	15% carbon fiber		"	"	"	204	800	244	<5,000	<10.5		453
LNP FP-VC-1003	15% carbon fiber		"	"	"	260	800	244	<5,000	<10.5		453
LNP FP-VC-1003	15% carbon fiber		carbon steel; surface finish: 12-16 µin; 18-20 Rockwell C	half cylindrical bearing		23	10	3	15000	31.5		453
LNP FP-VC-1003	15% carbon fiber		"	"		23	100	30.5	11000	23.1		453
LNP FP-VC-1003	15% carbon fiber		"	"		23	1000	304.8	<5,000	<10.5		453
LNP FP-VC-1003	15% carbon fiber		cold rolled steel; surface finish: 12-16 µin; 22 Rockwell C	journal bearing	apparatus: Faville-LeVally LFW5	204	100	30.5	<5,000	<10.5		453
LNP FP-VC-1003	15% carbon fiber		"	"	"	260	100	30.5	<5,000	<10.5		453
LNP FP-VC-1003	15% carbon fiber		"	"	"	23	100	30.5	11000	23.1		453
NYLON rubbing against												
RTP 200H	high impact	**steel**	C1018 steel; surface finish: 14-17 µin; 15-25 Rockwell C	thrust washer	apparatus: Falex Model No. 6	23	100	30.5	2000	4.2		457
RTP 200H TFE 20	20% PTFE, high impact		"	"	"	23	100	30.5	12000	25.2		457
RTP 205H	30% glass fiber, high impact		"	"	"	23	100	30.5	10000	21		457
RTP 205H TFE 15	15% PTFE, 30% glass fiber, high impact		"	"	"	23	100	30.5	18000	37.8		457
Unspecified grade	unfilled								4000	8.4	maximum contact temperature: 217°C; guideline values for reference only; actual values depend on test conditions	344

Material		Mating Surface		Test Method	Test Conditions Note	Temp.	Speed		Limiting PV		Note	Source
Supplier / Grade	Material Note	Mating Surface Material	Mating Surface Note			(°C)	(fpm)	(m/min)	(psi x fpm)	(MPa x m/min)		
NYLON, AMORPHOUS rubbing against												
LNP XC-1006	30% carbon fiber	steel	carbon steel; surface finish: 12-16 µin; 18-20 Rockwell C	half cylindrical bearing		23	10	3	10000	21		453
LNP XC-1006	30% carbon fiber		"	"		23	100	30.5	11000	23.1		453
LNP XC-1006	30% carbon fiber		"	"		23	1000	304.8	6000	12.6		453
NYLON 11 rubbing against												
RTP 200C TFE 20	20% PTFE	steel	C1018 steel; surface finish: 14-17 µin; 15-25 Rockwell C	thrust washer	apparatus: Falex Model No. 6	23	100	30.5	12000	25.2		457
RTP 203C TFE 20	20% PTFE, 20% glass fiber		"	"	"	23	100	30.5	15000	31.5		457
NYLON 12 rubbing against												
RTP 205F TFE 20	20% PTFE, 30% glass fiber	steel	C1018 steel; surface finish: 14-17 µin; 15-25 Rockwell C	thrust washer	apparatus: Falex Model No. 6	23	100	30.5	16000	33.6		457
RTP 20OF TFE 20	20% PTFE		"	"	"	23	100	30.5	12000	25.2		457
NYLON 6 rubbing against												
Bay Resins Lubriplas PA-211	unmodified resin	steel		half journal bearing			1000	304.8	<2000	<4.2		435
Bay Resins Lubriplas PA-211	unmodified resin			"			10	3	2500	5.3		435
Bay Resins Lubriplas PA-211	unmodified resin			"			100	30.5	2000	4.2		435
Bay Resins Lubriplas PA-211G30TF15	30% glass fiber, 15% PTFE			"			10	3	18000	37.8		435
Bay Resins Lubriplas PA-211G30TF15	30% glass fiber, 15% PTFE			"			100	30.5	20500	43.1		435
Bay Resins Lubriplas PA-211G30TF15	30% glass fiber, 15% PTFE			"			1000	304.8	13500	28.4		435
Bay Resins Lubriplas PA-211G33	33% glass fiber			"			100	30.5	8500	17.9		435
Bay Resins Lubriplas PA-211G33	33% glass fiber			"			1000	304.8	6000	12.6		435
Bay Resins Lubriplas PA-211G33	33% glass fiber			"			10	3	10000	21		435
Bay Resins Lubriplas PA-211TF20	20% PTFE			"			10	3	12500	26.3		435
Bay Resins Lubriplas PA-211TF20	20% PTFE			"			100	30.5	20000	42		435
Bay Resins Lubriplas PA-211TF20	20% PTFE			"			1000	304.8	6000	12.6		435
Ensinger Vekton 6PAL	oil impregnated, cast						50	15.2	18,000-20,000	37.8-42		449
LNP P-1000	unmodified		carbon steel; surface finish: 12-16 µin; 18-20 Rockwell C	half cylindrical bearing		23	10	3	2500	5.3		453

Material		Mating Surface		Test Method	Test Conditions Note	Temp.	Speed		Limiting PV		Note	Source
Supplier / Grade	Material Note	Mating Surface Material	Mating Surface Note			(°C)	(fpm)	(m/min)	(psi x fpm)	(MPa x m/min)		
NYLON 6 rubbing against												
LNP P-1000	unmodified	steel	carbon steel; surface finish: 12-16 µin; 18-20 Rockwell C	half cylindrical bearing		23	100	30.5	2000	4.2		453
LNP P-1000	unmodified		"	"		23	1000	304.8	2000	4.2		453
LNP PC-1006	30% carbon fiber		"	"		23	100	30.5	22000	46.2		453
LNP PC-1006	30% carbon fiber		"	"		23	1000	304.8	7500	15.8		453
LNP PC-1006	30% carbon fiber		"	"		23	10	3	18000	37.8		453
LNP PF-1006	30% glass fiber		"	"		23	10	3	10000	21		453
LNP PF-1006	30% glass fiber		"	"		23	100	30.5	8500	17.9		453
LNP PF-1006	30% glass fiber		"	"		23	1000	304.8	6000	12.6		453
LNP PFL-4036	30% glass fiber, 15% PTFE		"	"		23	100	30.5	20000	42		453
LNP PFL-4036	30% glass fiber, 15% PTFE		"	"		23	1000	304.8	13000	27.3		453
LNP PFL-4036	30% glass fiber, 15% PTFE		"	"		23	10	3	17500	36.8		453
LNP PFL-4536	30% glass fiber, 13% PTFE, 2% silicone		"	"		23	10	3	20000	42		453
LNP PFL-4536	30% glass fiber, 13% PTFE, 2% silicone		"	"		23	100	30.5	15500	32.6		453
LNP PFL-4536	30% glass fiber, 13% PTFE, 2% silicone		"	"		23	1000	304.8	15000	31.5		453
LNP PL-4040	20% PTFE		"	"		23	100	30.5	20500	43.1		453
LNP PL-4040	20% PTFE		"	"		23	1000	304.8	6500	13.7		453
LNP PL-4040	20% PTFE		"	"		23	10	3	12500	26.3		453
LNP PL-4410	2% silicone		"	"		23	10	3	3500	7.4		453
LNP PL-4410	2% silicone		"	"		23	100	30.5	4000	8.4		453
LNP PL-4410	2% silicone		"	"		23	1000	304.8	7500	15.8		453
LNP PL-4540	18% PTFE, 2% silicone		"	"		23	100	30.5	24500	51.5		453
LNP PL-4540	18% PTFE, 2% silicone		"	"		23	1000	304.8	10500	22.1		453
LNP PL-4540	18% PTFE, 2% silicone		"	"		23	10	3	12500	26.3		453
Polymer Nylatron GSM	cast, MoS_2 modified		unlubricated				100	30.5	3000	6.3		441
Polymer Nylatron MC901	cast, general purpose, high heat		"				100	30.5	3000	6.3		441
Polymer Nylatron NSM	w/ additives to improve bearing properties		"				100	30.5	15000	31.5		441
RTP 200A			C1018 steel; surface finish: 14-17 µin; 15-25 Rockwell C	thrust washer	apparatus: Falex Model No. 6	23	100	30.5	2000	4.2		457
RTP 200A TFF 20	20% PTFE		"	"	"	23	100	30.5	16000	33.6		457
RTP 205A	30% glass fiber		"	"	"	23	100	30.5	10000	21		457

Material		Mating Surface		Test Method	Test Conditions Note	Temp.	Speed		Limiting PV		Note	Source
Supplier / Grade	Material Note	Mating Surface Material	Mating Surface Note			(°C)	(fpm)	(m/min)	(psi x fpm)	(MPa x m/min)		
NYLON 6 rubbing against												
RTP 205A TFE 15	15% PTFE, 30% glass fiber	**steel**	C1018 steel; surface finish: 14-17 µin; 15-25 Rockwell C	thrust washer	apparatus: Falex Model No. 6	23	100	30.5	20000	42		457
Thermofil N-15FG-0100	15% glass fiber					23	100	30.5	2500	5.3		459
Thermofil N-30FG-0100	30% glass fiber					23	100	30.5	9500	20		459
Thermofil N-30FG-0214	30% glass fiber, PTFE lubricated					23	100	30.5	20000	42		459
Thermofil N-30NF-0100	30% graphite fiber					23	100	30.5	23000	48.3		459
Thermofil N-40-MF-0100	40% mineral					23	100	30.5	3000	6.3		459
Thermofil N-40BG-0100	40% glass bead					23	100	30.5	2800	5.9		459
Thermofil N-40FM-0100	40% glass/ mineral					23	100	30.5	9000	18.9		459
Unspecified grade	unmodified		carbon steel; surface finish: 16 µin; 18-22 Rockwell C	thrust washer	apparatus: Faville-LeVally	23	100	30.5	2000	4.2		458
NYLON 610 rubbing against												
LNP OL-4540	18% PTFE, 2% silicone	**steel**	carbon steel; surface finish: 12-16 µin; 18-20 Rockwell C	half cylindrical bearing		23	100	30.5	20500	43.1		453
LNP OL-4540	18% PTFE, 2% silicone		"	"		23	1000	304.8	9000	18.9		453
LNP OL-4540	18% PTFE, 2% silicone		"	"		23	10	3	8500	17.9		453
LNP Q-1000	unmodified		"	"		23	10	3	2500	5.3		453
LNP Q-1000	unmodified		"	"		23	100	30.5	2000	4.2		453
LNP Q-1000	unmodified		"	"		23	1000	304.8	<2000	<4.2		453
LNP QC-1006	30% carbon fiber		"	"		23	100	30.5	21000	44.1		453
LNP QC-1006	30% carbon fiber		"	"		23	1000	304.8	7500	15.8		453
LNP QC-1006	30% carbon fiber		"	"		23	10	3	18000	37.8		453
LNP QF-1006	30% glass fiber		"	"		23	10	3	10000	21		453
LNP QF-1006	30% glass fiber		"	"		23	100	30.5	8500	17.9		453
LNP QF-1006	30% glass fiber		"	"		23	1000	304.8	5500	11.6		453
LNP QFL-4036	30% glass fiber, 15% PTFE		"	"		23	1000	304.8	12000	25.2		453
LNP QFL-4036	30% glass fiber, 15% PTFE		"	"		23	10	3	20000	42		453
LNP QFL-4036	30% glass fiber, 15% PTFE		"	"		23	100	30.5	15000	31.5		453
LNP QFL-4536	30% glass fiber, 13% PTFE, 2% silicone		"	"		23	10	3	20000	42		453
LNP QFL-4536	30% glass fiber, 13% PTFE, 2% silicone		"	"		23	100	30.5	16000	33.6		453
LNP QFL-4536	30% glass fiber, 13% PTFE, 2% silicone		"	"		23	1000	304.8	13500	28.4		453

Material		Mating Surface		Test Method	Test Conditions Note	Temp.	Speed		Limiting PV		Note	Source
Supplier / Grade	Material Note	Mating Surface Material	Mating Surface Note			(°C)	(fpm)	(m/min)	(psi x fpm)	(MPa x m/min)		
NYLON 610 rubbing against												
LNP QL-4040	20% PTFE	steel	carbon steel; surface finish: 12-16 µin; 18-20 Rockwell C	half cylindrical bearing		23	100	30.5	17500	36.8		453
LNP QL-4040	20% PTFE		"	"		23	1000	304.8	6000	12.6		453
LNP QL-4040	20% PTFE		"	"		23	10	3	8500	17.9		453
LNP QL-4410	2% silicone		"	"		23	10	3	3000	6.3		453
LNP QL-4410	2% silicone		"	"		23	100	30.5	4000	8.4		453
LNP QL-4410	2% silicone		"	"		23	1000	304.8	7000	14.7		453
RTP 200B			C1018 steel; surface finish: 14-17 µin; 15-25 Rockwell C	thrust washer	apparatus: Falex Model No. 6	23	100	30.5	2000	4.2		457
RTP 205B	30% glass fiber		"	"	"	23	100	30.5	9000	18.9		457
RTP 205B TFE 15	15% PTFE, 30% glass fiber		"	"	"	23	100	30.5	20000	42		457
NYLON 612 rubbing against												
Akzo J-4/30/TF/15	30% glass fiber, 15% PTFE	steel	carbon steel; surface finish: 16 µin; 18-22 Rockwell C	thrust washer	apparatus: Faville-LeVally	23	100	30.5	20000	42		458
Akzo J-4/CF/30/TF/10	30% carbon fiber, 10% PTFE		"	"	"	23	100	30.5	36000	75.6		458
Akzo J-4/CF/30/TF/13/SI/2	30% carbon fiber, 13% PTFE, 2% silicone		"	"	"	23	100	30.5	43000	90.3		458
Akzo NY-4/TF/10	10% PTFE		"	"	"	23	100	30.5	14000	29.4		458
LNP I-1000	unmodified		carbon steel; surface finish: 12-16 µin; 18-20 Rockwell C	half cylindrical bearing		23	100	30.5	2000	4.2		453
LNP I-1000	unmodified		"	"		23	1000	304.8	<2,000	<4.2		453
LNP I-1000	unmodified		"	"		23	10	3	2500	5.3		453
LNP IC-1006	30% carbon fiber		"	"		23	10	3	18000	37.8		453
LNP IC-1006	30% carbon fiber		"	"		23	100	30.5	20	0		453
LNP IC-1006	30% carbon fiber		"	"		23	1000	304.8	17000	35.7		453
LNP IF-1006	30% glass fiber		"	"		23	100	30.5	8000	16.8		453
LNP IF-1006	30% glass fiber		"	"		23	1000	304.8	5000	10.5		453
LNP IF-1006	30% glass fiber		"	"		23	10	3	10000	21		453
LNP IFL-4036	30% glass fiber, 15% PTFE		"	"		23	10	3	20000	42		453
LNP IFL-4036	30% glass fiber, 15% PTFE		"	"		23	100	30.5	15000	31.5		453
LNP IFL-4036	30% glass fiber, 15% PTFE		"	"		23	1000	304.8	12000	25.2		453
LNP IFL-4536	30% glass fiber, 13% PTFE, 2% silicone		"	"		23	1000	304.8	13000	27.3		453

Material		Mating Surface		Test Method	Test Conditions Note	Temp.	Speed		Limiting PV		Note	Source
Supplier / Grade	Material Note	Mating Surface Material	Mating Surface Note			(°C)	(fpm)	(m/min)	(psi x fpm)	(MPa x m/min)		
NYLON 612 rubbing against												
LNP IFL-4536	30% glass fiber, 13% PTFE, 2% silicone	**steel**	carbon steel; surface finish: 12-16 µin; 18-20 Rockwell C	half cylindrical bearing		23	10	3	20000	42		453
LNP IFL-4536	30% glass fiber, 13% PTFE, 2% silicone		"	"		23	100	30.5	15000	31.5		453
LNP IL-4040	20% PTFE		"	"		23	10	3	9000	18.9		453
LNP IL-4040	20% PTFE		"	"		23	100	30.5	18000	37.8		453
LNP IL-4040	20% PTFE		"	"		23	1000	304.8	6000	12.6		453
LNP IL-4410	2% silicone		"	"		23	100	30.5	4000	8.4		453
LNP IL-4410	2% silicone		"	"		23	1000	304.8	7000	14.7		453
LNP IL-4410	2% silicone		"	"		23	10	3	3000	6.3		453
LNP IL-4540	18% PTFE, 2% silicone		"	"		23	10	3	9000	18.9		453
LNP IL-4540	18% PTFE, 2% silicone		"	"		23	100	30.5	20000	42		453
LNP IL-4540	18% PTFE, 2% silicone		"	"		23	1000	304.8	9000	18.9		453
RTP 200D			C1018 steel; surface finish: 14-17 µin; 15-25 Rockwell C	thrust washer	apparatus: Falex Model No. 6	23	100	30.5	2000	4.2		457
RTP 200D TFE 20	20% PTFE		"	"	"	23	100	30.5	17000	35.7		457
RTP 205D	30% glass fiber		"	"	"	23	100	30.5	9000	18.9		457
RTP 205D TFE 15	15% PTFE, 30% glass fiber		"	"	"	23	100	30.5	20000	42		457
Thermofil N6-30FG-0100	30% glass fiber					23	100	30.5	8000	16.8		459
Thermofil N6-30FG-0214	30% glass fiber, PTFE lubricated					23	100	30.5	15000	31.5		459
Thermofil N6-30FG-0282	30% glass fiber, lubricated					23	100	30.5	19000	39.9		459
Thermofil N6-30FG-0500	30% glass fiber, flame retardant					23	100	30.5	5000	10.5		459
Thermofil N6-30NF-0100	30% graphite fiber					23	100	30.5	21000	44.1		459
Thermofil N6-9900-0500	flame retardant					23	100	30.5	1000	2.1		459
Thermofil R-20NF-0214	20% graphite fiber, PTFE lubricated					23	100	30.5	28000	58.8		459
Thermofil R-30FG-0100	30% glass fiber					23	100	30.5	25000	52.5		459
Thermofil R-30FG-0214	30% glass fiber, PTFE lubricated					23	100	30.5	20000	42		459
Thermofil R-40NF-0100	40% graphite fiber					23	100	30.5	22000	46.2		459
Thermofil R-9900-0200	lubricated					23	100	30.5	2000	4.2		459
Thermofil R-9900-0214	PTFE lubricated					23	100	30.5	2800	5.9		459
Unspecified grade	unmodified		carbon steel; surface finish: 16 µin; 18-22 Rockwell C	thrust washer	apparatus: Faville-LeVally	23	100	30.5	2000	4.2		458

Material		Mating Surface		Test Method	Test Conditions Note	Temp.	Speed		Limiting PV		Note	Source
Supplier / Grade	Material Note	Mating Surface Material	Mating Surface Note			(°C)	(fpm)	(m/min)	(psi x fpm)	(MPa x m/min)		
NYLON 66 rubbing against												
Akzo G-1/30/MS/5	30% long glass fiber, 5% MoS_2	**steel**	carbon steel; surface finish: 16 µin; 18-22 Rockwell C	thrust washer	apparatus: Faville-LeVally	23	100	30.5	15000	31.5		458
Akzo G-1/30/TF/15	30% long glass fiber,15% PTFE		"	"	"	23	100	30.5	20000	42		458
Akzo J-1/30/MS/5	30% glass fiber, 5% MoS_2		"	"	"	23	100	30.5	15000	31.5		458
Akzo J-1/30/SI/3	30% glass fiber, 3% silicone		"	"	"	23	100	30.5	12000	25.2		458
Akzo J-1/30/TF/15	30% glass fiber,15% PTFE		"	"	"	23	100	30.5	20000	42		458
Akzo J-1/33/TF/13/SI/2	33% glass fiber,13% PTFE, 2% silicone		"	"	"	23	100	30.5	20000	42		458
Akzo J-1/CF/15/TF/20	15% carbon fiber, 20% PTFE		"	"	"	23	100	30.5	30000	63		458
Akzo J-1/CF/30/TF/13/SI/2	30% carbon fiber, 13% PTFE, 2% silicone		"	"	"	23	100	30.5	43000	90.3		458
Akzo J-1/CF/30/TF/15	30% carbon fiber, 15% PTFE		"	"	"	23	100	30.5	42000	88.2		458
Akzo NY-1/MS/5	5% MoS_2		"	"	"	23	100	30.5	7000	14.7		458
Akzo NY-1/MS/5/TF/30	5% MoS_2, 30% PTFE		"	"	"	23	100	30.5	9000	18.9		458
Akzo NY-1/SI/5	5% silicone		"	"	"	23	100	30.5	6000	12.6		458
Akzo NY-1/TF/10	10% PTFE		"	"	"	23	100	30.5	9000	18.9		458
Akzo NY-1/TF/15	15% PTFE		"	"	"	23	100	30.5	12000	25.2		458
Akzo NY-1/TF/30	30% PTFE		"	"	"	23	100	30.5	12000	25.2		458
Bay Resins Lubriplas PA-111	unmodified resin			half journal bearing			10	3	3000	6.3		435
Bay Resins Lubriplas PA-111	unmodified resin			"			100	30.5	2500	5.3		435
Bay Resins Lubriplas PA-111	unmodified resin			"			1000	304.8	<2500	<5.3		435
Bay Resins Lubriplas PA-111CF30	30% carbon fiber			"			1000	304.8	8500	17.9		435
Bay Resins Lubriplas PA-111CF30	30% carbon fiber			"			10	3	21500	45.2		435
Bay Resins Lubriplas PA-111CF30	30% carbon fiber			"			100	30.5	27000	56.7		435
Bay Resins Lubriplas PA-111G30TF15	30% glass fiber, 15% PTFE			"			10	3	17500	36.8		435
Bay Resins Lubriplas PA-111G30TF15	30% glass fiber, 15% PTFE			"			100	30.5	20000	42		435
Bay Resins Lubriplas PA-111G30TF15	30% glass fiber, 15% PTFE			"			1000	304.8	17500	36.8		435
Bay Resins Lubriplas PA-111G33	33% glass fiber			"			1000	304.8	7500	15.8		435
Bay Resins Lubriplas PA-111G33	33% glass fiber			"			10	3	12500	26.3		435
Bay Resins Lubriplas PA-111G33	33% glass fiber			"			100	30.5	10000	21		435
Bay Resins Lubriplas PA-111TF20	20% PTFE			"			10	3	14000	29.4		435
Bay Resins Lubriplas PA-111TF20	20% PTFE			"			100	30.5	17500	36.8		435

Material		Mating Surface		Test Method	Test Conditions Note	Temp.	Speed		Limiting PV		Note	Source
Supplier / Grade	Material Note	Mating Surface Material	Mating Surface Note			(°C)	(fpm)	(m/min)	(psi x fpm)	(MPa x m/min)		
NYLON 66 rubbing against												
Bay Resins Lubriplas PA-111TF20	20% PTFE	steel		half journal bearing			1000	304.8	8000	16.8		435
DuPont Maranyl A108	tribological properties, formerly by LNP; w/ MoS_2, w/ graphite		carbon EN8, dry	pad on ring (Amsler wear tester)		20	100	30.5	102816	216	maximum load: 50 kg	338
DuPont Maranyl A108	tribological properties, formerly by LNP; w/ MoS_2, w/ graphite		"	"		20	600	183	33796	71	maximum load: 3 kg	338
DuPont Maranyl A198	formerly by LNP, tribological properties; glass reinforced; w/ graphite		"	"		20	100	30.5	141848	298	maximum load: 110 kg	338
DuPont Maranyl A198	formerly by LNP, tribological properties; glass reinforced; w/ graphite		"	"		20	600	183	123760	260	maximum load: 10 kg	338
DuPont Maranyl A198	formerly by LNP, tribological properties; glass reinforced; w/ graphite		"	"		200	600	183	47124	99	maximum load: 5 kg	338
LNP R-1000	unmodified		carbon steel; surface finish: 12-16 µin; 18-20 Rockwell C	half cylindrical bearing		23	10	3	3000	6.3		453
LNP R-1000	unmodified		"	"		23	100	30.5	2500	5.3		453
LNP R-1000	unmodified		"	"		23	1000	304.8	2500	5.3		453
LNP RC-1004	20% carbon fiber		"	"		23	100	30.5	25000	52.5		453
LNP RC-1004	20% carbon fiber		"	"		23	1000	304.8	7000	14.7		453
LNP RC-1004	20% carbon fiber		"	"		23	10	3	19000	39.9		453
LNP RC-1006	30% carbon fiber		"	"		23	10	3	21000	44.1		453
LNP RC-1006	30% carbon fiber		"	"		23	100	30.5	27000	56.7		453
LNP RC-1006	30% carbon fiber		"	"		23	1000	304.8	8000	16.8		453
LNP RC-1008	40% carbon fiber		"	"		23	100	30.5	27500	57.8		453
LNP RC-1008	40% carbon fiber		"	"		23	1000	304.8	8500	17.9		453
LNP RC-1008	40% carbon fiber		"	"		23	10	3	22000	46.2		453
LNP RCL-4036	30% carbon fiber, 15% PTFE		"	"		23	10	3	29000	60.9		453
LNP RCL-4036	30% carbon fiber, 15% PTFE		"	"		23	100	30.5	42000	88.2		453
LNP RCL-4036	30% carbon fiber, 15% PTFE		"	"		23	1000	304.8	19000	39.9		453
LNP RCL-4536	30% carbon fiber, 13% PTFE, 2% silicone		"	"		23	100	30.5	43000	90.3		453
LNP RCL-4536	30% carbon fiber, 13% PTFE, 2% silicone		"	"		23	1000	304.8	20000	42		453
LNP RCL-4536	30% carbon fiber, 13% PTFE, 2% silicone		"	"		23	10	3	29000	60.9		453
LNP RF-1006 HS	30% glass fiber		"	"		23	10	3	12500	26.3		453
LNP RF-1006 HS	30% glass fiber		"	"		23	100	30.5	10000	21		453
LNP RF-1006 HS	30% glass fiber		"	"		23	1000	304.8	7500	15.8		453
LNP RFL-4036	30% glass fiber, 15% PTFE		"	"		23	100	30.5	20000	42		453

Material		Mating Surface		Test Method	Test Conditions Note	Temp.	Speed		Limiting PV		Note	Source
Supplier / Grade	Material Note	Mating Surface Material	Mating Surface Note			(°C)	(fpm)	(m/min)	(psi x fpm)	(MPa x m/min)		
NYLON 66 rubbing against												
LNP RFL-4036	30% glass fiber, 15% PTFE	**steel**	carbon steel; surface finish: 12-16 µin; 18-20 Rockwell C	half cylindrical bearing		23	1000	304.8	17500	36.8		453
LNP RFL-4036	30% glass fiber, 15% PTFE		"	"		23	10	3	17500	36.8		453
LNP RFL-4036	30% glass fiber, 15% PTFE		cold rolled steel; surface finish: 12-16 µin; 22 Rockwell C	journal bearing	apparatus: Faville-LeVally LFW5	260	100	30.5	19000	39.9		453
LNP RFL-4036	30% glass fiber, 15% PTFE		"	"	"	23	100	30.5	20000	42		453
LNP RFL-4036	30% glass fiber, 15% PTFE		"	"	"	93	100	30.5	20000	42		453
LNP RFL-4036	30% glass fiber, 15% PTFE		"	"	"	149	100	30.5	20000	42		453
LNP RFL-4036	30% glass fiber, 15% PTFE		"	"	"	204	100	30.5	20000	42		453
LNP RFL-4036	30% glass fiber, 15% PTFE		"	"	"	260	800	244	13000	27.3		453
LNP RFL-4036	30% glass fiber, 15% PTFE		"	"	"	204	800	244	15000	31.5		453
LNP RFL-4036	30% glass fiber, 15% PTFE		"	"	"	23	800	244	17500	36.8		453
LNP RFL-4036	30% glass fiber, 15% PTFE		"	"	"	93	800	244	17500	36.8		453
LNP RFL-4036	30% glass fiber, 15% PTFE		"	"	"	149	800	244	17500	36.8		453
LNP RFL-4536	30% glass fiber, 13% PTFE, 2% silicone		carbon steel; surface finish: 12-16 µin; 18-20 Rockwell C	half cylindrical bearing		23	10	3	17000	35.7		453
LNP RFL-4536	30% glass fiber, 13% PTFE, 2% silicone		"	"		23	100	30.5	20000	42		453
LNP RFL-4536	30% glass fiber, 13% PTFE, 2% silicone		"	"		23	1000	304.8	19000	39.9		453
LNP RL-4040	20% PTFE		"	"		23	100	30.5	17500	36.8		453
LNP RL-4040	20% PTFE		"	"		23	1000	304.8	8000	16.8		453
LNP RL-4040	20% PTFE		"	"		23	10	3	14000	29.4		453
LNP RL-4410	2% silicone		"	"		23	10	3	3000	6.3		453
LNP RL-4410	2% silicone		"	"		23	100	30.5	6000	12.6		453
LNP RL-4410	2% silicone		"	"		23	1000	304.8	9000	18.9		453
LNP RL-4540	18% PTFE, 2% silicone		"	"		23	100	30.5	30000	63		453
LNP RL-4540	18% PTFE, 2% silicone		"	"		23	1000	304.8	12000	25.2		453
LNP RL-4540	18% PTFE, 2% silicone		"	"		23	10	3	14000	29.4		453
Polymer Nylatron NS	w/ additives to improve bearing properties		unlubricated				100	30.5	11000	23.1		441
RTP 200	unmodified		C1018 steel; surface finish: 14-17 µin; 15-25 Rockwell C	thrust washer	apparatus: Falex Model No. 6	23	100	30.5	3000	6.3		457
RTP 200 AR 15	15% aramid fiber		"	"	"	23	100	30.5	15000	31.5		457
RTP 200 AR 15 TFE 15	15% PTFE, 15% aramid fiber		"	"	"	23	100	30.5	25000	52.5		457

Material		Mating Surface		Test Method	Test Conditions Note	Temp.	Speed		Limiting PV		Note	Source
Supplier / Grade	Material Note	Mating Surface Material	Mating Surface Note			(°C)	(fpm)	(m/min)	(psi x fpm)	(MPa x m/min)		
NYLON 66 rubbing against												
RTP 200 TFE 10	10% PTFE	**steel**	C1018 steel; surface finish: 14-17 µin; 15-25 Rockwell C	thrust washer	apparatus: Falex Model No. 6	23	100	30.5	14000	29.4		457
RTP 200 TFE 20	20% PTFE		"	"	"	23	100	30.5	17000	35.7		457
RTP 200 TFE 20 SI	20% PTFE, 0.5% silicone		"	"	"	23	100	30.5	22000	46.2		457
RTP 201 TFE 5 SI	5% PTFE, 0.5% silicone, 10% glass fiber		"	"	"	23	100	30.5	13000	27.3		457
RTP 203 TFE 10	10% PTFE, 20% glass fiber		"	"	"	23	100	30.5	15000	31.5		457
RTP 203 TFE 15	15% PTFE, 20% glass fiber		"	"	"	23	100	30.5	17000	35.7		457
RTP 203 TFE 20	20% PTFE, 20% glass fiber		"	"	"	23	100	30.5	18000	37.8		457
RTP 205	30% glass fiber		"	"	"	23	100	30.5	14000	29.4		457
RTP 205 SI 2	2% silicone, 30% glass fiber		"	"	"	23	100	30.5	13000	27.3		457
RTP 205 TFE 13 SI 2	13% PTFE, 2% silicone, 30% glass fiber		"	"	"	23	100	30.5	21000	44.1		457
RTP 205 TFE 15	30% glass fiber		"	"	"	23	100	30.5	20000	42		457
RTP 205 TFE 20	20% PTFE, 30% glass fiber		"	"	"	23	100	30.5	21000	44.1		457
RTP 205 TFE 5	5% PTFE, 30% glass fiber		"	"	"	23	100	30.5	15000	31.5		457
RTP 281 TFE 20	20% PTFE, 10% carbon fiber		"	"	"	23	100	30.5	25000	52.5		457
RTP 283 TFE 1 0	10% PTFE, 20% carbon fiber		"	"	"	23	100	30.5	25000	52.5		457
RTP 285	30% carbon fiber		"	"	"	23	100	30.5	25000	52.5		457
RTP 285 TFE 13 SI 2	13% PTFE, 2% silicone, 30% carbon fiber		"	"	"	23	100	30.5	41000	86.1		457
RTP 285 TFE 15	15% PTFE, 30% carbon fiber		"	"	"	23	100	30.5	40000	84		457
RTP 287 TFE 10	10% PTFE, 40% carbon fiber		"	"	"	23	100	30.5	42000	88.2		457
Thermofil N3-13FG-0100	13% glass fiber					23	100	30.5	5000	10.5		459
Thermofil N3-13FG-0700	13% glass fiber, impact modified					23	100	30.5	4000	8.4		459
Thermofil N3-15G-0560	15% glass fiber, flame retardant					23	100	30.5	3000	6.3		459
Thermofil N3-20NF-0100	20% graphite fiber					23	100	30.5	22000	46.2		459
Thermofil N3-30FG-0214	30% glass fiber, PTFE lubricated					23	100	30.5	20000	42		459
Thermofil N3-30FG-0231	30% glass fiber, MoS_2 lubricated					23	100	30.5	16500	34.7		459
Thermofil N3-30FG-0282	30% glass fiber, lubricated					23	100	30.5	20000	42		459
Thermofil N3-30FG-0560	30% glass fiber, flame retardant					23	100	30.5	10000	21		459
Thermofil N3-30NF-0214	30% graphite fiber, PTFE lubricated					23	100	30.5	43000	90.3		459
Thermofil N3-33FG-0100	33% glass fiber					23	100	30.5	10000	21		459
Thermofil N3-40NF-0100	40% graphite fiber					23	100	30.5	31000	65.1		459

Material		Mating Surface		Test Method	Test Conditions Note	Temp.	Speed		Limiting PV		Note	Source
Supplier / Grade	Material Note	Mating Surface Material	Mating Surface Note			(°C)	(fpm)	(m/min)	(psi x fpm)	(MPa x m/min)		
NYLON 66 rubbing against												
Thermofil N3-43FG-0100	43% glass fiber	**steel**				23	100	30.5	12000	25.2		459
Thermofil N3-9900-0200	lubricated					23	100	30.5	7000	14.7		459
Thermofil N3-9900-0214	PTFE lubricated					23	100	30.5	5000	10.5		459
Thermofil N3-9900-0231	MoS_2 lubricated					23	100	30.5	7500	15.8		459
Thermofil N3-9900-0279	40% carbon fiber					23	100	30.5	25000	52.5		459
Thermofil N3-9900-0560	flame retardant					23	100	30.5	2800	5.9		459
Unspecified grade	general purpose		unlubricated				100	30.5	2750	5.8		441
Unspecified grade	unmodified		carbon steel; surface finish: 16 µin; 18-22 Rockwell C	thrust washer	apparatus: Faville-LeVally	23	100	30.5	2500	5.3		458
POLYPHTHALAMIDE rubbing against												
RTP 4005 TFE 15	15% PTFE, 30% glass fiber	**steel**	C1018 steel; surface finish: 14-17 µin; 15-25 Rockwell C	thrust washer	apparatus: Falex Model No. 6	23	100	30.5	30000	63		457
RTP 4083 TFE 15 S12	15% PTFE, 2% silicone, 20% carbon fiber		"	"	"	23	100	30.5	40000	84		457
POLYCARBONATE rubbing against												
Akzo G-50/20/TF/15	20% long glass fiber, 15% PTFE	**steel**	carbon steel; surface finish: 16 µin; 18-22 Rockwell C	thrust washer	apparatus: Faville-LeVally	23	100	30.5	25000	52.5		458
Akzo J-50/20/TF/15	20% glass fiber, 15% PTFE		"	"	"	23	100	30.5	30000	63		458
Akzo J-50/20/TF/I 0	20% glass fiber, 10% PTFE		"	"	"	23	100	30.5	25000	52.5		458
Akzo J-50/3O/TF/15	30% glass fiber, 15% PTFE		"	"	"	23	100	30.5	30000	63		458
Akzo PC-50/TF/10	10% PTFE		"	"	"	23	100	30.5	18000	37.8		458
Akzo PC-50/TF/15	15% PTFE		"	"	"	23	100	30.5	20000	42		458
Bay Resins Lubriplas PC-1100	unmodified resin			half journal bearing			10	3	750	1.6		435
Bay Resins Lubriplas PC-1100	unmodified resin			"			100	30.5	600	1.3		435
Bay Resins Lubriplas PC-1100	unmodified resin			"			1000	304.8	500	1.1		435
Bay Resins Lubriplas PC-1100CF30	30% carbon fiber			"			100	30.5	8500	17.9		435
Bay Resins Lubriplas PC-1100CF30	30% carbon fiber			"			1000	304.8	5500	11.6		435
Bay Resins Lubriplas PC-1100CF30	30% carbon fiber			"			10	3	8000	16.8		435
Bay Resins Lubriplas PC-1100G30TF15	30% glass fiber, 15% PTFE			"			10	3	28000	58.8		435

Material		Mating Surface		Test Method	Test Conditions Note	Temp.	Speed		Limiting PV		Note	Source
Supplier / Grade	Material Note	Mating Surface Material	Mating Surface Note			(°C)	(fpm)	(m/min)	(psi x fpm)	(MPa x m/min)		
POLYCARBONATE rubbing against												
Bay Resins Lubriplas PC-1100G30TF15	30% glass fiber, 15% PTFE	steel		half journal bearing			100	30.5	30000	63		435
Bay Resins Lubriplas PC-1100G30TF15	30% glass fiber, 15% PTFE			"			1000	304.8	14000	29.4		435
Bay Resins Lubriplas PC-1100TF15	15% PTFE			"			1000	304.8	11000	23.1		435
Bay Resins Lubriplas PC-1100TF15	15% PTFE			"			10	3	15500	32.6		435
Bay Resins Lubriplas PC-1100TF15	15% PTFE			"			100	30.5	21000	44.1		435
Bay Resins Lubriplas PC-1100TF20	20% PTFE			"			10	3	16000	33.6		435
Bay Resins Lubriplas PC-1100TF20	20% PTFE			"			100	30.5	22000	46.2		435
Bay Resins Lubriplas PC-1100TF20	20% PTFE			"			1000	304.8	12000	25.2		435
LNP D-1000	unmodified		carbon steel; surface finish: 12-16 µin; 18-20 Rockwell C	half cylindrical bearing		23	100	30.5	500	1.1		453
LNP D-1000	unmodified		"	"		23	1000	304.8	<500	<1.1		453
LNP D-1000	unmodified		"	"		23	10	3	750	1.6		453
LNP DC-1006	30% carbon fiber		"	"		23	10	3	8000	16.8		453
LNP DC-1006	30% carbon fiber		"	"		23	100	30.5	8500	17.9		453
LNP DC-1006	30% carbon fiber		"	"		23	1000	304.8	5500	11.6		453
LNP DFL-4034	20% glass fiber, 15% PTFE		"	"		23	100	30.5	25000	52.5		453
LNP DFL-4034	20% glass fiber, 15% PTFE		"	"		23	1000	304.8	9000	18.9		453
LNP DFL-4034	20% glass fiber, 15% PTFE		"	"		23	10	3	20000	42		453
LNP DFL-4036	30% glass fiber, 15% PTFE		"	"		23	10	3	27500	57.8		453
LNP DFL-4036	30% glass fiber, 15% PTFE		"	"		23	100	30.5	30000	63		453
LNP DFL-4036	30% glass fiber, 15% PTFE		"	"		23	1000	304.8	14000	29.4		453
LNP DFL-4536	30% glass fiber, 13% PTFE, 2% silicone		"	"		23	100	30.5	30000	63		453
LNP DFL-4536	30% glass fiber, 13% PTFE, 2% silicone		"	"		23	1000	304.8	16000	33.6		453
LNP DFL-4536	30% glass fiber, 13% PTFE, 2% silicone		"	"		23	10	3	27500	57.8		453
LNP DL 4040	20% PTFE		"	"		23	10	3	16000	33.6		453
LNP DL 4040	20% PTFE		"	"		23	100	30.5	22000	46.2		453
LNP DL 4040	20% PTFE		"	"		23	1000	304.8	12000	25.2		453
LNP DL-4010	5% PTFE		"	"		23	100	30.5	9000	18.9		453

Material		Mating Surface		Test Method	Test Conditions Note	Temp.	Speed		Limiting PV		Note	Source
Supplier / Grade	Material Note	Mating Surface Material	Mating Surface Note			(°C)	(fpm)	(m/min)	(psi x fpm)	(MPa x m/min)		
POLYCARBONATE rubbing against												
LNP DL-4010	5% PTFE	steel	carbon steel; surface finish: 12-16 µin; 18-20 Rockwell C	half cylindrical bearing		23	1000	304.8	4000	8.4		453
LNP DL-4010	5% PTFE		"	"		23	10	3	6000	12.6		453
LNP DL-4020	10% PTFE		"	"		23	10	3	14000	29.4		453
LNP DL-4020	10% PTFE		"	"		23	100	30.5	18000	37.8		453
LNP DL-4020	10% PTFE		"	"		23	1000	304.8	8000	16.8		453
LNP DL-4030	15% PTFE		"	"		23	100	30.5	20000	42		453
LNP DL-4030	15% PTFE		"	"		23	1000	304.8	10500	22.1		453
LNP DL-4030	15% PTFE		"	"		23	10	3	15000	31.5		453
LNP DL-4530	13% PTFE, 2% silicone		"	"		23	10	3	14000	29.4		453
LNP DL-4530	13% PTFE, 2% silicone		"	"		23	100	30.5	23000	48.3		453
LNP DL-4530	13% PTFE, 2% silicone		"	"		23	1000	304.8	13000	27.3		453
RTP 300			C1018 steel; surface finish: 14-17 µin; 15-25 Rockwell C	thrust washer	apparatus: Falex Model No. 6	23	100	30.5	1000	2.1		457
RTP 300 AR 15	15% aramid fiber		"	"	"	23	100	30.5	12000	25.2		457
RTP 300 AR 15 TFE 15	15% PTFE, 15% aramid fiber		"	"	"	23	100	30.5	20000	42		457
RTP 300 TFE 10	10% PTFE		"	"	"	23	100	30.5	18000	37.8		457
RTP 300 TFE 15	15% PTFE		"	"	"	23	100	30.5	20000	42		457
RTP 300 TFE 15 SI	15% PTFE, 0.5% silicone		"	"	"	23	100	30.5	22000	46.2		457
RTP 300 TFE 20	20% PTFE		"	"	"	23	100	30.5	22000	46.2		457
RTP 300 TFE 5	5% PTFE		"	"	"	23	100	30.5	10000	21		457
RTP 301 TFE 5	5% PTFE, 10% glass fiber		"	"	"	23	100	30.5	20000	42		457
RTP 302 TFE 10	10% PTFE, 15% glass fiber		"	"	"	23	100	30.5	22000	46.2		457
RTP 302 TFE 15	15% PTFE, 15% glass fiber		"	"	"	23	100	30.5	23000	48.3		457
RTP 303 TFE 1 0	10% PTFE, 20% glass fiber		"	"	"	23	100	30.5	24000	50.4		457
RTP 303 TFE 15	15% PTFE, 20% glass fiber		"	"	"	23	100	30.5	23000	48.3		457
RTP 303 TFE 20	20% PTFE, 20% glass fiber		"	"	"	23	100	30.5	24000	50.4		457
RTP 303 TFE 20 SI 2	20% PTFE, 2% silicone, 20% glass fiber		"	"	"	23	100	30.5	25000	52.5		457
RTP 305	30% glass fiber		"	"	"	23	100	30.5	6000	12.6		457
RTP 305 TFE 13 SI 2	13% PTFE, 2% silicone, 30% glass fiber		"	"	"	23	100	30.5	31000	65.1		457
RTP 305 TFE 15	15% PTFE, 30% glass fiber		"	"	"	23	100	30.5	30000	63		457

Material		Mating Surface		Test Method	Test Conditions Note	Temp.	Speed		Limiting PV		Note	Source
Supplier / Grade	Material Note	Mating Surface Material	Mating Surface Note			(°C)	(fpm)	(m/min)	(psi x fpm)	(MPa x m/min)		
POLYCARBONATE rubbing against												
RTP 306 TFE 20	35% glass fiber	steel	C1018 steel; surface finish: 14-17 µin; 15-25 Rockwell C	thrust washer	apparatus: Falex Model No. 6	23	100	30.5	31000	65.1		457
RTP 385	30% carbon fiber		"	"	"	23	100	30.5	10000	21		457
RTP 385 TFE 13 SI 2	13% PTFE, 2% silicone, 30% carbon fiber		"	"	"	23	100	30.5	35000	73.5		457
Unspecified grade	unmodified		carbon steel; surface finish: 16 µin; 18-22 Rockwell C	"	apparatus: Faville-LeVally	23	100	30.5	500	1.1		458
POLYESTER, PBT rubbing against												
Bay Resins Lubriplas PBT-1100CF30	30% carbon fiber	steel		half journal bearing			1000	304.8	10500	22.1		435
Bay Resins Lubriplas PBT-1100CF30	30% carbon fiber			"			10	3	18500	38.9		435
Bay Resins Lubriplas PBT-1100CF30	30% carbon fiber			"			100	30.5	22000	46.2		435
Bay Resins Lubriplas PBT-1100G30TF15	30% glass fiber, 15% PTFE			"			10	3	20000	42		435
Bay Resins Lubriplas PBT-1100G30TF15	30% glass fiber, 15% PTFE			"			100	30.5	23000	48.3		435
Bay Resins Lubriplas PBT-1100G30TF15	30% glass fiber, 15% PTFE			"			1000	304.8	11000	23.1		435
Bay Resins Lubriplas PBT-1100TF20	20% PTFE			"			1000	304.8	7500	15.8		435
Bay Resins Lubriplas PBT-1100TF20	20% PTFE			"			10	3	12500	26.3		435
Bay Resins Lubriplas PBT-1100TF20	20% PTFE			"			100	30.5	16000	33.6		435
LNP WC-1006	30% carbon fiber		carbon steel; surface finish: 12-16 µin; 18-20 Rockwell C	half cylindrical bearing		23	100	30.5	22000	46.2		453
LNP WC-1006	30% carbon fiber		"	"		23	1000	304.8	10000	21		453
LNP WC-1006	30% carbon fiber		"	"		23	10	3	18000	37.8		453
LNP WFL-4036	30% glass fiber, 15% PTFE		"	"		23	10	3	20000	42		453
LNP WFL-4036	30% glass fiber, 15% PTFE		"	"		23	100	30.5	22000	46.2		453
LNP WFL-4036	30% glass fiber, 15% PTFE		"	"		23	1000	304.8	10000	21		453
LNP WFL-4536	30% glass fiber, 13% PTFE, 2% silicone		"	"		23	100	30.5	24000	50.4		453
LNP WFL-4536	30% glass fiber, 13% PTFE, 2% silicone		"	"		23	1000	304.8	13000	27.3		453
LNP WFL-4536	30% glass fiber, 13% PTFE, 2% silicone		"	"		23	10	3	20000	42		453
LNP WL-4040	20% PTFE		"	"		23	10	3	12500	26.3		453
LNP WL-4040	20% PTFE		"	"		23	100	30.5	15500	32.6		453

Material		Mating Surface		Test Method	Test Conditions Note	Temp.	Speed		Limiting PV		Note	Source
Supplier / Grade	Material Note	Mating Surface Material	Mating Surface Note			(°C)	(fpm)	(m/min)	(psi x fpm)	(MPa x m/min)		
POLYESTER, PBT rubbing against												
LNP WL-4040	20% PTFE	**steel**	carbon steel; surface finish: 12-16 µin; 18-20 Rockwell C	half cylindrical bearing		23	1000	304.8	7000	14.7		453
RTP 1000			C1018 steel; surface finish: 14-17 µin; 15-25 Rockwell C	thrust washer	apparatus: Falex Model No. 6	23	100	30.5	3000	6.3		457
RTP 1002 TFE 15	15% PTFE, 15% glass fiber		"	"	"	23	100	30.5	18000	37.8		457
RTP 1005	30% glass fiber		"	"	"	23	100	30.5	10000	21		457
RTP 1005 TFE 15	15% PTFE, 30% glass fiber		"	"	"	23	100	30.5	20000	42		457
RTP 1085	30% carbon fiber		"	"	"	23	100	30.5	15000	31.5		457
RTP 1085 TFE 15	15% PTFE, 30% carbon fiber		"	"	"	23	100	30.5	35000	73.5		457
Thermofil E-30FG-0287	30% glass fiber, lubricated					23	100	30.5	21000	44.1		459
Thermofil E-30NF-0100	30% graphite fiber					23	100	30.5	24000	50.4		459
POLYESTER, PET rubbing against												
RTP 1107 TFE 10	10% PTFE, 40% glass fiber	**steel**	C1018 steel; surface finish: 14-17 µin; 15-25 Rockwell C	thrust washer	apparatus: Falex Model No. 6	23	100	30.5	20000	42		457
Thermofil E2-30FG-7100	30% glass fiber, lubricated					23	100	30.5	23000	48.3		459
POLYETHERIMIDE rubbing against												
GE Ultem 4000	reinforced, internal lubricant, 85 Rockwell M	**steel**					100	30.5	60000	126		51
GE Ultem 4001	internal lubricant, 110 Rockwell M						100	30.5	80000	168		51
RTP 2100 AR 15	15% aramid fiber		C1018 steel; surface finish: 14-17 µin; 15-25 Rockwell C	thrust washer	apparatus: Falex Model No. 6	23	100	30.5	15000	31.5		457
RTP 2100 AR 15 TFE 15	15% PTFE, 15% aramid fiber		"	"	"	23	100	30.5	25000	52.5		457
RTP 2105 TFE 15	15% PTFE, 30% glass fiber		"	"	"	23	100	30.5	35000	73.5		457
RTP 2185 TFE 13 SI2	13% PTFE, 2% silicone, 30% carbon fiber		"	"	"	23	100	30.5	40000	84		457
RTP 2185 TFE 15	15% PTFE, 30% carbon fiber		"	"	"	23	100	30.5	40000	84		457
Thermofil W-10FG-0100	10% glass fiber					23	100	30.5	4000	8.4		459
Thermofil W-30NF-0100	30% graphite fiber					23	100	30.5	30000	63		459
Thermofil W-30NF-0214	30% graphite fiber, PTFE lubricated					23	100	30.5	27000	56.7		459

Material		Mating Surface		Test Method	Test Conditions Note	Temp.	Speed		Limiting PV		Note	Source
Supplier / Grade	Material Note	Mating Surface Material	Mating Surface Note			(°C)	(fpm)	(m/min)	(psi x fpm)	(MPa x m/min)		
POLYETHERETHERKETONE rubbing against		→										
LNP LCL-4033EM	15% carbon fiber, 15% PTFE	**steel**	cold rolled steel; surface finish: 12-16 µin; 22 Rockwell C	journal bearing	apparatus: Faville-LeVally LFW5	260	100	30.5	35000	73.5		453
LNP LCL-4033EM	15% carbon fiber, 15% PTFE		"	"	"	204	100	30.5	38000	79.8		453
LNP LCL-4033EM	15% carbon fiber, 15% PTFE		"	"	"	23	100	30.5	40000	84		453
LNP LCL-4033EM	15% carbon fiber, 15% PTFE		"	"	"	93	100	30.5	40000	84		453
LNP LCL-4033EM	15% carbon fiber, 15% PTFE		"	"	"	149	100	30.5	40000	84		453
LNP LCL-4033EM	15% carbon fiber, 15% PTFE		"	"	"	260	800	244	21000	44.1		453
LNP LCL-4033EM	15% carbon fiber, 15% PTFE		"	"	"	204	800	244	23000	48.3		453
LNP LCL-4033EM	15% carbon fiber, 15% PTFE		"	"	"	23	800	244	24000	50.4		453
LNP LCL-4033EM	15% carbon fiber, 15% PTFE		"	"	"	93	800	244	24000	50.4		453
LNP LCL-4033EM	15% carbon fiber, 15% PTFE		"	"	"	149	800	244	24000	50.4		453
LNP LCL-4033EM	15% carbon fiber, 15% PTFE		carbon steel; surface finish: 12-16 µin; 18-20 Rockwell C	half cylindrical bearing		23	10	3	42000	88.2		453
LNP LCL-4033EM	15% carbon fiber, 15% PTFE		"	"		23	100	30.5	40000	84		453
LNP LCL-4033EM	15% carbon fiber, 15% PTFE		"	"		23	1000	304.8	22000	46.2		453
RTP 2205 TFE	15% PTFE, 15% silicone, 30% glass fiber		C1018 steel; surface finish: 14-17 µin; 15-25 Rockwell C	thrust washer	apparatus: Falex Model No. 6	23	100	30.5	40000	84		457
RTP 2285TFE	15% PTFE, 15% silicone, 30% carbon fiber		"	"	"	23	100	30.5	50000	105		457
Thermofil K2-30NF-0100	30% graphite fiber					23	100	30.5	20000	42		459
Victrex PEEK 15OFC30	proprietary filler		carbon steel; surface finish: 12-16 µin; 18-20 Rockwell C	half cylindrical bearing		23	10	3	40000	84		453
Victrex PEEK 15OFC30	proprietary filler		"	"		23	100	30.5	45000	94.5		453
Victrex PEEK 15OFC30	proprietary filler		"	"		23	1000	304.8	28000	58.8		453
Victrex PEEK 450CA30	124 Rockwell R, 107 Rockwell M; 30% carbon fiber		carbon EN8, dry	pad on ring (Amsler wear tester)		200	100	30.5	237048	498	maximum load: 120 kg	338
Victrex PEEK 450CA30	124 Rockwell R, 107 Rockwell M; 30% carbon fiber		"	"		20	100	30.5	386988	813	maximum load: 160 kg	338
Victrex PEEK 450CA30	124 Rockwell R, 107 Rockwell M; 30% carbon fiber		"	"		200	600	183	211820	445	maximum load: 13 kg	338
Victrex PEEK 450CA30	124 Rockwell R, 107 Rockwell M; 30% carbon fiber		"	"		20	600	183	178976	376	maximum load: 22 kg	338
Victrex PEEK 450FC30	was Victrex D450HF30, tribological properties, bearing grade; 30% graphite/ carbon/ PTFE		"	"		200	100	30.5	366996	771	maximum load: 170 kg	338
Victrex PEEK 450FC30	was Victrex D450HF30, tribological properties, bearing grade; 30% graphite/ carbon/ PTFE		"	"		20	100	30.5	515984	1084	maximum load: 210 kg	338
Victrex PEEK 450FC30	was Victrex D450HF30, tribological properties, bearing grade; 30% graphite/ carbon/ PTFE		"	"		200	600	183	296072	622	maximum load: 40 kg	338

Material		Mating Surface		Test Method	Test Conditions Note	Temp.	Speed		Limiting PV		Note	Source
Supplier / Grade	Material Note	Mating Surface Material	Mating Surface Note			(°C)	(fpm)	(m/min)	(psi x fpm)	(MPa x m/min)		
POLYETHERETHERKETONE rubbing against												
Victrex PEEK 450FC30	was Victrex D450HF30, tribological properties, bearing grade; 30% graphite/ carbon/ PTFE	**steel**	carbon EN8, dry	pad on ring (Amsler wear tester)		20	600	183	377944	794	maximum load: 40 kg	338
Victrex PEEK 450G	gen. purp. grade, 126 Rockwell R, 99 Rockwell M		"	"		20	100	30.5	169932	357	maximum load: 70 kg	338
Victrex PEEK 450G	gen. purp. grade, 126 Rockwell R, 99 Rockwell M		"	"		200	100	30.5	187068	393	maximum load: 70 kg	338
Victrex PEEK 450G	gen. purp. grade, 126 Rockwell R, 99 Rockwell M		"	"		200	600	183	69972	147	maximum load: 8 kg	338
Victrex PEEK 450G	gen. purp. grade, 126 Rockwell R, 99 Rockwell M		"	"		20	600	183	69020	145	maximum load: 8 kg	338
Victrex PEEK D450HT15	no longer available, tribological properties, bearing grade		"	"		200	100	30.5	519792	1092	maximum load: 210 kg	338
Victrex PEEK D450HT15	no longer available, tribological properties, bearing grade		"	"		20	100	30.5	212772	447	maximum load: 110 kg	338
Victrex PEEK D450HT15	no longer available, tribological properties, bearing grade		"	"		20	600	183	365092	767	maximum load: 40 kg	338
Victrex PEEK D450HT15	no longer available, tribological properties, bearing grade		"	"		200	600	183	>416,000	>857	maximum load: 40 kg	338
POLYPROPYLENE rubbing against												
LNP MFL-4034HS	15% PTFE, 20% silicone	**steel**	carbon steel; surface finish: 12-16 µin; 18-20 Rockwell C	half cylindrical bearing		23	10	3	14000	29.4		453
LNP MFL-4034HS	15% PTFE, 20% silicone		"	"		23	100	30.5	12000	25.2		453
LNP MFL-4034HS	15% PTFE, 20% silicone		"	"		23	1000	304.8	7500	15.8		453
LNP ML-404OHS	20% PTFE		"	"		23	100	30.5	5000	10.5		453
LNP ML-404OHS	20% PTFE		"	"		23	1000	304.8	3000	6.3		453
LNP ML-404OHS	20% PTFE		"	"		23	10	3	7000	14.7		453
MODIFIED PPE rubbing against												
LNP Z-1000	unmodified	**steel**	carbon steel; surface finish: 12-16 µin; 18-20 Rockwell C	half cylindrical bearing		23	10	3	750	1.6		453
LNP Z-1000	unmodified		"	"		23	100	30.5	500	1.1		453
LNP Z-1000	unmodified		"	"		23	1000	304.8	<500	<1.1		453
LNP ZFL-4036	30% glass fiber, 15% PTFE		"	"		23	100	30.5	22000	46.2		453
LNP ZFL-4036	30% glass fiber, 15% PTFE		"	"		23	1000	304.8	9000	18.9		453
LNP ZFL-4036	30% glass fiber, 15% PTFE		"	"		23	10	3	18000	37.8		453

Material		Mating Surface		Test Method	Test Conditions Note	Temp.	Speed		Limiting PV		Note	Source
Supplier / Grade	Material Note	Mating Surface Material	Mating Surface Note			(°C)	(fpm)	(m/min)	(psi x fpm)	(MPa x m/min)		
POLYPHENYLENE SULFIDE rubbing against →												
Akzo J-1300/30/TF/15	30% glass fiber, 15% PTFE	**steel**	carbon steel; surface finish: 16 µin; 18-22 Rockwell C	thrust washer	apparatus: Faville-LeVally	23	100	30.5	27000	56.7		458
Akzo J-1300/CF/30/TF/15	30% carbon fiber, 15% PTFE		"	"	"	23	100	30.5	28000	58.8		458
GE Supec W331	116 Rockwell R; 30% glass fiber, 15% PTFE				GE method		433	132	35000	73.5		429
LNP Lubricomp O-BG	proprietary filler		carbon steel; surface finish: 12-16 µin; 18-20 Rockwell C	half cylindrical bearing		23	100	30.5	70000	147		453
LNP Lubricomp O-BG	proprietary filler		cold rolled steel; surface finish: 12-16 µin; 22 Rockwell C	journal bearing	apparatus: Faville-LeVally LFW5	260	100	30.5	63500	133.4		453
LNP Lubricomp O-BG	proprietary filler		"	"	"	204	100	30.5	67000	140.7		453
LNP Lubricomp O-BG	proprietary filler		"	"	"	23	100	30.5	70000	147		453
LNP O-1000	unmodified		carbon steel; surface finish: 12-16 µin; 18-20 Rockwell C	half cylindrical bearing		23	100	30.5	3000	6.3		453
LNP O-1000	unmodified		"	"		23	1000	304.8	4000	8.4		453
LNP O-1000	unmodified		"	"		23	10	3	2500	5.3		453
LNP OC-1006	30% carbon fiber		"	"		23	10	3	12000	25.2		453
LNP OC-1006	30% carbon fiber		"	"		23	100	30.5	20000	42		453
LNP OC-1006	30% carbon fiber		"	"		23	1000	304.8	10000	21		453
LNP OC-1006	30% carbon fiber		cold rolled steel; surface finish: 12-16 µin; 22 Rockwell C	journal bearing	apparatus: Faville-LeVally LFW5	260	100	30.5	15000	31.5		453
LNP OC-1006	30% carbon fiber		"	"	"	204	100	30.5	19000	39.9		453
LNP OC-1006	30% carbon fiber		"	"	"	23	100	30.5	20000	42		453
LNP OC-1006	30% carbon fiber		"	"	"	93	100	30.5	20000	42		453
LNP OC-1006	30% carbon fiber		"	"	"	149	100	30.5	20000	42		453
LNP OC-1006	30% carbon fiber		"	"	"	23	800	244	11000	23.1		453
LNP OC-1006	30% carbon fiber		"	"	"	93	800	244	11000	23.1		453
LNP OC-1006	30% carbon fiber		"	"	"	149	800	244	11000	23.1		453
LNP OC-1006	30% carbon fiber		"	"	"	204	800	244	11000	23.1		453
LNP OC-1006	30% carbon fiber		"	"	"	260	800	244	11000	23.1		453
LNP OCL-4036	30% carbon fiber, 15% PTFE		"	"	"	23	100	30.5	25000	52.5		453
LNP OCL-4036	30% carbon fiber, 15% PTFE		"	"	"	260	100	30.5	30000	63		453
LNP OCL-4036	30% carbon fiber, 15% PTFE		"	"	"	204	100	30.5	33000	69.3		453
LNP OCL-4036	30% carbon fiber, 15% PTFE		"	"	"	93	100	30.5	35000	73.5		453

Material		Mating Surface		Test Method	Test Conditions Note	Temp.	Speed		Limiting PV		Note	Source
Supplier / Grade	Material Note	Mating Surface Material	Mating Surface Note			(°C)	(fpm)	(m/min)	(psi x fpm)	(MPa x m/min)		
POLYPHENYLENE SULFIDE rubbing against												
LNP OCL-4036	30% carbon fiber, 15% PTFE	**steel**	cold rolled steel; surface finish: 12-16 µin; 22 Rockwell C	journal bearing	apparatus: Faville-LeVally LFW5	149	100	30.5	35000	73.5		453
LNP OCL-4036	30% carbon fiber, 15% PTFE		"	"	"	260	800	244	20000	42		453
LNP OCL-4036	30% carbon fiber, 15% PTFE		"	"	"	204	800	244	22000	46.2		453
LNP OCL-4036	30% carbon fiber, 15% PTFE		"	"	"	149	800	244	28000	58.8		453
LNP OCL-4036	30% carbon fiber, 15% PTFE		"	"	"	93	800	244	30000	63		453
LNP OCL-4036	30% carbon fiber, 15% PTFE		"	"	"	23	800	244	31000	65.1		453
LNP OCL-4036	30% carbon fiber, 15% PTFE		carbon steel; surface finish: 12-16 µin; 18-20 Rockwell C	half cylindrical bearing		23	100	30.5	35000	73.5		453
LNP OCL-4036	30% carbon fiber, 15% PTFE		"	"		23	1000	304.8	30000	63		453
LNP OCL-4036	30% carbon fiber, 15% PTFE		"	"		23	10	3	27000	56.7		453
LNP OF-1008	40% glass fiber		"	"		23	10	3	13000	27.3		453
LNP OF-1008	40% glass fiber		"	"		23	100	30.5	16000	33.6		453
LNP OF-1008	40% glass fiber		"	"		23	1000	304.8	14000	29.4		453
LNP OFL-4036	30% glass fiber, 15% PTFE		"	"		23	100	30.5	35000	73.5		453
LNP OFL-4036	30% glass fiber, 15% PTFE		"	"		23	1000	304.8	30000	63		453
LNP OFL-4036	30% glass fiber, 15% PTFE		"	"		23	10	3	27000	56.7		453
LNP OFL-4036	30% glass fiber, 15% PTFE		cold rolled steel; surface finish: 12-16 µin; 22 Rockwell C	journal bearing	apparatus: Faville-LeVally LFW5	260	100	30.5	28000	58.8		453
LNP OFL-4036	30% glass fiber, 15% PTFE		"	"	"	204	100	30.5	29000	60.9		453
LNP OFL-4036	30% glass fiber, 15% PTFE		"	"	"	23	100	30.5	30000	63		453
LNP OFL-4036	30% glass fiber, 15% PTFE		"	"	"	93	100	30.5	30000	63		453
LNP OFL-4036	30% glass fiber, 15% PTFE		"	"	"	149	100	30.5	30000	63		453
LNP OFL-4036	30% glass fiber, 15% PTFE		"	"	"	260	800	244	17000	35.7		453
LNP OFL-4036	30% glass fiber, 15% PTFE		"	"	"	204	800	244	19000	39.9		453
LNP OFL-4036	30% glass fiber, 15% PTFE		"	"	"	149	800	244	25000	52.5		453
LNP OFL-4036	30% glass fiber, 15% PTFE		"	"	"	23	800	244	30000	63		453
LNP OFL-4036	30% glass fiber, 15% PTFE		"	"	"	93	800	244	30000	63		453
LNP OFL-4536	30% glass fiber, 13% PTFE, 2% silicone		carbon steel; surface finish: 12-16 µin; 18-20 Rockwell C	half cylindrical bearing		23	10	3	27500	57.8		453
LNP OFL-4536	30% glass fiber, 13% PTFE, 2% silicone		"	"		23	100	30.5	30000	63		453
LNP OFL-4536	30% glass fiber, 13% PTFE, 2% silicone		"	"		23	1000	304.8	30000	63		453

Material		Mating Surface		Test Method	Test Conditions Note	Temp.	Speed		Limiting PV		Note	Source
Supplier / Grade	Material Note	Mating Surface Material	Mating Surface Note			(°C)	(fpm)	(m/min)	(psi x fpm)	(MPa x m/min)		
POLYPHENYLENE SULFIDE rubbing against												
RTP 1300		**steel**	C1018 steel; surface finish: 14-17 μin; 15-25 Rockwell C	thrust washer	apparatus: Falex Model No. 6	23	100	30.5	3000	6.3		457
RTP 1300 AR 15	15% aramid fiber		"	"	"	23	100	30.5	18000	37.8		457
RTP 1300 AR 15 TFE 15	15% PTFE, 15% aramid fiber		"	"	"	23	100	30.5	30000	63		457
RTP 1302 TFE 10	10% PTFE, 15% glass fiber		"	"	"	23	100	30.5	24000	50.4		457
RTP 1303 TFE 20	20% PTFE, 20% glass fiber		"	"	"	23	100	30.5	28000	58.8		457
RTP 1305	30% glass fiber		"	"	"	23	100	30.5	15000	31.5		457
RTP 1307 TFE 10	10% PTFE, 40% glass fiber		"	"	"	23	100	30.5	30000	63		457
RTP 1378	15% PTFE, 30% glass fiber		"	"	"	23	100	30.5	30000	63		457
RTP 1385 TFE 15	15% PTFE, 30% carbon fiber		"	"	"	23	100	30.5	40000	84		457
RTP 1387 TFE 10	10% PTFE, 40% carbon fiber		"	"	"	23	100	30.5	40000	84		457
Thermofil T-20NF-0100	20% graphite fiber					23	100	30.5	16000	33.6		459
Thermofil T-30FG-0214	30% glass fiber, PTFE lubricated					23	100	30.5	30000	63		459
Thermofil T-30NF-0214	30% graphite fiber, PTFE lubricated					23	100	30.5	45000	94.5		459
Thermofil T-40FG-0100	40% glass fiber					23	100	30.5	17000	35.7		459
Thermofil T-40NF-0100	40% graphite fiber					23	100	30.5	25000	52.5		459
	unmodified		carbon steel; surface finish: 16 μin; 18-22 Rockwell C	thrust washer	apparatus: Faville-LeVally	23	100	30.5	3000	6.3		458
POLYSULFONE rubbing against												
LNP G-1000	unmodified	**steel**	carbon steel; surface finish: 12-16 μin; 18-20 Rockwell C	half cylindrical bearing		23	100	30.5	5000	10.5		453
LNP G-1000	unmodified		"	"		23	1000	304.8	3000	6.3		453
LNP G-1000	unmodified		"	"		23	10	3	5000	10.5		453
LNP GC-1006	30% carbon fiber		"	"		23	10	3	8500	17.9		453
LNP GC-1006	30% carbon fiber		"	"		23	100	30.5	8500	17.9		453
LNP GC-1006	30% carbon fiber		"	"		23	1000	304.8	6000	12.6		453
LNP GFL-4036	30% glass fiber, 15% PTFE		"	"		23	100	30.5	35000	73.5		453
LNP GFL-4036	30% glass fiber, 15% PTFE		"	"		23	1000	304.8	15000	31.5		453
LNP GFL-4036	30% glass fiber, 15% PTFE		"	"		23	10	3	20000	42		453
RTP 900 TFE 15	15% PTFE		C1018 steel; surface finish: 14-17 μin; 15-25 Rockwell C	thrust washer	apparatus: Falex Model No. 6	23	100	30.5	22000	46.2		457

Material		Mating Surface		Test Method	Test Conditions Note	Temp.	Speed		Limiting PV		Note	Source
Supplier / Grade	Material Note	Mating Surface Material	Mating Surface Note			(°C)	(fpm)	(m/min)	(psi x fpm)	(MPa x m/min)		
POLYSULFONE rubbing against												
RTP 900 TFE 20	20% PTFE	**steel**	C1018 steel; surface finish: 14-17 µin; 15-25 Rockwell C	thrust washer	apparatus: Falex Model No. 6	23	100	30.5	24000	50.4		457
RTP 905 TFE 15	15% PTFE, 30% glass fiber		"	"	"	23	100	30.5	35000	73.5		457
Thermofil S-30FG-0214	30% glass fiber, PTFE lubricated					23	100	30.5	35000	73.5		459
Thermofil S-30NF-0100	30% graphite fiber					23	100	30.5	30000	63		459
POLYETHERSULFONE rubbing against												
LNP JFL-4036	30% glass fiber, 15% PTFE	**steel**	carbon steel; surface finish: 12-16 µin; 18-20 Rockwell C	half cylindrical bearing		23	10	3	18000	37.8		453
LNP JFL-4036	30% glass fiber, 15% PTFE		"	"		23	100	30.5	30000	63		453
LNP JFL-4036	30% glass fiber, 15% PTFE		"	"		23	1000	304.8	17000	35.7		453
RTP 1405 TFE 15	15% PTFE, 30% glass fiber		C1018 steel; surface finish: 14-17 µin; 15-25 Rockwell C	thrust washer	apparatus: Falex Model No. 6	23	100	30.5	35000	73.5		457
RTP 1485 TFE 15	15% PTFE, 30% carbon fiber		"	"	"	23	100	30.5	40000	84		457
Thermofil K-20FG-0100	20% glass fiber					23	100	30.5	7000	14.7		459
Thermofil K-30FG-0100	30% glass fiber					23	100	30.5	10000	21		459
Thermofil K-30FG-0214	30% glass fiber, PTFE lubricated					23	100	30.5	32000	67.2		459
Thermofil K-30NF-0100	30% graphite fiber					23	100	30.5	29000	60.9		459
Victrex PES 4800G	unmodified		carbon steel; surface finish: 12-16 µin; 18-20 Rockwell C	half cylindrical bearing		23	10	3	7000	14.7		453
Victrex PES 4800G	unmodified		"	"		23	100	30.5	7000	14.7		453
Victrex PES 4800G	unmodified		"	"		23	1000	304.8	4000	8.4		453
POLYURETHANE, RIGID rubbing against												
LNP T-1000	unmodified	**steel**	carbon steel; surface finish: 12-16 µin; 18-20 Rockwell C	half cylindrical bearing		23	1000	304.8	<1500	<3.2		453
LNP T-1000	unmodified		"	"		23	10	3	2000	4.2		453
LNP T-1000	unmodified		"	"		23	100	30.5	1500	3.2		453
LNP TFL-4036	30% glass fiber, 15% PTFE		"	"		23	10	3	7500	15.8		453
LNP TFL-4036	30% glass fiber, 15% PTFE		"	"		23	100	30.5	10000	21		453
LNP TFL-4036	30% glass fiber, 15% PTFE		"	"		23	1000	304.8	5500	11.6		453

Material		Mating Surface		Test Method	Test Conditions Note	Temp.	Speed		Limiting PV		Note	Source
Supplier / Grade	Material Note	Mating Surface Material	Mating Surface Note			(°C)	(fpm)	(m/min)	(psi x fpm)	(MPa x m/min)		
ABS rubbing against												
LNP AL-4030	15% PTFE	**steel**	carbon steel; surface finish: 12-16 µin; 18-20 Rockwell C	half cylindrical bearing		23	100	30.5	4000	8.4		453
LNP AL-4030	15% PTFE		"	"		23	1000	304.8	<2,000	<4.2		453
LNP AL-4030	15% PTFE		"	"		23	10	3	18000	37.8		453
RTP 600			C1018 steel; surface finish: 14-17 µin; 15-25 Rockwell C	thrust washer	apparatus: Falex Model No. 6	23	100	30.5	1000	2.1		457
RTP 600 SI 2	2% silicone		"	"	"	23	100	30.5	4000	8.4		457
Thermofil G-20NF-0100	20% graphite fiber					23	100	30.5	12000	25.2		459
Thermofil G-30FG-0100	30% glass fiber					23	100	30.5	8500	17.9		459
POLYSTYRENE rubbing against												
LNP C-1000	unmodified	**steel**	carbon steel; surface finish: 12-16 µin; 18-20 Rockwell C	half cylindrical bearing		23	10	3	750	1.6		453
LNP C-1000	unmodified		"	"		23	100	30.5	1500	3.2		453
LNP C-1000	unmodified		"	"		23	1000	304.8	500	1.1		453
LNP CL-4410	2% silicone		"	"		23	100	30.5	9000	18.9		453
LNP CL-4410	2% silicone		"	"		23	1000	304.8	1000	2.1		453
LNP CL-4410	2% silicone		"	"		23	10	3	4000	8.4		453
POLYSTYRENE , GP rubbing against												
Thermofil A-20FG-0100	20% glass fiber	**steel**				23	100	30.5	4000	8.4		459
SAN rubbing against												
LNP BFL-4036	30% glass fiber, 15% PTFE	**steel**	carbon steel; surface finish: 12-16 µin; 18-20 Rockwell C	half cylindrical bearing		23	10	3	17500	36.8		453
LNP BFL-4036	30% glass fiber, 15% PTFE		"	"		23	100	30.5	10000	21		453
LNP BFL-4036	30% glass fiber, 15% PTFE		"	"		23	1000	304.8	10000	21		453
LNP BL-4030	15% PTFE		"	"		23	100	30.5	5000	10.5		453
LNP BL-4030	15% PTFE		"	"		23	1000	304.8	3000	6.3		453
LNP BL-4030	15% PTFE		"	"		23	10	3	20000	42		453

Material		Mating Surface		Test Method	Test Conditions Note	Temp.	Speed		Limiting PV		Note	Source
Supplier / Grade	Material Note	Mating Surface Material	Mating Surface Note			(°C)	(fpm)	(m/min)	(psi x fpm)	(MPa x m/min)		
POLYMIDE rubbing against												
DuPont Vespel SP-21	15% graphite	**steel**	carbon EN8, dry	pad on ring (Amsler wear tester)		200	100	30.5	539784	1134	maximum load: 210 kg	338
DuPont Vespel SP-21	15% graphite		"	"		20	100	30.5	595952	1252	maximum load: 210 kg	338
DuPont Vespel SP-21	15% graphite		"	"		200	600	183	318920	670	maximum load: 20 kg	338
DuPont Vespel SP-21	15% graphite		"	"		20	600	183	426020	895	maximum load: 30 kg	338
Ube Upimol R	gen. purp. grade			S45C (ANSI 1045)	load: 50 N		98	30	809	1.7		123
Ube Upimol S	high heat grade			"	"		98	30	1047	2.2		123
DuPont Vespel SP-21	15% graphite								300000	630	maximum contact temperature: 393°C; guideline values for reference only; actual values depend on test conditions	344
DuPont Vespel SP-211	15% graphite, 10% PTFE modified								100000	210	maximum contact temperature: 260°C; guideline values for reference only; actual values depend on test conditions	344
DuPont Vespel SP-22	40 wt% graphite								300000	630	maximum contact temperature: 393°C; guideline values for reference only; actual values depend on test conditions	344

Appendix IV

Abrasion Resistance Data

Test Name	Material Supplier/ Grade	Material Note	Test Conditions							Source
			Test Method	Wheel / Abrader Type	Load (g)	Test Note	Taber Abrasion (mg/1000 cycles)	NBS Abrasion Index (%)	Test Result Note	

Acetal

Test Name	Material Supplier/ Grade	Material Note	Test Method	Wheel / Abrader Type	Load (g)	Test Note	Taber Abrasion (mg/1000 cycles)	NBS Abrasion Index (%)	Test Result Note	Source
Taber Abrasion	Unspecified grade		ASTM D1044	CS 10-F	250		4			78
	Unspecified grade		ASTM D1044	CS 17	1000		13			330
	Unspecified grade	PTFE modified	ASTM D1044	CS 17	1000		13			330

Acetal Copolymer

Test Name	Material Supplier/ Grade	Material Note	Test Method	Wheel / Abrader Type	Load (g)	Test Note	Taber Abrasion (mg/1000 cycles)	NBS Abrasion Index (%)	Test Result Note	Source
Taber Abrasion	Hoechst Cel. Celcon LW90	low wear, 80 Rockwell M; 9.0 g/10 min. MFI	ASTM D1044	CS 17	1000		8			321
	Hoechst Cel. Celcon LW90S2	low wear, 75 Rockwell M, low wear; 2% silicone; 9.0 g/10 min. MFI	ASTM D1044	CS 17	1000		14			321
	Hoechst Cel. Celcon M25	high molecular weight, 78 Rockwell M; 2.5 g/10 min. MFI	ASTM D1044	CS 17	1000		14			210
	Hoechst Cel. Celcon M270	low molecular weight, high flow, 80 Rockwell M; 27.0 g/10 min. MFI	ASTM D1044	CS 17	1000		14			210
	Hoechst Cel. Celcon M450	high flow, 80 Rockwell M; 45 g/10 min. MFI	ASTM D1044	CS 17	1000		14			210
	Hoechst Cel. Celcon M90	gen. purp. grade, 80 Rockwell M; 9.0 g/10 min. MFI	ASTM D1044	CS 17	1000		14			210
	Hoechst Cel. Celcon M90	80 Rockwell M; 9.0 g/10 min. MFI	ASTM D1044	CS 17	1000		8			321
	Hoechst Cel. Celcon M25	high molecular weight, 78 Rockwell M; 2.5 g/10 min. MFI	ASTM D1044	CS 17F	1000		6			210
	Hoechst Cel. Celcon M270	low molecular weight, high flow, 80 Rockwell M; 27.0 g/10 min. MFI	ASTM D1044	CS 17F	1000		6			210
	Hoechst Cel. Celcon M450	high flow, 80 Rockwell M; 45 g/10 min. MFI	ASTM D1044	CS 17F	1000		6			210
	Hoechst Cel. Celcon M90	gen. purp. grade, 80 Rockwell M; 9.0 g/10 min. MFI	ASTM D1044	CS 17F	1000		6			210

Fluoropolymer, FEP

Test Name	Material Supplier/ Grade	Material Note	Test Method	Wheel / Abrader Type	Load (g)	Test Note	Taber Abrasion (mg/1000 cycles)	NBS Abrasion Index (%)	Test Result Note	Source
Armstrong abrasion test	Unspecified grade	96 Shore A, 59 Shore D, 25 Rockwell R	ASTM D1242-56	No. 320 abrasive	6800	200 cycles; 100 minutes			0.052 g/cm² average weight loss	339
Tape abrasion test	Unspecified grade	96 Shore A, 59 Shore D, 25 Rockwell R	MIL-T-5438	400 grit tape	454	coating thickness of 0.015 in.			272 cm tape length required to abrade through wire coating	339
Taber Abrasion	Unspecified grade	96 Shore A, 59 Shore D, 25 Rockwell R	ASTM D1044	CS 17F	1000		7.5			339

Fluoropolymer, TFE

Test Name	Material Supplier/ Grade	Material Note	Test Method	Wheel / Abrader Type	Load (g)	Test Note	Taber Abrasion (mg/1000 cycles)	NBS Abrasion Index (%)	Test Result Note	Source
Armstrong abrasion test	Unspecified grade	98 Shore A, 58 Rockwell R, 52 Shore D	ASTM D1242-56	No. 320 abrasive	6800	200 cycles; 100 minutes			0.027 g/cm² average weight loss	339
Tape abrasion test	Unspecified grade	98 Shore A, 58 Rockwell R, 52 Shore D	MIL-T-5438	400 grit tape	454	coating thickness of 0.015 in.			193 cm tape length required to abrade through wire coating	339

Test Name	Material Supplier/ Grade	Material Note	Test Conditions							Source
			Test Method	Wheel / Abrader Type	Load (g)	Test Note	Taber Abrasion (mg/1000 cycles)	NBS Abrasion Index (%)	Test Result Note	

Fluoropolymer, TFE

Test Name	Material Supplier/ Grade	Material Note	Test Method	Wheel / Abrader Type	Load (g)	Test Note	Taber Abrasion (mg/1000 cycles)	NBS Abrasion Index (%)	Test Result Note	Source
Taber Abrasion	Unspecified grade	98 Shore A, 58 Rockwell R, 52 Shore D	ASTM D1044	CS 17F	1000		8.9			339

Fluoropolymer, PVDF

Test Name	Material Supplier/ Grade	Material Note	Test Method	Wheel / Abrader Type	Load (g)	Test Note	Taber Abrasion (mg/1000 cycles)	NBS Abrasion Index (%)	Test Result Note	Source
Taber Abrasion	Solvay Solef 1008	79 Shore D, injection molding	ASTM D1044	CS 10	1000		5-10			444
	Solvay Solef 1010	77 Shore D, extrusion	ASTM D1044	CS 10	1000		5-10			444
	Solvay Solef 1012	79 Shore D, semi-finished products	ASTM D1044	CS 10	1000		5-10			444
	Solvay Solef 3108	82 Shore D, anti-static	ASTM D1044	CS 10	1000		5-10			444
	Solvay Solef 3208	78 Shore D, low friction	ASTM D1044	CS 10	1000		5-10			444
	Solvay Solef 5708	79 Shore D, rotational molding	ASTM D1044	CS 10	1000		5-10			444
	Solvay Solef 6010	77 Shore D, compression molding	ASTM D1044	CS 10	1000		5-10			444
	Solvay Solef 8808	82 Shore D, carbon fiber reinforced	ASTM D1044	CS 10	1000		5-10			444
	Solvay Solef 8908	81 Shore D, mica reinforced	ASTM D1044	CS 10	1000		18			444
	Unspecified grade			CS 17	1000		16			437

Ionomer

Test Name	Material Supplier/ Grade	Material Note	Test Method	Wheel / Abrader Type	Load (g)	Test Note	Taber Abrasion (mg/1000 cycles)	NBS Abrasion Index (%)	Test Result Note	Source
NBS Abrasion Index	DuPont Surlyn 8020	transparent, sodium ion, 56 Shore D; 1.0 g/10 min. MFI	ASTM D1630					150		353
	DuPont Surlyn 8528	transparent, sodium ion, 60 Shore D; 1.3 g/10 min. MFI	ASTM D1630					600		353
	DuPont Surlyn 8550	sodium ion, 60 Shore D; 3.9 g/10 min. MFI	ASTM D1630					214		353
	DuPont Surlyn 8660	sodium ion, 62 Shore D; 10 g/10 min. MFI	ASTM D1630					170		353
	DuPont Surlyn 8920	transparent, sodium ion, 66 Shore D; 0.9 g/10 min. MFI	ASTM D1630					640		353
	DuPont Surlyn 8940	transparent, sodium ion, 65 Shore D; 2.8 g/10 min. MFI	ASTM D1630					370		353
	DuPont Surlyn 9020	transparent, zinc ion, 55 Shore D; 1.0 g/10 min. MFI	ASTM D1630					220		353
	DuPont Surlyn 9450	zinc ion, 54 Shore D; 5.5 g/10 min. MFI	ASTM D1630					170		353
	DuPont Surlyn 9520	zinc ion, 60 Shore D; 1.1 g/10 min. MFI	ASTM D1630					290		353
	DuPont Surlyn 9650	zinc ion, 63 Shore D; 5.0 g/10 min. MFI	ASTM D1630					270		353
	DuPont Surlyn 9720	zinc ion, 61 Shore D; 1.0 g/10 min. MFI	ASTM D1630					410		353
	DuPont Surlyn 9721	zinc ion, 61 Shore D; 1.0 g/10 min. MFI	ASTM D1630					410		353
	DuPont Surlyn 9730	zinc ion, 63 Shore D; 1.6 g/10 min. MFI	ASTM D1630					360		353
	DuPont Surlyn 9910	transparent, zinc ion, 64 Shore D; 0.7 g/10 min. MFI	ASTM D1630					610		353
	DuPont Surlyn 9950	transparent, zinc ion, 62 Shore D; 5.5 g/10 min. MFI	ASTM D1630					130		353
	DuPont Surlyn 9970	transparent, zinc ion, 62 Shore D; 14.0 g/10 min. MFI	ASTM D1630					120		353
Taber Abrasion	DuPont Surlyn 9520	zinc ion, 60 Shore D; 1.1 g/10 min. MFI		CS 17	1000		49			354
	Unspecified grade		ASTM C501	CS 17	1000	5000 cycles	2.4			427

Test Name	Material Supplier/ Grade	Material Note	Test Conditions							Source
			Test Method	Wheel / Abrader Type	Load (g)	Test Note	Taber Abrasion (mg/1000 cycles)	NBS Abrasion Index (%)	Test Result Note	
Nylon, Amorphous										
Taber Abrasion	Huls Trogamid T		ASTM D1044	CS 17	1000		21			447
Nylon 6										
Taber Abrasion	Unspecified grade		ASTM C501	CS 17	1000	5000 cycles	20.8			427
	Unspecified grade	30% glass fiber	ASTM D1044	CS 17	1000	DAM	28			236
Nylon 610										
Taber Abrasion	Unspecified grade		ASTM C501	CS 17	1000	5000 cycles	3.2			427
Nylon 612										
Taber Abrasion	DuPont Zytel 158L	lubricated, gen. purp. grade, 108 (50% RH) Rockwell R	ASTM D1044	CS 17	1000	50% RH	6			351
Nylon 66										
Taber Abrasion	Unspecified grade		ASTM D1044	CS 10-F	250		1.6			78
	DuPont Minlon 10B	mineral filled		CS 17	1000		14.1			68
	DuPont Minlon 11C	low warp grade, high impact, chrome platable; , mineral filled		CS 17	1000		22			68
	DuPont Minlon 12T	low warp grade, high impact; , mineral filled		CS 17	1000		21			68
	DuPont Minlon 20B	glass/ mineral reinforced		CS 17	1000		23.5			68
	DuPont Zytel 101	gen. purp. grade, unlubricated, 82 (50% RH) Shore D, 108 (50% RH) Rockwell R, 59 (50% RH) Rockwell M	ASTM D1044	CS 17	1000	50% RH	7			351
	DuPont Zytel 101L	lubricated, gen. purp. grade, 108 (50% RH) Rockwell R, 59 (50% RH) Rockwell M, 82 (50% RH) Shore D	ASTM D1044	CS 17	1000	50% RH	7			351
	DuPont Zytel 42A	108 (50% RH) Rockwell R, 60 (50% RH) Rockwell M, low flow; extrus.	ASTM D1044	CS 17	1000	50% RH	4			351
	DuPont Zytel 70G13L	lubricated, gen. purp. grade, 113 (50% RH) Rockwell R, 84 (50% RH) Rockwell M; 13% glass fiber	ASTM D1044	CS 17	1000	50% RH	12			351
	DuPont Zytel 70G33L	lubricated, gen. purp. grade; 33% glass fiber	ASTM D1044	CS 17	1000	50% RH	14			351
	DuPont Zytel 71G13L	lubricated, impact modified, 110 (50% RH) Rockwell R, 66 (50% RH) Rockwell M; 13% glass fiber	ASTM D1044	CS 17	1000	50% RH	34			351
	DuPont Zytel 71G33L	lubricated, impact modified, 90 (50% RH) Rockwell M, 118 (50% RH) Rockwell R; 33% glass fiber	ASTM D1044	CS 17	1000	50% RH	36			351
	DuPont Zytel ST801	high impact, 89 (50% RH) Rockwell R	ASTM D1044	CS 17	1000	50% RH	7			351
	DuPont Zytel ST801HS	high impact, heat stabilized, 89 (50% RH) Rockwell R	ASTM D1044	CS 17	1000	50% RH	7			351
	Hoechst Cel. Nylon 1500	33% glass fiber	ASTM D1044	CS 17	1000	DAM	20			317
	Hoechst Cel. Nylon 1500	33% glass fiber	ASTM D1044	CS 17	1000	50% RH	16			317

Nylon 66

Test Name	Material Supplier/ Grade	Material Note	Test Method	Wheel / Abrader Type	Load (g)	Test Note	Taber Abrasion (mg/1000 cycles)	NBS Abrasion Index (%)	Test Result Note	Source
Taber Abrasion	Unspecified grade	30% glass fiber	ASTM D1044	CS 17	1000	DAM	30			236
	Unspecified grade		ASTM D1044	CS 17	1000		7			330
	Unspecified grade	impact modified	ASTM D1044	CS 17	1000		7			330
	Unspecified grade		ASTM C501	CS 17	1000	5000 cycles	11.6			427

Nylon MXD6

Test Name	Material Supplier/ Grade	Material Note	Test Method	Wheel / Abrader Type	Load (g)	Test Note	Taber Abrasion (mg/1000 cycles)	NBS Abrasion Index (%)	Test Result Note	Source
Taber Abrasion	Mitsubishi Reny 1002	30% glass fiber	ASTM D1044	CS 17	1000	DAM	23			236
	Mitsubishi Reny 1022	12 g/10 min. MFI	ASTM D1044	CS 17	1000	DAM	16			236
	Mitsubishi RENY 1032	60% glass fiber reinforced	ASTM D1044	CS 17	1000	DAM	23			236
	Mitsubishi RENY 1313	40% glass fiber reinforced, high impact	ASTM D1044	CS 17	1000	DAM	15			236
	Mitsubishi RENY 1501A	30% glass fiber reinforced, flame retardant	ASTM D1044	CS 17	1000	DAM	36			236
	Mitsubishi RENY 1511A	40% glass fiber reinforced, flame retardant	ASTM D1044	CS 17	1000	DAM	38			236
	Mitsubishi RENY 1521A	50% glass fiber reinforced, flame retardant	ASTM D1044	CS 17	1000	DAM	38			236
	Mitsubishi RENY 1722	50% glass fiber reinforced, coating	ASTM D1044	CS 17	1000	DAM	45			236
	Mitsubishi RENY 2031	30% glass fiber/ mineral reinforced	ASTM D1044	CS 17	1000	DAM	39			236
	Mitsubishi RENY 2501	15% glass fiber/ mineral reinforced, flame retardant	ASTM D1044	CS 17	1000	DAM	40			236
	Mitsubishi RENY 2502	20% glass fiber/ mineral reinforced, flame retardant	ASTM D1044	CS 17	1000	DAM	35			236
	Mitsubishi RENY 2620	20% glass fiber/ mineral reinforced, low warpage	ASTM D1044	CS 17	1000	DAM	52			236
	Mitsubishi RENY 2714	20% glass fiber/ mineral reinforced, low warpage, coating	ASTM D1044	CS 17	1000	DAM	63			236
	Mitsubishi RENY 6002	general purpose	ASTM D1044	CS 17	1000	DAM	19			236
	Mitsubishi RENY 6301	high impact	ASTM D1044	CS 17	1000	DAM	21			236
	Mitsubishi RENY 9040	40% glass fiber reinforced	ASTM D1044	CS 17	1000	DAM	60			236
	Mitsubishi RENY 9115	15% glass fiber/ mineral reinforced	ASTM D1044	CS 17	1000	DAM	75			236

Polyarylamide

Test Name	Material Supplier/ Grade	Material Note	Test Method	Wheel / Abrader Type	Load (g)	Test Note	Taber Abrasion (mg/1000 cycles)	NBS Abrasion Index (%)	Test Result Note	Source
Taber Abrasion	Solvay Ixef 1022	50% glass fiber		CS 17	1000		16			135
	Solvay Ixef 1022	50% glass fiber		H 22	1000		53			135

Polycarbonate

Test Name	Material Supplier/ Grade	Material Note	Test Method	Wheel / Abrader Type	Load (g)	Test Note	Taber Abrasion (mg/1000 cycles)	NBS Abrasion Index (%)	Test Result Note	Source
Taber Abrasion	Dow Calibre 300-10	transparent, gen. purp. grade, 73 Rockwell M; 10 g/10 min. MFI	ASTM D1044	CS 10-F	250	1000 cycles			3 mm^3 volume loss; 4.5% change in light transmission	78
	Dow Calibre 300-10	transparent, gen. purp. grade, 73 Rockwell M; 10 g/10 min. MFI	ASTM D1044	CS 10-F	250	100 cycles			2.9% change in light transmission	78

Test Name	Material Supplier/ Grade	Material Note	Test Conditions							Source
			Test Method	Wheel / Abrader Type	Load (g)	Test Note	Taber Abrasion (mg/1000 cycles)	NBS Abrasion Index (%)	Test Result Note	

Polycarbonate

Test Name	Material Supplier/ Grade	Material Note	Test Method	Wheel / Abrader Type	Load (g)	Test Note	Taber Abrasion (mg/1000 cycles)	NBS Abrasion Index (%)	Test Result Note	Source
Taber Abrasion	Dow Calibre 300-15	transparent, gen. purp. grade, 72 Rockwell M; 15 g/10 min. MFI	ASTM D1044	CS 10-F	250	1000 cycles			3 mm³ volume loss; 4.4% change in light transmission	78
	Dow Calibre 300-15	transparent, gen. purp. grade, 72 Rockwell M; 15 g/10 min. MFI	ASTM D1044	CS 10-F	250	100 cycles			2.6% change in light transmission	78
	Dow Calibre 300-4	transparent, gen. purp. grade, 74 Rockwell M; 4 g/10 min. MFI	ASTM D1044	CS 10-F	250	1000 cycles			3 mm³ volume loss; 4.1% change in light transmission	78
	Dow Calibre 300-4	transparent, gen. purp. grade, 74 Rockwell M; 4 g/10 min. MFI	ASTM D1044	CS 10-F	250	100 cycles			2.8% change in light transmission	78
	Dow Calibre 301-10	transparent; 10 g/10 min. MFI; w/ mold release	ASTM D1044	CS 10-F	250	1000 cycles			3 mm³ volume loss; 4.6% change in light transmission	78
	Dow Calibre 301-10	transparent; 10 g/10 min. MFI; w/ mold release	ASTM D1044	CS 10-F	250	100 cycles			2.5% change in light transmission	78
	Dow Calibre 301-4	transparent, 74 Rockwell M; 4 g/10 min. MFI; w/ mold release	ASTM D1044	CS 10-F	250	1000 cycles			2 mm³ volume loss; 4% change in light transmission	78
	Dow Calibre 301-4	transparent, 74 Rockwell M; 4 g/10 min. MFI; w/ mold release	ASTM D1044	CS 10-F	250	100 cycles			2.5% change in light transmission	78
	Dow Calibre 302-10	UV stabilized, transparent, 73 Rockwell M; 10 g/10 min. MFI	ASTM D1044	CS 10-F	250	1000 cycles			4 mm³ volume loss; 4.4% change in light transmission	78
	Dow Calibre 302-10	UV stabilized, transparent, 73 Rockwell M; 10 g/10 min. MFI	ASTM D1044	CS 10-F	250	100 cycles			2.5% change in light transmission	78
	Unspecified grade	transparent	ASTM D1044	CS 10-F	250		4.3			78
	Dow Calibre	applies to unfilled grasdes	ASTM D1044	CS 10F	1000	500 cycles			45% change in haze	254
	Dow Calibre Megarad	gamma radiation stabilized	ASTM D1044	CS 10F	1000	500 cycles			35% change in haze	254

Test Name	Material Supplier/ Grade	Material Note	Test Conditions							Source
			Test Method	Wheel / Abrader Type	Load (g)	Test Note	Taber Abrasion (mg/1000 cycles)	NBS Abrasion Index (%)	Test Result Note	

Polycarbonate

Test Name	Material Supplier/ Grade	Material Note	Test Method	Wheel / Abrader Type	Load (g)	Test Note	Taber Abrasion (mg/1000 cycles)	NBS Abrasion Index (%)	Test Result Note	Source
Taber Abrasion	Bayer Makrolon	applies to unfilled grasdes	ASTM D1044	CS 17	1000		15			289
	GE Lexan 121	high flow, general purpose, transparent	ASTM D1044 modified	CS 17	1000		10			431
	GE Lexan 131	low flow, high MW, extrusion	ASTM D1044 modified	CS 17	1000		10			431
	GE Lexan 141	moderate flow, general purpose, transparent	ASTM D1044 modified	CS 17	1000		10			431
	GE Lexan 3412R	20% glass reinforced	ASTM D1044 modified	CS 17	1000		17			431
	GE Lexan 3414R	40% glass reinforced	ASTM D1044 modified	CS 17	1000		32			431
	GE Lexan 500R	glass reinforced, flame retardant	ASTM D1044 modified	CS 17	1000		11			431
	GE Lexan 940	flame retardant	ASTM D1044 modified	CS 17	1000		10			431
	GE Lexan HF 1130R	high flow, mold release, UV stable	ASTM D1044 modified	CS 17	1000		11			431
	GE Lexan HF1110	high flow, thin walls	ASTM D1044 modified	CS 17	1000		32			431
	Mitsubishi Iupilon GS-2010M	124 Rockwell R, 92 Rockwell M, 10% glass fiber	ASTM D1044	CS 17	1000		24			422
	Mitsubishi Iupilon GS-2020M	124 Rockwell R, 98 Rockwell M, 20% glass fiber	ASTM D1044	CS 17	1000		25			422
	Mitsubishi Iupilon GS-2030M	122 Rockwell R, 98 Rockwell M, 30% glass fiber	ASTM D1044	CS 17	1000		40			422
	LNP Lubricomp DFL-4036	30% glass fiber, 15% PTFE	ASTM D1044	CS-17	1000		40			455
	LNP Lubricomp DL-4030	15% PTFE	ASTM D1044	CS-17	1000		25			455
	LNP Stat-Kon D-FR	statically conductive	ASTM D1044	CS-17	1000		45			455
	LNP Stat-Kon DC-1003FR	15% carbon fiber, statically conductive	ASTM D1044	CS-17	1000		35			455
	LNP Stat-Kon DC-1006	30% carbon fiber, statically conductive	ASTM D1044	CS-17	1000		40			455
	LNP Stat-Kon DX-7	statically dissipative	ASTM D1044	CS-17	1000		48			455
	LNP Thermocomp DF-1004	20% glass fiber	ASTM D1044	CS-17	1000		30			455
	LNP Thermocomp DF-1006	30% glass fiber	ASTM D1044	CS-17	1000		35			455
	LNP Thermocomp DF-1008	40% glass fiber	ASTM D1044	CS-17	1000		40			455
	LNP Thermocomp DFA-113		ASTM D1044	CS-17	1000		15			455
		unmodified	ASTM D1044	CS-17	1000		11			455

Polyester, PBT

Test Name	Material Supplier/ Grade	Material Note	Test Method	Wheel / Abrader Type	Load (g)	Test Note	Taber Abrasion (mg/1000 cycles)	NBS Abrasion Index (%)	Test Result Note	Source
Taber Abrasion	Hoechst Cel. Celanex 2000	high flow, gen. purp. grade, 75 Rockwell M	ASTM D1044		1000		13			315
	Hoechst Cel. Celanex 2000K	moderate flow, 75 Rockwell M, keycap grade	ASTM D1044		1000		13			315
	Hoechst Cel. Celanex 2002	moderate flow, gen. purp. grade, 78 Rockwell M	ASTM D1044		1000		14			315
	Hoechst Cel. Celanex 2003	gen. purp. grade, mod.-high flow	ASTM D1044		1000		10			315
	Hoechst Cel. Celanex 2003K	keycap grade, mod.-high flow	ASTM D1044		1000		10			315
	Hoechst Cel. Celanex 3112	flame retardant, 88 Rockwell M; 13% glass fiber	ASTM D1044		1000		22			315
	Hoechst Cel. Celanex 3200	gen. purp. grade, 90 Rockwell M; 15% glass fiber	ASTM D1044		1000		24			315

Test Name	Material Supplier/ Grade	Material Note	Test Conditions							Source
			Test Method	Wheel / Abrader Type	Load (g)	Test Note	Taber Abrasion (mg/1000 cycles)	NBS Abrasion Index (%)	Test Result Note	

Polyester, PBT

Test Name	Material Supplier/ Grade	Material Note	Test Method	Wheel / Abrader Type	Load (g)	Test Note	Taber Abrasion (mg/1000 cycles)	NBS Abrasion Index (%)	Test Result Note	Source
Taber Abrasion	Hoechst Cel. Celanex 3210	flame retardant, 90 Rockwell M; 18% glass fiber	ASTM D1044		1000		30			315
	Hoechst Cel. Celanex 3300	gen. purp. grade, 90 Rockwell M; 30% glass fiber	ASTM D1044		1000		40			315
	Hoechst Cel. Celanex 3310	flame retardant, 90 Rockwell M; 30% glass fiber	ASTM D1044		1000		40			315
	Hoechst Cel. Celanex 3400	gen. purp. grade, 93 Rockwell M; 40% glass fiber	ASTM D1044		1000		18			315
	Hoechst Cel. Celanex 4300	high impact, 91 Rockwell M; 30% glass fiber	ASTM D1044		1000		29			315
	Hoechst Cel. Celanex 5300	good surface, 93 Rockwell M; 30% glass fiber	ASTM D1044		1000		17			315
	Hoechst Cel. Celanex 6400	low warp grade, good surface, 86 Rockwell M; 40% mineral/ glass	ASTM D1044		1000		25			315
	Bayer Pocan B1505	gen. purp. grade, unfilled	ASTM D1044	CS 17	1000		11			227
	Bayer Pocan B3235	natural resin; 30% glass fiber	ASTM D1044	CS 17	1000		16			227
	Bayer Pocan B4235	flame retardant; 30% glass fiber	ASTM D1044	CS 17	1000		26			227
	GE Valox 310SEO	general purpose, flame retardant	ASTM D1044 modified	CS 17	1000		19			431
	GE Valox 325	general purpose	ASTM D1044 modified	CS 17	1000		9			431
	GE Valox 357	high impact, flame retardant	ASTM D1044 modified	CS 17	1000		33			431
	GE Valox 359	high impact, flame retardant	ASTM D1044 modified	CS 17	1000		6			431
	GE Valox 420	30% glass	ASTM D1044 modified	CS 17	1000		19			431
	GE Valox 420SEO	30% glass, flame retardant	ASTM D1044 modified	CS 17	1000		22			431
	GE Valox 735	40% glass/ mineral	ASTM D1044 modified	CS 17	1000		64			431
	GE Valox 745	30% glass/ mineral	ASTM D1044 modified	CS 17	1000		26			431
	GE Valox 760	25% mineral	ASTM D1044 modified	CS 17	1000		37			431
	GE Valox 815	15% glass	ASTM D1044 modified	CS 17	1000		17			431
	GE Valox 855	15% glass, flame retardant	ASTM D1044 modified	CS 17	1000		17			431
	Rhone Pou. Techster T20001	80 Shore D, 105 Rockwell M, flame ret.	ASTM D1044	CS 17	1000		24			421
	Rhone Pou. Techster T2000V20	82 Shore D, 105 Rockwell M, 20% glass fiber, flame ret.	ASTM D1044	CS 17	1000		35			421
	Rhone Pou. Techster T2000V30	83 Shore D, 105 Rockwell M, 30% glass fiber, flame ret.	ASTM D1044	CS 17	1000		45			421
	Rhone Pou. Techster T29000V20	81 Shore D, 105 Rockwell M, 20% glass fiber	ASTM D1044	CS 17	1000		35			421
	Rhone Pou. Techster T29000V30	81 Shore D, 105 Rockwell M, 30% glass fiber	ASTM D1044	CS 17	1000		45			421
	Rhone Pou. Techster T29001	79 Shore D, 104 Rockwell M, gen. purp.	ASTM D1044	CS 17	1000		15-18			421
	Rhone Pou. Techster T29101	79 Shore D, 104 Rockwell M, high impact	ASTM D1044	CS 17	1000		18-24			421
		30% glass fiber	ASTM D1044	CS 17	1000	DAM	58			236

Test Name	Material Supplier/ Grade	Material Note	Test Conditions							Source
			Test Method	Wheel / Abrader Type	Load (g)	Test Note	Taber Abrasion (mg/1000 cycles)	NBS Abrasion Index (%)	Test Result Note	
Polyester, PET										
Taber Abrasion	DuPont Rynite 415HP	impact modified, 58 Rockwell M, 111 Rockwell R; 15% glass fiber	ASTM D1044	CS 17	1000		35			356
	DuPont Rynite 530	100 Rockwell M, 120 Rockwell R; 30% glass fiber	ASTM D1044	CS 17	1000		30			356
	DuPont Rynite 545	100 Rockwell M, 120 Rockwell R; 45% glass fiber	ASTM D1044	CS 17	1000		44			356
	DuPont Rynite 940	low warp grade, 118 Rockwell R; 40% mica/ glass	ASTM D1044	CS 17	1000		81			356
	DuPont Rynite FR515	flame retardant, 88 Rockwell M, 120 Rockwell R; 15% glass fiber	ASTM D1044	CS 17	1000		88			356
	DuPont Rynite FR530	flame retardant, 120 Rockwell R, 100 Rockwell M; 30% glass fiber	ASTM D1044	CS 17	1000		38			356
	DuPont Rynite FR543	flame retardant, 122 Rockwell R, 102 Rockwell M; 43% glass fiber	ASTM D1044	CS 17	1000		69			356
	DuPont Rynite FR945	low warp grade, flame retardant, 120 Rockwell R, 92 Rockwell M; 45% mica/ glass	ASTM D1044	CS 17	1000		81			356
	DuPont Rynite SST35	impact modified, high impact, 62 Rockwell M, 107 Rockwell R; 35% glass fiber	ASTM D1044	CS 17	1000		82			356
	Rhone Pou. Techster T20021V35	85 Shore D, 109 Rockwell M, 30% glass fiber, flame ret.	ASTM D1044	CS 17	1000		48			421
		30% glass fiber	ASTM D1044	CS 17	1000	DAM	33			236
Polyester Copolymer										
Taber Abrasion	Eastman PCTA 6761	amorphous, 105 Rockwell R, 54 Rockwell L	ASTM D1044	CS 10	1000		44.3			448
	Eastman PCTA 6761	amorphous, 105 Rockwell R, 54 Rockwell L	ASTM D1044	CS 17	1000		36.3			448
	Eastman PCTA 6761	amorphous, 105 Rockwell R, 54 Rockwell L	ASTM D1044	H 18	1000		197			448
Liquid Crystal Polymer										
Taber Abrasion	DuPont Zenite 6130	high heat grade, 3.2 mm thick, 63 Rockwell M, 110 Rockwell R; 30% glass fiber		CS 17	1000		63			354
	Hoechst AG Vectra A230	80 Rockwell M; Vectra A950 matrix, 30% carbon fiber	ASTM D1044	H 18	1000		121			333
	Hoechst AG Vectra A130	80 Rockwell M; 30% glass fiber, Vectra A950 matrix	ASTM D1044				7.4			70
	Hoechst AG Vectra A230	80 Rockwell M; Vectra A950 matrix, 30% carbon fiber	ASTM D1044				5.8			70
	Hoechst AG Vectra A515	50 Rockwell M; 15% mineral filler, Vectra A950 matrix	ASTM D1044				11			70
	Hoechst AG Vectra A950		ASTM D1044				56			70
Polyetherimide										
Taber Abrasion	GE Ultem 1000	unmodified, 109 Rockwell M	ASTM D1044	CS 17	1000		10			51
	GE Ultem 1010	high flow, 109 Rockwell M	ASTM D1044	CS 17	1000		10			51

Test Name	Material Supplier/ Grade	Material Note	Test Conditions							Source
			Test Method	Wheel / Abrader Type	Load (g)	Test Note	Taber Abrasion (mg/1000 cycles)	NBS Abrasion Index (%)	Test Result Note	

Polyetherimide

Test Name	Material Supplier/ Grade	Material Note	Test Method	Wheel / Abrader Type	Load (g)	Test Note	Taber Abrasion (mg/1000 cycles)	NBS Abrasion Index (%)	Test Result Note	Source
Taber Abrasion	GE Ultem 2100	10% glass fiber, 114 Rockwell M	ASTM D1044	CS 17	1000		15			51
	GE Ultem 2200	20% glass fiber, 114 Rockwell M	ASTM D1044	CS 17	1000		17			51
	GE Ultem 2300	30% glass fiber, 114 Rockwell M	ASTM D1044	CS 17	1000		20			51
	GE Ultem 2400	40% glass fiber, 114 Rockwell M	ASTM D1044	CS 17	1000		20			51
	GE Ultem 4000	reinforced, internal lubricant, 85 Rockwell M	ASTM D1044	CS 17	1000		33			51
	GE Ultem 4001	internal lubricant, 110 Rockwell M	ASTM D1044	CS 17	1000		2			51
	GE Ultem 6000	heat resistant, 110 Rockwell M	ASTM D1044	CS 17	1000		8			51
	GE Ultem 6100	10% glass fiber, heat resistant, 114 Rockwell M	ASTM D1044	CS 17	1000		17			51
	GE Ultem 6200	20% glass fiber, heat resistant, 114 Rockwell M	ASTM D1044	CS 17	1000		20			51
	GE Ultem 6202	20% mineral, heat resistant, 110 Rockwell M	ASTM D1044	CS 17	1000		11			51

Polyetheretherketone

Test Name	Material Supplier/ Grade	Material Note	Test Method	Wheel / Abrader Type	Load (g)	Test Note	Taber Abrasion (mg/1000 cycles)	NBS Abrasion Index (%)	Test Result Note	Source
Taber Abrasion	Victrex USA Victrex PEEK	natural, unfilled	ASTM D1044	CS 17	1000		.97			423
	Victrex USA Victrex PEEK	natural, unfilled	ASTM D1044	H 10	1000		.27			423
	Victrex USA Victrex PEEK	60% carborundum	ASTM D1044	H 10	1000		7.1			423

Polyethylene

Test Name	Material Supplier/ Grade	Material Note	Test Method	Wheel / Abrader Type	Load (g)	Test Note	Taber Abrasion (mg/1000 cycles)	NBS Abrasion Index (%)	Test Result Note	Source
Taber Abrasion	Unspecified grade		ASTM D1044	CS 10-F	250		4.5			78

Polyethylene, HDPE

Test Name	Material Supplier/ Grade	Material Note	Test Method	Wheel / Abrader Type	Load (g)	Test Note	Taber Abrasion (mg/1000 cycles)	NBS Abrasion Index (%)	Test Result Note	Source
Taber Abrasion	Unspecified grade	3 g/10 min. MFI	ASTM D1044	CS 17	1000		5.5			330

Polyethylene, UHMWPE

Test Name	Material Supplier/ Grade	Material Note	Test Method	Wheel / Abrader Type	Load (g)	Test Note	Taber Abrasion (mg/1000 cycles)	NBS Abrasion Index (%)	Test Result Note	Source
Taber Abrasion	Hoechst Cel. Hostalen GUR 5121	0.93 g/cm³ density; inj. mold. grade, 63 Shore D	ASTM D1044	CS 17	1000		.7			329
	Hoechst Cel. Hostalen GUR 5121	0.93 g/cm³ density; inj. mold. grade, 63 Shore D	ASTM D1044	CS 17	1000	3000 cycles			5.9 mg weight loss	329
	Hoechst Cel. Hostalen GUR 5121	0.93 g/cm³ density; inj. mold. grade, 63 Shore D	ASTM D1044	CS 17	1000	5000 cycles			9 mg weight loss	329
	Hoechst Cel. Hostalen GUR 5150	0.93 g/cm³ density; inj. mold. grade, 61 Shore D	ASTM D1044	CS 17	1000		1.1			329
	Hoechst Cel. Hostalen GUR 5150	0.93 g/cm³ density; inj. mold. grade, 61 Shore D	ASTM D1044	CS 17	1000	3000 cycles			3.2 mg weight loss	329
	Hoechst Cel. Hostalen GUR 5150	0.93 g/cm³ density; inj. mold. grade, 61 Shore D	ASTM D1044	CS 17	1000	5000 cycles			5.7 mg weight loss	329

Test Name	Material Supplier/ Grade	Material Note	Test Conditions							Source
			Test Method	Wheel / Abrader Type	Load (g)	Test Note	Taber Abrasion (mg/1000 cycles)	NBS Abrasion Index (%)	Test Result Note	

Polypropylene

Test Name	Material Supplier/ Grade	Material Note	Test Method	Wheel / Abrader Type	Load (g)	Test Note	Taber Abrasion (mg/1000 cycles)	NBS Abrasion Index (%)	Test Result Note	Source
Taber Abrasion	Unspecified grade		ASTM D1044	CS 10-F	250		4.3			78

Modified PPE

Test Name	Material Supplier/ Grade	Material Note	Test Method	Wheel / Abrader Type	Load (g)	Test Note	Taber Abrasion (mg/1000 cycles)	NBS Abrasion Index (%)	Test Result Note	Source
Taber Abrasion	GE Noryl 731	general purpose	ASTM D1044 modified	CS 17	1000		20			431
	GE Noryl GFN1	10% glass	ASTM D1044 modified	CS 17	1000		35			431
	GE Noryl GFN1-SE1	10% glass, flame retardant	ASTM D1044 modified	CS 17	1000		35			431
	GE Noryl GFN2	20% glass	ASTM D1044 modified	CS 17	1000		35			431
	GE Noryl GFN2-SE1	20% glass, flame retardant	ASTM D1044 modified	CS 17	1000		35			431
	GE Noryl GFN3	30% glass	ASTM D1044 modified	CS 17	1000		35			431
	GE Noryl GFN3-SE1	30% glass, flame retardant	ASTM D1044 modified	CS 17	1000		35			431
	GE Noryl N180	flame retardant	ASTM D1044 modified	CS 17	1000		20			431
	GE Noryl N180	general purpose, flame retardant	ASTM D1044 modified	CS 17	1000		20			431
	GE Noryl N190	general purpose, flame retardant	ASTM D1044 modified	CS 17	1000		20			431
	GE Noryl SE1	high heat, flame retardant, electrical properties	ASTM D1044 modified	CS 17	1000		20			431
	GE Noryl SE100	high heat, flame retardant	ASTM D1044 modified	CS 17	1000		20			431
	Mitsubishi Iupiace AH40	116 Rockwell R, UL94 HB	ASTM D1044	CS 17	1000		55			422
	Mitsubishi Iupiace AH50	118 Rockwell R, UL94 HB	ASTM D1044	CS 17	1000		50			422
	Mitsubishi Iupiace AH60	119 Rockwell R, UL94 HB	ASTM D1044	CS 17	1000		51			422
	Mitsubishi Iupiace AH70	121 Rockwell R, UL94 HB	ASTM D1044	CS 17	1000		48			422
	Mitsubishi Iupiace AH80	123 Rockwell R, UL94 HB	ASTM D1044	CS 17	1000		45			422
	Mitsubishi Iupiace AH90	124 Rockwell R, UL94 HB	ASTM D1044	CS 17	1000		39			422
	Mitsubishi Iupiace AHF6005	123 Rockwell R, low COF	ASTM D1044	CS 17	1000		40			422
	Mitsubishi Iupiace AHF6010	122 Rockwell R, low COF	ASTM D1044	CS 17	1000		35			422
	Mitsubishi Iupiace AHF6015	121 Rockwell R, low COF	ASTM D1044	CS 17	1000		30			422
	Mitsubishi Iupiace AN20	119 Rockwell R, UL94 V-0	ASTM D1044	CS 17	1000		51			422
	Mitsubishi Iupiace AN30	120 Rockwell R, UL94 V-0	ASTM D1044	CS 17	1000		51			422
	Mitsubishi Iupiace AN45	122 Rockwell R, UL94 V-0	ASTM D1044	CS 17	1000		48			422
	Mitsubishi Iupiace AN60	124 Rockwell R, UL94 V-0	ASTM D1044	CS 17	1000		39			422
	Mitsubishi Iupiace AN91	125 Rockwell R, UL94 V-0	ASTM D1044	CS 17	1000		39			422
	Mitsubishi Iupiace AV20	118 Rockwell R, UL94 V-1	ASTM D1044	CS 17	1000		60			422
	Mitsubishi Iupiace AV30	119 Rockwell R, UL94 V-1	ASTM D1044	CS 17	1000		55			422
	Mitsubishi Iupiace AV40	119 Rockwell R, UL94 V-1	ASTM D1044	CS 17	1000		59			422
	Mitsubishi Iupiace AV60	122 Rockwell R, UL94 V-1	ASTM D1044	CS 17	1000		39			422
	Mitsubishi Iupiace AV90	124 Rockwell R, UL94 V-1	ASTM D1044	CS 17	1000		38			422
	Mitsubishi Iupiace AVF6010	123 Rockwell R, low COF	ASTM D1044	CS 17	1000		35			422
	Mitsubishi Iupiace EV05	122 Rockwell R, 5% carbon fiber	ASTM D1044	CS 17	1000		40			422
	Mitsubishi Iupiace EV08	122 Rockwell R, 8% carbon fiber	ASTM D1044	CS 17	1000		35			422
	Mitsubishi Iupiace EV12	124 Rockwell R, 12% carbon fiber	ASTM D1044	CS 17	1000		30			422

Test Name	Material Supplier/ Grade	Material Note	Test Conditions							Source
			Test Method	Wheel / Abrader Type	Load (g)	Test Note	Taber Abrasion (mg/1000 cycles)	NBS Abrasion Index (%)	Test Result Note	

Modified PPE

Test Name	Material Supplier/ Grade	Material Note	Test Method	Wheel / Abrader Type	Load (g)	Test Note	Taber Abrasion (mg/1000 cycles)	NBS Abrasion Index (%)	Test Result Note	Source
Taber Abrasion	Mitsubishi Iupiace EV20	126 Rockwell R, 20% carbon fiber	ASTM D1044	CS 17	1000		25			422
	Mitsubishi Iupiace GHF3005	122 Rockwell R, low COF	ASTM D1044	CS 17	1000		35			422
	Mitsubishi Iupiace GHF3010	121 Rockwell R, low COF	ASTM D1044	CS 17	1000		30			422
	Mitsubishi Iupiace GHF3015	120 Rockwell R, low COF	ASTM D1044	CS 17	1000		30			422

Polyphenylene Sulfide

Test Name	Material Supplier/ Grade	Material Note	Test Method	Wheel / Abrader Type	Load (g)	Test Note	Taber Abrasion (mg/1000 cycles)	NBS Abrasion Index (%)	Test Result Note	Source
Taber Abrasion	Ryton R4	123 Rockwell R; 40% glass fiber	ASTM D1044	CS 10	1000		70			102
	GE Supec G401	123 Rockwell R; 40% glass fiber	ASTM D1044	CS 17	1000		51			429
	GE Supec G402	high flow, 123 Rockwell R; 40% glass fiber	ASTM D1044	CS 17	1000		51			429
	GE Supec W331	116 Rockwell R; 30% glass fiber, 15% PTFE	ASTM D1044	CS 17	1000		14			429
	Phillips Ryton A200	120 Rockwell R; 40% glass fiber	ASTM D1044	CS 17	1000		23			102
	Ryton R4	123 Rockwell R; 40% glass fiber	ASTM D1044	CS 17	1000		34			102
	Phillips Ryton R7	121 Rockwell R; 60% glass fiber/ mineral	ASTM D1044	CS 17	1000		68			102
	Unspecified grade	40% glass fiber reinforced	ASTM D1044	CS 17	1000	DAM	56			236

Polysulfone

Test Name	Material Supplier/ Grade	Material Note	Test Method	Wheel / Abrader Type	Load (g)	Test Note	Taber Abrasion (mg/1000 cycles)	NBS Abrasion Index (%)	Test Result Note	Source
Taber Abrasion	Unspecified grade	transparent, amber tint		CS 17	1000		20			15
	LNP Thermocomp GF-1004	20% glass fiber	ASTM D1044	CS-17	1000		35-40			456
	LNP Thermocomp GF-1006	30% glass fiber	ASTM D1044	CS-17	1000		35-40			456
	LNP Thermocomp GF-1008	40% glass fiber	ASTM D1044	CS-17	1000		35-40			456
	Unspecified grade	unmodified	ASTM D1044	CS-17	1000		20			456

Polyethersulfone

Test Name	Material Supplier/ Grade	Material Note	Test Method	Wheel / Abrader Type	Load (g)	Test Note	Taber Abrasion (mg/1000 cycles)	NBS Abrasion Index (%)	Test Result Note	Source
Taber Abrasion	Victrex USA Victrex PES	natural, unfilled	ASTM D1044	CS 17	1000		15			423
	Victrex USA Victrex PES	natural, unfilled	ASTM D1044	H 10	1000		56			423

Polyurethane, Rigid

Test Name	Material Supplier/ Grade	Material Note	Test Method	Wheel / Abrader Type	Load (g)	Test Note	Taber Abrasion (mg/1000 cycles)	NBS Abrasion Index (%)	Test Result Note	Source
Taber Abrasion	Dow Isoplast 101		ASTM D1044	CS 17	1000		11			434
	Dow Isoplast 102		ASTM D1044	CS 17	1000		12			434

ASA

Test Name	Material Supplier/ Grade	Material Note	Test Method	Wheel / Abrader Type	Load (g)	Test Note	Taber Abrasion (mg/1000 cycles)	NBS Abrasion Index (%)	Test Result Note	Source
Taber Abrasion	Dow Rovel 401	weatherable, no longer available, 100 Rockwell R		CS 17	1000		47			342

Polystyrene, IPS

Test Name	Material Supplier/ Grade	Material Note	Test Method	Wheel / Abrader Type	Load (g)	Test Note	Taber Abrasion (mg/1000 cycles)	NBS Abrasion Index (%)	Test Result Note	Source
Taber Abrasion	Unspecified grade		ASTM C501	CS 17	1000	5000 cycles	109			427

Test Name	Material Supplier/ Grade	Material Note	Test Conditions							Source
			Test Method	Wheel / Abrader Type	Load (g)	Test Note	Taber Abrasion (mg/1000 cycles)	NBS Abrasion Index (%)	Test Result Note	
Polyvinyl Chloride										
Taber Abrasion	Unspecified grade	high impact	ASTM C501	CS 17	1000	5000 cycles	17.8			427
Acrylic/PVC Alloy										
Taber Abrasion	Kleerdex Kydex 100		ASTM D1044	CS 10	1000		15			443
	Kleerdex Kydex 100	77 Shore D; sheet	ASTM D1044	CS 10	1000		.015	500		141
Polyolefin Alloy										
Taber Abrasion	Hoechst Cel. Hostalloy 731	0.95 g/cm³ density; 68 Shore D	ASTM D1044	CS 17	1000		1.7			330
PC/Polyester PBT Alloy										
Taber Abrasion	GE Xenoy 6120	low modulus, high impact	ASTM D1044 modified	CS 17	1000		37			431
	GE Xenoy CL200	semi-crystalline, high impact	ASTM D1044 modified	CS 17	1000		21			431
PPE/Nylon Alloy										
Taber Abrasion	GE Noryl GTX 900	general purpose, automotive	ASTM D1044 modified	CS 17	1000		14			431
	GE Noryl GTX 910	automotive, on-line painting	ASTM D1044 modified	CS 17	1000		14			431
TPE, Olefinic										
DIN 53516	Evode Forprene 630	55 Shore A	DIN 53516						240 mm³ volume loss	161
	Evode Forprene 631	60 Shore A	DIN 53516						220 mm³ volume loss	161
	Evode Forprene 632	65 Shore A	DIN 53516						200 mm³ volume loss	161
	Evode Forprene 633	70 Shore A	DIN 53516						150 mm³ volume loss	161
	Evode Forprene 634	75 Shore A	DIN 53516						120 mm³ volume loss	161
	Evode Forprene 635	80 Shore A	DIN 53516						115 mm³ volume loss	161
	Evode Forprene 636	85 Shore A	DIN 53516						110 mm³ volume loss	161
	Evode Forprene 637	90 Shore A	DIN 53516						100 mm³ volume loss	161
	Evode Forprene 638	40 Shore D	DIN 53516						95 mm³ volume loss	161
	Evode Forprene 639	45 Shore D	DIN 53516						95 mm³ volume loss	161
	Evode Forprene 640	50 Shore D	DIN 53516						95 mm³ volume loss	161
	Evode Forprene 641	55 Shore D	DIN 53516						95 mm³ volume loss	161

TPE, Olefinic

Test Name	Material Supplier/ Grade	Material Note	Test Conditions							Source
			Test Method	Wheel / Abrader Type	Load (g)	Test Note	Taber Abrasion (mg/1000 cycles)	NBS Abrasion Index (%)	Test Result Note	
DIN 53516	Evode Forprene 642	60 Shore D	DIN 53516						90 mm^3 volume loss	161
	Evode Forprene 643	65 Shore D	DIN 53516							161
NBS Abrasion Index	DuPont Alcryn 1201-B60	60 Shore A, black	ASTM D1630					grabs		424
	DuPont Alcryn 1201-B70	70 Shore A, black	ASTM D1630					115		424
	DuPont Alcryn 1201-B80	80 Shore A, black	ASTM D1630					98		424
	DuPont Alcryn ALR-6370	75 Shore A, black	ASTM D1630					100		424
Taber Abrasion	DSM Sarlink 3180	80 Shore A	ASTM D1044	CS 17	1000		10.5			442
	DSM Sarlink 3190	92 Shore A	ASTM D1044	CS 17	1000		5.9			442
	DuPont Alcryn 1060BK	black, 1.9 mm thick, 61 Shore A; compression molded	ASTM D1044	CS 17	1000		7			346
	DuPont Alcryn 1070BK	black, 1.9 mm thick, 69 Shore A; compression molded	ASTM D1044	CS 17	1000		7			346
	DuPont Alcryn 1080BK	black, 77 Shore A, 1.9 mm thick; compression molded	ASTM D1044	CS 17	1000		20			346
	DuPont Alcryn 1201-B60	60 Shore A, black	ASTM D1044	CS 17	1000		7			424
	DuPont Alcryn 1201-B70	70 Shore A, black	ASTM D1044	CS 17	1000		7			424
	DuPont Alcryn 1201-B80	80 Shore A, black	ASTM D1044	CS 17	1000		20			424
	DuPont Alcryn 2060BK	black, 58 Shore A, 1.9 mm thick; compression molded	ASTM D1044	CS 17	1000		5			346
	DuPont Alcryn 2060NC	natural resin, 1.9 mm thick, 59 Shore A; compression molded	ASTM D1044	CS 17	1000		5			346
	DuPont Alcryn 2070BK	black, 70 Shore A, 1.9 mm thick; compression molded	ASTM D1044	CS 17	1000		5			346
	DuPont Alcryn 2070NC	natural resin, 68 Shore A, 1.9 mm thick; compression molded	ASTM D1044	CS 17	1000		9			346
	DuPont Alcryn 2080BK	black, 75 Shore A, 1.9 mm thick; compression molded	ASTM D1044	CS 17	1000		3			346
	DuPont Alcryn 2080NC	natural resin, 73 Shore A, 1.9 mm thick; compression molded	ASTM D1044	CS 17	1000		10			346
	DuPont Alcryn 3055NC	natural resin, 57 Shore A, 1.9 mm thick; compression molded	ASTM D1044	CS 17	1000		0			346
	DuPont Alcryn 3065NC	natural resin, 64 Shore A, 1.9 mm thick; compression molded	ASTM D1044	CS 17	1000		.6			346
	DuPont Alcryn 3075NC	natural resin, 75 Shore A, 1.9 mm thick; compression molded	ASTM D1044	CS 17	1000		.6			346
	DuPont Alcryn ALR-6370	75 Shore A, black	ASTM D1044	CS 17	1000		9			424
	DuPont Alcryn 1201-B60	60 Shore A, black	ASTM D1044	H 18	1000		142			424
	DuPont Alcryn 1201-B70	70 Shore A, black	ASTM D1044	H 18	1000		146			424

TPE, Polyamide

Test Name	Material Supplier/ Grade	Material Note	Test Method	Wheel / Abrader Type	Load (g)	Test Note	Taber Abrasion (mg/1000 cycles)	NBS Abrasion Index (%)	Test Result Note	Source
Taber Abrasion	Atochem Pebax 2533	75 Shore A, 25 Shore D	DIN 53516	CS 17	1000		46			287
	Atochem Pebax 3533	85 Shore A, 35 Shore D	DIN 53516	CS 17	1000		25			287

Test Name	Material Supplier/ Grade	Material Note	Test Conditions							Source
			Test Method	Wheel / Abrader Type	Load (g)	Test Note	Taber Abrasion (mg/1000 cycles)	NBS Abrasion Index (%)	Test Result Note	

TPE, Polyamide

Test Name	Material Supplier/ Grade	Material Note	Test Method	Wheel / Abrader Type	Load (g)	Test Note	Taber Abrasion (mg/1000 cycles)	NBS Abrasion Index (%)	Test Result Note	Source
Taber Abrasion	Atochem Pebax 4033	90 Shore A, 40 Shore D	DIN 53516	CS 17	1000		14			287
	Atochem Pebax 5512	55 Shore D	DIN 53516	CS 17	1000		15			287
	Atochem Pebax 5533	55 Shore D	DIN 53516	CS 17	1000		11			287
	Atochem Pebax 6312	63 Shore D	DIN 53516	CS 17	1000		13			287
	Atochem Pebax 6333	63 Shore D	DIN 53516	CS 17	1000		12			287
	Atochem Pebax 2533	75 Shore A, 25 Shore D	DIN 53516	H 18	1000		94			287
	Atochem Pebax 2533	25 Shore D	ASTM D1044	H 18	1000		161			438
	Atochem Pebax 3533	85 Shore A, 35 Shore D	DIN 53516	H 18	1000		81			287
	Atochem Pebax 3533	35 Shore D	ASTM D1044	H 18	1000		104			438
	Atochem Pebax 4033	40 Shore D	ASTM D1044	H 18	1000		94			438
	Atochem Pebax 4033	90 Shore A, 40 Shore D	DIN 53516	H 18	1000		70			287
	Atochem Pebax 5512	55 Shore D	DIN 53516	H 18	1000		255			287
	Atochem Pebax 5533	55 Shore D	ASTM D1044	H 18	1000		93			438
	Atochem Pebax 5533	55 Shore D	DIN 53516	H 18	1000		65			287
	Atochem Pebax 6312	63 Shore D	DIN 53516	H 18	1000		88			287
	Atochem Pebax 6333	63 Shore D	ASTM D1044	H 18	1000		84			438
	Atochem Pebax 6333	63 Shore D	DIN 53516	H 18	1000		46			287
	Atochem Pebax 7033	69 Shore D	ASTM D1044	H 18	1000		57			438

TPE, Polyester

Test Name	Material Supplier/ Grade	Material Note	Test Method	Wheel / Abrader Type	Load (g)	Test Note	Taber Abrasion (mg/1000 cycles)	NBS Abrasion Index (%)	Test Result Note	Source
NBS Abrasion Index	DuPont Hytrel 4056	40 Shore D	ASTM D1630					590		347
	DuPont Hytrel 5556	55 Shore D	ASTM D1630					2250		347
	DuPont Hytrel 6346	63 Shore D	ASTM D1630					2340		347
	DuPont Hytrel 7246	72 Shore D	ASTM D1630					2620		347
	DuPont Hytrel G4075	40 Shore D	ASTM D1630					300		347
Taber Abrasion	DuPont Hytrel 3078	low modulus elastomer, w/ color stable antioxidants, specialty grade, 30 Shore D	ASTM D1044	CS 17	1000	modified test	2			348
	DuPont Hytrel 4056	40 Shore D	ASTM D1044	CS 17	1000		8			347
	DuPont Hytrel 4056	40 Shore D, w/ color stable antioxidants, high perform.	ASTM D1044	CS 17	1000	modified test	3			348
	DuPont Hytrel 4069	40 Shore D, w/ color stable antioxidants, high perform.	ASTM D1044	CS 17	1000	modified test	15			348
	DuPont Hytrel 4556	45 Shore D, w/ color stable antioxidants, high perform.	ASTM D1044	CS 17	1000	modified test	3			348
	DuPont Hytrel 5526	55 Shore D, high perform., w/ color stable antioxidants	ASTM D1044	CS 17	1000	modified test	7			348
	DuPont Hytrel 5556	55 Shore D	ASTM D1044	CS 17	1000		6			347
	DuPont Hytrel 5556	55 Shore D, w/ color stable antioxidants, high perform.	ASTM D1044	CS 17	1000	modified test	6			348
	DuPont Hytrel 6346	63 Shore D	ASTM D1044	CS 17	1000		15			347

TPE, Polyester

Test Name	Material Supplier/ Grade	Material Note	Test Conditions							Source
			Test Method	Wheel / Abrader Type	Load (g)	Test Note	Taber Abrasion (mg/1000 cycles)	NBS Abrasion Index (%)	Test Result Note	
Taber Abrasion	DuPont Hytrel 6356	63 Shore D, w/ color stable antioxidants, high perform.	ASTM D1044	CS 17	1000	modified test	7			348
	DuPont Hytrel 7246	72 Shore D	ASTM D1044	CS 17	1000		15			347
	DuPont Hytrel 7246	72 Shore D, w/ color stable antioxidants, high perform.	ASTM D1044	CS 17	1000	modified test	13			348
	DuPont Hytrel 8238	w/ color stable antioxidants, 82 Shore D, 104 Rockwell R, high perform.	ASTM D1044	CS 17	1000	modified test	9			348
	DuPont Hytrel G3548W	UV stabilized, color stable antioxidants, 35 Shore D, high productivity	ASTM D1044	CS 17	1000	modified test	30			348
	DuPont Hytrel G4074	40 Shore D, high productivity, w/ discoloring antioxidant	ASTM D1044	CS 17	1000	modified test	9			348
	DuPont Hytrel G4075	40 Shore D	ASTM D1044	CS 17	1000		10			347
	DuPont Hytrel G4078W	UV stabilized, 40 Shore D, high productivity, color stable antioxidants	ASTM D1044	CS 17	1000	modified test	20			348
	DuPont Hytrel G4774	high productivity, 47 Shore D, w/ discoloring antioxidant	ASTM D1044	CS 17	1000	modified test	13			348
	DuPont Hytrel G4778	high productivity, color stable antioxidants, 47 Shore D	ASTM D1044	CS 17	1000	modified test	12			348
	DuPont Hytrel G5544	55 Shore D, high productivity, w/ discoloring antioxidant	ASTM D1044	CS 17	1000	modified test	9			348
	DuPont Hytrel HTR4275BK	black, 55 Shore D, specialty grade	ASTM D1044	CS 17	1000	modified test	20			348
	DuPont Hytrel HTR5612BK	black, 50 Shore D, specialty grade	ASTM D1044	CS 17	1000	modified test	38			348
	DuPont Hytrel HTR6108	barrier prop., 60 Shore D	ASTM D1044	CS 17	1000	modified test	9			348
	DuPont Hytrel HTR8068	flame retardant, 46 Shore D	ASTM D1044	CS 17	1000	modified test	25			348
	DuPont Hytrel HTR8139LV	black, surface lubricity, specialty grade, fatigue resist., 46 Shore D	ASTM D1044	CS 17	1000	modified test	4			348
	DuPont Hytrel HTR8171	high MVTR, specialty grade, 32 Shore D, w/ color stable antioxidants	ASTM D1044	CS 17	1000	modified test	85			348
	DuPont Hytrel HTR8206	45 Shore D, high MVTR, w/ color stable antioxidants, specialty grade	ASTM D1044	CS 17	1000	modified test	0			348
	GE Lomod A1220	automotive, 60 Shore D	ASTM D1044 modified	CS 17	1000		31			431
	GE Lomod B0150	flame retardant, 40 Shore D	ASTM D1044 modified	CS 17	1000		171			431
	GE Lomod B0200	general purpose, 44 Shore D	ASTM D1044 modified	CS 17	1000		102			431
	GE Lomod B0220	high heat, 47 Shore D	ASTM D1044 modified	CS 17	1000		225			431
	GE Lomod B0250	flame retardant, 57 Shore D	ASTM D1044 modified	CS 17	1000		96			431
	GE Lomod B0320	high heat, 57 Shore D	ASTM D1044 modified	CS 17	1000		180			431
	GE Lomod B0520	high heat, 70 Shore D	ASTM D1044 modified	CS 17	1000		130			431
	GE Lomod B0800	general purpose, 70 Shore D	ASTM D1044 modified	CS 17	1000		105			431
	GE Lomod B0852	flame retardant, 69 Shore D	ASTM D1044 modified	CS 17	1000		106			431
	GE Lomod FR30125A	76 Shore D, flame retardant	ASTM D1044 modified	CS 17	1000		72			432
	GE Lomod FR5020A	flame retardant, 49 Shore D	ASTM D1044 modified	CS 17	1000		216			432
	GE Lomod FR5030A	flame retardant, 57 Shore D	ASTM D1044 modified	CS 17	1000		96			432
	GE Lomod ST3090A	32 Shore D, 90 Shore A	ASTM D1044 modified	CS 17	1000		235			432

TPE, Polyester

Test Name	Material Supplier/ Grade	Material Note	Test Conditions							Source
			Test Method	Wheel / Abrader Type	Load (g)	Test Note	Taber Abrasion (mg/1000 cycles)	NBS Abrasion Index (%)	Test Result Note	
Taber Abrasion	GE Lomod ST5090A	31 Shore D, 90 Shore A	ASTM D1044 modified	CS 17	1000		115			432
	GE Lomod TE3040A	43 Shore D, high heat	ASTM D1044 modified	CS 17	1000		185			432
	GE Lomod TE3045A	47 Shore D, high heat	ASTM D1044 modified	CS 17	1000		130			432
	GE Lomod TE3055A	54 Shore D	ASTM D1044 modified	CS 17	1000		120			432
	GE Lomod TE3055B	54 Shore D, improved heat aging	ASTM D1044 modified	CS 17	1000		120			432
	GE Lomod TE3060A	60 Shore D	ASTM D1044 modified	CS 17	1000		93			432
	GE Lomod TE3070A	71 Shore D	ASTM D1044 modified	CS 17	1000		67			432
	GE Lomod TE5040A	38 Shore D	ASTM D1044 modified	CS 17	1000		141			432
	GE Lomod TE5050A	51 Shore D	ASTM D1044 modified	CS 17	1000		86			432
	DuPont Hytrel 3078	low modulus elastomer, w/ color stable antioxidants, specialty grade, 30 Shore D	ASTM D1044	H 18	1000	modified test	90			348
	DuPont Hytrel 4056	40 Shore D, w/ color stable antioxidants, high perform.	ASTM D1044	H 18	1000	modified test	100			348
	DuPont Hytrel 4056	40 Shore D	ASTM D1044	H 18	1000		109			347
	DuPont Hytrel 4069	40 Shore D, w/ color stable antioxidants, high perform.	ASTM D1044	H 18	1000	modified test	80			348
	DuPont Hytrel 4556	45 Shore D, w/ color stable antioxidants, high perform.	ASTM D1044	H 18	1000	modified test	72			348
	DuPont Hytrel 5526	55 Shore D, high perform., w/ color stable antioxidants	ASTM D1044	H 18	1000	modified test	70			348
	DuPont Hytrel 5555HS	55 Shore D, specialty grade, w/ discoloring antioxidant	ASTM D1044	H 18	1000	modified test	112			348
	DuPont Hytrel 5556	55 Shore D, w/ color stable antioxidants, high perform.	ASTM D1044	H 18	1000	modified test	64			348
	DuPont Hytrel 5556	55 Shore D	ASTM D1044	H 18	1000		97			347
	DuPont Hytrel 6346	63 Shore D	ASTM D1044	H 18	1000		109			347
	DuPont Hytrel 6356	63 Shore D, w/ color stable antioxidants, high perform.	ASTM D1044	H 18	1000	modified test	77			348
	DuPont Hytrel 7246	72 Shore D, w/ color stable antioxidants, high perform.	ASTM D1044	H 18	1000	modified test	47			348
	DuPont Hytrel 7246	72 Shore D	ASTM D1044	H 18	1000		75			347
	DuPont Hytrel 8238	w/ color stable antioxidants, 82 Shore D, 104 Rockwell R, high perform.	ASTM D1044	H 18	1000	modified test	20			348
	DuPont Hytrel G3548W	UV stabilized, color stable antioxidants, 35 Shore D, high productivity	ASTM D1044	H 18	1000	modified test	310			348
	DuPont Hytrel G4074	40 Shore D, high productivity, w/ discoloring antioxidant	ASTM D1044	H 18	1000	modified test	193			348
	DuPont Hytrel G4075	40 Shore D	ASTM D1044	H 18	1000		223			347
	DuPont Hytrel G4078W	UV stabilized, 40 Shore D, high productivity, color stable antioxidants	ASTM D1044	H 18	1000	modified test	260			348
	DuPont Hytrel G4774	high productivity, 47 Shore D, w/ discoloring antioxidant	ASTM D1044	H 18	1000	modified test	168			348
	DuPont Hytrel G4778	high productivity, color stable antioxidants, 47 Shore D	ASTM D1044	H 18	1000	modified test	162			348

TPE, Polyester

Test Name	Material Supplier/ Grade	Material Note	Test Method	Wheel / Abrader Type	Load (g)	Test Note	Taber Abrasion (mg/1000 cycles)	NBS Abrasion Index (%)	Test Result Note	Source
Taber Abrasion	DuPont Hytrel G5544	55 Shore D, high productivity, w/ discoloring antioxidant	ASTM D1044	H 18	1000	modified test	116			348
	DuPont Hytrel HTR4275BK	black, 55 Shore D, specialty grade	ASTM D1044	H 18	1000	modified test	227			348
	DuPont Hytrel HTR5612BK	black, 50 Shore D, specialty grade	ASTM D1044	H 18	1000	modified test	186			348
	DuPont Hytrel HTR6108	barrier prop., 60 Shore D	ASTM D1044	H 18	1000	modified test	116			348
	DuPont Hytrel HTR8139LV	black, surface lubricity, specialty grade, fatigue resist., 46 Shore D	ASTM D1044	H 18	1000	modified test	65			348
	DuPont Hytrel HTR8171	high MVTR, specialty grade, 32 Shore D, w/ color stable antioxidants	ASTM D1044	H 18	1000	modified test	240			348
	DuPont Hytrel HTR8206	45 Shore D, high MVTR, w/ color stable antioxidants, specialty grade	ASTM D1044	H 18	1000	modified test	65			348
	EniChem Pibiflex 35M	35 Shore D	ASTM D1044	H 18	1000		90			451
	EniChem Pibiflex 40EM	38 Shore D	ASTM D1044	H 18	1000		87			451
	EniChem Pibiflex 40M	40 Shore D	ASTM D1044	H 18	1000		90			451
	EniChem Pibiflex 46EM	46 Shore D	ASTM D1044	H 18	1000		60			451
	EniChem Pibiflex 46M	46 Shore D	ASTM D1044	H 18	1000		85			451
	EniChem Pibiflex 47EM/UV	UV stabilized, 44 Shore D	ASTM D1044	H 18	1000		96			451
	EniChem Pibiflex 52BE	52 Shore D	ASTM D1044	H 18	1000		45			451
	EniChem Pibiflex 56M	56 Shore D	ASTM D1044	H 18	1000		73			451
	EniChem Pibiflex 56MWR	hydrolysis resistant, 56 Shore D	ASTM D1044	H 18	1000		98			451
	EniChem Pibiflex 59M	59 Shore D	ASTM D1044	H 18	1000		60			451
	EniChem Pibiflex B46MWR	hydrolysis resistant, 44 Shore D, blow molding grade	ASTM D1044	H 18	1000		68			451
	EniChem Pibiflex E40M	extrusion grade, 40 Shore D	ASTM D1044	H 18	1000		90			451
	EniChem Pibiflex E46M	extrusion grade, 46 Shore D	ASTM D1044	H 18	1000		90			451
	EniChem Pibiflex E59M	extrusion grade, 59 Shore D	ASTM D1044	H 18	1000		50			451
	EniChem Pibiflex PD25M	27 Shore D, powder grade	ASTM D1044	H 18	1000		120			451
	EniChem Pibiflex PD46LP	powder grade, 44 Shore D	ASTM D1044	H 18	1000		95			451
	EniChem Pibiflex PD53M	52 Shore D, powder grade	ASTM D1044	H 18	1000		70			451
	EniChem Pibiflex PD56E	53 Shore D, powder grade	ASTM D1044	H 18	1000		75			451
	EniChem Pibiflex PD59MAE	flame retardant, 56 Shore D, powder grade	ASTM D1044	H 18	1000		100			451
	EniChem Pibiflex PD63M	powder grade, 62 Shore D	ASTM D1044	H 18	1000		70			451
	EniChem Pibiflex PD72M	68 Shore D, powder grade	ASTM D1044	H 18	1000		65			451
	EniChem Pibiflex PDB40MWR	hydrolysis resistant, 40 Shore D, powder grade	ASTM D1044	H 18	1000		90			451
	EniChem Pibiflex PDB42MWR	hydrolysis resistant, 42 Shore D, powder grade	ASTM D1044	H 18	1000		87			451
	EniChem Pibiflex PDB5050MWR	hydrolysis resistant, 43 Shore D, blow molding grade, powder grade	ASTM D1044	H 18	1000		59			451
	EniChem Pibiflex PDB53MWR	hydrolysis resistant, blow molding grade, powder grade, 54 Shore D	ASTM D1044	H 18	1000		75			451
	EniChem Pibiflex PDE56E	extrusion grade, powder grade, 53 Shore D	ASTM D1044	H 18	1000		75			451

Test Name	Material Supplier/ Grade	Material Note	Test Conditions							Source
			Test Method	Wheel / Abrader Type	Load (g)	Test Note	Taber Abrasion (mg/1000 cycles)	NBS Abrasion Index (%)	Test Result Note	

TPE, Polyester

Test Name	Material Supplier/ Grade	Material Note	Test Method	Wheel / Abrader Type	Load (g)	Test Note	Taber Abrasion (mg/1000 cycles)	NBS Abrasion Index (%)	Test Result Note	Source
Taber Abrasion	EniChem Pibiflex PDE56M	extrusion grade, 56 Shore D, powder grade	ASTM D1044	H 18	1000		73			451
	GE Lomod B0100	general purpose, 35 Shore D	ASTM D1044 modified	H 18	1000		120			431
	Hoechst Cel. Riteflex 640	40 Shore D	ASTM D1044	H 18	1000		90			333
	Hoechst Cel. Riteflex 647	47 Shore D	ASTM D1044	H 18	1000		67			333
	Hoechst Cel. Riteflex 655	55 Shore D	ASTM D1044	H 18	1000		85			333
	Hoechst Cel. Riteflex 663	63 Shore D	ASTM D1044	H 18	1000		62			333
	Hoechst Cel. Riteflex 672	72 Shore D	ASTM D1044	H 18	1000		30			333
	Hoechst Cel. Riteflex BP8929	68 Shore D	ASTM D1044	H 18	1000		150			334
	Hoechst Cel. Riteflex BP9056	60 Shore D	ASTM D1044	H 18	1000		205			334
	Hoechst Cel. Riteflex BP9057	50 Shore D	ASTM D1044	H 18	1000		250			334
	Hoechst Cel. Riteflex BP9086	62 Shore D	ASTM D1044	H 18	1000		190			334

TPE, Urethane

Test Name	Material Supplier/ Grade	Material Note	Test Method	Wheel / Abrader Type	Load (g)	Test Note	Taber Abrasion (mg/1000 cycles)	NBS Abrasion Index (%)	Test Result Note	Source
Taber Abrasion	Dow Pellethane 2354-45DGA	46 Shore D, automotive grade	ASTM D1044	H 22	1000		10			254
	Dow Pellethane 2354-45DGA	56 Shore D, automotive grade	ASTM D1044	H 22	1000		40			254

TPE, Urethane (TPAU)

Test Name	Material Supplier/ Grade	Material Note	Test Method	Wheel / Abrader Type	Load (g)	Test Note	Taber Abrasion (mg/1000 cycles)	NBS Abrasion Index (%)	Test Result Note	Source
NBS Abrasion Index	Bayer Texin 355D	55 Shore D	ASTM D1630					1975		427
	Bayer Texin 480A	high heat grade, flame retardant, 123 Rockwell R, 83 Shore A, 100 Rockwell M; 30% glass fiber	ASTM D1630					190		427
	Bayer Texin 591A	91 Shore A	ASTM D1630					4600		427
Taber Abrasion	Bayer Texin 355D	55 Shore D	ASTM C501	CS 17	1000	5000 cycles	.64			427
	Bayer Texin 480A	high heat grade, flame retardant, 123 Rockwell R, 83 Shore A, 100 Rockwell M; 30% glass fiber	ASTM C501	CS 17	1000	5000 cycles	.54			427
	Bayer Texin 591A	91 Shore A	ASTM C501	CS 17	1000	5000 cycles	.08			427
	Dow Pellethane 2355-95AE	94 Shore A, polyester polyadipate	ASTM D1044	CS 17	1000		4			254
	Dow Pellethane 2355-95AESP	94 Shore A, polyester polyadipate	ASTM D1044	CS 17	1000		4			254
	Bayer Texin 355D	55 Shore D	ASTM C501	H 18	1000		55			427
	Bayer Texin 480A	high heat grade, flame retardant, 123 Rockwell R, 83 Shore A, 100 Rockwell M; 30% glass fiber	ASTM C501	H 18	1000		49			427
	Bayer Texin 591A	91 Shore A	ASTM C501	H 18	1000		39			427
	Dow Pellethane 2355-75A	83 Shore A, polyester polyadipate	ASTM D1044	H 18	1000		28			254
	Dow Pellethane 2102-55D	58 ShoreD, polyester polycaprolactone	ASTM D1044	H 22	1000		20			254
	Dow Pellethane 2102-65D	65 Shore D, polyester polycaprolactone	ASTM D1044	H 22	1000		20			254
	Dow Pellethane 2102-75A	77 Shore A, polyester polycaprolactone	ASTM D1044	H 22	1000		20			254
	Dow Pellethane 2102-80A	83 Shore A, polyester polycaprolactone	ASTM D1044	H 22	1000		20			254
	Dow Pellethane 2102-85A	88 Shore A, polyester polycaprolactone, UV stable	ASTM D1044	H 22	1000		20			254

Test Name	Material Supplier/ Grade	Material Note	Test Conditions							Source
			Test Method	Wheel / Abrader Type	Load (g)	Test Note	Taber Abrasion (mg/1000 cycles)	NBS Abrasion Index (%)	Test Result Note	
TPE, Urethane (TPAU)										
Taber Abrasion	Dow Pellethane 2102-90A	93 Shore A, polyester polycaprolactone	ASTM D1044	H 22	1000		20			254
	Dow Pellethane 2102-90AE	93 Shore A, polyester polycaprolactone	ASTM D1044	H 22	1000		20			254
	Dow Pellethane 2102-90AR	94 Shore A, polyester polycaprolactone	ASTM D1044	H 22	1000		20			254
	Dow Pellethane 2355-80AE	83 Shore A, polyester polyadipate	ASTM D1044	H 22	1000		10			254
	Dow Pellethane 2355-85ABR	87 Shore A, polyester polyadipate, FDA	ASTM D1044	H 22	1000		15			254
TPE, Urethane (TPEU)										
Taber Abrasion	Dow Pellethane 2103-80PF	83 Shore A, polytetramethylene glycol ether	ASTM D1044	CS 17	1000		5			254
	Dow Pellethane 2103-55D	55 Shore D, 96Shore A, polytetramethylene glycol ether	ASTM D1044	H 22	1000		80			254
	Dow Pellethane 2103-65D	63 Shore D, polytetramethylene glycol ether	ASTM D1044	H 22	1000		90			254
	Dow Pellethane 2103-70A	72 Shore A, polytetramethylene glycol ether	ASTM D1044	H 22	1000		3			254
	Dow Pellethane 2103-80AE	82 Shore A, polytetramethylene glycol ether	ASTM D1044	H 22	1000		20			254
	Dow Pellethane 2103-80AEF	83 Shore A, polytetramethylene glycol ether	ASTM D1044	H 22	1000		20			254
	Dow Pellethane 2103-80AEN	83 Shore A, polytetramethylene glycol ether	ASTM D1044	H 22	1000		20			254
	Dow Pellethane 2103-85AE	88 Shore A, polytetramethylene glycol ether	ASTM D1044	H 22	1000		5			254
	Dow Pellethane 2103-90A	48 Shore D, 92 Shore A, polytetramethylene glycol ether	ASTM D1044	H 22	1000		10			254
	Dow Pellethane 2103-90AE	91 Shore A, polytetramethylene glycol ether	ASTM D1044	H 22	1000		50			254
	Dow Pellethane 2103-90AEF	90 Shore A, 47 Shore D, polytetramethylene glycol ether	ASTM D1044	H 22	1000		50			254
	Dow Pellethane 2103-90AS	91 Shore A, polytetramethylene glycol ether	ASTM D1044	H 22	1000		10			254
TPE, Styrenic										
DIN 53516	Evode Evoprene Super G 931	60 Shore A	DIN 53516						307 mm^3 volume loss	160
	Evode Evoprene Super G 932	70 Shore A	DIN 53516						288 mm^3 volume loss	160
	Evode Evoprene Super G 946	26 Shore A	DIN 53516						1055 mm^3 volume loss	160
	Evode Evoprene Super G 947	37 Shore A	DIN 53516						435 mm^3 volume loss	160
	Evode Evoprene Super G 948	44 Shore A	DIN 53516						500 mm^3 volume loss	160
	Evode Evoprene Super G 949	54 Shore A	DIN 53516						310 mm^3 volume loss	160

Test Name	Material Supplier/ Grade	Material Note	Test Conditions							Source
			Test Method	Wheel / Abrader Type	Load (g)	Test Note	Taber Abrasion (mg/1000 cycles)	NBS Abrasion Index (%)	Test Result Note	
TPE, Styrenic										
Taber Abrasion	J-Von Hercuprene 1000-35	SBS, 35 Shore A	ASTM D1044	H 18	1000		650			446
	J-Von Hercuprene 1000-42	SBS, 42 Shore A	ASTM D1044	H 18	1000		620			446
	J-Von Hercuprene 1000-54	SBS, 54 Shore A	ASTM D1044	H 18	1000		450			446
	J-Von Hercuprene 1000-64	SBS, 64 Shore A	ASTM D1044	H 18	1000		400			446
	J-Von Hercuprene 1000-71	SBS, 71 Shore A	ASTM D1044	H 18	1000		375			446
	J-Von Hercuprene 1000-80	SBS, 80 Shore A	ASTM D1044	H 18	1000		325			446
	J-Von Hercuprene 1210-33	SBS, 33 Shore A, transparent	ASTM D1044	H 18	1000		400			446
	J-Von Hercuprene 1221-46	SBS, 46 Shore A, FDA grade	ASTM D1044	H 18	1000		480			446
	J-Von Hercuprene 1241-54	SBS, 54 Shore A, FDA grade	ASTM D1044	H 18	1000		480			446
	J-Von Hercuprene 3000-33	SEBS, 33 Shore A	ASTM D1044	H 18	1000		230			446
	J-Von Hercuprene 3000-45	SEBS, 45 Shore A	ASTM D1044	H 18	1000		80			446
	J-Von Hercuprene 3000-60	SEBS, 60 Shore A	ASTM D1044	H 18	1000		0			446
	J-Von Hercuprene 3000-70	SEBS, 70 Shore A	ASTM D1044	H 18	1000		0			446
	J-Von Hercuprene 3000-83	SEBS, 83 Shore A	ASTM D1044	H 18	1000		0			446
	J-Von Hercuprene 3001-33	SEBS, 33 Shore A, FDA grade	ASTM D1044	H 18	1000		250			446
	J-Von Hercuprene 3001-45	SEBS, 45 Shore A, FDA grade	ASTM D1044	H 18	1000		80			446
	J-Von Hercuprene 3001-60	SEBS, 60 Shore A, FDA grade	ASTM D1044	H 18	1000		0			446
	J-Von Hercuprene 3001-70	SEBS, 70 Shore A, FDA grade	ASTM D1044	H 18	1000		0			446
	J-Von Hercuprene 3001-83	SEBS, 83 Shore A, FDA grade	ASTM D1044	H 18	1000		0			446
	J-Von Hercuprene 3100-40	SEBS, 40 Shore A	ASTM D1044	H 18	1000		1600			446
	J-Von Hercuprene 3100-44	SEBS, 44 Shore A	ASTM D1044	H 18	1000		1535			446
	J-Von Hercuprene 3100-52	SEBS, 52 Shore A	ASTM D1044	H 18	1000		1400			446
	J-Von Hercuprene 3100-60	SEBS, 60 Shore A	ASTM D1044	H 18	1000		850			446
	J-Von Hercuprene 3100-73	SEBS, 73 Shore A	ASTM D1044	H 18	1000		310			446
	J-Von Hercuprene 3100-82	SEBS, 82 Shore A	ASTM D1044	H 18	1000		160			446
	J-Von Hercuprene 3100-90	SEBS, 92 Shore A	ASTM D1044	H 18	1000		150			446
Rubber, Chlorosulf. PE (CSM)										
NBS Abrasion Index	DuPont Hypalon 20	30 min. @ 153°C cure; 67 Shore A; 100 phr Hypalon 20, 0.5 phr MBTS, 2 phr Tetrone A, 30 phr aromatic process oil, 55 phr SRF black, 27.8 phr sublimed litharge; sublimed litharge (90% active) used, SRF N762 carbon black was used, Sundex 790 process oil						175		349

Test Name	Material Supplier/ Grade	Material Note	Test Conditions							Source
			Test Method	Wheel / Abrader Type	Load (g)	Test Note	Taber Abrasion (mg/1000 cycles)	NBS Abrasion Index (%)	Test Result Note	

Rubber, Chlorosulf. PE (CSM)

Test Name	Material Supplier/ Grade	Material Note	Test Method	Wheel / Abrader Type	Load (g)	Test Note	Taber Abrasion (mg/1000 cycles)	NBS Abrasion Index (%)	Test Result Note	Source
NBS Abrasion Index	DuPont Hypalon 20	30 min. @ 153°C cure; 63 Shore A; 100 phr Hypalon 20, 3 phr Pentaerythritol 200, 2 phr Tetrone A, 4 phr magnesia, 80 phr hard clay, 30 phr aromatic process oil; Suprex Clay, Sundex 790 process oil, Hercules PE-200 pentaerythritol used, Maglite D magnesia used						70		349
	DuPont Hypalon 30	30 min. @ 153°C cure; 89 Shore A; 0.5 phr MBTS, 2 phr Tetrone A, 27.8 phr sublimed litharge, 30 phr aromatic process oil, 100 phr Hypalon 30, 55 phr SRF black; Sundex 790 process oil, SRF N762 carbon black was used, sublimed litharge (90% active) used						220		349
	DuPont Hypalon 30	30 min. @ 153°C cure; 82 Shore A; 3 phr Pentaerythritol 200, 2 phr Tetrone A, 4 phr magnesia, 80 phr hard clay, 30 phr aromatic process oil, 100 phr Hypalon 30; Suprex Clay, Maglite D magnesia used, Hercules PE-200 pentaerythritol used, Sundex 790 process oil						52		349
	DuPont Hypalon 40	30 min. @ 153°C cure; 67 Shore A; 0.5 phr MBTS, 100 phr Hypalon 40, 2 phr Tetrone A, 27.8 phr sublimed litharge, 55 phr SRF black, 30 phr aromatic process oil; Sundex 790 process oil, sublimed litharge (90% active) used, SRF N762 carbon black was used						380		349
	DuPont Hypalon 40	30 min. @ 153°C cure; 67 Shore A; 100 phr Hypalon 40, 3 phr Pentaerythritol 200, 2 phr Tetrone A, 4 phr magnesia, 80 phr hard clay, 30 phr aromatic process oil; Maglite D magnesia used, Suprex Clay, Sundex 790 process oil, Hercules PE-200 pentaerythrit						146		349
	DuPont Hypalon 4085	30 min. @ 153°C cure; 67 Shore A; 0.5 phr MBTS, 2 phr Tetrone A, 100 phr Hypalon 4085, 55 phr SRF black, 30 phr aromatic process oil, 27.8 phr sublimed litharge; sublimed litharge (90% active) used, SRF N762 carbon black was used, Sundex 790 process oi						530		349
	DuPont Hypalon 4085	30 min. @ 153°C cure; 69 Shore A; 3 phr Pentaerythritol 200, 2 phr Tetrone A, 4 phr magnesia, 80 phr hard clay, 100 phr Hypalon 4085, 30 phr aromatic process oil; Sundex 790 process oil, Maglite D magnesia used, Suprex Clay, Hercules PE-200 pentaerythritol used						203		349
	DuPont Hypalon 40S	30 min. @ 153°C cure; 66 Shore A; 0.5 phr MBTS, 2 phr Tetrone A, 100 phr Hypalon 40S, 27.8 phr sublimed litharge, 30 phr aromatic process oil, 55 phr SRF black; SRF N762 carbon black was used, Sundex 790 process oil, sublimed litharge (90% active) used						340		349

Rubber, Chlorosulf. PE (CSM)

Test Name	Material Supplier/ Grade	Material Note	Test Conditions							Source
			Test Method	Wheel / Abrader Type	Load (g)	Test Note	Taber Abrasion (mg/1000 cycles)	NBS Abrasion Index (%)	Test Result Note	
NBS Abrasion Index	DuPont Hypalon 40S	30 min. @ 153°C cure; 63 Shore A; 3 phr Pentaerythritol 200, 2 phr Tetrone A, 4 phr magnesia, 80 phr hard clay, 30 phr aromatic process oil, 100 phr Hypalon 40S; Hercules PE-200 pentaerythritol used, Suprex Clay, Maglite D magnesia used, Sundex 790 process oil						108		349
	DuPont Hypalon 45	30 min. @ 153°C cure; 80 Shore A; 100 phr Hypalon 45, 0.5 phr MBTS, 2 phr Tetrone A, 55 phr SRF black, 30 phr aromatic process oil, 27.8 phr sublimed litharge; Sundex 790 process oil, sublimed litharge (90% active) used, SRF N762 carbon black was used						310		349
	DuPont Hypalon 45	30 min. @ 153°C cure; 81 Shore A; 100 phr Hypalon 45, 3 phr Pentaerythritol 200, 2 phr Tetrone A, 4 phr magnesia, 80 phr hard clay, 30 phr aromatic process oil; Hercules PE-200 pentaerythritol used, Sundex 790 process oil, Maglite D magnesia used						70		349
	DuPont Hypalon 48	30 min. @ 153°C cure; 77 Shore A; 0.5 phr MBTS, 2 phr Tetrone A, 27.8 phr sublimed litharge, 30 phr aromatic process oil, 55 phr SRF black, 100 phr Hypalon 48; Sundex 790 process oil, sublimed litharge (90% active) used, SRF N762 carbon black was used						220		349
	DuPont Hypalon 48	30 min. @ 153°C cure; 72 Shore A; 3 phr Pentaerythritol 200, 2 phr Tetrone A, 4 phr magnesia, 80 phr hard clay, 100 phr Hypalon 48, 30 phr aromatic process oil; Maglite D magnesia used, Suprex Clay, Hercules PE-200 pentaerythritol used, Sundex 790 process oil						73		349
	DuPont Hypalon 610	30 min. @ 153°C cure; 70 Shore A; 0.5 phr MBTS, 2 phr Tetrone A, 30 phr aromatic process oil, 27.8 phr sublimed litharge, 55 phr SRF black, 100 phr Hypalon 610; SRF N762 carbon black was used, sublimed litharge (90% active) used, Sundex 790 process oil						390		349
	DuPont Hypalon 610	30 min. @ 153°C cure; 70 Shore A; 3 phr Pentaerythritol 200, 2 phr Tetrone A, 4 phr magnesia, 80 phr hard clay, 30 phr aromatic process oil, 100 phr Hypalon 610; Hercules PE-200 pentaerythritol used, Suprex Clay, Maglite D magnesia used, Sundex 790 process oil						172		349
	DuPont Hypalon 623	30 min. @ 153°C cure; 80 Shore A; 0.5 phr MBTS, 2 phr Tetrone A, 30 phr aromatic process oil, 27.8 phr sublimed litharge, 100 phr Hypalon 623, 55 phr SRF black; sublimed litharge (90% active) used, Sundex 790 process oil, SRF N762 carbon black was used						230		349

Rubber, Chlorosulf. PE (CSM)

Test Name	Material Supplier/ Grade	Material Note	Test Conditions							Source
			Test Method	Wheel / Abrader Type	Load (g)	Test Note	Taber Abrasion (mg/1000 cycles)	NBS Abrasion Index (%)	Test Result Note	
NBS Abrasion Index	DuPont Hypalon 623	30 min. @ 153°C cure; 75 Shore A; 3 phr Pentaerythritol 200, 2 phr Tetrone A, 4 phr magnesia, 80 phr hard clay, 30 phr aromatic process oil, 100 phr Hypalon 623; Maglite D magnesia used, Sundex 790 process oil, Hercules PE-200 pentaerythritol used, Suprex Clay						60		349
	DuPont Hypalon 6525	30 min. @ 153°C cure; 65 Shore A; 0.5 phr MBTS, 2 phr Tetrone A, 27.8 phr sublimed litharge, 100 phr Hypalon 6525, 55 phr SRF black, 30 phr aromatic process oil; Sundex 790 process oil, SRF N762 carbon black was used, sublimed litharge (90% active) used						270		349
	DuPont Hypalon 6525	30 min. @ 153°C cure; 70 Shore A; 3 phr Pentaerythritol 200, 2 phr Tetrone A, 4 phr magnesia, 80 phr hard clay, 30 phr aromatic process oil, 100 phr Hypalon 6525; Suprex Clay, Sundex 790 process oil, Maglite D magnesia used, Hercules PE-200 pentaerythritol used						247		349
	DuPont Hypalon LD-999	30 min. @ 153°C cure; 66 Shore A; 0.5 phr MBTS, 2 phr Tetrone A, 100 phr Hypalon LD-999, 27.8 phr sublimed litharge, 55 phr SRF black, 30 phr aromatic process oil; sublimed litharge (90% active) used, SRF N762 carbon black was used, Sundex 790 process oil						260		349
	DuPont Hypalon LD-999	30 min. @ 153°C cure; 63 Shore A; 3 phr Pentaerythritol 200, 2 phr Tetrone A, 4 phr magnesia, 80 phr hard clay, 30 phr aromatic process oil, 100 phr Hypalon LD-999; Suprex Clay, Sundex 790 process oil, Maglite D magnesia used, Hercules PE-200 pentaerythritol used						93		349
Taber Abrasion	DuPont Acsium HPR6367	69 Shore A; 5 phr magnesia, 1 phr MBTS, 1 phr TMTD, 1 phr NBC, 0.5 phr sulfur, 55 phr N-774 black, 5 phr T(HRL) D-90, 20 phr Sundex 790; , 68 Mooney (ML 1+4 @ 100°C), 3 phr Carbowax 3350, 100 phr Acsium HPR6367	ASTM D1044		1000		.862			358
	DuPont Acsium HPR6367	76 DIN hardness; 3 phr Pentaerythritol 200, 4 phr magnesia, 1 phr MBTS, 0.5 phr sulfur, 35 phr N762 SRF black, 1 phr HVA2, 1 phr TMTD, 2 phr paraffin, 3 phr low MW PE; , 100 phr Acsium HPR6367, 70 Mooney (ML 1+4 @ 100°C)	ASTM D1044		1000		65.2			358
	DuPont Acsium HPR6367	80 DIN hardness; 3 phr Pentaerythritol 200, 4 phr magnesia, 1 phr MBTS, 70 phr N990 MT black, 1 phr TMTD, 0.5 phr sulfur, 3 phr low MW PE, 1 phr HVA2, 2 phr paraffin; , 100 phr Acsium HPR6367, 78 Mooney (ML 1+4 @ 100°C)	ASTM D1044		1000		120.8			358

Test Name	Material Supplier/ Grade	Material Note	Test Conditions							Source
			Test Method	Wheel / Abrader Type	Load (g)	Test Note	Taber Abrasion (mg/1000 cycles)	NBS Abrasion Index (%)	Test Result Note	
Rubber, Chlorosulf. PE (CSM)										
Taber Abrasion	DuPont Acsium HPR6367	67 DIN hardness; 3 phr Pentaerythritol 200, 4 phr magnesia, 1 phr MBTS, 1 phr HVA2, 1 phr TMTD, 0.5 phr sulfur, 25 phr N330 HAF black, 2 phr paraffin, 3 phr low MW PE; , 67 Mooney (ML 1+4 @ 100°C), 100 phr Acsium HPR6367	ASTM D1044		1000		60.6			358
	DuPont Acsium HPR6367	63 DIN hardness; 3 phr Pentaerythritol 200, 4 phr magnesia, 1 phr MBTS, 25 phr N220 ISAF black, 3 phr low MW PE, 0.5 phr sulfur, 1 phr TMTD, 2 phr paraffin, 1 phr HVA2; , 63 Mooney (ML 1+4 @ 100°C), 100 phr Acsium HPR6367	ASTM D1044		1000		59.1			358
	DuPont Acsium HPR6367	77 DIN hardness; 3 phr Pentaerythritol 200, 4 phr magnesia, 1 phr MBTS, 3 phr low MW PE, 2 phr paraffin, 1 phr TMTD, 0.5 phr sulfur, 30 phr N660 GPF black, 1 phr HVA2; , 71 Mooney (ML 1+4 @ 100°C), 100 phr Acsium HPR6367	ASTM D1044		1000		79.1			358
	DuPont Acsium HPR6367	76 DIN hardness, 77 DIN hardness; 3 phr Pentaerythritol 200, 4 phr magnesia, 1 phr MBTS, 3 phr low MW PE, 30 phr N550 FEF black, 0.5 phr sulfur, 1 phr TMTD, 1 phr HVA2, 2 phr paraffin; , 74 Mooney (ML 1+4 @ 100°C), 100 phr Acsium HPR6367	ASTM D1044		1000		98.1			358
	DuPont Hypalon 40	70 Shore A, 69 Shore A; 100 phr Hypalon 40, 5 phr magnesia, 1 phr MBTS, 1 phr NBC, 1 phr TMTD, 55 phr N-774 black, 20 phr Sundex 790, 5 phr T(HRL) D-90, 0.5 phr sulfur; , 3 phr Carbowax 3350, 63 Mooney (ML 1+4 @ 100°C)	ASTM D1044		1000		1.011			358
Fluoroelastomer, FFKM										
NBS Abrasion Index	DuPont Kalrez 1018	steam resistant, hot water resistant, 76-86 Shore A; w/ carb. bl.	ASTM D1630					121		425
Rubber, Latex (NR)										
Taber Abrasion	Unspecified grade	tread formulation	ASTM C501	CS 17	1000	5000 cycles	29.2			427
Polyurethane										
NBS Abrasion Index	Unspecified grade	55 Shore A; cast	ASTM D1630					100		406
	Unspecified grade	65 Shore A; cast	ASTM D1630					120		406
	Unspecified grade	73 Shore A, 25 Shore D; cast	ASTM D1630					140		406
	Unspecified grade	80 Shore A, 30 Shore D; cast	ASTM D1630					150		406
	Unspecified grade	85 Shore A, 35 Shore D; cast	ASTM D1630					250		406
	Unspecified grade	85 Shore A, 35 Shore D; cast	ASTM D1630					160		406
	Unspecified grade	90 Shore A, 40 Shore D; cast	ASTM D1630					175		406
	Unspecified grade	50 Shore D, 95 Shore A; cast	ASTM D1630					300		406
	Unspecified grade	60 Shore D; cast	ASTM D1630					370		406

100 in/min. The volume loss in cm^3 is calculated after 1000 rotation or revolution cycles by measuring the weight of the specimen before and after the test and dividing the difference by its density. Also called ASTM D1242-56 (1981).

ASTM D1242-56 (1981) See *ASTM D1242.*

ASTM D1708 An American Society for Testing of Materials (ASTM) standard method for determining tensile properties of plastics using microtensile specimens with maximum thickness 3.2 mm and minimum length 38.1 mm, including thin films. Tensile properties include yield strength, tensile strength, tensile strength at break, elongation at break, etc. determined per ASTM D638.

ASTM D1709 An American Society for Testing of Materials (ASTM) standard test method for determining resistance of polyethylene film to impact by the free-falling dart. The impact resistance is measured as the energy that causes 50% failure rate of the film. The energy is calculated as the product of dart weight and dropping height. There are 2 test methods (A and B) using darts with different diameters of their hemispherical head and different dropping heights.

ASTM D1894 An American Society for Testing of Materials (ASTM) standard test method for determining coefficients of starting and sliding friction (static and kinetic coefficients, respectively) of plastic film and sheeting when sliding over itself or other substances under specified conditions.

ASTM D1922 An American Society for Testing of Materials (ASTM) standard test method for determining the resistance of flexible plastic film or sheeting to tear propagation. The resistance is measured as the average force, in grams, required to propagate tearing from a precut slit through a specified length, using an Elmendorf-type pendulum tester and 2 specimens, a rectangular type and one with a constant radius testing length.

ASTM D2240 An American Society for Testing of Materials (ASTM) standard method for determining the hardness of materials ranging from soft rubbers to some rigid plastics by measuring the penetration of a blunt (type A) or sharp (type D) indenter of a durometer at a specified force. The blunt indenter is used for softer materials and the sharp indenter - for more rigid materials.

ASTM D3702 An American Society for Testing of Materials (ASTM) standard test method for assessing abrasive wear rate of self-lubricating plastics intended for bearing applications in which wear debris essentially remain in the contact zone. The specimen of thrust washer configuration is mounted on a rotary upper holder of a thrust washer tester and pressed against a steel washer placed on a stationary lower holder. The specimen is rotated at a speed of 36-900 rpm under 1-200 lb load for a specified test duration. The wear rate is calculated as an average specimen thickness decrease rate in in/h. Also called thrust washer friction and wear test method.

ASTM D3841 An American Society for Testing of Materials (ASTM) standard specification for glass fiber-reinforced polyester construction panels. The specification covers classification, inspection, certification, dimensions, weight, appearance, light transmission, weatherability, expansion, impact resistance, flammability and load-deflection properties of panels and their methods of testing.

ASTM D4637 An American Society for Testing of Materials (ASTM) standard specification for unreinforced or fabric-reinforced vulcanized rubber sheet made from EPDM or chloroprene rubber and used as single-ply roof membranes. The specification specifies grades, dimensions, mechanical properties, weatherability, resistance to ozone and heat aging, appearance, and test methods. The mechanical properties tested include tensile strength, set and elongation, seam strength, tear resistance and tearing strength. The exposure tests include water absorption. Also called ASTM DS D4637.

ASTM DS D4637 See *ASTM D4637.*

atm See *atmosphere.*

B

bar A metric unit of measurement of pressure equal to 1.0E+06 dynes/cm^2 or 1.0E+05 pascals. It has a dimension of unit of force per unit of area.Used to denote the pressure of gases, vapors and liquids.

bending properties See *flexural properties.*

bending strength See *flexural strength.*

bending stress See *flexural stress.*

bisphenol A diglycidyl ether The principal monomer used in the preparation of epoxy resins, comprising two benzene rings linked via isopropylidine bridge. Each ring is substituted with an epoxy group in the para position. Highly reactive. Polymerizes to form thermosetting epoxy resins. Also called diglycidyl ether of bisphenol A, DGEBA.

bisphenol A polyester A thermoset unsaturated polyester based on bisphenol A and fumaric acid.

blown film A plastic film produced by extrusion blowing, wherein an extruded plastic tube is continuously inflated by internal air pressure, cooled, collapsed by rolls and wound up. The thickness of the film is controlled by air pressure and rate of extrusion.

breaking elongation See *elongation.*

C

CA See *cellulose acetate.*

CAB See *cellulose acetate butyrate.*

carboxyl-terminated butadiene-acrylonitrile See *carboxyl-terminated butadiene-acrylonitrile copolymer.*

carboxyl-terminated butadiene-acrylonitrile copolymer Liquid nitrile rubber containing reactive carboxy end groups, used as a toughener or impact modifier in plastics. It cures at room or elevated temperatures to solid elastomeric block copolymers with of matrix resins. It improves crack resistance of epoxy and polyester compositions and enhances fatigue resistance of glass fiber-reinforced plastics. It has good electric and wetting properties. Used in epoxy coatings and structural adhesives, potting, encapsulation, and cable fillers. Can be processed by rotomolding. Also called CTBN, carboxyl-terminated butadiene-acrylonitrile.

cast film Film produced by pouring or spreading resin solution or melt over a suitable temporary substrate, followed by curing via solvent evaporation or melt cooling and removing the cured film from the substrate.

cellulose acetate Thermoplastic esters of cellulose with acetic acid. Have good toughness, gloss, clarity, processability, stiffness, hardness, and dielectric properties, but poor chemical, fire and water resistance and compressive strength. Processed by injection and blow molding and extrusion. Used for appliance cases, steering wheels, pens, handles, containers, eyeglass frames, brushes, and sheeting. Also called CA.

cellulose acetate butyrate Thermoplastic mixed esters of cellulose with acetic and butyric acids. Have good toughness, gloss, clarity, processability, dimensional stability, weatherability, and dielectric properties, but poor chemical, fire and water resistance and compressive strength. Processed by injection and blow molding and extrusion. Used for appliance cases, steering wheels, pens, handles, containers, eyeglass frames, brushes, and sheeting. Also called CAB.

cellulose propionate Thermoplastic esters of cellulose with propionic acid. Have good toughness, gloss, clarity, processability, dimensional stability, weatherability, and dielectric properties, but poor chemical, fire and water resistance and compressive strength. Processed by injection and blow molding and extrusion. Used for appliance cases, steering wheels, pens, handles, containers, eyeglass frames, brushes, and sheeting. Also called CP.

cellulosic plastics Thermoplastic cellulose esters and ethers. Have good toughness, gloss, clarity, processability, and dielectric properties, but poor chemical, fire and water resistance and compressive strength. Processed by injection and blow molding and extrusion. Used for appliance cases, steering wheels, pens, handles, containers, eyeglass frames, brushes, and sheeting.

chlorendic polyester A chlorendic anhydride-based unsaturated polyester.

chlorinated polyvinyl chloride Thermoplastic produced by chlorination of polyvinyl chloride. Has increased glass transition temperature, chemical and fire resistance, rigidity, tensile strength, and weatherability as compared to PVC. Processed by extrusion, injection molding, casting, and calendering. Used for pipes, auto parts, waste disposal devices, and outdoor applications. Also called CPVC.

chlorohydrins Halohydrins with chlorine as a halogen atom. One of the most reactive of halohydrins. Dichlorohydrins are used in the preparation of epichlorohydrins, important monomers in the manufacture of epoxy resins. Most chlorohydrins are reactive colorless liquids, soluble in polar solvents such as alcohols. **Note:** Chlorohydrins are a class of organic compounds, not to be mixed with a specific member of this class, 1-chloropropane-2,3-diol sometimes called chlorohydrin.

chlorosulfonated polyethylene rubber Thermosetting elastomers containing 20- 40% chlorine. Have good weatherability and heat and chemical resistance. Used for hoses, tubes, sheets, footwear soles, and inflatable boats.

coefficient of friction See *kinetic coefficient of friction.*

coefficient of friction See *kinetic coefficient of friction.*

coefficient of friction, kinetic See *kinetic coefficient of friction.*

coefficient of friction, kinetic See *kinetic coefficient of friction.*

coefficient of friction, static See *static coefficient of friction.*

coefficient of friction, static See *static coefficient of friction.*

conditioning Process of bringing the material or apparatus to a certain condition, e.g., moisture content or temperature, prior to further processing, treatment, etc. Also called conditioning cycle.

conditioning cycle See *conditioning.*

covulcanization Simultaneous vulcanization of a blend of two or more different rubbers to enhance their individual properties such as ozone resistance. Rubbers are often modified to improve covulcanization.

CP See *cellulose propionate.*

crack driving force See *stress-intensity factor range.*

crack growth See *fatigue crack growth.*

crack growth rate See *fatigue crack growth rate.*

crack propagation See *fatigue crack growth.*

crack propagation rate See *fatigue crack growth rate.*

cracking Appearance of external and/or internal cracks in the material as a result of stress that exceeds the strength of the material. The stress can be external and/or internal and can be caused by a variety of adverse conditions: structural defects, impact, aging, corrosion, etc. or a combination of thereof. Also called cracks. See also *processing defects.*

cracks See *cracking.*

crazes See *crazing.*

crazing Appearance of thin cracks on the surface of the material or, sometimes, minute frost-like internal cracks, as a result of stress that exceeds the strength of the material, impact, temperature changes, degradation, etc. Also called crazes.

crystal polystyrene See *general purpose polystyrene.*

CTBN See *carboxyl-terminated butadiene-acrylonitrile copolymer.*

CTFE See *polychlorotrifluoroethylene.*

cycles to failure See *fatigue life.*

D

da/dN See *fatigue crack growth rate.*

DAP See *diallyl phthalate resins.*

degradation Loss or undesirable change in the properties, such as color, of a material as a result of aging, chemical reaction, wear, exposure, etc. See also *stability.*

delta E See *color difference.*

delta K See *stress-intensity factor range.*

DGEBA See *bisphenol A diglycidyl ether.*

diallyl phthalate resins Thermosets supplied as diallyl phthalate prepolymer or monomer. Have high chemical, heat and water resistance, dimensional stability, and strength. Shrink during peroxide curing. Processed by injection, compression and transfer molding. Used in glass-reinforced tubing, auto parts, and electrical components. Also called DAP.

diglycidyl ether of bisphenol A See *bisphenol A diglycidyl ether.*

DIN 53453 A German Standards Institute (DIN) standard specifying conditions for the flexural impact testing of molded or laminated plastics. The bar specimens are either unnotched or notched on one side, mounted on two-point support and struck in the middle (on the unnotched side for notched specimens) by a hammer of the pendulum impact machine. Impact strength of the specimen is calculated relative to the cross-sectional area of the specimen as the energy required to break the specimen equal to the difference between the energy in the pendulum at the instant of impact and the energy remaining after complete fracture of the specimen. Also called DIN 53453 impact test.

DIN 53453 impact test See *DIN 53453.*

DIN 53456 A German Standards Institute (Deutsches Institut fuer Normen, DIN) standard test method for determining ball indentation hardness of plastics. The indentor is forced into the specimen under the action of the major load, the position of the indentor having been fixed beforehand as a zero point by the application of a minor load. The hardness is calculated as the ratio of the major load to the area of indentation.

durometer A hardness See *Shore hardness.*

durometer hardness Indentation hardness of a material as determined by either the depth of an indentation made with an indentor under specified load or the indentor load required to produced specified indentation depth. The tool used to measure indentation hardness of polymeric materials is called durometer, e.g., Shore-type durometer.

E

ECTFE See *ethylene chlorotrifluoroethylene copolymer.*

Elmendorf tear strength The resistance of flexible plastic film or sheeting to tear propagation. It is measured, according to ASTM D1922, as the average force, in grams, required to propagate tearing from a pre-cut slit through a specified length, using an Elmendorf-type pendulum tester and 2 specimens, a rectangular type and one with a constant radius testing length.

elongation The increase in gauge length of a specimen in tension, measured at or after the fracture, depending on the viscoelastic properties of the material. **Note:** Elongation is usually expressed as a percentage of the original gauge length. Also called tensile elongation, elongation at break, ultimate elongation, breaking elongation, elongation at rupture. See also *tensile strain.*

elongation at break See *elongation.*

elongation at rupture See *elongation.*

EMAC See *ethylene methyl acrylate copolymer.*

embrittlement A reduction or loss of ductility or toughness in materials such as plastics resulting from chemical or physical damage.

endurance limit The maximum stress below which a material can endure an infinite number of loading-unloading cycles of specified type without failure or, in practice, a very large number of cycles. Also called fatigue endurance limit.

EPDM See *EPDM rubber.*

EPDM rubber Sulfur-vulcanizable thermosetting elastomers produced from ethylene, propylene, and a small amount of nonconjugated diene such as hexadiene. Have good weatherability and chemical and heat resistance. Used as impact modifiers and for weather stripping, auto parts, cable insulation, conveyor belts, hoses, and tubing. Also called EPDM.

epoxides Organic compounds containing three-membered cyclic group(s) in which two carbon atoms are linked with an oxygen atom as in an ether. This group is called an epoxy group and is quite reactive, allowing the use of epoxides as intermediates in preparation of certain fluorocarbons and cellulose derivatives and as monomers in preparation of epoxy resins. Also called epoxy compounds.

epoxies See *epoxy resins.*

epoxy compounds See *epoxides.*

epoxy resins Thermosetting polyethers containing crosslinkable glycidyl groups. Usually prepared by polymerization of bisphenol A and epichlorohydrin or reacting phenolic novolaks with epichlorohydrin. Can be made unsaturated by acrylation. Unmodified varieties are cured at room or elevated temperatures with polyamines or anhydrides. Bisphenol A epoxy resins have excellent adhesion and very low shrinkage during curing. Cured novolak epoxies have good UV stability and dielectric properties. Cured acrylated epoxies have high strength and chemical resistance. Processed by molding, casting, coating, and lamination. Used as protective coatings, adhesives, potting compounds, and binders in laminates and composites. Also called epoxies.

epoxyethane See *ethylene oxide.*

EPR See *ethylene propene rubber.*

ETFE See *ethylene tetrafluoroethylene copolymer.*

ethylene acrylic rubber Copolymers of ethylene and acrylic esters. Have good toughness, low temperature properties, and resistance to heat, oil, and water. Used in auto and heavy equipment parts.

ethylene copolymers See *ethylene polymers.*

ethylene methyl acrylate copolymer Thermoplastic copolymers of ethylene with <40% methyl acrylate. Have good dielectric properties, toughness, thermal stability, stress crack resistance, and compatibility with other polyolefins. Transparency decreases with increasing content of acrylate. Processed by blown film extrusion and blow and injection molding. Used in heat-sealable films, disposable gloves, and packaging. Some grades are FDA-approved for food packaging. Also called EMAC.

ethylene polymers Ethylene polymers include ethylene homopolymers and copolymers with other unsaturated monomers, most importantly olefins such as propylene and polar substances such as vinyl acetate. The properties and uses of ethylene polymers depend on the molecular structure and weight. Also called ethylene copolymers.

ethylene propene rubber Stereospecific copolymers of ethylene with propylene. Used as impact modifiers for plastics. Also called EPR.

ethylene tetrafluoroethylene copolymer Thermoplastic alternating copolymer of ethylene and tetrafluoroethylene. Has good impact strength, abrasion and chemical resistance, weatherability, and dielectric properties. Processed by molding, extrusion, and powder coating. Used in tubing, cables, pump parts, and tower packing in a wide temperature range. Also called ETFE.

ethylene vinyl alcohol copolymer Thermoplastics prepared by hydrolysis of ethylene-vinyl acetate polymers. Have good barrier properties, mechanical strength, gloss, elasticity, weatherability, clarity, and abrasion resistance. Barrier properties and processibility improve with increasing content of ethylene due to lower absorption of moisture. Ethylene content of high barrier grades range from 32 40 44 mole %. Processed by extrusion, coating, blow and blow film molding, and thermoforming. Used as packaging films and container liners. Also called EVOH.

ethylene-acrylic acid copolymer A flexible thermoplastic with water and chemical resistance and barrier properties similar to those of low-density polyethylene and enhanced adhesion, optics, toughness, and hot tack properties, compared to the latter. Contains 3-20% acrylic acid, with density and adhesion to polar substrates increasing with increasing acrylic acid content. FDA-approved for direct contact with food. Processed by extrusion, blow and film methods and extrusion molding, and extrusion coating. Used in rubberlike small parts like pipe caps, hoses, gaskets, gloves, hospital sheeting, diaper liners, and packaging film.

EVOH See *ethylene vinyl alcohol copolymer.*

extenders Relatively inexpensive resin, plasticizer or filler such as carbonate used to reduce cost and/or to improve processing of plastics, rubbers or nonmetallic coatings.

F

falling dart impact See *falling weight impact energy.*

falling dart impact energy See *dart impact energy.*

falling dart impact strength See *falling weight impact energy.*

falling sand abrasion test A test for determining abrasion resistance of coatings by the amount of abrasive sand required to wear through a unit thickness of the coating, when the sand falls against it at a specified angle from a specified height through a guide tube. Also called falling sand test method.

falling sand test method See *falling sand abrasion test.*

falling weight impact See *falling weight impact energy.*

falling weight impact energy The mean energy of a free-falling dart or weight (tup) that will cause 50% failures after 50 tests to a directly or indirectly stricken specimen. The energy is calculated by multiplying dart mass, gravitational acceleration and drop height. Also called falling weight impact strength, falling weight impact, falling dart impact energy, falling dart impact strength, falling dart impact, drop dart impact energy, drop dart impact strength.

falling weight impact strength See *falling weight impact energy.*

fatigue crack growth Crack extension caused by constant-amplitude fatigue loading of material specimen. The initial crack is often introduced by artificial notching or cutting. Also called fatigue crack propagation, crack growth, crack propagation.

fatigue crack growth rate The rate of crack extension caused by constant-amplitude fatigue loading, expressed in terms of crack extension per cycle. Also called crack propagation rate, da/dN, crack growth rate.

fatigue crack propagation See *fatigue crack growth.*

fatigue endurance limit See *endurance limit.*

fatigue life Number of loading-unloading cycles of a specified type that material specimen can endure before failing in a fatigue test. Also called cycles to failure.

Faville-LeValley Falex 6 See *thrust-washer testing machine.*

FEP See *fluorinated ethylene propylene copolymer.*

flexural fatigue Progressive localized permanent structural change occurring in a material subjected to cyclic flexural stress that may culminate in cracks or complete fracture after a sufficient number of cycles.

flexural properties Properties describing the reaction of physical systems to flexural stress and strain. Also called bending properties.

flexural strength The maximum stress in the extreme fiber of a specimen loaded to failure in bending. **Note:** Flexural strength is calculated as a function of load, support span and specimen geometry. Also called modulus of rupture in bending, modulus of rupture, bending strength.

flexural stress The maximum stress in the extreme fiber of a specimen in bending. **Note:** Flexural stress is calculated as a function of load at a given strain or at failure, support span and specimen geometry. Also called bending stress.

fluorinated ethylene propylene copolymer Thermoplastic copolymer of tetrafluoroethylene and hexafluoropropylene. Has decreased tensile strength and wear and creep resistance, but good weatherability, dielectric properties, fire and chemical resistance, and friction. Decomposes above 204°C (400°F), releasing toxic products. Processed by molding, extrusion, and powder coating. Used in chemical apparatus liners, pipes, containers, bearings, films, coatings, and cables. Also called FEP.

fluoro rubber See *fluoroelastomers.*

fluoroelastomers Fluorine-containing synthetic rubber with good chemical and heat resistance. Used in underhood applications such as fuel lines, oil and coolant seals, and fuel pumps, and as a flow additive for polyolefins. Also called fluoro rubber.

fluoroplastics See *fluoropolymers.*

fluoropolymers Polymers prepared from unsaturated fluorine-containing hydrocarbons. Have good chemical resistance, weatherability, thermal stability, antiadhesive properties and low friction and flammability, but low creep resistance and strength and poor processibility. The properties vary with the fluorine content. Processed by extrusion and molding. Used as liners in chemical apparatus, in bearings, films, coatings, and containers. Also called fluoroplastics.

fluorosilicones Polymers with chains of alternating silicon and oxygen atoms and trifluoropropyl pendant groups. Most are rubbers.

FMQ See *methylfluorosilicones.*

fractional melt index resin Thermoplastics having a low melt index of <1. These resins have higher molecular weights and are harder to extrude because of lower rate and greater force requirements compared to the lower molecular weight resins. They are mainly used for heavy duty applications such as pipe.

fracture mechanics A method of fracture analysis that can determine the stress required to induce fracture instability in a structure containing a crack of known size and shape. Also called linear elastic fracture mechanics.

G

general purpose polystyrene General purpose polystyrene is an amorphous thermoplastic prepared by homopolymerization of styrene. It has good tensile and flexural strengths, high light transmission and adequate resistance to water, detergents and inorganic chemicals. It is attached by hydrocarbons and has a relatively low impact resistance. Processed by injection molding and foam extrusion. Used to manufacture containers, health care items such as pipettes, kitchen and bathroom housewares, stereo and camera parts and foam sheets for food packaging. Also called crystal polystyrene.

glycol modified polycyclohexylenedimethylene terephthalate Thermoplastic polyester prepared from glycol, cyclohexylenedimethanol, and terephthalic acid. Has good impact strength and other mechanical properties, chemical resistance, and clarity. Processed by injection molding and extrusion. Can be blended with polycarbonate. Also called PCTG.

Graves tear strength A force required to tear completely across a specially designed nicked rubber test specimen, or right-angled test specimen, by elongating it at a specified rate using a power-driven tensile testing machine (Graves machine) as described in the ASTM D624. Expressed in units of force per thickness of specimen.

H

halogen compounds A class of organic compounds containing halogen atoms such as chlorine. A simple example is halocarbons but many other subclasses with various functional groups and of different molecular structure exist as well.

halohydrins Halogen compounds that contain a halogen atom(s) and a hydroxy (OH) group(s) attached to a carbon chain or ring. Can be prepared by reaction of halogens with alkenes in the presence of water or by reaction of halogens with triols. Halohydrins can be easily dehydrochlorinated in the presence of a base to give an epoxy compound.

HDPE See *high density polyethylene.*

HDT See *heat deflection temperature.*

heat deflection point See *heat deflection temperature.*

heat deflection temperature The temperature at which a material specimen (standard bar) is deflected by a certain degree under specified load. Also called heat distortion temperature, heat distortion point, heat deflection point, deflection temperature under load, tensile heat distortion temperature, HDT.

heat distortion point See *heat deflection temperature.*

heat distortion temperature See *heat deflection temperature.*

heterocyclic See *heterocyclic compounds.*

heterocyclic compounds A class of cyclic compounds containing rings with some carbon atoms replaced by other atoms such as oxygen, sulfur and nitrogen. Also called heterocyclic.

high density polyethylene A linear polyethylene with density 0.94-0.97 g/cm^3. Has good toughness at low temperatures, chemical resistance, and dielectric properties and high softening temperature, but poor weatherability. Processed by extrusion, blow and injection molding, and powder coating. Used in houseware, containers, food packaging, liners, cable insulation, pipes, bottles, and toys. Also called HDPE.

high impact polystyrene See *impact polystyrene.*

high molecular weight low density polyethylene Thermoplastic with improved abrasion and stress crack resistance and impact strength, but poor processibility and reduced tensile strength. Also called HMWLDPE.

HIPS See *impact polystyrene.*

HMWLDPE See *high molecular weight low density polyethylene.*

I

impact energy The energy required to break a specimen, equal to the difference between the energy in the striking member of the impact apparatus at the instant of impact and the energy remaining after complete fracture of the specimen. Also called impact strength. See also *ASTM D256, ASTM D3763.*

impact polystyrene Impact polystyrene is a thermoplastic produced by polymerizing styrene dissolved in butadiene rubber. Impact polystyrene has good dimensional stability, high rigidity and good low temperature impact strength, but poor barrier properties, grease resistance and heat resistance. Processed by extrusion, injection molding, thermoforming and structural foam molding. Used in food packaging, kitchen housewares, toys, small appliances, personal care items and audio products. Also called IPS, high impact polystyrene, HIPS, impact PS.

impact property tests Names and designations of the methods for impact testing of materials. Also called impact tests. See also *impact toughness.*

impact PS See *impact polystyrene.*

impact strength See *impact energy.*

impact tests See *impact property tests.*

impact toughness Property of a material indicating its ability to absorb energy of a high-speed impact by plastic deformation rather than crack or fracture. See also *impact property tests.*

initial tear resistance The force required to initiate tearing of a flexible plastic film or thin sheeting at very low rates of loading, measured as maximum stress usually found at the onset of tearing. Also called tear resistance, initial.

ionomers Thermoplastics containing a relatively small amount of pendant ionized acid groups. Have good flexibility and impact strength in a wide temperature range, puncture and chemical resistance, adhesion, and dielectric properties, but poor weatherability, fire resistance, and thermal stability. Processed by injection, blow and rotational molding, blown film extrusion, and extrusion coating. Used in food packaging, auto bumpers, sporting goods, and foam sheets.

IPS See *impact polystyrene.*

ISO 2039-2 An International Organization for Standardization (ISO) standard test method for determination of indentation hardness of plastics by Rockwell tester using Rockwell M, L, and R hardness scales. The hardness number is derived from the net increase in the depth of impression as the load on a ball indenter is increased from a fixed minor load (98.07 N) to a major load and then returned to the minor load. This number consists of the number of scale divisions (each corresponding to 0.002 mm vertical movement of the indentor) and scale symbol. Rockwell scale vary depending on the diameter of the indentor and the major load. For example, scale R corresponds to the ball diameter 12.7 mm and major load 588.4 N. Also called ISO 2039-B.

ISO 2039-B See *ISO 2039-2.*

isophthalate polyester An unsaturated polyester based on isophthalic acid.

Izod See *Izod impact energy.*

Izod impact See *Izod impact energy.*

Izod impact energy The energy required to break a specimen equal to the difference between the energy in the striking member of the Izod-type impact apparatus at the instant of impact and the energy remaining after complete fracture of the specimen. Also called Izod impact, Izod impact strength, Izod.

Izod impact strength See *Izod impact energy.*

J

J See *joule.*

JIS P8116 A Japanese Standards Association (Nippon Kikaku Kyokai) standard test for determining the resistance of flexible plastic film or sheeting to tear propagation. The resistance is measured as the average force, in grams, required to propagate tearing from a precut slit through a specified length, using an Elmendorf-type pendulum tester.

joule A unit of energy in SI system that is equal to the work done when the point of application of a force of one newton (N) is displaced through distance of one meter (m) in the direction of the force. The dimension of joule is N m. Also called J.

K

K See *wear factor.*

K factor See *wear factor.*

kinetic coefficient of friction The ratio of tangential force, which is required to sustain motion without acceleration of one surface with respect to another, to the normal force, which presses the two surfaces together. Also called coefficient of friction, coefficient of friction, kinetic.

kinetic strip test An ozone resistance test for rubbers that involves a strip-shaped specimen stretched to 23% and relax to 0 at a rate of 30 cycles per minute, while subjected to ozone attack in the test chamber. The results of the test are reported with 2 digits separated with a virgule. The number before the virgule indicates the number of quarters of the test strip which showed the cracks. The number after the virgule indicates the size of the cracks in length perpendicular to the length of the strip.

L

labile crack Crack in the process of growth or propagation. Unstable crack that can readily begin to grow as a result of internal or external processes in the material, such application of stress.

LCP See *liquid crystal polymers.*

LDPE See *low density polyethylene.*

limiting pressure-velocity value The value of the product between the load applied to the specimen normal to its friction surface and the sliding speed of this surface against a countersurface in wear testing of plastics at which friction-generated temperatures reach melting or softening points of a plastic or at which the wear rate begins to increase rapidly. Limiting PV is usually expressed in the unit of pressure or stress such as kg(f)/cm^2 times the unit of speed such as cm/s. Also called limiting PV, LPV, PV, PV limit.

limiting PV See *limiting pressure-velocity value.*

linear elastic fracture mechanics See *fracture mechanics.*

linear low density polyethylene Linear polyethylenes with density 0.91-0.94 g/cm^3. Has better tensile, tear, and impact strength and crack resistance properties, but poorer haze and gloss than branched low-density polyethylene. Processed by extrusion at increased pressure and higher melt temperatures compared to branched low-density polyethylene, and by molding. Used to manufacture film, sheet, pipe, electrical insulation, liners, bags and food wraps. Also called LLDPE, LLDPE resin.

linear polyethylenes Linear polyethylenes are polyolefins with linear carbon chains. They are prepared by copolymerization of ethylene with small amounts of higher alfa-olefins such as 1-butene. Linear polyethylenes are stiff, tough and have good resistance to environmental cracking and low temperatures. Processed by extrusion and molding. Used to manufacture film, bags, containers, liners, profiles and pipe.

liquid crystal polymers Thermoplastic aromatic copolyesters with highly ordered structure. Have good tensile and flexural properties at high temperatures, chemical, radiation and fire resistance, and weatherability. Processed by sintering and injection molding. Used to substitute ceramics and metals in electrical components, electronics, chemical apparatus, and aerospace and auto parts. Also called LCP.

LLDPE See *linear low density polyethylene.*

LLDPE resin See *linear low density polyethylene.*

low density polyethylene A branched-chain thermoplastic with density 0.91-0.94 g/cm^3. Has good impact strength, flexibility, transparency, chemical resistance, dielectric properties, and low water permeability and brittleness temperature, but poor heat, stress cracking and fire resistance and weatherability. Processed by extrusion coating, injection and blow molding, and film extrusion. Can be crosslinked. Used in packaging and shrink films, toys, bottle caps, cable insulation, and coatings. Also called LDPE.

LPV See *limiting pressure-velocity value.*

M

macroscopic properties See *thermodynamic properties.*

mechanical properties Properties describing the reaction of physical systems to stress and strain.

methylfluorosilicones Silicone rubbers containing pendant fluorine and methyl groups. Have good chemical and heat resistance. Used in gasoline lines, gaskets, and seals. Also called FMQ.

methylphenylsilicones Silicone rubbers containing pendant phenyl and methyl groups. Have good resistance to heat, oxidation, and radiation, and compatibility with plastics.

methylsilicone Silicone rubbers containing pendant methyl groups. Have good heat and oxidation resistance. Used in electrical insulation and coatings. Also called MQ.

methylvinylfluorosilicone Silicone rubbers containing pendant vinyl, methyl, and fluorine groups. Can be additionally crosslinked via vinyl groups. Have good resistance to petroleum products at elevated temperatures.

methylvinylsilicone Silicone rubbers containing pendant methyl and vinyl groups. Can be additionally crosslinked via vinyl groups. vulcanized to high degrees of crosslinking. Used in sealants, adhesives, coatings, cables, gaskets, tubing, and electrical tape.

micron A unit of length equal to 1E-6 meter. Its symbol is Greek small letter mu or mum.

microtensile specimen A small specimen as specified in ASTM D1708 for determining tensile properties of plastics. It has maximum thickness 3.2 mm and minimum length 38.1 mm. Tensile properties determined with this specimen include yield strength, tensile strength, tensile strength at break and elongation at break.

MIL-T-5438 See *Armstron abrasion test.*

modified polyphenylene ether Thermoplastic polyphenylene ether alloys with impact polystyrene. Have good impact strength, resistance to heat and fire, but poor resistance to solvents. Processed by injection and structural foam molding and extrusion. Used in auto parts, appliances, and telecommunication devices. Also called MPE, MPO, modified polyphenylene oxide.

modified polyphenylene oxide See *modified polyphenylene ether.*

modulus of rupture See *flexural strength.*

modulus of rupture in bending See *flexural strength.*

molecular weight The sum of the atomic weights of all atoms in a molecule. Also called MW.

molybdenum disulfide Molybdenum disulfide (MoS_2) is a crystalline filler used as an external lubricant in plastics such as polystyrene, nylons, and fluoropolymers to improve their wear resistance.

MPE See *modified polyphenylene ether.*

MPO See *modified polyphenylene ether.*

MQ See *methylsilicone.*

MW See *molecular weight.*

N

NBS Abrasion Index See *Abrasive Index.*

neoprene rubber Polychloroprene rubbers with good resistance to petroleum products, heat, and ozone, weatherability, and toughness.

nitrile rubber Rubbers prepared by free-radical polymerization of acrylonitrile with butadiene. Have good resistance to petroleum products, heat, and abrasion. Used in fuel hoses, shoe soles, gaskets, oil seals, and adhesives.

nonelastomeric thermoplastic polyurethanes See *rigid thermoplastic polyurethanes.*

nonelastomeric thermosetting polyurethane Curable mixtures of isocyanate prepolymers or monomers. Have good abrasion resistance and low-temperature stability, but poor heat, fire, and solvent resistance and weatherability. Processed by reaction injection and structural foam molding, casting, potting, encapsulation, and coating. Used in heat insulation, auto panels and trim, and housings for electronic devices.

notch effect The effect of the presence of specimen notch or its geometry on the outcome of a test such as an impact strength test of plastics. Notching results in local stresses and accelerates failure in both static and cycling testing (mechanical, ozone cracking, etc.).

notched Izod See *notched Izod impact energy.*

notched Izod impact See *notched Izod impact energy.*

notched Izod impact energy The energy required to break a notched specimen equal to the difference between the energy in the striking member of the Izod-type impact apparatus at the instant of impact and the energy remaining after complete fracture of the specimen. **Note:** Energy depends on geometry (e.g., width, depth, shape) of the notch, on the cross-sectional area of the specimen and on the place of impact (on the side of the notch or on the opposite side). In some tests notch is made on both sides of the specimen Also called notched Izod impact strength, notched Izod impact, notched Izod.

notched Izod impact strength See *notched Izod impact energy.*

nylon Thermoplastic polyamides often prepared by ring-opening polymerization of lactam. Have good resistance to most chemicals, abrasion, and creep, good impact and tensile strengths, barrier properties, and low friction, but poor resistance to moisture and light. Have high mold shrinkage. Processed by injection, blow, and rotational molding, extrusion, and powder coating. Used in fibers, auto parts, electrical devices, gears, pumps, appliance housings, cable jacketing, pipes, and films.

nylon 11 Thermoplastic polymer of 11-aminoundecanoic acid having good impact strength, hardness, abrasion resistance, processability, and dimensional stability. Processed by powder coating, rotational molding, extrusion, and injection molding. Used in electric insulation, tubing, profiles, bearings, and coatings.

nylon 12 Thermoplastic polymer of lauric lactam having good impact strength, hardness, abrasion resistance, and dimensional stability. Processed by powder coating, rotational molding, extrusion, and injection molding. Used in sporting goods and auto parts.

nylcn 46 Thermoplastic copolymer of 2-pyrrolidone and caprolactam.

nylon 6 Thermoplastic polymer of caprolactam. Has good weldability and mechanical properties but rapidly picks up moisture which results in strength losses. Processed by injection, blow, and rotational molding and extrusion. Used in fibers, tire cord, and machine parts.

nylon 610 Thermoplastic polymer of hexamethylenediamine and sebacic acid having decreased melting point and water absorption and good retention of mechanical properties. Processed by injection molding and extrusion. Used in fibers and machine parts.

nylon 612 Thermoplastic polymer of 1,12-dodecanedioic acid and hexamethylenediamine having good dimensional stability, low moisture absorption, and good retention of mechanical properties. Processed by injection molding and extrusion. Used in wire jackets, cable sheath, packaging film, fibers, bushings, and housings.

nylon 66 Thermoplastic polymer of adipic acid and hexamethylenediamine having good tensile strength, elasticity, toughness, heat resistance, abrasion resistance, and solvent resistance but low weatherability and color resistance. Processed by injection molding and extrusion. Used in fibers, bearings, gears, rollers, and wire jackets.

nylon 666 Thermoplastic polymer of adipic acid, caprolactam, and hexamethylenediamine having good strength, toughness, abrasion and fatigue resistance, and low friction but high moisture absorption and low dimensional stability. Processed by injection molding and extrusion. Used in electrical devices and auto and mechanical parts.

nylon MXD6 Thermoplastic polymer of m-xylyleneadipamide having good flexural strength and chemical resistance but decreased tensile strength.

olefin resins See *polyolefins.*

olefinic resins See *polyolefins.*

olefinic thermoplastic elastomers Blends of EPDM or EP rubbers with polypropylene or polyethylene, optionally crosslinked. Have low density, good dielectric and mechanical properties, and processibility but low oil resistance and high flammability. Processed by extrusion, injection and blow molding, thermoforming, and calendering. Used in auto parts, construction, wire jackets, and sporting goods. Also called TPO.

one-side notched specimen See *single-edge notched specimen.*

OPP See *oriented polypropylene.*

organic compounds See *halogen compounds.* Also called organic substances.

organic substances See *organic compounds.*

orientation A process of drawing or stretching of as-spun synthetic fibers or hot thermoplastic films to orient polymer molecules in the direction of stretching. The fibers are drawn uniaxially and the films are stretched either uniaxially or biaxially (usually longitudinally or longitudinally and transversely, respectively). Oriented fibers and films have enhanced mechanical properties. The films will shrink in the direction of stretching, when reheated to the temperature of stretching.

oriented polypropylene A grade of polypropylene film hot stretched uniaxially or biaxially (usually longitudinally or longitudinally and transversely, respectively) to orient polymer molecules in the direction of stretching. Oriented films have enhanced mechanical properties. They will shrink in the direction of stretching when reheated, e.g., during heat sealing. Also called OPP.

Pa See *pascal.*

PABM See *polyaminobismaleimide resins.*

paraffinic Containing, being derived from, or belonging to a class of liquid or solid long-chain alkanes or paraffins, like in paraffinic oils. The molecules of paraffins are linear or branched hydrocarbon chains ($-CH_2-$) that are fully saturated, i.e., contain no double or triple bonds between carbon atoms.

paraffinic plasticizer Plasticizers for plastics comprising liquid or solid long-chain alkanes or paraffins (saturated linear or branched hydrocarbons).

Paris plot A plot of fatigue crack propagation rate da/dN (crack extension per cycle) vs. stress-intensity factor range delta K.

parts per hundred A relative unit of concentration, parts of one substance per 100 parts of another. Parts can be measured by weight, volume, count or any other suitable unit of measure. Used often to denote composition of a blend or mixture, such as plastic, in terms of the parts of a minor ingredient, such as plasticizer, per 100 parts of a major, such as resin. Also called phr.

parts per hundred million A relative unit of concentration, parts of one substance per 100 million parts of another. Parts can be measured by weight, volume, count or any other suitable unit of measure. Used often to denote very small concentration of a substance, such as impurity or toxin, in a medium, such as air. Also called pphm.

pascal An SI unit of measurement of pressure equal to the pressure resulting from a force of one newton acting uniformly over an area of one square meter. Used to denote the pressure of gases, vapors or liquids and the strength of solids. Also called Pa.

PBI See *polybenzimidazoles.*

PBT See *polybutylene terephthalate.*

PC See *polycarbonates.*

PCT See *polycyclohexylenedimethylene terephthalate.*

PCTG See *glycol modified polycyclohexylenedimethylene terephthalate.*

PE copolymer See *polyethylene copolymer.*

PEEK See *polyetheretherketone.*

PEI See *polyetherimides.*

PEK See *polyetherketone.*

pentaerythritol A polyol, $C(CH_2OH)_4$, prepared by reaction of acetaldehyde with an excess formaldehyde in alkaline medium. Used as plasticizer and as monomer in alkyd resins.

perfluoroalkoxy resins Thermoplastic polymers of perfluoroalkoxyethylenes having good creep, heat, and chemical resistance and processibility but low compressive and tensile strengths. Processed by molding, extrusion, rotational molding, and powder coating. Used in films, coatings, pipes, containers, and chemical apparatus linings. Also called PFA.

PES See *polyethersulfone.*

PET See *polyethylene terephthalate.*

PETG See *polycyclohexylenedimethylene ethylene terephthalate.*

PFA See *perfluoroalkoxy resins.*

phase transition See *phase transition properties.*

phase transition point The temperature at which a phase transition occurs in a physical system such as material. **Note:** An example of phase transition is glass transition. Also called phase transition temperature, transition point, transition temperature.

phase transition properties Properties of physical systems such as materials associated with their transition from one phase to another, e.g., from liquid to solid phase. Also called phase transition.

phase transition temperature See *phase transition point.*

phenolic resins Thermoset polymers of phenols with excess or deficiency of aldehydes, mainly formaldehyde, to give resole or novolak resins, respectively. Heat-cured resins have good dielectric properties, hardness, thermal stability, rigidity, and compressive strength but poor chemical resistance and dark color. Processed by coating, potting, compression, transfer, or injection molding and extrusion. Used in coatings, adhesives, potting compounds, handles, electrical devices, and auto parts.

phr See *parts per hundred.*

PI See *polyimides.*

plasticizer A substance incorporated into a material such as plastic or rubber to increase its softness, processability and flexibility via solvent or lubricating action or by lowering its molecular weight. Plasticizers can lower melt viscosity, improve flow and increase low-temperature resilience of material. Most plasticizers are nonvolatile organic liquids or low-melting-point solids, such as dioctyl phthalate or stearic acid. They have to be non-bleeding, nontoxic and compatible with material. Sometimes plasticizers play a dual role as stabilizers or crosslinkers.

plastics See *polymers.*

PMMA See *polymethyl methacrylate.*

PMP See *polymethylpentene.*

polyacrylates See *acrylic resins.*

polyallomer Crystalline thermoplastic block copolymers of ethylene, propylene, and other olefins. Have good impact strength and flex life and low density.

polyamide thermoplastic elastomers Copolymers containing soft polyether and hard polyamide blocks having good chemical, abrasion, and heat resistance, impact strength, and tensile properties. Processed by extrusion and injection and blow molding. Used in sporting goods, auto parts, and electrical devices. Also called polyamide TPE.

polyamide TPE See *polyamide thermoplastic elastomers.*

polyamides Thermoplastic aromatic or aliphatic polymers of dicarboxylic acids and diamines, of amino acids, or of lactams. Have good mechanical properties, chemical resistance, and antifriction properties. Processed by extrusion and molding. Used in fibers and molded parts. Also called PA.

polyaminobismaleimide resins Thermoset polymers of aromatic diamines and bismaleimides having good flow and thermochemical properties and flame and radiation resistance. Processed by casting and compression molding. Used in aircraft parts and electrical devices. Also called PABM.

polyarylamides Thermoplastic crystalline polymers of aromatic diamines and aromatic dicarboxylic anhydrides having good heat, fire, and chemical resistance, property retention at high temperatures, dielectric and mechanical properties, and stiffness but poor light resistance and processibility. Processed by solution casting, molding, and extrusion. Used in films, fibers, and molded parts.

polyarylsulfone Thermoplastic aromatic polyether-polysulfone having good heat, fire, and chemical resistance, impact strength, resistance to environmental stress cracking, dielectric properties, and rigidity. Processed by injection and compression molding and extrusion. Used in circuit boards, lamp housings, piping, and auto parts.

polybenzimidazoles Mainly polymers of 3,3',4,4'-tetraminonbiphenyl (diaminobenzidine) and diphenyl isophthalate. Have good heat, fire, and chemical resistance. Used as coatings and fibers in aerospace and other high-temperature applications. Also called PBI.

polybutylene terephthalate Thermoplastic polymer of dimethyl terephthalate and butanediol having good tensile strength, dielectric properties, and chemical and water resistance, but poor impact strength and heat resistance. Processed by injection and blow molding, extrusion, and thermoforming. Used in auto body parts, electrical devices, appliances, and housings. Also called PBT.

polycarbodiimide Polymers containing -N=C=N- linkages in the main chain, typically formed by catalyzed polycondensation of polyisocyanates. They are used to prepare open-celled foams with superior thermal stability. Sterically hindered polycarbodiimides are used as hydrolytic stabilizers for polyester-based urethane elastomers.

polycarbonate See *polycarbonates.*

polycarbonate polyester alloys High-performance thermoplastics processed by injection and blow molding. Used in auto parts.

polycarbonate resins See *polycarbonates.*

polycarbonates Polycarbonates are thermoplastics prepared by either phosgenation of dihydric aromatic alcohols such as bisphenol A or by transesterification of these alcohols with carbonates, e.g., diphenyl carbonate. Polycarbonates consist of chains with repeating carbonyldioxy groups and can be aliphatic or aromatic. They have very good mechanical properties, especially impact strength, low moisture absorption and good thermal and oxidative stability. They are self-extinguishing and some grades are transparent. Polycarbonates have relatively low chemical resistance and resistance to stress cracking. Processed by injection and blow molding, extrusion, thermoforming at relatively high processing temperatures. Used in telephone parts, dentures, business machine housings, safety equipment, nonstaining dinnerware, food packaging, etc. Also called polycarbonate, PC, polycarbonate resins.

polychlorotrifluoroethylene Thermoplastic polymer of chlorotrifluoroethylene having good transparency, barrier properties, tensile strength, and creep resistance, modest dielectric properties and solvent resistance, and poor processibility. Processed by extrusion, injection and compression molding, and coating. Used in chemical apparatus, low-temperature seals, films, and internal lubricants. Also called CTFE.

polycyclohexylenedimethylene ethylene terephthalate Thermoplastic polymer of cyclohexylenedimethylenediol, ethylene glycol, and terephthalic acid. Has good clarity, stiffness, hardness, and low-temperature toughness. Processed by injection and blow molding and extrusion. Used in containers for cosmetics and foods, packaging film, medical devices, machine guards, and toys. Also called PETG.

polycyclohexylenedimethylene terephthalate Thermoplastic polymer of cyclohexylenedimethylenediol and terephthalic acid having good heat resistance. Processed by molding and extrusion. Also called PCT.

polyester resins See *polyesters.*

polyester thermoplastic elastomers Copolymers containing soft polyether and hard polyester blocks having good dielectric strength, chemical and creep resistance, dynamic performance, appearance, and retention of properties in a wide temperature range but poor light resistance. Processed by injection, blow, and rotational molding, extrusion casting, and film blowing. Used in electrical insulation, medical products, auto parts, and business equipment. Also called polyester TPE.

polyester TPE See *polyester thermoplastic elastomers.*

polyesters A broad class of polymers usually made by condensation of a diol with dicarboxylic acid or anhydride. Polyesters consist of chains with repeating carbonyloxy group and can be aliphatic or aromatic. There are thermosetting polyesters, such as alkyd resins and unsaturated polyesters, and thermoplastic polyesters such as PET. The properties, processing methods and applications of polyesters vary widely. Also called polyester resins.

polyetheretherketone Semi-crystalline thermoplastic aromatic polymer having good chemical, heat, fire, and radiation resistance, toughness, rigidity, bearing strength, and processibility. Processed by injection molding, spinning, cold forming, and extrusion. Used in fibers, films, auto engine parts, aerospace composites, and electrical insulation. Also called PEEK.

polyetherimides Thermoplastic cyclized polymers of aromatic diether dianhydrides and aromatic diamine. Have good chemical, creep, and heat resistance and dielectric properties. Processed by extrusion, thermoforming, and compression, injection, and blow molding. Used in auto parts, jet engines, surgical instruments, industrial apparatus, food packaging, cookware, and computer disks. Also called PEI.

polyetherketone Thermoplastic having good heat and chemical resistance. Thermal stability. Used in advanced composites, wire coating, filters, integrated circuit boards, and bearings. Also called PEK.

polyethersulfone Thermoplastic aromatic polymer having good heat and fire resistance, transparency, dielectric properties, dimensional stability, rigidity, and toughness, but poor solvent and stress cracking resistance, processibility, and weatherability. Processed by injection, blow, and compression molding and extrusion. Used in high temperature applications electrical devices, medical devices, housings, and aircraft and auto parts. Also called PES.

polyethylene copolymer Thermoplastics polymers of ethylene with other olefins such as propylene. Processed by molding and extrusion. Also called PE copolymer.

polyimides Thermoplastic aromatic cyclized polymers of trimellitic anhydride and aromatic diamine. Have good tensile strength, dimensional stability, dielectric and barrier properties, and creep, impact, heat, and fire resistance, but poor processibility. Processed by compression and injection molding, powder sintering, film casting, and solution coating. Thermoset uncyclized polymers are heat curable and have good processability. Processed by transfer and injection molding, lamination, and coating. Used in jet engines, compressors, sealing coatings, auto parts, and business machines. Also called PI.

polymer chain unsaturation See *chemical unsaturation.*

polymers Polymers are high-molecular-weight organic or inorganic compounds the molecules of which comprise linear, branched, crosslinked or otherwise shaped chains of repeating molecular groups. Synthetic polymers are prepared by polymerization of one or more monomers. The monomers are low-molecular-weight substances with one or more reactive bonds or functional groups. Also called resins, plastics.

polymethylpentene Thermoplastic stereoregular polyolefin obtained by polymerizing 4-methyl-1-pentene based on dimerization of propylene; having low density, good transparency, rigidity, dielectric and tensile properties, and heat and chemical resistance. Processed by injection and blow molding and extrusion. Used in laboratory ware, coated paper, light fixtures, auto parts, and electrical insulation. Also called PMP.

polyolefin resins See *polyolefins.*

polyolefins Polyolefins are a broad class of hydrocarbon-chain elastomers or thermoplastics usually prepared by addition (co)polymerization of alkenes such as ethylene. There are branched and linear polyolefins and some are chemically or physically modified. Unmodified polyolefins have relatively low thermal stability and a nonporous, nonpolar surface with poor adhesive properties. Processed by extrusion, injection molding, blow molding and rotational molding. Polyolefins are used more and have more applications than any other polymers. Also called olefinic resins, olefin resins, polyolefin resins.

polyphenylene ether nylon alloys Thermoplastics having improved heat and chemical resistance and toughness. Processed by molding and extrusion. Used in auto body parts.

polyphenylene sulfide High-performance engineering thermoplastic having good chemical, water, fire, and radiation resistance, dimensional stability, and dielectric properties, but decreased impact strength and poor processibility. Processed by injection, compression, and transfer molding and extrusion. Used in hydraulic components, bearings, electronic parts, appliances, and auto parts. Also called PPS.

polyphenylene sulfide sulfone Thermoplastic having good heat, fire, creep, and chemical resistance and dielectric properties. Processed by injection molding. Used in electrical devices. Also called PPSS.

polyphthalamide Thermoplastic polymer of aromatic diamine and phthalic anhydride. Has good heat, chemical, and fire resistance, impact strength, retention of properties at high temperatures, dielectric properties, and stiffness, but decreased light resistance and poor processibility. Processed by solution casting, molding, and extrusion. Used in films, fibers, and molded parts. Also called PPA.

polypropylene Thermoplastic polymer of propylene having low density and good flexibility and resistance to chemicals, abrasion, moisture, and stress cracking, but decreased dimensional stability, mechanical strength, and light, fire, and heat resistance. Processed by injection molding, spinning, and extrusion. Used in fibers and films for adhesive tapes and packaging. Also called PP.

polypyrrole A polymer of pyrrole, a five-membered heterocyclic substance with one nitrogen and four carbon atoms and with two double bonds. The polymer can be prepared via electrochemical polymerization. Polymers thus prepared are doped by electrolyte anion and are electrically conductive. Polypyrrole is used in lightweight secondary batteries, as electromagnetic interference shielding, anodic coatings, photoconductors, solar cells, and transistors.

polystyrene Polystyrenes are thermoplastics produced by polymerization of styrene with or without modification (e.g., by copolymerization or blending) to make impact resistant or expandable grades. They have good rigidity, high dimensional stability, low moisture absorption, optical clarity, high gloss and good dielectric properties. Unmodified polystyrenes have poor impact strength and resistance to solvents, heat and UV radiation. Processed by injection molding, extrusion, compression molding, and foam molding. Used widely in medical devices, housewares, food packaging, electronics and foam insulation. Also called polystyrenes, PS, polystyrol.

polystyrenes See *polystyrene.*

polystyrol See *polystyrene.*

polysulfones Thermoplastics, often aromatic and with ether linkages, having good heat, fire, and creep resistance, dielectric properties, transparency, but poor weatherability, processibility, and stress cracking resistance. Processed by injection, compression, and blow molding and extrusion. Used in appliances, electronic devices, auto parts, and electric insulators. Also called PSO.

polytetrafluoroethylene Thermoplastic polymer of tetrafluoroethylene having good dielectric properties, chemical, heat, abrasion, and fire resistance, antiadhesive properties, impact strength, and weatherability, but decreased strength, processibility, barrier properties, and creep resistance. Processed by sinter molding and powder coating. Used in nonstick coatings, chemical apparatus, electrical devices, bearings, and containers. Also called PTFE.

polyurethane resins See *polyurethanes.*

polyurethanes Polyurethanes (PUs) are a broad class of polymers consisting of chains with a repeating urethane group, prepared by condensation of polyisocyanates with polyols, e.g., polyester or polyether diols. PUs may be thermoplastic or thermosetting, elastomeric or rigid, cellular or solid, and offer a wide range of properties depending on composition and molecular structure. Many PUs have high abrasion resistance, good retention of properties at low temperatures and good foamability. Some have poor heat resistance, weatherability and resistance to solvents. PUs are flammable and can release toxic substances. Thermoplastic PUs are not crosslinked and are processed by injection molding and extrusion. Thermosetting PUs can be cured at relatively low temperatures and give foams with good heat insulating properties. They are processed by reaction injection molding, rigid and flexible foam methods, casting and coating. PUs are used in load bearing rollers and wheels, acoustic clamping materials, sporting goods, seals and gaskets, heat insulation, potting and encapsulation. Also called PUR, PU, urethane polymers, urethane resins, urethanes, polyurethane resins.

polyvinyl chloride Thermoplastic polymer of vinyl chloride, available in rigid and flexible forms. Has good dimensional stability, fire resistance, and weatherability, but decreased heat and solvent resistance and high density. Processed by injection and blow molding, calendering, extrusion, and powder coating. Used in films, fabric coatings, wire insulation, toys, bottles, and pipes. Also called PVC.

polyvinyl fluoride Crystalline thermoplastic polymer of vinyl fluoride having good toughness, flexibility, weatherability, and low-temperature and abrasion resistance. Processed by film techniques. Used in packaging, glazing, and electrical devices. Also called PVF.

polyvinylidene chloride Stereoregular thermoplastic polymer of vinylidene chloride having good abrasion and chemical resistance and barrier properties. Vinylidene chloride (VDC) content always exceeds 50%. Processed by molding and extrusion. Used in food packaging films, bag liners, pipes, upholstery, fibers, and coatings. Also called PVDC.

polyvinylidene fluoride Thermoplastic polymer of vinylidene fluoride having good strength, processibility, wear, fire, solvent, and creep resistance, and weatherability, but decreased dielectric properties and heat resistance. Processed by extrusion, injection and transfer molding, and powder coating. Used in electrical insulation, pipes, chemical apparatus, coatings, films, containers, and fibers. Also called PVDF.

PP See *polypropylene.*

PPA See *polyphthalamide.*

pphm See *parts per hundred million.*

ppm A unit for measuring small concentrations of material or substance as the number of its parts (arbitrary quantity) per million parts of medium consisting of another material or substance.

PPS See *polyphenylene sulfide.*

PPSS See *polyphenylene sulfide sulfone.*

pressure Stress exerted equally in all directions. See *processing pressure*

prevulcanization See *scorching.*

process characteristics See *processing parameters.*

process conditions See *processing parameters.*

process media See *processing agents.*

process parameters See *processing parameters.*

process pressure See *processing pressure.*

process rate See *processing rate.*

process speed See *processing rate.*

process time See *processing time.*

process velocity See *processing rate.*

processing additives See *processing agents.*

processing agents Agents or media used in the manufacture, preparation and treatment of a material or article to improve its processing or properties. The agents often become a part of the material. Also called process media, processing aids, processing additives.

processing aids See *processing agents.*

processing defects Structural and other defects in material or article caused inadvertently during manufacturing, preparation and treatment processes by using wrong tooling, process parameters, ingredients, part design, etc. Usually preventable. Also called processing flaw, defects, flaw. See also *cracking.*

processing flaw See *processing defects.*

processing methods Method names and designations for material or article manufacturing, preparation and treatment processes. **Note:** Both common and standardized names are used. Also called processing procedures.

processing parameters Measurable parameters such as temperature prescribed or maintained during material or article manufacture, preparation and treatment processes. Also called process characteristics, process conditions, process parameters.

processing pressure Pressure maintained in an apparatus during material or article manufacture, preparation and treatment processes. Also called process pressure. See also *pressure.*

processing procedures See *processing methods.*

processing rate Speed of the process in manufacture, preparation and treatment of a material or article. It usually denotes the change in a process parameter per unit of time or the throughput speed of material in a unit of weight, volume, etc. per unit of time. Also called process speed, process velocity, process rate.

processing time Time required for the completion of a process in the manufacture, preparation and treatment of a material or article. Also called process time, cycle time. See also *time.*

propene See *propylene.*

propylene An alkene (unsaturated aliphatic hydrocarbon) with three carbon atoms, $CH_2=CHCH_3$. A colorless, highly flammable gas. Autoignition temperature 497°C. Derived by thermal cracking of ethylene or from naphtha. Used as monomer in polymer and organic synthesis. Also called propene.

PS See *polystyrene.*

PSO See *polysulfones.*

PTFE See *polytetrafluoroethylene.*

PU See *polyurethanes.*

PUR See *polyurethanes.*

PV See *limiting pressure-velocity value.*

PV limit See *limiting pressure-velocity value.*

PVC See *polyvinyl chloride.*

PVDC See *polyvinylidene chloride.*

PVDF See *polyvinylidene fluoride.*

PVF See *polyvinyl fluoride.*

PVT relationship Pressure (P)-volume (V)-temperature (T) relationship of Boyle's law stating that the product of the volume of a gas times its pressure is a constant at a given temperature, PV/T=R, where R is Boltzmann constant.

R

Ra See *roughness average.*

reaction injection molding system Liquid compositions, mostly polyurethane-based, of thermosetting resins, prepolymers, monomers, or their mixtures. Have good processibility, dimensional stability, and flexibility. Processed by foam molding with in-mold curing at high temperatures. Used in auto parts and office furniture. Also called RIM.

resins See *polymers.*

resistance to flex cut growth A measure of vulcanized rubber deterioration on repeated bending in a suitable flexing machine, such as Ross or De Mattia machine. A pierced or unpierced specimen is bent repeatedly at a specified angle. The resistance is expressed as the average number of cycles for each 100% increase in cut size. **Note:** ASTM D3708 Also called resistance to flex cut growth, Ross-pierced, resistance to flex cut growth, Ross-unpierced, resistance to flex cut growth, De Mattia-pierced.

resistance to flex cut growth, De Mattia-pierced See *resistance to flex cut growth.*

resistance to flex cut growth, Ross-pierced See *resistance to flex cut growth.*

resistance to flex cut growth, Ross-unpierced See *resistance to flex cut growth.*

resorcinol modified phenolic resins Thermosetting polymers of phenol, formaldehyde, and resorcinol having good heat and creep resistance and dimensional stability.

RIM See *reaction injection molding system.*

Rockwell A See *Rockwell hardness.*

Rockwell E See *Rockwell hardness.*

Rockwell hardness A number derived from the net increase in the depth of impression as the load on an indenter is increased from a fixed minor load (10 kgf) to a major load and then returned to the minor load. This number consists of the number of scale divisions (each corresponding to 0.002 mm vertical movement of the indentor) and scale symbol. Rockwell scales, designated by a single capital letter of English alphabet, vary depending on the diameter of the indentor and the major load. For example, scale A indicates the use of a diamond indentor and major load 60 kgf, E - 1/8" ball indentor and 100 kgf, K - same ball and 150 kgf, M - 1/4" ball and 100 kgf, R - 1/2" ball and 60 kgf. The hardness increases in the order of R, M, K, E, and A scales. Also called Rockwell A, Rockwell E, Rockwell K, Rockwell M, Rockwell R.

Rockwell K See *Rockwell hardness.*

Rockwell M See *Rockwell hardness.*

Rockwell R See *Rockwell hardness.*

roughness average A height parameter of surface roughness equal to the average absolute deviation of surface profile from the mean line, calculated as the integrated area of peaks and valleys above and below the mean line, respectively, divided by the length of this line. Also called Ra.

S

S-N curve A plot showing the relationship of stress (S) to the number of cycles (N) before fracture in fatigue testing of materials. Also called Wohler curve, S-N diagram, S-N plot.

S-N diagram See *S-N curve.*

S-N plot See *S-N curve.*

SAN See *styrene acrylonitrile copolymer.*

SAN copolymer See *styrene acrylonitrile copolymer.*

SAN resin See *styrene acrylonitrile copolymer.*

SEN See *single-edge notched specimen.*

service life The period of time required for the specified properties of the material to deteriorate under normal use conditions to the minimum allowable level with material retaining its overall usability.

shelf life Time during which a physical system, such as a material, retains its storage stability under specified conditions. Also called storage life.

Shore A See *Shore hardness.*

Shore D See *Shore hardness.*

Shore hardness Indentation hardness of a material as determined by the depth of an indentation made with an indentor of the Shore-type durometer. The scale reading on this durometer is from 0, corresponding to 0.100" depth, to 100 for zero depth. The Shore A indenter has a sharp point, is spring-loaded to 822 gf, and is used for softer plastics. The Shore B indenter has a blunt point, is spring-loaded to 10 lbf, and is used for harder plastics. Also called Shore D, Shore A, durometer A hardness.

silicone There are rigid thermoplastic and liquid silicones and silicone rubbers consisting of alternating silicone and oxygen atom chains with organic pendant groups, prepared by hydrolytic condensation of chlorosilanes, followed by crosslinking. Silicone rubbers have good adhesion, flexibility, dielectric properties, weatherability, barrier properties, and heat and fire resistance, but decreased strength. Rigid silicones have good flexibility, weatherability, soil repelling properties, dimensional stability, but poor solvent resistance. Processed by coating, casting, and injection compression, and transfer molding. Used in coatings, electronic devises, diaphragms, medical products, adhesives, and sealants. Also called siloxane.

single-edge notched specimen A specimen, such as bar specimen used in impact testing of plastics, that has a notch or groove with a roughly triangular profile of various sharpness machined transversely across the entire width of its one side or edge. The notch is intended to simulate the crack and ensures the maximum test stress in the notched cross-section of the specimen. Also called SEN, one-side notched specimen.

sinusoidal wave form Something that resembles a sine wave which is a wave whose amplitude varies as the sine of a linear function of time. For example, in fatigue testing of materials, the applied stress amplitude can be varied sinusoidally, i.e., increasing gradually to a maximum value and then decreasing to a minimum value.

sliding velocity The relative speed of movement of one body against the surface of another body (counterbody) without the loss of contact as in a sliding motion during wear and friction testing of materials. In the sliding motion, the velocity vectors of the body and the counterbody remain parallel and should be unequal if they have the same direction.

slip factor A property that characterizes the lubricity of a material such as plastic sliding in contact with another material that is reciprocal of the friction coefficient.

SMA See *styrene maleic anhydride copolymer.*

SMA PTB alloy See *styrene maleic anhydride copolymer PBT alloy.*

specific wear rate See *wear factor.*

square wave form Something that resembles a square wave which is a wave whose amplitude shows periodic discontinuities between two values, remaining constant between jumps. For example, in fatigue testing of materials, the applied stress amplitude can be varied by increasing stress rapidly to a maximum value, keeping it constant for a period of time, then decreasing it rapidly to a minimum value and keeping it constant for a period of time before increasing it again.

stability The ability of a physical system, such as a material, to resist a change or degradation under exposure to outside forces, including mechanical force, heat and weather. See also *degradation.*

static coefficient of friction The ratio of the force that is required to start the friction motion of one surface against another to the force, usually gravitational, acting perpendicular to the two surfaces in contact. Also called coefficient of friction, static.

storage life See *shelf life.*

storage stability The resistance of a physical system, such as a material, to decomposition, deterioration of properties or any type of degradation in storage under specified conditions.

strain The per unit change, due to force, in the size or shape of a body referred to its original size or shape. **Note:** Strain is nondimensional but is often expressed in unit of length per unit of length or percent.

stress amplitude One-half the algebraic difference between the maximum and minimum stresses in one cycle of a repetitively varying stress as in fatigue testing of materials.

stress cracking Appearance of external and/or internal cracks in the material as a result of stress that is lower than its short-term strength.

stress cycle frequency Number of loading-unloading cycles per unit time in fatigue testing of materials.

stress pattern Distribution of applied or residual stress in a specimen, usually throughout its bulk. Applied stress is a stress induced by an outside force, e.g., by loading. Residual stress or stress memory may be a result of processing or exposure. The stress pattern can be made visible in transparent materials by polarized light.

stress-intensity factor The magnitude of the ideal crack tip stress field for a particulate mode of fracture in a homogeneous, linearly elastic body. It can be calculated as a limit of a power function of stress and a distance directly forward from the crack tip to a location where the stress is calculated in fatigue testing of materials.

stress-intensity factor range The difference between maximum and minimum values of stress-intensity factor in a cycle during fatigue testing of materials. Also called crack driving force, delta K.

styrene acrylonitrile copolymer SAN resins are thermoplastic copolymers of about 70% styrene and 30% acrylonitrile with higher strength, rigidity and chemical resistance than polystyrene. Characterized by transparency, high heat deflection properties, excellent gloss, hardness and dimensional stability. Have low continuous service temperature and impact strength. Processed by injection molding, extrusion, injection-blow molding and compression molding. Used in appliances, housewares, instrument lenses for automobiles, medical devices, and electronics. Also called styrene-acrylonitrile copolymer, SAN, SAN resin, SAN copolymer.

styrene butadiene block copolymer Thermoplastic amorphous block polymer of butadiene and styrene having good impact strength, rigidity, gloss, compatibility with other styrenic resins, water resistance, and processibility. Used in food and display containers, toys, and shrink wrap.

styrene butadiene copolymer Thermoplastic polymers of butadiene and >50% styrene having good transparency, toughness, and processibility. Processed by extrusion, injection and blow molding, and thermoforming. Used in film wraps, disposable packaging, medical devices, toys, display racks, and office supplies.

styrene maleic anhydride copolymer Thermoplastic copolymer of styrene with maleic anhydride having good thermal stability and adhesion, but decreased chemical and light resistance. Processed by injection and foam molding and extrusion. Used in auto parts, appliances, door panels, pumps, and business machines. Also called SMA.

styrene maleic anhydride copolymer PBT alloy Thermoplastic alloy of styrene maleic anhydride copolymer and polybutylene terephthalate having improved dimensional stability and tensile strength. Processed by injection molding. Also called SMA PTB alloy.

styrene plastics See *styrenic resins.*

styrene polymers See *styrenic resins.*

styrene resins See *styrenic resins.*

styrene-acrylonitrile copolymer See *styrene acrylonitrile copolymer.*

styrenic resins Styrenic resins are thermoplastics prepared by free-radical polymerization of styrene alone or with other unsaturated monomers. The properties of styrenic resins vary widely with molecular structure, attaining the high performance level of engineering plastics. Processed by blow and injection molding, extrusion, thermoforming, film techniques and structural foam molding. Used heavily for the manufacture of automotive parts, household goods, packaging, films, tools, containers and pipes. Also called styrene resins, styrene polymers, styrene plastics.

styrenic thermoplastic elastomers Linear or branched copolymers containing polystyrene end blocks and elastomer (e.g., isoprene rubber) middle blocks. Have a wide range of hardnesses, tensile strength, and elongation, and good low-temperature flexibility, dielectric properties, and hydrolytic stability. Processed by injection and blow molding and extrusion. Used in coatings, sealants, impact modifiers, shoe soles, medical devices, tubing, electrical insulation, and auto parts. Also called TES.

surface roughness Relatively fine spaced surface irregularities, the heights, widths and directions of which establish the predominant surface pattern.

surface tack Stickiness of a surface of a material such as wet paint when touched.

T

Taber abrasion See *Taber abrasion resistance.*

Taber abrasion resistance The weight loss of a plastic or other material specimen after it was subjected to abrasion in Taber abraser for a prescribed number of specimen disk rotations, usually 1000. Taber abraser consists of an idling abrasive wheel, designated depending on the type and grit of the abrasive used as CS-10F, H 22, etc., and a rotary disk with the specimen mounted on it. The load is applied to the wheel. The produced motion simulates that of rolling with slip. Also called Taber abrasion.

tape abrasion test See *Armstron abrasion test.*

tear propagation resistance The force required to propagate a slit in a flexible plastic film or thin sheeting at a constant rate of loading, calculated as an average between the initial and the maximum tear-propagation forces. Also called tear resistance, propagated.

tear resistance, initial See *initial tear resistance.*

tear resistance, propagated See *tear propagation resistance.*

tearing energy Tearing energy is a function of strain energy density and crack length, often expressed in kN/m. Plots of tearing energy vs. fatigue crack growth rate are used to characterize the kinetics of fatigue crack extension in rubbers, which do not obey the classical theory of elasticity. Also called tearing energy parameter.

tearing energy parameter See *tearing energy.*

temperature Property which determines the direction of heat flow between objects. **Note:** The heat flows from the object with higher temperature to that with lower.

tensile elongation See *elongation.*

tensile fatigue Progressive localized permanent structural change occurring in a material subjected to cyclic tensile stress that may culminate in cracks or complete fracture after a sufficient number of cycles.

tensile properties Properties describing the reaction of physical systems to tensile stress and strain. See also *tensile property tests.*

tensile property tests Names and designations of the methods for tensile testing of materials. Also called tensile tests. See also *tensile properties.*

tensile strain The relative length deformation exhibited by a specimen in tension. See also *elongation.*

tensile strength The maximum tensile stress that a specimen can sustain in a test carried to failure. **Note:** The maximum stress can be measured at or after the failure or reached before the fracture, depending on the viscoelastic behavior of the material. Also called tensile ultimate strength, ultimate tensile strength, UTS, tensile strength at break, ultimate tensile stress. See also *ASTM D638.*

tensile strength at break See *tensile strength.*

tensile stress The stress is perpendicular and directed to the opposite plane on which the forces act.

tensile tests See *tensile property tests.*

tensile ultimate strength See *tensile strength.*

terephthalate polyester Thermoset unsaturated polymer of terephthalic anhydride.

TES See *styrenic thermoplastic elastomers.*

test methods Names and designations of material test methods. Also called testing methods.

test variables Terms related to the testing of materials such as test method names.

testing methods See *test methods.*

tetrachloroethylene Also called perchloroethylene.

tetrafluoroethylene propylene copolymer Thermosetting elastomeric polymer of tetrafluoroethylene and propylene having good chemical and heat resistance and flexibility. Used in auto parts.

thermal properties Properties related to the effects of heat on physical systems such as materials and heat transport. The effects of heat include the effects on structure, geometry, performance, aging, stress-strain behavior, etc.

thermal stability The resistance of a physical system, such as a material, to decomposition, deterioration of properties or any type of degradation in storage under specified conditions.

thermodynamic properties A quantity that is either an attribute of the entire system or is a function of position, which is continuous and does not vary rapidly over microscopic distances, except possibility for abrupt changes at boundaries between phases of the system. Also called macroscopic properties.

thermoplastic polyesters A class of polyesters that can be repeatedly made soft and pliable on heating and hard (flexible or rigid) on subsequent cooling.

thermoplastic polyurethanes A class of polyurethanes including rigid and elastomeric polymers that can be repeatedly made soft and pliable on heating and hard (flexible or rigid) on subsequent cooling. Also called thermoplastic urethanes, TPUR, TPU.

thermoplastic urethanes See *thermoplastic polyurethanes.*

thrust washer friction and wear test method See *ASTM D3702.*

thrust washer test apparatus See *thrust-washer testing machine.*

thrust washer test machine See *thrust-washer testing machine.*

thrust-washer testing machine A machine, such as Falex Multispecimen Test Machine, used for assessing abrasive wear rate of of self-lubricating plastics intended for bearing applications in which wear debris essentially remain in the contact zone. The specimen of thrust washer configuration is mounted on a rotary upper holder and pressed against a steel washer placed on a stationary lower holder. The pressure load is applied to the lower holder via a level arm with dead weights. The upper holder is driven via a spindle by a d.c. electric motor at speeds 14-4420 rpm. The speed is controlled by a potentiometer and can be monitored by a tachometer. The machine also has an elapsed time indicator and a torque measurement device. Loose particles may be introduced in the contact zone either in a continuous stream or dispressed in a carrier fluid. Also called thrust washer test machine, Faville-LeValley Falex 6, thrust washer test apparatus.

time One of basic dimensions of the universe designating the duration and order of events at a given place. See also *processing time.*

toughness Property of a material indicating its ability to absorb energy by plastic deformation rather than crack or fracture.

TPO See *olefinic thermoplastic elastomers.*

TPU See *thermoplastic polyurethanes.*

TPUR See *thermoplastic polyurethanes.*

tribological Something having to do with friction, like in tribological behavior of material.

U

UHMWPE See *ultrahigh molecular weight polyethylene.*

ultimate elongation See *elongation.*

ultimate seal strength Maximum force that a heat-sealed thermoplastic film can sustain in a tensile test without seal failure per unit length of the seal.

ultimate tensile strength See *tensile strength.*

ultimate tensile stress See *tensile strength.*

ultrahigh molecular weight polyethylene Thermoplastic linear polymer of ethylene with molecular weight in the millions. Has good wear and chemical resistance, toughness, and antifriction properties, but poor processibility. Processed by compression molding and ram extrusion. Used in bearings, gears, and sliding surfaces. Also called UHMWPE.

uniaxially oriented A state of material such as polymeric film or composite characterized by the permanent orientation of its components such as polymer molecules or reinforcing fibers in one direction. The orientation is achieved by a number of different processes, e.g., stretching, and is intended to improve the mechanical properties of the material.

units See *units of measurement.*

units of measurement Systematic and non-systematic units for measuring physical quantities, including metric and US pound-inch systems. Also called units.

urea resins Thermosetting polymers of formaldehyde and urea having good clarity, colorability, scratch, fire, and solvent resistance, rigidity, dielectric properties, and tensile strength, but decreased impact strength and chemical, heat, and moisture resistance. Must be filled for molding. Processed by compression and injection molding, impregnation, and coating. Used in cosmetic containers, housings, tableware, electrical insulators, countertop laminates, adhesives, and coatings.

urethane polymers See *polyurethanes.*

urethane resins See *polyurethanes.*

urethane thermoplastic elastomers Block polyether or polyester polyurethanes containing soft and hard segments. Have good tensile strength, elongation, adhesion, and a broad hardness and service temperature ranges, but decreased moisture resistance and processibility. Processed by extrusion, injection molding, film blowing, and coating. Used in tubing, packaging film, adhesives, medical devices, conveyor belts, auto parts, and cable jackets. Also called TPU.

urethanes See *polyurethanes.*

UTS See *tensile strength.*

V

vinyl ester resins Thermosetting acrylated epoxy resins containing styrene reactive diluent. Cured by catalyzed polymerization of vinyl groups and crosslinking of hydroxy groups at room or elevated temperatures. Have good chemical, solvent, and heat resistance, toughness, and flexibility, but shrink during cure. Processed by filament winding, transfer molding, pultrusion, coating, and lamination. Used in structural composites, coatings, sheet molding compounds, and chemical apparatus.

vinyl resins Thermoplastics polymers of vinyl compounds such as vinyl chloride or vinyl acetate. Have good weatherability, barrier properties, and flexibility, but decreased solvent and heat resistance. Processed by molding, extrusion, and coating. Used in films and packaging.

vinyl thermoplastic elastomers Vinyl resin alloys having good fire and aging resistance, flexibility, dielectric properties, and toughness. Processed by extrusion. Used in cable jackets and wire insulation.

vinylidene fluoride hexafluoropropylene copolymer Thermoplastic polymer of vinylidene fluoride and hexafluoropropylene having good antistick, dielectric, and antifriction properties and chemical and heat resistance, but decreased mechanical strength and creep resistance and poor processibility. Processed by molding, extrusion, and coating. Used in chemical apparatus, containers, films, and coatings.

vinylidene fluoride hexafluoropropylene tetrafluoroethylene terpolymer Thermosetting elastomeric polymer of vinylidene fluoride, hexafluoropropylene, and tetrafluoroethylene having good chemical and heat resistance and flexibility. Used in auto parts.

volumetric wear rate See *wear factor.*

vulcanizate Rubber that had been irreversibly transformed from predominantly plastic to predominantly elastic material by vulcanization (chemical curing or crosslinking) using heat, vulcanization agents, accelerants, etc.

vulcanizate crosslinks Chemical bonds formed between polymeric chains in rubber as a result of vulcanization.

W

wear coefficient See *wear factor.*

wear factor The ratio of the wear volume or volume loss, caused by the abrasive wear of a specimen, to the product of the sliding distance that the specimen travels against the counterbody, and the load applied to the specimen. Also called K, K factor, wear coefficient, abrasion factor, specific wear rate, volumetric wear rate.

weight The gravitational force with which the earth attracts a body.

Wohler curve See *S-N curve.*

(15) *Udel Polysulfone Design Engineering Handbook*, supplier design guide (F-47178) - Amoco Performance Products, Inc, 1988.

(20) *Torlon Engineering Polymers / Design Manual*, supplier design guide (F-49893) - Amoco Performance Products.

(27) *Ultrapek Product Line, Properties, Processing*, supplier design guide (B 607 e/10.92) - BASF Aktiengesellschaft, 1992.

(28) *Ultrason E, Ultrason S Product Line, Properties, Processing*, supplier design guide (B 602 e/10.92) - BASF Aktiengesellschaft, 1992.

(30) *Luran Product Line, Properties, Processing*, supplier design guide (B 565 e/10.83) - BASF Aktiengesellschaft, 1983.

(34) Rakus, J., Brantley, T., Szabo, E., *Environmental Concerns With Vinyl Medical Devices, ANTEC, 1991*, conference proceedings - Society of Plastics Engineers, 1991.

(37) *Exploring Vespel Territory - The Properties of Du Pont Vespel Parts*, supplier design guide (E-26800) - Du Pont Company.

(39) *Handbook Of Properties For Teflon PFA*, supplier design guide (E-96679) - Du Pont Company, 1987.

(51) *Ultem Design Guide*, supplier design guide (ULT-201G (6/90) RTB) - General Electric Company, 1990.

(68) *Design Handbook For Du Pont Engineering Plastics - Module II*, supplier design guide (E-42267) - Du Pont Engineering Polymers.

(70) *Vectra Polymer Materials*, supplier design guide (B 121 BR E 9102/014) - Hoechst AG, 1991.

(71) *Physical Properties Of Unfilled And Filled Fluon Polytetrafluoroethylene*, supplier design guide (Technical Service Note F12/13) - ICI PLC, 1981.

(75) *Victrex PEK Properties And Processing*, supplier design guide (VP2/ October, 1987) - ICI Advanced Materials, 1987.

(76) *Victrex PES Data For Design Unreinforced Grades*, supplier design guide - ICI Advanced Materials, 1987.

(78) *Calibre Engineering Thermoplastics Basic Design Manual*, supplier design guide (301-1040-1288) - Dow Chemical Company, 1988.

(92) *Kinel Polyimide Compounds, Properties, Applications*, supplier design guide - Rhone Poulenc.

(93) *Ultramid Nylon Resins Product Line, Properties, Processing*, supplier design guide (B 568/1e/4.91) - BASF Corporation, 1991.

(94) *Hostalen Polymer Materials*, supplier design guide (HDKR 101 E 9050/022) - Hoechst AG.

(96) *Kel-F 81 PCTFE Engineering Manual*, supplier design guide (98-0211-5944-1 (120.5) DPI) - 3M Industrial Chemical Products Division, 1990.

Reference numbers correspond to our assigned source document number, if you wish additional information, please contact Plastics Design Library.

(101) *Engineering Properties Of Marlex Resins*, supplier design guide (TSM-243) - Phillips 66 Company, 1983.

(102) *Ryton Polyphenylene Sulfide Resins Engineering Properties Guide*, supplier design guide (1065(a)-89 A 02) - Phillips 66 Company, 1989.

(122) *Celanex Thermoplastic Polyester Properties and Proecessing (CX-1A)*, supplier design guide (HCER 91-343/10M/692) - Hoechst Celanese Corporation, 1992.

(123) *Upimol Polyimide Shape*, supplier technical report - Ube Industries.

135) *IXEF Reinforced Polyarylamide Based Thermoplastic Compounds Technical Manual*, supplier design guide (Br 1409c-B-2-1190) - Solvay, 1990.

(140) *Ryton Polyphenylene Sulfide Compounds Engineering Properties*, supplier design guide (TSM-266) - Phillips Chemical Company, 1983.

(141) *Physical Properties Kydex 100 Acrylic PVC Alloy Sheet*, supplier technical report (KC-89-03) - Kleerdex Company, 1989.

(143) *Luran S Acrylonitrile Styrene Acrylate Product Line, Properties, Processing*, supplier design guide (B 566 e / 10.83) - BASF Aktiengesellschaft, 1983.

(160) *Evoprene Super G Thermoplastic Elastomer Compounds*, supplier marketing literature (RDS 050/9240) - Evode Plastics.

(161) *Forprene By S.O.F.TER.*, supplier marketing literature (RDS 049/9240) - Evode Plastics.

(180) *Ultradur Polybutylene Terephthalate (PBT) Product Line, Properties, Processing*, supplier design guide (B 575/1e - (819) 4.91) - BASF Aktiengesellschaft, 1991.

(181) *Ultraform Polyacetal (POM) Product Line, Properties, Processing*, supplier design guide (B 563/1e - (888) 4.91) - BASF Aktiengesellschaft, 1991.

(185) *Ultramid T Polyamid 6/6T (PA) Product Line, Properties, Processing*, supplier design guide (B 605 e / 3.93) - BASF Aktiengesellschaft, 1993.

(189) *Engineering Design Guide To Rigid Geon Custom Injection Molding Vinyl Compounds*, supplier design guide (CIM-020) - BFGoodrich Geon Vinyl Division, 1989.

(190) *BFGoodrich Fiberloc Polymer Composites Engineering Design Data*, supplier design guide (FL-0101) - BFGoodrich Geon Vinyl Division, 1989.

(200) *Rynite Design Handbook For Du Pont Engineering Plastics*, supplier design guide (E-62620) - Du Pont Company, 1987.

(201) *Delrin Design Handbook For Du Pont Engineering Plastics*, supplier design guide (E-62619) - Du Pont Company, 1987.

(205) *Tefzel Fluoropolymer Design Handbook*, supplier design guide (E-31301-1) - Du Pont Company, 1973.

(210) *Celcon Acetal Copolymer*, supplier design guide (90-350 7.5M/490) - Hoechst Celanese Corporation, 1990.

(227) *Pocan Thermoplastic PBT Polyester - A General Reference Manual*, supplier design guide (55-B635(7.5)J) - Mobay Corporation, 1985.

(229) Stein, Harvey L, *Ultrahigh Molecular Weight Polyethylenes (UHMWPE), Engineered Materials Handbook, Vol. 2, Engineering Plastics*, reference book - ASM International, 1988.

(236) *RENY Property And Design Guide*, supplier design guide - Mitsubishi Gas Chemical Company, Inc.

(238) *Ube Nylon Technical Brochure*, supplier design guide (1989.8.1000) - Ube Industries, Ltd, 1989.

(254) *621 Ways To Succeed -, 1993-1994 Materials Selection Guide*, supplier technical report (304-00286-1292X SMG) - Dow Chemical Company, 1992.

(262) *Styron Polystyrene Resins For Applications Requiring Impact Resistance*, supplier design guide (301-471-1281) - Dow Chemical Company, 1981.

(263) *General Purpose Styron*, supplier design guide (301-678-1085) - Dow Chemical Company, 1985.

(282) *Santoprene Thermoplastic Elastomer Physical Properties*, supplier technical report (AES-1015) - Advanced Elastomer Systems, 1990.

(287) *Pebax*, supplier design guide - Atochem, 1987.

(289) *Guide Data Makrolon (Mechanical, Thermal, Electrical and Other Properties)*, supplier technical report (KU 46.100a/e) - Bayer AG, 1992.

(315) *Celanex Thermoplastic Polyester Short Term Properties (CX-4)*, supplier technical report (90-334 7.5M/490) - Hoechst Celanese Corporation, 1990.

(316) *Celanex Polybutylene Terephthalate (PBT)*, supplier design guide (BYKR 123 E 9070/014) - Hoechst AG, 1990.

(317) *Nylon Glass Reinforced Nylon Bulletin NY-B1*, supplier design guide (TNYBI/5M/1188) - Hoechst Celanese Corporation, 1988.

(318) Willis, E.N. Wang, L.C., *Ultra-High Performance Melt Processable Polybenzimidazole Parts, SAE Technical Paper Series - International Congress & Exposition*, conference proceedings (920818) - SAE International, 1992.

(320) *Celazole Data Sheets*, supplier technical report (ENW-91-41, CTD-106-1/92, CTD-108-2/92) - Hoechst Celanese Corporation, 1992.

(321) *Celcon Acetal Copolymer Specialty Grades Tailored To Meet the Needs of a More Demanding Marketplace*, supplier technical report (10M/8-85, CS-29) - Hoechst Celanese Corporation, 1985.

(323) *Engineering Plastics Calculations, Design, Applications - A.2.2 Calculation Principles, Characteristic Values, Calculation Examples Hostaform*, supplier design guide (HBKP 0395/A.2.2 E-UK-9021/022) - Hoechst AG, 1991.

(324) *Technical Plastics Calculations, Design, Applications - A.1.1 Grades and Properties Hostaform*, supplier design guide (HBKP 395/A.1.1 E-8127/022) - Hoechst AG, 1987.

(325) *Technical Plastics Calculations, Design, Applications - A.2.3 Calculation Principles, Characteristic Values, Calculation Examples Reinforced Hostalen PP*, supplier design guide (HBKP 395/A.2.3 E/FFM-8076/022) - Hoechst AG, 1986.

(326) *Hostacom Reinforced Polypropylene*, supplier design guide (B115BRE9072/046) - Hoechst AG, 1992.

(327) *Hoechst Plastics Hostalen GUR*, supplier design guide (HKR112E8102/14) - Hoechst AG.

(328) *Hostalen GUR Ultra High Molecular Weight Polymer Product Guide*, supplier technical report - Hoechst Celanese, 1983.

(329) *Hostalen GUR UHMW Polymer Data Sheets*, supplier technical report - Hoechst Celanese, 1990.

(330) *Hostalloy Engineering Resins*, supplier technical report - Hoechst Celanese, 1993.

(333) *Riteflex Thermoplastic Polyester Elastomer Short Term Properties (RF-4)*, supplier data sheets (HCER 93-301/2.5M/193) - Hoechst Celanese, 1993.

(334) *Riteflex BP Thermoplastic Polyester Elastomer Data Sheets*, supplier data sheets (THCER 89-384/IM/1289) - Hoechst Celanese, 1989.

(336) *Typical Physical Properties of Pellethane Thermoplastic Polyurethane Elastomers*, supplier data sheets (306-183-1293X SMG) - Dow Chemical Company, 1993.

(337) *Victrex PEEK The High Temperature Engineering Thermoplastic - Data For Design*, supplier design guide (VK4/2/1290) - ICI Advanced Materials, 1990.

(338) *Victrex PEEK For Bearing Applications*, supplier design guide (VKT1/0986) - ICI Advanced Materials, 1986.

(339) *Teflon Fluorocarbon Resin Mechanical Design Data*, supplier design guide (E-05561-3) - DuPont Company.

(340) *Teflon/ Tefzel Fluoropolymer Resin Product Information*, supplier marketing literature (E-96678-2) - DuPont Company, 1993.

(341) *Styron XL-8035 MFD High Impact Polystyrene Resin for Microfloppy Diskettes*, supplier marketing literature (301-1607-791X SMG) - Dow Chemical Company, 1991.

(342) *Rovel 401 Weatherable Polymer for Sheet Extrusion*, supplier marketing literature (301-1201-688 RJD) - Dow Chemical Company, 1988.

(343) *VESPEL - A Publication on Du Pont's high-performance Vespel Polyimide Parts and Shapes*, supplier marketing literature (H-36046) - DuPont Company, 1991.

(344) *VESPEL - Using Vespel Bearings Design and Technical Data*, supplier design guide (E-61500 5-88) - DuPont Company, 1988.

(345) *VESPEL Polyimide Parts Properties Summary*, supplier technical report (H-44018) - DuPont Company, 1992.

(346) *Alcryn Melt-Processible Rubber Product and Properties Guide*, supplier technical report (H-44599) - DuPont Company, 1992.

(347) *Hytrel Polyester Elastomer Design Handbook*, supplier design guide (E-52083-1) - DuPont Company, 1988.

(348) *Hytrel Engineering Thermoplastic Elastomer Product and Properties Guide*, supplier technical report (H-14888-2) - DuPont Company, 1993.

(349) *Hypalon Synthetic Rubber Types, Properties and Uses Guide - HP-210.1 (R3)*, supplier technical report (H-42029) - DuPont Company, 1992.

(350) *Delrin Acetal Resin Product and Properties Guide*, review report (H-41651-2) - DuPont Company, 1993.

(352) *General Electric Plastics Engineering Design Database*, computer database (EDD) - General Electric Company, 1995.

(351) *Zytel Nylon Resin Product and Properties Guide*, supplier data sheets (H-57463) - DuPont Company, 1994.

(353) *Product Guide - Surlyn Ionomer Resins*, review report (H-0473) - DuPont Company.

(354) *Zenite Liquid Crystal Polymer Resins Product and Properties Guide*, supplier data sheets (H-49658-1) - DuPont Company, 1994.

(355) Gotham, K.V. (RAPRA), *Fatigue And Long-Term Strength Of Thermoplastics, Developments in Plastics Technology-3*, reference book - Elsevier Applied Science Publishers, 1986.

(356) *Rynite PET Product and Properties Guide*, supplier data sheets (E-62735-1) - DuPont Company, 1993.

(358) *Acsium HPR-6367 Synthetic Rubber Product Information*, supplier data sheets (H-36563-1) - DuPont Company, 1991.

(359) *Amodel PPA Resins*, supplier technical report (AM-F-49992) - Amoco Performance Products.

(360) *Kinel Polyimide Compounds, Applications, Properties*, supplier design guide - Rhone-Poulenc.

(362) *Polyphenylene Ether (PPE) Resin - Iupiace*, supplier design guide (M.G.C.91042000P.A.) - Mitsubishi Gas Chemical Company, 1992.

(363) *Iupiace Polyphenylene Ether Resin Physical Properties Edition*, supplier design guide - Mitsubishi Gas Chemical Company.

(364) *Iupital Acetal Copolymer*, supplier design guide (MGC.MC88.2.1.1000A) - Mitsubishi Gas Chemical Company, 1988.

(365) O'Toole, J.L., *Selecting Plastics For Fatigue Resistance, Modern Plastics Encyclopedia -, 1986-1987*, reference book (ISSN 0085-3518/86) - McGraw Hill Inc, 1986.

(366) Poloso, A. Bradley, M.B., *High-Impact Polystyrenes, Engineered Materials Handbook, Vol. 2, Engineering Plastics*, reference book - ASM International, 1988.

(368) Riddell, M.N., Koo, G.P., O'Toole, J.L., *Fatigue Mechanisms of Thermoplastics, Polymer Engineering and Science*, technical journal, 1966.

(369) Crawford, R.J., Benham, P.P., *Polymer*, technical journal (Vol 16), 1975.

(370) Lesser, Alan J., *High-Cycle Fatigue Behavior Of Engineering Thermoplastics, ANTEC, 1995*, conference proceedings - Society of Plastics Engineers, 1995.

(371) Seibel, S.R. (Case Western), Moet, A. (Case Western), Bank, D.H. (Dow), Nichols, K. (Dow), *Effect Of Filler On the Fatigue Performance of a 65 PC/ 35 ABS Blend, ANTEC, 1993*, conference proceedings - Society of Plastics Engineers, 1993.

(372) Boukhili, R. (Ecole Polytechnique de Montreal), Gauvin, R. (Ecole Polytechnique de Montreal), Gosselin, M. (Ecole Polytechnique de Montreal), *Fatigue Behavior of Injection Moded Polycarbonate and Polystyrene Containing Weld Lines, ANTEC, 1989*, conference proceedings - Society of Plastics Engineers, 1989.

(374) Benabdallah, H. (Ecole Polytechnique de Montreal), Fisa, B. (Ecole Polytechnique de Montreal), *Static Friction Characteristics of Some Thermoplastics, ANTEC, 1989*, conference proceedings - Society of Plastics Engineers, 1989.

(375) Tanrattanakul, V. (Case Western Reserve University), Moet, A. (Case Western Reserve University), Bank, D. (Dow), *Fatigue Behavior of Rigid Thermoplastic Polyurethanes, ANTEC, 1993*, conference proceedings - Society of Plastics Engineers, 1993.

(376) Duvall, D.E. (L.J. Broutman), So, P.K. (L.J. Broutman), Broutman, L.J. (L.J. Broutman, *Effect of Molecular Weight on Slow Crack Growth in High Density Polyethylene, ANTEC, 1992*, conference proceedings - Society of Plastics Engineers, 1992.

(378) Young, D.G. (Exxon), Doyle, M.J. (Exxon), *Fracture Mechanics Characterization of Rubbber Compounds for Fatigue Crack Growth Resistance, ANTEC, 1992*, conference proceedings - Society of Plastics Engineers, 1992.

(380) Benabdallah, S.H. (Royal Military College of Canada), *Mechanical and Tribological Properties of PEHD/PA 6 Blends, ANTEC, 1992*, conference proceedings - Society of Plastics Engineers, 1992.

(381) Seeler, K.A. (Lafayette College), Kumar, V. (University of Washington), *Tension-Tension Fatigue of Microcellular Polycarbonate: Inital Results, ANTEC, 1992*, conference proceedings - Society of Plastics Engineers, 1992.

(382) Bolvari, Anne (LNP), *Analysis of Dry Sliding Wear and Friction Data of Internally Lubricated Composites, ANTEC, 1994*, conference proceedings - Society of Plastics Engineers, 1994.

(384) Yang, S.G. (University of Florida), Bennett, D. (University of Florida), Beatty, C. (University of Florida), *Flexural Fatigue Resistance of Recycled HDPE, ANTEC, 1994*, conference proceedings - Society of Plastics Engineers, 1994.

(385) Li, Z. (University of Florida), El-Rahman, M. (University of Florida), Beatty, C. (University of Florida), *Wear/ Abrasion Studies of Styrene-Butadiene Block Copolymer Containing Recycled Rubber, ANTEC, 1994*, conference proceedings - Society of Plastics Engineers, 1994.

(386) Seeler, K.A. (Lafayette College), Kumar, V. (University of Washington), *Effect of Carbon Dioxide Saturation and Desorption on the Notch Sensitivity of Polycarbonate Fatigue, ANTEC, 1994*, conference proceedings - Society of Plastics Engineers, 1994.

(388) Mimura, K. (The University of Michigan), Hristov, H. (The University of Michigan), Yee, A.F. (The University of Michigan), *Effect of Compatibilizer on Fatigue Properties of PC/PS Blend, ANTEC, 1994*, conference proceedings - Society of Plastics Engineers, 1994.

(390) Azimi, H.R. (Lehigh University), Pearson, R.A. (Lehigh University), Hertzberg, R.W. (Lehigh University), *A Mechanistic Understanding of Fatigue Crack Propagation Behavior of Rubber-Modified Epoxy Polymers, ANTEC, 1995*, conference proceedings - Society of Plastics Engineers, 1995.

(391) Grove, D.A. (LNP), Kim, H.C. (LNP), *Effect of Constituents on the Fatigue Behavior of Long Fiber Reinforced Thermoplastics, ANTEC, 1995*, conference proceedings - Society of Plastics Engineers, 1995.

(392) *Victrex PES Bearing Applications*, supplier design guide (VST5/0286) - ICI Advanced Materials, 1986.

(393) *Victrex PES Data For Design*, supplier design guide (VS4/1089) - ICI Advanced Materials, 1989.

(394) *LNP Engineering Graphical Database*, computer database - LNP Engineering Plastics, 1995.

(395) Newby, G.B. (LNP), Theberge, J.E. (LNP), *Long-Term Behavior of Reinforced Thermoplastics, Machine Design*, trade journal - Penton Publishing, 1984.

(397) Barron, D.L. (Dow), Kelley, D.H. (Dow), Blankenship, L.T. (Dow), *Fracture Mechanics Approach Differentiates Fatigue Performance of Thermoset Resin Systems, 44th Annual Conference, SPI Composites Institute*, conference proceedings - Society of the Plastics Industry, 1989.

(398) Sandberg, L.B. (Michicgan Technological University), Rintala, A.E. (Rowe Engineering), *Resistance of Structural Silicones to Creep Rupture and Fatigue, Building Sealants: Materials, Properties and Performance - ASTM STP 1069 - Symposium Proceedings*, conference proceedings - American Society for Testing and Materials, 1990.

(399) Royo, J., *Fatigue Testing of Rubber Materials and Articles, Polymer Testing 11 (1992)*, technical journal - Elsevier Science Publishers Ltd, 1992.

(400) Bank, D.H. (Dow), Seibel, S. (Case Western Reserve University), Moet, A. (Case Western Reserve University), *Fatigue and Fracture Toughness Evaluation of Mineral Filled Polycarbonate/ Polyethylene Terephthalate Blends, ANTEC, 1992*, conference proceedings - Society of Plastics Engineers, 1992.

(401) Wyzgoski, M.G. (General Motors), Novak, G.E. (General Motors), *Fatigue-Resistant Nylon Alloys, Journal of Applied Polymer Science (1994)*, technical journal (Vol. 51; CCC 0021-8995/94/050873-13) - John Wiley & Sons, Inc, 1994.

(402) *Lupolen Ultrahigh Molecular Weight Polyethylene (UHMWPE) Product Line, Properties, Processing*, supplier technical report (B598e/10.89) - BASF, 1989.

(403) *Huls Vestodur Polybutylene Terephthalate*, supplier technical report (42.01.003e/505.93/bu) - Huls Aktiengesellschaft, 1993.

(404) *Rogers Phenolic Powertrain Components*, supplier marketing literature (6299-093-3.0W) - Rogers Corporation, 1993.

(405) *Mearthane Products Corp. - The Polyurethane Leader Technical Guide*, supplier technical report - Mearthane Products Corp, 1993.

(406) *Design and Application Guide Cast & Molded Components of Polyurethane Elastomers*, supplier technical report - Gallagher Corporation, 1994.

(407) *Properties of Toyolac*, supplier technical report - Toray Industries, 1987.

(408) Prevorsek, D.C. (Allied-Signal), Kwon, Y.D. (Allied-Signal), Beringer, C.W. (Allied-Signal), Feldstein, M. (Allied-Signal), *Mechanics of Short Fiber Reinforced Elastomers (paper 71), ACS Rubber Division, 1990 Fall Meeting - Washington D.C.*, conference proceedings - American Chemical Society, 1990.

(409) Lauretti, E. (Enichem Elastomeri), Miani, B. (Enichem Elastomeri), Mistrali, F. (Enichem Elastomeri), *Improving Fatigue Resistance with Neodimium Polybutadiene (paper 62), ACS Rubber Division 144th Meeting - Orlando, Florida*, conference proceedings - , 1993.

(410) Omura, N. (Shin-Etsu), Takahashi, M. (Shin-Etsu), Nakamura, T. (Shin-Etsu), *Silicone Rubber and its Fatigue Properties, Rubber & Plastics News*, trade journal, 1991.

(411) Young, D.G. (Exxon), Danik, J.A. (Exxon), *Fatigue Tests Can Provide Useful Information, Rubber & Plastics News*, trade journal, 1994.

(412) Roland, C.M. (Naval Research Laboratory), Sobieski, J.W. (Geo-Centers, Inc.), *Anomalous Fatigue Behavior in Polyisoprene (paper 35), ACS Rubber Division 134th Meeting - Cincinnati, Ohio*, conference proceedings - American Chemical Society, 1988.

(413) Adams, M.E., Buckley, D.J., Colborn, R.E., England, W.P., Schissel, D.N., *Acrylonitrile Butadiene Styrene Polymers - Rapra Review Report 70*, review report - RAPRA Technology Ltd, 1993.

(414) Wright, D. (RAPRA), *Knowledge Based System for Plastics - SPE ANTEC '95*, presentation - RAPRA Technology Ltd, 1995.

(415) *Huls Trogamid Polyamide 6-3-T*, supplier design guide (42.01.027e) - Huls Aktiengesellschaft, 1992.

(416) *Makrolon Polycarbonate Design Guide*, supplier design guide (55-A840(10)L) - Bayer, 1988.

(417) O'Toole, J.L. (Allied Chemical), Radosta, J.A. (Allied Chemical), Moylan, J.J. (Allied Chemical), *Thermosets vs. New Thermoplastics - Elevated Temperature Behavior*, review report - Allied Chemical Corporation, 1980.

(418) *Primef Polyphenylene Sulphides - High Performance Engineering Polymers*, supplier design guide (Br-1444c-B-4-0490) - Solvay, 1990.

(419) *Stanyl 46 Nylon General Information*, supplier design guide (MBC-PP-492-5M) - DSM, 1992.

(420) *Utilizing Inherent Hinge Properties of Tenite Polypropylene and Tenite Polyallomer*, supplier technical report (TR-14E) - Eastman Chemical, 1987.

(421) *Techster Thermoplastic Polyester for Electronic Components*, supplier technical report (TR/MA/AN/0687/012) - Rhone Poulenc, 1987.

(422) *Mitsubishi Gas Chemical Company, Inc. Engineering Plastics List of Properties*, supplier technical report (01 05 5000 DPR2) - Mitsubishi Gas Chemical, 1991.

(423) *Victrex PEEK Aromatic Polymer Friction and Wear Properties*, supplier technical report - ICI, 1983.

(424) *Alcryn Melt Processible Rubber General Technical Guide*, supplier technical report (E-90901) - DuPont Company, 1990.

(425) *Kalrez Perfluoroelastomer Parts Physical and Compound Comparisons*, supplier technical report (E-94428-2) - DuPont Company, 1994.

(426) *Fortron Polyphenylene Sulfide*, supplier design guide (BYKR 120 E-9110/014) - Hoechst AG, 1991.

(427) *Texin Urethane Elastomer - An Engineering Handbook*, supplier design guide - Bayer, 1993.

(428) *Tedur Polyphenylene Sulfide Design Guide*, supplier design guide (KU-C1005(3.5)) - Bayer, 1991.

(429) *Supec Design Guide*, supplier design guide (SUP-120A (4/90) RTB) - General Electric Company, 1990.

(430) *Noryl Design Guide*, supplier design guide (CDX-83D (11/86) RTB) - General Electric Company, 1986.

(431) *GE Plastics Engineering Materials Profile*, supplier technical report - General Electric Company.

(432) *Lomod Product Data Sheets*, supplier technical report (Eris) - General Electric Company, 1991.

(433) *Santoprene Thermoplastic Rubber General Product Bulletin*, supplier technical report (TPE-02-11) - Advanced Elastomer Systems.

434) *Isoplast Engineering Thermoplastic Resin*, supplier technical report (301-120*-688 SMG) - Dow Chemical Company, 1988.

(435) *Lubriplas Molding Compounds Product Information*, supplier technical report (94-01) - Bay Resins, 1994.

(436) *Solidur UHMW-PE Product Selection Guide*, supplier marketing literature - Solidur Plastics Company, 1988.

(437) *Foraflon Polyvinylidene Fluoride*, supplier design guide - Elf Atochem, 1987.

(438) *Pebax Resins 33 Series Property Comparison*, supplier design guide - Elf Atochem, 1994.

(439) *Halon TFE Filled Compounds Product Line and Properties*, supplier design guide - Ausimont.

(440) *Mechanical Properties of Halar Fluoropolymer*, supplier design guide - Ausimont.

(441) *Engineering Plastic Products*, supplier marketing literature - The Polymer Corporation, 1993.

(442) *Sarlink 3000 Series Data Sheets*, supplier marketing literature - DSM Thermoplastic Elastomers, 1994.

(443) *Kydex Wallcovering*, supplier marketing literature - Kleerdex Company.

(444) *Solef PVDF Product Line*, supplier marketing literature - Solvay, 1985.

(445) *Millathane 76 - A Millable Urethane Elastomer*, supplier design guide - TSE Industries, 1993.

(446) *J-Von Introduces Hercuprene*, supplier marketing literature (01-0192) - J-Von Limited Partnership, 1992.

(447) *Trogamid T Mechanical , electrical and Thermal Properties*, supplier design guide (KR 67/2.87/4000/5081) - Dynamit Nobel Chemie, 1987.

(448) *Molding Properties PCTA 6761 Amorphous Copolyester*, supplier design guide - Eastman Performance Products.

449) *Standard Values of Ensinger Reinforced Engineering Plastics*, supplier design guide (300-5/93) - Ensinger Inc, 1993.

(451) *Pibiflex Elastomeric Copolyesters*, supplier design guide - ECP EniChem Polimeri, 1992.

(452) *Ferro Plastics Europe*, supplier design guide (F-1000-E) - Ferro Plastics Europe, 1993.

(453) *Lubricomp Internally Lubricated Reinforced Thermoplastics and Fluoropolymer Composites*, supplier design guide (254-691) - LNP Engineering Plastics, 1991.

(454) Wu, Y.T., *How Short Aramid Fiber Improves Wear Resistance, Modern Plastics*, trade journal - McGraw-Hill, 1988.

(455) *Polycarbonate Composites*, supplier marketing literature (202-192) - LNP Engineering Plastics, 1992.

(456) *LNP Glass Reinforced Polysulfone Thermocomp GF Series*, supplier marketing literature (206-1191) - LNP Engineering Plastics, 1991.

(457) *Wear Resistant Thermoplastics*, supplier technical report - RTP Company, 1991.

(458) *Plaslube Internally Lubricated and Reinforced Engineering Resins*, supplier technical report (PLAS 1289 5M) - Akzo Engineering Plastics, 1989.

(459) *Reinforced Thermoplastics Engineering Manual*, supplier design guide - Thermofil Engineering Thermoplastics.

(460) *RTP Polyphthalamide Data Sheets*, supplier marketing literature - RTP Company, 1992.

(461) Hawley, S.W., *Physical Testing of Thermoplastics - Rapra Review Report 60*, review report - RAPRA Technology Ltd, 1992.

(462) Moet, Abdelsamie (Case Western) Aglan, Heshmat (Case Western), *Fatigue Failure, Engineered Materials Handbook, Vol. 2, Engineering Plastics*, reference book - ASM International, 1988.

(463) Yelle, Henri J.M. Estabrook, F. Reed Jr., *Fatigue Loading, Engineered Materials Handbook, Vol. 2, Engineering Plastics*, reference book - ASM International, 1988.

(464) Turner, S. (Queen Mary College), *Mechanical Testing, Engineered Materials Handbook, Vol. 2, Engineering Plastics*, reference book - ASM International, 1988.

(465) Brick, Robert M. Pense, Alan W. Gordon, Robert B., *Engineered Materials Handbook, Vol. 2, Engineering Plastics*, reference book - ASM International, 1988.

Trade Name Index

Appendix I - Coefficient of Friction

Appendix II - Wear Factor (K) and Wear Rate

Appendix III - Limiting PV Values

Appendix IV - Abrasion Resistance Data

Fatigue and tribological properties of plastics & elastomers